AF352865

Improving Athletes' Performance and Avoiding Health Issues

Improving Athletes' Performance and Avoiding Health Issues

Editors

João Paulo Brito
Rafael Oliveira

Basel • Beijing • Wuhan • Barcelona • Belgrade • Novi Sad • Cluj • Manchester

Editors

João Paulo Brito	Rafael Oliveira
Sports Science School	Sport Sciences School
of Rio Maior	of Rio Maior
Polytechnic Institute of Santarém	Polytechnic Institute of Santarem
Rio Maior	Rio Maior
Portugal	Portugal

Editorial Office
MDPI
St. Alban-Anlage 66
4052 Basel, Switzerland

This is a reprint of articles from the Special Issue published online in the open access journal *Healthcare* (ISSN 2227-9032) (available at: www.mdpi.com/journal/healthcare/special_issues/athletes_performance).

For citation purposes, cite each article independently as indicated on the article page online and as indicated below:

Lastname, A.A.; Lastname, B.B. Article Title. *Journal Name* **Year**, *Volume Number*, Page Range.

ISBN 978-3-0365-8971-8 (Hbk)
ISBN 978-3-0365-8970-1 (PDF)
doi.org/10.3390/books978-3-0365-8970-1

Contents

About the Editors

João Paulo Brito

João Paulo Moreira de Brito is a full professor at Escola Superior de Desporto de Rio Maior–Instituto Politécnico de Santarém (Sport Sciences School of Rio Maior–Polytechnic Institute of Santarem), an integrated member of the Life Quality Research Centre, and a collaborator at the Research Centre in Sport Sciences, Health Sciences and Human Development. He obtained his PhD in Sports Sciences at the University of Lleida, Lleida, Spain.

His research activities focus on exercise physiology, physiological assessment of athletes from different sports, determination of physical and physiological profiles of elite athletes, subjective and objective load monitoring/quantification, and strength training to improve performance in male and female athletes. Currently, he is an author of more than 80 published scientific articles in indexed journals and about 17 books or book chapters. He has received several awards for achieving a good citation rate in scientific publication in recent years.

Rafael Oliveira

Rafael Franco Soares Oliveira is an adjunct professor at Escola Superior de Desporto de Rio Maior–Instituto Politécnico de Santarém (Sport Sciences School of Rio Maior–Polytechnic Institute of Santarem), an integrated member of the Life Quality Research Centre, and a collaborator at the Research Centre in Sport Sciences, Health Sciences and Human Development. He obtained his PhD in Sports Sciences at the University of Beira Interior, Covilhã, Portugal.

His research activities focus on exercise physiology, testing, control and prescription, load monitoring/quantification, strength and conditioning and sports training methodology. Currently, he is an author of more than 100 published scientific articles (indexed in Journal of Citation Reports or Scientific Journal Ranking) and about 20 books or book chapters.

In 2022 and 2023, he was awarded with the 1st place for his contribution to research and development by the Polytechnic Institute of Santarem. He was also awarded with an honorable mention as a researcher with the best citation rate in scientific publication in 2022 and the 1st place as a researcher with the best citation rate in scientific publication in 2023 by the Life Quality Research Centre.

Preface

In several elite sports, training and match loads are assessed. However, only clubs with the most resources and money can apply more valid and reliable tools. For instance, at the regional level, clubs and their coaches do not have sufficient resources to allow them to precisely monitor the training/match loads. As an example, elite football generates the most media and commercial attention, but it is at the regional level that the greatest numbers of clubs, players, and coaches are found. Meanwhile, studies that investigate load monitoring among teams at the regional level are sparse.

The planning and quantification of intensities are important for optimizing athletes' physical fitness to guarantee wellbeing and a better performance in male and female athletes.

Considering other sports, such as futsal, basketball, volleyball, rugby, hockey, and handball, resources could be equally distributed to apply proper load management. Conducting studies at the regional level enables the development of knowledge about athlete populations, while promoting research on the evolution of sports at all competitive levels will also contribute to avoiding any kind of healthcare issues, such as injury or illness.

Therefore, the aim of this Special Issue, which now constitutes a reprint, was to compile and present information on load and competition monitoring, especially in less resourceful sports, and the effects of gender, as well as to demonstrate the use of various tools for monitoring load to avoid healthcare issues among athletes.

Any coach or practitioner of any sport who is interested in improving their understanding of the relationship between intensity control and injury prevention or overtraining would be interested in reading this reprint.

João Paulo Brito and Rafael Oliveira
Editors

Editorial

Load Monitoring and Its Relationship with Healthcare in Sports

Rafael Oliveira [1,2,3,*] and João Paulo Brito [1,2,3]

[1] Sports Science School of Rio Maior, Polytechnic Institute of Santarém, 2040-413 Rio Maior, Portugal; jbrito@esdrm.ipsantarem.pt
[2] Life Quality Research Centre, 2040-413 Rio Maior, Portugal
[3] Research Centre in Sport Sciences, Health Sciences and Human Development, 5001-801 Vila Real, Portugal
*   Correspondence: rafaeloliveira@esdrm.ipsantarem.pt

Citation: Oliveira, R.; Brito, J.P. Load Monitoring and Its Relationship with Healthcare in Sports. *Healthcare* **2023**, *11*, 2330. https://doi.org/10.3390/healthcare11162330

Received: 3 August 2023
Accepted: 10 August 2023
Published: 18 August 2023

Load monitoring consists of training/match demand quantification as well as wellness and readiness to maximize the likelihood of optimal athletic performance [1]. The literature divides load into two dimensions: internal and external. Internal load is associated with psychophysiological demands that can be objectively and subjectively measured (e.g., heart rate and rating of perceived exertion, respectively) [2,3]. External load is associated with mechanical/locomotor demands, usually collected by global positioning systems, global navigation satellite systems, local positioning systems, and inertial measurement units that belong to micro-electro-mechanical systems (which provide a combination of 3D accelerometers, 3D gyroscopes, and 3D magnetometers). Despite different technologies, they provide external load measures, such as distances covered at various running speeds, accelerations, decelerations, player load, and others [2–5].

Indeed, there are other types of wearable technology that were considered to be among the top worldwide fitness trends in 2016 and 2017 [6]. Such technology (i.e., smartwatches and mobiles) allows for the quantification of different physical variables such as step counts, metabolic work, or power [7].

Another relevant dimension to monitor is the wellness/well-being of athletes, which is regularly collected by questionnaires that include different categories such as fatigue, quality of sleep, muscle soreness, mood, and stress [8,9]. For instance, a systematic review showed several relationships between wellness and training load measures that ranged from no association to a very large association [10].

The monitoring of different dimensions is useful in sports to optimize training adaptation, which can consequently improve performance and reduce injury risk [11]. Still, inappropriate load management can be a significant risk factor for acute illness and overtraining syndrome [12]. Therefore, all quantification can contribute towards better healthcare for recreational or elite sport athletes. However, there are few research papers that combine load monitoring and its relationship with healthcare in sports.

The present Special Issue contributed to the field with 35 articles. Of those, 16 articles involved only soccer athletes. For instance, one study was a systematic review that summarized studies about external and internal training load monitoring to provide range values for the main measures in young male soccer players [13]. Another study compared the external load between official and friendly matches and between the first and second halves of professional soccer players [14], while another analyzed differences among playing positions: whether playing home/away matches and if playing in the first or second part of the championship influence the external load of amateur soccer [15]. Moreover, external load was compared between starters and non-starters [16] and among the playing positions [17] of professional soccer players based on different parts of a full season. A sub-analysis of a specific type of training exercise (i.e., small-sided games) was performed to analyze the between-session and within-player variability of heart rates and external load of young male soccer players [18]. While the previous literature included male athletes, the analysis of female professional soccer players was also conducted by quantifying external and internal load as well as the wellness profile of a typical microcycle [19]. Another study tested objective and subjective external and internal loads in primary education students [20].

Other investigations included the analysis of physical fitness and competitive performance in different age categories (8.0–9.9, 10.0–11.9, and 12.0–13.9 years) [21] and in professional soccer [22]. In this regard, different intervention protocols were applied to professional soccer players [23,24] and archers [25] to improve several fitness characteristics. In the same way, fitness and technical skill were compared among young soccer players [26]. Additionally, the relationship between different inter-limb jumping asymmetries and performance measures in male senior and professional soccer players was analyzed [27]. Furthermore, the effectiveness of different training programs on the reactive strength index was compared in a systematic review with a meta-analysis [28], isotonic and isometric exercise interventions were reviewed to analyze the strength and flexibility of the hamstring muscles [29], and a dose-response meta-analysis was conducted to assess the velocity loss effects on strength development and related training efficiency [30]. In fact, other authors analyzed different sports training protocols. For instance, aquatic and bicycling training was applied to improve leg function and range of motion in the intermediate stage of rehabilitation in amateur athletes that underwent meniscal allograft transplantation [31].

Other negative psychological variables, such as stress, injury, anxiety, and depression in adult soccer players, were also analyzed [32]. In this regard, one study applied a mindfulness program to address levels of impulsivity, mood, and pre-competition anxiety in samples of athletics, tennis, swimming, basketball, handball, volleyball, and soccer athletes [33]. Coping strategies were another topic analyzed in professional soccer during Ramadan fasting [34], as well as in other sports than soccer, such as handball, martial arts, rugby, basketball, athletics, aerobic and artistic gymnastics, volleyball, tennis, and swimming during the COVID-19 pandemic [35]. Lastly, the personality and resilience of competitive drivers were another topic of research [36].

In basketball, one study compared the redox, hormonal, metabolic, and lipid profiles between adult male and female athletes and sedentary controls [37], while others characterized the salivary proteome and metabolome of highly trained female and male young basketball players [38].

Body composition and rapid weight loss were another topic of analysis for combat athletes [39]. Moreover, not only body composition but also physiology and morphology were compared between male and female Olympic-distance triathletes [40,41]. Another study analyzed the effects of low-intensity aerobic training combined with blood flow restriction on body composition, physical fitness, and vascular responses in recreational runners [42]. An analysis of the vitamin D receptor (VDR), the rs2228570 polymorphism, and its effect on elite athletes' performance was also compared between track and field athletes with non-athletes (controls with a physical activity record) [43].

In tennis, high-intensity interval training effects in athletes with and without cognitive load were analyzed on accuracy, critical flicker fusion threshold, and rating of perceived exertion [44].

In Olympic weightlifting, a comparison of the fatigue prompted by the "Clean and Jerk", and "the Snatch" and their derivative exercises among male and female participants was performed [45].

A case study about the influence of swimming training on an athlete with active Chron's disease, where scarce research exists, was conducted [46]. Finally, another case study analyzed four athletes who participated in a 768 km ultra-trail race for 11 days to address bone turnover alterations [47].

This Special Issue provides relevant information to update the state of the art in this field. It addresses several sports, including young, professional, recreational, male, and female athletes. Moreover, it addresses some gaps in the literature (e.g., Chron's disease, Olympic weightlifting, or velocity speed loss). Notwithstanding, this Special Issue, along with its included studies, contributes information to improve load monitoring (of training and competition) and healthcare through direct or indirect research.

Author Contributions: Conceptualization: R.O. and J.P.B.; writing—original draft: R.O. and J.P.B.; writing—review and editing: R.O. and J.P.B.; project administration: R.O. and J.P.B. All authors have read and agreed to the published version of the manuscript.

Conflicts of Interest: The authors declare no conflict of interest.

References

1. Gabbett, T.J.; Nassis, G.P.; Oetter, E.; Pretorius, J.; Johnston, N.; Medina, D.; Rodas, G.; Myslinski, T.; Howells, D.; Beard, A.; et al. The Athlete Monitoring Cycle: A Practical Guide to Interpreting and Applying Training Monitoring Data. *Br. J. Sports Med.* **2017**, *51*, 1451–1452. [CrossRef]
2. Bourdon, P.C.; Cardinale, M.; Murray, A.; Gastin, P.; Kellmann, M.; Varley, M.C.; Gabbett, T.J.; Coutts, A.J.; Burgess, D.J.; Gregson, W.; et al. Monitoring Athlete Training Loads: Consensus Statement. *Int. J. Sports Physiol. Perform.* **2017**, *12*, 161–170. [CrossRef]
3. Miguel, M.; Oliveira, R.; Loureiro, N.; García-Rubio, J.; Ibáñez, S.J. Load Measures in Training/Match Monitoring in Soccer: A Systematic Review. *Int. J. Environ. Res. Public Health* **2021**, *18*, 2721. [CrossRef]
4. Impellizzeri, F.M.; Marcora, S.M.; Coutts, A.J. Internal and External Training Load: 15 Years on. *Int. J. Sports Physiol. Perform.* **2019**, *14*, 270–273. [CrossRef]
5. Helwig, J.; Diels, J.; Röll, M.; Mahler, H.; Gollhofer, A.; Roecker, K.; Willwacher, S. Relationships between External, Wearable Sensor-Based, and Internal Parameters: A Systematic Review. *Sensors* **2023**, *23*, 827. [CrossRef] [PubMed]
6. Bunn, J.A.; Navalta, J.W.; Fountaine, C.J.; Reece, J.D. Current State of Commercial Wearable Technology in Physical Activity Monitoring 2015–2017. *Int. J. Exerc. Sci.* **2018**, *11*, 503–515.
7. Henriksen, A.; Haugen Mikalsen, M.; Woldaregay, A.Z.; Muzny, M.; Hartvigsen, G.; Hopstock, L.A.; Grimsgaard, S. Using Fitness Trackers and Smartwatches to Measure Physical Activity in Research: Analysis of Consumer Wrist-Worn Wearables. *J. Med. Internet Res.* **2018**, *20*, e110. [CrossRef] [PubMed]
8. Hooper, S.L.; Mackinnon, L.T. Monitoring Overtraining in Athletes: Recommendations. *Sport. Med.* **1995**, *20*, 321–327. [CrossRef]
9. McLean, B.D.; Coutts, A.J.; Kelly, V.; McGuigan, M.R.; Cormack, S.J. Neuromuscular, Endocrine, and Perceptual Fatigue Responses during Different Length between-Match Microcycles in Professional Rugby League Players. *Int. J. Sports Physiol. Perform.* **2010**, *5*, 367–383. [CrossRef]
10. Duignan, C.; Doherty, C.; Caulfield, B.; Blake, C. Single-Item Self-Report Measures of Team-Sport Athlete Wellbeing and Their Relationship with Training Load: A Systematic Review. *J. Athl. Train.* **2020**, *55*, 944–953. [CrossRef] [PubMed]
11. Halson, S.L. Monitoring Training Load to Understand Fatigue in Athletes. *Sport. Med.* **2014**, *44*, 139–147. [CrossRef]
12. Schwellnus, M.; Soligard, T.; Alonso, J.M.; Bahr, R.; Clarsen, B.; Dijkstra, H.P.; Gabbett, T.J.; Gleeson, M.; Hägglund, M.; Hutchinson, M.R.; et al. How Much Is Too Much? (Part 2) International Olympic Committee Consensus Statement on Load in Sport and Risk of Illness. *Br. J. Sports Med.* **2016**, *50*, 1043–1052. [CrossRef]
13. Oliveira, R.; Brito, J.P.; Moreno-Villanueva, A.; Nalha, M.; Rico-González, M.; Clemente, F.M. Reference Values for External and Internal Training Intensity Monitoring in Young Male Soccer Players: A Systematic Review. *Healthcare* **2021**, *9*, 1567. [CrossRef] [PubMed]
14. Nobari, H.; Brito, J.P.; Pérez-Gómez, J.; Oliveira, R. Variability of External Intensity Comparisons between Official and Friendly Soccer Matches in Professional Male Players. *Healthcare* **2021**, *9*, 1708. [CrossRef] [PubMed]
15. Miguel, M.; Oliveira, R.; Brito, J.P.; Loureiro, N.; García-Rubio, J.; Ibáñez, S.J. External Match Load in Amateur Soccer: The Influence of Match Location and Championship Phase. *Healthcare* **2022**, *10*, 594. [CrossRef]
16. Ghollzadeh, R.; Nobari, H.; Bolboli, L.; Siahkouhian, M.; Brito, J.P. Comparison of Measurements of External Load between Professional Soccer Players. *Healthcare* **2022**, *10*, 1116. [CrossRef]
17. Nobari, H.; Ramachandran, A.K.; Brito, J.P.; Oliveira, R. Quantification of Pre-Season and In-Season Training Intensity across an Entire Competitive Season of Asian Professional Soccer Players. *Healthcare* **2022**, *10*, 1367. [CrossRef]
18. Silva, A.F.; González-Fernández, F.T.; Aquino, R.; Akyildiz, Z.; Vieira, L.P.; Yıldız, M.; Birlik, S.; Nobari, H.; Praça, G.; Clemente, F.M. Analyzing the within and between Players Variability of Heart Rate and Locomotor Responses in Small-Sided Soccer Games Performed Repeatedly over a Week. *Healthcare* **2022**, *10*, 1412. [CrossRef] [PubMed]
19. Fernandes, R.; Ibrahim, H.; Clemente, F.M.; Martins, A.D.; Nobari, H.; Reis, V.M.; Oliveira, R. In-Season Microcycle Quantification of Professional Women Soccer Players—External, Internal and Wellness Measures. *Healthcare* **2022**, *10*, 695. [CrossRef]
20. García-Ceberino, J.M.; Gamero, M.G.; Ibáñez, S.J.; Feu, S. Are Subjective Intensities Indicators of Player Load and Heart Rate in Physical Education? *Healthcare* **2022**, *10*, 428. [CrossRef]
21. Irurtia, A.; Torres-Mestre, V.M.; Cebrián-Ponce, Á.; Carrasco-Marginet, M.; Altarriba-Bartés, A.; Vives-Usón, M.; Cos, F.; Castizo-Olier, J. Physical Fitness and Performance in Talented & Untalented Young Chinese Soccer Players. *Healthcare* **2022**, *10*, 98. [CrossRef] [PubMed]
22. Mijatovic, D.; Krivokapic, D.; Versic, S.; Dimitric, G.; Zenic, N. Change of Direction Speed and Reactive Agility in Prediction of Injury in Football; Prospective Analysis over One Half-Season. *Healthcare* **2022**, *10*, 440. [CrossRef] [PubMed]

23. Mohammadi Nia Samakosh, H.; Brito, J.P.; Shojaedin, S.S.; Hadadnezhad, M.; Oliveira, R. What Does Provide Better Effects on Balance, Strength, and Lower Extremity Muscle Function in Professional Male Soccer Players with Chronic Ankle Instability? Hopping or a Balance Plus Strength Intervention? A Randomized Control Study. *Healthcare* **2022**, *10*, 1822. [CrossRef] [PubMed]

24. de Oliveira-Sousa, S.L.; León-Garzón, M.C.; Gacto-Sánchez, M.; Ibáñez-Vera, A.J.; Espejo-Antúnez, L.; León-Morillas, F. Does Inspiratory Muscle Training Affect Static Balance in Soccer Players? A Pilot Randomized Controlled Clinical Trial. *Healthcare* **2023**, *11*, 262. [CrossRef]

25. Liao, C.-N.; Fan, C.-H.; Hsu, W.-H.; Chang, C.-F.; Yu, P.-A.; Kuo, L.-T.; Lu, B.-L.; Hsu, R.W.-W. Twelve-Week Lower Trapezius-Centred Muscular Training Regimen in University Archers. *Healthcare* **2022**, *10*, 171. [CrossRef] [PubMed]

26. Yapici, H.; Soylu, Y.; Gulu, M.; Kutlu, M.; Ayan, S.; Muluk, N.B.; Aldhahi, M.I.; AL-Mhanna, S.B. Agility Skills, Speed, Balance and CMJ Performance in Soccer: A Comparison of Players with and without a Hearing Impairment. *Healthcare* **2023**, *11*, 247. [CrossRef]

27. Espada, M.C.; Jardim, M.; Assunção, R.; Estaca, A.; Ferreira, C.C.; Pessôa Filho, D.M.; Verardi, C.E.L.; Gamonales, J.M.; Santos, F.J. Lower Limb Unilateral and Bilateral Strength Asymmetry in High-Level Male Senior and Professional Football Players. *Healthcare* **2023**, *11*, 1579. [CrossRef]

28. Rebelo, A.; Pereira, J.R.; Martinho, D.V.; Duarte, J.P.; Coelho-e-Silva, M.J.; Valente-dos-Santos, J. How to Improve the Reactive Strength Index among Male Athletes? A Systematic Review with Meta-Analysis. *Healthcare* **2022**, *10*, 593. [CrossRef]

29. Widodo, A.F.; Tien, C.-W.; Chen, C.-W.; Lai, S.-C. Isotonic and Isometric Exercise Interventions Improve the Hamstring Muscles’ Strength and Flexibility: A Narrative Review. *Healthcare* **2022**, *10*, 811. [CrossRef]

30. Zhang, X.; Feng, S.; Li, H. The Effect of Velocity Loss on Strength Development and Related Training Efficiency: A Dose–Response Meta–Analysis. *Healthcare* **2023**, *11*, 337. [CrossRef]

31. Chen, Y.; Kim, Y.; Choi, M. Effects of Aquatic Training and Bicycling Training on Leg Function and Range of Motion in Amateur Athletes with Meniscal Allograft Transplantation during Intermediate-Stage Rehabilitation. *Healthcare* **2022**, *10*, 1090. [CrossRef]

32. Zafra, A.O.; Martins, B.; Ponseti-Verdaguer, F.J.; Ruiz-Barquín, R.; García-Mas, A. It Is Not Just Stress: A Bayesian Approach to the Shape of the Negative Psychological Features Associated with Sport Injuries. *Healthcare* **2022**, *10*, 236. [CrossRef]

33. Sánchez-Sánchez, L.C.; Franco, C.; Amutio, A.; García-Silva, J.; González-Hernández, J. Influence of Mindfulness on Levels of Impulsiveness, Moods and Pre-Competition Anxiety in Athletes of Different Sports. *Healthcare* **2023**, *11*, 898. [CrossRef] [PubMed]

34. Hajji, J.; Sabah, A.; Aljaberi, M.A.; Lin, C.-Y.; Huang, L.-Y. The Effect of Ramadan Fasting on the Coping Strategies Used by Male Footballers Affiliated with the Tunisian First Professional League. *Healthcare* **2023**, *11*, 1053. [CrossRef] [PubMed]

35. Makarowski, R.; Predoiu, R.; Piotrowski, A.; Görner, K.; Predoiu, A.; Oliveira, R.; Pelin, R.A.; Moanță, A.D.; Boe, O.; Rawat, S.; et al. Coping Strategies and Perceiving Stress among Athletes during Different Waves of the COVID-19 Pandemic–Data from Poland, Romania, and Slovakia. *Healthcare* **2022**, *10*, 1770. [CrossRef]

36. Rawat, S.; Deshpande, A.P.; Predoiu, R.; Piotrowski, A.; Malinauskas, R.; Predoiu, A.; Vazne, Z.; Oliveira, R.; Makarowski, R.; Görner, K.; et al. The Personality and Resilience of Competitive Athletes as BMW Drivers–Data from India, Latvia, Lithuania, Poland, Romania, Slovakia, and Spain. *Healthcare* **2023**, *11*, 811. [CrossRef]

37. Pinto, G.; Militello, R.; Amoresano, A.; Modesti, P.A.; Modesti, A.; Luti, S. Relationships between Sex and Adaptation to Physical Exercise in Young Athletes: A Pilot Study. *Healthcare* **2022**, *10*, 358. [CrossRef] [PubMed]

38. Luti, S.; Militello, R.; Pinto, G.; Illiano, A.; Amoresano, A.; Chiappetta, G.; Marzocchini, R.; Modesti, P.A.; Pratesi, S.; Pazzagli, L.; et al. Chronic Training Induces Metabolic and Proteomic Response in Male and Female Basketball Players: Salivary Modifications during In-Season Training Programs. *Healthcare* **2023**, *11*, 241. [CrossRef]

39. Baranauskas, M.; Kupčiūnaitė, I.; Stukas, R. The Association between Rapid Weight Loss and Body Composition in Elite Combat Sports Athletes. *Healthcare* **2022**, *10*, 665. [CrossRef] [PubMed]

40. Puccinelli, P.J.; de Lira, C.A.B.; Vancini, R.L.; Nikolaidis, P.T.; Knechtle, B.; Rosemann, T.; Andrade, M.S. The Performance, Physiology and Morphology of Female and Male Olympic-Distance Triathletes. *Healthcare* **2022**, *10*, 797. [CrossRef]

41. Barbosa, J.G.; de Lira, C.A.B.; Vancini, R.L.; dos Anjos, V.R.; Vivan, L.; Seffrin, A.; Forte, P.; Weiss, K.; Knechtle, B.; Andrade, M.S. Physiological Features of Olympic-Distance Amateur Triathletes, as Well as Their Associations with Performance in Women and Men: A Cross–Sectional Study. *Healthcare* **2023**, *11*, 622. [CrossRef] [PubMed]

42. Beak, H.J.; Park, W.; Yang, J.H.; Kim, J. Effect of Low-Intensity Aerobic Training Combined with Blood Flow Restriction on Body Composition, Physical Fitness, and Vascular Responses in Recreational Runners. *Healthcare* **2022**, *10*, 1789. [CrossRef]

43. Bulgay, C.; Bayraktar, I.; Kazan, H.H.; Yıldırım, D.S.; Zorba, E.; Akman, O.; Ergun, M.A.; Cerit, M.; Ulucan, K.; Eken, Ö.; et al. Evaluation of the Association of VDR Rs2228570 Polymorphism with Elite Track and Field Athletes’ Competitive Performance. *Healthcare* **2023**, *11*, 681. [CrossRef] [PubMed]

44. Clemente-Suárez, V.J.; Villafaina, S.; García-Calvo, T.; Fuentes-García, J.P. Impact of HIIT Sessions with and without Cognitive Load on Cortical Arousal, Accuracy and Perceived Exertion in Amateur Tennis Players. *Healthcare* **2022**, *10*, 767. [CrossRef] [PubMed]

45. Antunes, J.P.; Oliveira, R.; Reis, V.M.; Romero, F.; Moutão, J.; Brito, J.P. Comparison between Olympic Weightlifting Lifts and Derivatives for External Load and Fatigue Monitoring. *Healthcare* **2022**, *10*, 2499. [CrossRef] [PubMed]

46. Papadimitriou, K. The Influence of Aerobic Type Exercise on Active Crohn’s Disease Patients: The Incidence of an Elite Athlete. *Healthcare* **2022**, *10*, 713. [CrossRef] [PubMed]
47. Castellar-Otín, C.; Lecina, M.; Pradas, F. Bone Turnover Alterations after Completing a Multistage Ultra-Trail: A Case Study. *Healthcare* **2022**, *10*, 798. [CrossRef]

Systematic Review

Reference Values for External and Internal Training Intensity Monitoring in Young Male Soccer Players: A Systematic Review

Rafael Oliveira [1,2,3,*], João Paulo Brito [1,2,3], Adrián Moreno-Villanueva [4], Matilde Nalha [1], Markel Rico-González [5] and Filipe Manuel Clemente [6,7]

1   Sports Science School of Rio Maior, Polytechnic Institute of Santarém, 2140-413 Rio Maior, Portugal; jbrito@esdrm.ipsantarem.pt (J.P.B.); matildenalha@gmail.com (M.N.)
2   Research Center in Sport Sciences, Health Sciences and Human Development, 5001-801 Vila Real, Portugal
3   Life Quality Research Centre, 2140-413 Rio Maior, Portugal
4   Department of Physical Activity and Sport Sciences, International Excellence Campus "Mare Nostrum", Faculty of Sports Sciences, University of Murcia, 30720 San Javier, Spain; more_adri@hotmail.com
5   Department of Didactics of Musical, Plastic and Corporal Expression, University of the Basque Country, UPV-EHU, 48940 Leioa, Spain; markeluniv@gmail.com
6   Escola Superior Desporto e Lazer, Instituto Politécnico de Viana do Castelo, Rua Escola Industrial e Comercial de Nun'Álvares, 4900-347 Viana do Castelo, Portugal; filipe.clemente5@gmail.com
7   Instituto de Telecomunicações, Delegação da Covilhã, 1049-001 Lisboa, Portugal
*   Correspondence: rafaeloliveira@esdrm.ipsantarem.pt

Abstract: Training intensity monitoring is a daily practice in soccer which allows soccer academies to assess the efficacy of its developmental interventions and management strategies. The current systematic review's purpose is to: (1) identify and summarize studies that have examined external and internal training intensity monitoring, and to (2) provide references values for the main measures for young male soccer players. A systematic review of EBSCO, PubMed, Scielo, Scopus, SPORTDiscus, and Web of Science databases was performed according to the Preferred Reporting Items for Systematic Reviews and Meta-Analyses (PRISMA) guidelines. From the 2404 studies initially identified, 8 were fully reviewed, and their outcome measures were extracted and analyzed. From them, the following range intervals were found for training: rated perceived exertion (RPE) 2.3–6.3 au; session-RPE, 156–394 au; total distance, 3964.5–6500 m and; distance >18 km/h, 11.8–250 m. Additionally, a general tendency to decrease the intensity in the day before the match was Found. This study allowed to provide reference values of professional young male players for the main internal and external measures. All together, they can be used by coaches, their staff, or practitioners in order to better adjust training intensity.

Keywords: football; male; training; youth; RPE; GPS; running; high-speed running; sprint

Citation: Oliveira, R.; Brito, J.P.; Moreno-Villanueva, A.; Nalha, M.; Rico-González, M.; Clemente, F.M. Reference Values for External and Internal Training Intensity Monitoring in Young Male Soccer Players: A Systematic Review. *Healthcare* **2021**, *9*, 1567. https://doi.org/10.3390/healthcare9111567

Academic Editor: Parisi Attilio

Received: 20 October 2021
Accepted: 11 November 2021
Published: 17 November 2021

Publisher's Note: MDPI stays neutral with regard to jurisdictional claims in published maps and institutional affiliations.

1. Introduction

Training intensity monitoring is a daily practice within the soccer training context [1]. Training intensity describes how hard a player is exercising [2]. Understanding the impact of training stimulus on the players and identify the variations between players contributes to an adjustment of planning and for adjusting specific recovery or managing strategies [3]. For that reason, coaches expect that training intensity monitoring guides them for planning better for an improvement in performance and reduction in injury risk [1]. Quantifying aspects of young players development and performance (e.g., physical abilities, physiological capacities, technical and tactical skill) allows soccer academies to assess the efficacy of its developmental interventions and management strategies [4]. By collecting and analyzing regular inputs related to player development and performance, organizations create a feedback loop in which their planning and training interventions can be objectively assessed.

The monitoring is commonly organized in two dimensions [5]: (i) internal intensity; and (ii) external intensity. The internal intensity represents the psychophysiological impact of external intensity on the player's organism while the external intensity represents the mechanical and physical impact of training drills on the players [6]. Different instruments can be used to assess these dimensions, however in soccer, the most common instruments for measuring internal intensity are heart rate monitors and rated perceived exertion (RPE) scale, while for measuring the external intensity the most common are microelectromechanical systems (e.g., global positioning system, local positioning system, inertial measurement units) [7].

From such instruments, the typical variables extracted are heart rate-bases scores, RPE-based scores and distances covered at different speed thresholds, or changes of velocity (accelerations and decelerations). The measures are daily quantified in training and match scenarios. Using absolute intensity per session and per week, the sports scientist has some possibilities of understanding the within- and between-weeks variations [7].

The analysis of intensity impact on players is highly relevant. One of the reasons is justified by the gap between coaches perception and the actual intensity imposed on the players [8]. Based on that, employing intensity monitoring in youth teams became a typical strategy used by coaches [9]. Thus, a growing number of researches have been published in training monitoring in youth soccer players [10–12].

Although the increase in intensity monitoring reports in young soccer, most of the articles are from the same team. This reduces the possibility of generalization of the evidence. However, is also known that is not expectable to reach different teams at the same time over long periods of the season. Thus, one of the strategies can be to summarize the available evidence about intensity values of young soccer players by conducting a systematic review. Although a possible idea, no systematic review was found about that. Although there is pertinence of looking into both males and females, summarizing both within the same article would not be useful since the huge differences related with a number of publications and a logical difference between sexes produces a different approach to the consequences for the article logic. Thus, a more focused contribution as a systematic review may help coaches to compare values of training intensity in young male soccer and provide some benchmarks. Thus, the aim of this systematic review is to identify and summarize studies that have examined external and internal training monitoring and to provide references values for the main measures for young male soccer players.

2. Materials and Methods

The preferred reporting items for systematic reviews and meta-analyses (PRISMA) guidelines were followed to write this systematic review [13] and guidelines for performing systematic reviews in sport sciences [14]. The protocol of the systematic review was a priori registered in the International Platform of Registered Systematic Review and Meta Analysis Protocols with the number INPLASY202180055 and the DOI number 10.37766/inplasy2021.8.0055.

2.1. Eligibility Criteria

The inclusion and exclusion criteria can be found in Table 1.

The screening process related to analysis of the title, abstract and reference list of each study to locate potentially relevant studies was independently executed by two of the authors (A.M.V. and M.R.G.). Moreover, both authors also reviewed the full version of the included papers in detail to identify which article met the inclusion criteria. Additionally, a search within the reference lists of the included records was performed to add additional relevant studies. In the cases of discrepancies, a discussion was performed with the participation of a third author (R.O.). Possible errata for the included articles were considered.

Table 1. Eligibility criteria.

PICOS	Inclusion Criteria	Exclusion Criteria
Population	Healthy young (under eighteen) male soccer players.	Studies conducted with professional or amateur players or in other sports, or with female populations.
Intervention/Exposure	Exposure to entire training sessions for number of weeks and sessions included (minimum one week).	No exposure to entire training sessions (e.g., specific exercises only reported; only matches; only simulated matches).
Comparator	Not required. Eventually, comparisons between playing positions and/or competitive levels within the same age-group and/or age-groups.	No study will be excluded based on comparators.
Outcomes	Presents at least of one measure among the included in internal (e.g., heart rate, rate of perceived exertion) and/or external intensity (e.g., distances covered at different speed thresholds, acceleration-based measures) in absolute values. Whenever relative values allow to calculate absolute values, the study will be included.	Absence of data characterizing the intensity during the training sessions (e.g., wellness variables, readiness parameters) and or only reports the data in relative values without allowing the calculation of absolute values. Data from workload calculations will also be excluded (e.g., accumulated weekly intensity, training monotony, strain, ACWR, EWMA).
Study design	No restrictions imposed on study design.	No study was excluded on the basis of study design.
Others	Only original and full-text studies written in English.	Written in other language than English. Other article types than original (e.g., reviews, letters to editors, trial registrations, proposals for protocols, editorials, book chapters and conference abstracts).

PICOS: (P) population; (I) intervention/exposure; (C) comparator; (O) outcomes; (S) study design.

2.2. Information Sources

The following electronic databases were used to search for relevant publication on the 3 July 2021, after protocol registration: FECYT (MEDLINE, Scielo, and Web of Science), PubMed, and Scopus. A manual search was also conducted after search in electronic databases to retrieve additional studies that could fit our eligibility criteria.

2.3. Search Strategy

Keywords and synonyms were entered in various combinations in the title, abstract or keywords which means that the following research content was applied: ("soccer" OR "football") AND ("internal load" OR "external load" OR "workload" OR "training load" OR "load monitoring"). Search results were exported to EndNote 20.0.1 for Mac (Clarivate Analytics). No filters or limits were applied.

2.4. Data Extraction

A specific spreadsheet was designed in Microsoft Excel (Microsoft Corporation, Readmon, WA, USA) for process the data extraction. The design followed the recommendations of the Cochrane Consumers and Communication Review Group's data extraction template [15]. In this spreadsheet the information about inclusion and exclusion requirements and reasons was detailed. The selection of the articles was made independently by two authors (A.M.-V. and M.R.-G.). In the cases of discrepancies, a discussion was performed with the participation of a third author (R.O.).

2.5. Methodological Assessment

The methodological quality was assessed using the methodological index for non-randomized studies (MINORS) by two authors (A.M.-V. and M.R.-G.) [16]. MINORS consists of twelve items, four of which are only applicable to comparative studies which was not the case of the included studies. Thus, only eight items were applied. Each item is rated as 0 when the criterion is not reported in the article, 1 if reported but not sufficiently fulfilled, or 2 when adequately met. Higher scores indicate good methodological quality of the article and a low risk of bias. The highest possible score is 16 for non-comparative studies. MINORS has yielded acceptable inter- and intra-rater reliability, internal consistency, content validity, and discriminative validity [16,17].

3. Results

A total of 2404 (i.e., FECYT: 1481; PubMed: 806; Scopus: 117) original articles were initially retrieved from the mentioned databases, of which 834 were duplicates. Thus, a total of 1570 original articles were found. After this, a total of 1558 articles checked by title and abstract were excluded. The remaining 12 articles were checked in full text, leading to the exclusion of 3 articles according to criterion n° 1, and 1 according to criterion n° 2. Additionally, 1 article was included from additional sources. A total of 8 articles met all the inclusion criteria and were finally included in the qualitative synthesis. All the steps followed for the selection of the articles are available in Figure 1.

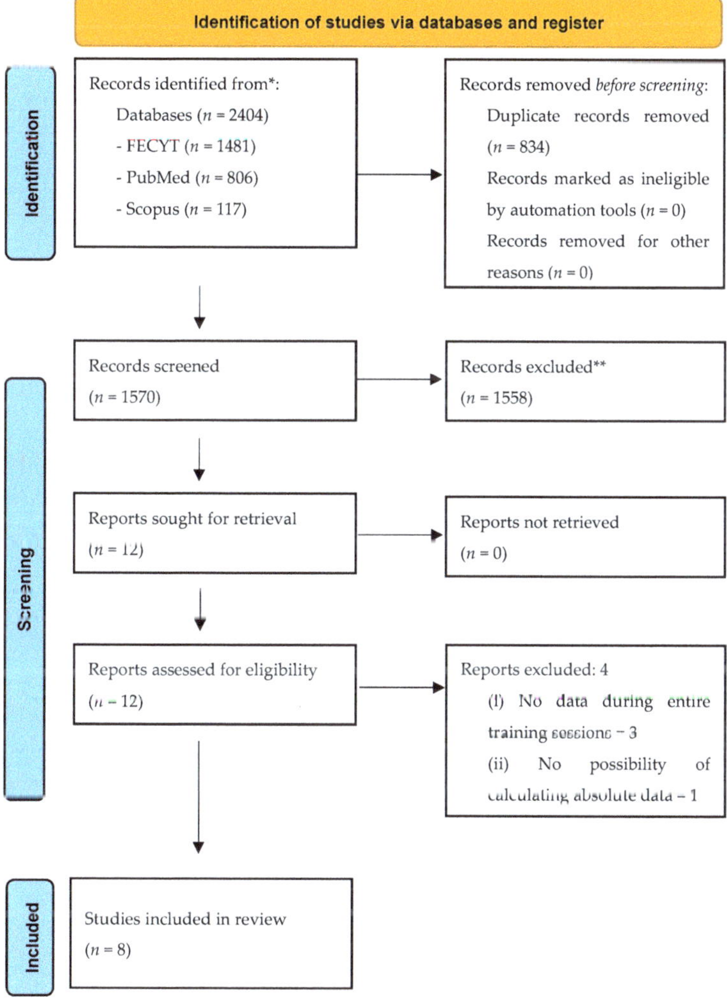

Figure 1. Preferred reporting item for systematic reviews (PRISMA) flow diagram.

3.1. Methodological Quality

The overall methodological quality of the cross-sectional studies can be found in Table 2.

Table 2. Methodological assessment using MINORS checklist.

Study	1	2	3	4	5	6	7	8	Score
[18]	2	2	1	1	1	2	2	1	12/16
[19]	2	2	2	2	0	1	2	1	12/16
[20]	2	2	1	1	1	2	2	2	13/16
[21]	2	2	2	2	1	2	2	2	15/16
[22]	2	2	2	2	1	1	2	2	14/16
[23]	2	2	1	2	0	2	2	1	12/16
[24]	2	2	2	2	1	2	2	2	15/16
[25]	2	2	2	1	0	1	2	1	11/16

Note: The MINORS checklist asks the following information (2 = High quality; 1 = Medium quality; 0 = Low quality): 1. Clearly defined objective. 2. Inclusion of patients consecutively. 3. Information collected retrospectively. 4. Assessments adjusted to objective. 5. Evaluations carried out in a neutral way. 6. Follow-up phase consistent with the objective. 7. Dropout rate during follow-up less than 5%. 8. Appropriate statistical analysis.

3.2. Results of the Studies

Table 3 presents the characteristics of the studies regarding their sample size, their age of the sample, their duration, as well as training duration, internal and external measures, and instruments used. From the eight studies included, the study of Hannon et al. was the only to analyze the categories of under12 (U12), U13 [21]. Two studies analyzed U14 category [21,25]. Four studies analyzed U15 category [18,20,21,24]. Two studies analyzed the category of U16 [21,25]. Three studies analyzed the category of U17 [18,23,24]. Two studies analyzed the category of U18 [21,25]. Finally, three studies analyzed the category of U19 [18,22,24].

Table 3. Study characteristics.

Study	Sample	Age	Study and Training Duration	Internal Measures/ Instruments	External Measures/ Instruments
[18]	U15 N:56 U17 N:66 U19 N:29	14.0 ± 0.2 15.8 ± 0.4 17.8 ± 0.6	9 weeks Training duration: 90 min	1 Hz Polar Team System, Polar, FI: HR < 75%; HR 75–84.9%; HR 85% to 89.9%; HR $\geq$ 90%	15 Hz GPS (SPI-Pro X II, GPSports, Canberra, Australia): Total distance Distance 0–6.9 km/h Distance 7.0–9.9 km/h Distance 10.0–12.9 km/h Distance 13–15.9 km/h Distance 16–17.9 km/h Distance $\geq$ 18.0 km/h Distance $\geq$ 18.0 km/h NR Impacts NR
[19]	U17 N:18	15.7 ± 0.5	38 training sessions Training duration ND	1 HZ Polar Team Pro tracking system: HR	10 Hz Polar Team Pro tracking System: Total distance, Distance 11–14.9 km/h Distance 15.0–18.9 km/h Distance $\geq$ 19.00 km/h Acceleration $\geq$ 2.0 m.s^{-2} NR
[20]	U15 N:22	14.5 ± 0.3	4 weeks Training duration ND	s-RPE (CR-10)	-
[21]	U12 N:15 U13 N:13 U14 N:12 U15 N:10 U16 N:11 U18 N:15	11.7 ± 0.2 12.6 ± 0.3 13.7 ± 0.2 14.5 ± 0.3 15.5 ± 0.2 17.0 ± 0.4	1 in-season Training duration: described in Table 5	-	10 Hz GPS (Apex, STATSports, Northern Ireland): Total distance Distance 19.8–25.2 km/h Distance > 25 km/h

Table 3. *Cont.*

Study	Sample	Age	Study and Training Duration	Internal Measures/ Instruments	External Measures/ Instruments
[22]	U19 N:9	17.6 ± 0.6	1 in-season Training duration 79 min	-	10 Hz GPS OptimEye X4 (Catapult Sports, Melbourne, Australia): Total distance Distance 12–15 km/h Distance 15–20 km/h Distance 20–25 km/h Distance > 25 km/h
[23]	U17 N: 21	16.1 ± 0.2	36 weeks Training duration ND	s-RPE (CR-10)	-
[24]	U15 N:20 U17 N:20 U19 N:20	13.2 ± 0.5 15.4 ± 0.5 17.39 ± 0.55	2 Weeks Training duration 90 min	RPE (CR-20) s-RPE (CR-20)	18 Hz GPS STATSports Apex® (Newry, Northern Ireland): Total distance Maximal speed HMLD (>25.5 $W \cdot kg^{-1}$) Distance > 25 km/h Distance > 25 km/h NR Accelerations ≥ 3 m.s^{-2} Decelerations ≤ -3 m.s ^{2}
[25]	U14 N:8 U16 N:8 U18 N:8	13 ± 1 15 ± 1 17 ± 1	2 Weeks Field training duration: U14—90 min U16—102 min U18—104 min Gym training duration: 30–35 min	5 Hz Polar Team System®, Kempele, Finland): HR s-RPE (CR10)	-

U: under; min: minutes; HR: heart rate; NR: number; ND: non described; HMLD: high metabolic load distance; RPE: rated perceived exertion; s-RPE: session rated of perceived exertion.

In addition, three studies analyzed both internal and external measures [18,19,24], three studies only analyzed internal measures [20,23,25] and two studies only analyzed external measures [21,22].

Table 4 presents the results for external measures in which four studies were included [18,19,22,24]. The table was organized according to the age categories.

Only one study presented data for starters and non-starters. Since non-starter presented lower values than starters, we provided a range interval with lower values related to non-starters and higher values related to starters [19].

The last line of Table 4 presents the range intervals for the external measures most used.

Table 5 presents the results for external intensity by match-day minus (MD-) and by players positions (last two lines). The approach of MD- was used by two studies [21,24], while player positions was analyzed by one study [24]. The last line of Table 5, before player positions data, presents the range intervals according to the MD- approach for the measures most used.

Table 6 presents results for internal intensity in which one study included 1-match week and 2 matches-week analysis [20], one study analyzed player positions [23], one study analyzed player status (starters vs. non-starters) [19], three studies analyzed the average overall team [24–26] and two studies used MD approach [23,24]. The session-rated of perceived exertion (s-RPE) was the variable most used to quantify internal training intensity. Finally, one study presented internal intensity for gym training [25]. The main results showed a range interval range of 2.3 to 6.3 arbitrary units (au) for RPE between U14, U15, U16, and U18 [20,25] and 156–394 au for s-RPE between U15 and U17 [20,23].

Table 4. Results for external training by overall team.

Study	Measures	U12	U13	U14	U15	U16	U17	U18	U19
[18]	Total distance (m)	-	-	-	3964.5 ± 725.4	-	4648.3 ± 831.9	-	4212.5 ± 935.4
	Distance $\geq$ 18.0 km/h (NR)				10.9 ± 6.3		16.4 ± 8.2		11.8 ± 7.9
	Distance $\geq$ 18.0 km/h (m)				12.1 ± 4.9		13.0 ± 5.3		11.8 ± 6.7
	Impacts (NR)				490.8 ± 309.5		584.0 ± 363.5		613.1 ± 329.4
[19]	Distance 15.0–18.9 km/h (m)	-	-	-	-	-	~380–400	-	-
	Distance $\geq$ 19.00 km/h (m)						~180–200		
	Acceleration $\geq$ 2.0 m.s^{-2} (NR)						~65–75		
[21]	Duration (min)	~83–90	~85–100	~83–100	~85–110	~87–117	-	~70–115	-
	Total Distance (m)	~4200–5200	~4600–5100	~4500–5100	~4800–6000	~4800–6100		~3800–6500	
	Distance 19.8–25.2 km/h (m)	~20–40	~10–40	~40–120	~60–180	~80–250		~60–210	
	Distance > 25 km/h (m)	~0–2	~0–1	~0–2	~0–23	~2–30		~2–25	
[22]	Total distance (m)	-	-	-	-	-	-	-	~5611.3
	Distance 12–15 km/h (m)								~541.9
	Distance 15–20 km/h (m)								~427.8
	Distance 20–25 km/h (m)								~149.7
	Distance > 25 km/h (m)								~28.5
[24]	Total Distance (m)	-	-	-	5316.2 ± 1354.5	-	6021.4 ± 1675.6	-	4750.4 ± 1593.5
	HMLD (>25.5 W·kg^{-1}) (m)				489.1 ± 228.4		730.6 ± 483.4		524.9 ± 291.4
	Distance > 25 km/h (m)				28.1 ± 41.7		130.4 ± 462.6		40.2 ± 50.4
	Distance > 25 km/h (NR)				1.9 ± 2.5		4.8 ± 4.8		3.1 ± 2.9
	Acceleration $\geq$ 3 m.s^{-2} (NR)				33.6 ± 18.8		53.8 ± 20.6		49.9 ± 20.2
	Deceleration $\leq$ −3 m.s^{-2} (NR)				30.3 ± 19.8		49.8 ± 25.1		44.0 ± 22.5
Reference values	**Duration (min)**				**All teams**				
	Total distance (m)				**79–117**				
	Distance > 18 km/h (m)				**3964.5–6500**				
	Distance > 25 km/h (m)				**11.8–250**				
					0–30				

U: under; min: minutes; NR: number; HMLD: high metabolic load distance.

Table 5. Results for external training intensity by match-day minus and by players positions.

Study	Team and Measures	MD-5	MD-4	MD-3	MD-2	MD-1
	U12					-
	Duration (min)	~85	~83	~88	~90	
[21]	Total Distance (m)	~4800	~5200	~4400	~4200	
	Distance 19.8–25.2 km/h (m)	~30	~40	~30	~20	
	Distance > 25 km/h (m)	~0	~0	~2	~1	
	U13					-
	Duration (min)	~100	~85	~85	~87	
[21]	Total Distance (m)	~5100	~4900	~4700	~4600	
	Distance 19.8–25.2 km/h (m)	~10	~40	~20	~30	
	Distance > 25 km/h (m)	~0	~0	~0	~1	
	U14					-
	Duration (min)	~100	~85	~83	~90	
[21]	Total Distance (m)	~5100	~5100	~4500	~4800	
	Distance 19.8–25.2 km/h (m)	~60	~120	~40	~60	
	Distance > 25 km/h (m)	~1	~2	~0	~0	
	U15			-		
	Duration (min)	~98	~110		~90	~85
[21]	Total Distance (m)	~5100	~6000		~5100	~4800
	Distance 19.8–25.2 km/h (m)	~60	~180		~110	~90
	Distance > 25 km/h (m)	~0	~23		~3	~1
	U16			-		
	Duration (min)	~93	~117		~90	~87
[21]	Total Distance (m)	~5000	~6100		~4900	~4800
	Distance 19.8–25.2 km/h (m)	~80	~250		~120	~80
	Distance > 25 km/h (m)	~2	~30		~10	~2
	U18			-		
	Duration (min)	~90	~115		~100	~70
[21]	Total Distance (m)	~4900	~6500		~6000	~3800
	Distance 19.8–25.2 km/h (m)	~105	~210		~200	~60
	Distance > 25 km/h (m)	~3	~25		~23	~2

Table 5. *Cont.*

Study	Team and Measures	MD-5	MD-4	MD-3	MD-2	MD-1
	U15–U17–U19	-	-			
	Total Distance (m)			5372.0 ± 1452.1	5796.0 ± 1773.3	4728.0 ± 1618.6
	HMLD (m)			591.2 ± 284.9	568.2 ± 287.7	488.8 ± 293.6
[24]	Distance > 25 km/h (m)			39.7 ± 49.1	40.4 ± 51.1	58.1 ± 76.5
	Distance > 25 km/h (NR)			3.1 ± 2.9	2.9 ± 3.7	3.8 ± 4.7
	Accelerations ≥ 3 m.s^{-2} (NR)			48.9 ± 22.8	43.6 ± 20.5	43.2 ± 19.9
	Decelerations ≤ -3 m.s^{-2} (NR)			46.0 ± 25.6	40.3 ± 20.8	34.4 ± 21.8
Reference values	**Total distance (m)**	**4800–5100**	**4900–6500**	**4400–5372**	**4200–6000**	**3800–4800**
	Distance > 18 km/h (m)	**10–105**	**40–250**	**20–40**	**20–200**	**60–90**
	Distance > 25 km/h (m)	**0–3**	**0–30**	**0–39.7**	**0–40**	**1–58**

Study	Team and Measures	Central Defender	Wide Defender	Central Midfielder	Wide Midfielder	Forward
	U15–U17–U19					
	Total Distance (m)	5282.3 ± 1407.5	5275.9 ± 1774.6	5456.9 ± 1565.9	5370.1 ± 1692.6	5156.9 ± 1820.9
	HMLD (m)	541.3 ± 243.7	548.5 ± 282.1	602.2 ± 275.4	562.2 ± 275.4	529.5 ± 360.6
[24]	Distance > 25 km/h (m)	44.3 ± 56.9	51.2 ± 26.6	49.1 ± 57.1	38.8 ± 48.9	56.6 ± 77.8
	Distance > 25 km/h (NR)	3.2 ± 3.3	2.7 ± 3.1	3.4 ± 3.6	3.1 ± 3.4	4.1 ± 5.1
	Accelerations ≥ 3 m.s^{-2} (NR)	44.6 ± 19.4	45.6 ± 20.0	47.8 ± 21.9	46.6 ± 24.0	45.3 ± 23.8
	Decelerations ≤ -3 m.s^{-2} (NR)	39.6 ± 18.7	40.2 ± 19.5	43.3 ± 22.2	41.2 ± 26.0	43.8 ± 34.4

U: under; min: minutes; HR: heart rate; NR: number; HMLD: high metabolic load distance; MD-: match-day minus.

Table 6. Results for internal training intensity in one or two matches per week, by match-day minus and by player positions.

Study	Measures	Positions/Status/Team	Training with 1-Match Week			Training with 2 Matches-Week		
			Session 1	Session 2	Session 3	Session 1	Session 2	Session 3
[20]	RPE (CR-10, au)	Overall U15	3.9 ± 0.6	3.5 ± 0.9	3.5 ± 0.6	3.4 ± 0.6	3.3 ± 0.9	2.3 ± 0.5
	s-RPE (CR-10, au)	Overall U15	157.1 ± 42.2	275.7 ± 62.0	283.0 ± 47.8	313.2 ± 43.9	293.9 ± 71.1	167.2 ± 26.4

Table 6. *Cont.*

Study	Measures	Positions/Status/Team	Training with 1-Match Week			Training with 2 Matches-Week	
			U17				
			MD-5	MD-4	MD-3	MD-2	MD-1
[23]	s-RPE (CR-10, au)	Wide defenders	~240	~360	~385	-	~150
		Central defenders	~245	~348	~390	-	~160
		Central midfielders	~243	~375	~400	-	~158
		Wide midfielders	~242	~380	~400	-	~158
		Forwards/strikers	~215	~355	~395	-	~152
		Overall team	~237	~364	~394	-	~156
[24]	RPE (CR-20, au)	Overall U15-U17-U19	-	-	13.3 ± 2.4	12.5 ± 1.7	13.3 ± 2.3
	s-RPE (CR-20, au)	Overall U15-U17-U19	-	-	1196.1 ± 211.2	1158.1 ± 221.2	1194.4 ± 205.2

Study			Central Defender	Wide Defender	Central Midfielder	Wide Midfielder	Forward
[23]	s-RPE (CR-10, au)	Overall U17	~160–390	~150–385	~158–400	~158–400	~152–395
[24]	RPE (CR-20, au)	Overall U15-U17-U19	44.6	44.6	47.7	46.6	45.3
	s-RPE (CR-20, au)	Overall U15-U17-U19	261.2	230.5	265.3	238.1	255.9

			Player status				
[19]	Banister TRIMP (au)	U17 Starters	~105	-	-	-	-
		U17 Non-starters	~110	-	-	-	-

			Average training per overall team				
[24]	RPE (CR-20, au)	Overall U15	13.7 ± 1.9	U17	15.5 ± 1.8	U19	12.5 ± 2.5
	s-RPE (CR-20, au)	Overall U15	1235.3 ± 171.9	U17	1215.5 ± 158.7	U19	1120.2 ± 224.7
[25]	RPE (CR-10, au)	U18 Gym training	5.5 ± 0.3	U18 Field Training	6.6 ± 0.6	-	-
	RPE (CR-10, au)	U16 Gym training	5.8 ± 0.4	U16 Field Training	6.3 ± 0.4	-	-
	RPE (CR-10, au)	U14 Gym training	6.3 ± 0.4	U14 Field Training	6.2 ± 0.2	-	-

U: uncer; min: minutes; MD-: match-day minus; au: arbitrary units; RPE: rated perceived exertion; s-RPE: session rated perceived exertion.

4. Discussion

The aim of this systematic review was to identify and summarize studies that have examined external and internal training intensity monitoring in young male soccer players and to provide references values for the main training intensity measures. The main results showed the following range intervals by overall teams that include (U12 to U19):

training duration of 79–117 min [18,19,21,22,24];
total distance of 364.5–6500 m [18,19,21,22,24];
distance > 18 km/h of 11.8–250 m [18,19,21,22];
distance > 25 km/h of 0–30 m [18,19,21,22,24].

The importance of prescribing appropriate training intensities to improve player performance is well recognized in the current literature [27,28]. However, there remains a lack of clarity regarding the training intensity values most likely to promote improvements in young players performance.

4.1. Reference Values Depending on Age Group

When analyzing the mean of the total distance per training session for overall team of the selected studies, there was a pattern of increasing distance until older ages [18,21,22,24] which was also corroborated by increase in the number of body impacts [18]. However, Dalen and Lorås reported a decrease in the number of accelerations from U15 to U17, respectively, [19], which was in line with Teixeira et al. that found the same pattern from U17 to U19 [24].

When considering, the accumulated weekly, total distance did not show that pattern from under U12 (~18.6 Km), U13 (~19.3 km), U14 (~19.5 km), U15 (~21.0 km), U16 (~20.8 km), U17 (~16.0 km), U18 (~21.2 km) to U19 (~16.0 km) age-groups [18,19,21,22,24,25]. Considering only the study of Hannon et al. that quantifying the training and match volume in male players from an English Premier League academy during two in-season microcycles, it was found an increase in the accumulative total distance from U12/13 (38.3 ± 5.1 km), to U15 (53.7 ± 4.5 km) and to U18 (54.4 ± 7.1 km) [21]. Additionally, Abade et al. reported a higher total distance covered for U17 players than for U19 [18].

Perhaps technical-tactical methodologies, namely game model, can explain these data. There was an age-related increase in the training intensity and to a greater extent in the training volume [25]. Due to this fact, associated with a more conscious pacing strategy and better game interpretation with age, it was possible to also increase the exercise economics [24,29]. Even more interaction effects were found between inter-day and age [21,24], confirming an increase in the pacing strategy in the aging progression. Training periodization also seemed to influence the external intensity, concerning the training day and weekly microcycle [24].

The sprints number reported by the studies showed a high dispersion. In the study by Abade et al. [18] 11 ± 6 sprints (distance = 12 ± 5 m) were reported in U15 but in Teixeira et al. [24] only 2 ± 3 sprints (sprint distance = 28 ± 42 m), however the distance is twice that reported by Abade et al. [18]. In U17, the sprint number and sprint distance data were, respectively, between 16.4 ± 8.2 (sprint distance = 13.0 ± 5.3 m) [18] and 4.8 ± 4.8 (sprint distance = 130.4 ± 462.6 m) [24]. Teixeira et al. (2021) study evidenced the lowest intensity in U15 players' training sessions regarding high-speed run, average sprint distance, number of sprints [24]. The high standard deviation expresses the dispersion of results, which makes it difficult to standardize training intensity patterns related to the sprint number and distance at youth age.

Regarding internal training intensity, the main results showed a range interval range of 2.3 to 6.3 au for RPE between U14, U15, U16, and U18 [20,25] and 156 to 394 au for s-RPE between U15 and U17 [20,23]. Additionally, after conducting the research analysis of the present systematic review, a new study in U17 soccer players that analyze RPE and s-RPE measures was published [30]. That study [30] is in line with the interval range for RPE but higher values of 640 and 595 au were found for s-RPE during pre- and in-season with training durations around 96 and 95 min, respectively, which may justify the higher values.

Moreover, Wrigley et al. [25] noted a higher weekly RPE in the older age group (i.e., U18). However, the authors Teixeira et al. verified a higher training volume in younger players (i.e., U15 vs. U19) [24]. It is reasonable to argue that coaching team tends to code training programs with more volume and less intensity when it comes to younger players [24,31,32]. Furthermore, a focus on the basic tactical principles and technical skills using constrained training tasks was reported in younger age groups [25]. Nevertheless, the time spent at high-intensity zones and normalizing the session duration may affect the perceived exertion [33].

There were only one study analyzed Banister TRIMP and player status and found no differences between starters and non-starters [19] which was also corroborated by Martins et al. [30]. Nonetheless, it is important to reinforce that the period of the season and microcycle can influence result interpretations. For instance, the comparison of RPE between starters and non-starters showed significant differences in the following day after the match due to the recovery session for starters. Additionally, some differences were found during some mesocycles of the in-season [30].

4.2. Training Intensity by Match Day Minus

The main findings showed that young players usually training between 3 to 4 days per week. The higher intensity was found on MD-4, MD-3, and MD-2 while the lower intensity was found on MD-1, although only two studies used this approach for data analysis [21,24]. The previous findings were similar to adult players [34–39] which seems to be convergent in a tapering strategy based on a gradual reduction until the last day before the match [40].

The decrease in high-speed running distance and sprint distance in training sessions before match day was also evident [21] but it was not confirmed in sprint distance by Teixeira et al. [24]. Some studies reported that some coaches use sprinting and acceleration exercises for neuromuscular activation as a pre-match activation methodology [24,40].

Regarding internal intensity analysis, two studies analyzed s-RPE with different scales, CR-10 [23] vs. CR-20 [24] and different training schedules, 4 [23] vs. 3 training sessions [24]. Even so, Nobari et al. found lower intensity in the following day of the match, higher intensity in mid-week, and the lowest intensities in the day before the match [23]. Although the study of Martins et al. [30] was in line with these findings, no differences between training days were found in Teixeira et al., study [24]. By contrast, previous studies reported an decreasing intensity phase in young players concerning RPE values [25,41] and Wrigley et al. (2012) evidenced a tapering in U18 players [25].

4.3. Player Positions

In youth soccer, the positional role has been analyzed in constrained training tasks [32,42,43], however, the present systematic review only found one study that analyzed intensity by position [24]. The study by Teixeira et al. analyzed three age-groups studied (U15, U17, U19) and reported that the greatest total distance was performed by the central midfielder (5456.9 ± 1565.9 m), followed by the wide midfielder (5370.1 ± 1692.6 m) [24]. The same authors only found significant differences between central defenders vs. forwards in high-speed running and sprint distance (minimum to moderate effect). Additionally, the internal training intensity presented significant differences between wide midfielder and forwards players (minimum effect). The same authors also reported an interaction effect between age, week, training day and playing position for deceleration [24]

The influence of playing position on physical and physiological performance during competition is well documented [44–46], while in training it seems to have a minimal effect for young players. In contrast, the influence of playing position in the adult training football has been well documented but it was non-conclusive [36,47–49]. On one hand, it was found greater training intensity for wide defenders and wide midfielders with respect to high-speed running and number of sprints when compared with the other positions [48]. In the same line, it was found in several mesocycles during an in-season, that central midfielders and wide midfielders displayed higher training intensity than central defenders with respect to total distance and high-speed running (>19 km.h^{-1}) [49]. Other study found that midfielders had the highest training weekly acute load of high metabolic load distance (6901 AU), while central defenders had the lowest (4986 AU) [47]. On the other hand, it was found no differences between players positions for total distance, high-speed running (>19 km.h^{-1}) [36] in mesocycle and microcycle analysis during an in-season. The results of the present review seems unclear regarding the influence of player positions on training intensity. A possible explanation could be related to the training tasks, which may not be representative of the positional role specificity as referred by Ferraz et al. [29]. In future studies, the weekly training intensity quantification should consider the game model and representative game-based situations to promote playing position specificity. Furthermore, speed and acceleration thresholds in the studies of this review were based on elite gold-standard guidelines. Future research should focus on the adjustment thresholds for elite and sub-elite youth football.

In our literature review, only two studies reported values of the internal training intensity between playing positions [23,24]. Teixeira et al. Reported significant differences [24], however Nobari et al. found no differences in the comparisons within weekdays between the playing positions [23]. In addition, and Gjaka et al. also found no differences in accumulated weekly intensity between player positions [20]. If data from adult training were considered, the results seem to be in line because, it was found in several mesocycles during an in-season, that central defenders displayed higher training intensity than strikers or central midfielders with respect to s-RPE [49], while higher values of s-RPE were reported for midfielders than other positions [50]. On the other hand, no differences were between players positions for the same variable [36] in mesocycle and microcycle analysis during an in-season. This information revealed that a non-consensus still remain which suggest future studies to confirm the results.

4.4. Study Limitations and Future Directions

This study presents some limitations that influenced the results and the reference values provided. First, the small number of studies proves that much more research is needed in young players, even more considering the different age-group categories. Additionally, the studies included only analyze one soccer team which limit their results. Consequently, few studies analyzed contextual variables, such as player status, player positions, and MD- approach. In addition, few studies included full seasons. Moreover, the different speed thresholds and scales used makes difficult to generalize the results and compare them between studies. In addition, training intensity and reference values provided came from general training, but this information was not revealed by the studies. Only one study identify data from the field and from the gym training [25]. Finally, the present systematic review did not consider some possible inter-country variability or the context of each study which should be considered in future studies.

Despite the limitations, the present study constitutes a relevant tool in the field of training intensity quantification of young male soccer players that can be used for coaches, their staff and practitioners as a reference for future studies in order to replicate such values or even to increase the numbers presented.

In future studies, the comparison between elite and sub-elite soccer academies is an important research gap which should be considered. Additionally, information on what type of training and exercises could provide further knowledge on training intensity. At last, future studies can analyze values of training intensity for injury prevention, since the general aims of the studies in this field are to improve performance and avoid injuries.

5. Conclusions

The standard microcycle from U12 to U16 included three training sessions, while U17 and U19 included four training sessions per week. The duration of training sessions for all age-groups was between ~70 and 117 min.

Specifically, the following range intervals were found for training: RPE 2.3–6.3 au; s-RPE, 156–394 au; total distance, 3964.5–6500 m and; distance > 18 km/h, 11.8–250 m.

Even though the internal training intensity did not present significant differences for weekly inter-day analysis, there was an inverted U-shaped curve in the distribution of the weekly external training intensity. In the older age-group players, tapering is evidenced. On the other hand, the U14 to U16 players seems present relatively high training intensities across the weekly microcycle. Indeed, it could be suggested that coaches opt for different tapering strategies when the age and competition focus increases.

The playing position seems to have a minimal effect on the weekly training intensity of young players.

Author Contributions: Conceptualization, R.O. and F.M.C.; methodology, R.O., A.M.-V., M.R.-G. and F.M.C.; software, R.O., A.M.-V.; validation, R.O., J.P.B. and F.M.C.; formal analysis, R.O. and M.N.; investigation, R.O., J.P.B., A.M.-V., M.N. and F.M.C.; resources, R.O., J.P.B., A.M.-V., M.N. and F.M.C.; writing—original draft preparation, R.O., J.P.B., A.M.-V., M.N. and F.M.C.; writing—review and editing, R.O., J.P.B., A.M.-V., M.N. and F.M.C.; visualization, R.O., M.R.-G. and F.M.C.; supervision, R.O. and J.P.B.; project administration, R.O. All authors have read and agreed to the published version of the manuscript.

Funding: This research was funded by Portuguese Foundation for Science and Technology, I.P., grant number UIDP/04748/2020. The funders had no role in the design of the study, in the collection, analyses, or interpretation of data, in the writing of the manuscript, or in the decision to publish the results.

Institutional Review Board Statement: Not applicable.

Informed Consent Statement: Not applicable.

Data Availability Statement: Not applicable.

Conflicts of Interest: The authors declare no conflict of interest.

References

1. Weston, M. Training load monitoring in elite English soccer: A comparison of practices and perceptions between coaches and practitioners. *Sci. Med. Footb.* **2018**, *2*, 216–224. [CrossRef]
2. Staunton, C.A.; Abt, G.; Weaving, D.; Wundersitz, D.W.T. Misuse of the term 'load' in sport and exercise science. *J. Sci. Med. Sport* **2021**. [CrossRef]
3. Rebelo, A.; Brito, J.; Seabra, A.; Oliveira, J.; Drust, B.; Krustrup, P. A New Tool to Measure Training Load in Soccer Training and Match Play. *Int. J. Sports Med.* **2012**, *33*, 297–304. [CrossRef] [PubMed]
4. Williams, A.M.; Reilly, T. Talent identification and development in soccer. *J. Sports Sci.* **2000**, *18*, 657–667. [CrossRef] [PubMed]
5. Impellizzeri, F.M.; Marcora, S.M.; Coutts, A.J. Internal and External Training Load: 15 Years On. *Int. J. Sports Physiol. Perform.* **2019**, *14*, 270–273. [CrossRef]
6. Halson, S.L. Monitoring Training Load to Understand Fatigue in Athletes. *Sport Med* **2014**, *44*, 139–147. [CrossRef] [PubMed]
7. Miguel, M.; Oliveira, R.; Loureiro, N.; García-Rubio, J.; Ibáñez, S.J. Load measures in training/match monitoring in soccer: A systematic review. *Int. J. Environ. Res. Public Health* **2021**, *18*, 2721. [CrossRef]
8. Scantlebury, S.; Till, K.; Sawczuk, T.; Weakley, J.; Jones, B. Understanding the Relationship Between Coach and Athlete Perceptions of Training Intensity in Youth Sport. *J. Strength Cond. Res.* **2018**, *32*, 3239–3245. [CrossRef]
9. Salter, J.; De Ste Croix, M.B.A.; Hughes, J.D.; Weston, M.; Towlson, C. Monitoring Practices of Training Load and Biological Maturity in UK Soccer Academies. *Int. J. Sports Physiol. Perform.* **2021**, *16*, 395–406. [CrossRef]
10. Decker, M.; Sperlich, B.; Zinner, C.; Achtzehn, S. Intra-Individual and Seasonal Variation of Selected Biomarkers for Internal Load Monitoring in U-19 Soccer Players. *Front. Physiol.* **2020**, *11*, 838. [CrossRef]
11. Maughan, P.C.; MacFarlane, N.G.; Swinton, P.A. Relationship Between Subjective and External Training Load Variables in Youth Soccer Players. *Int. J. Sports Physiol. Perform.* **2021**, *16*, 1127–1133. [CrossRef]
12. Akubat, I.; Patel, E.; Barrett, S.; Abt, G. Methods of Monitoring the Training and Match Load and Their Relationship to Changes in Fitness in Professional Youth Soccer Players. *J. Sports Sci.* **2012**, *30*, 1473–1480. [CrossRef] [PubMed]
13. Page, M.J.; McKenzie, J.E.; Bossuyt, P.M.; Boutron, I.; Hoffmann, T.C.; Mulrow, C.D.; Shamseer, L.; Tetzlaff, J.M.; Akl, E.A.; Brennan, S.E.; et al. The PRISMA 2020 statement: An updated guideline for reporting systematic reviews. *BMJ* **2021**, *372*. [CrossRef]
14. Rico-González, M.; Pino-Ortega, J.; Clemente, F.; Los Arcos, A. Guidelines for performing systematic reviews in sports science. *Biol. Sport* **2021**, *39*, 463–471. [CrossRef]

15. Ryan, R.; Synnot, A.; Prictor, M.; Hill, S. Cochrane Consumers and Communication Group Data Extraction Template for Included Studies. Available online: http://cccrg.cochrane.org/author-resources (accessed on 27 July 2021).
16. Slim, K.; Nini, E.; Forestier, D.; Kwiatkowski, F.; Panis, Y.; Chipponi, J. Methodological index for non-randomized studies (MINORS): Development and validation of a new instrument. *ANZ J. Surg.* **2003**, *73*, 712–716. [CrossRef] [PubMed]
17. Kim, S.Y.; Park, J.E.; Lee, Y.J.; Seo, H.-J.; Sheen, S.-S.; Hahn, S.; Jang, B.-H.; Son, H.-J. Testing a tool for assessing the risk of bias for nonrandomized studies showed moderate reliability and promising validity. *J. Clin. Epidemiol.* **2013**, *66*, 408–414. [CrossRef]
18. Abade, E.A.; Gonçalves, B.V.; Leite, N.M.; Sampaio, J.E. Time-motion and physiological profile of football training sessions performed by under-15, under-17, and under-19 elite portuguese players. *Int. J. Sports Physiol. Perform.* **2014**, *9*, 463–470. [CrossRef]
19. Dalen, T.; Lorås, H. Monitoring Training and Match Physical Load in Junior Soccer Players: Starters versus Substitutes. *Sports* **2019**, *7*, 70. [CrossRef]
20. Gjaka, M.; Tschan, H.; Francioni, F.M.; Tishkuaj, F.; Tessitore, A. Monitoring of Loads and Recovery Perceived During Weeks With Different Schedule in Young Soccer Players Spremljanje Obremenitev in Okrevanja Mladih Nogometašev V Več Tednih Z Različnimi Programi Treniranja. *Orig. Artic. Kinesiol. Slov.* **2016**, *22*, 16–26.
21. Hannon, M.P.; Coleman, N.M.; Parker, L.J.F.; McKeown, J.; Unnithan, V.B.; Close, G.L.; Drust, B.; Morton, J.P. Seasonal training and match load and micro-cycle periodization in male Premier League academy soccer players. *J. Sports Sci.* **2021**, *39*, 1838–1849. [CrossRef]
22. Houtmeyers, K.C.; Jaspers, A.; Brink, M.S.; Vanrenterghem, J.; Varley, M.C.; Helsen, W.F. External load differences between elite youth and professional football players: Ready for take-off? *Sci. Med. Footb.* **2020**, *5*, 1–5. [CrossRef]
23. Nobari, H.; Vahabidelshad, R.; Pérez-Gómez, J.; Ardigò, L.P. Variations of Training Workload in Micro- and Meso-Cycles Based on Position in Elite Young Soccer Players: A Competition Season Study. *Front. Physiol.* **2021**, *12*, 1–10. [CrossRef]
24. Teixeira, J.E.; Forte, P.; Ferraz, R.; Leal, M.; Ribeiro, J.; Silva, A.J.; Barbosa, T.M.; Monteiro, A.M. Quantifying sub-elite youth football weekly training load and recovery variation. *Appl. Sci.* **2021**, *11*, 4871. [CrossRef]
25. Wrigley, R.; Drust, B.; Stratton, G.; Scott, M.; Gregson, W. Quantification of the typical weekly in-season training load in elite junior soccer players. *J. Sports Sci.* **2012**, *30*, 1573–1580. [CrossRef] [PubMed]
26. Silva, P.; Dos Santos, E.; Grishin, M.; Rocha, J.M. Validity of heart rate-based indices to measure training load and intensity in elite football players. *J. Strength Cond. Res.* **2018**, *32*, 2340–2347. [CrossRef]
27. Bourdon, P.C.; Cardinale, M.; Murray, A.; Gastin, P.; Kellmann, M.; Varley, M.C.; Gabbett, T.J.; Coutts, A.J.; Burgess, D.J.; Gregson, W.; et al. Monitoring athlete training loads: Consensus statement. *Int. J. Sports Physiol. Perform.* **2017**, *12*, 161–170. [CrossRef]
28. McLaren, S.J.; Macpherson, T.W.; Coutts, A.J.; Hurst, C.; Spears, I.R.; Weston, M. The Relationships Between Internal and External Measures of Training Load and Intensity in Team Sports: A Meta-Analysis. *Sport. Med.* **2018**, *48*, 641–658. [CrossRef]
29. Ferraz, R.; Gonçalves, B.; Coutinho, D.; Oliveira, R.; Travassos, B.; Sampaio, J.; Marques, M.C. Effects of knowing the task's duration on soccer players' positioning and pacing behaviour during small-sided games. *Int. J. Environ. Res. Public Health* **2020**, *17*, 3843. [CrossRef]
30. Martins, A.D.; Oliveira, R.; Brito, J.P.; Loureiro, N.; Querido, S.M.; Nobari, H. Intra-season variations in workload parameters in europe's elite young soccer players: A comparative pilot study between starters and non-starters. *Healthcare* **2021**, *9*, 977. [CrossRef]
31. Coutinho, D.; Gonçalves, B.; Figueira, B.; Abade, E.; Marcelino, R.; Sampaio, J. Typical weekly workload of under 15, under 17, and under 19 elite Portuguese football players. *J. Sports Sci.* **2015**, *33*, 1229–1237. [CrossRef]
32. Beenham, M.; Barron, D.J.; Fry, J.; Hurst, H.H.; Figueirdo, A.; Atkins, S. A Comparison of GPS Workload Demands in Match Play and Small-Sided Games by the Positional Role in Youth Soccer. *J. Hum. Kinet.* **2017**, *57*, 129–137. [CrossRef]
33. Haddad, M.; Stylianides, G.; Djaoui, L.; Dellal, A.; Chamari, K. Session-RPE method for training load monitoring: Validity, ecological usefulness, and influencing factors. *Front. Neurosci.* **2017**, *11*, 612. [CrossRef] [PubMed]
34. Clemente, F.M.; Owen, A.; Serra-Olivares, J.; Nikolaidis, P.T.; Van Der Linden, C.M.I.; Mendes, B. Characterization of the Weekly External Load Profile of Professional Soccer Teams from Portugal and the Netherlands. *J. Hum. Kinet.* **2019**, *66*, 155–164. [CrossRef]
35. Los Arcos, A.; Mendez-Villanueva, A.; Martínez-Santos, R. In-season training periodization of professional soccer players. *Biol. Sport* **2017**, *34*, 149–155. [CrossRef]
36. Oliveira, R.; Brito, J.P.; Martins, A.; Mendes, B.; Marinho, D.A.; Ferraz, R.; Marques, M.C. In-season internal and external training load quantification of an elite European soccer team. *PLoS ONE* **2019**, *14*, e0209393. [CrossRef] [PubMed]
37. Sanchez-Sanchez, J.; Hernández, D.; Martin, V.; Sanchez, M.; Casamichana, D.; Rodriguez-Fernandez, A.; Ramirez-Campillo, R.; Nakamura, F.Y. Assessment of the external load of amateur soccer players during four consecutive training microcycles in relation to the external load during the official match. *Motriz. Rev. Educ. Fis.* **2019**, *25*, e101938. [CrossRef]
38. Anderson, L.; Orme, P.; Di Michele, R.; Close, G.L.; Morgans, R.; Drust, B.; Morton, J.P. Quantification of training load during one-, two- and three-game week schedules in professional soccer players from the English Premier League: Implications for carbohydrate periodisation. *J. Sport. Sci.* **2016**, *34*, 1250–1259. [CrossRef]

39. Oliveira, R.; Brito, J.; Martins, A.; Mendes, B.; Calvete, F.; Carriço, S.; Ferraz, R.; Marques, M.C. In-season training load quantification of one-, two- and three-game week schedules in a top European professional soccer team. *Physiol. Behav.* **2019**, *201*, 146–156. [CrossRef] [PubMed]

40. Brito, J.; Hertzog, M.; George, N. Do match-related contextual variables Influence training load in highly trained Soccer players? *J. Strength Cond. Res.* **2016**, *30*, 393–399. [CrossRef]

41. Impellizzeri, F.M.; Rampinini, E.; Coutts, A.J.; Sassi, A.; Marcora, S.M. Use of RPE-based training load in soccer. *Med. Sci. Sports Exerc.* **2004**, *36*, 1042–1047. [CrossRef]

42. Barnabé, L.; Volossovitch, A.; Duarte, R.; Ferreira, A.P.; Davids, K. Age-related effects of practice experience on collective behaviours of football players in small-sided games. *Hum. Mov. Sci.* **2016**, *48*, 74–81. [CrossRef]

43. Coutinho, D.; Gonçalves, B.; Santos, S.; Travassos, B.; Wong, D.P.; Sampaio, J. Effects of the pitch configuration design on players' physical performance and movement behaviour during soccer small-sided games. *Res. Sport. Med.* **2019**, *27*, 298–313. [CrossRef]

44. Pettersen, S.A.; Brenn, T. Activity Profiles by Position in Youth Elite Soccer Players in Official Matches. *Sport. Med. Int. Open* **2019**, *3*, E19–E24. [CrossRef]

45. Saward, C.; Morris, J.G.; Nevill, M.E.; Sunderland, C. The effect of playing status, maturity status, and playing position on the development of match skills in elite youth football players aged 11–18 years: A mixed-longitudinal study. *Eur. J. Sport Sci.* **2019**, *19*, 315–326. [CrossRef] [PubMed]

46. Aslan, A.; Açikada, C.; Güvenç, A.; Gören, H.; Hazir, T.; Özkara, A. Metabolic demands of match performance in young soccer players. *J. Sport. Sci. Med.* **2012**, *11*, 170–179.

47. Clemente, F.M.; Silva, R.; Ramirez-Campillo, R.; Afonso, J.; Mendes, B.; Chen, Y.S. Accelerometry-based variables in professional soccer players: Comparisons between periods of the season and playing positions. *Biol. Sport* **2020**, *37*, 389–403. [CrossRef] [PubMed]

48. Clemente, F.M.; Silva, R.; Castillo, D.; Arcos, A.L.; Mendes, B.; Afonso, J. Weekly load variations of distance-based variables in professional soccer players: A full-season study. *Int. J. Environ. Res. Public Health* **2020**, *17*, 3300. [CrossRef] [PubMed]

49. Oliveira, R.; Martins, A.; Nobari, H.; Nalha, M.; Mendes, B.; Clemente, F.M.; Brito, J.P. In-season monotony, strain and acute/chronic workload of perceived exertion, global positioning system running based variables between player positions of a top elite soccer team. *BMC Sports Sci. Med. Rehabil.* **2021**, *13*, 126. [CrossRef]

50. Clemente, F.M.; Mendes, B.; Nikolaidis, P.T.; Calvete, F.; Carriço, S.; Owen, A.L. Internal training load and its longitudinal relationship with seasonal player wellness in elite professional soccer. *Physiol. Behav.* **2017**, *179*, 262–267. [CrossRef]

 *healthcare*

Article

Variability of External Intensity Comparisons between Official and Friendly Soccer Matches in Professional Male Players

Hadi Nobari [1,2,3,4,*], João Paulo Brito [5,6,7], Jorge Pérez-Gómez [8] and Rafael Oliveira [5,6,7,*]

1 Department of Physical Education and Sports, University of Granada, 18010 Granada, Spain
2 Department of Physiology, School of Sport Sciences, University of Extremadura, 10003 Cáceres, Spain
3 Sports Scientist, Sepahan Football Club, Isfahan 81887-78473, Iran
4 Department of Exercise Physiology, Faculty of Educational Sciences and Psychology, University of Mohaghegh Ardabili, Ardabil 56199-11367, Iran
5 Sports Science School of Rio Maior, Polytechnic Institute of Santarém, 2140-413 Rio Maior, Portugal; jbrito@esdrm.ipsantarem.pt
6 Research Center in Sport Sciences, Health Sciences and Human Development, 5001-801 Vila Real, Portugal
7 Life Quality Research Centre, 2140-413 Rio Maior, Portugal
8 HEME Research Group, Faculty of Sport Sciences, University of Extremadura, 10003 Cáceres, Spain; jorgepg100@gmail.com
* Correspondence: hadi.nobari1@gmail.com (H.N.); rafaeloliveira@esdrm.ipsantarem.pt (R.O.)

Abstract: The aims of this study were to compare the external intensity between official (OMs) and friendly matches (FMs), and between first and second halves in the Iranian Premier League. Twelve players participated in this study (age, 28.6 ± 2.7 years; height, 182.1 ± 8.6 cm; body mass, 75.3 ± 8.2 kg). External intensity was measured by total duration, total distance, average speed, high-speed running distance, sprint distance, maximal speed and body load. In general, there was higher intensity in OMs compared with FMs for all variables. The first half showed higher intensities than the second half, regardless of the type of the match. Specifically, OMs showed higher values for total sprint distance ($p = 0.012$, ES = 0.59) and maximal speed ($p < 0.001$, ES = 0.27) but lower value for body load ($p = 0.038$, ES = -0.42) compared to FMs. The first half of FMs only showed lower value for body load ($p = 0.004$, ES = -0.38) than FMs, while in the second half of OMs, only total distance showed a higher value than FMs ($p = 0.013$, ES = 0.96). OMs showed higher demands of high intensity, questioning the original assumption of FMs demands. Depending on the period of the season that FMs are applied, coaches may consider requesting higher demands from their teams.

Keywords: performance; load monitoring; high-speed running; match load; player load; sprint

Citation: Nobari, H.; Brito, J.P.; Pérez-Gómez, J.; Oliveira, R. Variability of External Intensity Comparisons between Official and Friendly Soccer Matches in Professional Male Players. *Healthcare* **2021**, *9*, 1708. https://doi.org/10.3390/healthcare9121708

Academic Editor: Parisi Attilio

Received: 10 November 2021
Accepted: 6 December 2021
Published: 8 December 2021

Publisher's Note: MDPI stays neutral with regard to jurisdictional claims in published maps and institutional affiliations.

1. Introduction

An alert was recently launched for the soccer sports community about the need to discuss the global phenomenon of pre-season soccer matches as friendly matches (FMs) [1]. Pre-season is a specific period where the main objective is the acquisition of individual and collective adaptations that allow starting the competition adequately [2]. Usually, pre-season lasts four to six weeks and is of critical importance to develop high-level performance in soccer [3], which is supposed to be conducted with the aim of maximizing players' participation in team training sessions to develop technical, psychological, physical and tactical performance [4,5]. From a conditioning point of view, the pre-season is characterized by a high volume of training and a gradually increasing intensity [1,6]; however, the improvement of strategical and tactical training should also be a major aim in this period [7].

According to the requirements of the field position, to reach a high level of physical performance at the elite competitive level, the analysis of match demands is the only feasible way to establish physical conditioning standards to be implemented in players. Some studies analysed the demand patterns during elite-level soccer match play [4,8,9].

For instance, they found that, during competitive matches, several intermittent periods of high-intensity activity such as high-speed running distance (HSRD) interspersed with low intensity periods occurred. Despite being a common practice, to use small-sided games during the pre-season in professional and semi-professional soccer, they do not reflect the intensity of the competitive game, and therefore coaches make use of FMs [8–10] in an attempt to improve physical fitness and skill development of the team [8]. This has been one of the aspects that have led to a greater use of FMs for the squads' preparation [1,8].

The vast majority of FMs take place in the pre-season, where commercial imperatives are increasingly present in the team's performance during this period, which can condition the training intensity and the pressure placed on players [1,11,12]. Consequently, these matches have been scheduled increasingly closer to the beginning of the pre-season, causing an extra physical and psychological demand to the players and coaching staff. This has had a major impact on strength and conditioning processes, which need to be much faster, which therefore require skipping important phases in the training process to apply higher intensity values than expected for the period [1,11]. It will probably generate unwanted consequences at different levels. Moreover, the "need to win" all matches contributed as an additional stressing factor [1,12]. In addition, an inappropriate training in the pre-season can be associated with a higher number of injuries through the in-season [13]. Beyond the training process, the higher number of matches played with a similar intensity to the official matches [14,15] may be associated with a higher injury rate [16]. It had been shown that high levels of training and matches are associated with higher risk of illness and injury [17–20].

Some studies reported that the pre-season is the period with the highest training intensity [1,21]. For instance, Jeong et al. [21] showed higher values of mean heart rate at 124 $\pm$ 7 beats/min and session-rated perceived exertion at 4343 $\pm$ 329 arbitrary units (AU) in pre-season, while lower values were revealed (heart rate, 112$\pm$ 7 beats/min; session-rated perceived exertion 1703 $\pm$ 173 AU) during in-season [21].

The pre-season is currently considered by most teams as the period with the highest intensity and a high risk of non-traumatic injuries [1,6,22]. Conversely, Coppalle et al. [18] reported that this period is not associated with performance of the team because there are many more factors such as technical and tactical levels, opponents and environment that can influence the performance. Nonetheless and due to the commercial commitments, the pressure to win all matches, which also includes FMs, has increased. In this sense and according to Calleja-Gonzalez et al. [1], it was suggested to analyse external measures between official matches (OMs) and FMs.

Currently, tracking systems are used in professional clubs for better external intensity management [22]. The most used technologies are micro-sensors devices, usually known as global positioning systems (GPS) [23–25]. They have shown to provide reliable and valid measures of the physical activity profile of team sports [26–28]. They allow to quantify several running distance speeds and accelerometery-based measures which are associated with the physical demands performed (e.g., training or matches). Usually, the external monitoring is known as external load [29–31]. However, due to the misuse of the load concept, "intensity" is used instead of "load" [29].

The development of high levels of physical performance is essential for performance in soccer. For that reason, it is necessary to develop a specific range of physical qualities [26] in order to best each the physical and physiological demands for the matches [8]. In this sense, some investigations have analysed the differences between the physical demands of training sessions or small-sided games in OMs [8,30] or FMs [11,12,32].

Another determinant aspect in modern soccer is the ability of players to maintain high levels of intensity during both halves of the match, and although some studies reported a decline in the total distances covered and HSRD in the second half compared to the first half [31,32], some studies in soccer and other sports have reported no differences between halves [4,33]. These fluctuations in intensity running can be caused by a variety of factors, including tactical alterations, the quality of the match, or the players' level of preparation [31].

Based on previous literature, our working hypothesis was to verify if there is a decrement on players' performance toward the end of match and between first and second halves for OMs and FMs. Understanding soccer players' match-related demands and fatigue profiles likely helps with developing conditioning programs that increase team performance [1,34]. Despite match physical demands being a frequent study topic during the last years, the novelty of this research relies on the comparison for the determination of physical FMs and OMs demands. Therefore, the aims of this study were: to compare the running distances variables and body load between OMs and FMs; to compare all variables between first and second halves for OMs and FMs, respectively. It was hypothesised that some GPS measures would present higher values in FMs than in OMs, and that the first half of the matches would display higher intensity levels than the second half.

2. Materials and Methods

2.1. Design

A cohort study was conducted to identify differences between FMs and OMs during the season through GPS-derived variables. This professional soccer team had participated in the highest level of the Iranian Premier League called the Persian Gulf. In this league, teams were allowed to use GPS in competitions. All external monitoring and receiving information were performed by GPS with model GPSPORTS systems Pty Ltd., and SPI High-Performance Unit (HPU), Canberra, Australia. Finally, for the present study, 27 OMs and 10 FMs were analysed. The characteristics of the weeks and matches are presented in Table 1.

Table 1. Characterization of the weeks and matches included for analysis.

Weeks	Type of Matches
1–2	Not included
3–5	Friendly
6–11	Official
12	Not included
13–16	Official
17	Friendly
18–21	Official
22	Non included
23–24	Official
25	Not included
26	Official
27	Not included
28	Friendly
29	Not included
30–31	Friendly
32–33	Official
34	Friendly
35–37	Official
38	Friendly
39	Official
40	Not included *
41–44	Official
45	Not included
46	Not included *
47	Friendly
48	Official

* weeks with two official matches.

2.2. Participants

Twelve professional players were selected according to the inclusion criteria (age, 28.6 ± 2.7 years; height, 182.1 ± 8.6 cm; body mass, 75.3 ± 8.2 kg; body mass index, 22.6 ± 0.7 kg/m^2) and consisted of participating in at least three consecutive matches and training with the team at least three training sessions a week. The anthropometric measurements were performed by specialists at the Iran Football Medical Assessment and Rehabilitation Center (http://ifmarc.ir/) (8 August 2021). In order to measure height and weight, the participants stood without shoes and with only shorts. For both measurements, a portable stadiometer (accuracy of ± 5 mm) and balance weighting scales (accuracy of ± 0.1 kg) (Seca model 207, Germany) was used. Body mass index (kg/m^2) was calculated through the formula: weight/height^2. The exclusion criteria of this study were players who did not attend training for more than two weeks, who were excluded from the study for any reason. In addition, the goalkeepers were omitted from the study. After coordination and obtaining official permission from the club director, an introductory session with the players as well as the team staff was conducted for the experimental approach of this study and individual consent was obtained from the players. This study was approved by the ethics committee of the University of Isfahan and Mohaghegh Ardabili University. During the study, the Helsinki Declaration was also considered for human studies.

2.3. Monitoring External Measures

GPS receiver specifications. During the season, all workouts and match sessions were monitored using GPSPORTS systems Pty Ltd. (Model: SPI High Performance Unit, Canberra, Australia) for professional athletes, which includes a 15 Hz position GPS and a tri-axial accelerometer to collect body load data. According to a previous study, this device has a high validity and inter-unit reliability within $\pm 2\%$ based on root mean square error [35]. There were no reported adverse weather conditions to affect data collection.

Prior to the start of the match, belts were worn, and after the cooldown session post-match, they were collected. Then, GPS was placed in the dock system that allow downloading the information to save it through the Team AMS software. The same steps were described in a previous study [36].

According to the aims of the present study, match duration, total distance, average speed, HSRD (18–23 km$\cdot$h^{-1}), total sprint distance (>23 km$\cdot$h^{-1}), maximal speed (MS) and body load were measured.

2.4. Statistical Analysis

SPSS version 22.0 (SPSS Inc., Chicago, IL, USA) was used to analyse GPS data. First, the participants and GPS measures were described through descriptive statistics. Second, to verify the assumption normality and homoscedasticity of the several measures, Shapiro-Wilk and Levene's tests were applied, respectively.

In order to accomplish the study aims of comparing OMs vs. FMs and 1st vs. 2nd halves, t tests with 95% confidence interval (CI) were conducted. A $p \leq 0.05$ was considered for statistical significance. In addition, t test family sample power was calculated for a post hoc compute achieve power (α level = 0.05, effect size = 0.8 and n = 12) by the G-Power [37]. There was an actual power of 83% for the present analysis and sample.

The last step consisted of the effect size (ES) calculation with CI (95%) to determine the magnitude of effects which was then analysed considering the range intervals. <0.2 = trivial, $0.2 > 0.6$ = small effect, $0.6 > 1.2$ = moderate effect, $1.2 > 2.0$ = large effect and >2.0 = very large [38].

3. Results

Table 2 presents descriptive results, comparisons between first and second halves and full data between OMs and FMs.

Table 2. Comparison of full match-day, 1st and 2nd halves between official matches and friendly matches, mean ± standard deviation and CI (95%).

Full-Match	Official Matches	Friendly Matches	p	CI (95%)	Effect Size
Duration (min)	87.9 ± 11.6 (80.5–95.3)	85.8 ± 4.1 (83.2–88.4)	0.514	−4.7, 8.9	0.24 (−57, 1.04)
Total Distance (m)	9424.7 ± 1224.5 (8646.7–10,202.7)	9125.7 ± 1224.5 (8697.8–9553.5)	0.435	−514.0, 1112.1	0.24 (−57, 1.04)
Average speed (m/min)	107.9 ± 11.4 (100.6–115.2)	106.8 ± 11.4 (99.5–114.0)	0.585	−3.4, 5.7	0.10 (−0.71, 0.89)
HSRD (m)	241.7 ± 82.2 (189.4–293.9)	220.5 ± 74.7 (173.1–268.0)	0.274	−19.3, 61.5	0.27 (−0.54, 1.06)
Total sprint distance (m)	28.4 ± 9.5 (22.3–34.4)	21.8 ± 12.7 (13.7–29.9)	0.012 *	1.8, 11.3	0.59 (−0.25, 1.38)
Maximal speed ($km \cdot h^{-1}$)	29.0 ± 1.2 (28.2–29.8)	28.7 ± 1.0 (28.0–29.3)	<0.001 *	−0.1, 0.8	0.27 (−0.54, 1.07)
Body Load (au)	157.5 ± 38.9 (132.9–182.2)	179.8 ± 63.5 (139.5–220.2)	0.038 *	−43.1, −1.5	−0.42 (−1.22, 0.40)

1st Half	Official Matches	Friendly Matches	p	CI (95%)	Effect Size
Duration (min)	47.1 ± 1.8 (45.9–48.2)	47.2 ± 2.7 (45.5–48.9)	0.872	−1.9, 1.6	−4.40 (−5.69, −2.81)
Total Distance (m)	5181.5 ± 412.7 (4919.2–5443.7)	5123.1 ± 375.5 (4884.5–5361.7)	0.494	−123.3, 240.0	0.15 (−0.66, 0.94)
Average speed (m/min)	110.2 ± 9.5 (104.2–116.3)	108.9 ± 10.0 (102.5–115.2)	0.595	−4.1, 6.8	0.13 (−0.67, 0.93)
HSRD (m),	137.7 ± 58.9 (100.3–175.2)	108.7 ± 39.3 (83.7–133.7)	0.012 *	7.8, 50.3	0.58 (−0.26, 1.37)
Total sprint distance (m)	15.1 ± 7.5 (10.3–19.8)	12.1 ± 7.4 (7.5–16.8)	0.031 *	0.3, 5.6	0.40 (−0.42, 1.20)
Maximal speed ($km \cdot h^{-1}$)	29.2 ± 1.4 (28.3–30.1)	28.4 ± 1.4 (27.5–29.3)	0.031 *	0.1, 1.5	0.57 (−0.26, 1.37)
Body Load (au)	88.8 ± 28.0 (71.0–106.6)	100.5 ± 33.4 (79.2–121.7)	0.004 *	−18.7, −4.6	−0.38 (−1.17, 0.44)

2nd Half	Official Matches	Friendly Matches	p	CI (95%)	Effect Size
Duration (min)	43.4 ± 6.2 (39.4–47.3)	38.6 ± 4.2 (35.9–41.3)	0.016 *	1.1, 8.4	0.91 (0.04, 1.71)
Total Distance (m)	4531.9 ± 638.8 (4126.0–4937.8)	4002.6 ± 442.3 (3721.6–4283.6)	0.013 *	136.0, 922.7	0.96 (0.09, 1.77)
Average speed (m/min)	105.6 ± 14.3 (96.4–114.7)	104.9 ± 16.4 (94.5–115.3)	0.815	−5.3, 6.6	0.05 (−0.76, 0.84)
HSRD (m)	115.1 ± 33.9 (93.6–136.7)	111.9 ± 38.5 (87.4–136.4)	0.742	−17.9, 24.4	0.09 (−0.72, 0.89)
Total sprint distance (m)	15.4 ± 4.6 (12.5–18.4)	11.6 ± 6.5 (7.5–15.8)	0.145	−1.5, 9.2	0.67 (−0.17, 1.47)
Maximal speed ($km \cdot h^{-1}$)	28.9 ± 1.2 (28.1–29.7)	28.9 ± 1.1 (28.2–29.6)	0.974	−0.6, 0.7	0.00 (−0.80, 0.80)
Body Load (au)	75.1 ± 15.1 (65.5–84.7)	79.4 ± 32.1 (59.0–99.8)	0.520	−18.6, 10.0	−0.17 (−0.97, 0.64)

au, arbitrary units; m, meters; HSRD, high-speed running distance; AvS, average speed; TSD, total sprint distance; MS, maximal speed.
* significant differences between official match vs. friendly match, $p < 0.05$.

The comparisons between first half from OMs vs. FMs showed no significant differences in duration, but HSRD, sprint distance, and MS showed higher values in OMs, while body load presented higher values in FMs (all, $p < 0.05$, small effect size, which means a low power in considering the statistical power).

The comparisons between the second half from OMs vs. FMs showed higher values for duration and total distance (all, $p < 0.05$, moderate effect size), but the other variables did not present differences.

Considering full-match comparisons, total sprint distance and MS showed higher values in OMs than FMs, while body load showed higher values in FMs than OMs (all, $p < 0.05$ with small effect size, which mean a low statistical power). The other variables did not present significant differences.

Comparisons between first and second halves for OMs and FMs, respectively, are presented in Table 3. Regarding OMs, there were higher duration, total distance and body load in the first half of OMs (all, $p < 0.05$, moderate to large effect size). Regarding FMs, there were higher duration, total distance and body load in the first half of FMs (all, $p < 0.05$, moderate effect size).

Table 3. Comparison of first and second halves data for official matches and friendly matches, respectively.

Official Matches	p (1st Half vs. 2nd Half)	Confidence Interval (95%)	Effect Size
Duration (min)	0.034 *	0.3, 7.1	-1.38 (-2.22, -0.45)
Total Distance (m)	0.005 *	237.0, 1062.1	1.21 (0.30, 2.03)
Average speed (m/min)	0.079	-0.6, 10.0	0.38 (-0.44, 1.17)
HSRD (m)	0.057	-0.8, 46.0	0.47 (-0.36, 1.26)
Total sprint distance (m)	0.846	-4.4, 3.7	-0.05 (-0.85, 0.75)
Maximal speed (km·h^{-1})	0.322	-0.3, 1.0	0.23 (-0.58, 1.02)
Body Load (au)	0.021 *	2.5, 25.0	0.61 (-0.23, 1.41)
Friendly Matches	**p (1st Half vs. 2nd Half)**	**Confidence Interval (95%)**	**Effect Size**
Duration (min)	<0.001 *	4.9, 12.3	2.44 (1.31, 3.39)
Total Distance (m)	<0.001 *	822.7, 1418.4	2.73 (1.54, 3.73)
Average speed (m/min)	0.377	-5.5, 13.4	0.29 (-0.52, 1.09)
HSRD (m)	0.622	-17.2, 10.8	-0.08 (-0.88, 0.72)
Total sprint distance (m)	0.851	-5.1, 6.1	0.07 (-0.73, 0.87)
Maximal speed (km·h^{-1})	0.316	-1.4, 0.5	-0.40 (-1.19, 0.42)
Body Load (au)	0.001 *	11.0, 31.1	0.64 (-0.20, 1.044)

au, arbitrary units; m, meters; HSRD, high-speed running distance; AvS, average speed; TSD, total sprint distance; MS, maximal speed.
* denotes difference from 2nd half. all $p < 0.05$.

4. Discussion

The study' aims were: to compare the running distances variables and body load between OMs and FMs; to compare all variables between first and second halves for OMs and FMs, respectively. The impact of the current phenomenon of the high-level of FMs which is related to its commercial imperatives and the additional stressing factor of the "need to win" [12] matches was the rationale for the present study. The need to win matches implies to understand soccer players' match-related demands and fatigue profiles that will likely help in developing conditioning programs to increase team performance, wellness and to reduce injuries, illnesses [34].

In general, the results indicated some differences between OMs and FMs throughout the season; however, in the full-match data, there was no difference in OMs and FMs in duration, total distance, average speed and HSRD. The major findings were found in sprint distance and MS, where higher values were found in OMs, while body load showed higher values in FMs. According to Akenhead et al. [39] and Ade et al [40] the external intensity imposed by OMs is often high because of the large amount of high-intensity activity demands required, such as accelerations and high-speed running. However, in the present study, despite the sprint distance and the maximal speed being higher in the OMs, the body load values are higher in the FMs, which can eventually be understood due to the coach's tactical options, such as a more tactical positioning of the team and in the greater pressure to regain ball' possession, due to the "need not to lose" the game. Thus, a greater number of impact/tackle/collision actions for ball recovery can cause a higher BL [41].

Usually, body load was used [41,42] to access the physiological demands of different sports. However, Gomez-Piriz et al. [43] analysed body load validity through the analysis of relationship with session-rated perceived exertion during training session and reported that the relation between session-rated perceived exertion and body load was weak and non-linear despite being significant. For this reason, it was suggested that body load may not be a valid measure to assess intensity in soccer. Since it uses an algorithm that calculates the total measure body load, it could be limited on this ability to globally quantify soccer-specific intensity. This measure also known as player load is not rigorous in identifying the influence that mode of motion and ball actions have on the energy expenditure during different actions in soccer [43]. Thus, Reilly and Bowen [44] reported that modes of displacement, such as running backward, running sideways and changing direction, can accentuate the metabolic charge. The combination of accelerometery, magnetometry and GPS software with match recordings may provide more insight into categorization of forces/accelerations received/exerted during the many contact elements within the game.

Some studies [10,45] recognized that accumulated measures of accelerometery can provide a different construct of the training process versus internal physiological intensity (i.e., rated perceived exertion, heart rate and blood lactate concentration). Nonetheless, they constitute valid measures to quantify the physical demands of the players [41].

Regarding the analysis of the lowest total distance observed in the FMs, there may possibly be a consequence of the fact that in these matches sometimes the number of interruptions is higher for information/substitutions.

Another factor that may explain the obtained results is that during the in-season, coaches tended to reduce training intensity [46] to allow the players to recover and reach the match at optimum fitness levels.

However, pre-season is characterized by having a high weekly intensity, both due to the intensity assigned in the training sessions and the number of FMs, which can eventually condition fatigue. This possibly reflects the low priority of the coaches to prepare their teams before FMs during pre-season and can consequently contribute to higher accumulated fatigue level [8]. Campos-Vázquez et al. [8] reinforced the finding of FMs being the session with highest intensity during pre-season compared with data from training. Such results highlight the importance of playing FMs during the pre-season and/or to compensate the OMs absence during some periods of the in-season. Nevertheless, further investigations should aim to clarify if FMs actually reproduce the demands of OMs.

When comparing the first and second halves between OMs and FMs, higher values were found in the first half of OMs in the HSDR, total sprint distance and maximal speed variables. Once again, body load is higher in FMs. It can be inferred that the decisive actions of the game that usually underlie high intensity activities occur with greater magnitude in the first half of the OMs. In the second half, there were lower values of duration and total distance covered in FMs.

In the present study, a decrease of total distance was observed in the second half in both type of matches. However, there seemed to be no change in the intensity of the matches, although Mortimer et al. [47] reported that accumulated fatigue may contribute to a reduction in match intensity during the second half of a soccer match [48]. However, the reduction of total distance could be associated with the progressive use of glycogen during the game, which decreases performance in the second half [49].

As a conclusion of the analysis of our results, it is verified that the OMs present higher values in the maximal speed and in total sprint distance, which indicates that the intensity is higher in the displacements that underlie the decisive actions of the game.

Regarding the concern recently expressed by Calleja-Gonzalez et al. [1] that FMs are becoming less and less friendly, we found that in all friendly games performed, the intensity was lower than OMs.

The present study points out, as a main limitation, the small sample size. Only players with at least three consecutive matches participated, which did not allow to provide further insights into the players with lower match participation, known in other studies as non-starters. Furthermore, because all players were from the same team, it is unclear whether the results obtained would be generalizable to other teams and competitive levels. Nevertheless, the present study represents the actual training and competition environment from athletes.

5. Conclusions

The results of this study provided evidence for the difference in activity patterns between OMs and FMs in male professional soccer players. Specifically, OMs showed higher demands in the high-intensity domain, questioning the original assumption of FMs demands.

Furthermore, and because of this study, the use of FMs within the pre-season phase or during the in-season should warrant additional care when planned between high-intensity and high-volume training. For instance, pre-season FMs should be prepared to progressively increase the intensity in training program, which theoretically means that lower intensity should be applied in an early phase of the pre-season period when compared to the final phase of in-season periods. In addition, during in-season, FMs should be performed in the weeks without OMs in order to keep a day in the week with higher intensity, once matches constitute the most demanding intensity to players.

Despite all findings from this study, the results should be carefully interpreted due to the small sample size. Therefore, it is suggested to conduct more studies with identical design to confirm the present findings.

Author Contributions: Conceptualization, H.N., R.O., and J.P.-G., methodology, H.N., R.O., J.P.B., and J.P.-G., data collection, H.N., analysis, H.N., R.O., and J.P.-G., writing—original draft preparation, H.N., J.P.B., R.O., and J.P.-G., writing—review and editing, H.N., J.P.B., R.O., and J.P.-G. All authors have read and agreed to the published version of the manuscript.

Funding: This study was funded by the Portuguese Foundation for Science and Technology, I.P., grant/award number UIDP/04748/2020.

Institutional Review Board Statement: The training coaches of the club, after obtaining permission from the relevant authorities and the head coach of the club, conducted this research. Before commencing the study, approval of the research ethics committee from the University of Isfahan and Mohaghegh Ardabili University was received.

Informed Consent Statement: All players were informed of the purpose of the study before completing the informed consent. All stages of this study were carried out based on the ethical principles in the Helsinki Declaration. Written informed consent has been obtained from the participants to publish this paper.

Data Availability Statement: Data are available through the corresponding authors: hadi.nobari1@gmail.com (Hadi Nobari) and rafaeloliveira@esdrm.ipsantarem.pt (Rafael Oliveira).

Acknowledgments: The authors would like to thank the team's coaches and players for their cooperation during all data collection procedures.

Conflicts of Interest: The authors declare no conflict of interest. The funders had no role in the design of the study; in the collection, analyses, or interpretation of data, in the writing of the manuscript, or in the decision to publish the results.

References

1. Calleja-Gonzalez, J.; Lalín, C.; Cos, F.; Marques-Jimenez, D.; Alcaraz, P.E.; Gómez-Díaz, A.J.; Freitas, T.T.; Mielgo Ayuso, J.; Loturco, I.; Peirau, X.; et al. SOS to the Soccer World. Each Time the Preseason Games Are Less Friendly. *Front. Sports Act. Living* **2020**, *2*, 210. [CrossRef] [PubMed]
2. Windt, J.; Gabbett, T.J.; Ferris, D.; Khan, K.M. Training load-Injury paradox: Is greater preseason participation associated with lower in-season injury risk in elite rugby league players? *Br. J. Sports Med.* **2017**, *51*, 645–650. [CrossRef] [PubMed]
3. Loturco, I.; Pereira, L.A.; Reis, V.P.; Bishop, C.; Zanetti, V.; Alcaraz, P.E.; Freitas, T.T.; Mcguigan, M.R. Power training in elite young soccer players: Effects of using loads above or below the optimum power zone. *J. Sports Sci.* **2020**, *38*, 1416–1422. [CrossRef] [PubMed]
4. Di Salvo, V.; Baron, R.; Tschan, H.; Calderon Montero, F.J.; Bachl, N.; Pigozzi, F. Performance characteristics according to playing position in elite soccer. *Int. J. Sports Med.* **2007**, *28*, 222–227. [CrossRef]
5. Fessi, M.S.; Nouira, S.; Dellal, A.; Owen, A.; Elloumi, M.; Moalla, W. Changes of the psychophysical state and feeling of wellness of professional soccer players during pre-season and in-season periods. *Res. Sports Med.* **2016**, *24*, 375–386. [CrossRef] [PubMed]

6. Androulakis, N.E.; Koundourakis, N.E.; Nioti, E.; Spatharaki, P.; Hatzisymeon, D.; Miminas, I.; Alexandrakis, M.G. Preseason preparation training and endothelial function in elite professional soccer players. *Vasc. Health Risk Manag.* **2015**, *11*, 595–599. [CrossRef] [PubMed]

7. Carey, D.L.; Crow, J.; Ong, K.L.; Blanch, P.; Morris, M.E.; Dascombe, B.J.; Crossley, K.M. Optimizing preseason training loads in Australian football. *Int. J. Sports Physiol. Perform.* **2018**, *13*, 194–199. [CrossRef]

8. Campos-Vázquez, M.Á.; Castellano, J.; Toscano-Bendala, F.J.; Owen, A. Comparison of the physical and physiological demands of friendly matches and different types of preseason training sessions in professional soccer players. *Rev. Int. Cienc. Deport.* **2019**, *58*, 339–352. [CrossRef]

9. Castellano, J.; Casamichana, D. Differences in the number of accelerations between small-sided games and friendly matches in soccer. *J. Sports Sci. Med.* **2013**, *12*, 209–210.

10. Casamichana, D.; Castellano, J.; Castagna, C. Comparing The Physical Demands Of Friendly Matches And Small-Sided Games In Semiprofessional Soccer Players. *Strength Cond.* **2012**, *23*, 837–843. [CrossRef] [PubMed]

11. Rabelo, F.N.; Pasquarelli, B.N.; Gonçalves, B.; Matzenbacher, F.; Campos, F.; Sampaio, J.; Nakamura, F. Monitoring the intended and perceived training load of a professional futsal team over 45 weeks, a case study. *J. Strength Cond. Res.* **2016**, *16*, 134–140. [CrossRef] [PubMed]

12. Lee, A.J.; Garraway, W.M.; Arneil, D.W. Influence of preseason training, fitness, and existing injury on subsequent rugby injury. *Br. J. Sports Med.* **2001**, *35*, 412–417. [CrossRef]

13. Oliveira, R.; Brito, J.; Martins, A.; Mendes, B.; Calvete, F.; Carriço, S.; Ferraz, R.; Marques, M.C. In-season training load quantification of one-, two- and three-game week schedules in a top European professional soccer team. *Physiol. Behav.* **2019**, *201*, 146–156. [CrossRef]

14. Eliakim, E.; Doron, O.; Meckel, Y.; Nemet, D.; Eliakim, A. Pre-season Fitness Level and Injury Rate in Professional Soccer—A Prospective Study. *Sports Med. Int. Open* **2018**, *2*, 84–90. [CrossRef]

15. Clemente, F.M.; Nikolaidis, P.T.; Linden, C.M.I. (Niels) van der; Silva, B. Effects of small-sided soccer games on internal and external load and lower limb power: A pilot study in collegiate players. *Hum. Mov.* **2017**, *18*, 50–57. [CrossRef]

16. Clemente, F.M.; Mendes, B.; Nikolaidis, P.T.; Calvete, F.; Carriço, S.; Owen, A.L. Internal training load and its longitudinal relationship with seasonal player wellness in elite professional soccer. *Physiol. Behav.* **2017**, *179*, 262–267. [CrossRef]

17. Foster, C. Monitoring training in athletes with reference to overtraining syndrome. *Med. Sci. Sports Exerc* **1998**, *30*, 1164–1168. [CrossRef] [PubMed]

18. Coppalle, S.; Rave, G.; Ben Abderrahman, A.; Ali, A.; Salhi, I.; Zouita, S.; Zouita, A.; Brughelli, M.; Granacher, U. Relationship of Pre-season Training Load With In-Season Biochemical Markers, Injuries and Performance in Professional Soccer Players. *Front. Physiol.* **2019**, *10*, 409. [CrossRef] [PubMed]

19. Owen, A.L.; Forsyth, J.J.; Wong, D.P.; Dellal, A.; Connelly, S.P.; Chamari, K. Heart rate–based training intensity and its impact on injury incidence among elite-level professional soccer players. *J. Strength Cond. Res.* **2015**, *29*, 1705–1712. [CrossRef]

20. Hawkins, R.D.; Hulse, M.A.; Wilkinson, C.; Hodson, A.; Gibson, M.; Way, N.; Road, H.; Care, P. The association football medical research programme: An audit of injuries in professional football. *Br. J. Sports Med.* **2001**, *35*, 43–47. [CrossRef]

21. Jeong, T.S.; Reilly, T.; Morton, J.; Bae, S.W.; Drust, B. Quantification of the physiological loading of one week of "pre-season" and one week of "in-season" training in professional soccer players. *J Sports Sci* **2011**, *29*, 1161–1166. [CrossRef]

22. Jones, C.M.; Griffiths, P.C.; Mellalieu, S.D. Training Load and Fatigue Marker Associations with Injury and Illness: A Systematic Review of Longitudinal Studies. *Sports Med.* **2016**, *47*, 943–974. [CrossRef]

23. Aquino, R.; Gonçalves, L.G.; Galgaro, M.; Maria, T.S.; Rostaiser, E.; Pastor, A.; Nobari, H.; Garcia, G.R.; Moraes-Neto, M.V.; Nakamura, F.Y. Match running performance in Brazilian professional soccer players: Comparisons between successful and unsuccessful teams. *BMC Sports Sci. Med. Rehabil.* **2021**, *13*, 1–10. [CrossRef]

24. Nobari, H.; Oliveira, R.; Siahkouhian, M.; Pérez-Gómez, J.; Cazan, F.; Ardigò, L. Variations of accelerometer and metabolic power global positioning system variables across a soccer season: A within-group study for starters and non-starters. *Appl. Sci.* **2021**, *11*, 6747. [CrossRef]

25. Nobari, H.; Sögüt, M.; Oliveira, R.; Pérez-Gómez, J.; Suzuki, K.; Zouhal, H. Wearable inertial measurement unit to accelerometer-based training monotony and strain during a soccer season: A within-group study for starters and non-starters. *Int. J. Environ. Res. Public Health* **2021**, *18*, 8007. [CrossRef] [PubMed]

26. Bourdon, P.C.; Cardinale, M.; Murray, A.; Gastin, P.; Kellmann, M.; Varley, M.C.; Gabbett, T.J.; Coutts, A.J.; Burgess, D.J.; Gregson, W.; et al. Monitoring athlete training loads: Consensus statement. *Int. J. Sports Physiol. Perform.* **2017**, *12*, 161–170. [CrossRef]

27. Scott, M.T.U.; Scott, T.J.; Kelly, V.G. The Validity and Reliability of Global Positioning Systems in Team Sport. *J. Strength Cond. Res.* **2016**, *30*, 1470–1490. [CrossRef]

28. Coutts, A.J.; Duffield, R. Validity and reliability of GPS devices for measuring movement demands of team sports. *J. Sci. Med. Sport* **2010**, *13*, 133–135. [CrossRef] [PubMed]

29. Staunton, C.A.; Abt, G.; Weaving, D.; Wundersitz, D.W.T. Misuse of the term 'load' in sport and exercise science. *J. Sci. Med. Sport* **2021**, in press. [CrossRef]

30. Desgorces, F.D.; Senegas, X.; Garcia, J.; Decker, L.; Noirez, P. Methods to quantify intermittent exercises. *App. Physiol. Nutr. Metabol.* **2007**, *769*, 762–769. [CrossRef] [PubMed]

31. Casamichana, D.; Castellano, J. Demandas físicas en jugadores semiprofesionales de fútbol: ¿se entrena igual que se com-pite? *Rev. Cult. Cienc. Deport.* **2011**, *17*, 121–127. [CrossRef]
32. Salvo, D.; Gregson, W.; Atkinson, G.; Tordoff, P.; Drust, B. Analysis of High Intensity Activity in Premier League Soccer. *Int. J. Sports Med.* **2009**, *30*, 205–212. [CrossRef]
33. Bradley, P.S.; Sheldon, W.; Wooster, B.; Olsen, P.; Boanas, P.; Krustrup, P. High-intensity running in English FA Premier League soccer matches. *J. Sports Sci* **2009**, *27*, 159–168. [CrossRef] [PubMed]
34. Torreño, N.; Munguía-Izquierdo, D.; Coutts, A.; De Villarreal, E.S.; Asian-Clemente, J.; Suarez-Arrones, L. Relationship Between External and Internal Load of Professional Soccer Players During Full-Matches in Official Games Using GPS and Heart Rate Technology. *Int. J. Sports Physiol. Perform.* **2016**, *11*, 940–946. [CrossRef]
35. Tessaro, E.; Williams, J.H. Validity and Reliability of a 15 Hz GPS Device for Court-Based Sports Movements. *Sport Perform. Sci. Rep.* **2018**, *29*, 1–4.
36. Nobari, H.; Khalili, S.M.; Oliveira, R.; Castillo-Rodríguez, A.; Jorge, P.; Paolo, L. Comparison of Official and Friendly Matches through Acceleration, Deceleration and Metabolic Power Measures: A Full-Season Study in Professional Soccer Players. *Int. J. Environ. Res. Public Health* **2021**, *18*, 5980. [CrossRef] [PubMed]
37. Faul, F.; Erdfelder, E.; Lang, A.-G.; Buchner, A. G* power 3: A flexible statistical power analysis program for the social, Behavioral and biomedical sciences. *Behav. Res. Methods* **2007**, *39*, 175–191. [CrossRef]
38. Hopkins, W.G.; Marshall, S.W.; Batterham, A.M.; Hanin, J. Progressive statistics for studies in sports medicine and exercise science. *Med. Sci. Sports Exerc.* **2009**, *41*, 3–13. [CrossRef]
39. Akenhead, R.; Harley, J.A.; Tweddle, S.P. Examining the external training load of an English Premier League football team with special reference to acceleration. *J. Strength Cond. Res.* **2016**, *30*, 2424–2432. [CrossRef]
40. Ade, J.; Fitzpatrick, J.; Bradley, P.S.; Ade, J.; Fitzpatrick, J.; High-intensity, P.S.B. High-intensity efforts in elite soccer matches and associated movement patterns, technical skills and tactical actions. Information for position- specific training drills. *J. Sports Sci.* **2016**, *34*, 2205–2214. [CrossRef]
41. Vanrenterghem, J.; Nedergaard, N.J.; Robinson, M.A.; Drust, B. Training Load Monitoring in Team Sports: A Novel Framework Separating Physiological and Biomechanical Load-Adaptation Pathways. *Sports Med.* **2017**, *47*, 2135–2142. [CrossRef]
42. Ehrmann, F.E.; Duncan, C.S.; Sindhusake, D.; Franzsen, W.N.; Greene, D.A. GPS and injury prevention in professional soccer. *J. Strength Cond. Res.* **2016**, *30*, 360–367. [CrossRef] [PubMed]
43. Gomez-Piriz, P.T.; Jiménez-Reyes, P.; Ruiz-Ruiz, C. Relation between total body load and Session-RPE in professional soc-cer players. *J. Strength Cond. Res.* **2011**, *25*, 2100–2103. [CrossRef] [PubMed]
44. Reilly, T.; Bowen, T. Exertional costs of changes in directional modes of running. *Percept. Mot. Ski.* **1984**, *58*, 149–150. [CrossRef]
45. Scott, B.R.; Lockie, R.G.; Knight, T.J.; Clark, A.C.; De Jonge, X.A.K.J. A comparison of methods to quantify the in-season training load of professional soccer players. *Int. J. Sports Physiol. Perform.* **2013**, *8*, 195–202. [CrossRef]
46. Malone, J.; Di Michele, R.; Morgans, R.; Burgess, D.; Morton, J.; Drust, B. Seasonal Training-Load Quantification in Elite English Premier League Soccer Players. *Int. J. Sports Physiol. Perform.* **2015**, *10*, 489–497. [CrossRef]
47. Mortimer, L.A.C.F.; Condessa, L.; Rodrigues, V.; Coelho, D.; Soares, D.; Silami-Garcia, E. Comparison between the effort intensity of young soccer players in the first and second halves of the soccer game. *Rev. Port. Cien. Desp.* **2006**, *6*, 154–159.
48. Rienzi, E.; Drust, B.; Reilly, T.; Carter, J.E.; Martin, A. Investigation of anthropometric and work-rate profiles of elite South American interna-tional soccer players. *J. Sports Med. Phys. Fitness.* **2000**, *40*, 162–169. [PubMed]
49. Bangsbo, J. The physiology of soccer, with special reference to intense intermittent exercise. *Acta Physiol. Scan.* **1994**, *619*, 1–155.

healthcare

Article

External Match Load in Amateur Soccer: The Influence of Match Location and Championship Phase

Mauro Miguel [1,2,3,*], Rafael Oliveira [2,3,4], João Paulo Brito [2,3,4], Nuno Loureiro [2,3], Javier García-Rubio [1] and Sergio Jose Ibáñez [1]

[1] Training Optimization and Sports Performance Research Group (GOERD), Sport Science Faculty, University of Extremadura, 10003 Cáceres, Spain; jagaru@unex.es (J.G.-R.); sibanez@unex.es (S.J.I.)

[2] Sport Sciences School of Rio Maior, Polytechnic Institute of Santarém, 2040-413 Rio Maior, Portugal; rafaeloliveira@esdrm.ipsantarem.pt (R.O.); jbrito@esdrm.ipsantarem.pt (J.P.B.); nunoloureiro@esdrm.ipsantarem.pt (N.L.)

[3] Life Quality Research Centre (CIEQV), Polytechnic Institute of Santarém, 2001-902 Santarém, Portugal

[4] Research Centre in Sport Sciences, Health Sciences and Human Development, 5001-801 Vila Real, Portugal

[*] Correspondence: mauromiguel@esdrm.ipsantarem.pt

Abstract: Assessment of the physical dimension implicit in the soccer match is crucial for the improvement and individualization of training load management. This study aims to: (a) describe the external match load at the amateur level, (b) analyze the differences between playing positions, (c) verify whether the home/away matches and if (d) the phase (first or second) of the championship influence the external load. Twenty amateur soccer players (21.5 ± 1.9 years) were monitored using the global positioning system. The external load was assessed in 23 matches, where 13 were part of the first phase of the competition (seven home and six away matches) and the other 10 matches belonged to the second (and final) phase of the championship (five home and five away matches). A total of 173 individual match observations were analyzed. The results showed significant differences between playing positions for all the external load measures ($p < 0.001$). There were higher values observed in the total distance covered for central defenders ($p = 0.037$; ES = 0.70) and in high-intensity decelerations for forwards ($p = 0.022$; ES = 1.77) in home matches than in away matches. There were higher values observed in the total distance ($p = 0.026$; ES = 0.76), relative distance ($p = 0.016$; ES = 0.85), and moderate-intensity accelerations ($p = 0.008$; ES = 0.93) for central defenders, in very high-speed running distance for forwards ($p = 0.011$; ES = 1.97), and in high-intensity accelerations ($p = 0.036$; ES = 0.89) and moderate-intensity decelerations ($p = 0.006$; ES = 1.11) for wide midfielders in the first phase than in the second phase of the championship. Match location and championship phase do not appear to be major contributing factors to influence the external load while the playing position should be used as the major reference for planning the external training load.

Keywords: external load; contextual variables; soccer; amateur; home match; away match

Citation: Miguel, M.; Oliveira, R.; Brito, J.P.; Loureiro, N.; García-Rubio, J.; Ibáñez, S.J. External Match Load in Amateur Soccer: The Influence of Match Location and Championship Phase. *Healthcare* **2022**, *10*, 594. https://doi.org/10.3390/healthcare10040594

Academic Editors: Parisi Attilio and Jitendra Singh

Received: 20 January 2022
Accepted: 18 March 2022
Published: 22 March 2022

Publisher's Note: MDPI stays neutral with regard to jurisdictional claims in published maps and institutional affiliations.

1. Introduction

The assessment and knowledge of the physical dimension implicit in the soccer match is crucial in the improvement and individualization of planning structures since specific protocols can be designed in accordance with these demands [1]. In this regard, Bourdon et al. [2] assert that the load monitoring process should assist the coaches' decision-making regarding the players' availability to train and compete in order to achieve the main objectives of performance and injury prevention [3–5]. As manifested by Zurutuza et al. [6], it is essential to individualize training as much as possible in order to strengthen collective training and thus optimize competition performance.

Over the years, several methods have been used to determine the physical profile of soccer players [7]. In this sense, the global positioning system (GPS) and inertial sensors in wearable devices are widely used to measure external loads [8–14], which are objective measures of work performed by an athlete during training or competition [2]. Based on the perspective that measuring loads relative to competition demands could be an advantageous strategy that coaches use within training periodization models [12], the influence of the playing position and contextual factors (e.g., home/away, opponent's standard positions, match period) on the match load has been a subject of particular attention [12,13,15–17] and their effects are well-reported.

In elite soccer, match loads have been being assessed for several years [8–14]. However, at the amateur level, clubs and their coaches do not have sufficient resources to allow them to precisely monitor the training/match load. Based on the identified differences between amateurs and professionals, Dellal et al. [18] suggest an adequation of training for these athlete populations. At the amateur level where athletes associate sports practice with another professional activity, the use of instruments that provide accurate data, even if used in spaced-out periods of the sports season, will allow the collection of data that help coaches to carry out adequate management of loads.

As stated by García-Rubio et al. [19], identifying the interactive effects of contextual variables on performance indicators can enable better team preparation. In this respect, Springham et al. [20] exposed that, of the analyzed contextual variables, only the playing position and goal deficit were identified as predictors of match physical performance. Additionally, some studies [21,22] describe that match location can exert a confounding effect on match physical performance, where home matches place greater physical demands on players compared with playing away. The study by Oliveira et al. [23] confirms that match location can influence internal and external load data preceding home and away matches. Lastly, Reche-Soto et al. [13] consider that physical, technical, tactical, and psychological training should be planned in relation to match location and level of the opponent. Amateur soccer teams usually have a lower frequency of training sessions, which requires physical, technical, tactical, and psychological training to be planned with great thoughtfulness with regard to match location [24]. According to the contextual factors (home/away, first/second phase), it may be necessary to adjust the microcycle to induce optimal physiological and performance recovery before the match [25,26]. In this sense, it is essential to analyze the effect that match location and championship phase can have on effort intensity of players on amateur teams.

Thus, to better identify and understand the current physical match demands placed on players in different playing positions at the amateur level as well as the impacts of contextual variables, this study aims to (a) describe the external match load at the amateur level, (b) analyze the differences in the external match load between playing positions, (c) verify whether the home/away contextual factor impacts the external match load, and (d) determine whether there are differences in match load between the first and the second phases of the championship.

2. Materials and Methods

2.1. Experimental Approach to the Problem

This investigation follows an associative strategy [27] where an attributive variable is utilized and differences between groups are examined. It is a longitudinal observational study performed with amateur soccer players participating in an official Portuguese regional competition.

Match data were collected over the 2018/2019 competitive season from October to June. The standard competitive microcycle included a match (Sunday) and three training sessions (Tuesday, MD–5; Thursday, MD–3; and Friday, MD–2—according to Malone et al. [28], training sessions are classified in relation to the number of days before the next competitive match). Twenty-four matches were observed throughout the data collection period, one of which was excluded from the analysis because the match was interrupted due to weather conditions and completed days later. Out of the 23 analyzed matches, 13 were part of the first phase of the competition (seven home and six away matches) while the other 10 matches belonged to the second (and final) phase of the championship (five home and five away matches). In the first phase, eight teams competed for the qualification to the final phase (the top three to classify would get access to this phase) where they would play for promotion (the analyzed team classified in the first place). In the final phase, six teams competed for the top three places giving access to the higher division (the analyzed team classified in the second place).

Assessment of the demands of a soccer match at the amateur level will allow increasing the knowledge about this little studied context (in terms of training/match monitoring). Evaluating the match load according to match location and championship phase will allow assessing the influence of these variables on the game and the players, and through this type of analysis, planning structures can be properly adapted to the specificities and particularities of the competitive period.

2.2. Participants

Twenty amateur soccer players (age: 21.5 ± 1.9 years old; height: 174.5 ± 7.9 cm; body weight: 71.2 ± 7.6 kg; fat mass: $17.5 \pm 3.9\%$) from the same team that participated in the Portuguese men's soccer championship (regional level) were included in this analysis. Given the preliminary nature of our study, we applied a stringent inclusion criterion. Players were only included in the analysis if they participated in at least four full matches during the data collection period. All the players and coaches were informed about the research protocol, requisites, benefits, and risks, and their written consent was obtained before the start of the study. The study protocol was approved by the ethics committee of the local university (No. 67/2017) and performed according to the ethical standards of the Declaration of Helsinki [29].

The analyzed team played in the 1:4:3:3 formation throughout the season, with two defensive midfielders and one offensive midfielder (these three players are hereinafter referred to as central midfielders). All the analyses were conducted according to playing positions (goalkeeper, central defender, full-back, central midfielder, wide midfielder, and forward). Goalkeeper's data were only used to describe the match load for this playing position and were excluded from the comparative analysis. A total of 173 individual match observations were analyzed: goalkeeper (GK; $n = 3$ players, $n = 18$ cases), central defender (CD; $n = 4$ players, $n = 35$ cases), full-back (FB; $n = 4$ players, $n = 38$ cases), central midfielder (CM; $n = 5$ players, $n = 44$ cases), wide midfielder (WM; $n = 5$ players, $n = 28$ cases), and forward (F; $n = 3$ players, $n = 10$ cases).

2.3. External Match Load

The data from the external match load were collected using portable 10 Hz GPS devices (PlayerTek, Catapult Innovations, Melbourne, Australia); each of them also incorporated a triaxial 100 Hz accelerometer. This type of GPS devices seems to be the most valid and reliable for use in team sports [30].

The PlayerTek inertial devices were turned on and placed in a specific customized vest pocket located on the posterior side of the upper torso fitted tightly to the body, as is typically used in matches. These devices were turned on 10 min before the start of the warm-up period. During the monitoring period, the GPS devices would always be placed and checked by the same coach of the team, and each player would always use the same device [31].

The running variables obtained from the GPS were the total distance covered (TDC, m), the relative distance covered (RDC, m/min), and the distance covered (m) at five different speed thresholds: walking/jogging distance (WJD), 0.0 to 3.0 m/s; running-speed distance (RSD), 3.0 to 4.0 m/s; high-speed running distance (HSRD), 4.0 to 5.5 m/s; very high-speed running distance (VHSRD), 5.5 to 7.0 m/s; and sprint distance (SpD), a speed greater than 7.0 m/s [32]. The total number of accelerations and decelerations in three zones was also analyzed: low-intensity (LI Acc./LI Dec.), 0.0 to 2.0 m/s^2; moderate-intensity (MI Acc./MI Dec.), 2.0 to 4.0 m/s^2; and high-intensity (HI Acc./HI Dec.), greater than 4.0 m/s^2 [9].

Moreover, player load (PL) was also included as a global load indicator in volume (AU) and intensity (AU/min).

2.4. Statiscal Analyses

All the statistical analyses were conducted using SPSS for Windows statistical software package version 22.0 (SPSS Inc., Chicago, IL, USA). Initially, descriptive statistics were used to describe and characterize the sample. Shapiro–Wilk and Levene's tests were conducted to determine normality and homoscedasticity, respectively. One-way ANOVA was used with Scheffe's post-hoc method. One-way analyses of variance were used to compare all the dependent variables (external match load measures) across the playing positions. Student's t-test was also used to compare data by match location (home/away contextual factor) and championship phase (first and second phases). The effect size with 95% confidence interval (ES 95% CI) statistic was calculated to determine the magnitude of effects. Furthermore, Hopkins' thresholds for the effect size statistics were used as follows: $\leq$0.2, trivial; >0.2, small; >0.6, moderate; >1.2, large; >2.0, very large; and >4.0, nearly perfect [33]. Alpha was set at $p \leq 0.05$.

3. Results

3.1. Description of the External Match Load by Playing Position

Description of the 15 dependent variables by playing position is presented in Table 1. The central midfielders were the players who showed the greatest total distance covered (11,020 $\pm$ 720 m) and relative distance covered (112.7 $\pm$ 7.5 m/min). The forwards were those who showed greater very high-speed running distance (667 $\pm$ 158 m) and sprint distance (299 $\pm$ 96.6 m). The forwards and the wide midfielders exhibited the largest number of high-intensity accelerations, 36 $\pm$ 7 and 35 $\pm$ 8, respectively. The forwards were also the ones with the largest number of high-intensity decelerations (49 $\pm$ 7). The central midfielders were those who presented with greater player load, both absolute and relative (476 $\pm$ 31.2 AU and 4.9 $\pm$ 0.3 AU/min, respectively).

Table 1. External match load by playing position (mean $\pm$ SD).

	Goalkeeper	Central Defender	Full-Back	Central Midfielder	Wide Midfielder	Forward	Team (a)
TDC (m)	4852 $\pm$ 592	9443 $\pm$ 547	10,129 $\pm$ 704	11,020 $\pm$ 720	10,003 $\pm$ 1004	10,906 $\pm$ 844	10,265 $\pm$ 963
RDC (m/min)	49.4 $\pm$ 5.7	96.5 $\pm$ 5.2	103.8 $\pm$ 7.2	112.7 $\pm$ 7.5	101.9 $\pm$ 9.8	110.9 $\pm$ 9.3	104.6 $\pm$ 9.8
WSJ (m)	4317 $\pm$ 545	6350 $\pm$ 375	5750 $\pm$ 373	6044 $\pm$ 438	5892 $\pm$ 218	3963 $\pm$ 2712	5914 $\pm$ 879
RSD (m)	324 $\pm$ 65.8	1717 $\pm$ 224	1954 $\pm$ 277	2463 $\pm$ 437	1855 $\pm$ 498	4184 $\pm$ 2705	2179 $\pm$ 957
HSRD (m)	178 $\pm$ 64.5	1048 $\pm$ 154	1551 $\pm$ 413	1832 $\pm$ 336	1501 $\pm$ 414	1494 $\pm$ 636	1505 $\pm$ 461
VHSRD (m)	31.9 $\pm$ 32.6	268 $\pm$ 74.7	636 $\pm$ 188	515 $\pm$ 166	579 $\pm$ 132	667 $\pm$ 158	503 $\pm$ 198
SpD (m)	0.9 $\pm$ 2.3	59.6 $\pm$ 41.8	239 $\pm$ 115	97.5 $\pm$ 77.3	190 $\pm$ 54.7	299 $\pm$ 96.6	148 $\pm$ 111
LI Acc. (n)	104 $\pm$ 23	210 $\pm$ 23	200 $\pm$ 23	244 $\pm$ 41	170 $\pm$ 39	206 $\pm$ 30	212 $\pm$ 40
MI Acc. (n)	73 $\pm$ 17	179 $\pm$ 23	231 $\pm$ 39	244 $\pm$ 38	212 $\pm$ 43	237 $\pm$ 39	221 $\pm$ 44
HI Acc. (n)	10 $\pm$ 6	16 $\pm$ 6	32 $\pm$ 7	29 $\pm$ 9	35 $\pm$ 8	36 $\pm$ 7	29 $\pm$ 10
LI Dec. (n)	94 $\pm$ 24	226 $\pm$ 40	214 $\pm$ 25	228 $\pm$ 33	192 $\pm$ 42	219 $\pm$ 39	219 $\pm$ 36
MI Dec. (n)	68 $\pm$ 17	147 $\pm$ 17	186 $\pm$ 38	218 $\pm$ 29	175 $\pm$ 33	205 $\pm$ 22	186 $\pm$ 40
HI Dec. (n)	10 $\pm$ 4	21 $\pm$ 7	39 $\pm$ 8	44 $\pm$ 12	41 $\pm$ 11	49 $\pm$ 7	37 $\pm$ 13
PL (AU)	207 $\pm$ 17.4	380 $\pm$ 22.3	420 $\pm$ 34.7	476 $\pm$ 31.2	399 $\pm$ 42.5	435 $\pm$ 22.5	425 $\pm$ 49.0
PL (AU/min)	2.1 $\pm$ 0.2	3.9 $\pm$ 0.2	4.3 $\pm$ 0.4	4.9 $\pm$ 0.3	4.1 $\pm$ 0.4	4.4 $\pm$ 0.3	4.3 $\pm$ 0.5

SD = standard deviation; TDC = total distance covered; RDC = relative distance covered; WJD = walking/jogging distance (0.0 to 3.0 m/s); RSD = running-speed distance (3.0 to 4.0 m/s); HSRD = high-speed running distance (4.0 to 5.5 m/s); VHSRD = very high-speed running distance (5.5 to 7.0 m/s); SpD = sprint distance (>7.0 m/s); LI Acc. = low-intensity accelerations (0.0 to 2.0 m/s^2); MI Acc. = moderate-intensity accelerations (2.0 to 4.0 m/s^2); HI Acc. = high-intensity accelerations (>4.0 m/s^2); LI Dec. = low-intensity decelerations (0.0 to -2.0 m/s^2); MI Dec. = moderate-intensity decelerations (-2.0 to -4.0 m/s^2); HI Dec. = high-intensity decelerations (>-4.0 m/s^2); PL = player load; AU = arbitrary units; m = meters; min = minutes; (a) all playing positions, excluding goalkeeper's data.

3.2. External Match Load—Comparison between Playing Positions

Significant differences were found between playing positions for all the external match load measures ($p \leq 0.001$) and can be observed in detail in Table 2. Regarding the total distance covered, the central defenders (CD) presented with a significantly smaller distance compared to the other playing positions ($p = 0.006$; ES = −1.07 to −2.40), except for the wide midfielders (W), $p = 0.075$. The central midfielders showed a larger total distance than the other playing positions ($p = 0.000$; ES = 1.20 to 2.40), except for the forwards ($p = 0.996$). Regarding the relative distance covered, it appears that the central midfielders had a greater relative distance covered than the other playing positions ($p = 0.000$; ES = 1.20 to 2.44), with the exception of the forwards ($p = 0.980$); in relation to the distance covered at different speed zones, the central defenders and the full-backs showed significant differences in all the zones ($p = 0.025$; ES = −2.51 to 1.59), with the exception of the running-speed distance ($p = 0.774$). A tendency towards the higher-intensity speed zone was noted for the full-backs, while the walking/jogging distance had higher values among the central defenders.

Table 2. External match load—comparison between playing positions, CI 95%, and ES.

	CD vs. FB	CD vs. CM	CD vs. WM	CD vs. F	FB vs. CM	FB vs. WM	FB vs. F	CM vs. WM	CM vs. F	WM vs. F
TDC (m)	−1234.6 to −138.3 **/moderate	−2107.0 to −1047.1 ***/very large	−1153.8 to 32.7	−2302.5 to −624.6 ***/very large	−1408.8 to −372.5 ***/large	−456.8 to 708.7	−1608.7 to 54.5	450.9 to 1582.2 ***/large	−706.2 to 933.2	−1765.0 to −41.0 */moderate
RDC (m/min)	−12.8 to −1.8 **/moderate	−21.5 to −10.8 ***/very large	−11.4 to 0.6	−22.9 to −5.9 ***/very large	−14.1 to −3.6 ***/large	−3.9 to 7.8	−15.5 to 1.3	5.1 to 16.5 ***/large	−6.5 to 10.0	−17.8 to −0.3 */moderate
WSJ (m)	48.5 to 1152.7 */large	−227.3 to 840.2	−139.3 to 1055.7	1542.0 to 3232.0 ***/large	−816.0 to 227.8	−729.3 to 444.6	948.9 to 2624.0 ***/large	−418.0 to 721.4	1254.9 to 2906.1 ***/large	1060.7 to 2797.0 ***/large
RSD (m)	−787.7 to 314.8	−1278.6 to −212.8 **/moderate	−734.4 to 458.7	−3610.4 to −1923.1 ***/very large	−1030.3 to 11.8	−487.4 to 684.6	−3366.5 to −1694.1 ***/very large	39.0 to 1176.7 */large	−2845.3 to −1196.8 ***/large	−3495.7 to −1762.1 ***/large
HSRD (m)	−771.7 to −234.9 ***/large	−1043.4 to −524.5 ***/large	−743.4 to −162.5 ***/large	−857.0 to −35.5 */large	−534.4 to −27.0 */moderate	−235.0 to 335.6	−350.1 to 464.2	54.0 to 607.9 */moderate	−63.6 to 739.1	−415.3 to 428.8
VHSRD (m)	−477.3 to −257.9 ***/very large	−352.6 to −140.5 ***/very large	−429.3 to −191.9 ***/very large	−566.4 to −230.7 ***/very large	17.4 to 224.8 */moderate	−59.6 to 173.7	−197.3 to 135.5	−177.3 to 49.1	−316.1 to 12.0	−260.5 to 84.5
SpD (m)	−240.1 to −119.3 ***/very large	−96.3 to 20.5	−195.8 to −65.0 ***/very large	−331.5 to −146.5 ***/nearly perfect	84.7 to 198.9 ***/large	−14.9 to 113.6	−151.0 to 32.4	−154.8 to −30.1 ***/large	−291.5 to −110.7 ***/moderate	−203.7 to −13.6 */large
LI Acc. (n)	−14 to 33	−56 to −6 */moderate	15 to 66 ***/large	−32 to 41	−66 to −21 ***/large	6 to 56 */moderate	−41 to 31	50 to 99 ***/large	3 to 74 */moderate	−73 to 1
MI Acc. (n)	−79 to −25 ***/large	−91 to −39 ***/very large	−62 to −5 */moderate	−99 to −17 **/very large	−38 to 12	−10 to 47	−46 to 35	4 to 59 */moderate	−33 to 47	−66 to 18
HI Acc. (n)	−21 to −10 ***/very large	−18 to −7 ***/large	−24 to −13 ***/very large	−28 to −12 ***/very large	−2 to 9	−8 to 3	−12 to 4	−12 to 0 */moderate	−16 to 1	−10 to 7
LI Dec. (n)	−14 to 37	−27 to 23	6 to 61 */moderate	−32 to 46	−38 to 10	−5 to 49	−44 to 34	9 to 62 **/moderate	−29 to 47	−67 to 14
MI Dec. (n)	−60 to −17 ***/large	−92 to −50 ***/very large	−52 to −4 */moderate	−91 to −24 ***/very large	−53 to −12 ***/moderate	−13 to 34	−52 to 14	21 to 66 ***/large	−19 to 46	−64 to 4
HI Dec. (n)	−25 to −11 ***/very large	−29 to −16 ***/very large	−28 to −12 ***/very large	−38 to −17 ***/very large	−11 to 2	−9 to 6	−20 to 1	−5 to 10	−16 to 5	−19 to 4

Table 2. *Cont.*

	CD vs. FB	CD vs. CM	CD vs. WM	CD vs. F	FB vs. CM	FB vs. WM	FB vs. F	CM vs. WM	CM vs. F	WM vs. F
PL (AU)	−63.9 to −16.7 ***/large	−119.5 to −73.8 ***/very large	−45.4 to 5.7	−91.5 to −19.2 ***/very large	−78.7 to −34.0 ***/large	−4.7 to 45.5	−50.9 to 20.8	52.4 to 101.1 ***/very large	6.0 to 76.6 **/large	−72.6 to 1.7
PL (AU/min)	−0.7 to −0.2 ***/large	−1.2 to −0.8 ***/very large	−0.5 to 0.1	−0.9 to −0.2 ***/very large	−0.8 to −0.3 ***/large	−0.3 to 0.5	−0.5 to 0.3	0.5 to 1.1 ***/very large	0.1 to 0.8 **/large	−0.7 to 0.0

CD = central defender; FB = full-back; CM = central midfielder; WM = wide midfielder; F = forward; TDC = total distance covered; RDC = relative distance covered; WJD = walking/jogging distance (0.0 to 3.0 m/s); RSD = running-speed distance (3.0 to 4.0 m/s); HSRD = high-speed running distance (4.0 to 5.5 m/s); VHSRD = very high-speed running distance (5.5 to 7.0 m/s); SpD = sprint distance (>7.0 m/s); LI Acc. = low-intensity accelerations (0.0 to 2.0 m/s^2); MI Acc. = moderate-intensity accelerations (2.0 to 4.0 m/s^2); HI Acc. = high-intensity accelerations (>4.0 m/s^2); LI Dec. = low-intensity decelerations (0.0 to −2.0 m/s^2); MI Dec. = moderate-intensity decelerations (−2.0 to −4.0 m/s^2); HI Dec. = high-intensity decelerations (>−4.0 m/s^2); PL = player load; AU = arbitrary units; m = meters; min = minutes; CI 95% = 95% confidence interval; * $p < 0.05$; ** $p < 0.005$; *** $p < 0.001$. Effect size: ≤ 0.2, trivial; > 0.2, small; > 0.6, moderate; > 1.2, large; > 2.0, very large; and > 4.0, nearly perfect.

The central defenders and the forwards presented with significant differences in all the speed zones ($p = 0.025$; ES = −4.07 to 1.83). The forwards presented with higher values for running categories than the central defenders. Despite the forwards presenting with the highest average values in very high-speed running distance and sprint distance, there were no significant differences compared to the full-backs ($p = 0.400$) who presented with slightly lower values. In the three "acceleration zones", the central midfielders showed higher values than both the central defenders ($p \leq 0.001$; ES = 0.98 to 2.00) and the wide midfielders ($p = 0.030$; ES = −0.69 to 1.82). The central defenders presented with significant differences with all the other playing positions in moderate- ($p = 0.013$; ES = −0.98 to −2.10) and high-intensity accelerations ($p = 0.000$; ES = −1.65 to −3.16) and in moderate- ($p \leq 0.011$; ES = −1.09 to −3.14) and high-intensity decelerations ($p = 0.000$; ES = −2.20 to −3.93).

Concerning the player load (AU), the central defenders presented with significantly lower values than all the other playing positions ($p = 0.000$; ES = −1.34 to −3.44), except for the wide midfielders ($p = 0.213$). The central midfielders presented with significantly higher values than all the other playing positions ($p = 0.012$; ES = 1.35 to 3.44). The wide midfielders only had significant differences with the central midfielders ($p = 0.000$; ES = −0.97). The same results were verified for the relative player load (AU/min).

3.3. External Match Load—Home vs. Away and First vs. Second Championship Phase

Variations in the external match load by playing position, between home and away matches, and between the first and the second championship phases are presented in Tables 3 and 4, respectively. In the match location, significant differences were observed in the total distance covered for the central defenders ($p = 0.037$; ES = 0.70) and in high-intensity decelerations for the forwards ($p = 0.022$; ES = 1.77)—higher values in home matches. In the championship phase, significant differences were observed in the total distance covered ($p = 0.026$; ES = 0.76), relative distance covered ($p = 0.016$; ES = 0.85), and moderate-intensity accelerations ($p = 0.008$; ES = 0.93) for the central defenders, in very high-speed running distance for the forwards ($p = 0.011$; ES = 1.97), and in high-intensity accelerations ($p = 0.036$; ES = 0.89) and moderate-intensity decelerations ($p = 0.006$; ES = 1.11) for the wide midfielders—higher values in the first phase of the championship.

Table 3. External match load—comparison between match location (home/away), mean ± SD, CI 95%, and ES.

	Central Defender		Full-Back		Central Midfielder		Wide Midfielder		Forward	
	Home	Away	Home	Away	Home	Away	Home	Away	Home	Away
TDC (m)	9646 ± 431	9271 ± 585	10,333 ± 583	9925 ± 769	11,163 ± 650	10,863 ± 775	10,214 ± 965	9793 ± 1032	10,948 ± 914	10,865 ± 875
	23.9 to 725.2/*/moderate		−40.9 to 857.1		−134.0 to 733.6		−355.1 to 1197.1		−1221.7 to 1387.3	
RDC (m/min)	98.0 ± 4.3	95.3 ± 5.6	105.3 ± 6.3	102.4 ± 8.0	113.2 ± 7.5	112.0 ± 7.6	103.1 ± 9.8	100.6 ± 10.0	110.2 ± 10.0	111.6 ± 9.7
	−0.9 to 6.1		−1.8 to 7.6		−3.4 to 5.8		−5.2 to 10.2		−15.8 to 13.0	
WSJ (m)	6465 ± 358	6253 ± 369	5771 ± 368	5729 ± 387	5973 ± 552	6120 ± 256	5921 ± 214	5863 ± 227	3852 ± 2768	4074 ± 2975
	−39.3 to 463.7		−206.5 to 290.5		−413.9 to 118.2		−113.9 to 228.7		−4412.6 to 3968.6	
RSD (m)	1760 ± 193	1681 ± 246	2033 ± 237	1874 ± 297	2482 ± 478	2442 ± 399	1956 ± 439	1755 ± 549	4542 ± 2925	4426 ± 2813
	−75.7 to 233.1		−17.9 to 335.3		−229.3 to 309.2		−184.9 to 587.4		−4068.3 to 4301.1	
HSRD (m)	1073 ± 165	1026 ± 146	1637 ± 375	1465 ± 440	1906 ± 289	1750 ± 370	1598 ± 313	1403 ± 487	1569 ± 175	1419 ± 609
	−59.2 to 154.6		−96.6 to 441.7		−44.9 to 357.5		−122.6 to 513.8		−826.1 to 1126.2	
VHSRD (m)	281 ± 86.4	257 ± 63.5	653 ± 201	619 ± 178	555 ± 179	470 ± 141	590 ± 142	567 ± 127	677 ± 174	657 ± 160
	−27.3 to 75.9		−90.4 to 159.2		−13.4 to 184.3		−81.3 to 127.5		−223.9 to 263.8	
SpD (m)	65.9 ± 49.6	54.3 ± 34.5	240 ± 110	239 ± 123	113 ± 98.8	80.2 ± 65.0	175 ± 39.7	205 ± 64.4	308 ± 132	289 ± 58.1
	−17.4 to 40.7		−76.3 to 77.4		−18.4 to 84.4		−71.8 to 11.3		−130.1 to 167.1	
LI Acc. (n)	212 ± 24	208 ± 23	202 ± 27	199 ± 19	240 ± 41	248 ± 41	175 ± 42	164 ± 38	211 ± 35	200 ± 27
	−13 to 19		−12 to 18		−33 to 17		−20 to 42		−34 to 57	
MI Acc. (n)	184 ± 19	175 ± 26	238 ± 41	223 ± 35	242 ± 36	246 ± 41	217 ± 38	207 ± 49	246 ± 35	227 ± 44
	−6 to 25		−10 to 40		−28 to 19		−24 to 44		−39 to 77	
HI Acc. (n)	14 ± 6	18 ± 5	33 ± 7	31 ± 7	31 ± 8	27 ± 9	33 ± 9	37 ± 8	35 ± 9	38 ± 4
	−7 to 0		−3 to 7		−2 to 9		−10 to 3		−13 to 7	
LI Dec. (n)	227 ± 48	225 ± 32	216 ± 24	213 ± 27	223 ± 38	234 ± 25	199 ± 40	185 ± 45	210 ± 41	228 ± 40
	−26 to 30		−14 to 19		−31 to 9		−19 to 47		−77 to 41	
MI Dec. (n)	149 ± 16	145 ± 18	194 ± 40	177 ± 34	216 ± 28	221 ± 31	179 ± 31	171 ± 36	215 ± 21	195 ± 21
	−9 to 15		−8 to 41		−22 to 14		−18 to 34		−11 to 49	
HI Dec. (n)	21 ± 7	21 ± 7	41 ± 8	38 ± 8	44 ± 11	43 ± 13	42 ± 11	41 ± 11	54 ± 4	44 ± 6
	−5 to 5		−2 to 8		−6 to 9		−7 to 10		2 to 18/*/large	
PL (AU)	384 ± 18.7	376 ± 24.8	427 ± 32.4	414 ± 36.5	479 ± 28.3	473 ± 34.5	409 ± 42.4	390 ± 42.0	436 ± 25.8	435 ± 21.8
	−7.1 to 23.7		−9.8 to 35.6		−13.0 to 25.3		−13.8 to 51.7		−33.7 to 36.0	
PL (AU/min)	3.9 ± 0.2	3.9 ± 0.2	4.3 ± 0.3	4.3 ± 0.4	4.9 ± 0.3	4.9 ± 0.3	4.1 ± 0.4	4.0 ± 0.4	4.4 ± 0.3	4.5 ± 0.2
	−0.1 to 0.2		−0.2 to 0.3		−0.2 to 0.2		−0.2 to 0.5		−0.5 to 0.3	

TDC = total distance covered; RDC = relative distance covered; WJD = walking/jogging distance (0.0 to 3.0 m/s); RSD = running-speed distance (3.0 to 4.0 m/s); HSRD = high-speed running distance (4.0 to 5.5 m/s); VHSRD = very high-speed running distance (5.5 to 7.0 m/s); SpD = sprint distance (>7.0 m/s); LI Acc. = low-intensity accelerations (0.0 to 2.0 m/s2); MI Acc. = moderate-intensity accelerations (2.0 to 4.0 m/s2); HI Acc. = high-intensity accelerations (>4.0 m/s2); LI Dec. = low-intensity decelerations (0.0 to −2.0 m/s2); MI Dec. = moderate-intensity decelerations (−2.0 to −4.0 m/s2); HI Dec. = high-intensity decelerations (>−4.0 m/s2); PL = player load; AU = arbitrary units; m = meters; min = minutes; CI 95% = 95% confidence interval; * $p < 0.05$. Effect size: ≤0.2, trivial; >0.2, small; >0.6, moderate; >1.2, large; >2.0, very large; and >4.0, nearly perfect.

Table 4. External match load—comparison between the championship phases (first phase/second phase), mean ± SD, CI 95%, and ES.

	Central Defender		Full-Back		Central Midfielder		Wide Midfielder		Forward	
	First Phase	Second Phase	First Phase	Second Phase	First Phase	Second Phase	First Phase	Second Phase	First Phase	Second Phase
TDC (m)	9603 ± 550	9201 ± 461	10,052 ± 650	10,260 ± 797	11,067 ± 728	10,951 ± 724	10,090 ± 1102	9887 ± 889	11,266 ± 926	10,752 ± 832
	52.3 to 753.7/*/moderate		−690.1 to 273.3		−332.8 to 565.7		−596.0 to 1001.0		−848.6 to 1875.8	
RDC (m/min)	98.2 ± 5.0	94.0 ± 4.5	103.1 ± 6.9	105.1 ± 7.7	113.3 ± 7.7	111.8 ± 7.3	102.8 ± 10.9	100.6 ± 8.4	112.7 ± 9.9	110.2 ± 9.8
	0.8 to 7.5/*/moderate		−7.0 to 2.8		−3.1 to 6.2		−5.6 to 10.0		−13.0 to 18.2	
WSJ (m)	6439 ± 367	6216 ± 357	5777 ± 370	5703 ± 388	5958 ± 521	6167 ± 242	5833 ± 210	5970 ± 212	6008 ± 206	3087 ± 2833
	−32.3 to 477.8		−183.1 to 330.5		−475.9 to 56.9		−302.7 to 28.5		−987.3 to 6829.1	
RSD (m)	1772 ± 236	1636 ± 182	1911 ± 269	2028 ± 283	2463 ± 469	2462 ± 400	1988 ± 506	1679 ± 449	2251 ± 628	5441 ± 2700
	−16.3 to 288.0		−304.2 to 70.1		−273.6 to 273.9		−69.6 to 687.8		−6944.3 to 564.1	
HSRD (m)	1081 ± 158	998 ± 141	1507 ± 406	1626 ± 429	1901 ± 334	1732 ± 321	1551 ± 405	1433 ± 434	1918 ± 312	1312 ± 668
	−23.4 to 188.8		−401.7 to 163.5		−34.5 to 372.6		−209.2 to 446.3		−347.5 to 1558.7	
VHSRD (m)	256 ± 71.2	285 ± 79.0	622 ± 171	660 ± 219	531 ± 166	491 ± 168	559 ± 144	605 ± 116	837 ± 81.8	593 ± 120
	−81.2 to 23.2		−167.7 to 90.8		−63.1 to 143.3		−150.0 to 58.5		82.1 to 405.8/*/large	
SpD (m)	55.9 ± 43.8	65.2 ± 39.6	237 ± 94.5	244 ± 148	97.1 ± 79.1	98.0 ± 95.4	182 ± 47.0	201 ± 64.1	252 ± 31.9	319 ± 110
	−38.9 to 20.3		−87.1 to 72.1		−54.2 to 52.4		−62.4 to 23.8		−220.3 to 86.9	
LI Acc. (n)	215 ± 25	203 ± 18	197 ± 21	207 ± 25	245 ± 42	242 ± 41	180 ± 42	155 ± 31	204 ± 36	206 ± 30
	−4 to 28		−25 to 6		−22 to 28		−5 to 55		−53 to 38	
MI Acc. (n)	187 ± 21	167 ± 21	231 ± 32	230 ± 49	251 ± 38	233 ± 36	223 ± 46	199 ± 37	268 ± 36	223 ± 33
	6 to 35/*/moderate		−25 to 28		−5 to 41		−10 to 57		−10 to 99	
HI Acc. (n)	17 ± 7	16 ± 4	32 ± 7	33 ± 8	30 ± 9	28 ± 9	38 ± 8	31 ± 7	41 ± 8	35 ± 6
	−3 to 5		−6 to 4		−4 to 7		1 to 13/*/moderate		−4 to 16	
LI Dec. (n)	234 ± 44	214 ± 29	213 ± 22	216 ± 31	230 ± 39	225 ± 22	201 ± 47	181 ± 34	218 ± 58	219 ± 35
	−8 to 47		−20 to 15		−15 to 26		−13 to 52		−68 to 65	
MI Dec. (n)	150 ± 17	143 ± 18	182 ± 38	191 ± 39	224 ± 30	211 ± 28	189 ± 31	156 ± 26	222 ± 19	197 ± 20
	−6 to 19		−35 to 17		−5 to 30		10 to 55/*/moderate		−6 to 56	
HI Dec. (n)	21 ± 8	22 ± 6	39 ± 9	40 ± 7	44 ± 13	43 ± 10	42 ± 10	40 ± 12	47 ± 9	49 ± 7
	−6 to 4		−6 to 5		−6 to 9		−7 to 10		−14 to 10	
PL (AU)	385 ± 22.9	372 ± 19.6	416 ± 31.0	427 ± 40.5	478 ± 34.9	475 ± 25.8	405 ± 44.4	393 ± 40.9	441 ± 32.6	433 ± 19.5
	−2.2 to 28.3		−34.2 to 12.2		−16.8 to 22.3		−22.4 to 43.0		−28.4 to 46.2	
PL (AU/min)	3.9 ± 0.2	3.8 ± 0.2	4.3 ± 0.3	4.4 ± 0.4	4.9 ± 0.4	4.8 ± 0.3	4.1 ± 0.4	4.0 ± 0.4	4.4 ± 0.4	4.4 ± 0.2
	−0.1 to 0.3		−0.3 to 0.1		−0.2 to 0.3		−0.2 to 0.5		−0.5 to 0.4	

TDC = total distance covered; RDC = relative distance covered; WJD = walking/jogging distance (0.0 to 3.0 m/s); RSD = running-speed distance (3.0 to 4.0 m/s); HSRD = high-speed running distance (4.0 to 5.5 m/s); VHSRD = very high-speed running distance (5.5 to 7.0 m/s); SpD = sprint distance (>7.0 m/s); LI Acc. = low-intensity accelerations (0.0 to 2.0 m/s^2); MI Acc. = moderate-intensity accelerations (2.0 to 4.0 m/s^2); HI Acc. = high-intensity accelerations (>4.0 m/s^2); LI Dec. = low-intensity decelerations (0.0 to −2.0 m/s^2); MI Dec. = moderate-intensity decelerations (−2.0 to −4.0 m/s^2); HI Dec. = high-intensity decelerations (>−4.0 m/s^2); PL = player load; AU = arbitrary units; m = meters; min = minutes; CI 95% = 95% confidence interval; * $p < 0.05$. Effect size: ≤0.2, trivial; >0.2, small; >0.6, moderate; >1.2, large; >2.0, very large; and >4.0, nearly perfect.

4. Discussion

The four aims of the present study were as follows: (a) to describe the external match load at the amateur level, (b) to analyze the differences in the external match load between playing positions, (c) to verify whether the home/away contextual factor influences the external match load, and (d) to determine whether there are differences in match load between the first and the second phases of the championship. While the first aim was merely descriptive (shown in Table 1), another analysis allowed noting relevant findings. The second purpose of the study showed significant differences between playing positions in several load measures. Finally, the third and the fourth purposes of the study found a tendency towards higher values in home matches than in away matches and in the first phase of the championship in comparison with the second phase, respectively, although mostly insignificant.

In the external load measures that are possible to compare, the results presented some peculiarities in comparison with other recent studies in professional/elite teams [9,14,21,25,34–41]. Based on the comparison of our results with those studies, greater differences were observed in the distance covered in higher-speed zones (developed in the following paragraphs). If the TDC seems identical, this could demonstrate that at higher competitive levels, one of the differences is in the physical requirement involved in a match. Therefore, examining the high-intensity activity provides a valid insight into physical performance and its strong relationship with the training status [42,43].

4.1. External Match Load—Description and Comparisons between Playing Positions

In relation to playing positions, the results showed that the central midfielders were the players who would cover the longest distance during a match, which is in line with recent literature findings [9,31,36,41,43–45], followed by the forwards, the full-backs, and the wide midfielders. As in other studies [34,38,41], the forwards were the ones who had the longest SpD, followed by the full-backs and the wide midfielders. Other features have been presented in other studies [36,46]. For example, Ingebrigtsen et al. [47] registered that players in lateral playing positions sprint the longest distances and more often compared to centrally playing players, while Paraskevas et al. [48] report that full-backs cover more SpD and VHSRD compared to all the other positions.

Regarding accelerations and decelerations, our data indicate that the central midfielders were those that presented with the greatest quantity, mainly of low and moderate intensity. The forwards were the ones that presented with higher values in HI Acc. and HI Dec. Similar results were found by Modric et al. [44] who reported that central midfielders were the players who performed more accelerations and decelerations (in a greater quantity in tactical formations with three defensive players in comparison with tactical formations with four such players), but the most intense accelerations and decelerations were carried out by forwards and full-backs. In contrast, Ingebrigtsen et al. [47] found that players in lateral positions accelerated more than those in central positions.

In recent studies [13,45], central midfielders were the players who showed the highest PL, followed by forwards, full-backs, wide midfielders, and central defenders. It has also been observed that the position of a central defender is the one that presents an external match load profile with more significant differences compared to the other playing positions. Modric [36] asserted that this is understandable knowing that their technical roles (e.g., aerial duels, tackles, positioning, and interception of the balls passed to the attackers) are generally more focused on reactions or accelerations and high-speed running. On the other hand, full-backs and wide midfielders are the playing positions that have more similarities (full-backs and forwards also have an identical profile). Wide midfielders and forwards showed a similar profile in terms of accelerations, decelerations, and PL.

Based on the results obtained and comparing them with other studies [9,36,41,44,45,47], it is possible to determine that the differences between the competitive levels (amateur, semi-professional, and professional) are more visible in the most demanding external load measures (e.g., SpD), as well as in the existence of variability in the external load profiles presented by the different playing positions. Through this perspective, coaches and technical staff must continually evaluate their own methodologies, strategies, game styles, player characteristics, among other factors [43], because although there are references of the external load for the team (depending on the competitive level) and the players (depending on the position occupied), the search for the best possible performance implies the customization and individualization of the approaches to be developed with the athletes. Thus, it is essential that coaches know the requirements of their own way of playing [8] and, with this, plan and organize the external load to be applied during the microcycle, because the one-size-fits-all approach could provide tactically constrained physical data for selected positions that are challenging to interpret given the lack of contextualization [42]. At the same time, the identification of differences and similarities in the external load profiles between the different positions will allow the conception and selection of training exercises in which each group of players will participate. This statement is corroborated by Modric [36], who affirmed that training prescriptions in soccer should be based on the established requirements specific to the playing positions, thereby ensuring that players are more able to fulfill their game duties and tactical responsibilities throughout the competition. Therefore, all training exercises must be characterized knowing the physical demands of each exercise as well as the external load applied on each playing position [31,49]. Complementary analytical training exercises (no ball) can also be developed to ensure that everyone achieves the desired external load, which was also supported by Modric [48], who suggested that when using small-side games which do not require running at high speeds, coaches should include specific running drills that entail high-intensity running (e.g., high-speed running and sprinting) in the training sessions.

4.2. External Match Load—Home vs. Away Matches

In the present study, there was a clear trend towards higher values in the TDC, RDC, distance covered by speed zone, and PL for home matches; however, only the TDC for the central defenders was significant, with a moderate effect size. While Castellano [50], Lago [51], and Gonçalves [52] also found that the external load was higher when playing at home, Gonçalves [17] verified that the match load was not influenced by the match location. In the meantime, Teixeira [43] suggested that the quality of opposition and match outcome have a greater influence than match location. Paraskevas et al. [48] reported that a greater TDC was covered during home matches against weak opponents compared to home matches against strong opponents and the opposite during away matches.

The same tendency is observed with regard to accelerations, decelerations (only HI Dec. was significant for the forwards). Otherwise, although not significantly, the central midfielders had smaller numbers of accelerations and decelerations at low and moderate intensity in home matches. Aquino [21] and Lago-Peñas [22] found that home matches place greater physical demands on players, owing to the combined effects of crowd, travel, familiarity, referee bias, territoriality, specific tactics, and psychological factors [22]. Contrary to that, Reche-Soto [13] described that the external load was higher when the team was playing away, but clarified that the effect of the match location was not clear. Chena [53] advised that this finding could be more related to the emotional variables than to the requirement of training sessions during that week. In other scope and in line with our results, Gonçalves [17] found that external loads were not influenced by match location and asserted that each competition may present its own idiosyncrasies, which is the reason why data generalization should be performed with caution.

The trend towards higher external load values in home matches can be explained by the offensive and defensive strategies adopted by both teams [43,48,52,54] as well as the size of the field. The analyzed team played at a stadium with natural grass and large dimensions (meets the requirements for international competitions) in home matches, which is different from the experience of most amateur clubs in Portugal (artificial grass and medium dimensions, some of which do not meet the requirements for national competitions). According to Almeida et al. [54], the defensive strategies used by better teams imply more intense and organized collective processes in order to recover the ball directly from the opposing team. Gonçalves et al. [52] reported that the counterattacking/transitional styles may result in greater high-intensity activities while the possession style may increase the distance covered in lower-speed zones. While, on the one hand, home matches increase the playing area available for opposing counterattacks (they demand a greater radius of action from central defenders), on the other hand, away matches reduce the area available for demarcations and moves by forwards.

Despite the well-proven knowledge that the size of the playing areas influences the load imposed on the players in small-side games [55–57], in amateur soccer, there is a diversity of fields (type of grass and dimensions) that can impact the physical demands of the game itself.

In our opinion, despite the match location presents an unclear effect on the external load, this contextual variable should be a subject of attention from coaches [17,43]. Existing changes, even if only in some specific metrics and for specific playing positions, may require reflection about and consideration of the need to adjust the planning, meeting the competition's requests for each playing position (individual approach) [42,47]. Our results should not be used to generalize the dynamics of the external load in home/away matches—once again, we suggest that a continuous assessment be made of the team itself and that the approaches be centered on the analysis of the data themselves.

4.3. External Match Load—First vs. Second Championship Phase

The data suggest that there was a clear trend towards higher values in the TDC, RDC, accelerations, decelerations, and PL in matches of the first championship phase. However only the TDC and RDC for the central defenders were significant, with a moderate effect size, as well as MI Acc. for the central defenders with a moderate effect and for the wide midfielders in HI Acc. and MI Dec., also with a moderate effect. Curiously, in relation to the distance covered by speed zone, an inverse trend was observed. Lower values of VHSRD and SpD in the matches of the first phase of the championship (however, the forwards had a significantly higher VHSRD, with a large effect), which is contrary to Ingebrigtsen [47] results, who observed a shift towards more walking and fewer high-intensity locomotor activities during a match towards the end compared to the start of the season. Although not significant, the trend towards higher values of SpD and VHSRD in the second phase of the championship also contradicts the results by Springham [38] who claimed that the most notable decreases in performance were observed in sprint performance indices for which the greatest reductions were observed in full-backs, central midfielders, and wide midfielders. Additionally, Teixeira [43] suggested that contextual factors (quality of opposition and match outcome) and their changes seem to differ between playing positions.

In our study, the analysis of the championship phase required a careful reflection because of a simultaneous association between different opponents' levels (in the second phase the teams had a similar level, with no weak teams) and periods of the competitive season. If, on the one hand, the confrontation with opponents of a higher qualitative level could cause a greater physical demand [48,52], on the other hand, the course of the competitive period itself leads to the loss of physical availability. According to Springham [38] results, all the physical performance indices decreased across the season; decreases in match physical performance indices in all the positions were observed. This author claimed that the cross-season decreases in match physical performance observed might be explained by longitudinal fatigue. Additionally, the absence of quality of opposition as a predictor variable for match physical performance is somewhat surprising as players are reported to complete more high-intensity activity and high-speed running when playing against high-quality opposition as opposed to low-quality opposition [20]. Conversely, Aquino [21] found that matches against weak opposition place greater physical demands on players.

In our opinion, the general and significant decreases observed may be associated simultaneously with the period of the season, as well as the type of opposition faced in the first and the second phases of the championship. Firstly, at the level (amateur) where the existing resources to help with physical recovery are scarce, the accumulation of trainings and matches throughout the competitive season can cause a decrease in physical freshness of the players. Finally, the offensive and defensive constraints posed by the opponents can assume co-responsibility for these results (mainly on the trend towards increasing SpD values) because a confrontation with opponents of greater value implies more moments of defensive and offensive transitions, moments of the game that demand fast and intense displacements on the part of all the players [52]. More specifically, while in the first phase the team dominated the games, continuously installing itself in the offensive midfield, in the second phase, this domain was not so evident, and the game assumed a more balanced nature.

4.4. Limitations

The main limitation of this study is the fact that only one team was observed, which is a very common obstacle in studies with soccer players [8,58]. In addition, the small sample size and the existence of other contextual factors such as the match result or the quality of opponents were not considered in the analysis and could provide better insights. Secondly, we only included the players who participated in full matches, which excludes the contribution of the other players who entered during in the game. All the matches played home were on natural grass, and of the 11 matches played away, 10 were on artificial grass and one—on natural grass. Finally, for us, it is difficult to associate the trend towards the increase in the external load in home matches only with the location when there are other variables that change with the match location.

5. Conclusions

These findings are novel and provide relevant information about the external match load in amateur soccer that can promote the reflection of coaches about the periodization and organization of training sessions at this competitive level:

- Match location and championship phase are not major contributing factors influencing the external load. Although there are several considerations regarding the influence of contextual variables on the match load, these should not be generalized and require their own evaluation and adoption of strategies based on them.
- The position occupied by players prevails as the most important factor in determining the load imposed by the competition, which is the reason why coaches must evaluate and characterize their game model in order to determine the physical demand imposed on the team and each player. Through the analysis of collective and individual match loads, it is possible to customize and individualize methodological approaches with regard to the management and regulation of the external training load.

Author Contributions: Conceptualization, M.M. and N.L.; methodology, R.O. and J.P.B.; validation, N.L., J.G.-R., and S.J.I.; formal analysis, M.M. and N.L.; investigation, M.M.; resources, N.L.; data curation, M.M.; writing—original draft preparation, M.M.; writing—review and editing, R.O., J.P.B., and N.L.; visualization, J.G.-R.; supervision, S.J.I.; project administration, M.M., J.G.-R., and S.J.I.; funding acquisition, S.J.I. All authors have read and agreed to the published version of the manuscript.

Funding: This study has been partially subsidized by the Aid for Research Groups (GR21149) from the Regional Government of Extremadura (Department of Economy, Science and Digital Agenda), with a contribution from the European Union from the European Funds for Regional Development, and by the Portuguese Foundation for Science and Technology, I.P., grant No. UIDP/04748/2020.

Institutional Review Board Statement: The study was conducted according to the guidelines of the Declaration of Helsinki and approved by the Ethics Committee of the University of Extremadura (No. 67/2017).

Informed Consent Statement: Informed consent was obtained from all the subjects involved in the study.

Data Availability Statement: Data are available upon request to the contact author.

Acknowledgments: The authors thank the players and staff at Rio Maior Sport Clube for their cooperation and participation in this study.

Conflicts of Interest: The authors declare no conflict of interest. The funders had no role in the design of the study; in the collection, analyses, or interpretation of data; in the writing of the manuscript, or in the decision to publish the results.

References

1. Suarez-Arrones, L.J.; Torreño, N.; Requena, B.; de Villarreal, E.S.; Casamichana, D.; Carlos, J.; Barbero-Alvarez, D.M. Match-Play Activity Profile in Professional Soccer Players during Official Games and the Relationshop between External and Internal Load. *J. Sports Med. Phys. Fit.* **2015**, *55*, 1417–1422.
2. Bourdon, P.C.; Cardinale, M.; Murray, A.; Gastin, P.; Kellmann, M.; Varley, M.C.; Gabbett, T.J.; Coutts, A.J.; Burgess, D.J.; Gregson, W.; et al. Monitoring Athlete Training Loads: Consensus Statement. *Int. J. Sports Physiol. Perform.* **2017**, *12*, 161–170. [CrossRef] [PubMed]
3. Akenhead, R.; Nassis, G.P. Training Load and Player Monitoring in High-Level Football: Current Practice and Perceptions. *Int. J. Sports Physiol. Perform.* **2016**, *11*, 587–593. [CrossRef] [PubMed]
4. Gabbett, T.J.; Nassis, G.P.; Oetter, E.; Pretorius, J.; Johnston, N.; Medina, D.; Rodas, G.; Myslinski, T.; Howells, D.; Beard, A.; et al. The Athlete Monitoring Cycle: A Practical Guide to Interpreting and Applying Training Monitoring Data. *Br. J. Sports Med.* **2017**, *51*, 1451–1452. [CrossRef]
5. Vanrenterghem, J.; Nedergaard, N.J.; Robinson, M.A.; Drust, B. Training Load Monitoring in Team Sports: A Novel Framework Separating Physiological and Biomechanical Load-Adaptation Pathways. *Sports Med.* **2017**, *47*, 2135–2142. [CrossRef]
6. Zurutuza, U.; Castellano, J.; Echeazarra, I.; Casamichana, D. Absolute and Relative Training Load and Its Relation to Fatigue in Football. *Front. Psychol.* **2017**, *8*, 878. [CrossRef]
7. Casamichana, D.; Castellano, J.; Castagna, C. Comparing the Physical Demands of Friendly Matches and Small-Sided Games in Semiprofessional Soccer Players. *J. Strength Cond. Res.* **2012**, *26*, 837–843. [CrossRef]
8. Clemente, F.M.; Owen, A.; Serra-Olivares, J.; Nikolaidis, P.T.; van der Linden, C.M.I.; Mendes, B. Characterization of the Weekly External Load Profile of Professional Soccer Teams From Portugal and the Netherlands. *J. Hum. Kinet.* **2019**, *66*, 155–164. [CrossRef]
9. Curtis, R.M.; Huggins, R.A.; Looney, D.P.; West, C.A.; Fortunati, A.; Fontaine, G.J.; Casa, D.J. Match Demands of National Collegiate Athletic Association Division I Men's Soccer. *J. Strength Cond. Res.* **2018**, *32*, 2907–2917. [CrossRef]
10. Howle, K.; Waterson, A.; Duffield, R. Recovery Profiles Following Single and Multiple Matches per Week in Professional Football. *Eur. J. Sport Sci.* **2019**, *19*, 1303–1311. [CrossRef]
11. Jones, R.N.; Greig, M.; Mawéné, Y.; Barrow, J.; Page, R.M. The Influence of Short-Term Fixture Congestion on Position Specific Match Running Performance and External Loading Patterns in English Professional Soccer. *J. Sports Sci.* **2019**, *37*, 1338–1346. [CrossRef] [PubMed]
12. Martín-García, A.; Gómez Díaz, A.; Bradley, P.S.; Morera, F.; Casamichana, D. Quantification of a Professional Football Team's External Load Using a Microcycle Structure. *J. Strength Cond. Res.* **2018**, *32*, 3511–3518. [CrossRef] [PubMed]
13. Reche-Soto, P.; Cardona-Nieto, D.; Diaz-Suarez, A.; Bastida-Castillo, A.; Gomez-Carmona, C.; Garcia-Rubio, J.; Pino-Ortega, J. Player Load and Metabolic Power Dynamics as Load Quantifiers in Soccer. *J. Hum. Kinet.* **2019**, *69*, 259–269. [CrossRef]
14. Torreño, N.; Munguía-Izquierdo, D.; Coutts, A.; de Villarreal, E.S.; Asian-Clemente, J.; Suarez-Arrones, L. Relationship between External and Internal Loads of Professional Soccer Players during Full Matches in Official Games Using Global Positioning Systems and Heart-Rate Technology. *Int. J. Sports Physiol. Perform.* **2016**, *11*, 940–946. [CrossRef]
15. Barrett, S.; McLaren, S.; Spears, I.; Ward, P.; Weston, M. The Influence of Playing Position and Contextual Factors on Soccer Players' Match Differential Ratings of Perceived Exertion: A Preliminary Investigation. *Sports* **2018**, *6*, 13. [CrossRef] [PubMed]
16. Brito, Â.; Roriz, P.; Silva, P.; Duarte, R.; Garganta, J. Effects of Pitch Surface and Playing Position on External Load Activity Profiles and Technical Demands of Young Soccer Players in Match Play. *Int. J. Perform. Anal. Sport* **2017**, *17*, 902–918. [CrossRef]
17. Gonçalves, L.G.C.; Kalva-Filho, C.A.; Nakamura, F.Y.; Rago, V.; Afonso, J.; de Souza Bedo, B.L.; Aquino, R. Effects of Match-Related Contextual Factors on Weekly Load Responses in Professional Brazilian Soccer Players. *Int. J. Environ. Res. Public Health* **2020**, *17*, 5163. [CrossRef]
18. Dellal, A.; Hill-Haas, S.; Lago-Penas, C.; Chamari, K. Small-Sided Games in Soccer: Amateur vs. Professional Players' Physiological Responses, Physical, and Technical Activities. *J. Strength Cond. Res.* **2011**, *25*, 2371–2381. [CrossRef]
19. García-Rubio, J.; Gómez, M.Á.; Lago-Peñas, C.; Ibáñez, J.S. Effect of Match Venue, Scoring First and Quality of Opposition on Match Outcome in the UEFA Champions League. *Int. J. Perform. Anal. Sport* **2015**, *15*, 527–539. [CrossRef]
20. Springham, M.; Williams, S.; Waldron, M.; Strudwick, A.J.; Mclellan, C.; Newton, R.U. Prior Workload Has Moderate Effects on High-Intensity Match Performance in Elite-Level Professional Football Players When Controlling for Situational and Contextual Variables. *J. Sports Sci.* **2020**, *38*, 2279–2290. [CrossRef]
21. Aquino, R.; Carling, C.; Vieira, L.H.P.; Martins, G.; Jabor, G.; Machado, J.O.; Santiago, P.; Garganta, J.L.; Puggina, E. Influence of Situation Variables, Team Formation, and Playing Position on Match Running Performance and Social Network Analysis in Brazilian Professional Soccer Players. *J. Strength Cond. Res.* **2018**, *34*, 808–817. [CrossRef] [PubMed]
22. Lago-Peñas, C. The Role of Situational Variables in Analysing Physical Performance in Soccer. *J. Hum. Kinet.* **2012**, *35*, 89–95. [CrossRef] [PubMed]

23. Oliveira, R.; Brito, J.P.; Loureiro, N.; Padinha, V.; Nobari, H.; Mendes, B. Will Next Match Location Influence External and Internal Training Load of a Top-Class Elite Professional European Soccer Team? *Int. J. Environ. Res. Public Health* **2021**, *18*, 5229. [CrossRef] [PubMed]
24. Sanchez-Sanchez, J.; Hernández, D.; Martin, V.; Sanchez, M.; Casamichana, D.; Rodriguez-Fernandez, A.; Ramirez-Campillo, R.; Nakamura, F.Y. Assessment of the External Load of Amateur Soccer Players during Four Consecutive Training Microcycles in Relation to the External Load during the Official Match. *Motriz Rev. Educ. Fis.* **2019**, *25*, e101938. [CrossRef]
25. Stevens, T.G.A.; de Ruiter, C.J.; Twisk, J.W.R.; Savelsbergh, G.J.P.; Beek, P.J. Quantification of In-Season Training Load Relative to Match Load in Professional Dutch Eredivisie Football Players. *Sci. Med. Footb.* **2017**, *1*, 117–125. [CrossRef]
26. Owen, A.L.; Lago-Peñas, C.; Gómez, M.-Á.; Mendes, B.; Dellal, A. Analysis of a Training Mesocycle and Positional Quantification in Elite European Soccer Players. *Int. J. Sports Sci. Coach.* **2017**, *12*, 665–676. [CrossRef]
27. Ato, M.; López-García, J.J.; Benavente, A. Un sistema de clasificación de los diseños de investigación en psicología. *Analesps* **2013**, *29*, 1038–1059. [CrossRef]
28. Malone, J.J.; Murtagh, C.F.; Morgans, R.; Burgess, D.J.; Morton, J.P.; Drust, B. Countermovement Jump Performance Is Not Affected during an In-Season Training Microcycle in Elite Youth Soccer Players. *J. Strength Cond. Res.* **2015**, *29*, 752–757. [CrossRef]
29. World Medical Association. World Medical Association Declaration of Helsinki: Ethical Principles for Medical Research Involving Human Subjects. *JAMA* **2013**, *310*, 2191. [CrossRef]
30. Scott, M.T.U.; Scott, T.J.; Kelly, V.G. The Validity and Reliability of Global Positioning Systems in Team Sport: A Brief Review. *J. Strength Cond. Res.* **2016**, *30*, 1470–1490. [CrossRef]
31. Ravé, G.; Granacher, U.; Boullosa, D.; Hackney, A.C.; Zouhal, H. How to Use Global Positioning Systems (GPS) Data to Monitor Training Load in the "Real World" of Elite Soccer. *Front. Physiol.* **2020**, *11*, 944. [CrossRef] [PubMed]
32. Miguel, M.; Oliveira, R.; Loureiro, N.; García-Rubio, J.; Ibáñez, S.J. Load Measures in Training/Match Monitoring in Soccer: A Systematic Review. *Int. J. Environ. Res. Public Health* **2021**, *18*, 2721. [CrossRef] [PubMed]
33. Hopkins, W.G.; Marshall, S.W.; Batterham, A.M.; Hanin, J. Progressive Statistics for Studies in Sports Medicine and Exercise Science. *Med. Sci. Sports Exerc.* **2009**, *41*, 3–12. [CrossRef] [PubMed]
34. Carling, C.; Bradley, P.; McCall, A.; Dupont, G. Match-to-Match Variability in High-Speed Running Activity in a Professional Soccer Team. *J. Sports Sci.* **2016**, *34*, 2215–2223. [CrossRef] [PubMed]
35. Iacono, A.D.; Martone, D.; Cular, D.; Milic, M.; Padulo, J. Game-Profile-Based Training in Soccer: A New Field Approach. *J. Strength Cond. Res.* **2017**, *31*, 3333–3342. [CrossRef]
36. Modric, T.; Versic, S.; Sekulic, D.; Liposek, S. Analysis of the Association between Running Performance and Game Performance Indicators in Professional Soccer Players. *Int. J. Environ. Res. Public Health* **2019**, *16*, 4032. [CrossRef]
37. Oliveira, R.; Brito, J.; Martins, A.; Mendes, B.; Calvete, F.; Carriço, S.; Ferraz, R.; Marques, M.C. In-Season Training Load Quantification of One-, Two- and Three-Game Week Schedules in a Top European Professional Soccer Team. *Physiol. Behav.* **2019**, *201*, 146–156. [CrossRef]
38. Springham, M.; Williams, S.; Waldron, M.; Burgess, D.; Newton, R.U. Large Reductions in Match Play Physical Performance Variables across a Professional Football Season with Control for Situational and Contextual Variables. *Front. Sports Act. Living* **2020**, *2*, 570937. [CrossRef]
39. Swallow, W.E.; Skidmore, N.; Page, R.M.; Malone, J.J. An Examination of In-Season External Training Load in Semi-Professional Soccer Players: Considerations of One and Two Match Weekly Microcycles. *Int. J. Sports Sci. Coach.* **2021**, *16*, 192–199. [CrossRef]
40. Tuo, Q.; Wang, L.; Huang, G.; Zhang, H.; Liu, H. Running Performance of Soccer Players during Matches in the 2018 FIFA World Cup: Differences among Confederations. *Front. Psychol.* **2019**, *10*, 1044. [CrossRef]
41. Varley, M.C.; Di Salvo, V.; Modonutti, M.; Gregson, W.; Mendez-Villanueva, A. The Influence of Successive Matches on Match-Running Performance during an under-23 International Soccer Tournament: The Necessity of Individual Analysis. *J. Sports Sci.* **2018**, *36*, 585–591. [CrossRef] [PubMed]
42. Bradley, P.S.; Ade, J.D. Are Current Physical Match Performance Metrics in Elite Soccer Fit for Purpose or Is the Adoption of an Integrated Approach Needed? *Int. J. Sports Physiol. Perform.* **2018**, *13*, 656–664. [CrossRef] [PubMed]
43. Teixeira, J.E.; Leal, M.; Ferraz, R.; Ribeiro, J.; Cachada, J.M.; Barbosa, T.M.; Monteiro, A.M.; Forte, P. Effects of Match Location, Quality of Opposition and Match Outcome on Match Running Performance in a Portuguese Professional Football Team. *Entropy* **2021**, *23*, 973. [CrossRef] [PubMed]
44. Modric, T.; Versic, S.; Sekulic, D. Position Specific Running Performances in Professional Football (Soccer): Influence of Different Tactical Formations. *Sports* **2020**, *8*, 161. [CrossRef]
45. Wiig, H.; Raastad, T.; Luteberget, L.S.; Ims, I.; Spencer, M. External Load Variables Affect Recovery Markers up to 72 h After Semiprofessional Football Matches. *Front. Physiol.* **2019**, *10*, 689. [CrossRef]
46. Baptista, I.; Johansen, D.; Figueiredo, P.; Rebelo, A.; Pettersen, S.A. Positional Differences in Peak- and Accumulated- Training Load Relative to Match Load in Elite Football. *Sports* **2019**, *8*, 1. [CrossRef]
47. Ingebrigtsen, J.; Dalen, T.; Hjelde, G.H.; Drust, B.; Wisløff, U. Acceleration and Sprint Profiles of a Professional Elite Football Team in Match Play. *Eur. J. Sport Sci.* **2015**, *15*, 101–110. [CrossRef]
48. Paraskevas, G.; Smilios, I.; Hadjicharalambous, M. Effect of Opposition Quality and Match Location on the Positional Demands of the 4-2-3-1 Formation in Elite Soccer. *J. Exerc. Sci. Fit.* **2020**, *18*, 40–45. [CrossRef]

49. Morgans, R.; Orme, P.; Anderson, L.; Drust, B. Principles and Practices of Training for Soccer. *J. Sport Health Sci.* **2014**, *3*, 251–257. [CrossRef]
50. Castellano, J.; Blanco-Villaseñor, A.; Álvarez, D. Contextual Variables and Time-Motion Analysis in Soccer. *Int. J. Sports Med.* **2011**, *32*, 415–421. [CrossRef]
51. Lago, C.; Casais, L.; Dominguez, E.; Sampaio, J. The Effects of Situational Variables on Distance Covered at Various Speeds in Elite Soccer. *Eur. J. Sport Sci.* **2010**, *10*, 103–109. [CrossRef]
52. Gonçalves, L.G.C.; Clemente, F.; Palucci Vieira, L.H.; Bedo, B.; Puggina, E.F.; Moura, F.; Mesquita, F.; Pereira Santiago, P.R.; Almeida, R.; Aquino, R. Effects of Match Location, Quality of Opposition, Match Outcome, and Playing Position on Load Parameters and Players' Prominence during Official Matches in Professional Soccer Players. *Hum. Mov.* **2021**, *22*, 35–44. [CrossRef]
53. Chena, M.; Morcillo, J.A.; Rodríguez-Hernández, M.L.; Zapardiel, J.C.; Owen, A.; Lozano, D. The Effect of Weekly Training Load across a Competitive Microcycle on Contextual Variables in Professional Soccer. *Int. J. Environ. Res. Public Health* **2021**, *18*, 5091. [CrossRef]
54. Almeida, C.H.; Ferreira, A.P.; Volossovitch, A. Effects of Match Location, Match Status and Quality of Opposition on Regaining Possession in UEFA Champions League. *J. Hum. Kinet.* **2014**, *41*, 203–214. [CrossRef]
55. Modena, R.; Togni, A.; Fanchini, M.; Pellegrini, B.; Schena, F. Influence of Pitch Size and Goalkeepers on External and Internal Load during Small-Sided Games in Amateur Soccer Players. *Sport Sci. Health* **2021**, *17*, 797–805. [CrossRef]
56. Santos, F.J.; Verardi, C.E.L.; de Moraes, M.G.; Filho, D.M.P.; Macedo, A.G.; Figueiredo, T.P.; Ferreira, C.C.; Borba, R.P.; Espada, M.C. Effects of Pitch Size and Goalkeeper Participation on Physical Load Measures during Small-Sided Games in Sub-Elite Professional Soccer Players. *Appl. Sci.* **2021**, *11*, 8024. [CrossRef]
57. Casamichana, D.; Bradley, P.S.; Castellano, J. Influence of the Varied Pitch Shape on Soccer Players Physiological Responses and Time-Motion Characteristics during Small-Sided Games. *J. Hum. Kinet.* **2018**, *64*, 171–180. [CrossRef]
58. Modric, T.; Jelicic, M.; Sekulic, D. Relative Training Load and Match Outcome: Are Professional Soccer Players Actually Undertrained during the In-Season? *Sports* **2021**, *9*, 139. [CrossRef]

 *healthcare* 

Article

Comparison of Measurements of External Load between Professional Soccer Players

Roghayyeh Gholizadeh [1,2], Hadi Nobari [1,3,4,*], Lotfali Bolboli [1], Marefat Siahkouhian [1] and João Paulo Brito [5,6,7,8,*]

1 Department of Exercise Physiology, Faculty of Educational Sciences and Psychology, University of Mohaghegh Ardabili, Ardabil 56199-11367, Iran; r.gholizadeh60@yahoo.com (R.G.); l.bolboli@uma.ac.ir (L.B.); m_siahkohian@uma.ac.ir (M.S.)
2 Department of Physical Education and Sport, Faculty of Sport Science, University of Granada, 18010 Granada, Spain
3 Department of Motor Performance, Faculty of Physical Education and Mountain Sports, Transilvania University of Braşov, 500068 Braşov, Romania
4 Faculty of Sport Sciences, University of Extremadura, 10003 Cáceres, Spain
5 Sepahan Football Club, Isfahan 81887-78473, Iran
6 Sports Science School of Rio Maior, Polytechnic Institute of Santarém, 2040-413 Rio Maior, Portugal
7 Life Quality Research Centre, 2040-413 Rio Maior, Portugal
8 Research Centre in Sport Sciences, Health Sciences and Human Development, 5001-801 Vila Real, Portugal
* Correspondence: hadi.nobari1@gmail.com or nobari.hadi@unitbv.ro (H.N.); jbrito@esdrm.ipsantarem.pt (J.P.B.)

Abstract: Background: The excessive and rapid increases in training load (TL) may be responsible for most non-contact injuries in soccer. This study's aims were to describe, week(w)-by-week, the acute (AW), chronic (CW), acute:chronic workload ratio (wACWR), total distance (wTD), duration training (wDT), sprint total distance (wSTD), repeat sprint (wRS), and maximum speed (wMS) between starter and non-starter professional soccer players based on different periods (i.e., pre-, early-, mid-, and end-season) of a full-season (Persian Gulf Pro League, 2019–2020). Methods: Nineteen players were divided according to their starting status: starters ($n = 10$) or non-starters ($n = 9$). External workload was monitored for 43 weeks: pre- from w1–w4; early- from w5–w17; mid- from w18–w30, and end-season from w31–w43. Results: In starters, AW, CW, and wACWR were greater than non-starters ($p < 0.05$) throughout the periods of early- (CW, $p \leq 0.0001$), mid- (AW, $p = 0.008$; CW, $p \leq 0.0001$; wACWR, $p = 0.043$), or end-season (AW, $p = 0.035$; CW, $p = 0.017$; wACWR, $p = 0.010$). Starters had a greater wTD ($p \leq 0.0001$), wSTD ($p \leq 0.0001$ to 0.003), wDT ($p \leq 0.0001$ to 0.023), wRS ($p \leq 0.0001$ to 0.018), and wMS ($p \leq 0.0001$) than non-starters during early-, mid-, and end-season. Conclusion: Starters experienced more CW and AW during the season than non-starters, which underlines the need to design tailored training programs accounting for the differences between playing status.

Keywords: external training load; technology; soccer; performance; wearable inertial measurement units

Citation: Gholizadeh, R.; Nobari, H.; Bolboli, L.; Siahkouhian, M.; Brito, J.P. Comparison of Measurements of External Load between Professional Soccer Players. *Healthcare* **2022**, *10*, 1116. https://doi.org/10.3390/healthcare10061116

Academic Editor: Daniele Giansanti

Received: 1 May 2022
Accepted: 11 June 2022
Published: 15 June 2022

Publisher's Note: MDPI stays neutral with regard to jurisdictional claims in published maps and institutional affiliations.

1. Introduction

Soccer is a sport with unstructured movement patterns, in which high-intensity performance-related energy systems are significantly involved [1]. During a competitive game, the total distances and sprints performed by professional soccer players are higher when compared to semi-professional soccer players. This reflects the different physical demands between soccer categories [2]. It is very important for coaches to pay attention to individual differences, such as physical fitness factors, workload responses, training intensity types, playing position, etc. [3,4], in order to improve the quality of teams' activities and to reduce injuries [5]. Hence, the non-functional overreaching syndrome (NFOR) that could reduce team and player performance and increase club costs justifies the recent attention of researchers to avoid injuries [6–9]. Birrer reported (2013) that the

NFOR/OTS career prevalence rate of Swiss elite athletes can be estimated at approximately 30%. In reviews of adult studies, using self-report questionnaires, prevalence rates for OTS/NFOR have been reported between 10–60%, with a predominance reported between 25–30% [10,11].

Soccer as a sport also requires variety of ballistic concentric and eccentric muscle contractions to perform a variety of movements, such as acceleration, deceleration and change of direction, tackling, jumping, and shooting, which increases the need for physiological and metabolic stress to support these movements. Soccer players accumulate fatigue from training and weekly matches during a competitive season, and when training/competition volume increases with inadequate recovery, players may enter NFOR mode, which is characterized by decreased performance [11]. Dietary and ergogenic studies have been reported as an intervention in athletes' programs to achieve faster adaptation and manage exercise-induced fatigue. Sports coaches and researchers have the opportunity to reduce the risk of injury in athletes by manipulating some of these risk factors [8]. Training load (TL) is one modifiable risk factor that has recently received a lot of research attention. TL includes training intensity and duration, and research has shown that a higher TL increases the level of workload parameters, such as the acute:chronic workload ratio (ACWR), which leads to more non-contact injury occurrences and greater injury severity [12–15]. Some evidence has reported that many non-contact injuries occur when an athlete has a large amount of TL, in which case, any training or competition has the potential for athletic injury, indicating that an unfit workload can increase injury risk [16–19].

Lu et al. reported that professional soccer players had more total and relative exposure in the three weeks prior to injury [20]. However, an another perspective contends that TLs that underprepared players for the demands of match play could put them at risk of injury, and impede results [21]. Therefore, optimal training stimulus is crucial to maximize efficiency by using an acceptable TL while also limiting the detrimental effects of undertraining and overtraining [22]. TL can be considered as an internal or external load. The objectively assessed work performed by the athlete during training and/or competition, irrespective of internal workloads, is referred to as external TL [23]. Today, with the advancement of science, up-to-date and various tools are available to increase the efficiency of teams. Among these tools, we can see the reason for training monitoring. Devices such as GPS provide extensive information, such as acceleration and deceleration of speed, total distance in training, maximum heart rate, maximum velocity, etc., to coaches and team analysts [19]. Speed, acceleration, and distance covered are some of the most common external TL assessments. Internal TL represents an individual athlete's response to training, and can be quantified by the intensity and duration of the physiological stress imposed on the athlete. Despite the advantages of both internal and external TL, it has been proposed that combining the two approaches could be the most efficient way to control TL [13]. Using this information, coaches can monitor player progress in training sessions, assess perceived training pressure levels on players, fatigue, and recovery, and use this information to design training sessions for future sessions. They can also use this information to identify the best players in each position who have the best physical condition during training, and place them in the starting line-up. The acute:chronic workload ratio (ACWR), a relative measure derived from Hulin et al., is a significant recent advancement in exploring TL [24]. Hulin et al. suggested the ACWR, which separates an athlete's recent TL (i.e., acute workload: AW) by their TL over a long period of time (i.e., chronic workload: CW). This metric is suggested as a tool to help practitioners control TL within clear limits. The original idea was to hold ACWR within an arbitrary "optimal range" of 0.8–1.5 by preventing abrupt changes in TL. Understanding the workload–injury relationship is crucial to improve player availability and results. There is a contradiction of investigation on the association between workload and damage in professional soccer. The International Olympic Committee (2016) published a consensus statement that suggests that the use of the ACWR approach is effective in preventing non-contact injuries [25]. Nobari et al. have been reported that sprint variables (i.e., total distance, repeat sprint, or sprint total distance) may be related to the odds ratio of non-contact injuries in professional soccer players [14]. However, Impellizzeri et al. [26] concluded that there is no evidence that ACWR is used to manage TL and reduce the risk of injury. On the other hand, other researchers believe that excessive and rapid increases in TL may be responsible for most non-contact injuries [19,27]. Therefore, despite its increasing popularity as a load monitoring system, the ACWR and the injury risks associated with it need to be investigated further. Thus, the aims of this analysis were to: describe, week(w)-by-week, the AW, CW, and wACWR between starter and non-starter professional soccer players using the body load (BL); analyze differences in the AW, CW, and wACWR on a weekly basis between starter and non-starter professional soccer players based on different periods of the full-season (pre-, early-, mid-, and end-season); and analyze weekly total distance (wTD), weekly duration training (wDT), sprint total distance (wSTD), and repeat sprint (wRS) by season periods between starter versus non-starter professional soccer players. We hypothesized that starters would record greater workload and sprint variables during all periods of a soccer season compared to non-starters.

2. Materials and Methods

2.1. Study Design

This study included a full season of a professional football team for 43 weeks in the Persian Gulf Premier League and knockout tournament during the 2019–2020 season. The 43 weeks of the full season were divided into 4 periods: pre-season (W1 to W4), early-season (W5 to W17), mid-season (W18 to W30), and end-season (W31 to W43). The three periods of the in-season have been considered as exactly the same number of weeks [28]. Forty-eight matches and 229 training sessions were held during the full season, which includes 11 congested weeks (i.e., two matches within 7 days) and 26 non-congested weeks (i.e., one match within 7 days). These periods were used to analyze the differences between starters and non-starters for AW, CW, wACWR, wTD, wSTD, wRS, wDT, and weekly maximum speed (wMS). Nineteen players were divided into two groups according to their starting status: starters (n = 10) and non-starters (n = 9). This criterion, based on a previous study, defined starter and non-starter players according to the time of participation in matches and training week-by-week [29]. Moreover, due to the outbreak of the COVID-19 pandemic, the team's training between the 34th and 35th weeks was closed for four months from 28 February 2020 to 28 June 2020, and we have not considered this course due to the closure of training. On the other hand, due to the club's decisions to maintain the level of physical fitness of the players during the lockdown from the COVID-19 pandemic, a training program video was provided for each player, and daily reports of training performance were sent by the players.

2.2. Participants

The present study, as a longitudinal descriptive study, was conducted throughout 2019–2020 in the Iranian Premier League. Nineteen professional soccer players of one club (28 ± 4.6 years; 181.6 ± 5.8 cm, 74.5 ± 5.6 kg) took part in the study during the full season, which was 43 weeks long. This research has been done by the club's training coaches after coordination with the club managers and the head coach. The selection of players to participate in this study includes: continuous participation in competitions and training sessions during the season (when a player did not play in the weekly competition, they performed an alternative training session such as high-intensity interval training or a small side game); players could not be cross-trained during the study period; and players were not injured for more than two weeks. Based on previous studies [30,31], goalkeepers were not included in the report because of physiological variations in preparation and competition. The players participating in this study were divided into two groups, starter and non-starter. Starter and non-starter players were defined on a weekly basis. If a player suffered an injury and was unable to play and train weekly, he would be removed from the starting line-up in the same week. Before starting the study, players were introduced to the objectives of the study, and the consent forms were signed. The project was approved with the ethical code IR.ARUMS.REC.1399.546 by the Ardabil University of Medical Sciences, and is in compliance with the Declaration of Helsinki for human subjects [32].

2.3. External Training Load

2.3.1. GPS

Players' daily physical activity was monitored using a GPS (GPS SPORTS systems Pty Ltd., Model: SPI High-Performance Unit (HPU), made in Canberra, Australia). This GPS model's features include the following: 15 Hz position GPS, distance, and speed measurement; accelerometer: 100 Hz, 16 G Tri-Axial-Track impacts, accelerations, and decelerations, as well as data source body load (BL); Mag: 50 Hz, Tri-Axial; dimensions: smallest device on the market (74 mm $\times$ 42 mm $\times$ 16 mm); robustness; SPI high-performance unit based on Mining/Industrial Strength Electronics design; water resistance; and data transmission: infra-red and weighs 56 g [33]. Previous studies have shown that GPS can provide valid and reliable estimates of instantaneous and constant velocity movements during linear, multidirectional, and soccer-specific activities [24].

2.3.2. Data Collection by GPS

The GPS information was used to measure external TL speed activities in all training and competition sessions. For every training session, all players used the same GPS to avoid mid-season takes between units. To allow access to satellite signals, and synchronize the GPS clock with the satellite atomic clock, all the devices were triggered 30 min before data collection. The information was downloaded to a computer after the recording, and analyzed using the software kit (GPS software, SPI High-Performance Unit (HPU), Canberra, Australia). GPS data were intended for the main team session (i.e., the beginning of the warm-up to the end of the last organized drill). In this study, baseline regions of SPI IQ absolute values were defined, and then, variables were assessed for each of the four time periods. 1. TD, 2. BL, 3 RS, 4.MS, 5. STD, and 6. BL reflects the quantity and intensity of acceleration events calculated from the accelerometer data. BL also acts as an integrated load variable used as an alternative TL marker (BL) for the original GPSports BL variable, and as a task speed marker (BL/min) as a TL criterion.

2.3.3. Calculate Training Load

BL is calculated from accelerometer data from GPSports devices taken from 3-axis accelerometers (planes X, Y, Z), and reflects both the amount and intensity of acceleration performed by the player. The following steps were performed in the BL calculation for each acceleration level: Initialize the BL counter to 0; Calculate the magnitude of the acceleration vector (V) of the current acceleration (V = ax2 + ay2 + az2); Normalize the magnitude vector (NV) by subtracting the country's 1 G (NV = V 1.0 G). The unscaled BL (USBL) was then calculated using the formula USBLC = NV + (NV3) [19]. Next, the accelerometer recording speed (100 Hz) and load factor (EF) (SBLC = USBLC/100/EF) were used to calculate the scaled BL (SBLC) [34]. Finally, the final BL (BL = BL + SBLC) was determined. BL was replacing the original GPS port. Regular TL data were analyzed to report weekly load adjustments during the match season (AW, CW, and ACWR). The CW represents the rolling exponential of average accumulated training load of training session experience in the previous three weeks [35]. The AW represents the total of the training load experienced in the previous seven days [36,37]. ACWR represents the used uncoupled formula [38], which is defined in the equation below. All variables were calculated in each week of the experimental period, which is defined in the equation below [39].

$$\text{Uncoupled ACWR4} = \frac{\text{AW4}}{0.333 \times (\text{AW1} + \text{AW2} + \text{AW3})} \tag{1}$$

ACWR was not computed for the pre-season period, and CW started from the third week according to the above formula. However, cumulative loads were used to calculate the ratio of CW and ACWR during the first few weeks of the season.

2.4. Statistical Analysis

Statistical analysis data were analyzed using IBM SPSS Statistics for Windows, Version 22.0 (IBM Corp., Armonk, NY, USA) statistical software package. The Shapiro–Wilk and Leven's tests were used separately to evaluate the normality and homogeneity of variances to analyze data. The results were reported as mean $\pm$ standard deviation, with a 95% confidence interval (CI). Changes between the four in-season periods were assessed using a repeated-measures analysis of variance (ANOVA), followed by a Bonferroni with adjustment *post-hoc* test for pairwise comparisons. Partial eta-square ($\eta2p$) was calculated as the effect size of the repeated-measures ANOVA, and if the variable was not statistically normal, the Kruskal–Wallis H test was used to analyze the intergroup differences. Moreover, the effect size of Hedge's g (95% CI) was reported. The Hopkins threshold was used to calculate the effect size [40] as follows: <0.2 = trivial, 0.2 to 0.6 = small, >0.6 to 1.2 = medium, >1.2 to 2.0 = large, >2.0 to 4.0 = very large, and >4.0, almost perfect. The significance level was considered at $p \leq 0.05$.

3. Results

Figure 1 shows the changes in the AW variations over a full season with different periods (pre-, early-, mid-, and end-season) for starters and non-starters. The highest load in these variables was in w 35 (1303.9) arbitrary units (A.U. for starters). Besides, the highest values for non-starters in AW were in w 34 (1379.3 A.U.), whereas the lowest load was observed in w 29 (28.0 A.U.).

Figure 2 shows changes in the CW variations during the full season based on periods (pre-, early-, mid-, and end-season) and between weekly coefficient of variation (%CV) for this variable based on groups (starters and non-starters). The highest load in these variables was in w 38 (1073.7 A.U.); while the lowest load was observed in w 24 (509.0 A.U.) for starters. Besides, the highest values for non-starter CW were in w 5 (968.4 A.U.), whereas the lowest load was observed in w 29 (160.0 A.U.).

Figure 3 shows an overall vision of ACWR variations across the full season and its different periods (pre-season, early-season, mid-season, and end-season) for starter and non-starter players. Overall, the highest ACWR occurred in w 35 (2.3 A.U.) and w 21 (4.8 A.U.) for starters and non-starters, respectively; however, the lowest ACWR was found in w 11 and 33 (0.6 A.U.) for starters and w 30 (0.1 A.U.) for non-starters, respectively (Figure 3B).

The results of the repeated-measures ANOVA revealed differences between season periods in AW starters ($p < 0.001$, $\eta2p = 0.886$), AW non-starters ($p < 0.001$, $\eta2p = 0.944$), CW starters ($p < 0.004$, $\eta2p = 0.833$) CW non-starters ($p < 0.001$, $\eta2p = 0.957$), ACWR starters ($p < 0.003$, $\eta2p = 0.771$), and ACWR non-starters ($p < 0.04$, $\eta2p = 0.582$). Theresults revealed significant greater weekly AW and CW and ACWR of starters compared to non-starters during the mid-season season (AW: $p = 0.008$, $g = 2.36$; CW: $p \leq 0.0001$; $g = 2.17$; wACWR: $p = 0.043$, $g = -0.95$) and end-season (AW: $p = 0.035$, $g = 0.91$; CW: $p = 0.017$; $g = 1.0$; wACWR: $p = 0.010$, $g = -0.92$). Moreover, there was a significant difference in CW ($p \leq 0.0001$, $g = 2.18$) between starters and non-starters in the early season. However, non-significant differences between groups were found for AW and CW during the pre-season, but not for AW and wACWR during the early season (Table 1).

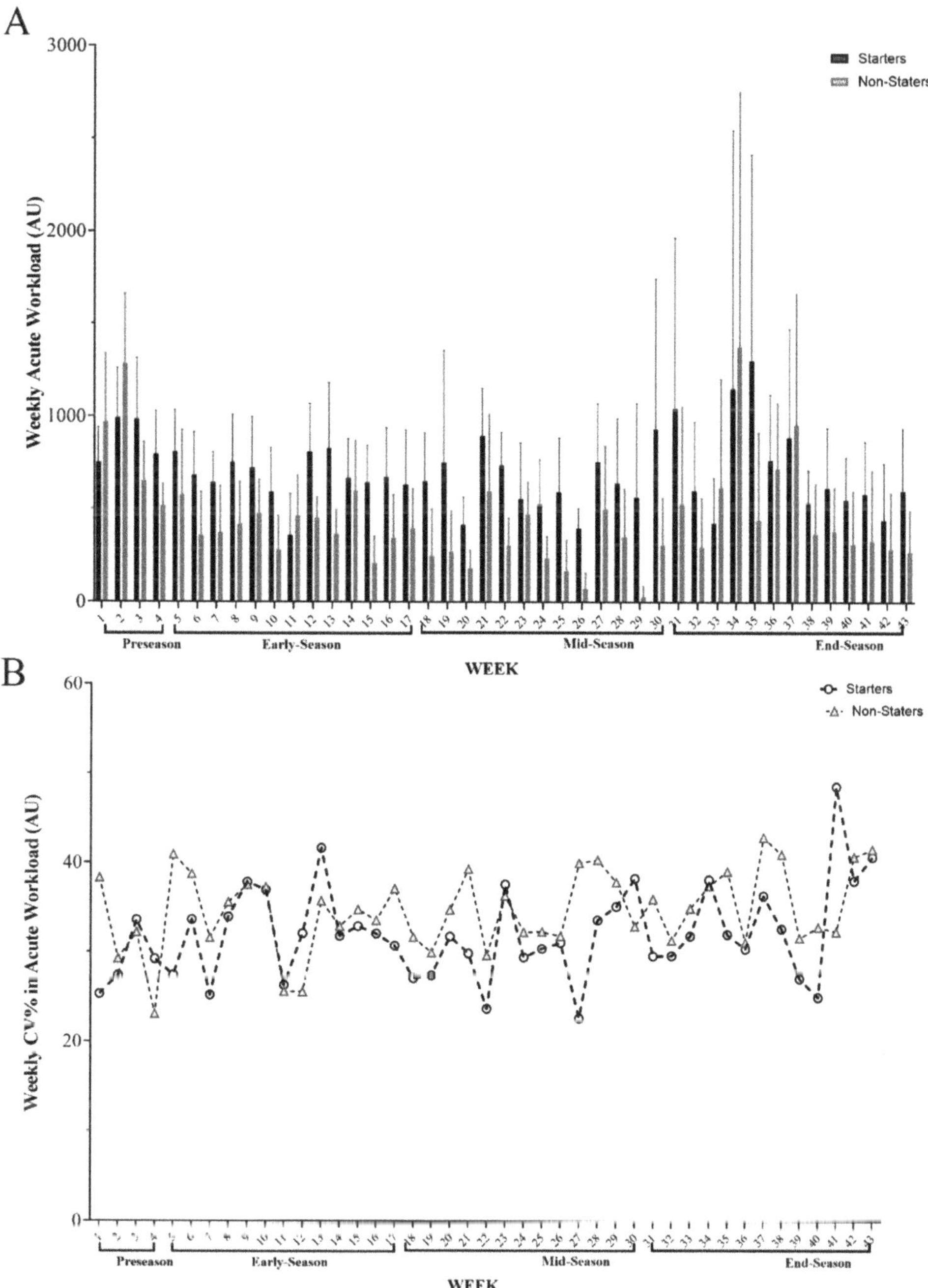

Figure 1. (**A**) Descriptive variations across AW (weekly acute workload) during the full season based on periods and (**B**) between weekly coefficient of variation (%CV) for this variable based on groups.

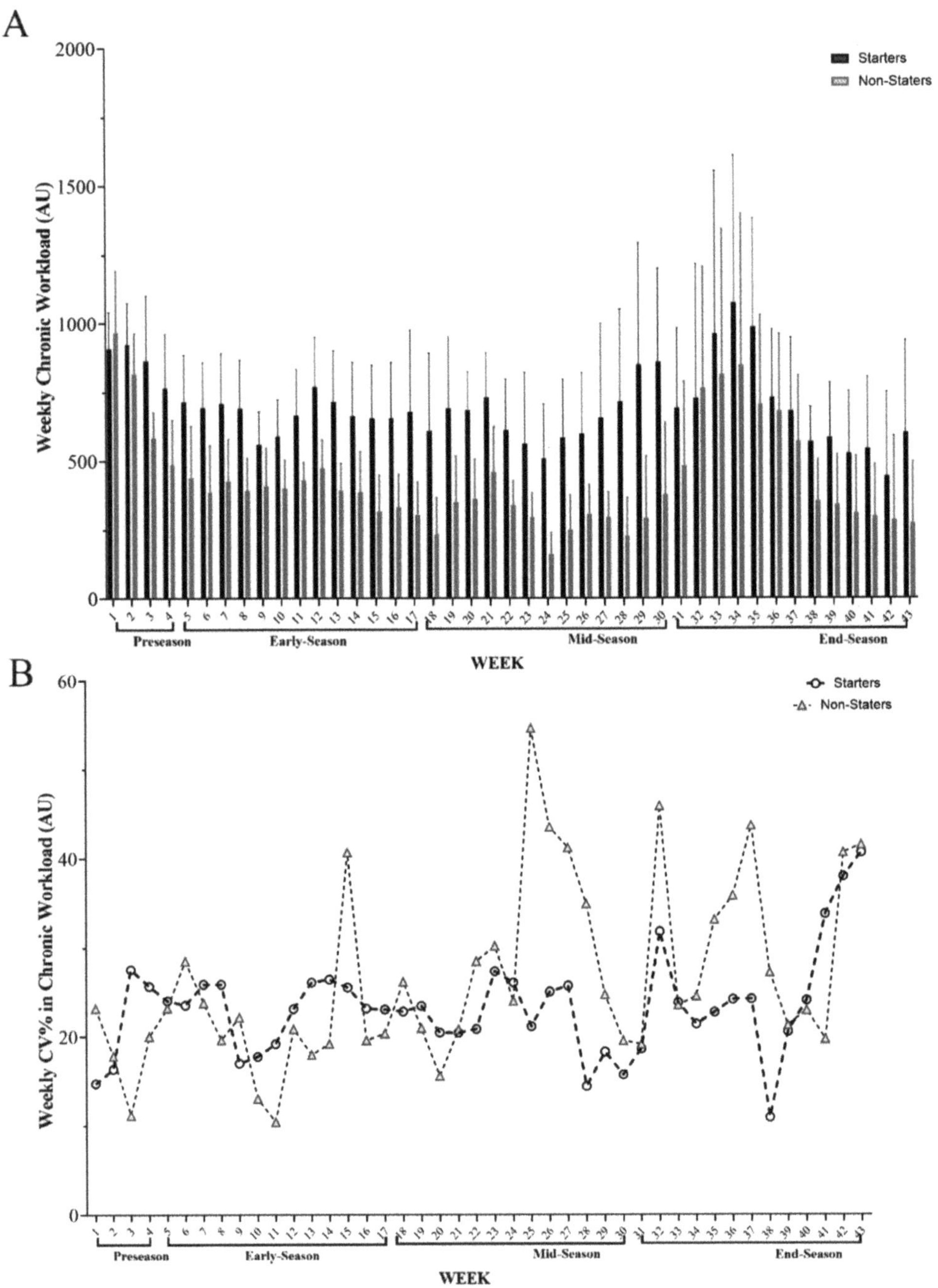

Figure 2. (**A**) Descriptive variations across CW (weekly chronic workload) during the full season based on periods and (**B**) between weekly coefficient of variation (%CV) for this variable based on groups.

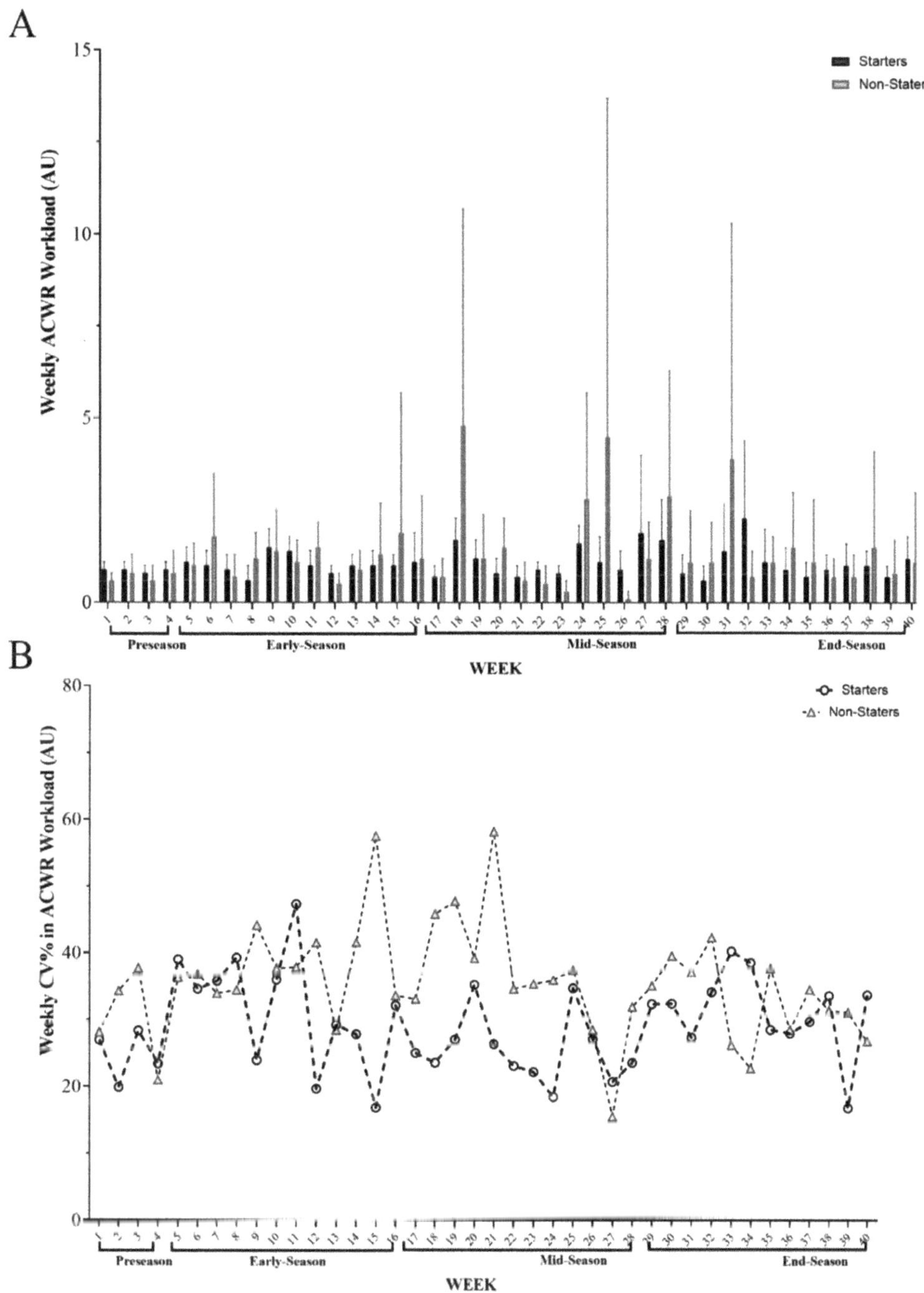

Figure 3. (**A**) Descriptive variations across wACWR (weekly acute/chronic workload) during the full season based on periods and (**B**) between weekly coefficient of variation (%CV) for this variable based on groups.

Table 1. Differences between starters and non-starters for AW, CW, and wACWR related to body load in the different periods of the season.

Variables	Season Period	Group		% Difference (Non-Starters vs. Starters)	p	Hedges's g (95% CI) (Non-Starters vs. Starters)	Inference
		Starters	Non-Starters				
AW (A.U.)	Pre-Season	880.9 (129.8)	856.0 (183.6)	2.4 (−12.7 to 7.7)	0.307	0.1 (−0.7 to 1.0)	small
	Early-Season	680.6 (151.9)	411.4 (56.1)	2.6 (1.5 to 3.8)	0.227	2.1 (1.0 to 3.3)	very large
	Mid-Season	649.8 (190.9)	288.5 (66.2)	3.6 (2.2 to 5.0)	0.008 *	2.3 (1.1 to 3.5)	very large
	End-Season	732.5 (258.6)	531.1 (137.9)	4.0 (−275.5 to 20.1)	0.035 *	0.9 (−0.0 to 1.8)	moderate to large
CW (A.U.)	Pre-Season	916.4 (132.6)	893.5 (182.8)	2.2 (−1.3 to 1.7)	0.756	0.1 (−0.7 to 1.0)	small
	Early-Season	696.8 (155.6)	425.4 (51.8)	2.7 (1.5 to 3.8)	<0.001 *	2.1 (1.0 to 3.3)	very large
	Mid-Season	637.3 (194.2)	301.3 (60.4)	3.3 (1.9 to 4.7)	<0.001 *	2.1 (1.0 to 3.3)	very large
	End-Season	752.7 (250.5)	526.3 (148.6)	2.2 (2.3 to 4.2)	0.017 *	1.0 (0.0 to 1.9)	large to very large
wACWR (A.U.)	Early-Season	1.0 (0.0)	1.0 (0.1)	−0.5 (−12 to 13)	0.133	0.0 (−0.9 to 0.9)	trivial
	Mid-Season	1.1 (0.2)	1.6 (0.7)	−5.0 (−9.8 to −1.4)	0.043 *	−0.9 (−1.9 to −0.0)	trivial
	End-Season	1.10 (0.16)	1.4 (0.38)	−3.0 (−6.0 to 0.1)	0.010 *	−0.9 (−1.8 to 0.0)	trivial

Abbreviations: AU, arbitrary units; AW, weekly acute workload in AU; CW, weekly chronic workload in AU; wACWR, weekly acute:chronic workload ratio in AU; p, p-value at alpha level 0.05; Hedges's g (95% CI), Hedges's g effect size magnitude with 95% confidence interval. * Significant differences for $p \leq 0.05$. Hedge's g was interpreted as: <0.2 = trivial, 0.2 to 0.6 = small, >0.6 to 1.2 = medium, >1.2 to 2.0 = large, >2.0 to 4.0 = very large, and > 4.0, almost perfect.

Results of repeated-measures ANOVA revealed differences between season periods in starters wTD starters ($p < 0.001$, η2p = 0.966), wTD non-starters ($p < 0.001$, η2p = 0.982), wSTD starters ($p < 0.003$, η2p = 0.683), wSTD non-starters ($p < 0.001$, η2p = 0.953), wDT starters ($p < 0.001$, η2p = 0.965) and wDT non-starters ($p < 0.001$, η2p = 0.993), wMS starters ($p < 0.001$, η2p = 0.978) and wMS non-starters ($p < 0.001$, η2p = 0.992), and wRS starters ($p < 0.02$, η2p = 0.713) and wRS non-starters ($p < 0.03$, η2p = 0.749). The between-group comparisons for derived-GPS variables of distance and sprint in the different periods of the season are displayed in Table 2. Overall, the results showed that there was a significant difference between groups starters and non-starters for wTD (average value) ($p \leq 0.0001$, g = 3.49 to 2.01), wSTD ($p \leq 0.0001$ to 0.003; g = 2.88 to 1.55), wDT (average value) ($p \leq 0.0001$ to 0.023; g = 3.34 to 1.09), wMS (average value) ($p \leq 0.0001$; g = 3.48 to 2.07), and wRS ($p \leq 0.0001$ to 0.018; g = 2.17 to 1.15) in the early-, mid-, and end-season. However, there were no differences between starters and non-starters for wTD, wSTD, wDT, wMS, wRS during the pre-season.

Table 2. Differences between starters and non-starters for derived GPS variables of distance and sprint in the different periods of the season.

Variables	Season Period	Group		%Difference (Non-Starters vs. Starters)	p	Hedges's g (95% CI) (Non-Starters vs. Starters)	Inference
		Starters	Non-Starters				
wTD **(m)**	*Pre-Season*	4995.8 (399.9)	4755.8 (356.3)	240 (−1.2 to 6)	0.187	0.60 (−0.31 to 1.52)	trivial
	Early-Season	4745.9 (498.9)	2974.2 (466.7)	17.7 (1.3 to 2.2)	<0.001 *	3.49 (2.06 to 4.92)	very large to nearly perfect
	Mid-Season	4568.3 (593.2)	2385.7 (559.2)	218.2 (1.6 to 2.7)	<0.001 *	3.61 (2.15 to 5.06)	very large to nearly perfect
	End-Season	4116.3 (668)	2696.7 (679.9)	141.9 (7.6 to 2)	<0.001 *	2.01 (0.90 to 3.11)	very large
wSTD **(m)**	*Pre-Season*	2541.4 (479.7)	2546.6 (571.3)	−520 (−5.1 to 5.0)	0.883	−0.00 (−0.91 to 0.89)	trivial
	Early-Season	1869.3 (282.8)	1123.7 (199.4)	7.4 (5.0 to 9.8)	<0.000 *	2.88 (1.59 to 4.16)	very large
	Mid-Season	1782.5 (402.8)	854.8 (220.4)	9.2 (6.0 to 1.2)	<0.000 *	2.68 (1.44 to 3.92)	very large
	End-Season	1876.4 (425.9)	1194.7 (413.0)	6.8 (2.7 to 1.0)	<0.003 *	1.55 (0.52 to 2.57)	large
wDT **(min)**	*Pre-Season*	76.8 (5.8)	76.1 (4)	70 (−4.1 to 5.5)	0.704	0.13 (−0.76 to 1.04)	trivial
	Early-Season	67.2 (5.4)	48.9 (7.4)	1.8 (1.2 to 2.4)	<0.001 *	3.34 (1.94 to 4.73)	very large
	Mid-Season	59 (7.5)	37.7 (6.7)	2.1 (1.4 to 2.8)	<0.001 *	4.25 (2.63 to 5.88)	nearly perfect
	End-Season	51.8 (10.3)	39.3 (11.5)	1.2 (1.9 to 2.3)	<0.017 *	1.09 (0.127 to 2.058)	small to moderate
wMS **(km·h^{1})**	*Pre-Season*	26.3 (2.5)	23.9 (2.3)	240 (0 to 4.7)	0.052	0.91 (−0.03 to 1.85)	Moderate to large
	Early-Season	28.5 (2.4)	18.8 (2.4)	97 (7.3 to 12)	<0.001 *	3.48 (2.05 to 4.91)	very large
	Mid-Season	26.1 (3.3)	17.1 (1.7)	90 (6.4 to 11.5)	<0.001 *	3.34 (1.954 to 4.74)	very large
	End-Season	23.9 (3.2)	18.5 (2)	54 (2.7 to 8)	<0.001 *	2.07 (0.95 to 3.18)	very large
wRS **(number)**	*Pre-Season*	88.3 (23.6)	75.4 (26.1)	1.2 (−1.1 to 3.7)	0.271	0.49 (−0.41 to 1.41)	small
	Early-Season	62.5 (9.5)	43.2 (7.2)	1.9 (1.1.0 to 2.7)	<0.001 *	2.17 (1.03 to 3.30)	very large
	Mid-Season	59.5 (10.0)	36.5 (8.9)	2.3 (1.3 to 3.2)	<0.001 *	2.31 (1.15 to 3.47)	very large
	End-Season	61.3 (13.3)	44.1 (15.3)	1.7 (0.3 to 3.1)	<0.016 *	1.15 (0.17 to 2.12)	large

Abbreviations: wTD, weekly total distance in meters; wSTD, weekly sprint total distance in meters; wDT, weekly duration training in minutes; wMS, the average accumulated of the maximum sprint is calculated and reported each week in kilometres per hour; wRS, weekly repeat sprint in meters; p, p-value at alpha level 0.05; Hedges's g (95% CI), Hedges's g effect size magnitude with 95% confidence interval. * Significant differences for $p \leq 0.05$. Hedge's g was interpreted as: <0.2 = trivial, 0.2 to 0.6 = small, >0.6 to 1.2 = medium, >1.2 to 2.0 = large, >2.0 to 4.0 = very large, and > 4.0, almost perfect.

4. Discussion

The goals of this study were to: (i) identify, week-by-week, the AW, CW, and wACWR between starter and non-starter professionals soccer players using the BL; (ii) examine the aforementioned details related to the importance of quantifying the external TL over the season while attending to the starting status of the players; (iii) compare the gaps in AW, CW, and wACWR in BL, as well as the weekly average of distance and sprint variables, between starters and non-starters over the four season periods (pre-, early-, mid-, and end-season). On the other hand, in the present study, with the closure of the Persian Gulf League due to the COVID-19 epidemic, matches and group training were closed during this period, and training continued individually at home. Therefore, this time period was excluded from the present study.

The results of the BL analysis showed that both groups of players achieved the highest amount of AW and CW during the pre- and end-season. In non-starter players, the highest amount of wACWR was observed in the end-season, whereas in starter players, the highest amount was obtained in the mid-season. The fluctuations of AW and CW changes were consistent in the starter and non-starter players during the competitive season. However, the intergroup comparison showed that starters experienced more CW and AW during the season than the non-starters. Throughout the season, AW values were higher among non-starter players. Since the starter players are the starters of the matches, and for most of the playing time, these players are present in the match, it can be justified that these values are higher compared to non-starters, and also, this result could be related to the different levels of physical fitness of the players [41]. Other reasons include additional physiological stressors (high-speed running and number of sprints) [42] or psychological stains (rating of perceived exertion) [43] during a competitive season, which leads the starter group to increase its weekly training load and have a higher weekly workload than non-starters. On the other hand, perhaps more training can be attributed to the common strategy of the team's technical staff to compensate, after the match or immediately after the match-play, for the participation of players who played for less than 45 min in an official game [30]. According to the contradictory results of recent research, the concept of ACWR, which indicates the possibility of injury, is questioned [26].

Changes in ACWR in the early-season were close to each other and there was no significant difference between groups, whereas in the mid- and end-season, there was a significant difference between groups. Some studies have shown an increased risk of injury with greater workloads [35,44] because exposure to a higher load is inevitable. Besides, Bowen et al. pointed out that an acute, excessive, and rapid increase in loads compared to a higher chronic load may be responsible for a large proportion of non-contact injuries [45]. Thus, non-starters should be prepared to experience greater stress during periods when they are not participating as starters in matches in order to reduce their injury risk [46]. Curtis et al. demonstrated that non-starters had higher training workloads, as measured from total covering distance during weekly training sessions [47]. Researchers attribute the workload difference between starters and non-starters to a variety of factors, such as age, competitive environment, tactic strategy of the league, length of the season, and league fixture [30]. However, in another part of the results, in non-starters, ACWR in the mid- or early-season were at 1.4 and 1.6, respectively. Bowne et al. reported a significant increase in injury risk (relative risk = 2.6) when CW was low (<938 m) and when ACWR was high at high-speeds (1.4–1.9 A.U.). These results, though seemingly not generalized, suggest that ACWR surveillance can be an important strategy to prevent injury in professional soccer [17]. Most studies on the relationship between workload and injuries have removed contact injuries, as they appear to be unavoidable, but many researchers have suggested that ACWR is a good variable of the risk of non-contact injury in players [45]. Recent research has questioned the concept of ACWR as a marker to identify possible injuries [48,49]. No one criticizes the true value of understanding changes in TL progression or changes over the course of a few weeks. Interestingly, training heterogeneity can lead to ACWRs that differ from play position differences during the season, as they are stagnant or show sensitivity for regular periods. Of course, the difference between these changes may depend on the type of measurement considered.

The results of the present study revealed that starters registered greater wTD, wSTD, wDT, wRS, and wMS during the early-, mid-, and end-season in comparison with non-starters in every season period. Anderson et al. and Oliveria et al. did not report the TD difference between starters and non-starters; however, the authors concluded that the high-intensity season loading pattern depended on the starting position of the players in favour of the starters in the English Premier League [46,50]. In line with previous studies, our results showed greater wDT and wTD in starters compared to non-starters during the season. Consistent with our results, the highest TD values were reported by researchers during the pre- and early-season compared to the mid- and end-season. Oliveria et al. attributed the great TD in the pre-season to the coaches' emphasis on physical conditioning, which reduces the number of other courses compared to the pre-season. In addition to examining the external TL, some articles have also analyzed the internal TL over a season [50]. Researchers related the difference between starter and non-starter players to the starting position of the players, coach's decisions for designing drills, play styles, and competition demands across leagues, which should be considered by the technical staff so that a suitable training strategy can be considered for each player [19,51]. It seems that the differences in GPS variables between starters and non-starters are understandable because only 11 players participate in each competition, so the other players are not exposed the load of the match, and so, these reported differences are justified. Managing starter and non-starter players is very important for the team manager and staff of the professional teams present during the tournament schedule. External load monitoring provides good reference information about the physical impact and physiological stresses during an annual cycle. It seems that awareness of training intensity, recovery status, and substitution strategies in systematic planning in a competitive season to prevent injury, overtraining syndrome, and NFOR is too important.

Limitations of Study

Regarding the limitations of the present study, we can mention the lack of internal TL criteria. This study examines a team in the Persian Gulf League that has faced a shutdown due to the expansion of COVID-19. Due to the fact that we only had access to one of the 16 teams in the Premier League, so the number of samples was small, we could not report the sample size calculation [32]. A lack of physiological analysis in the training load portion was one of the limitations of the present study. Another limitation of this study is the fact that players trained at home during the COVID-19 league closure. Simultaneous examination of internal and external load may have been more appropriate. It is also possible in future studies to compare AW and CW orientation strategies for TL management in higher statistical samples.

5. Conclusions

According to the results, starters' AW, CW, and ACWR were greater than non-starters. Moreover, the ACWR value of the starters in the mid- and end-season showed high numbers that should be considered by the technical staff to review the TL strategies. In the next section, the results of starters' wTD, wSTD, wDT, wRS, and wMS variables in comparison to non-starters showed a higher mean, and also in both groups, compared to the pre-season, the average of all variables was reduced.

According to the results of this study, it seems that training monitoring using the parameters measured in this study can be a good way to check the external loads during training and competition. Moreover, coaches and professional soccer players can use the information obtained in this way to check the progress during the training season, review the performance during the match, and design appropriate training sessions.

Author Contributions: Conceptualization, R.G., H.N., L.B. and M.S.; Data curation, R.G. and H.N.; Formal analysis, R.G., H.N. and J.P.B.; Methodology, R.G. and H.N.; Writing—original draft, R.G., H.N., L.B. and J.P.B.; Writing—review and editing, R.G., H.N., L.B., M.S. and J.P.B. All authors have read and agreed to the published version of the manuscript.

Funding: This research was funded by the Portuguese Foundation for Science and Technology, I.P., Grant/Award Number UIDP/04748/2020.

Institutional Review Board Statement: The study was conducted according to the guidelines of the Declaration of Helsinki. The project was approved with the ethical code IR.ARUMS.REC.1399.546 by the Ardabil University of Medical Sciences.

Informed Consent Statement: Informed consent was obtained from all subjects and their parents involved in the study.

Data Availability Statement: The datasets used and/or analyzed during the current study are available from the corresponding author on reasonable request.

Acknowledgments: The authors would like to thank the team's coaches and players for their cooperation during all data collection procedures.

Conflicts of Interest: The authors declare no conflict of interest.

References

1. Draganidis, D.; Chatzinikolaou, A.; Jamurtas, T.; Barbero, J.C.; Tsoukas, D.; Theodorou, A.; Margonis, K.; Michailidis, Y.; Avloniti, A.; Theodorou, A.; et al. The time-frame of acute resistance exercise effects on football skill performance: The impact of exercise intensity. *J. Sports Sci.* **2013**, *31*, 714–722. [CrossRef] [PubMed]
2. Swallow, W.E.; Skidmore, N.; Page, R.M.; Malone, J.J. An examination of in-season external training load in semi-professional soccer players: Considerations of one and two match weekly microcycles. *Int. J. Sports Sci. Coach.* **2021**, *16*, 192–199. [CrossRef]
3. Nobari, H.; Oliveira, R.; Clemente, F.; Pérez-Gómez, J.; Pardos-Mainer, E.; Ardigò, L. Somatotype, Accumulated Workload, and Fitness Parameters in Elite Youth Players: Associations with Playing Position. *Children* **2021**, *8*, 375. [CrossRef] [PubMed]
4. Nobari, H.; Vahabidelshad, R.; Pérez-Gómez, J.; Ardigò, L.P. Variations of training workload in micro-and meso-cycles based on position in elite young soccer players: A competition season study. *Front. Physiol.* **2021**, *12*, 529. [CrossRef] [PubMed]
5. Rice, S.M.; Purcell, R.; De Silva, S.; Mawren, D.; McGorry, P.D.; Parker, A. The Mental Health of Elite Athletes: A Narrative Systematic Review. *Sports Med.* **2016**, *46*, 1333–1353. [CrossRef]
6. Eckard, T.G.; Padua, D.A.; Hearn, D.W.; Pexa, B.S.; Frank, B.S. The Relationship Between Training Load and Injury in Athletes: A Systematic Review. *Sports Med.* **2018**, *48*, 1929–1961. [CrossRef] [PubMed]
7. Drew, M.K.; Finch, C.F. The Relationship Between Training Load and Injury, Illness and Soreness: A Systematic and Literature Review. *Sports Med.* **2016**, *46*, 861–883. [CrossRef]
8. Nobari, H.; Kargarfard, M.; Minasian, V.; Cholewa, J.M.; Pérez-Gómez, J. The effects of 14-week betaine supplementation on endocrine markers, body composition and anthropometrics in professional youth soccer players: A double blind, randomized, placebo-controlled trial. *J. Int. Soc. Sports Nutr.* **2021**, *18*, 20. [CrossRef]
9. Schmikli, S.L.; de Vries, W.R.; Brink, M.S.; Backx, F.J. Monitoring performance, pituitary–adrenal hormones and mood profiles: How to diagnose non-functional over-reaching in male elite junior soccer players. *Br. J. Sports Med.* **2012**, *46*, 1019–1023. [CrossRef]
10. Meeusen, R.; Duclos, M.; Foster, C.; Fry, A.; Gleeson, M.; Nieman, D.; Raglin, J.; Rietjens, G.; Steinacker, J.; Urhausen, A. Prevention, diagnosis and treatment of the overtraining syndrome: Joint consensus statement of the European College of Sport Science (ECSS) and the American College of Sports Medicine (ACSM). *Eur. J. Sport Sci.* **2013**, *13*, 1–24. [CrossRef]
11. Schmikli, S.; Brink, M.; De Vries, W.; Backx, F. Can we detect non-functional overreaching in young elite soccer players and middle-long distance runners using field performance tests? *Br. J. Sports Med.* **2011**, *45*, 631–636. [CrossRef] [PubMed]
12. Braun, H.; Von Andrian-Werburg, J.; Schänzer, W.; Thevis, M. Nutrition status of young elite female German football players. *Pediatric Exerc. Sci.* **2018**, *30*, 157–167. [CrossRef] [PubMed]
13. Griffin, A.; Kenny, I.C.; Comyns, T.M.; Lyons, M. The association between the acute: Chronic workload ratio and injury and its application in team sports: A systematic review. *Sports Med.* **2020**, *50*, 561–580. [CrossRef] [PubMed]
14. Nobari, H.; Mainer-Pardos, E.; Zamorano, A.D.; Bowman, T.G.; Clemente, F.M.; Pérez-Gómez, J. Sprint variables are associated with the odds ratios of non-contact injuries in professional soccer players. *Int. J. Environ. Res. Public Health* **2021**, *18*, 10417. [CrossRef]
15. Ball, S.; Halaki, M.; Sharp, T.; Orr, R. Injury patterns, physiological profile, and performance in university rugby union. *Int. J. Sports Physiol. Perform.* **2018**, *13*, 69–74. [CrossRef]
16. Rogalski, B.; Dawson, B.; Heasman, J.; Gabbett, T.J. Training and game loads and injury risk in elite Australian footballers. *J. Sci. Med. Sport* **2013**, *16*, 499–503. [CrossRef]
17. Bowen, L.; Gross, A.S.; Gimpel, M.; Li, F.-X. Accumulated workloads and the acute: Chronic workload ratio relate to injury risk in elite youth football players. *Br. J. Sports Med.* **2017**, *51*, 452–459. [CrossRef]
18. Nobari, H.; Aquino, R.; Clemente, F.M.; Khalafi, M.; Adsuar, J.C.; Pérez-Gómez, J. Description of acute and chronic load, training monotony and strain over a season and its relationships with well-being status: A study in elite under-16 soccer players. *Physiol. Behav.* **2020**, *225*, 113117. [CrossRef]

19. Nobari, H.; Khalili, S.M.; Zamorano, A.D.; Bowman, T.G.; Adsuar, J.C.; Perez-Gomez, J.; Granacher, U. Workload is Associated with the Occurrence of Non-Contact Injuries in Professional Male Soccer Players: A Pilot Study. *Preprint* **2021**. [CrossRef]

20. Lu, D.; Howle, K.; Waterson, A.; Duncan, C.; Duffield, R. Workload profiles prior to injury in professional soccer players. *Sci. Med. Footb.* **2017**, *1*, 237–243. [CrossRef]

21. Barnes, C.; Archer, D.T.; Hogg, B.; Bush, M.; Bradley, P.S. The evolution of physical and technical performance parameters in the English Premier League. *Int. J. Sports Med.* **2014**, *35*, 1095–1100. [CrossRef]

22. Morton, R. Modelling training and overtraining. *J. Sports Sci.* **1997**, *15*, 335–340. [CrossRef] [PubMed]

23. Rico-González, M.; Los Arcos, A.; Rojas-Valverde, D.; Clemente, F.M.; Pino-Ortega, J. A survey to assess the quality of the data obtained by radio-frequency technologies and microelectromechanical systems to measure external workload and collective behavior variables in team sports. *Sensors* **2020**, *20*, 2271. [CrossRef] [PubMed]

24. Hulin, B.T.; Gabbett, T.J.; Caputi, P.; Lawson, D.W.; Sampson, J.A. Low chronic workload and the acute: Chronic workload ratio are more predictive of injury than between-match recovery time: A two-season prospective cohort study in elite rugby league players. *Br. J. Sports Med.* **2016**, *50*, 1008–1012. [CrossRef]

25. Schwellnus, M.; Soligard, T.; Alonso, J.-M.; Bahr, R.; Clarsen, B.; Dijkstra, H.P.; Gabbett, T.J.; Gleeson, M.; Hägglund, M.; Hutchinson, M.R.; et al. How much is too much? (Part 2) International Olympic Committee consensus statement on load in sport and risk of illness. *Br. J. Sports Med.* **2016**, *50*, 1043–1052. [CrossRef] [PubMed]

26. Impellizzeri, F.M.; Tenan, M.S.; Kempton, T.; Novak, A.; Coutts, A.J. Acute: Chronic workload ratio: Conceptual issues and fundamental pitfalls. *Int. J. Sports Physiol. Perform.* **2020**, *15*, 907–913. [CrossRef]

27. Gabbett, T.J. The training—Injury prevention paradox: Should athletes be training smarter and harder? *Br. J. Sports Med.* **2016**, *50*, 273–280. [CrossRef] [PubMed]

28. Nobari, H.; Kharatzadeh, M.; Khalili, S.M.; Pérez-Gómez, J.; Ardigò, L.P. Fluctuations of Training Load Variables in Elite Soccer Players U-14 throughout the Competition Season. *Healthcare* **2021**, *9*, 1418. [CrossRef]

29. Nobari, H.; Oliveira, R.; Clemente, F.M.; Adsuar, J.C.; Pérez-Gómez, J.; Carlos-Vivas, J.; Brito, J.P. Comparisons of Accelerometer Variables Training Monotony and Strain of Starters and Non-Starters: A Full-Season Study in Professional Soccer Players. *Int. J. Environ. Res. Public Health* **2020**, *17*, 6547. [CrossRef]

30. Nobari, H.; Castillo, D.; Clemente, F.M.; Carlos-Vivas, J.; Pérez-Gómez, J. Acute, chronic and acute/chronic ratio between starters and non-starters professional soccer players across a competitive season. *Proc. Inst. Mech. Eng. Part P J. Sports Eng. Technol.* **2021**, 17543371211016594. [CrossRef]

31. Nobari, H.; Oliveira, R.; Brito, J.; Pérez-Gómez, J.; Clemente, F.; Ardigò, L. Comparison of running distance variables and body load in competitions based on their results: A full-season study of professional soccer players. *Int. J. Environ. Res. Public Health* **2021**, *18*, 2077. [CrossRef]

32. World Medical Association. *Declaración de Helsinki de la AMM-Principios Eticos Para las Investigaciones Médicas en Seres Humanos*; World Medical Association: Ferney-Voltaire, France, 2019.

33. Williams, J.; Tessaro, E. Validity and reliability of a 15 Hz GPS Device for court-based sports movements. *J. Strength Cond. Res.* **2018**, *28*. [CrossRef]

34. Nobari, H.; Banoocy, N.K.; Oliveira, R.; Pérez-Gómez, J. Win, Draw, or Lose? Global Positioning System-Based Variables' Effect on the Match Outcome: A Full-Season Study on an Iranian Professional Soccer Team. *Sensors* **2021**, *21*, 5695. [CrossRef] [PubMed]

35. Malone, S.; Owen, A.; Newton, M.; Mendes, B.; Collins, K.D.; Gabbett, T.J. The acute: Chronic workload ratio in relation to injury risk in professional soccer. *J. Sci. Med. Sport* **2017**, *20*, 561–565. [CrossRef] [PubMed]

36. Clemente, F.M.; Silva, A.F.; Clark, C.C.; Conte, D.; Ribeiro, J.; Mendes, B.; Lima, R. Analyzing the seasonal changes and relationships in training load and wellness in elite volleyball players. *Int. J. Sports Physiol. Perform.* **2020**, *15*, 731–740. [CrossRef]

37. Hulin, B.T.; Gabbett, T.J.; Lawson, D.W.; Caputi, P.; Sampson, J.A. The acute: Chronic workload ratio predicts injury: High chronic workload may decrease injury risk in elite rugby league players. *Br. J. Sports Med.* **2016**, *50*, 231–236. [CrossRef]

38. Hedges, L.V.; Olkin, I. *Statistical Methods for Meta-Analysis*; Academic Press: Cambridge, MA, USA, 2014.

39. Haddad, M.; Stylianides, G.; Djaoui, L.; Dellal, A.; Chamari, K. Session-RPE method for training load monitoring: Validity, ecological usefulness, and influencing factors. *Front. Neurosci.* **2017**, *11*, 612. [CrossRef]

40. Hopkins, W.G.; Marshall, S.W.; Batterham, A.M.; Hanin, J. Progressive statistics for studies in sports medicine and exercise science. *Med. Sci. Sports Exerc.* **2009**, *41*, 3. [CrossRef]

41. Silva, J.R.; Magalhães, J.F.; Ascensão, A.A.; Oliveira, E.M.; Seabra, A.F.; Rebelo, A.N. Individual match playing time during the season affects fitness-related parameters of male professional soccer players. *J. Strength Cond. Res.* **2011**, *25*, 2729–2739. [CrossRef]

42. Dalen, T.; Jørgen, I.; Gertjan, E.; Havard, H.G.; Ulrik, W. Player load, acceleration, and deceleration during forty-five competitive matches of elite soccer. *J. Strength Cond. Res.* **2016**, *30*, 351–359. [CrossRef]

43. Arcos, A.L.; Mendez-Villanueva, A.; Martinez-Santos, R. In-season training periodization of professional soccer players. *Biol. Sport* **2017**, *34*, 149. [CrossRef]

44. Gabbett, T.J.; Jenkins, D.G. Relationship between training load and injury in professional rugby league players. *J. Sci. Med. Sport* **2011**, *14*, 204–209. [CrossRef] [PubMed]

45. Bowen, L.; Gross, A.S.; Gimpel, M.; Bruce-Low, S.; Li, F.-X. Spikes in acute: Chronic workload ratio (ACWR) associated with a 5–7 times greater injury rate in English Premier League football players: A comprehensive 3-year study. *Br. J. Sports Med.* **2020**, *54*, 731–738. [CrossRef] [PubMed]

46. Anderson, L.; Orme, P.; Di Michele, R.; Close, G.L.; Milsom, J.; Morgans, R.; Drust, B.; Morton, J.P. Quantification of seasonal-long physical load in soccer players with different starting status from the English Premier League: Implications for maintaining squad physical fitness. *Int. J. Sports Physiol. Perform.* **2016**, *11*, 1038–1046. [CrossRef] [PubMed]
47. Curtis, R.M.; Huggins, R.A.; Benjamin, C.L.; Sekiguchi, Y.; Arent, S.M.; Armwald, B.C.; Pullara, J.M.; West, C.A.; Casa, D.J. Seasonal Accumulated Workloads in Collegiate Men's Soccer: A Comparison of Starters and Reserves. *J. Strength Cond. Res.* **2019**, *35*, 3184–3189. [CrossRef] [PubMed]
48. Fanchini, M.; Rampinini, E.; Riggio, M.; Coutts, A.J.; Pecci, C.; McCall, A. Despite association, the acute: Chronic work load ratio does not predict non-contact injury in elite footballers. *Sci. Med. Footb.* **2018**, *2*, 108–114. [CrossRef]
49. Enright, K.; Green, M.; Hay, G.; Malone, J.J. Workload and injury in professional soccer players: Role of injury tissue type and injury severity. *Int. J. Sports Med.* **2020**, *41*, 89–97. [CrossRef]
50. Oliveira, R.; Brito, J.P.; Martins, A.; Mendes, B.; Marinho, D.A.; Ferraz, R.; Marques, M.C. In-season internal and external training load quantification of an elite European soccer team. *PLoS ONE* **2019**, *14*, e0209393. [CrossRef]
51. Clemente, F.M.; Rabbani, A.; Conte, D.; Castillo, D.; Afonso, J.; Clark, C.C.T.; Nikolaidis, P.T.; Rosemann, T.; Knechtle, B. Training/match external load ratios in professional soccer players: A full-season study. *Int. J. Environ. Res. Public Health* **2019**, *16*, 3057. [CrossRef]

Article

Quantification of Pre-Season and In-Season Training Intensity across an Entire Competitive Season of Asian Professional Soccer Players

Hadi Nobari [1,2,3,*], Akhilesh Kumar Ramachandran [4], João Paulo Brito [5,6,7] and Rafael Oliveira [5,6,7,*]

1 Department of Motor Performance, Faculty of Physical Education and Mountain Sports, Transilvania University of Braşov, 500068 Braşov, Romania
2 HEME Research Group, Faculty of Sport Sciences, University of Extremadura, 10003 Cáceres, Spain
3 Sports Scientist, Sepahan Football Club, 81887-78473 Isfahan, Iran
4 Sports Dynamix Private Limited, Chennai 600006, India; akhil24.kumar@gmail.com
5 Sports Science School of Rio Maior—Polytechnic Institute of Santarém, 2040-413 Rio Maior, Portugal; jbrito@esdrm.ipsantarem.pt
6 Research Centre in Sport Sciences, Health Sciences and Human Development, Quinta de Prados, Edifício Ciências de Desporto, 5001-801 Vila Real, Portugal
7 Life Quality Research Centre, 2040-413 Rio Maior, Portugal
* Correspondence: hadi.nobari1@gmail.com (H.N.); rafaeloliveira@esdrm.ipsantarem.pt (R.O.)

Citation: Nobari, H.; Ramachandran, A.K.; Brito, J.P.; Oliveira, R. Quantification of Pre-Season and In-Season Training Intensity across an Entire Competitive Season of Asian Professional Soccer Players. *Healthcare* **2022**, *10*, 1367. https://doi.org/10.3390/healthcare10081367

Academic Editor: Herbert Löllgen

Received: 25 June 2022
Accepted: 21 July 2022
Published: 23 July 2022

Publisher's Note: MDPI stays neutral with regard to jurisdictional claims in published maps and institutional affiliations.

Abstract: The aim of this study was to quantify the training load in two microcycles (Ms) from pre- and another two from in-season and to analyze playing position influences on the load experienced by professional soccer players. Nineteen Asian athletes, including four central defenders, four wide defenders, six central midfielders, three wide midfielders, and two strikers participated in this study. The micro-electromechanical system was used to collect training duration, total distance, and data from Zone 1 (0–3.9 km·h^{-1}), Zone 2 (4–7.1 km·h^{-1}), Zone 3 (7.2–14.3 km·h^{-1}), Zone 4 (14.4–19.7 km·h^{-1}), and Zone 5 (>19.8 km·h^{-1}), heart rate maximum (HRmax), and average (HRavg). The load was reduced on the last day of the Ms, with the exception of Zone 5, in M1, where higher values were found on the last day. Significant differences were observed between central and wide defenders for distance covered in Zone 4 (effect-size: ES = −4.83) in M2 and M4 (ES = 4.96). Throughout all the Ms, a constant HRmax (165–188 bpm) and HRavg (119–145 bpm) were observed. There was a tendency to decrease the load on the last day of the Ms. In general, there were higher external training loads in Ms from the pre-season than in-season. Wide defenders and wide midfielders showed higher distances covered with high-intensity running.

Keywords: heart rate; high speed running; monitoring; sprint; sports technology; training load; load quantification; competition phases

1. Introduction

The increasing intensity of soccer matches has resulted in drastic changes in the physiological demands of the sport [1,2]. The soccer game involves brief bouts of linear high-intensity running, sprinting, accelerations, decelerations, and multi-directional activities or change-of-direction movements, interspersed by longer recovery periods of lower intensity activities [3]. For instance, players in the English Premier League have been found to cover 681 m at running speeds ranging from 19.5 to 25.1 km·h^{-1} [4]. Previously, professional soccer players have performed 150–200 intense actions per game [5] and covered very high-intensity distances every 72 s [6]. The findings from the available literature highlight the importance for players to maintain high physical fitness levels to withstand the increasing physical demands during the training session and competitive matches [7]. Therefore, monitoring/tracking each player's daily intensity has become a necessity for coaches and practitioners in order to maintain an optimal fitness level of the players throughout the competitive season [8], and also to reduce the risk of injuries [9].

The weekly intensity of soccer players varies according to the number of matches during the season and phase of the annual plan [10,11]. The aim during the pre-season is to rebuild the fitness levels of the players after the off-season [12]. However, the aim during the in-season is to maintain and if possible improve the specific fitness components developed during the pre-season [12]. Professional soccer players train between four to six sessions/week during the competitive season [10], while it might be as high as one or two sessions, five days a week during the pre-season [11]. Such variation in training patterns might drastically influence the physiological demands placed on players and might result in differences in the physiological stress [13]. Therefore, knowledge regarding the intensity, during the different phases of the season, can be helpful for coaches in designing and implementing periodized soccer-specific training routines.

In this sense, load monitoring includes both internal and external dimensions. For better clarity, we avoid the load term and replace it with intensity following a recent suggestion by Staunton et al. [14]. Thus, the external dimension constitutes the physical demands performed during a training session or match, while the internal dimension is related to biochemical and biomechanical responses to the external stressors [15]. In soccer, subjective scales such as the rating of perceived exertion (RPE) initially developed by the Borg scale [16] and later adapted by Foster [17], and s-RPE developed by Foster et al. [18,19] and Impellizzeri et al. [20] were initially used to quantify internal intensity. This was followed by using heart rate (HR) telemetry to monitor the cardiovascular response during each training session. On the other hand, the external intensity is measured in training and matches using micro-electro-mechanical systems (MEMS) such as location position systems, global positioning systems (GPS), or optical camera tracking systems, using metrics such as distance covered and intensities of different movements/actions [21]. Therefore, soccer teams in recent years employ a combination of the above-mentioned methods to quantify external and internal intensities.

Recent research in season intensity quantification has provided valuable information by players in various leagues across European soccer [22–27]. Further, several studies have attempted to quantify the in-season training intensity in a professional soccer team, which involves a comparison of training days within weekly mesocycles [26–28]. Based on the evidence available from the existing literature, the intensity has shown to have little variation during the competition phase [26,27]. Additionally, within the weekly microcycles (Ms), the intensity was generally similar within training days with a reduction in the intensity in the days prior to the competition [22,24,26]. However, a comparison of pre-season and in-season has been made by few studies [12,29]. Thereafter, there is still paucity in the available evidence, since these studies have only reported limited information on intensity and have considered only the duration of training and s-RPE for the internal intensity quantification, and not the external intensity data collected using MEMS. Additionally, there are currently not many available studies that have analyzed internal and external intensities at two different time points during the soccer pre- and in-season of the Asian Professional League.

A better understanding of the periodization practices in professional soccer will be helpful for performance and technical coaches to manipulate the training volume, in order to help players increase their physical levels during different phases of a soccer season [30]. However, due to the lack of clear evidence in professional players, the periodization practices in soccer are currently poorly known and are reliant on the coach's training philosophy based on years of coaching experience [31]. Therefore, it can be argued that there is still a lack of clarity with respect to periodization practices and whether these practices demonstrate the required variation in intensity. Since the coaching practices vary, further research is needed to broaden our understanding of this topic, and how it should be programmed across an entire season. Moreover, most of the research on intensity during training and competition during various Ms have predominantly focused on European soccer clubs [22,24]. As a result, these findings cannot be applied across all the soccer teams, mainly due to the variation in match demands, culture, the number of matches played, and tactical reasons [32]. To the author's knowledge, there are limited studies [32] that have quantified intensity from both soccer pre- and in-season in Asian Professional League. Therefore, the primary aim of this study was to quantify intensity in four different Ms from the pre- and in-season period (two Ms from each period) of soccer players competing in an Asian Professional League. The secondary aims of the study were to compare training days in each M as well as to compare playing positions.

2. Materials and Methods

2.1. Experimental Approach to the Problem

For this longitudinal study, training data were collected over a 48-week period during full season. For the purposes of the present study, all training sessions conducted during the main team sessions were considered which means that rehabilitation or recovery sessions were not used for analysis. The duration of the training sessions includes the warm-up, main and slow-down phases, plus stretching. All training programs were planned by the coach and staff, and the researchers only standardized the first 30 min and the final 30 min, (i.e., before and after each training session) regarding research procedures for external and internal data collection (described below in Sections 2.4 and 2.5).

The weeks were chosen based on similar characteristics in pre- and in-season periods which included pre- and in-season weeks with the exact same scheduling, (i.e., M1 and M4 had five training sessions while M2 and M3 had four training sessions). Such similarities would allow a proper comparison between accumulated data. The characterization of weekly training Ms analyzed in the present study is described in Table 1.

Table 1. Weekly training characterization of Ms from pre- and in-season.

M1, Pre-Season						
Day 1	Day 2	Day 3	Day 4	Day 5	Day 6	Day 7
		training sessions				day-off

M2, Pre-Season						
Day 1	Day 2	Day 3	Day 4	Day 5	Day 6	Day 7
		training sessions				day-off

M3, In-Season						
Day 1	Day 2	Day 3	Day 4	Day 5	Day 6	Day 7
		training sessions				match-day

M4, In-Season						
Day 1	Day 2	Day 3	Day 4	Day 5	Day 6	Day 7
day off			training sessions			match-day

2.2. Participants

Nineteen professional soccer players belonging to a soccer team in Asian Professional League participated in this study. The players were four central defenders (CD), four wide defenders (WD), six central midfielders (CM), three wide midfielders (WM), and two strikers (ST). They had an age of 28.4 ± 3.7 years, a height of 180.9 ± 7.1 cm, a body mass of 74.0 ± 7.9 kg, and a body mass index of 22.5 ± 1.1 kg/m^2.

All players participated in the full season (>60% of the total matches played in a season). Specifically, for the present study, the inclusion criteria were a training session participation of 100% regarding frequency and duration of the session while the exclusion criteria were players who participated in rehabilitation or recovery sessions. All participants were familiarized with the training protocols prior to investigation. This study was conducted according to the requirements of the Declaration of Helsinki and was approved by the University of Isfahan research ethics committee (IR.UI.REC., 1399.064 Isfahan, Iran).

2.3. Procedures

All training sessions were conducted on the natural grass field and in the afternoon. The players had the same nutrition every week for at least several days in camp and throughout the match days. All the information was provided by the researcher and global positioning system (GPS) manager of the team who had more than eight years of experience. Training data were collected over the course of four different 7-day Ms: from pre-season, M1 and M2 included five and six training sessions, respectively; from in-season, M3 and M4 included six and five training sessions, with one match (national league), respectively. All Ms analyzed included consecutively training sessions. Days off or matches only occurred on the last days or the first day of the week (Table 1). Although other weeks also fit the descriptions provided, all weeks selected met the criteria for participants, meaning that they completed all training sessions during the chosen weeks. A total number of 22 training sessions (418 individuals) were observed for this study. This study did not influence or alter the training sessions in any way. Training data were collected at the soccer club's outdoor training pitches. Data were analyzed per day of the week, (i.e., day 1, day 2, ... , and day 7).

2.4. External Intensity Monitoring—Training Data

A 15 Hz MEMS (GPSPORTS systems Pty Ltd., model: SPI high-performance units, Australia) system was used in all the training sessions. This MEMS presents a 100 Hz accelerometer, with 16 G Tri-Axial-Track impacts, accelerations, and decelerations and 50 Hz Tri-Axial magnetometers. The GPS size dimensions were (74 mm × 42 mm × 16 mm). Additionally, it is water-resistant and uses infra-red, and weighed 56 g for data transmission. The validity and reliability of the device have been confirmed by Tessaro and Williams [33] study. Thus, the following variables were selected: total duration of training session, total distance, and distance of different exercise intensity zones which were adapted from previous studies [22,34]: Zone 1 (0–3.9 $km \cdot h^{-1}$), Zone 2 (4–7.1 $km \cdot h^{-1}$), Zone 3 (7.2–14.3 $km \cdot h^{-1}$), Zone 4 (14.4–19.7 $km \cdot h^{-1}$) and Zone 5 (>19.8 $km \cdot h^{-1}$). Considering that several players did not achieve a higher speed of 24 $km \cdot h^{-1}$ (which was Zone 6), Zone 5 included all data to avoid zeros and a lower comparison power analysis between players. Weekly zones (1–5) and weekly total distance were calculated by the sum value for the entire week for each M, respectively.

2.5. Internal Intensity Monitoring—Training Data

A flashing red light was used to track HR. We placed each unit perpendicular to the bag and made sure the logo on the unit was facing backward and the RED light was on. High-performance units are designed to automatically collect athletes' accelerometer and HR data in one session. In addition to the GPS receiver, the SPI Pro X unit consists of a tri-axial accelerometer, to estimate the forces on the player and an integrated HR monitor. The following variables were selected: maximal HR (HRmax) and average HR (HRavg). Then, weekly HRmax (wHRmax) and weekly HRavg (wHRavg) were calculated by the average value for the entire week for each M, respectively. The way this information was recorded was similar to previous studies [35–38].

2.6. Statistical Analysis

Data were analyzed using the IBM SPSS Statistics for Windows, Version 22.0. (IBM Corp., Armonk, NY, USA) statistical software package. Initially, descriptive statistics were used to characterize the sample. Shapiro-Wilk and Levene tests were conducted to determine normality and homoscedasticity, respectively. Once variables obtained a normal distribution (Shapiro–Wilk>0.05), it was used a repeated measures ANOVA test and the Bonferroni post hoc correction test to compare variables for days of the week. This process was repeated to also allow a comparison between all Ms/weeks and player positions. The results are significant with a $p \leq 0.05$. Hedge's g effect size (ES) was also calculated to determine the magnitude of pairwise comparisons. The following criteria in absolute values was used: <0.2 = trivial, 0.2 to 0.6 = small effect, 0.7 to 1.2 = moderate effect, 1.3 to 2.0 = large effect, and >2.0 = very large [39].

Considering the main of this study and using G-Power [40], a post hoc calculation was performed for n = 19, p = 0.05, ES = 0.6, one group, four measurements (four Ms), for repeated measures ANOVA which displayed 99.9% of actual power.

3. Results

3.1. Day-to-Day Intensity Variations across the Four Ms

Training duration is presented in Table 2 while Tables 3 and 4 showed comparisons between days of the week for each M, respectively (all, p < 0.05).

Table 2. Training duration (minutes) during the 7-day testing period for squad average.

Periods	Day 1	Day 2	Day 3	Day 4	Day 5	Day 6	Day 7
M1, pre-season	83.3 ± 0.0 [a,b,c,d]	76.6 ± 2.0 [b,c,d]	93.7 ± 0.4 [c,d]	95.3 ± 1.9	92.4 ± 1.0	X	X
M2, pre-season	70.8 ± 3.1 [a,b,c,d,e]	98.9 ± 1.0 [b,e]	107.2 ± 0.0 [c,d,e]	98.7 ± 1.6 [e]	100.9 + 0.0 [e]	110.3 ± 0.0	X
M3, in-season	78.6 ± 3.3 [a,e]	97.6 ± 0.0 [b,c,d,e]	80.4 ± 0.1 [c,e]	73.4 ± 0.0 [d,e]	80.8 ± 0.5 [e]	60.5 ± 0.1	MD
M4, in-season	X	73.1 ± 5.9 [b]	97.0 ± 0.0 [e]	91.4 ± 2.9 [e]	87.2 ± 0.9 [e]	60.1 ± 0.0	MD

M, microcycle; X, day-off; MD = match-day; [a], denotes difference from day 2; [b], denotes difference from day 3; [c], denotes difference from day 4; [d], denotes difference from day 5; [e], denotes difference from day 6; all p < 0.05.

Table 3. Distances covered at different speed thresholds and heart rate variables (representative of squad average data) during the 7-day period in pre-season.

M1, Pre-Season	Day 1	Day 2	Day 3	Day 4	Day 5	Day 6	Day 7
Total Distance (m)	6951.5 ± 143.8 [a,b]	6000.1 ± 153.4 [b,d]	8135.8 ± 196.8 [c,d]	6422.4 ± 217.9 [d]	7117.0 ± 191.5	X	X
Zone1 (m)	1671.8 ± 109.1 [b,c,d]	1933.8 ± 104.4 [c,d]	2233.1 ± 72.2 [d]	2349.2 ± 63.0	2538.4 ± 66.0	X	X
Zone2 (m)	2600.1 ± 49.5 [a,b]	2155.6 ± 89.8 [b,d]	3174.4 ± 80.2 [c,d]	2224.1 ± 116.0	2587.8 ± 66.0	X	X
Zone3 (m)	2131.2 ± 83.0 [a,c,d]	1504.4 ± 79.6 [b]	2041.3 ± 108.8 [c,d]	1279.6 ± 60.8 [d]	1530.5 ± 88.3	X	X
Zone4 (m)	519.4 ± 53.0 [a]	359.8 ± 35.6 [b,c]	619.9 ± 31.2 [d]	524.1 ± 38.2 [d]	384.9 ± 45.0	X	X
Zone5 (m)	29.1 ± 3.4	46.5 ± 7.1	67.2 ± 10.7	45.3 ± 5.1	75.4 ± 16.9	X	X
HRmax (bpm)	187 ± 2	184 ± 4	188 ± 2	186 ± 2	186 ± 3	X	X
HRavg (bpm)	145 ± 3 [d]	141 ± 3	144 ± 1 [d]	138 ± 2	139 ± 2	X	X
M2, Pre-Season	Day 1	Day 2	Day 3	Day 4	Day 5	Day 6	Day 7
Total Distance (m)	5734.5 ± 216.4 [b,d]	6807.0 ± 223.1 [b,d]	8742.1 ± 136.2 [c,e]	6215.3 ± 150.7 [d]	8091.8 ± 250.6 [e]	6133.6 ± 142.5	X
Zone1 (m)	1999.4 ± 80.3 [a,b,d,e]	2608.2 ± 77.5 [b,c]	2910.2 ± 63.6 [c]	2285.6 ± 54.3 [d,e]	2914.3 ± 105.2	2670.5 ± 91.3	X
Zone2 (m)	1650.2 ± 98.6 [b,d]	1747.0 ± 61.3 [b,c,d]	2665.3 ± 80.8 [c,e]	2097.0 ± 72.6 [e]	2330.6 ± 94.8 [e]	1675.6 ± 36.4	X
Zone3 (m)	1143.9 ± 93.8 [a,b,d]	1760.7 ± 91.3 [b]	2097.0 ± 75.8 [c,d,e]	1377.9 ± 75.1	1554.5 ± 65.9	1387.8 ± 67.9	X
Zone4 (m)	809.8 ± 86.1 [c,d,e]	618.4 ± 37.1 [b,c,d,e]	762.3 ± 44.6 [c,d,e]	421.7 ± 34.4 [d]	1201.9 ± 6.8 [e]	358.9 ± 28.3	X
Zone5 (m)	131.1 ± 22.3 [b,c,e]	72.8 ± 12.9 [b]	307.2 ± 19.3 [c,d,e]	30.1 ± 5.2	90.5 ± 21.7	40.7 ± 4.8	X
HRmax (bpm)	170 ± 9	188 ± 2	187 ± 2	176 ± 5	186 ± 2 [e]	180 ± 2	X
HRavg (bpm)	128 ± 7	138 ± 3 [c,e]	139 ± 3 [c,e]	127 ± 3	135 ± 2 [e]	124 ± 2	X

M, microcycle; MD, match-day; bpm, beats per minute; m, meters; HRmax, heart rate maximum; HRavg, heart rate average; X, Day Off; [a], denotes difference from day 2; [b], denotes difference from day 3; [c], denotes difference from day 4; [d], denotes difference from day 5; [e], denotes difference from day 6; all p < 0.05.

3.1.1. M1, Pre-Season

Total distance showed significant differences in day 1 > day 2 (ES = 6.40) and <day 3 (ES = −6.87). Day 2 < day 3 (ES = −12.10) and <day 4 (ES = −6.44). Day 3 > day 4 (ES = 8.25) and >day 5 (ES = 5.25). Additionally, day 4 < day 5 (ES = −3.39)

Zone 1 showed significant differences in day 1 < day 3 (ES = −6.07), <day 4 (ES = −7.60), <day 5 (ES − 9.60). Day 2 < day 3 (ES − 4.02) and day 4 (ES = −6.92). Additionally, day 3 < day 5 (ES = −4.41).

Zone 2 showed significant differences in day 1 > day 2 (ES = 6.13), <day 3 (ES = −8.62). Day 2 < day 3 (ES = −11.67) and< day 5 (ES = −0.66). Day 3 > day 4 (ES = 9.53) and day (ES = 7.99).

Zone 3 showed significant differences in day 1 > day 2 (ES = 7.71), >day 4 (ES = 11.71), >day 5 (ES = 7.01). Day 2 < day 3 (ES = −5.63). Day 3 > day 4 (ES = 8.64) and >day 5 (ES = 5.16). Day 4 < day 5 (ES = −3.31).

Zone 4 showed significant differences in day 1 > day 2 (ES = 3.54). Day 2 < day 3 (ES = −7.77) and <day 4 (ES = −4.45). Day 3 > day 5 (ES = 6.07). Day 4 > day 6 (ES = 3.34).

HRavg showed significant differences in day 1 > day 5 (ES = 2.35) and day 3 > day 5 (ES = 3.16).

Table 4. Distances covered at different speed thresholds and heart rate variables (representative of squad average data) during the 7-day period in season.

M3, In-Season	Day 1	Day 2	Day 3	Day 4	Day 5	Day 6	Day 7
Total Distance (m)	4818.1 ± 370.9 [b]	5488.8 ± 189.1 [c,d,e]	6966.4 ± 415.3 [c,d,e]	4019.6 ± 96.3 [e]	3757.0 ± 174.7	3584.6 ± 60.0	MD
Zone1 (m)	1791.0 ± 129.0 [a,b]	2394.5 ± 83.6 [c,d,e]	2675.6 ± 140.8 [c,d,e]	1750.3 ± 38.2 [e]	1798.1 ± 126.4	1616.9 ± 35.9	MD
Zone2 (m)	1423.4 ± 178.0	1669.6 ± 91.4 [c,d,e]	1583.6 ± 98.2 [c,d,e]	904.6 ± 21.3 [e]	892.3 ± 50.2 [e]	1061.7 ± 20.5	MD
Zone3 (m)	1115.9 ± 142.6 [b]	937.9 ± 78.7 [b,e]	1953.5 ± 154.1 [c,d,e]	942.1 ± 37.7 [d,e]	682.6 ± 28.3	630.7 ± 31.4	MD
Zone4 (m)	405.3 ± 50.3	445.0 ± 25.8 [b,c,d,e]	651.4 ± 74.3 [c,d,e]	366.7 ± 28.5 [e]	305.7 ± 22.6	249.0 ± 8.3	MD
Zone5 (m)	82.5 ± 14.1 [e]	41.6 ± 9.2 [b]	102.3 ± 13.4 [c]	55.9 ± 10.4	78.4 ± 19.7	26.1 ± 3.3	MD
HRmax (bpm)	180 ± 3 [e]	174 ± 5	179 ± 6	172 ± 4	189 ± 2	167 ± 5	MD
HRavg (bpm)	129 ± 4	125 ± 3	139 ± 6	126 ± 2	121 ± 2	124 ± 2	MD
M4, In-Season	**Day 1**	**Day 2**	**Day 3**	**Day 4**	**Day 5**	**Day 6**	**Day 7**
Total Distance (m)	X	5739.9 ± 628.8 [e]	4883.3 ± 96.2 [c,d,e]	6768.5 ± 358.5 [d,e]	3988.4 ± 150.7	3486.8 ± 56.8	MD
Zone1 (m)	X	2042.9 ± 245.8	2091.3 ± 85.1 [e]	2424.9 ± 139.8 [e]	2085.9 ± 63.5 [e]	1509.6 ± 32.9	MD
Zone2 (m)	X	1867.4 ± 228.1 [d,e]	1535.6 ± 23.4 [d,e]	1821.0 ± 102.2 [d,e]	792.4 ± 66.3	1027.8 ± 33.5	MD
Zone3 (m)	X	1299.7 ± 222.4	1043.8 ± 115.5 [c,e]	1989.5 ± 153.2 [d,e]	758.9 ± 47.7	661.2 ± 45.3	MD
Zone4 (m)	X	396.0 ± 26.8 [b,e]	193.6 ± 17.6 [c]	493.8 ± 53.6 [d,e]	269.8 ± 26.3	249.9 ± 18.5	MD
Zone5 (m)	X	133.9 ± 11.4 [c,d,e]	18.9 ± 3.3 [d]	39.3 ± 5.2	81.3 ± 12.7	38.3 ± 13.3	MD
HRmax (bpm)	X	170 ± 6	166 ± 7	176 ± 5	170 ± 4	165 ± 8	MD
HRavg (bpm)	X	127 ± 4	120 ± 3 [c]	133 ± 3 [d]	119 ± 3	121 ± 5	MD

M, microcycle; MD, match-day; bpm, beats per minute; m, meters; HRmax, heart rate maximum; HRavg, heart rate average; X, Day Off; [a], denotes difference from day 2; [b], denotes difference from day 3; [c], denotes difference from day 4; [d], denotes difference from day 5; [e], denotes difference from day 6; all $p < 0.05$.

3.1.2. M2, Pre-Season

Total distance showed significant differences in day 1 < day 3 (ES = −16.63) and <day 5 (ES = −10.07). Day 2 < day 3 (ES = −10.47) and <day 5 (ES = −5.42). Day 3 > day 4 (ES = 17.59) and >day 6 (ES = 18.71). Additionally, day 4 < day 5 (ES = −9.08).

Zone 1 showed significant differences in day 1 < day 2 (ES = −7.71), <day 3 (ES = −12.57), <day 5 (ES = −9.78), <day 6 (ES = −7.81). Day 2 < day 3 (ES = −4.26) and >day 4 (ES = 4.82). Additionally, day 3 > day 4 (ES = 10.56). Additionally, day 4 > day 5 (ES = −7.51) and >day 6 (ES = −5.12).

Zone 2 showed significant differences in day 1 < day 2 (ES = −11.26), <day 4 (ES = −7.03). Day 2 < day 3 (ES = −12.80), <day 4 (ES = −5.21) and <day 5 (ES = −7.31). Day 3 > day 4 (ES = 7.40) and >day 6 (ES = 15.79). Day 4 and day 5 were higher than day 6 (ES = 7.34, ES = 9.12, respectively).

Zone 3 showed significant differences in day 1 < day 2 (ES = −6.66), <day 3 (ES = −11.18), <day 5 (ES = −5.07). Day 2 < day 3 (ES = −4.01). Day 3 < day 4 (ES = 9.53), <day 5 (ES = 7.64) and <day 6 (ES = 9.86).

Zone 4 showed significant differences in day 1 > day 4 (ES = 5.92), <day 5 (ES = −6.42), >day 6 (ES = 7.04). Day 2 < day 3 (ES = −3.51), >day 4 (ES = 5.50), <day 5 (ES = −21.88), > day 6 (ES = 7.86). Additionally, day 3 > day 4 (ES = 8.55), <day 5 (ES = −13.78), >day 6 (E = 10.80). Additionally, day 4 < day 5 (ES = 1.99). Additionally, day 5 > day 6 (ES = 40.96).

Zone 5 showed significant differences in day 1 > day 3 (ES = −8.44), >day 4 (ES = 6.24), >day 6 (ES = 5.60). Day 2 < day 3 (ES = −14.28). Additionally, day 3 > day 4 (ES = 19.61), > day 5 (ES = 10.55), >day 6 (ES = 18.95).

HRmax showed significant differences in day 5 > day 6 (ES = 3.00). HRavg showed significant differences in day 2 > day 4 (ES = 3.67) and >day 6 (ES = 5.49). Additionally, day 3 > day 4 (ES = 4.00) and >day 6 (ES = 5.88). Finally, day 5 > day 6 (ES = 5.50).

3.1.3. M3, In-Season

Total distance showed significant differences in day 1 < day 3 (ES = −5.46). Day 2 > day 4 (ES = 9.79), >day 5 (ES = 9.51) and >day 6 (ES = 13.57). Day 3 > day 4 (ES = 9.78), >day 5 (ES = 10.07) and >day 6 (ES = 11.40). Additionally, day 4 > day 6 (ES = 5.42).

Zone 1 showed significant differences in day 1 < day 2 (ES = −5.55), <day 3 (ES = −6.55). Day 2 > day 4 (ES = 9.91), >day 5 (ES = 5.57) and >day 6 (ES = 12.09). Additionally, day 3 > day 4 (ES = 8.97), >day 5 (ES = 6.56) and >day 6 (ES = 10.30). Additionally, day 4 > day 6 (ES = 3.60).

Zone 2 showed significant differences in day 2 > day 4 (ES = 11.53), >day 5 (ES = 10.54) and >day 6 (ES = 9.18). Day 3 > day 4 (ES = 9.56), >day 5 (ES = 8.86) and >day 6 (ES = 7.36). Day 4 and day 5 were higher than day 6 (ES = −7.52, ES = −4.42, respectively).

Zone 3 showed significant differences in day 1 < day 3 (ES = −5.64). Day 2 < day 3 (ES = −8.30) and >day 6 (ES = 5.12). Day 3 > day 4 (ES = 9.02), >day 5 (ES = 11.47) and >day 6 (ES = 11.90). Day 4 > day 5 (ES = 7.79) and >day 6 (ES = 8.98).

Zone 4 showed significant differences in day 2 < day 3 (ES = −3.71), >day 4 (ES = 2.88), >day 5 (ES = 5.74), >day 6 (ES = 10.23). Day 3 > day 4 (ES = 5.06), >day 5 (ES = 6.30), >day 6 (ES = 7.61). Additionally, day 4 > day 6 (ES = 5.61).

Zone 5 showed significant differences in day 1 > day 6 (ES = 5.51). Day 2 < day 3 (ES = −5.28). Additionally, day 3 > day 4 (ES = 3.87).

HRmax showed significant differences in day 1 > day 6 (ES − 3.15).

3.1.4. M4, In-Season

Total distance showed significant differences in day 2 > day 6 (ES = 5.05). Day 3 < day 4 (ES = −7.18), >day 5 (ES = 7.08) and >day 6 (ES = 17.68). Day 4 > day 5 (ES = 10.11), >day 6 (ES = 12.79).

Zone 1 showed significant differences in day 3 > day 6 (ES = 9.02). Day 4 > day 6 (ES = 9.01). Day 5 > day 6 (ES = 11.40).

Zone 2 showed significant differences in day 2 > day 5 (ES = 6.40), >day 6 (ES = 5.15). Day 3 > day 5 (ES = 14.95), >day 6 (ES = 17.57). Day 4 > day 5 (ES = 11.94) and day 6 (ES = 10.43).

Zone 3 showed significant differences in day 3 < day 4 (ES = −6.97) and >day 6 (ES = 3.22). Day 4 > day 5 (ES = 10.85) and >day 6 (ES = 11.76).

Zone 4 showed significant differences in day 2 > day 3 (ES = 8.93), >day 6 (ES = 6.34). Day 3 < day 4 (ES = −7.53). Additionally, day 4 > day 5 (ES = 5.31) and >day 6 (ES = 6.08).

Zone 5 showed significant differences in day 2 > day 4 (ES = 10.68), >day 5 (ES = 4.36), >day 6 (ES = 7.72). Additionally, day 3 < day 5 (ES = −6.73).

HRavg showed significant differences in day 3 < day 4 (ES = −4.33) and day 4 > day 5 (ES = 4.67).

3.2. Comparisons between Ms

Figures 1–3 displayed the differences between Ms. In Figure 1A, there were significant differences in the weekly total distance between M1 and M2 (ES = −14.72), M1 and M3 (ES − 9.26), M1 and M4 (ES − 17.42), M2 and M3 (ES − 23.09), M2 and M4 (ES = 36.22), M3 and M4 (ES − 5.93). Figure 1B showed that weekly Zone 1 presented significant differences between M1 and M2 (ES = −15.05), M2 and M3 (ES = 11.97), M2 and M4 (ES= 17.26), and M3 and M4 (ES = 6.50). Finally, Figure 1C showed that weekly Zone 2 presented significant differences between M1 and M2 (ES − 3.51), M1 and M3 (ES − 1.04), M1 and M4 (ES = 29.91), M2 and M3 (ES = 20.3), M2 and M4 (ES = 27.34).

In Figure 2A, there were significant differences for weekly Zone 3 between M1 and M2 (ES = 3.15), M1 and M3 (ES = 7.41), M1 and M4 (ES = 9.61), M2 and M3 (ES = 11.58), M2 and M4 (ES = 14.47). In Figure 2B, there were significant differences for weekly Zone 4 between M1 and M2 (ES = −12.20), M1 and M3 (ES = 6.33), M2 and M3 (ES = 13.10), M2 and M4 (ES = 24.02), M3 and M4 (ES = 7.17). Ultimately, in Figure 2C, there were significant differences for weekly Zone 5 between M1 and M2 (ES = −10.24), M1 and M3 (ES = −3.80), M2 and M3 (ES = 6.64), and M2 and M4 (ES = 9.27).

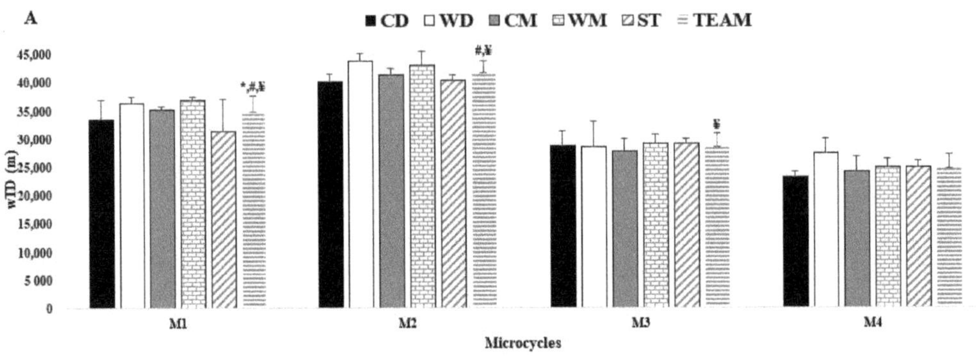

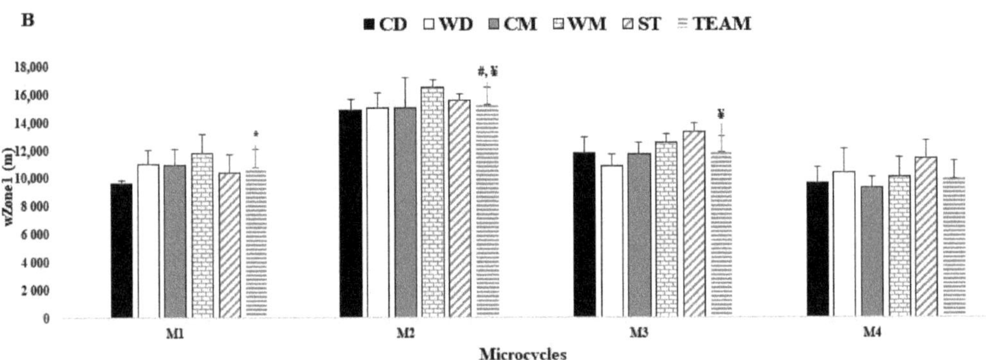

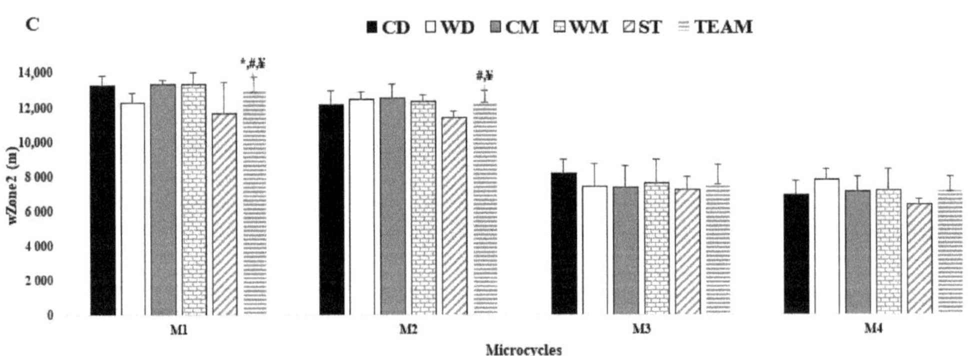

Figure 1. Comparisons between player positions for weekly total distance (wTD), Zone 1 (wZone1), and Zone 2 (wZone2) for each microcycle, (e.g., M1, M2, M3, and M4) and between Ms by team average. (**A**) wTD; (**B**) wZone1; (**C**) wZone2; * Denotes difference from M2; # denotes difference from M3; ¥ denotes difference from M4; all $p < 0.05$.

In Figure 3A, there were significant differences for HRmax between M1 and M3 (ES = 4.26) and M1 and M4 (ES = 4.92). In Figure 3B, there were significant differences for HRavg between M1 and M2 (ES = 4.00), M1 and M3 (ES = 6.50 and M1 and M4 (ES = −8.04), M2 and M3 (ES = 2.78), M2 and M4 (ES = −10.34), and M3 and M4 (ES = −11.75).

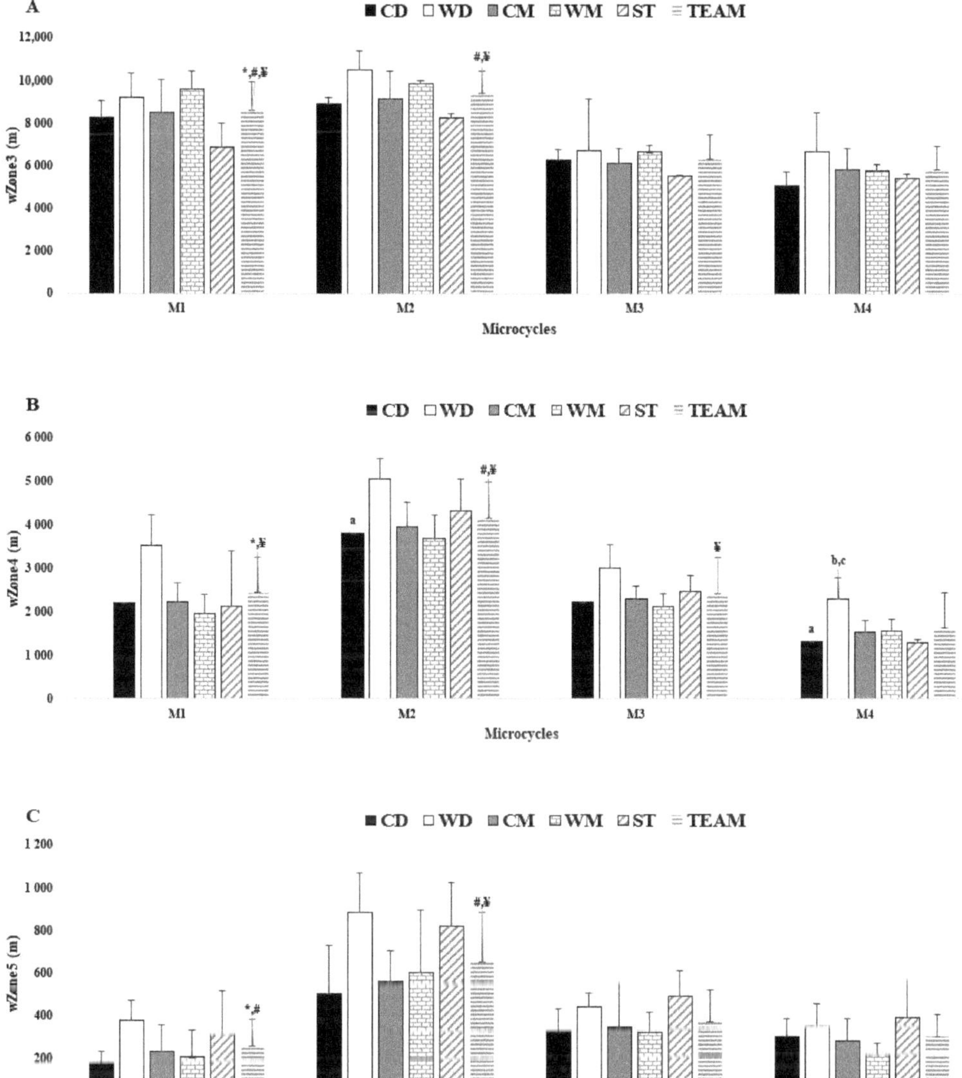

Figure 2. Comparisons between player positions for weekly Zones 3, 4, and 5 for each microcycle, (e.g., M1, M2, M3, and M4) and between Ms by team average. (**A**) wZone3; (**B**) wZone4; (**C**) wZone5; a denotes difference from WD, b denotes difference from WM, c denotes difference from ST, * denotes difference from M2, # denotes difference from M3, ¥ denotes difference from M4, all *p* < 0.05.

3.3. Comparisons between Player Positions

There were no differences between playing positions for all Ms and variables, except for Zone 4, which displayed some difference between playing positions, namely CD vs. WD (ES = −4.83) in M2 and M4 (ES = 4.96). Additionally, WD vs. WM (ES = −5.88) and WD vs. ST (ES = 5.38) in M4 (Figure 2B).

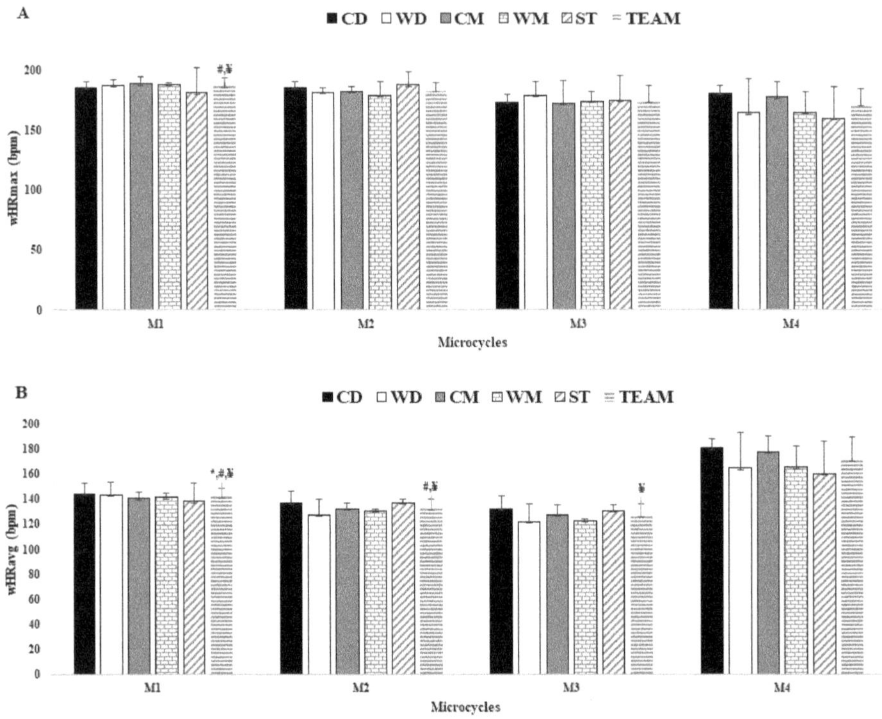

Figure 3. Comparisons between player positions for average HRmax and average HRavg for each microcycle, (e.g., M1, M2, M3, and M4) and between Ms by team average. (**A**) wHRmax; (**B**) wHRavg; * Denotes difference from M2; # denotes difference from M3; ¥ denotes difference from M4; all $p < 0.05$.

4. Discussion

The aim of this study was to quantify the training intensity in two Ms from pre-season and another two during in-season. The secondary aim was to analyze the influence of playing position on the intensity experienced by professional soccer players. To the best of our knowledge, this study provides the first report of daily external and internal intensity over four different Ms (two from pre-season and two from in-season) and for playing positions in Asian soccer players. This study could find significant differences in daily and accumulated intensities for the within and between match schedules.

4.1. Day-to-Day Intensity Variations across the Four Ms

The off-season is designed to recover, maintain and/or improve specific individual weaknesses from the in-season. However, sometimes a detraining process may occur and, in this sense, the emphasis during pre-season could be rebuilding fitness parameters after the off-season period. During the pre-season phase (M1), the training distance covered was highest on day 3, while the lowest was on day 2. The distance covered in Zone 1 was the highest on day 5, while the distance covered in Zone 2 and Zone 4 was highest on day 3. Further, the distance covered in Zone 3 was highest on day 1 while day 5 showed the highest distance covered in Zone 5. These results indicate that the high-speed running and the intensity of the training increased as the week progressed. The increase in physiological demands during the pre-season phase could be to provide optimal conditioning to the soccer players in order to prepare them for the competitive season [41].

There was no difference in the HRmax of the players throughout the week. The HR reported in our study was higher than that reported by Jeong et al. [12]. The HRmax was reported to be 124 ± 7 beats/min in professional Korean soccer players. This difference might be due to the external work performed by each team during their respective pre-season, individual differences in the player characteristics, and differences in the periodization plans specific to the team included in our study. This further highlights that the coaches maintained an overall consistent intensity during the entire week on the majority of the training days.

During M2, the highest training duration was on day 6. The highest distance covered was on day 3 and the lowest was on day 1. The most distance covered in Zone 1 was on day 5 and in Zone 2, 3, and 5 was on day 3. There might be a possibility that this could be due to the reduction in intensity during the start of the week and then a progressively increase as the week progressed. There was no difference in the HRmax between the days of the session, but the HRavg showed differences on days 2, 4, and 6. The HR during this period was also found to be higher than in the study by Jeong et al. [12]. It must be taken into account that training/match intensity is different today compared to more than 10 years ago.

During the in-season phase (M3 and M4), the training duration and the distance covered were reduced one day before the match. It can also be observed that the distances were increased progressively during the first three days of the week with the total distance starting to reduce from day 4. This has been associated with tapering strategies used to reduce intensity and volume on the day preceding the matches to improve player readiness [22,25,42]. Furthermore, a rest day was provided after the match during M4. These findings are in line with the previous studies in the existing literature. For instance, Kelly et al. [31] reported a recovery day after the match followed by a subsequent increase in intensity as the week progressed. It is interesting to note that the distance covered in Zone 5 was higher on M4 post-recovery day. One possible reason could be the difference in training duration, and the level of intensity applied by the coach [22]. For example, the duration of training post-match day in M4 was 73 min, whereas it was between 87–97 min during the other days except for the day before the match (day 6). This could be due to the high-intensity training approach, or simply due to the context and objectives related to the training intensity management [22]. The HRmax did not vary much between the sessions during M4. However, it was significantly different on day 1 and day 6 in M3. One possible explanation for this could be due to the increased distance covered in Zone 5 (82.5 m vs. 26.1 m) on both days. The distance covered at increased speed could have contributed to the higher HR. Additionally, the training duration was higher on day 1 compared to day 6 (78.6 vs. 60.5 min). These findings reflect the day-to-day intensity variation during different Ms, and it seems that there are some marked differences in the intensity during the 1-week Ms during pre-season and in-season.

4.2. Comparison between Ms

The weekly total distance and the distance covered in Zone 1–5 during both Ms of the pre-season was higher compared to the in-season. Previous studies showed that pre-season intensity is generally higher compared to the in-season [12,43]. The aim during the pre-season is to work on the fitness levels of soccer players, while the aim during the in-season is focused on the technical and tactical development, and to maintain the optimal level of fitness developed during the pre-season [26]. The pre-season training period is traditionally the time when the majority of the physical preparation work is completed by players, to enable them to fulfill the physiological requirements of the competitive season [41]. Although the pre-season varies according to specific purposes of training [12], it is generally accepted that the physiological demands of this phase of training are greater than at other times during in-season. As such, maximal tests can be applied both within the gym and on-field during the pre-season phase, with the results being used to prescribe general running-based conditioning that accompanies more specific methods of training such as simulated phases of play, small-sided games, and skill-based drills [44]. There was also a significant difference in the HRmax and HRavg during both Ms of the in-season compared to the pre-season. This variation could be due to the differences in training administered during both Ms.

In particular, there seem to be differences in the total distance covered during the pre-season. The distance covered during the M2 was higher than M1. Further, the distance covered in Zone 5 was high in M2 compared to M1. One reason could be that the intensity of the training sessions could have increased and could have involved more acceleration/deceleration drills just before the beginning of the in-season. During the in-season, the total distance covered during the first M was higher when compared to the latter one. This is in agreement with the results reported by Malone et al. [26] in which soccer players covered more distance during the initial phase of the in-season compared to the final M. One further reason for this could be associated with the emphasis on physical conditioning by coaches as a continuation of the pre-season phase. The data reported within the current investigation may show that coaches adopted a pre-competition reduction in intensity to protect against injury and prevent fatigue in players as the in-season progresses. The data from our study suggested that training intensity is periodically manipulated before the competition, in order to reduce the risk of injury and improve potential match performance as the competition progresses [45–47].

4.3. Comparisons between Player Positions

Our results revealed that there were significant differences in the distance covered in Zone 4, it varied significantly between CD, WD, WM and ST during M2 and M4. The WD was found to cover the most distance in Zone 4 during both Ms. Wide areas are generally less congested compared to the central ones, resulting in increased chances of achieving high speed [48]. Further, since CD operated in congested, and central areas of the pitch, this could have led them to cover less distance in high-speed running when compared to WD [49]. Previous research by Abbott et al. [48] also indicated that the high-speed running and sprinting distances covered by WD and WA are similar because they operate on the flanks of the pitch. However, our results contradict these findings as there was a significant difference observed between both playing positions during M4. When comparing WD and ST, the lesser involvement of the ST during defensive tasks could explain the lesser distance covered in high-intensity zones [48]. In addition, every position saw similar efforts during the other periods of the M and there was no significant difference in the distance covered in the other M.

4.4. Practical Applications, Limitations and Considerations for Future Studies

This study provides useful information regarding the intensity employed by professional soccer teams during the pre- and in-season. It provides further evidence of the value of using the combination of different measures of intensity to fully evaluate the patterns observed across the two periods of the season. For coaches and practitioners, the study generates reference values for professional players, which can be considered when planning training sessions, and also designing and implementing training programs considering the physiological demands of each playing position. Considering the importance of summarizing the main evidence about the influence of factors for optimizing the exercises and the potential practical applications for coaches, it is suggested that HR should be used with some parsimony in controlling the intensity of the training. In maximizing the physical fitness of players from different field positions, coaches can take into account that only speed zones higher than 14.4 km·h^{-1} denoted some significant differences for player positions.

Despite the previous practical applications, this study is based on one top professional Asian soccer team with only a small sample size and for those reasons, it cannot represent the usual training demands of other leagues/countries. Thus, future studies with larger sample sizes and from other leagues/countries should be considered. Moreover, playing position analysis was limited to the small number of participants in each position (2–6 players). Furthermore, it was previously suggested to conduct future studies, that included players that also completed exercise training sessions, without competitive matches [22]. Even so, we consider that data are still limited because there are other variables such as acceleration, deceleration, player load, and metabolic power as well as technical, tactical, or other physical and physiological variables or information about training content and periodization structure that could also give important knowledge and references for training which makes us suggest the use of covariates in future studies. In addition, future studies can consider training and match data to amplify science and confirm these results. Additionally, it is important to understand whether several contextual variables, such as the opponent level, the match result, or the time during the season, could affect the results. Lastly, the higher running speed zone included a threshold higher than 19.8 km·h^{-1} and future studies may want to consider other speed thresholds such as higher, (i.e., 24 km·h^{-1}) to provide additional information. However, in the present study, several players did not achieve such running speed as previously explained. This information should be considered by performance and technical coaches, especially those from the Iranian league.

5. Conclusions

This study quantifies the daily soccer training and accumulative weekly intensity of a team from the Asian Professional League in four different Ms schedules. It is important to note that customary external intensity did not exhibit a regular pattern for the analyzed Ms. However, there was a similar tendency of constant HRmax (range values: 165–188 bpm's) and HRavg (range values: 119–145 values) over the different Ms. Moreover, this study confirmed previous evidence that there was a general tendency to decrease intensity in the last day of the M.

In general, there was a higher external intensity in Ms from the pre-season than in-season. Finally, only speed zones higher than 14.4 km·h^{-1} denoted some significant differences for player positions. WD and WM showed higher distances covered with high intensity. In opposition, internal intensity does not vary from different playing positions.

Author Contributions: Conceptualization, H.N. and R.O.; methodology, H.N., J.P.B. and R.O.; software, H.N. and R.O.; validation, R.O. and H.N.; formal analysis, H.N., A.K.R. and R.O.; investigation, H.N., A.K.R., J.P.B. and R.O.; resources, H.N., A.K.R., J.P.B. and R.O.; data curation, R.O. and H.N.; writing—original draft preparation, H.N., A.K.R., J.P.B. and R.O.; writing—review and editing, H.N., A.K.R., J.P.B. and R.O.; visualization, H.N., A.K.R., J.P.B. and R.O.; supervision, H.N. and R.O.; project administration, H.N. and R.O.; funding acquisition, J.P.B. and R.O. All authors have read and agreed to the published version of the manuscript.

Funding: This research was funded by the Portuguese Foundation for Science and Technology, I.P., Grant/Award Number UIDP/04748/2020.

Institutional Review Board Statement: The study was conducted according to the guidelines of the Declaration of Helsinki and approved by the Ethics Committee of the University of Isfahan (IR.UI.REC. 1399.064).

Informed Consent Statement: Written informed consent was obtained from the participants to publish this paper.

Data Availability Statement: The data presented in this study are available on request from the corresponding author.

Acknowledgments: The authors would like to thank the team's coaches and players for their cooperation during all data collection procedures.

Conflicts of Interest: The authors declare no conflict of interest. The funders had no role in the design of the study; in the collection, analyses, or interpretation of data; in the writing of the manuscript, or in the decision to publish the results.

References

1. Carling, C.; Orhant, E.; Legall, F. Match Injuries in Professional Soccer: Inter-Seasonal Variation and Effects of Competition Type, Match Congestion and Positional Role. *Int. J. Sports Med.* **2010**, *31*, 271–276. [CrossRef] [PubMed]
2. Clemente, F.M.; Couceiro, M.S.; Martins, F.M.L.; Ivanova, M.O.; Mendes, R. Activity Profiles of Soccer Players during the 2010 World Cup. *J. Hum. Kinet.* **2013**, *38*, 201–211. [CrossRef] [PubMed]
3. Aughey, R.J.; Varley, M.C. Acceleration Profiles in Elite Australian Soccer. *Int. J. Sports Med.* **2013**, *34*, 282. [CrossRef] [PubMed]
4. Bradley, P.S.; Carling, C.; Gomez Diaz, A.; Hood, P.; Barnes, C.; Ade, J.; Boddy, M.; Krustrup, P.; Mohr, M. Match Performance and Physical Capacity of Players in the Top Three Competitive Standards of English Professional Soccer. *Hum. Mov. Sci.* **2013**, *32*, 808–821. [CrossRef]
5. Mohr, M.; Krustrup, P.; Bangsbo, J. Match Performance of High-Standard Soccer Players with Special Reference to Development of Fatigue. *J. Sports Sci.* **2003**, *21*, 519–528. [CrossRef]
6. Bradley, P.S.; Sheldon, W.; Wooster, B.; Olsen, P.; Boanas, P.; Krustrup, P. High-Intensity Running in English FA Premier League Soccer Matches. *J. Sports Sci.* **2009**, *27*, 159–168. [CrossRef]
7. Coppalle, S.; Rave, G.; Ben Abderrahman, A.; Ali, A.; Salhi, I.; Zouita, S.; Zouita, A.; Brughelli, M.; Granacher, U.; Zouhal, H. Relationship of Pre-Season Training Load with in-Season Biochemical Markers, Injuries and Performance in Professional Soccer Players. *Front. Physiol.* **2019**, *10*, 409. [CrossRef]
8. Borresen, J.; Ian Lambert, M. The Quantification of Training Load, the Training Response and the Effect on Performance. *Sport. Med.* **2009**, *39*, 779–795. [CrossRef]
9. Gabbett, T.J.; Nassis, G.P.; Oetter, E.; Pretorius, J.; Johnston, N.; Medina, D.; Rodas, G.; Myslinski, T.; Howells, D.; Beard, A.; et al. The Athlete Monitoring Cycle: A Practical Guide to Interpreting and Applying Training Monitoring Data. *Br. J. Sports Med.* **2017**, *51*, 1451–1452. [CrossRef]
10. Bangsbo, J.; Mohr, M.; Krustrup, P. Physical and Metabolic Demands of Training and Match-Play in the Elite Football Player. *J. Sports Sci.* **2006**, *24*, 665–674. [CrossRef]
11. Impellizzeri, F.M.; Marcora, S.M.; Castagna, C.; Reilly, T.; Sassi, A.; Iaia, F.M.; Rampinini, E. Physiological and Performance Effects of Generic versus Specific Aerobic Training in Soccer Players. *Int. J. Sports Med.* **2006**, *27*, 483–492. [CrossRef] [PubMed]
12. Jeong, T.S.; Reilly, T.; Morton, J.; Bae, S.W.; Drust, B. Quantification of the Physiological Loading of One Week of "Pre-Season" and One Week of "in-Season" Training in Professional Soccer Players. *J. Sports Sci.* **2011**, *29*, 1161–1166. [CrossRef] [PubMed]
13. Goto, K.; Ishii, N.; Mizuno, A.; Takamatsu, K. Enhancement of Fat Metabolism by Repeated Bouts of Moderate Endurance Exercise. *J. Appl. Physiol.* **2007**, *102*, 2158–2164. [CrossRef] [PubMed]
14. Staunton, C.A.; Abt, G.; Weaving, D.; Wundersitz, D.W.T. Misuse of the Term 'Load' in Sport and Exercise Science. *J. Sci. Med. Sport* **2021**, *25*, 439–444. [CrossRef]
15. Vanrenterghem, J.; Nedergaard, N.J.; Robinson, M.A.; Drust, B. Training Load Monitoring in Team Sports: A Novel Framework Separating Physiological and Biomechanical Load-Adaptation Pathways. *Sports Med.* **2017**, *47*, 2135–2142. [CrossRef]
16. Borg, G. Perceived Exertion as an Indicator of Somatic Stress. *Scand. J. Rehabil. Med.* **1970**, *2*, 92–98.
17. Foster, C. Monitoring Training in Athletes with Reference to Overtraining Syndrome. *Med. Sci. Sports Exerc.* **1998**, *30*, 1164–1168. [CrossRef]
18. Foster, C.; Hector, L.L.; Welsh, R.; Schrager, M.; Green, M.A.; Snyder, A.C. Effects of Specific versus Cross-Training on Running Performance. *Eur. J. Appl. Physiol. Occup. Physiol.* **1995**, *70*, 367–372. [CrossRef]
19. Foster, C.; Florhaug, J.A.; Franklin, J.; Gottschall, L.; Hrovatin, L.A.; Praker, S.; Doleshal, P.; Dodge, C. A New Approach to Monitoring Exercise Training. *J. Strength Cond. Res.* **2001**, *15*, 109–115. [CrossRef]
20. Impellizzeri, F.M.; Rampinini, E.; Coutts, A.J.; Sassi, A.; Marcora, S.M. Use of RPE-Based Training Load in Soccer. *Med. Sci. Sports Exerc.* **2004**, *36*, 1042–1047. [CrossRef]
21. Akenhead, R.; Nassis, G.P. Training Load and Player Monitoring in High-Level Football: Current Practice and Perceptions. *Int. J. Sports Physiol. Perform.* **2016**, *11*, 587–593. [CrossRef] [PubMed]
22. Oliveira, R.; Brito, J.; Martins, A.; Mendes, B.; Calvete, F.; Carriço, S.; Ferraz, R.; Marques, M.C. In-Season Training Load Quantification of One-, Two- and Three-Game Week Schedules in a Top European Professional Soccer Team. *Physiol. Behav.* **2019**, *201*, 146–156. [CrossRef] [PubMed]
23. Los Arcos, A.; Mendez-Villanueva, A.; Martínez-Santos, R. In-Season Training Periodization of Professional Soccer Players. *Biol. Sport* **2017**, *34*, 149–155. [CrossRef] [PubMed]
24. Martín-García, A.; Gómez Díaz, A.; Bradley, P.S.; Morera, F.; Casamichana, D. Quantification of a Professional Football Team's External Load Using a Microcycle Structure. *J. Strength Cond. Res.* **2018**, *32*, 3511–3518. [CrossRef]
25. Stevens, T.G.A.; de Ruiter, C.J.; Twisk, J.W.R.; Savelsbergh, G.J.P.; Beek, P.J. Quantification of In-Season Training Load Relative to Match Load in Professional Dutch Eredivisie Football Players. *Sci. Med. Footb.* **2017**, *1*, 117–125. [CrossRef]
26. Malone, J.J.; Di Michele, R.; Morgans, R.; Burgess, D.; Morton, J.P.; Drust, B. Seasonal Training-Load Quantification in Elite English Premier League Soccer Players. *Int. J. Sports Physiol. Perform.* **2015**, *10*, 489–497. [CrossRef]
27. Anderson, L.; Orme, P.; Di Michele, R.; Close, G.L.; Milsom, J.; Morgans, R.; Drust, B.; Morton, J.P. Quantification of Seasonal-Long Physical Load in Soccer Players with Different Starting Status from the English Premier League: Implications for Maintaining Squad Physical Fitness. *Int. J. Sports Physiol. Perform.* **2016**, *11*, 1038–1046. [CrossRef]

28. Akenhead, R.; Harley, J.A.; Tweddle, S.P. Examining the External Training Load of an English Premier League Football Team with Special Reference to Acceleration. *J. Strength Cond. Res.* **2016**, *30*, 2424–2432. [CrossRef]
29. Manzi, V.; Bovenzi, A.; Impellizzeri, M.F.; Carminati, I.; Castagna, C. Individual Training-Load and Aerobic-Fitness Variables in Premiership Soccer Players during the Precompetitive Season. *J. Strength Cond. Res.* **2013**, *27*, 631–636. [CrossRef]
30. Malone, S.; Collins, K.; McRobert, A.; Doran, D. Quantifying the Training and Match-Play External and Internal Load of Elite Gaelic Football Players. *Appl. Sci.* **2021**, *11*, 1756. [CrossRef]
31. Kelly, D.M.; Strudwick, A.J.; Atkinson, G.; Drust, B.; Gregson, W. Quantification of Training and Match-Load Distribution across a Season in Elite English Premier League Soccer Players. *Sci. Med. Footb.* **2020**, *4*, 59–67. [CrossRef]
32. Owen, A.L.; Wong, D.P.; Newton, M.; Weldon, A.; Koundourakis, N.E. Quantification, Tapering and Positional Analysis of across 9-Weekly Microcycles in a Professional Chinese Super League Soccer Team. *EC Orthop.* **2020**, *12*, 39–56.
33. Tessaro, E.; Williams, J.H. Validity and Reliability of a 15 Hz GPS Device for Court-Based Sports Movements. *Sport. Perform. Sci. Reports* **2018**, *29*, 1–4.
34. Ispyrlidis, I.; Gourgoulis, V.; Mantzouranis, N.; Gioftsidou, A.; Athanailidis, I. Match and Training Loads of Professional Soccer Players in Relation to Their Tactical Position. *J. Phys. Educ. Sport* **2020**, *20*, 2269–2276. [CrossRef]
35. Clemente, F.M.; Silva, R.; Chen, Y.S.; Aquino, R.; Praça, G.M.; Paulis, J.C.; Nobari, H.; Mendes, B.; Rosemann, T.; Knechtle, B. Accelerometry-Workload Indices Concerning Different Levels of Participation during Congested Fixture Periods in Professional Soccer: A Pilot Study Conducted over a Full Season. *Int. J. Environ. Res. Public Health* **2021**, *18*, 1137. [CrossRef]
36. Nobari, H.; Oliveira, R.; Brito, J.P.; Pérez-gómez, J.; Clemente, F.M.; Ardigò, L.P. Comparison of Running Distance Variables and Body Load in Competitions Based on Their Results: A Full-season Study of Professional Soccer Players. *Int. J. Environ. Res. Public Health* **2021**, *18*, 2077. [CrossRef] [PubMed]
37. Nobari, H.; Oliveira, R.; Clemente, F.M.; Adsuar, J.C.; Pérez-Gómez, J.; Carlos-Vivas, J.; Brito, J.P. Comparisons of Accelerometer Variables Training Monotony and Strain of Starters and Non-Starters: A Full-Season Study in Professional Soccer Players. *Int. J. Environ. Res. Public Health* **2020**, *17*, 6547. [CrossRef]
38. Nobari, H.; Praça, G.M.; Clemente, F.M.; Pérez-Gómez, J.; Carlos Vivas, J.; Ahmadi, M. Comparisons of New Body Load and Metabolic Power Average Workload Indices between Starters and Non-Starters: A Full-Season Study in Professional Soccer Players. *Proc. Inst. Mech. Eng. Part P J. Sport. Eng. Technol.* **2020**, *235*, 105–113. [CrossRef]
39. Hopkins, W.; Marshall, S.; Batterham, A.; Hanin, J. Progressive Statistics for Studies in Sports Medicine and Exercise Science. *Med. Sci. Sport. Exerc.* **2009**, *4*, 3. [CrossRef]
40. Faul, F.; Erdfelder, E.; Lang, A.G.; Buchner, A. G*Power 3: A Flexible Statistical Power Analysis Program for the Social, Behavioral, and Biomedical Sciences. *Behav. Res. Methods* **2007**, *39*, 175–191. [CrossRef]
41. Bangsbo, J. The Physiology of Soccer—With Special Reference to Intense Intermittent Exercise. *Acta Physiol. Scand. Suppl.* **1994**, *619*, 1–155. [PubMed]
42. Owen, A.L.; Wong, D.P.; Dunlop, G.; Groussard, C.; Kebsi, W.; Dellal, A.; Morgans, R.; Zouhal, H. High-Intensity Training and Salivary Immunoglobulin a Responses in Professional Top-Level Soccer Players: Effect of Training Intensity. *J. Strength Cond. Res.* **2016**, *30*, 2460–2469. [CrossRef] [PubMed]
43. Fessi, M.S.; Nouira, S.; Dellal, A.; Owen, A.; Elloumi, M.; Moalla, W. Changes of the Psychophysical State and Feeling of Wellness of Professional Soccer Players during Pre-Season and in-Season Periods. *Res. Sport. Med.* **2016**, *24*, 375–386. [CrossRef]
44. Reilly, T.; Collins, K. Science and the Gaelic Sports: Gaelic Football and Hurling. *Eur. J. Sport Sci.* **2008**, *8*, 231–240. [CrossRef]
45. Malone, S.; Roe, M.; Doran, D.A.; Gabbett, T.J.; Collins, K.D. Aerobic Fitness and Playing Experience Protect Against Spikes in Workload: The Role of the Acute:Chronic Workload Ratio on Injury Risk in Elite Gaelic Football. *Int. J.* **2017**, *14*, 156–162.
46. Ritchie, D.; Hopkins, W.; Buchheit, M.; Cordy, J.; Bartlett, J. Quantification of Training and Competition Load Across a Season in an Elite Australian Football Club. *Int. J. Sports Physiol. Perform.* **2016**, *11*, 474–479. [CrossRef] [PubMed]
47. Gabbett, T.J. The Training-Injury Prevention Paradox: Should Athletes Be Training Smarter and Harder? *Br. J. Sports Med.* **2016**, *50*, 273–280. [CrossRef]
48. Abbott, W.; Brickley, G.; Smeeton, N.J. Physical Demands of Playing Position within English Premier League Academy Soccer. *J. Hum. Sport Exerc.* **2018**, *13*, 285–295. [CrossRef]
49. Di Salvo, V.; Baron, R.; Tschan, H.; Calderon Montero, F.J.; Bachl, N.; Pigozzi, F. Performance Characteristics According to Playing Position in Elite Soccer. *Int. J. Sports Med.* **2007**, *28*, 222–227. [CrossRef]

 *healthcare*

Article

Analyzing the within and between Players Variability of Heart Rate and Locomotor Responses in Small-Sided Soccer Games Performed Repeatedly over a Week

Ana Filipa Silva [1,2,3], Francisco Tomás González-Fernández [4], Rodrigo Aquino [5], Zeki Akyildiz [6], Luiz Palucci Vieira [7], Mehmet Yıldız [8], Sabri Birlik [8], Hadi Nobari [9,10,11], Gibson Praça [12] and Filipe Manuel Clemente [1,2,13,*]

1   Escola Superior Desporto e Lazer, Instituto Politécnico de Viana do Castelo, Rua Escola Industrial e Comercial de Nun'Álvares, 4900-347 Viana do Castelo, Portugal; anafilsilva@gmail.com
2   Research Center in Sports Performance, Recreation, Innovation and Technology (SPRINT), 4960-320 Melgaço, Portugal
3   The Research Centre in Sports Sciences, Health Sciences and Human Development (CIDESD), 5001-801 Vila Real, Portugal
4   Department of Physical Education and Sport, Faculty of Education and Sport Sciences, Campus of Melilla, University of Granada, 52006 Melilla, Spain; ftonzalez@ugr.es
5   LabSport, Department of Sports, Center of Physical Education and Sports, Federal University of Espírito Santo, Vitória 29075-910, Brazil; aquino.rlq@gmail.com
6   Sports Science Department, Gazi University, Ankara 06500, Turkey; zekiakyldz@hotmail.com
7   MOVI-LAB Human Movement Research Laboratory, School of Sciences, Graduate Program in Movement Sciences, Physical Education Department, UNESP São Paulo State University, Av. Eng. Luís Edmundo Carrijo Coube, 2085-Nucleo Res. Pres. Geisel, Bauru 17033-360, Brazil; luiz.palucci@unesp.br
8   School of Physical Education and Sports, Afyon Kocatepe University, Afyonkarahisar 03204, Turkey; mehmetyildiz@aku.edu.tr (M.Y.); sabribirlik38@gmail.com (S.B.)
9   Department of Exercise Physiology, Faculty of Educational Sciences and Psychology, University of Mohaghegh Ardabili, Ardabil 5619911367, Iran; hadi.nobari1@gmail.com
10   Faculty of Sport Sciences, University of Extremadura, 10003 Cáceres, Spain
11   Department of Motor Performance, Faculty of Physical Education and Mountain Sports, Transilvania University of Braşov, 500068 Braşov, Romania
12   Sports Department, Universidade Federal de Minas Gerais, Belo Horizonte 31270-901, Brazil; gibson_moreira@yahoo.com.br
13   Instituto de Telecomunicações, Delegação da Covilhã, 1049-001 Lisboa, Portugal
*   Correspondence: filipe.clemente5@gmail.com

Citation: Silva, A.F.; González-Fernández, F.T.; Aquino, R.; Akyildiz, Z.; Vieira, L.P.; Yıldız, M.; Birlik, S.; Nobari, H.; Praça, G.; Clemente, F.M. Analyzing the within and between Players Variability of Heart Rate and Locomotor Responses in Small-Sided Soccer Games Performed Repeatedly over a Week. *Healthcare* **2022**, *10*, 1412. https://doi.org/10.3390/healthcare10081412

Academic Editor: Herbert Löllgen

Received: 30 June 2022
Accepted: 27 July 2022
Published: 28 July 2022

Publisher's Note: MDPI stays neutral with regard to jurisdictional claims in published maps and institutional affiliations.

Abstract: Background: Small-sided games (SSGs) are drill-based and constrained exercises designed to promote a technical/tactical and physiological/physical stimulus on players while preserving some dynamics of the real game. However, as a dynamic game, they can offer some variability making the prediction of the stimulus hardest for the coach. Aim: The purpose of this study was to analyze the between-session and within-player variability of heart rates and locomotor responses of young male soccer players in 3v3 and 5v5 small-sided game formats. Methods: This study followed a repeated-measures study design. Twenty soccer players were enrolled in a study design in which the SSG formats 3v3 and 5v5 were performed consecutively across four days. Twenty under-17 male youth soccer players (16.8 ± 0.4 years old) voluntarily participated in this study. Participants were monitored using a Polar Team Pro for measuring the heart rate mean and maximum, distances covered at different speed thresholds, and peak speed. Results: Between-players variability revealed that maximum heart rate was the outcome with a smaller coefficient of variation (3v3 format: 3.1% to 11.1%; 5v5 format: 6.6% to 15.2%), while the distance covered at Z5 (3v3 format: 82.5% to 289.8%; 5v5 format: 94.0% to 221.1%). The repeated measures ANOVA revealed that the four games tested were different in the within-player variability considering the maximum heart rate ($p = 0.032$), total distance ($p < 0.001$), and distances at zone 1, 2, and 5 of speed ($p < 0.001$). Conclusions: The smaller small-sided game tested promotes greater within-player variability in locomotor demands while promoting smaller within-player variability heart rate responses. Possibly, 5v5 is more recommended to stabilize the locomotor demands, while the 3v3 is recommended to stabilize the heart rate stimulus.

Keywords: football; exercise monitoring; athletic performance; reproducibility

---

1. Introduction

Small-sided games (SSGs) are a widespread form of soccer practice given their capability to simulate both physical and technical–tactical demands at least in part similar to those required in official matches [1–3]. In the context of youth soccer, SSGs have been used for various purposes including training prescription, monitoring/testing, and possible talent detection [4–6]. However, aside from a number of additional strengths (e.g., exercise specificity and player buy-in), potentially undesired variability may be among the prominent pitfalls of SSGs, particularly when a training program aims to reach specific intensities (i.e., individualisation) and progress across the time. Otherwise, it may be an important advantage of such a tool since offering unpredictable stimuli generally benefits technical and tactical development [7,8]. As a consequence, it is clear that there is a need to know the level of variability promoted by distinct SSGs formats [9] before applying them in order to attempt to match the delivered stimuli with the coaches' intentions.

Two common manners to check the consistency/inconsistency of SSG responses consist in determining their within- and between-session variability, and looking for capturing the possible day and seasonal oscillations through repeated measures, respectively. Importantly, the SSG variability data obtained in senior standards cannot be extrapolated to teenagers owing to the discrepancies previously detected [10,11]. This implies that the body of knowledge in this research area, notably in the former, i.e., more than half of existing studies (for a systematic review, see [8]), may not always apply to developing players. In addition to age, the time of day has some impact on young players' performance based on the assumption that a single age group may encompass players with distinct chronotype profiles [12]. Although previous studies controlled such variables when testing within- or between-session variability of SSGs in youth soccer (e.g., performed testing at the same time across different days [13–16]), it remains questionable whether these may be representative for whole samples. As such, it is still necessary to gather information about SSG variability considering distinct periods of the day.

Research conducted on soccer players [15] indicates that either 2v2 or 4v4 SSGs are reproducible in under-19 (mean aged 16.3 years) players on a day-to-day and between-day basis whilst increases in game format tended to decrease control over the SSG intensity. In a follow-up study, contrasting results were reported for the same population and a similar experimental approach with no such evident influence of game format on the physiological variability of SSGs across time was used [9].

It is recognized that combining data collection and analysis of player effort, skill outputs, and psychological aspects (e.g., respectively, heart rate responses, ball handling, and enjoyment) is important in promoting a holistic approach to understanding SSGs [14,17–19]. While preliminary studies are available as aforementioned, with no consensus regarding SSG format effects on the variability of physiological parameters [9,15], the examination of additional features of SSGs such as technical–tactical and individual engagement is lacking in assessing its within- and between-session variability considering youth soccer athletes in particular [8]. It is particularly important to control the within- and between-player variability since the stimulus and volume of the demands imposed can propitiate asymmetries in the dose provided to each player which may expose players to under or over-stimulus. A consistent under- or over-exposure to training load may enhance the risk of bad adaptations or ultimately increase the risk of injury. Understanding within and between players' variability in typical drills used in soccer training (as the SSGs) may help coaches to understand how to manage the implementation of SSGs and adjust them based on the players' needs. Therefore, the present study aimed to investigate the within/between-session variability of physiological and locomotor responses during small-sided games (3v3 and 5v5) in young male soccer players.

2. Materials and Methods

2.1. Experimental Approach to the Problem

This study followed a repeated-measures study design. Twenty soccer players were enrolled in a study design in which the SSGs formats 3v3 and 5v5 were performed consecutively across four days (Table 1). The study was performed early in the season, two weeks after the beginning of the season. The study occurred after 24-h of rest (regarding the latest match), and was performed in the following environmental conditions: 3v3, day 1, (time period: 17:00 h; temperature:17 °C; relative humidity: 65%); 3v3, day 2, (time period: 17:00 h; temperature: 15 °C; relative humidity: 65%); 3v3, day 3, (time period: 17:00 h; temperature: 18 °C; relative humidity: 50%); 3v3, day 4, (time period: 17:00 h; temperature: 19 °C; relative humidity: 48%); 5v5, day 1, (time period: 17:00 h; temperature: 21 °C; relative humidity: 65%); 5v5, day 2, (time period: 17:00 h; temperature: 16 °C; relative humidity: 53%); 5v5, day 3, (time period: 17:00 h; temperature: 18 °C; relative humidity: 65%); 5v5, day 4, (time period: 17:00 h; temperature: 20 °C; relative humidity: 55%).

Table 1. Study design.

		Day 1	Day 2	Day 3	Day 4				Day 1	Day 2	Day 3	Day 4
Match day	Day off	Warm-up FIFA 11+	Warm-up FIFA 11+	Warm-up FIFA 11+	Warm-up FIFA 11+	Day off	Match day	Day off	Warm-up FIFA 11+	Warm-up FIFA 11+	Warm-up FIFA 11+	Warm-up FIFA 11+
		Format: 5v5 Pitch: 50 × 31 m Area per player: 155 m² Length per width ratio: 1.6 Time: 5 min Small goal in the center No offside	Format: 5v5 Pitch: 40 × 25 m Area per player: 100 m² Length per width ratio: 1.6 Time: 5 min Small goal in the center No offside	Format: 5v5 Pitch: 50 × 31 m Area per player: 155 m² Length per width ratio: 1.6 Time: 5 min Small goal in the center No offside	Format: 5v5 Pitch: 40 × 25 m Area per player: 100 m² Length per width ratio: 1.6 Time: 5 min Small goal in the center No offside				Format: 3v3 Pitch: 39 × 24 m Area per player: 156 m² Length per width ratio: 1.6 Time: 3 min Small goal in the center No offside	Format: 3v3 Pitch: 32 × 19 m Area per player: 101 m² Length per width ratio: 1.7 Time: 3 min Small goal in the center No offside	Format: 3v3 Pitch: 39 × 24 m Area per player: 156 m² Length per width ratio: 1.6 Time: 3 min Small goal in the center No offside	Format: 3v3 Pitch: 32 × 19 m Area per player: 101 m² Length per width ratio: 1.7 Time: 3 min Small goal in the center No offside
		Recovery 3 min	Recovery 3 min	Recovery 3 min	Recovery 3 min				Recovery 3 min	Recovery 3 min	Recovery 3 min	Recovery 3 min
		Format: 5v5 Pitch: 40 × 25 m Area per player: 100 m² Length per width ratio: 1.6 Time: 5 min Small goal in the center No offside	Format: 5v5 Pitch: 50 × 31 m Area per player: 155 m² Length per width ratio: 1.6 Time: 5 min Small goal in the center No offside	Format: 5v5 Pitch: 40 × 25 m Area per player: 100 m² Length per width ratio: 1.6 Time: 5 min Small goal in the center No offside	Format: 5v5 Pitch: 50 × 31 m Area per player: 155 m² Length per width ratio: 1.6 Time: 5 min Small goal in the center No offside				Format: 3v3 Pitch: 32 × 19 m Area per player: 101 m² Length per width ratio: 1.7 Time: 3 min Small goal in the center No offside	Format: 3v3 Pitch: 39 × 24 m Area per player: 156 m² Length per width ratio: 1.6 Time: 3 min Small goal in the center No offside	Format: 3v3 Pitch: 32 × 19 m Area per player: 101 m² Length per width ratio: 1.7 Time: 3 min Small goal in the center No offside	Format: 3v3 Pitch: 39 × 24 m Area per player: 156 m² Length per width ratio: 1.6 Time: 3 min Small goal in the center No offside

2.2. Participants

Participants were enrolled based on convenience sampling. The sample size was calculated a priori in G*power (version 3.1.9.6.). For a power of 0.80 and an effect size of 0.5 (medium), the recommended total sample was 21. From the initial recruitment of twenty-four players, twenty male youth soccer players (16.8 ± 0.4 years old; 6.4 ± 0.7 years of experience; 167.9 ± 3.4 cm of stature; 65.4 ± 6.4 kg of body mass) belonging to the same team competing in the regional/national level voluntarily participated in this study. Four players from the initial twenty-four were excluded due to them being goalkeepers (n = 3) and missed one assessment (n = 1). The players were typically involved in five training sessions per week, plus one official match. Each training session lasted about 90 min. The eligibility criteria for being enrolled in the study were: (i) an outfield player; (ii) not injured in the last month before the experimental data collection and not present any signals or reports of injury or illness on the days of data collection; (iii) did not miss any of the data collection moments or repetitions performed; and (iv) did not take any drugs or energy drinks or changed any of their daily dietary routines.

2.3. Physical Fitness Assessment

The participants were preliminarily analyzed for their final locomotor profile. In the week before starting the observation, they were tested for the maximal sprint speed in a 30-m sprint test and for their ability to perform repeated efforts in the 30–15 Intermittent Fitness test. The data collection was preceded by a 24-h rest and occurred in the afternoon (5 pm). The players performed a standardized warm-up protocol (FIFA 11+). After that, they rested for 3 min and started the 30-m sprint speed assessment. They started with a split position (using the preferred leg in front), and the peak speed was assessed using a Polar Team Pro GPS (10 hz, Polar, Kempele, Finland) which was confirmed for the validity and reliability to estimate their peak speed in the previous study. After performing two trials of the 30-m sprint (interspaced by 5 min rest), they recovered for 3 min and followed the 30–15 Intermittent Fitness tests. The original protocol was implemented, and the final velocity completed was registered as the main outcome. For the case of maximal sprint speed, the best speed in the two trials was considered the main outcome.

2.4. Small-Sided Games

The formats 3v3 and 5v5 without a goalkeeper were implemented based on the characteristics presented in Table 1. Players were grouped into teams by their coaches. The same teams faced the same opponents at all times (bouts and different days) aiming to minimize the contextual variation influence. All the formats were preceded by the FIFA11+ warm-up protocol [20]. Four balls were positioned around the pitches aiming to quickly replace the ball in case of going out of boundaries. No verbal encouragement was conceded during the games.

2.5. Heart Rate (HR) and Locomotor Demands

The heart rate and locomotor demands were monitored using the Polar Team Pro (Polar, Finland) which is confirmed for its reliability to measure the demands analyzed [21,22]. The sensor was positioned in the center of the chest while using a band. Each player always used the same unit in order to avoid inter-unit variability. The following measures were obtained per game: minimum heart rate (HRmin); average heart rate (HRmean); peak heart rate (HRpeak); peak speed; average speed; distance covered per minute; distance covered at zone 1 (Z1: 3.00 to 6.99 km/h) per minute; distance covered at zone 2 (Z2: 7.00 to 10.99 km/h) per minute; distance covered at zone 3 (Z3: 11.00 to 14.99 km/h) per minute; distance covered at zone 4 (Z4: 15.00 to 18.99 km/h) per minute; distance covered at zone 5 (Z5: >19.00 km/h) per minute.

2.6. Statistical Procedures

The descriptive statistics were presented in form of mean and standard deviation. The coefficient of variation (expressed as a percentage) was calculated to analyze within-player variability (variability of the player across different days for the same condition) and between-player variability (heterogeneity of the responses for a specific game and condition). The normality of the sample was not verified in all the measures; however, since the data were greater than 30, we have assumed normality by the Central Limit Theorem.

A repeated measures ANOVA was performed to compare the within-players variability obtained in each game (format and pitch dimension). Moreover, for each format of play and pitch dimension, a repeated measures ANOVA was tested (aiming to analyze between training-days variation). Partial eta squared was used to calculate the effect size. Bonferroni was used as a post hoc test in the pairwise comparisons. The statistical procedures were executed in the SPSS (version 28.0.0.0, IBM, Chicago, IL, USA) for a $p < 0.05$.

3. Results

The physical fitness assessment revealed that players presented a maximal speed sprint of 28.7 ± 2.8 km/h and a final velocity in the 30–15 Intermittent Fitness test of 15.2 ± 1.3 km/h. Descriptive statistics of between-training sessions variability can be found in Table 2 for 3v3 and Table 3 for 5v5 formats. While the maximum heart rate presented in the smaller coefficient of variation in 3v3 played at 39×24 m (3.2% to 11.1%) and 32×19 m (3.1% to 8.9%), the distance covered at Z5 presented the greater coefficient of variation in 3v3 played at 39×24 m (82.5% to 289.8%) and 32×19 m (91.1% to 195.7%).

Similarly, the maximum heart rate presented a smaller coefficient of variation in 5v5 played at 50×31 m (6.6% to 16.9%) and 40×25 m (7.0% to 15.2%) and a greater coefficient of variation in distance covered at Z5 in 5v5 played at 50×31 m (103.7% to 221.1%) and 40×25 m (94.0% to 156.8%).

Table 4 presents the descriptive statistics of the within-player variability across the different training sessions performed for each format of play. Additionally, Figures 1–5 present the within-player variability in a graphical representation. The repeated measures ANOVA tested the differences in coefficient of variation between formats and pitch dimensions. The repeated measures ANOVA revealed that the four games tested were different in the within-player variability considering the maximum heart rate ($p = 0.032$; $\eta_p^2 = 0.172$), the total distance ($p < 0.001$; $\eta_p^2 = 0.441$), the distance at Z1 ($p < 0.001$; $\eta_p^2 = 0.745$), the distance at Z2 ($p < 0.001$; $\eta_p^2 = 0.444$), and the distance at Z5 ($p < 0.001$; $\eta_p^2 = 0.394$). The 3v3 played at 39×24 m presented the greater CV% in total distance (34.8 ± 10.5), distance at Z1 (60.3 ± 13.8), the distance at Z2 (52.8 ± 14.0), and the distance at Z5 (170.5 ± 27.6). On the other hand, the 5v5 played at 50×31m presented a smaller CV% in the total distance (18.4 ± 8.4), the distance at Z1 (19.6 ± 10.1), the distance at Z2 (32.2 ± 13.5), and the distance at Z5 (113.7 ± 38.5).

The between-session analysis for each game was performed using a repeated measures ANOVA. In the case of format 3v3 played at 39×24 m (Table 5), significant differences were found between sessions on HRmean ($p = 0.001$; $\eta_p^2 = 0.303$), HRpeak ($p = 0.023$; $\eta_p^2 = 0.189$), total distance ($p = 0.005$; $\eta_p^2 = 0.281$), distances at Z1 ($p = 0.001$; $\eta_p^2 = 0.422$), Z2 ($p = 0.007$; $\eta_p^2 = 0.255$), Z4 ($p = 0.006$; $\eta_p^2 = 0.196$), Z5 ($p = 0.002$; $\eta_p^2 = 0.324$), and peak speed ($p < 0.001$; $\eta_p^2 = 0.330$). In the case of 3v3 played at 32×19 m (Table 5), significant differences were found between sessions on the total distance ($p = 0.003$; $\eta_p^2 = 0.301$), distances at Z1 ($p = 0.007$; $\eta_p^2 = 0.259$), Z2 ($p = 0.001$; $\eta_p^2 = 0404$), Z4 ($p = 0.009$; $\eta_p^2 = 0.220$), and peak speed ($p = 0.037$; $\eta_p^2 = 0.180$).

The between-session analysis for each game was performed using a repeated measures ANOVA. In the case of format 5v5 played at 50×31 m (Table 6), no significant differences between sessions were found for any of the main outcomes. In the case of 5v5 played at 40×25 m (Table 6), significant differences were found between sessions on the total distance ($p < 0.001$; $\eta_p^2 = 0.291$), distances at Z1 ($p < 0.001$; $\eta_p^2 = 0.377$), Z4 ($p = 0.008$; $\eta_p^2 = 0.225$), Z5 ($p < 0.001$; $\eta_p^2 = 0.438$), and peak speed ($p = 0.012$; $\eta_p^2 = 0.174$).

Table 2. Between-session variability (CV%) in the 3v3 format.

	3v3 (39 × 24 m) Day1 BPV (CV%)	3v3 (39 × 24 m) Day2 BPV (CV%)	3v3 (39 × 24 m) Day3 BPV (CV%)	3v3 (39 × 24 m) Day4 BPV (CV%)	3v3 (32 × 19 m) Day1 BPV (CV%)	3v3 (32 × 19 m) Day 2 BPV (CV%)	3v3 (32 × 19 m) Day 3 BPV (CV%)	3v3 (32 × 19 m) Day 4 BPV (CV%)
HRmean (bpm)	11.0	7.1	5.8	4.6	9.7	6.8	9.2	9.4
HRpeak (bpm)	11.1	3.5	5.1	3.2	3.1	5.1	8.9	3.5
Total distance (m)	33.6	29.0	30.7	20.1	20.8	27.0	14.8	16.4
Distance at Z1 (m)	32.3	50.3	104.4	27.7	41.7	60.0	68.5	17.5
Distance at Z2 (m)	28.3	56.1	41.5	47.7	36.3	48.9	30.2	16.4
Distance at Z3 (m)	94.7	40.0	39.4	26.5	104.4	28.9	60.3	54.1
Distance at Z4 (m)	118.4	58.1	48.9	75.2	113.8	58.6	48.2	72.0
Distance at Z5 (m)	289.8	149.2	82.5	245.0	195.7	148.5	91.1	156.3
Peak speed (km/h)	29.1	16.1	17.5	12.8	34.7	16.2	22.0	17.9

BPV: between-player variability; CV%: coefficient of variation expressed as a percentage.

Table 3. Between-session variability (CV%) in 5v5 format.

	5v5 (50 × 31 m) Day1 BPV (CV%)	5v5 (50 × 31 m) Day2 BPV (CV%)	5v5 (50 × 31 m) Day3 BPV (CV%)	5v5 (50 × 31 m) Day4 BPV (CV%)	5v5 (40 × 25 m) Day1 BPV (CV%)	5v5 (40 × 25 m) Day 2 BPV (CV%)	5v5 (40 × 25 m) Day 3 BPV (CV%)	5v5 (40 × 25 m) Day 4 BPV (CV%)
HRmean (bpm)	10.5	7.4	19.7	8.3	8.5	16.5	13.6	8.5
HRpeak (bpm)	8.5	8.5	16.9	6.6	7.0	15.2	12.4	8.5
Total distance (m)	21.6	14.1	23.6	13.4	18.6	21.0	32.1	17.4
Distance at Z1 (m)	19.8	17.3	24.9	17.6	13.9	16.4	26.7	15.6
Distance at Z2 (m)	40.6	25.0	33.7	25.0	38.5	34.3	40.2	27.5
Distance at Z3 (m)	50.4	40.0	74.8	38.5	42.2	64.2	65.1	51.1
Distance at Z4 (m)	68.0	65.2	84.5	61.7	62.9	73.3	87.5	70.3
Distance at Z5 (m)	103.7	221.1	111.1	116.8	127.1	134.0	156.8	94.0
Peak speed (km/h)	11.9	13.3	30.6	16.6	13.8	24.6	19.3	11.8

BPV: between-player variability; CV%: coefficient of variation expressed as a percentage.

Table 4. Within-player variability (CV%) in both formats and pitch dimensions.

	3v3 (39 × 24 m) All WPV (CV%)	3v3 (32 × 19 m) All WPV (CV%)	5v5 (50 × 31 m) All WPV (CV%)	5v5 (40 × 25 m) All WPV (CV%)	Repeated Measures ANOVA $p \mid \eta_p^2$
HRmean (bpm)	7.9 ± 4.0	8.7 ± 4.1	11.9 ± 6.5	11.1 ± 7.1	0.089	0.126
HRpeak (bpm)	6.0 ± 4.3	6.5 ± 4.1	10.3 ± 6.7	10.4 ± 6.3	0.032	0.172
Total distance (m)	34.8 ± 10.5 [b,c,d]	23.0 ± 7.2 [a]	18.4 ± 8.4 [a,d]	26.0 ± 8.2 [a,c]	<0.001	0.441
Distance at Z1 (m)	60.3 ± 13.8 [c,d]	53.1 ± 19.1 [c,d]	19.6 ± 10.1 [a,b]	20.4 ± 7.7 [a,b]	<0.001	0.745
Distance at Z2 (m)	52.8 ± 14.0 [c,d]	46.3 ± 14.4 [c,d]	32.2 ± 13.5 [a,b]	32.7 ± 12.7 [a,b]	<0.001	0.444
Distance at Z3 (m)	51.2 ± 20.5	62.8 ± 20.9	51.1 ± 24.7	52.9 ± 21.9	0.208	0.076
Distance at Z4 (m)	77.5 ± 31.7	83.0 ± 29.1	68.3 ± 23.8	79.4 ± 33.9	0.457	0.044
Distance at Z5 (m)	170.5 ± 27.6 [c,d]	158.6 ± 32.3 [c]	113.7 ± 38.5 [a,b]	139.7 ± 33.9 [a]	<0.001	0.394
Peak speed (km/h)	22.1 ± 9.5	25.0 ± 9.9	17.0 ± 10.5	18.6 ± 8.8	0.052	0.146

WPV: within-player variability (average of within-players repeated measures across the days); CV%: coefficient of variation expressed as percentage; [a]: significantly different from 3v3 (39 × 24 m) at $p < 0.05$; [b]: significantly different from 3v3 (32 × 19 m) at $p < 0.05$; [c]: significantly different from 5v5 (50 × 31m) at $p < 0.05$; [d]: significantly different from 5v5 (40 × 25 m) at $p < 0.05$.

Table 5. Descriptive statistics (mean and standard deviation) of heart rate and locomotor demands between sessions in regard to the 3v3 format.

	3v3 (39 × 24 m) Day1	3v3 (39 × 24 m) Day2	3v3 (39 × 24 m) Day3	3v3 (39 × 24 m) Day4	Repeated Measures ANOVA $p \mid \eta_p^2$
HRmean (bpm)	157.0 ± 17.3 [c,d]	170.4 ± 12.1	172.8 ± 10.1 [a]	174.8 ± 8.0 [a]	0.001	0.303
Hrpeak (bpm)	176.2 ± 19.6	187.0 ± 6.5	184.2 ± 9.4	189.4 ± 6.1	0.023	0.189
Total distance (m)	324.3 ± 108.9 [c]	384.6 ± 111.7	489.5 ± 150.4 [a,d]	343.9 ± 69.1 [c]	0.005	0.281
Distance at Z1 (m)	79.0 ± 25.5 [c,d]	78.2 ± 39.3 [d]	41.1 ± 42.9 [a,d]	115.2 ± 31.8 [a,b,c]	0.001	0.422
Distance at Z2 (m)	153.6 ± 43.5 [d]	165.5 ± 92.8	215.9 ± 89.5 [d]	112.4 ± 53.6 [a,c]	0.007	0.255
Distance at Z3 (m)	50.5 ± 47.8	61.9 ± 24.7	78.1 ± 30.8	65.9 ± 17.5	0.092	0.116
Distance at Z4 (m)	20.7 ± 24.5 [c]	35.0 ± 20.3	46.7 ± 22.8 [a]	29.8 ± 22.4	0.006	0.196
Distance at Z5 (m)	14.3 ± 41.3 [c]	40.1 ± 59.8	104.1 ± 85.8 [a,d]	13.3 ± 32.5 [c]	0.002	0.324
Peak speed (km/h)	17.1 ± 5.0 [b,c]	22.0 ± 3.6 [a]	23.5 ± 4.1 [a,d]	20.0 ± 2.6 [c]	<0.001	0.330
	3v3 (32 × 19 m) Day1	3v3 (32 × 19 m) Day 2	3v3 (32 × 19 m) Day 3	3v3 (32 × 19 m) Day 4	Repeated Measures ANOVA $p \mid \eta_p^2$
Hrmean (bpm)	162.6 ± 15.8	169.0 ± 11.4	173.1 ± 15.9	168.2 ± 15.9	0.213	0.075
Hrpeak (bpm)	177.9 ± 17.6	186.7 ± 5.8	187.3 ± 9.6	182.7 ± 16.3	0.138	0.100
Total distance (m)	409.9 ± 85.4 [d]	435.8 ± 117.7 [d]	454.8 ± 67.4 [d]	331.3 ± 54.3 [a,b,c]	0.003	0.301
Distance at Z1 (m)	76.5 ± 31.9 [d]	76.8 ± 46.1	53.4 ± 36.6 [d]	103.1 ± 18.1 [a,c]	0.007	0.259
Distance at Z2 (m)	233.6 ± 84.9 [d]	172.3 ± 84.1	245.7 ± 74.2 [d]	123.0 ± 20.2 [a,c]	0.001	0.404
Distance at Z3 (m)	37.5 ± 39.2 [b]	84.9 ± 24.5 [a,c]	53.9 ± 32.5 [b]	71.2 ± 38.5	0.001	0.289
Distance at Z4 (m)	25.6 ± 29.1	43.0 ± 25.2 [d]	45.9 ± 22.1 [d]	20.6 ± 14.8 [b,c]	0.009	0.220
Distance at Z5 (m)	32.0 ± 62.6	51.2 ± 76.0	54.3 ± 49.5 [d]	4.6 ± 7.1 [c]	0.066	0.151
Peak speed (km/h)	17.8 ± 6.2 b	21.5 ± 3.5 [a,d]	22.1 ± 4.9 [d]	18.0 ± 3.2 [b,c]	0.037	0.180

[a]: significantly different from 3v3 (39 × 24 m) at $p < 0.05$; [b]: significantly different from 3v3 (32 × 19 m) at $p < 0.05$; [c]: significantly different from 5v5 (50 × 31 m) at $p < 0.05$; [d]: significantly different from 5v5 (40 × 25 m) at $p < 0.05$.

Table 6. Descriptive statistics (mean and standard deviation) of heart rate and locomotor demands between sessions in regard to the 5v5 format.

	5v5 (50 × 31 m) Day1	5v5 (50 × 31 m) Day2	5v5 (50 × 31 m) Day3	5v5 (50 × 31 m) Day4	Repeated Measures ANOVA $p \mid \eta_p^2$
HRmean (bpm)	164.6 ± 17.3	165.4 ± 12.2	154.9 ± 30.5	160.8 ± 13.3	0.357 I 0.053
Hrpeak (bpm)	182.3 ± 15.7	190.0 ± 16.1	170.9 ± 28.9	183.9 ± 12.2	0.062 I 0.136
Total distance (m)	486.2 ± 104.9	555.8 ± 78.3	529.8 ± 124.8	531.8 ± 71.0	0.197 I 0.078
Distance at Z1 (m)	199.8 ± 39.5	212.8 ± 36.8	207.5 ± 51.6	181.9 ± 32.1	0.128 I 0.100
Distance at Z2 (m)	125.6 ± 51.0	148.1 ± 43.0	148.5 ± 50.0	156.0 ± 38.9	0.187 I 0.080
Distance at Z3 (m)	84.6 ± 42.6	96.5 ± 38.6	86.5 ± 64.7	112.6 ± 43.4	0.263 I 0.067
Distance at Z4 (m)	35.3 ± 24.0	44.5 ± 29.0	45.7 ± 38.6	41.3 ± 25.5	0.703 I 0.024
Distance at Z5 (m)	14.4 ± 14.9	34.1 ± 75.3	28.4 ± 31.6	13.8 ± 16.1	0.323 I 0.056
Peak speed (km/h)	20.9 ± 2.5	21.5 ± 2.9	19.7 ± 6.0	21.5 ± 3.6	0.389 I 0.049
	5v5 (40 × 25 m) Day1	5 v5 (40 × 25 m) Day 2	5v5 (40 × 25 m) Day 3	5v5 (40 × 25 m) Day 4	Repeated Measures ANOVA $p \mid \eta_p^2$
Hrmean (bpm)	168.6 ± 14.3	160.8 ± 26.6	163.4 ± 22.3	166.4 ± 14.1	0.668 I 0.027
Hrpeak (bpm)	183.6 ± 12.9	179.7 ± 27.2	181.7 ± 22.4	190.1 ± 16.2	0.464 I 0.044
Total distance (m)	522.0 ± 97.0	522.6 ± 109.7	413.9 ± 132.9	596.8 ± 103.8	<0.001 I 0.291
Distance at Z1 (m)	207.1 ± 28.8	227.2 ± 37.4	174.4 ± 46.5	176.9 ± 27.6	<0.001 I 0.377
Distance at Z2 (m)	146.8 ± 56.5	137.5 ± 47.2	108.1 ± 43.5	134.0 ± 36.8	0.054 I 0.124
Distance at Z3 (m)	94.1 ± 39.7 [d]	82.6 ± 53.1	65.1 ± 42.4	64.4 ± 32.9 [a]	0.075 I 0.113
Distance at Z4 (m)	40.4 ± 25.4	39.8 ± 29.2	33.9 ± 29.6	78.0 ± 54.8	0.008 I 0.225
Distance at Z5 (m)	13.8 ± 17.5 [d]	19.2 ± 25.7 [d]	14.8 ± 23.2 [d]	127.6 ± 119.9 [a,b,c]	<0.001 I 0.438
Peak speed (km/h)	20.7 ± 2.9 [d]	20.0 ± 4.9 [d]	20.1 ± 3.9 [d]	23.7 ± 2.8 [a,b,c]	0.012 I 0.174

[a]: significantly different from 3v3 (39 × 24 m) at $p < 0.05$; [b]: significantly different from 3v3 (32 × 19 m) at $p < 0.05$; [c]: significantly different from 5v5 (50 × 31 m) at $p < 0.05$; [d]: significantly different from 5v5 (40 × 25 m) at $p < 0.05$.

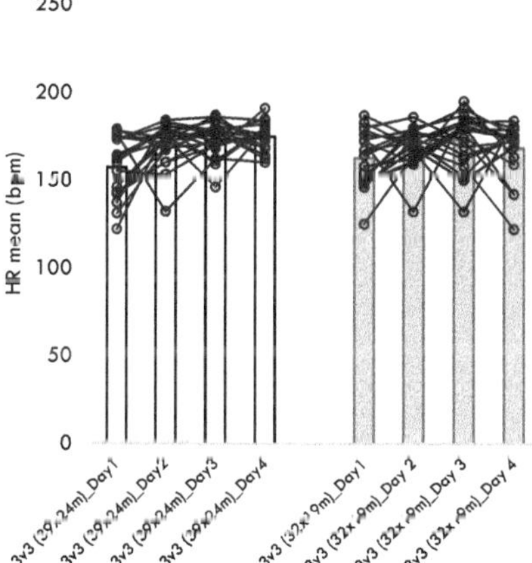
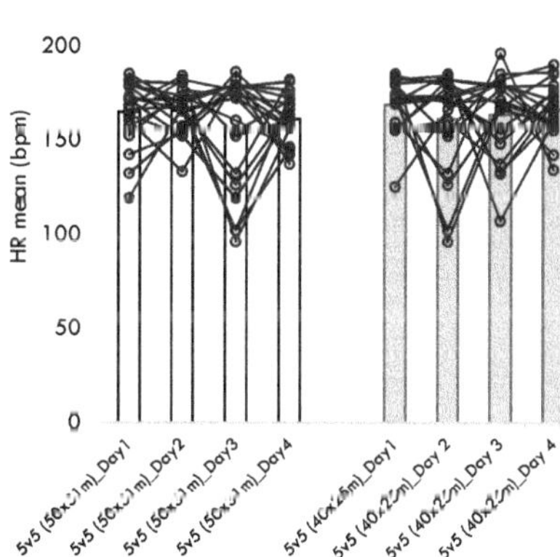

Figure 1. *Cont.*

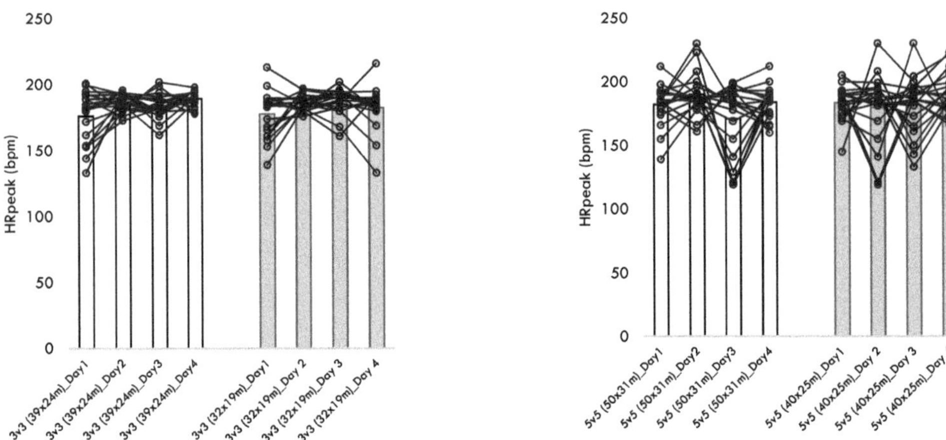

Figure 1. Descriptive statistics of heart rate (HR) responses over the different formats, pitch dimensions, and days.

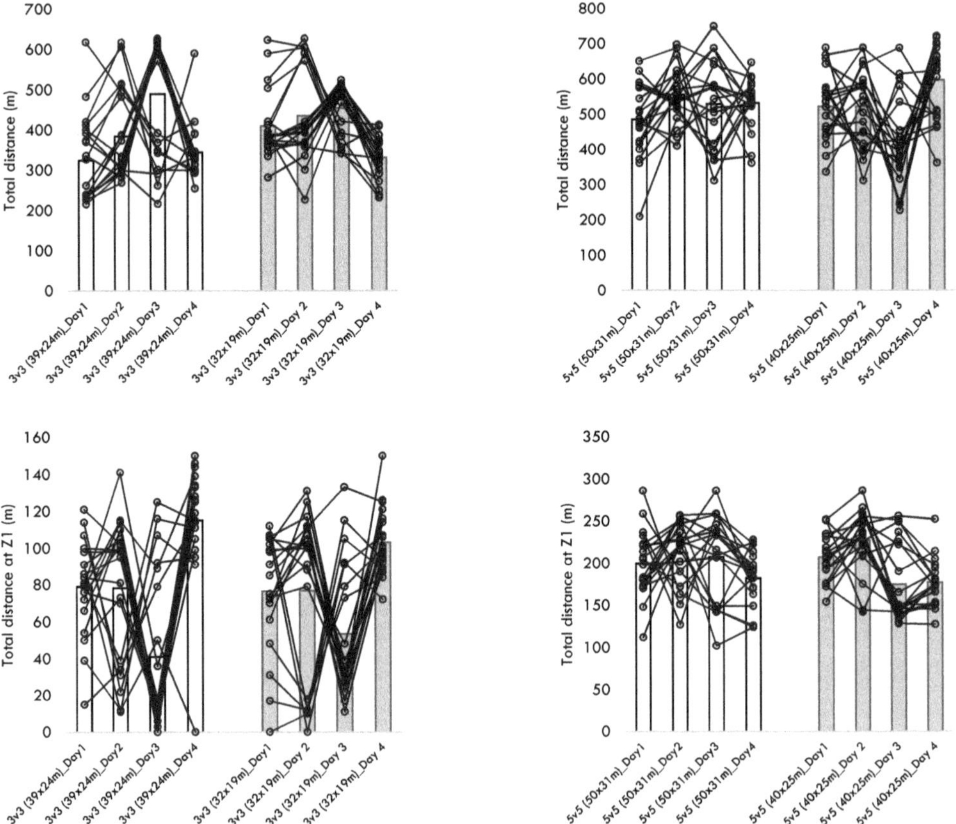

Figure 2. Descriptive statistics of total distance and distance covered at Z1 over the different formats, pitch dimensions, and days.

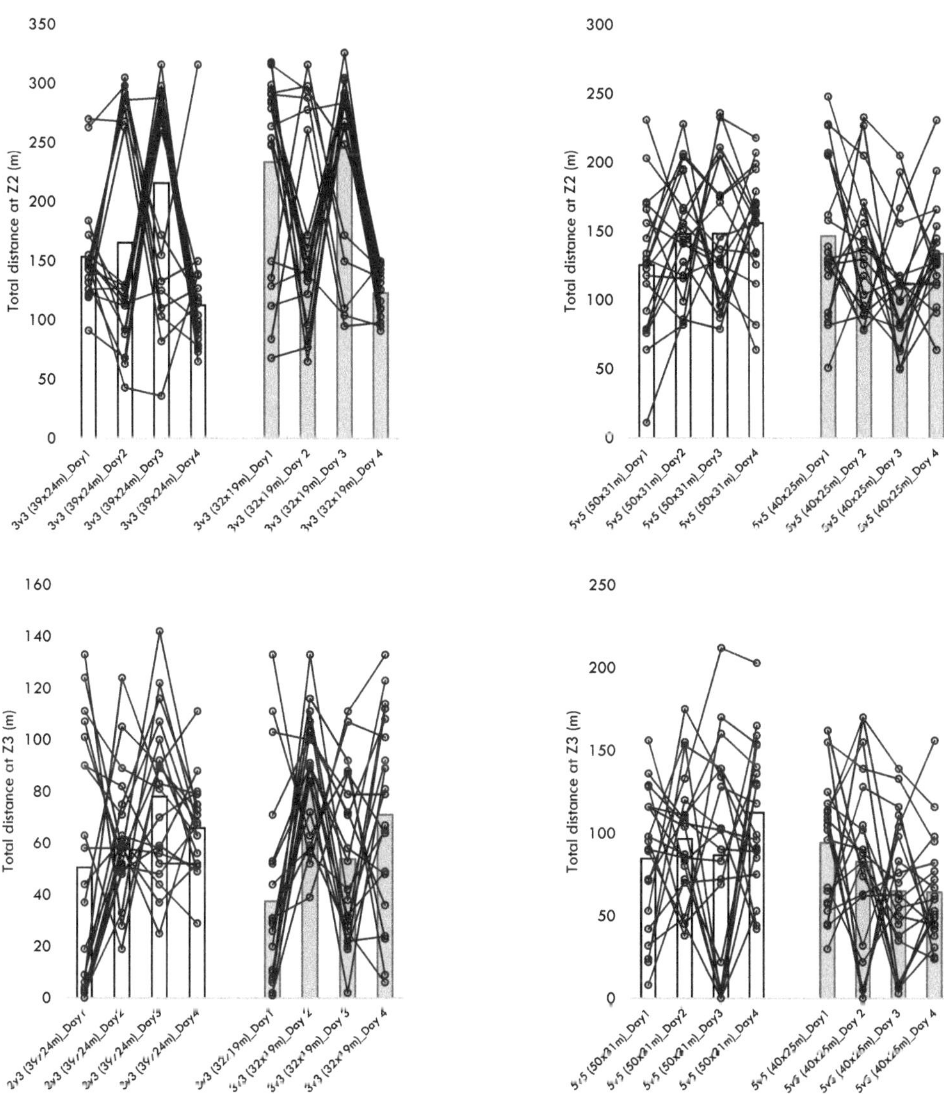

Figure 3. Descriptive statistics of distances covered at Z2 and Z3 over the different formats, pitch dimensions, and days.

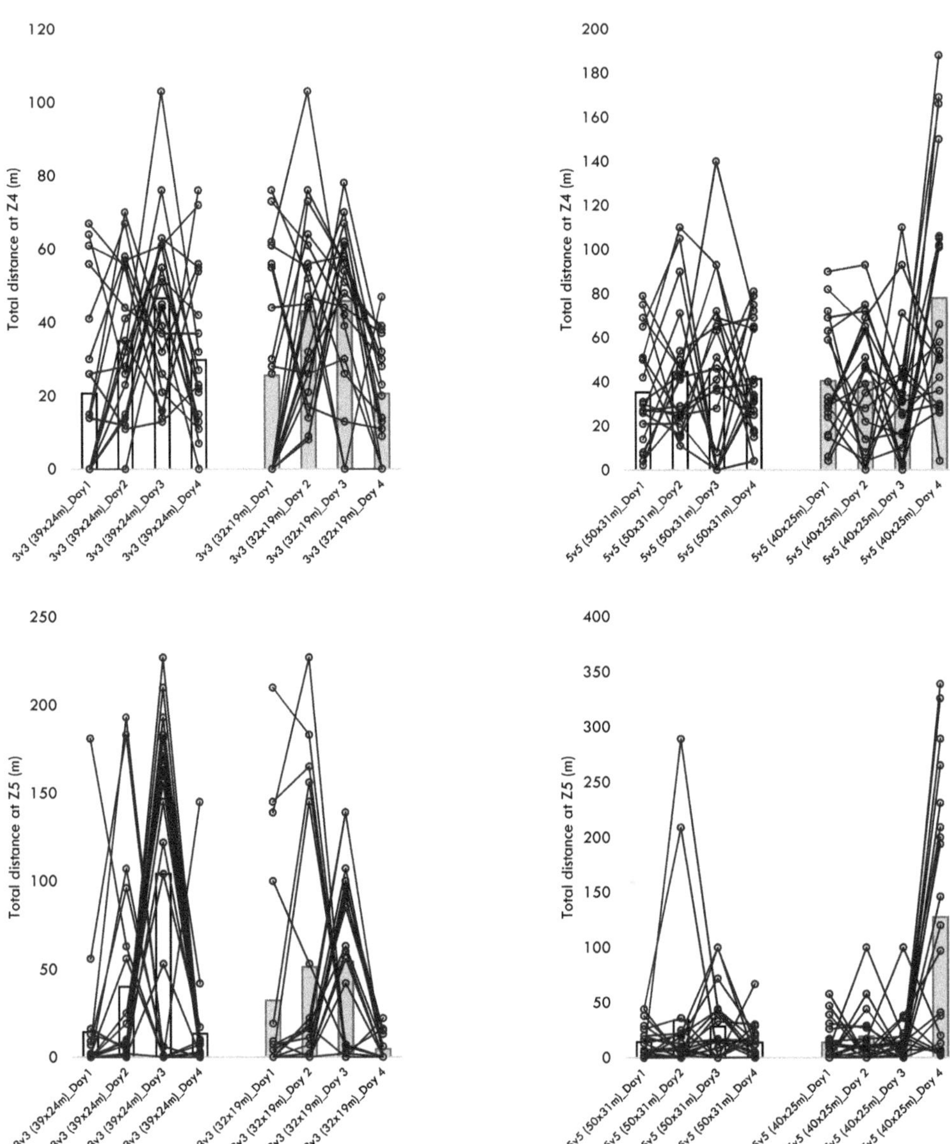

Figure 4. Descriptive statistics of distances covered at Z4 and Z5 over the different formats, pitch dimensions, and days.

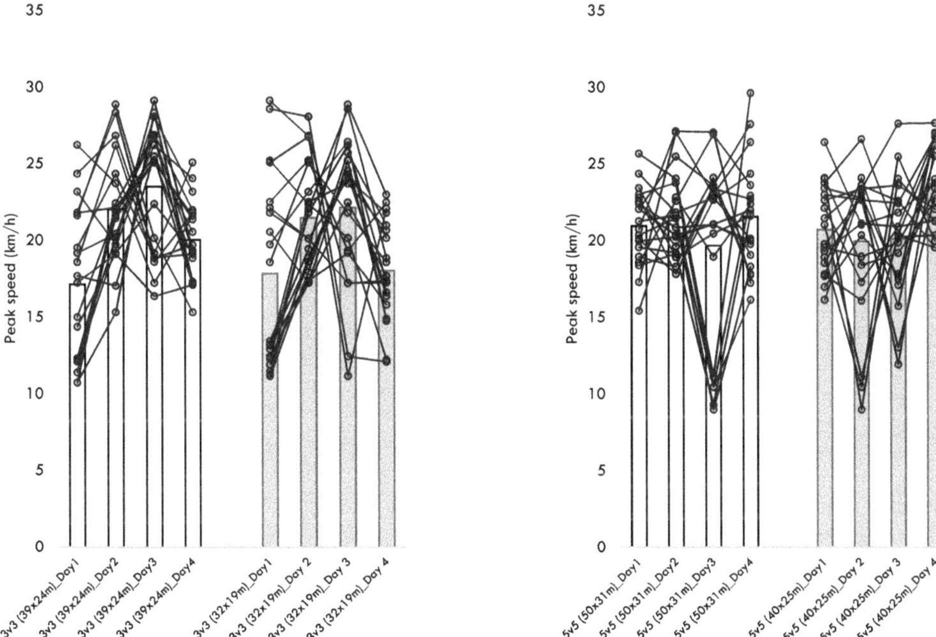

Figure 5. Descriptive statistics of peak speed over the different formats, pitch dimensions, and days.

4. Discussion

The current research reveals that locomotor demands present greater between-session and within-player variability than heart rate responses. Additionally, it was found that 3v3 small-sided game formats promote smaller within-player variability in heart rate responses, yet promote greater locomotor demand variability than the 5v5 format.

One of the concerns in training prescription is projecting the dose for the players. The act of designing small-sided games is complex since many task constraints can be used in combination and some of them can contribute to changing behaviors that are naturally associated with variation in players' responses [6]. In our study, we tried to identify how physiological and locomotor players' responses can vary across four consecutive days for the same games. The within-player variability revealed that heart rate responses are more stable, and the coefficients of variation are, on average, between 6.0 and 11.9% for maximum and mean heart rate. These values are slightly higher than those reported by previous studies on the variability of heart rate responses in the same small-sided games [4,15,23], yet they confirm that physiological responses are relatively reproducible across the days and conditions. The games played no effect on the within-players variability in heart rate mean, although significant differences were found between games on the maximum heart rate. On average, the 3v3 varied between 6.0% and 6.5% of the coefficient of variation, while 5v5 games ranged from 10.3% to 10.4%. The fact that smaller formats of play induce smaller variability can be associated with the time of repetition (3 min) and the fact that, in the 3v3 format, the levels of participation were more equal [24]. Interestingly, this finding may help coaches in the training prescription, since 3v3 induces values closer to those necessary for developing aerobic power [1], and these stable values presented by players across sessions can help coaches to trust in the 3v3 format to induce a similar physiological stimulus while prescribing small-sided games for targeting aerobic power [25].

On the other hand, locomotor demands presented great levels of within-players variability. While the total distance covered presented coefficients of variation between 18% and 35% in the games tested, the distances covered by different speed thresholds revealed a progressive increase of within-player variability. The distance covered at running intensity Z5 was about 113% and 171% in the games tested. These values and tendencies are aligned with previous studies [26,27] which suggested that distances covered with higher intensities are the same with greater within-player variability. Considering that the small-sided games shared the dynamics of the game and are dependent on the tactical behaviors and contextual factors [28], it is expected that intensities can vary based on the moment and tactical dimension of the game [29]. Interestingly, it was observed that the 5v5 format presented a significantly smaller coefficient of variations for within-player variability (e.g., total distance covered, distances covered at Z1, Z2, and Z5) which suggests that possibly bigger formats of play may offer a more stable pattern of locomotor profile. This can be associated with the playing position which will be more structured in formats with a greater number of players as in 5v5. In the case of 3v3, the players will not have a specific position (or, at least, not so structured) which will increase the participation from game to game.

A descriptive observation allowed also us to identify the heterogeneity level in which the game (between players) was smaller in heart rate responses and greater in locomotor demands. Using small-sided games as the format of exercise for prescribing physical stimulus is challenging since the game is dynamic. The heterogeneity between players is expectable based on the dynamic of the game as well. Between-player variability varied from 3.2% to 11.1% in the worst case for the heart rate responses in the 3v3 games performed, while it varied from 7.0% to 19.7% in the 5v5 format. This means that 5v5 may induce a greater inter-player variability. This can be caused by the greater number of players participating in the game and by the fact that tactical positions assumed by the players may interact to explain these varieties in the physiological stimulus. However, greater variabilities occur in locomotor demands as in the case of distance covered at Z5 in 3v3 in which variabilities between players achieved 245% of the coefficient of variation, while in the 5v5, dropped to a maximum of 221.1%. Coaches must be attentive to this between-player variability since locomotor demands are commonly considered for the identification of neuromuscular stimuli. Considering that Z5 corresponds to the sprint zone, and keeping in mind that sprint is one of the most important neuromuscular stimuli for players, some players can suffer in small-sided games by under or over-stimulation.

4.1. Study Limitations

This study presents some limitations. The small sample included is one of the limitations that should be considered as a potential bias. Another limitation is convenience sampling. All the players were part of the same team, which should be faced as a bias for generalization of the evidence. Thus, any conclusion should not be observed as an irrefutable finding and requires more observations to be confirmed. Moreover, this study tested repeated measures on consecutive days which, in fact, can interfere with the way players respond day-by-day. However, this is also an issue for the practice in which specific formats are applied in different contextual scenarios. Another limitation is that some thresholds (heart rate, speed thresholds) are associated with physical fitness, which makes it harder to detect stable values. For example, speed thresholds are player-dependent which means that some thresholds (exempla as sprint) may not start at the same velocity for all. Future research should consider increasing the number of weeks observed (to have a real repeated measure considering the one-game format played on the same day of the week). Moreover, increases in the number of players will favor the generalization of the findings. Finally, other measures that can better describe the anaerobic contribution as blood lactate concentrations must be considered in future research to identify the influence of the most intense SSGs.

4.2. Practical Applications

The current research may provide some practical applications. In this case, the 3v3 format must be used by coaches for an aerobic power stimulus, since it seems to be more stable in terms of stress on heart rate. On the other hand, in the case of aiming to focus on locomotor demands, possibly adjusting the format for medium-sided games (as in 5v5) can be better, since the values are less variable across the sessions.

5. Conclusions

The current research suggests that within- and between-player variability occurs in small-sided games; however, they are associated with the formats and task constraints implemented. The results indicate that heart rate responses are less variable in 3v3 formats, while locomotor demands are less variable in 5v5 formats. Additionally, it is observed that between-player variability is smaller in heart rate responses than in locomotor demands. The results come from a small sample and convenience sampling. Thus, these results are context-dependent which should be faced as a limitation for the generalization of the findings. Future studies should be conducted to confirm or refute the evidence. The next studies must include more players to increase the sample size. Additionally, different teams and populations must be included for a more robust analysis. Finally, consideration for moderators and mediators such as training status, skill level, and moment of the season must be also included.

Author Contributions: Conceptualization, A.F.S., F.M.C. and Z.A.; methodology, F.M.C. and Z.A.; formal analysis, A.F.S. and F.M.C.; investigation, Z.A., M.Y. and S.B.; data curation, F.M.C.; writing—original draft preparation, A.F.S., F.T.G.-F., R.A., Z.A., L.P.V., H.N., M.Y., S.B., G.P. and F.M.C.; writing—review and editing, A.F.S., F.T.G.-F., R.A., Z.A., L.P.V., H.N., M.Y., S.B., G.P. and F.M.C.; supervision, F.M.C. All authors have read and agreed to the published version of the manuscript.

Funding: This research received no external funding.

Institutional Review Board Statement: The study was conducted in accordance with the Declaration of Helsinki and approved by the Institutional Review Board (or Ethics Committee) of Afyon Kocatepe University (protocol code AKU-2021/2).

Informed Consent Statement: Informed consent was obtained from all subjects involved in the study.

Conflicts of Interest: The authors declare no conflict of interest.

References

1. Lacome, M.; Simpson, B.M.; Cholley, Y.; Lambert, P.; Buchheit, M. Small-Sided Games in Elite Soccer: Does One Size Fit All? *Int. J. Sports Physiol. Perform.* **2018**, *13*, 568–576. [CrossRef] [PubMed]
2. Olthof, S.B.H.; Frencken, W.G.P.; Lemmink, K.A.P.M. A Match-Derived Relative Pitch Area Facilitates the Tactical Representativeness of Small-Sided Games for the Official Soccer Match. *J. strength Cond. Res.* **2019**, *33*, 523–530. [CrossRef] [PubMed]
3. Riboli, A.; Esposito, F.; Coratella, G. Small-Sided Games in Elite Football: Practical Solutions to Replicate the 4-min Match-Derived Maximal Intensities. Available online: https://pubmed.ncbi.nlm.nih.gov/35333202/ (accessed on 1 June 2022).
4. Aquino, R.; Melli-Neto, B.; Ferrari, J.V.S.; Bedo, B.L.S.; Vieira, L.H.P.; Santiago, P.R.P.; Gonçalves, L.G.C.; Oliveira, L.P.; Puggina, E.F. Validity and reliability of a 6-a-side small-sided game as an indicator of match-related physical performance in elite youth Brazilian soccer players. *J. Sports Sci.* **2019**, *37*, 2639–2644. [CrossRef] [PubMed]
5. Bennett, K.J.M.; Novak, A.R.; Pluss, M.A.; Stevens, C.J.; Coutts, A.J.; Fransen, J. The use of small-sided games to assess skill proficiency in youth soccer players: A talent identification tool. *Sci. Med. Footb.* **2018**, *2*, 231–236. [CrossRef]
6. Clemente, F.M.; Afonso, J.; Sarmento, H.; Manuel, F.; Id, C.; Clemente, F.M.; Afonso, J.; Sarmento, H. Small-sided games: An umbrella review of systematic reviews and meta-analyses. *PLoS ONE* **2021**, *16*, e0247067. [CrossRef]
7. Clemente, F.M. The Threats of Small-Sided Soccer Games. *Strength Cond. J.* **2020**, *42*, 100–105. [CrossRef]
8. Clemente, F.; Aquino, R.; Praça, G.M.; Rico González, M.; Oliveira, R.; Filipa Silva, A.; Sarmento, H., Afonso, J. Variability of internal and external loads and technical/tactical outcomes during small-sided soccer games: A systematic review. *Biol. Sport* **2022**, *39*, 647–672. [CrossRef]
9. Hill-Haas, S.; Rowsell, G.; Coutts, A.; Dawson, B.; Rowsell, G.; Dawson, B. The Reproducibility of Physiological Responses and Performance Profiles of Youth Soccer Players in Small-Sided Games. *Int. J. Sports Physiol. Perform.* **2008**, *3*, 393–396. [CrossRef]
10. Dellal, A.; Drust, B.; Lago-Penas, C. Variation of Activity Demands in Small-Sided Soccer Games. *Int. J. Sports Med.* **2012**, *33*, 370–375. [CrossRef]
11. Köklü, Y.; Asçı, A.; Koçak, F.Ü.; Alemdaroglu, U.; Dündar, U.; Aşçi, A.; Koçak, F.Ü.; Alemdaroğlu, U.; Dündar, U. Comparison of the Physiological Responses to Different Small-Sided Games in Elite Young Soccer Players. *J. Strength Cond. Res.* **2011**, *25*, 1522–1528. [CrossRef]

12. Palucci Vieira, L.H.; Lastella, M.; da Silva, J.P.; Cesário, T.; Santinelli, F.B.; Moretto, G.F.; Santiago, P.R.P.; Barbieri, F.A. Low sleep quality and morningness-eveningness scale score may impair ball placement but not kicking velocity in youth academy soccer players. *Sci. Med. Footb.* **2021**, 1–11. [CrossRef]
13. Bredt, S.d.G.T.; Praça, G.M.; Figueiredo, L.S.; Paula, L.V.d.; Silva, P.C.R.; Andrade, A.G.P.d.; Greco, P.J.; Chagas, M.H. Reliability of physical, physiological and tactical measures in small-sided soccer Games with numerical equality and numerical superiority. *Brazilian J. Kinanthropometry Hum. Perform.* **2016**, *18*, 602.
14. Clemente, F.M.; Sarmento, H.; Costa, I.T.; Enes, A.R.; Lima, R. Variability of Technical Actions During Small-Sided Games in Young Soccer Players. *J. Hum. Kinet.* **2019**, *69*, 201–212. [CrossRef] [PubMed]
15. Hill-Haas, S.; Coutts, A.; Rowsell, G.; Dawson, B. Variability of acute physiological responses and performance profiles of youth soccer players in small-sided games. *J. Sci. Med. Sport* **2008**, *11*, 487–490. [CrossRef] [PubMed]
16. McLean, S.; Kerhervé, H.; Naughton, M.; Lovell, G.; Gorman, A.; Solomon, C. The Effect of Recovery Duration on Technical Proficiency during Small Sided Games of Football. *Sports* **2016**, *4*, 39. [CrossRef]
17. Arcos, A.L.; Vázquez, J.S.; Martín, J.; Lerga, J.; Sánchez, F.; Villagra, F.; Zulueta, J.J. Effects of Small-Sided Games vs. Interval Training in Aerobic Fitness and Physical Enjoyment in Young Elite Soccer Players. *PLoS ONE* **2015**, *10*, e0137224. [CrossRef]
18. Palucci Vieira, L.H.; Aquino, R.; Moura, F.A.; Barros, R.M.d.; Arpini, V.M.; Oliveira, L.P.; Bedo, B.L.; Santiago, P.R. Team Dynamics, Running, and Skill-Related Performances of Brazilian U11 to Professional Soccer Players During Official Matches. *J. Strength Cond. Res.* **2019**, *33*, 2202–2216. [CrossRef]
19. Sanchez-Sanchez, J.; Hernández, D.; Casamichana, D.; Martínez-Salazar, C.; Ramirez-Campillo, R.; Sampaio, J. Heart Rate, Technical Performance, and Session-RPE in Elite Youth Soccer Small-Sided Games Played With Wildcard Players. *J. Strength Cond. Res.* **2017**, *31*, 2678–2685. [CrossRef]
20. Bizzini, M.; Impellizzeri, F.M.; Dvorak, J.; Bortolan, L.; Schena, F.; Modena, R.; Junge, A. Physiological and performance responses to the "FIFA 11+" (part 1): Is it an appropriate warm-up? *J. Sports Sci.* **2013**, *31*, 1481–1490. [CrossRef]
21. Akyildiz, Z.; Yildiz, M.; Clemente, F.M. The Reliability and Accuracy of Polar Team Pro GPS Units. Available online: https://journals.sagepub.com/doi/abs/10.1177/1754337120976660 (accessed on 1 June 2022).
22. Sagiroglu, İ.; Akyildiz, Z.; Yildiz, M.; Clemente, F.M. Validity and Reliability of Polar Team Pro GPS Units for Assessing Maximum Sprint Speed in Soccer Players. Available online: https://journals.sagepub.com/doi/abs/10.1177/17543371211047224 (accessed on 1 June 2022).
23. Stevens, T.G.A.; De Ruiter, C.J.; Beek, P.J.; Savelsbergh, G.J.P. Validity and reliability of 6-a-side small-sided game locomotor performance in assessing physical fitness in football players. *J. Sports Sci.* **2016**, *34*, 527–534. [CrossRef]
24. Younesi, S.; Rabbani, A.; Clemente, F.M.; Sarmento, H.; Figueiredo, A.J. Session-to-session variations of internal load during different small-sided games: A study in professional soccer players. *Res. Sport. Med.* **2021**, *29*, 462–474. [CrossRef] [PubMed]
25. Clemente, F.M.F.M.; Lourenço Martins, F.M.; Mendes, R.S.R.S.; Martins, F.M.; Mendes, R.S.R.S. Developing Aerobic and Anaerobic Fitness Using Small-Sided Soccer Games: Methodological Proposals. *Strength Cond. J.* **2014**, *36*, 76–87. [CrossRef]
26. Younesi, S.; Rabbani, A.; Clemente, F.M.; Sarmento, H.; Figueiredo, A.J. Session-to-session variations in external load measures during small-sided games in professional soccer players. *Biol. Sport* **2021**, *38*, 185–193. [CrossRef] [PubMed]
27. Clemente, F.M.; Chen, Y.-S.; Bezerra, P.; Guiomar, J.; Lima, R. Between-format differences and variability of technical actions during small-sided soccer games played by young players. *Hum. Mov.* **2018**, *19*, 114–120. [CrossRef]
28. Clemente, F.M.; Afonso, J.; Castillo, D.; Arcos, A.L.; Silva, A.F.; Sarmento, H. The effects of small-sided soccer games on tactical behavior and collective dynamics: A systematic review. *Chaos Solitons Fractals* **2020**, *134*, 109710. [CrossRef]
29. Paul, D.J.; Bradley, P.S.; Nassis, G.P. Factors Affecting Match Running Performance of Elite Soccer Players: Shedding Some Light on the Complexity. *Int. J. Sports Physiol. Perform.* **2015**, *10*, 516–519. [CrossRef]

Article

In-Season Microcycle Quantification of Professional Women Soccer Players—External, Internal and Wellness Measures

Renato Fernandes [1,2,3,*], Halil İbrahim Ceylan [4], Filipe Manuel Clemente [5,6,7], João Paulo Brito [1,2,8], Alexandre Duarte Martins [1,2,9], Hadi Nobari [10,11,12], Victor Machado Reis [3,8] and Rafael Oliveira [1,2,8,*]

1   Sports Science School of Rio Maior—Polytechnic Institute of Santarém, 2040-413 Rio Maior, Portugal; jbrito@esdrm.ipsantarem.pt (J.P.B.); alexandremartins@esdrm.ipsantarem.pt (A.D.M.)
2   Life Quality Research Centre, 2040-413 Rio Maior, Portugal
3   Sport Sciences Department, University of Trás-os-Montes e Alto Douro, 5001-801 Vila Real, Portugal; victormachadoreis@gmail.com
4   Physical Education and Sports Teaching Department, Faculty of Kazim Karabekir Education, Ataturk University, Erzurum 25240, Turkey; halil.ibrahimceylan60@gmail.com
5   Escola Superior Desporto e Lazer, Instituto Politécnico de Viana do Castelo, Rua Escola Industrial e Comercial de Nun'Álvares, 4900-347 Viana do Castelo, Portugal; filipe.clemente5@gmail.com
6   Instituto de Telecomunicações, Delegação da Covilhã, 1049-001 Lisboa, Portugal
7   Research Center in Sports Performance, Recreation, Innovation and Technology (SPRINT), 4960-320 Melgaço, Portugal
8   Research Centre in Sport Sciences, Health Sciences and Human Development, 5001-801 Vila Real, Portugal
9   Comprehensive Health Research Centre (CHRC), Departamento de Desporto e Saúde, Escola de Saúde e Desenvolvimento Humano, Universidade de Évora, Largo dos Colegiais, 7004-516 Évora, Portugal
10   HEME Research Group, Faculty of Sport Sciences, University of Extremadura, 10003 Cáceres, Spain; hadi.nobari1@gmail.com
11   Department of Exercise Physiology, Faculty of Educational Sciences and Psychology, University of Mohaghegh Ardabili, 56199-11367 Ardabil, Iran
12   Sports Scientist, Sepahan Football Club, 81887-78473 Isfahan, Iran
*   Correspondence: rfernandes@esdrm.ipsantarem.pt (R.F.); rafaeloliveira@esdrm.ipsantarem.pt (R.O.)

Citation: Fernandes, R.; Ceylan, H.İ.; Clemente, F.M.; Brito, J.P.; Martins, A.D.; Nobari, H.; Reis, V.M.; Oliveira, R. In-Season Microcycle Quantification of Professional Women Soccer Players—External, Internal and Wellness Measures. *Healthcare* **2022**, *10*, 695. https://doi.org/10.3390/healthcare10040695

Academic Editor: Alessandro Sartorio

Received: 7 March 2022
Accepted: 6 April 2022
Published: 7 April 2022

Publisher's Note: MDPI stays neutral with regard to jurisdictional claims in published maps and institutional affiliations.

Abstract: Although data currently exists pertaining to the intensity in the women's football match, the knowledge about training is still scarce. Therefore, the aim of this study was to quantify external (locomotor activity) and internal (psychophysiological) intensities, as well as the wellness profile of the typical microcycle from professional female soccer players during the 2019/20 in-season. Ten players (24.6 ± 2.3 years) from an elite Portuguese women soccer team participated in this study. All variables were collected in 87 training session and 15 matches for analysis from the 2019–2020 in-season. Global positioning variables such total distance, high-speed running, acceleration, deceleration and player load were recorded as intensity while Rated Perceived Exertion (RPE) and session-RPE were recorded as internal measures. The Hooper Index (HI) was collected as a wellness parameter. The results showed that internal and external intensity measures were greater in matches compared to trainings during the week (match day minus [MD-], MD-5, MD-4, MD-2), $p < 0.05$ with very large effect size (ES). In the same line, higher internal and external intensity values were found in the beginning of the week while the lowest values were found in MD-2 ($p < 0.05$, with very large ES). Regarding wellness, there was no significant differences in the HI parameters between the training days and match days ($p > 0.05$). This study confirmed the highest intensity values during MD and the lowest on the training session before the MD (MD-2). Moreover, higher training intensities were found in the beginning of the training week sessions which were then reduced when the MD came close. Wellness parameters showed no variation when compared to intensity measures. This study confirmed the hypothesis regarding internal and external intensity but not regarding wellness.

Keywords: external load; female; football; Hooper Index; internal load; load; match; training load

1. Introduction

The number of women's soccer participants is growing [1–3] which contributes to the progressive raising of publications related to the topic (in the PubMed, a simple search by "women" AND "soccer" revealed an increment of publications from 75 in 2019 to 125 in 2021). This reveals the concerns of community science to offer more evidence to practical community. Currently, it is recognized that women soccer players covers about 10 to 11 km (km) per professional match [4]. From those km, high-intensity running (18–25 km/h) demands can vary between 718 [4] and 3000 m [5]. These range values are dependent from competitive level, playing position, age-group or event moment of the season and physical fitness [6,7].

Since soccer demands in women soccer are intermittent, this represents a mixed contribution of energetic systems to support the intensities presented in match [7]. Although a greater contribution of low-to-moderate activities, which is consistent with greater aerobic participation, anaerobic systems are also part of the process considering that blood lactate concentrations can vary between 2 and 7 mmol/L [8], and the average heart rate responses in adult women soccer players is around 87% of maximum heart rate [6].

In addition to a well-known knowledge regarding match demands, it is still needed some research to characterize the training process [2]. As an example, a study conducted in elite women soccer players [9] revealed that in the first part of the season a heavy intensity week (week 16) may present 357 arbitrary units (AU) in average per session, while a low intensity week (week 13) may be about 210 AU (values only included weeks with four training sessions, while matches were not included) (considering the multiplication of time of sessions in minutes by the score in a 10-point rate of perceived exertion scale). In the same study, it was found that in a heavy week women player may cover 5090 m per session, while in a low week the average drops to 3870 m [9].

Possibly, these week variations can be affected by the intensities imposed between days of the microcycle. As an example, a study conducted with senior international women players in which was found intra-week variations, in particular greater values of physical and physiological demands in match day (MD)-5 (five days before next match), followed by a progressive intensity decrease until MD-1 [10]. Considering the locomotor activity demands, these intra-week variations can also influence some wellness outcomes. In this sense, a systematic review [11] revealed that the level of a negative correlation between wellness measures (e.g., sleep quality, mood, fatigue, delayed onset muscle soreness, stress) and physiological/physical and locomotor activity demands was small-to-moderate (in general, when there were higher training intensity, the wellness decreased). Even so, the previous systematic review highlighted that not all research was able to find such relationships due to different study designs, methodologies and statistical analysis approaches [11]. Nonetheless, it can be expected to observe intra-week variations of training intensity as well as wellness.

Although there are some cases of studies characterising soccer intra-week variations regarding physiological and locomotor activity demands, and wellness in women soccer, is still need greater research which helps to characterize how the periodization occurs. This type of evidence (intra-week variation) is well-established for men in the last decade where it was observed a tapering on the physiological and locomotor activity demands in last two days before match [12–14]. In the case of women, more research is needed to provide possibilities for comparisons between scenarios and contexts. This may help to characterise the reality of training process in professional women soccer players.

Thus, the purpose of this study was to quantify external (locomotor activity) and internal (psychophysiological) intensities, as well as the wellness profile of the typical microcycle from professional women soccer players during the 2019/20 in-season. For this purpose, the match day minus (MD-) approach used in previous studies was applied for data analysis [14–16]. It was hypothesized that the training session intensity and Hooper Index values are lower on the training day closer to the next match and that match-day presents the highest intensity of the week.

2. Materials and Methods

2.1. Design

The observational period occurred during seven months, from September to March (early-to-mid-season) due to the COVID-19 pandemic, which provoked the disruption of training sessions and matches and the suspension of the season in March. Thus, the observational cohort study contemplated 87 training session and 15 matches for analysis from the 2019–2020 in-season.

The players belong to a team that participated in the BPI League, the women's first League in Portugal. A typical microcycle had three training session and one match per week. For better clarity, the training session occurred in MD-5, MD-4 and MD-2. During MD-3, MD-1 and in the day after the match, the athletes rested.

2.2. Participants

Similar to previous studies with small sample sizes [17–19], 10 elite women soccer players with a professional experience of 4.9 ± 2.1 years, an age of 24.6 ± 2.3 years, a height of 165 ± 6.0 cm (Seca 220, Hamburg, Germany), a body mass of 58.5 ± 9.3 kg (Seca 220, Hamburg, Germany), and body mass index of 22.3 ± 3.8 kg/m^2, participated in this study. Moreover, the power of the sample size was estimated through G-Power [20]. The analysis featured 99.2% of actual power, with a total of 10 participants with a $p < 0.05$ and effect-size for 0.6.

We adopted inclusion/exclusion criteria from our previous studies [14,16,21,22] where participants need to achieve a minimum of 80% of the training sessions and an average of 75 min from all matches, while the exclusion criteria were based on becoming injured, ill, sick for two consecutive weeks. Only defenders ($n = 3$), midfielders ($n = 4$), and strikers ($n = 3$) were included for analysis while goalkeepers were removed.

Before the beginning of the study, all explanations about the study design were provided to players. Then, a written informed consent was recorded from all participants. In addition, the study was approved by the research Ethics Committee of the Polytechnic Institute of Santarém, Santarém, Portugal (252020 Desporto) and it was developed according to the requirements of the Declaration of Helsinki.

To control more outside factors that could influence the intensity and wellness over the period of analysis, all participants were asked to maintain their normal diet throughout the study period. To confirm their habits, nutritional intake questionnaire was used to record a 24 h diet over seven days of the week. This procedure was applied in the first and last week of the analysed period following the procedures of our previous study [21]. The size of the food portions, supplements, and other aspects pertaining to an accurate recording of their energy intake were addressed and reviewed for macronutrient composition and total energy intake [23].

2.3. Internal Intensity Quantification

The CR10-point scale, adapted by Foster et al., was used 30 min after the end of each training/match session [24]. To avoid non-valid values, all players were previously familiarized with the scale and all answers were provided on google forms, through a tablet. Then, each training/match session value was multiplied by the, respectively, session duration to produce the s-RPE [24,25].

2.4. Wellness Quantification

The Hooper Index (HI, 1–7 scale) questionnaire was also used 30 min before each training/match session. The same procedure described in the previous point was used for data collection. This questionnaire has four questions: fatigue, stress, delayed onset muscle soreness (DOMS) (in which 1 is very, very low and 7 is very, very high), and the quality of sleep of the night that preceded the evaluation (in which 1 is very, very bad and 7 is very, very good). Moreover, to produce a final score of HI, data from all questions was summed [26].

2.5. External Intensity Quantification

A portable 10 Hz GPS device was used to collect external data (PlayerTek, Catapult Innovations, Melbourne, Australia), which also incorporates a tri-axial 100 Hz accelerometer. These types of GPS devices seem to be the most valid and reliable to use in team sports [27].

Ten minutes before each training session and match, PlayerTek devices were turned on and the players were asked to use them. The devices were turned on and placed in a specific customized vest pocket located on the posterior side of the upper torso fitted tightly to the body, as is typically used in matches. The devices were placed and checked always by the same coach of the team, and the players always used the same device [28].

The measures used for analysis were total distance, high-speed running distance (≥ 15 km/h) [29], number of accelerations (ACC, >1–2 m.s^{-2} [ACC1]; >2–3 m.s^{-2} [ACC2]; >3–4 m.s^{-2} [ACC3]; >4 m.s^{-2} [ACC4]) and decelerations (DEC, <-1–2 m.s^{-2} [DEC1]; <-2–3 m.s^{-2} [DEC2]; <-3–4 m.s^{-2} [DEC3]; <-4 m.s^{-2} [DEC4]) maximal speed, average speed and player load.

2.6. Statistical Analysis

Descriptive statistics (mean $\pm$ standard deviation, SD) were performed for all measures. All the variables were checked for normality and homoscedasticity, respectively, using the Shapiro–Wilk and Levene tests. Then, repeated measures ANOVA with the Bonferroni post hoc test was calculated to compare the training and match sessions. The p-value ≤ 0.05 was used as significant and all the data were analysed using SPSS version 22.0 (SPSS Inc., Chicago, IL, USA) for the Windows statistical software package.

Finally, the Hedges effect-size (ES) was performed to determine the effect magnitude through the difference of two means divided by the standard deviation from the data and the following criteria were used: <0.2 = trivial, 0.2 to 0.6 = small effect, 0.6 to 1.2 = moderate effect, 1.2 to 2.0 = large effect, and >2.0 = very large [30].

3. Results

Table 1 shows the MD- differences for duration and running distance variables between training and match days. With the exception for duration, where MD-4 displayed the highest value, all running-based variables showed to be significantly higher in MD with very large effect sizes.

Table 1. External Intensity by running-based variables during training and matches for squad average (mean $\pm$ SD).

Day	Duration (min)	Total Distance (m)	Average Speed (m/min)	Maximal Speed (km/h)	HSR (m)
MD-5	85.1 ± 2.8	5121.6 ± 82.2 [a,b,c,*]	60.6 ± 1.7 [a,b,c,*]	23.3 ± 0.5 [c,*]	306.5 ± 33.9 [c,*]
MD-4	90.9 ± 2.5 [b,*]	4638.6 ± 73.0 [b,c,*]	53.5 ± 1.8 [b,c,*]	23.5 ± 0.7 [c,*]	312.8 ± 38.2 [c,*]
MD-2	78.3 ± 2.1	3857.5 ± 73.0 [c,*]	47.8 ± 1.2 [c,*]	22.9 ± 0.5 [c,*]	311.3 ± 22.2 [c,*]
MD	87.2 ± 2.0	7616.1 ± 395.2	89.9 ± 5.4	26.7 ± 1.2	879.7 ± 102.2

MD, match-day; MD-, matchday minus (5, 4, 2); min, minutes; m, meters; HSR; high-speed running; [a] denotes difference from MD-4; [b] denotes difference from MD-2; [c] denotes difference from MD; all $p \leq 0.05$; * means a very large effect size for all differences (>2.0).

Table 2 shows the MD- differences for accelerometery-based variables, namely, ACC, DEC and player load between training and match days. With the exceptions of player load in MD-5 and ACC4, all variables showed to be significantly higher in MD with very large effect sizes.

Table 2. External Intensity by accelerometery-based variables during training and matches for squad average (mean ± SD).

MD	Player Load (AU)	ACC1	ACC2	ACC3	ACC4	DEC1	DEC2	DEC3	DEC4
MD-5	284.4 ± 11.7 [a,b,*]	138.6 ± 7.6 [b,c,*]	80.9 ± 4.3 [b,c,*]	29.9 ± 2.7 [a,b,*]	9.6 ± 1.4	126.3 ± 6.8 [b,c,*]	77.8 ± 4.6 [b,c,*]	28.0 ± 2.4 [b,c,*]	11.8 ± 1.6 [c,*]
MD-4	263.4 ± 10.5 [b,c,*]	134.8 ± 6.7 [b,c,*]	79.8 ± 3.5 [b,c,*]	26.7 ± 2.5 [b,c,*]	7.5 ± 1.2	121.9 ± 5.6 [b,c,*]	77.6 ± 3.6 [b,c,*]	27.9 ± 2.2 [b,c,*]	11.2 ± 1.4 [c,*]
MD-2	222.6 ± 7.2 [c,*]	100.7 ± 4.1 [c,*]	52.9 ± 1.7 [c,*]	21.3 ± 1.3 [c,*]	9.8 ± 1.0	89.6 ± 4.1 [c,*]	56.0 ± 2.3 [c,*]	21.2 ± 1.3 [c,*]	9.7 ± 1.1 [c,*]
MD	324.4 ± 14.0	177.3 ± 8.2	106.8 ± 6.4	34.9 ± 3.4	10.3 ± 1.8	169.0 ± 4.8	98.8 ± 5.9	39.0 ± 4.1	18.9 ± 2.9

MD, match-day; MD- = matchday minus (5, 4, 2); AU, Arbitrary Units; ACC, acceleration; DEC, deceleration. Both ACC and DEC were measured in number (counts); [a] denotes difference from MD-4; [b] denotes difference from MD-2; [c] denotes difference from MD; all $p \leq 0.05$; * means a very large effect size for all differences (>2.0).

Table 3 shows the MD- differences for internal intensity and wellness profile. While variables from HI did not show any significant difference, both RPE and s-RPE showed to be significantly higher in MD with very large effect sizes. There was an exception regarding MD-5 versus MD for s-RPE that showed a large effect size instead of a very large.

Table 3. Internal Intensity and Wellness Profile during training and matches for squad average (mean ± SD).

MD	RPE (AU)	s-RPE (AU)	Fatigue (AU)	Stress (AU)	DOMS (AU)	Sleep Quality (AU)	HI (AU)
MD-5	5.9 ± 0.3 [b,c,*]	508.3 ± 29.0 [b,#]	3.3 ± 0.2	3.6 ± 0.4	2.9 ± 0.3	3.6 ± 0.3	13.4 ± 0.8
MD-4	5.4 ± 0.2 [b,c,*]	473.7 ± 20.5 [b,c,*]	3.4 ± 0.2	3.1 ± 0.3	3.0 ± 0.3	3.4 ± 0.2	13.0 ± 0.7
MD-2	4.4 ± 0.3 [c,*]	353.5 ± 21.9 [c,*]	3.3 ± 0.2	3.2 ± 0.4	2.7 ± 0.2	3.3 ± 0.2	12.5 ± 0.7
MD	7.9 ± 0.3	604.7 ± 36.5	3.1 ± 0.1	3.0 ± 0.2	3.0 ± 0.2	3.3 ± 0.3	11.6 ± 0.8

MD, match-day; MD-, matchday minus (5, 4, 2); AU, Arbitrary Units; RPE, rated perceived exertion; s-RPE, session-RPE; DOMS, delayed onset muscle soreness; HI, Total Hooper Index; [a] denotes difference from MD-4. [b] denotes difference from MD-2. [c] denotes difference from MD; all $p \leq 0.05$; [#] means a large effect size (>1.2–2.0); * means a very large effect size for all differences (>2.0).

4. Discussion

The purpose of this study was to quantify external and internal intensities, as well as the wellness profile of a typical microcycle in professional women soccer players during the 2019/20 in-season. The present study indicated that internal and external intensity measures were greater in MD than in weekday training sessions (MD-5, MD-4, MD-2). Moreover, it was found that the wellness parameters (Hooper Index, HI) of the players did not differ significantly between training sessions and matches. During the week, it was observed that the internal and external intensity gradually decreased close to the match, and even in the last two days before the match, there was a serious decrease in the intensity measures so that the players could prepare for the match. Our results supported one part of the hypothesis of the present study that confirmed that the training session intensity values are lower on the training day closer to the next match and that match-day presents the highest intensity of the week. On the other hand, our results did not support that Hooper Index values are lower on the training day closer to the next match.

The present study revealed there was no significant difference in the HI parameters between the three training days and match days in professional women soccer players. Consistent with our results, recent studies reported no intra-week variations (between match day and training sessions) in wellness variables in elite soccer players [13,14]. In our study, it was seen that the wellness measures vary between 2 and 3 on average throughout the microcycle in women soccer players. This result was in agreement with the study carried out by Fernandes et al. [16]. Similarly, Clemente et al. [31] stated that basketball players showed similar health profiles in both training and matches during normal and congested weeks, and the health profiles of the players (very low DOMS, fatigue, and stress and very good sleep quality, around 2 on average) were quite well stated. This shows that both training sessions and matches have similar effects on players and do not sufficiently trigger stress factors. This result is also supported by Clemente et al. [31].

Concerning the RPE and s-RPE, our study demonstrated that MD-5 presented highest intensity than MD-4 (session with highest training duration). Both MD-5 and MD-4 presented higher values than MD-2 (lowest intensity), which revealed a pyramid weekly microcycle shape regarding internal intensity (lower intensity after and before in the training sessions close to the match and higher intensity in the middle week training sessions). Doyle et al. [10] found that the total training intensity in elite female football players increased from MD-7 to MD-5 compared to other training sessions and peaked at MD-5.

In the first and second training sessions of the week (MD-5 and MD-4), the players were exposed to the greatest psychophysiological intensity, which was supported by the study by Ispirlidis et al. [32] and that by Fernandes et al. [16], who noted that fatigue, DOMS and total HI were significantly higher in MD-4 (second weekly training day) as compared with other training and match days in soccer players. As in our study, Romero-Moraleda et al. [29] also reported that training sessions in the middle of the week (MD-4 and MD-3) demonstrated greater intensities in professional female soccer players. Similarly, Clemente et al. [31] stated that fatigue and stress were high due to the training intensity in MD-4 (1st training day of the week) during normal and congested weeks, and accordingly, sleep quality was quite low.

The present study indicated that the intensity of the training (RPE and s-RPE) gradually decreased as the match day approached. Our results were in agreement with Doyle et al. [10], who reported that the total training volume and intensity on both MD-3 and MD-1 decreased during the training sessions close to the match. Similarly, Romero-Moraleda et al. [29] stated that lower intensities implemented in training sessions (MD plus 1, [MD + 1] and MD-2) were closest to MD. Additionally, Malone et al. [15] noted that the training intensity decreased in MD-1 compared to MD-2 and MD-5. In contrast, Oliveira et al. [14] remarked that s-RPE value decreased in an imperfect order from MD-5 to MD-1 in elite male soccer players. In our study, HI scores did not show any change in the days before the match day, which means that during the week, whereas the decrease in RPE and s-RPE from MD-5 as the match day approaches, HI presented no changes. Previous study of Haddad et al. [33] was in line with the present findings. Haddad et al. [33] reported that HI parameters were not significant determinants of perceived effort during traditional soccer training without excessive training intensity. On the other hand, Clemente et al. [34] observed that the relationship between s-RPE and HI was significant and negative (small-medium) in the weeks when there were two official matches and there was no relationship between s-RPE and HI in the weeks when there was one match, as in our study.

Regarding the total distance covered and average speed, the total distance and average speed gradually decreased as the match day approached and this variable reached the highest values on the match day when compared with the training sessions. Consistent with our results, Trewin et al. [35] reported that the female soccer players exhibited lower values for the total distance covered (approx. 4000–5000 m) before the match compared to the other training sessions. Similarly, recent study notified that the total distance covered by female soccer players in the MD-2 (3024 ± 1220 m) was lower than in the MD-4 (4831 ± 860 m) and MD-3 (4975 ± 1318 m) in an one-week training program [29]. Additionally, Doyle et al. [10] noted that MD-5 and MD-2 were the most intense training sessions compared to other training sessions in terms of total distance covered (5933.5 and 5151.5 m, respectively), very high speed running (387.5 and 201 m, respectively) and sprint distance (187.5 and 49 m, respectively) in elite female soccer players. Our results were supported by some studies conducted on top elite male soccer players which indicated that the total distance covered and average speed before the match decreased in MD-1 (in our study, MD-2) compared to other training days (MD-5 and MD-4) [12,14,15,36].

The gradual decrease in internal and external intensity towards the match in the weekly program may be due to the content of the weekday training sessions. Indeed, Romero-Moraleda et al. [29] showed that MD-2 included skill and strategy exercises applied to reduce the training intensity before match day. The present study also used speed and ACC exercises in the execution of skills and goal achievement during MD-2.

Furthermore, the present study demonstrated that the accelerometer-based variables, namely ACC, DEC and player load (except ACC4), decreased from MD-5 to MD-2. In addition, the fact that these variables were higher in MD-5 can be explained by the intense application of small-sided games to players during this training session, and previous studies notified that ACC, DEC and player load were affected by pitch area, and small-sided games increased these variables more [9,37].

Nevertheless, our results for ACC are similar to the previous study that indicated that the number of ACC performed in MD-2 was significantly lower than in MD-5, and there was no difference in DEC between MD-2 and MD-5 in elite female soccer players [10]. Other studies showed that total number of ACC and DEC decreased from MD-4 to MD-2 in elite female soccer players [29], and also that these variables decreased from MD-4 to MD-1 in elite male soccer players [38]. A study of elite male soccer players found lower ACCs values before matches compared to other days [14]. Considering the above studies, Harper et al. [39] stated that greater volume of explosive eccentric actions such as ACCs and DECs during the match could cause greater perceived exertion, muscle damage, neuromuscular fatigue and, as a result, a higher risk of injury. However, the available evidence from the recent study emphasized that the volume of very high intensity running and sprint distance on MD-2 was decreased by 48% and 73%, respectively, showing us how the training session content was modified to optimize player preparation for competition in MD [10]. Additionally, the same authors asserted that the optimum taper period process for international women's matches are not clear enough in the literature and therefore the most effective periodization methods should be investigated to ensure optimal preparation for matchday performance [10].

In our study, it was seen that the volume and intensity were high on the 1st and 2nd (MD-5, MD-4) training days, and then the volume and intensity decreased towards MD-2. This shows that the tapering strategy (significant reduction in volume and intensity before the match) was used for the weekly training design such as in previous studies [12,35,36,40]. The tapering strategy seems to be a preferred method for reducing the residual fatigue accumulation in the last two days before the match, to optimize the performance and to prepare the players for the match [29]. In addition, previous studies found that the use of the tapering strategy throughout a microcycle created certain performance advantages on players. For instance, in one study, it was reported that reducing the training intensity of professional soccer players by ~25% during taper weeks during matches resulted in a 15% increase in intense and high intensity activities during matches [41]. In addition, previous studies showed that a reduction in intensity, in the last training session before a match leads to improvements in wellness variables (total HI, DOMS, fatigue, and sleep quality) on a matchday and prevent overtraining in female soccer players [16], and professional basketball players [31]

Soccer matches show the highest intensity in the weekly schedule [14]. The results from our study showed that when comparing MD with the training sessions, the highest RPE and s-RPE values were recorded in MD. This finding was compatible with the study conducted by Fernandes et al. [16] which showed the highest matchday RPE values in female soccer players. Subsequently, the same researchers stated that the match offered lower s-RPE values and interestingly there was no significant difference between MD-5, MD-4 and MD. This result is not in line with the results of present study. In another study, Romero-Moraleda et al. [29] observed that the internal intensity in professional female soccer players was higher in official matches (RPE: 8.4 AU, and s-RPE: 792 AU) compared to training sessions (RPE: 3.1–6.2 AU, and s-RPE: 167–579 AU).

Moreover, the present study revealed that HSR and maximal speed were found to be significantly higher in MD (HSR: 879.7 ± 102.2 m, maximal speed. 26.7 ± 1.2 km/h) compared to weekday training days (HSR: 306.5 ± 33.9–312.8 ± 38.2 m, maximal speed: 22.9 ± 0.5–23.5 ± 0.7 km/h). In addition, ACC and DEC were significantly higher in MD than weekday training days. Our results were supported by the study performed by Romero-Moraleda et al. [29].

Considering the results of the above studies, it was observed that women soccer players are exposed to higher internal and external intensities in matches compared to trainings during the week, intensely engage in high-intensity activities during the match and reach higher maximum speeds in activities. However, Teixeira et al. [42] reported that intensity differences between training and matches may be affected by contextual factors such as the type of weekly schedule, player's starting status, playing positions, age group, training mode, opponent's level, the location of matches.

The present study includes some limitations. Firstly, it was carried out with 10 professional adult women soccer players. Clemente et al. [12] found that intensity and tapering methods differed between teams in different countries, so it can be difficult to generalize current results to male or female amateur players of different age categories in different countries. At the same time, the sample size in the present study does not permit the generalization of the results for other teams. Furthermore, players from different positions are exposed to different internal and external intensities [15,29]. The neglect of player positions due to the sample size in our study could be considered the second limitation.

The weekly training design used in this study is valid for women's soccer teams using the same design. The present study showed that soccer matches generate very high external, internal and wellness measures on the players. Therefore, it is recommended to include different time-efficient training methods (high-speed straight runs, running involving directional changes, repeated short-medium-large sprint ability, time spent in game-based situations, with modifications of pitch dimensions) in training planning, taking into account the player positions in order to cope with the differences in both internal and external intensities between training and official matches [29]. In the literature, the number of studies on women soccer players related to weekly intensity design is still reduced compared to male soccer players. Therefore, the methodology of this study can be comprehensively replicated in women soccer players of different age categories in different countries, with the addition of evaluation of various biochemical parameters, especially after match day.

Increasing studies on women soccer players are important for the development and periodization of training programs. For instance, this study did not address the menstrual cycle. It has been revealed that there are hormonal variations during this cycle and those can affect cardiorespiratory [43] and neuromuscular performance [44], although results are not always consensual [45] and more research is needed. Such analysis would be relevant for future studies to analyse if there are variations of external and internal intensities with respect to the menstrual cycle.

Finally, we suggest that future studies could analyse if the wellness values of the day before influence the training intensity of the day after and if the training intensity of the day influence the HI values of the day after as previously conducted in the Silva et al. study [46].

From a practical point of view, the results of this study can express that one of the main focuses of technical teams is the balanced allocation of the microcycle intensity in order to induce the best performance in the players and simultaneously ensure that the effort management allows to avoid possible harmful effects that may arise from it. This fact was evident in two aspects: (1) in the wellness profiles evidenced by the players with low levels of DOMS, fatigue and stress, plus good sleep quality throughout the microcycle; and (2) the good management of intensity that revealed greater levels in the first and second sessions of the microcycle and then, a reduction until the match. Nowadays, professional technical teams have human resources, whether they are assistant coaches, physical trainers, physiologists, psychologists, or physiotherapists who, working as a team also have the mission of protecting their players and putting them at the highest physical and physiopsychological levels for the competition.

5. Conclusions

The present study revealed that the internal (RPE, s-RPE) and external measures (total distance, average speed, maximal speed, high-speed running distance, almost ACC and DEC) were higher on the match day than during the weekday training sessions (MD-5, MD-4, MD-2) throughout the in-season in elite female soccer players. In addition, it was observed that HI values did not differ significantly between training sessions and matches during the week. Finally, the present study showed that players generally reached the highest internal and external intensity in MD-5 and after that, intensity decreased until the MD.

This study confirmed the hypothesis regarding internal and external intensity but not regarding wellness. It is recommended to consider the results of the current study in the weekly intensity distribution in female soccer players and accordingly in the optimal adjustment of the relationship between intensity and rest.

Author Contributions: Conceptualization, R.F. and R.O.; methodology, R.F., J.P.B., A.D.M., H.N., F.M.C. and R.O., software, R.F., H.N. and R.O.; validation, V.M.R., R.O. and J.P.B.; formal analysis, R.F., R.O. and A.D.M.; investigation, R.O., H.İ.C., R.F., A.D.M., J.P.B. and F.M.C.; resources, R.O., R.F., H.İ.C., A.D.M., J.P.B., V.M.R. and F.M.C.; data curation, R.F., R.O., A.M and H.N.; writing—original draft preparation, R.O., J.P.B., H.İ.C., F.M.C. and R.F.; writing—review and editing, R.O., H.İ.C., R.F., A.D.M., J.P.B. and F.M.C.; visualization, R.O. and R.F.; supervision, R.O., V.M.R. and J.P.B.; project administration, R.O., J.P.B. and R.F.; funding acquisition, R.O. and J.P.B. All authors have read and agreed to the published version of the manuscript.

Funding: This research was funded by the Portuguese Foundation for Science and Technology, I.P., Grant/Award Number UIDP/04748/2020. Victor Machado Reis was funded by FCT–Fundação para a Ciência e Tecnologia (UID04045/2020). This work was also funded by Fundação para a Ciência e Tecnologia/Ministério da Ciência, Tecnologia e Ensino Superior through national funds and when applicable co-funded EU funds under the project UIDB/50008/2020.

Institutional Review Board Statement: The study was conducted according to the guidelines of the Declaration of Helsinki and approved by the Ethics Committee of Polytechnic Institute of Santarém (252020Desporto).

Informed Consent Statement: Written informed consent was obtained from the participants to publish this paper.

Data Availability Statement: The data presented in this study are available on request from the corresponding author.

Acknowledgments: The authors would like to thank the team's coaches and players for their cooperation during all data collection procedures.

Conflicts of Interest: The authors declare no conflict of interest. The funders had no role in the design of the study; in the collection, analyses, or interpretation of data; in the writing of the manuscript, or in the decision to publish the results.

References

1. Haugen, T.A.; Tønnessen, E.; Seiler, S. Speed and countermovement-jump characteristics of elite female soccer players, 1995–2010. *Int. J. Sports Physiol. Perform.* **2012**, *7*, 340–349. [CrossRef] [PubMed]
2. Milanović, Z.; Sporiš, G.; James, N.; Trajković, N.; Ignjatović, A.; Sarmento, H.; Trecroci, A.; Mendes, B. Physiological demands, morphological characteristics, physical abilities and injuries of female soccer players. *J. Hum. Kinet.* **2017**, *60*, 77–83. [CrossRef] [PubMed]
3. FIFA. *Women's Football Member Associations Survey Report*; FIFA: Zurich, Switzerland, 2019.
4. Bradley, P.S.; Vescovi, J.D. Velocity Thresholds for Women's Soccer Matches: Sex Specificity Dictates High-Speed-Running and Sprinting Thresholds—Female Athletes in Motion (FAiM). *Int. J. Sports Physiol. Perform.* **2015**, *10*, 112–116. [CrossRef] [PubMed]

5. Bradley, P.S.; Dellal, A.; Mohr, M.; Castellano, J.; Wilkie, A. Gender differences in match performance characteristics of soccer players competing in the UEFA Champions League. *Hum. Mov. Sci.* **2014**, *33*, 159–171. [CrossRef]

6. Griffin, J.; Larsen, B.; Horan, S.; Keogh, J.; Dodd, K.; Andreatta, M.; Minahan, C. Women's Football: An Examination of Factors That Influence Movement Patterns. *J. Strength Cond. Res.* **2020**, *34*, 2384–2393. [CrossRef]

7. Gonçalves, L.; Clemente, F.M.; Barrera, J.I.; Sarmento, H.; Praça, G.M.; de Andrade, A.G.P.; Figueiredo, A.J.; Silva, R.; Silva, A.F.; Carral, J.M.C. Associations between physical status and training load in women soccer players. *Int. J. Environ. Res. Public Health* **2021**, *18*, 10015. [CrossRef]

8. Krustrup, P.; Zebis, M.; Jensen, J.M.; Mohr, M. Game-Induced Fatigue Patterns in Elite Female Soccer. *J. Strength Cond. Res.* **2010**, *24*, 437–441. [CrossRef]

9. Douchet, T.; Humbertclaude, A.; Cometti, C.; Paizis, C.; Babault, N. Quantifying accelerations and decelerations in elite women soccer players during regular in-season training as an index of training load. *Sports* **2021**, *9*, 109. [CrossRef]

10. Doyle, B.; Browne, D.; Horan, D. Quantification of internal and external training load during a training camp in senior international female footballers. *Sci. Med. Footb.* **2021**, *6*, 7–14. [CrossRef]

11. Duignan, C.; Doherty, C.; Caulfield, B.; Blake, C. Single-item self-report measures of team-sport athlete wellbeing and their relationship with training load: A systematic review. *J. Athl. Train.* **2020**, *55*, 944–953. [CrossRef]

12. Clemente, F.M.; Owen, A.; Serra-Olivares, J.; Nikolaidis, P.T.; Van Der Linden, C.M.I.; Mendes, B. Characterization of the Weekly External Load Profile of Professional Soccer Teams from Portugal and the Netherlands. *J. Hum. Kinet.* **2019**, *66*, 155–164. [CrossRef]

13. Oliveira, R.; Brito, J.P.; Loureiro, N.; Padinha, V.; Ferreira, B.; Mendes, B. Does the distribution of the weekly training load account for the match results of elite professional soccer players? *Physiol. Behav.* **2020**, *225*, 113118. [CrossRef]

14. Oliveira, R.; Brito, J.P.J.P.; Martins, A.; Mendes, B.; Marinho, D.A.D.A.; Ferraz, R.; Marques, M.C.M.C. In-season internal and external training load quantification of an elite European soccer team. *PLoS ONE* **2019**, *14*, e0209393. [CrossRef]

15. Malone, J.; Di Michele, R.; Morgans, R.; Burgess, D.; Morton, J.; Drust, B. Seasonal Training-Load Quantification in Elite English Premier League Soccer Players. *Int. J. Sports Physiol. Perform.* **2015**, *10*, 489–497. [CrossRef]

16. Fernandes, R.; Oliveira, R.; Martins, A.; Paulo Brito, J. Internal training and match load quantification of one-match week schedules in female first league Portugal soccer. *Cuad. Psicol. Deport.* **2021**, *21*, 126–138. [CrossRef]

17. Oliveira, R.; Brito, J.; Martins, A.; Mendes, B.; Calvete, F.; Carriço, S.; Ferraz, R.; Marques, M.C. In-season training load quantification of one-, two- and three-game week schedules in a top European professional soccer team. *Physiol. Behav.* **2019**, *201*, 146–156. [CrossRef]

18. Nobari, H.; Brito, J.P.; Pérez-Gómez, J.; Oliveira, R. Variability of External Intensity Comparisons between Official and Friendly Soccer Matches in Professional Male Players. *Healthcare* **2021**, *9*, 1708. [CrossRef]

19. Nobari, H.; Khalili, S.M.; Oliveira, R.; Castillo-Rodríguez, A.; Pérez-Gomez, J.; Ardigò, L. Comparison of Official and Friendly Matches through Acceleration, Deceleration and Metabolic Power Measures: A Full-Season Study in Professional Soccer Players. *Int. J. Environ. Res. Public Health* **2021**, *18*, 5980. [CrossRef]

20. Faul, F.; Erdfelder, E.; Lang, A.-G.; Buchner, A. G* power 3: A flexible statistical power analysis program for the social, Behavioral and biomedical sciences. *Behav. Res. Methods* **2007**, *39*, 175–191. [CrossRef]

21. Oliveira, R.; Francisco, R.; Fernandes, R.; Martins, A.; Nobari, H.; Clemente, F.M. In-Season Body Composition Effects in Professional Women Soccer Players. *Int. J. Environ. Res. Public Health* **2021**, *18*, 12023. [CrossRef]

22. Fernandes, R.; Brito, J.P.; Vieira, L.H.P.; Martins, A.D.; Clemente, F.M.; Nobari, H.; Reis, V.M.; Oliveira, R. In-season internal load and wellness variations in professional women soccer players: Comparisons between playing positions and status. *Int. J. Environ. Res. Public Health* **2021**, *18*, 12817. [CrossRef] [PubMed]

23. Silva, A.M.; Matias, C.N.; Santos, D.A.; Rocha, P.M.; Minderico, C.S.; Sardinha, L.B. Increases in intracellular water explain strength and power improvements over a season. *Int. J. Sports Med.* **2014**, *35*, 1101–1105. [CrossRef] [PubMed]

24. Foster, C.; Florhaug, J.A.; Franklin, J.; Gottschall, L.; Hrovatin, L.A.; Parker, S.; Doleshal, P.; Dodge, C. A New Approach to Monitoring Exercise Training. *J. Strength Cond. Res.* **2001**, *15*, 109–115. [CrossRef] [PubMed]

25. Foster, C.; Hector, L.L.; Welsh, R.; Schrager, M.; Green, M.A.; Snyder, A.C. Effects of specific versus cross-training on running performance. *Eur. J. Appl. Physiol. Occup. Physiol.* **1995**, *70*, 367–372. [CrossRef]

26. Hooper, S.L.; Mackinnon, L.T. Monitoring Overtraining in Athletes: Recommendations. *Sport. Med.* **1995**, *20*, 321–327. [CrossRef]

27. Scott, M.T.U.; Scott, T.J.; Kelly, V.G. The Validity and Reliability of Global Positioning Systems in Team Sport. *J. Strength Cond. Res.* **2016**, *30*, 1470–1490. [CrossRef]

28. Ravé, G.; Granacher, U.; Boullosa, D.; Hackney, A.C. How to Use Global Positioning Systems (GPS) Data to Monitor Training Load in the " Real World " of Elite Soccer. *Front. Physiol.* **2020**, *11*, 944. [CrossRef]

29. Romero-moraleda, B.; Nedergaard, N.J.; Morencos, E.; Ramirez-campillo, R.; Vanrenterghem, J.; Nedergaard, N.J.; Morencos, E. External and internal loads during the competitive season in professional female soccer players according to their playing position: Differences between training and competition. *Res. Sport. Med.* **2021**, *29*, 449–461. [CrossRef]

30. Hopkins, W.G.; Marshall, S.W.; Batterham, A.M.; Hanin, J. Progressive statistics for studies in sports medicine and exercise science. *Med. Sci. Sports Exerc.* **2009**, *41*, 3–13. [CrossRef]

31. Clemente, F.M.; Mendes, B.; da Bredt, S.G.T.; Praça, G.M.; Silvério, A.; Carriço, S.; Duarte, E. Perceived Training Load, Muscle Soreness, Stress, Fatigue, and Sleep Quality in Professional Basketball: A Full Season Study. *J. Hum. Kinet.* **2019**, *67*, 199–207. [CrossRef]

32. Ispirlidis, I.; Fatouros, I.G.; Jamurtas, A.Z.; Nikolaidis, M.G.; Michailidis, I.; Douroudos, I.; Margonis, K.; Chatzinikolaou, A.; Kalistratos, E.; Katrabasas, I.; et al. Time-course of Changes in Inflammatory and Performance Responses Following a Soccer Game. *Clin. J. Sport Med.* **2008**, *18*, 423–431. [CrossRef]
33. Haddad, M.; Chaouachi, A.; Wong, D.P.; Castagna, C.; Hambli, M.; Hue, O.; Chamari, K. Influence of fatigue, stress, muscle soreness and sleep on perceived exertion during submaximal effort. *Physiol. Behav.* **2013**, *119*, 185–189. [CrossRef]
34. Clemente, F.M.; Mendes, B.; Nikolaidis, P.T.; Calvete, F.; Carriço, S.; Owen, A.L. Internal training load and its longitudinal relationship with seasonal player wellness in elite professional soccer. *Physiol. Behav.* **2017**, *179*, 262–267. [CrossRef]
35. Trewin, J.; Meylan, C.; Varley, M.C.; Cronin, J. The match-to-match variation of match-running in elite female soccer. *J. Sci. Med. Sport* **2018**, *21*, 196–201. [CrossRef]
36. Akenhead, R.; Harley, J.A.; Tweddle, S.P. Examining the External Training Load of an English Premier League Football Team With Special Reference to Acceleration. *J. Strength Cond. Res.* **2016**, *30*, 2424–2432. [CrossRef]
37. Owen, A.L.; Wong, D.P.; Mckenna, M.; Dellal, A. Heart rate responses and technical comparison between small-vs. large-sided games in elite professional soccer. *J. Strength Cond. Res.* **2011**, *25*, 2104–2110. [CrossRef]
38. Stevens, T.G.A.; de Ruiter, C.J.; Twisk, J.W.R.; Savelsbergh, G.J.P.; Beek, P.J. Quantification of in-season training load relative to match load in professional Dutch Eredivisie football players. *Sci. Med. Footb.* **2017**, *1*, 117–125. [CrossRef]
39. Harper, D.J.; Carling, C.; Kiely, J. High-Intensity Acceleration and Deceleration Demands in Elite Team Sports Competitive Match Play: A Systematic Review and Meta-Analysis of Observational Studies. *Sport. Med.* **2019**, *49*, 1923–1947. [CrossRef]
40. Manzi, V.; D'Ottavio, S.; Impellizzeri, F.M.; Chaouachi, A.; Chamari, K.; Castagna, C. Profile of Weekly Training Load in Elite Male Professional Basketball Players. *J. Strength Cond. Res.* **2010**, *24*, 1399–1406. [CrossRef]
41. Fessi, M.S.; Zarrouk, N.; Di Salvo, V.; Filetti, C.; Barker, A.R.; Moalla, W. Effects of tapering on physical match activities in professional soccer players. *J. Sports Sci.* **2016**, *34*, 2189–2194. [CrossRef]
42. Teixeira, J.E.; Forte, P.; Ferraz, R.; Leal, M.; Ribeiro, J.; Silva, A.J.; Barbosa, T.M.; Monteiro, A.M. Monitoring accumulated training and match load in football: A systematic review. *Int. J. Environ. Res. Public Health* **2021**, *18*, 3906. [CrossRef]
43. Samsudeen, N.; Rajagopalan, A. Effect of different phases of menstrual cycle on cardio-respiratory efficiency in normal, overweight and obese female undergraduate students. *J. Clin. Diagn. Res.* **2016**, *10*, CC01–CC04. [CrossRef]
44. Chidi-Ogbolu, N.; Baar, K. Effect of estrogen on musculoskeletal performance and injury risk. *Front. Physiol.* **2019**, *10*, 1834. [CrossRef]
45. McNulty, K.L.; Elliott-Sale, K.J.; Dolan, E.; Swinton, P.A.; Ansdell, P.; Goodall, S.; Thomas, K.; Hicks, K.M. The Effects of Menstrual Cycle Phase on Exercise Performance in Eumenorrheic Women: A Systematic Review and Meta-Analysis. *Sport. Med.* **2020**, *50*, 1813–1827. [CrossRef]
46. Silva, A.F.; Oliveira, R.; Akyildiz, Z.; Yıldız, M.; Ocak, Y.; Günay, M.; Sarmento, H.; Marques, A.; Badicu, G.; Clemente, F.M. Sleep Quality and Training Intensity in Soccer Players: Exploring Weekly Variations and Relationships. *Appl. Sci.* **2022**, *12*, 2791. [CrossRef]

healthcare

Are Subjective Intensities Indicators of Player Load and Heart Rate in Physical Education?

Juan M. García-Ceberino [1,2,*], María G. Gamero [1,3], Sergio J. Ibáñez [1,4] and Sebastián Feu [1,3]

1 Optimization of Training and Sports Performance Research Group (GOERD), University of Extremadura, 10003 Cáceres, Spain; mgamerob@alumnos.unex.es (M.G.G.); sibanez@unex.es (S.J.I.); sfeu@unex.es (S.F.)
2 Faculty of Humanities and Social Sciences, University of Isabel I, 09003 Burgos, Spain
3 Faculty of Education, University of Extremadura, 06006 Badajoz, Spain
4 Faculty of Sports Science, University of Extremadura, 10003 Cáceres, Spain
* Correspondence: juanmanuel.garcia.ceberino@ui1.es

Abstract: Physical education teachers need valid, low-cost, subjective techniques as an alternative to high-cost new technologies to monitor students' intensity monitoring. This study aimed to investigate the correlations between both objective and subjective external (eTL) and internal (iTL) intensities. A total of 95 primary education students participated in this study. In this regard, 40 played soccer, and 55 performed basketball tasks, recording a total of 3956 units of analysis. The intensities caused by the different soccer and basketball tasks were measured using objective techniques (inertial devices and heart rate monitors) and subjective techniques (a sheet of task analysis and ratings of perceived exertion). Matrix scatter plots were made to show the values of two variables for a dataset. In this regard, adjustment lines were plotted to determine the trend of the correlations. Then, Spearman's correlation was calculated to measure the association between two variables. Despite the low correlation levels obtained, the main results showed significant positive correlations between the intensities. This means that the high intensity values recorded by objective techniques also implied high intensity values recorded by subjective techniques, and vice versa. Negative correlations (r Rho = -0.19; $p = 0.00$) were only found between the following eTL variables: task eTL per minute (subjective technique) and player load per minute (objective technique). This negative correlation occurred when students played in the same 3 vs. 3 game situation without variability in subjective eTL (M ± SD, 28.00 ± 0.00). Therefore, subjective eTL and iTL techniques could be proposed as a suitable alternative for planning and monitoring the intensities supported by students in physical education classes. Moreover, these subjective techniques are easy to use in schools.

Keywords: correlation; external intensity; inertial device; internal intensity; perceived exertion; SIATE

Citation: García-Ceberino, J.M.; Gamero, M.G.; Ibáñez, S.J.; Feu, S. Are Subjective Intensities Indicators of Player Load and Heart Rate in Physical Education? *Healthcare* **2022**, *10*, 428. https://doi.org/10.3390/healthcare10030428

Academic Editors: João Paulo Brito and Rafael Oliveira

Received: 25 January 2022
Accepted: 23 February 2022
Published: 24 February 2022

Publisher's Note: MDPI stays neutral with regard to jurisdictional claims in published maps and institutional affiliations.

1. Introduction

The planning and quantification of intensities are important for optimizing students´ physical fitness, as well as for achieving teaching objectives [1,2]. However, it is a technique rarely used by physical education teachers in primary education [3]. This may be due to poor teacher training on intensity planning and monitoring in physical education classes [4]. In this regard, teachers usually aim for a higher time of motor commitment (useful time) for the learning tasks, but they do not take into account whether the demands of these tasks are high or low. Therefore, adequate learning-task planning, using different pedagogical, organizational (time-related), and subjective intensity variables, should be performed to achieve the recommended levels of physical fitness [5–7].

In physical education classes, students should spend at least 50% of their time recording moderate to vigorous physical activities for adequate cardiovascular work in order to prevent diseases such as weight gain and obesity [6]. In this regard, García-Ceberino et al. [7] proposed a list of recommendations for students to spend more than 50% of the class in moderate to vigorous physical activity. Similarly, the World Health Organization [8] recommends spending at least 60 min per day in moderate to intense physical activity, mainly aerobic, throughout the week (for populations aged 5 to 17 years). In primary schools, subjective techniques can help teachers analyze whether sessions achieve recommended physical activity levels.

Intensity monitoring (this replaces the term "workload" [9–11]) encompasses the psychological and biological demands (internal intensity, iTL) caused by tasks or competition (external intensity, eTL). For example, the modification of game spaces or number of participants (eTL) affect the iTL of players. Thus, intensity quantification can be obtained using the eTL and iTL dimensions [12,13]. The eTL is the mechanical and locomotor stress produced by a physical activity (physical demands) [14,15]. The iTL is the physiological reaction (heart rate, HR) and stress experienced by a stimulus (eTL) [16]. HR indicates the intensity of physical activity (physiological demands) [17]. The iTL is individual and specific to each subject. A learning task, planned with the same eTL, may cause different HR values in the subjects, and these HR values will be appropriate for some subjects and inappropriate for others [18].

There are different instruments available for measuring, objectively and subjectively, the eTL and iTL. In this regard, inertial motion devices that integrate a multitude of sensors (such as accelerometers, gyroscopes, magnetometers, and GPS, among others) are used to quantify objective eTL. Among the variables recorded by these inertial devices, the most used and predictive of eTL is the Player Load (PL) [14,19]. The PL is the vectorial sum of the accelerations in three axes (vertical, anteroposterior, and lateral), and it measures the neuromuscular eTL [20]. Currently, new technologies are not accessible to all sports professionals due to their high cost [13], making it difficult to quantify these intensities. Likewise, smartphones could be integrated as an objective technique. However, many Spanish state schools prohibit their use during school periods. (This decision is made by each school.)

Faced with these problems, Ibáñez et al. [21] proposed an Integral Analysis System of Training Tasks, SIATE (Spanish acronym). The SIATE is an observational and categorization sheet, which makes it possible to quantify the subjective eTL (actions prior to sports practice). It is obtained through the sum of the categorical-ordinal values (1 to 5) assigned to six variables when categorizing learning tasks: degree of opposition, task density, percentage of simultaneous performers, competitive load, game space, and cognitive involvement. Thus, the SIATE allows the sports professionals to discover the factors that affect the sports teaching process using pedagogical, organizational, and subjective external intensity variables.

In addition, the HR (objective iTL) is measured through HR monitors that are synchronized with inertial devices [13]. The use of HR as a measure to determine the intensity of physical activity is well defined, validated, and accepted because it is a simple and non-invasive technique [12]. Its economic cost can also be high. For this reason, other types of subjective iTL instruments have appeared.

As an alternative, there is the Rating of Perceived Exertion (RPE) (psycho-physiological demands, subjective iTL). In the sports field, the most commonly used instrument is Borg's RPE scale. The subjects indicate how tired they are by means of this scale, thus defining the intensity of the physical activity [22]. Eston and Parfitt [23] also designed a curvilinear pictorial scale (with graphic illustrations) representing the degree of perceived effort. These authors adapted Borg's RPE scale for the child population. Therefore, subjective techniques present an alternative because of their low cost, accessibility, and ease of use [24].

The different intensities and measuring instruments mentioned above [1] are detailed in Figure 1.

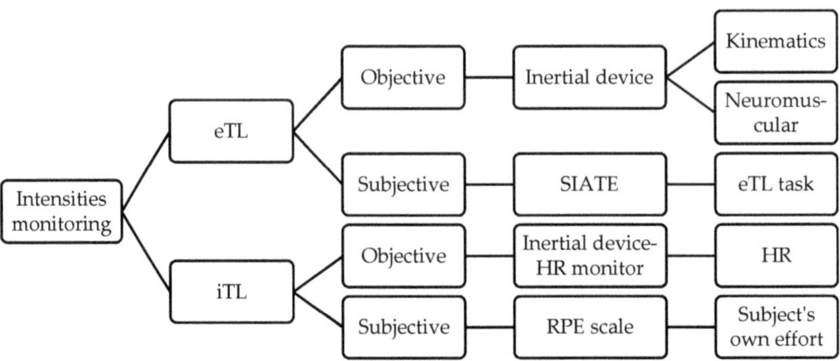

Figure 1. Classification of the intensities and measuring instruments according to García-Ceberino et al. [1]. Note: eTL = External Intensity; SIATE = Integral Analysis System of Training Tasks; iTL = Internal Intensity; HR = Heart Rate; RPE = Rating of Perceived Exertion.

After analyzing the scientific literature and instruments available to sports professionals to control and monitor sports practices, it is necessary to study their correlation because many of these instruments are not accessible to everyone. In women's basketball training, Reina et al. [12] investigated the quantification of intensities using three instruments (WIMU^{TM} inertial devices, GARMIN^{TM} HR monitors, and the SIATE sheet) to establish correlations among them. They confirmed a direct correlation between the intensities obtained by the subjective categorization of training tasks (SIATE) with the PL and HR intensities provided by objective techniques. In soccer training, Gómez-Carmona et al. [13] also confirmed that subjective intensity variables influence objective intensity variables, finding a strong correlation between them. In addition, a combination of objective eTL and iTL (HR) factors predicted RPE in rugby training [25].

To our knowledge, the correlation between the intensities recorded by objective and subjective techniques has not been studied in the school context. Therefore, this study aimed to investigate the correlations between both objective and subjective intensities (eTL and iTL) in order to analyze the reliability of subjective techniques with respect to objective techniques. We hypothesized that correlations would be found between the following variables: (A) the objective (inertial device) and subjective (SIATE sheet) eTL variables; (B) the objective (HR) and subjective (RPE) iTL variables; and (C) the objective and subjective intensities studied.

2. Materials and Methods

2.1. Study Design

A correlational study was designed to analyze the relationships between variables in order to identify a categorical variable [26]. In turn, it was subdivided into two studies: (1) Study 1 aimed to analyze the correlations between intensities caused by different types of tasks (i.e., without opposition, individual game, small-sided games—SSGs, and full games), such that each task type implied a variability in the subjective eTL (see Table 1). (2) Study 2 aimed to analyze the correlations between intensities caused by 3 vs. 3 game situations involving the same subjective eTL (see Table 1). Study 2 also differed from Study 1 because it measured the students' RPE. Consequently, the study was ecological and aimed to analyze the reliability of the subjective techniques with respect to the objective ones.

2.2. Sample

The basic unit of analysis was the record of students in each learning task type performed during nine soccer and basketball sessions (Study 1) and in two 3 vs. 3 sessions (Study 2), obtaining 3305 and 651 records, respectively. Table 1 describes the learning tasks for each study and sport. The subjective eTL of the learning tasks was always known to the teachers during the design period, i.e., before applying them in the physical education classes.

Table 1. Description of the learning tasks (by study and sport).

Study	Task Type	Example	Soccer		Basketball	
			%	eTL_{avg}	%	eTL_{avg}
Study 1 Tasks with different subjective eTL	Without Opposition	1 vs. 0 …	39.90	10.85	56.50	12.03
	Individual Game	1 vs. 1	28.30	17.83	15.70	19.52
	Inequality SSG	2 vs. 1 …	25.10	20.84	17.80	19.17
	Equality SSG	2 vs. 2 …	1.80	17.00	5.70	21.48
	Full Game	5 vs. 5	4.90	28.64	4.30	24.66
Study 2 3 vs. 3 with same subjective eTL	Equality SSG	3 vs. 3	100.00	28	100.00	28.00

Note: eTL = External Intensity; avg = Average; SSG = Small-Sided Game.

The learning tasks were played by a total of 95 primary education students from two state schools. The fifth-grade students (school 1) played soccer, while the sixth-grade students (school 2) played basketball. The soccer [27,28] and basketball [29,30] tasks are valid and reliable for application in physical education classes. The characteristics of the students are described in Table 2.

Table 2. Characteristics of the students participating (by study).

Demographic Data	Study 1 Tasks with Different Subjective eTL		Study 2 3 vs. 3 with Same Subjective eTL	
School, grade	School 1, 5th PE	School 2, 6th PE	School 1, 5th PE	School 2, 6th PE
Students, girls	40, 18 girls	55, 32 girls	33, 16 girls	48, 25 girls
Years (M ± SD)	10.65 ± 0.48	11.09 ± 0.29	10.67 ± 0.48	11.10 ± 0.31

Note: M = Mean; SD = Standard Deviation; eTL = External Intensity; PE = Primary Education.

All the students who participated in at least 80% of the soccer and basketball sessions were selected as the study sample in Study 1. Of these students, those who participated in the two soccer and basketball 3 vs. 3 sessions were included in Study 2. Parents or legal guardians were required to sign an informed consent form. Similarly, the study was conducted in accordance with the ethical guidelines of the Helsinki Declaration of 1975 (with modifications in subsequent years) and Organic Law 3/2018, of December 5, on the protection of personal research data and the guarantee of digital rights (BOE, 294, 6 December 2018), to fulfil the ethical considerations of scientific research with human beings.

2.3. Variables and Instruments

The variables analyzed in both studies were the intensities recorded in physical education classes, grouped into eTL and iTL variables:

-   Objective eTL variables: (1) PL; and (2) PL per minute (PL/min). These neuromuscular eTL variables were measured using WIMU Pro™ inertial devices (RealTrack System, Almería, Spain). PL was measured only during the time of motor commitment in order to eliminate distorting values, such as PL during rest periods.
-   Subjective eTL variables: (1) density of task, categorical–ordinal variable with five levels: 1-walking; 2-gentle pace; 3-intensity with rest; 4-intensity without rest; and 5-high intensity without rest. In Study 1, the variability of the learning tasks applied involved a different level for each task type. In contrast, the 3 vs. 3 game situations involved only one level in Study 2. (2) eTL, obtained by the sum of the values (1 to 5) given to six categorical–ordinal variables when categorizing tasks: degree of

opposition, task density, percentage of simultaneous performers, competitive load, game space, and cognitive involvement. Thus, the eTL value for each learning task ranges from 6 to 30. (3) eTL*minute (eTL*min): these variables were measured through the SIATE observation sheet [21].

- Objective iTL variables: (1) average HR (HR_{avg}); and (2) maximum HR (HR_{max}). These were measured with GARMIN^{TM} HR monitors (Garmin Ltd., Olathe, KS, USA), synchronized with the above-mentioned inertial devices through Ant+ technology [31]. HR was also measured only during the time of motor commitment.
- Only in study 2 was subjective iTL (psycho-physiological demands) measured using the curvilinear pictorial scale with graphic illustrations [23], which represents the RPE.

Table 3 summarizes the variables and measurement instruments applied.

Table 3. Summary of study variables and instruments.

Intensity	Variable	Unit	Description	Instrument
eTL (objective)	PL PL/min	Arbitrary units (per min)	Neuromuscular eTL resulting from accelerations	WIMU Pro^{TM}
eTL (subjective)	Task density	Scale 1 to 5	Intensity of the learning task	SIATE observation sheet
	Task eTL eTL*min	Number 6 to 30 (per min)	Intensity resulting from the sum of six categorical variables	
iTL (objective)	HR_{avg} HR_{max}	Beats per minute	Number (average/maximum) of beats per minute	GARMIN^{TM} monitors
iTL (subjective)	RPE	Scale 1 to 10	Perception of one's own effort	CPS (graphics)

Note: eTL = External Intensity; PL = Player Load; min = Minute; SIATE = Integral Analysis System of Training Tasks; iTL = Internal Intensity; HR = Heart Rate; avg = Average; max = Maximum; RPE = Ratings of Perceived Exertion; CPS = Curvilinear Pictorial Scale.

2.4. Procedure

First, it was necessary to obtain a series of authorizations: (1) approval from the University Bioethics Committee [Ref. 247/2019]; (2) authorization from the schools and physical education teachers; (3) approval from the school council to include the research in the curriculum of the schools; and (4) written informed consent from the parents or legal guardians.

Study 1. The students played soccer (school 1) and basketball (school 2). All learning tasks (n = 180) applied in both invasion sports ranged from simple (e.g., 1 vs. 0, 1 vs. 1, …) to more complex (e.g., 3 vs. 3, 4 vs. 4, …) [27,29]. These tasks were validated by experts in the field of Sport Pedagogy [28,30]. The teaching progressed based on the variability and difficulty of the tasks. The variability of the tasks implied that each one of them caused a different subjective eTL. When designing the learning tasks using the SIATE observation sheet [21], the physical education teachers were aware of their subjective eTLs before implementing them. The interventions lasted nine teaching sessions with one or two sessions per week (1 h per session). These were conducted on the school's outdoor sports field. In this regard, the school authorities indicated the days on which the teachers could teach these invasion sports in the physical education classes. The intensity data from the soccer and basketball sessions were pooled for Study 1.

Study 2. The students played 3 vs. 3 matches with three players per team (n = 120), performing a five-minute warm-up. Each of the teams played against each of the other teams in this game situation. All matches were played on mini-fields. A fair balance between the teams was sought, and the teams were mixed by gender and experience. In all matches, the 3 vs. 3 game situation involved the same subjective eTL known to the teachers.

A 3 vs. 3 session was performed for each sport, pre- and post-intervention (Study 1). In the soccer matches, there was no goalkeeper (with mini-goals) because their demands differ from those of field players [32]. The RPE was measured just after each match. For Study 2, the intensity data from all 3 vs. 3 matches were pooled.

In both studies, the physical education teachers (researchers) directed the soccer and basketball sessions, and then collected and analyzed the data. They had academic and professional training in these invasion sports. Inertial devices were placed using anatomical harnesses at the beginning of each session. In addition, HR monitoring with the GARMIN^{TM} device was chest-based. The data recorded from these devices were exported to the SPRO^{TM} software (RealTrack System, Almería, Spain) for quantification and statistical analysis. This methodological procedure was carried out by both researchers equally so as to guarantee the reliability and validity of the data, as suggested by Murillo-Lorente et al. [33].

Figure 2 presents the study procedure.

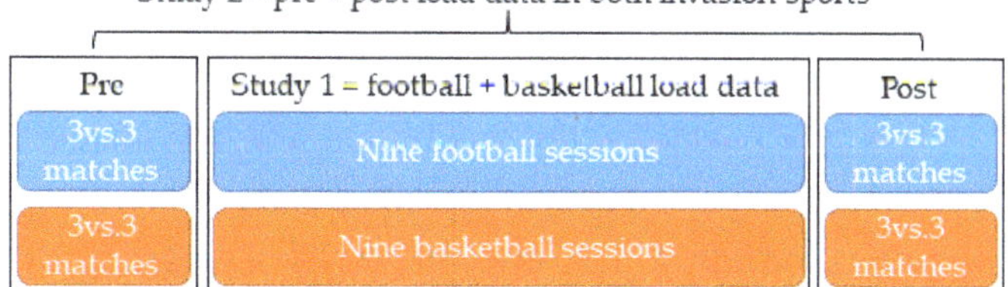

Figure 2. Procedure for both studies.

2.5. Statistical Analyses

First, the criterion assumption tests were performed to determine the use of parametric/non-parametric tests for hypothesis testing [34]. The Kolmogorov-Smirnov, Levene, and Rachas tests indicated the use of non-parametric tests.

Then, scatter plots were used to show the values of two variables for a dataset. Specifically, matrix scatter plots were used for each study. The scatter plot shows the correlation (non-causality) between the two variables. It can be positive (increase), negative (decrease) or null (the two variables are not correlated) [34]. In this regard, adjustment lines were plotted to determine the trend of the correlations.

Finally, Spearman's correlation (Spearman's Rho) for non-parametric tests was performed. It measures association between two variables and values from -1 to +1, indicating negative or positive associations, respectively. A positive association indicates that high values of one variable are correlated with high values of the other variable, and vice versa. On the other hand, a negative association indicates that high values of one variable are correlated with low values of the other variable, and vice versa [34]. The following interpretation scale determines the level of association: null = 0.00; between null and low = 0.00–0.25 (−0.25); low = 0.26 (−0.26)–0.50 (−0.50); between moderate and strong = 0.51 (−0.51)–0.75 (−0.75); and, between strong and perfect = 0.76 (−0.76)–1.00 (−1.00) [35].

The same scale was always used for each instrument. Therefore, Spearman's Rho was not affected by changes in the units of measurement [35].

Graphics and statistical analysis were performed with SPSS 25.0 (Released 2017. IBM SPSS Statistics for Windows, Version 25. IBM Corp., Armonk, NY, USA).

3. Results

Study 1. Figure 3 shows the matrix scatter plots between the intensities studied (in pairs). These indicate that the possible correlations follow a positive/increase trend.

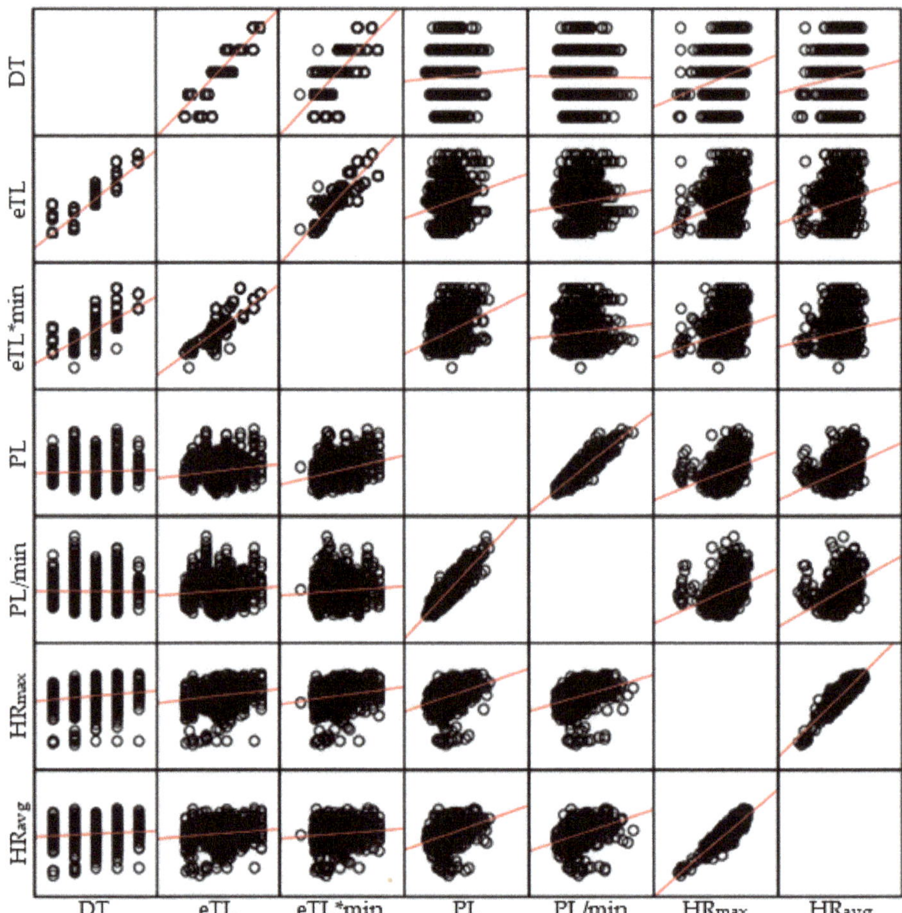

Figure 3. Matrix scatter plots between the intensities studied (in pairs) in the interventions. Note: eTL = External Intensity; TD = Task Density; PL = Player Load; min = Minute; iTL = Internal Intensity; HR = Heart Rate; avg = Average; max = Maximum.

Table 4 shows the correlation results between the objective eTL, the subjective eTL, and the objective iTL variables. The objective intensities (eTL and iTL) were recorded during the application of the intervention programs. In contrast, the design of the learning tasks prior to their application implies a variability of subjective eTL ($M \pm SD$, 17.15 ± 4.93, minimum 8–maximum 30). The main results show significant positive correlations between all the intensity variables studied, both objective and subjective. Therefore, high values of the subjective eTL variables also indicate high values of the objective eTL and iTL variables, and vice versa. Moreover, the moderate to strong correlation between the PL/min variable (eTL) and the HR_{avg} and HR_{max} variables (iTL) is noteworthy. This level of correlation is similar to that between PL and HR_{max} variables.

Study 2. The matrix scatter plots between the intensities studied (in pairs) are shown in Figure 4. These indicate that the possible correlations follow a positive/increase trend, except for the variables: eTL*min and PL/min (negative correlation).

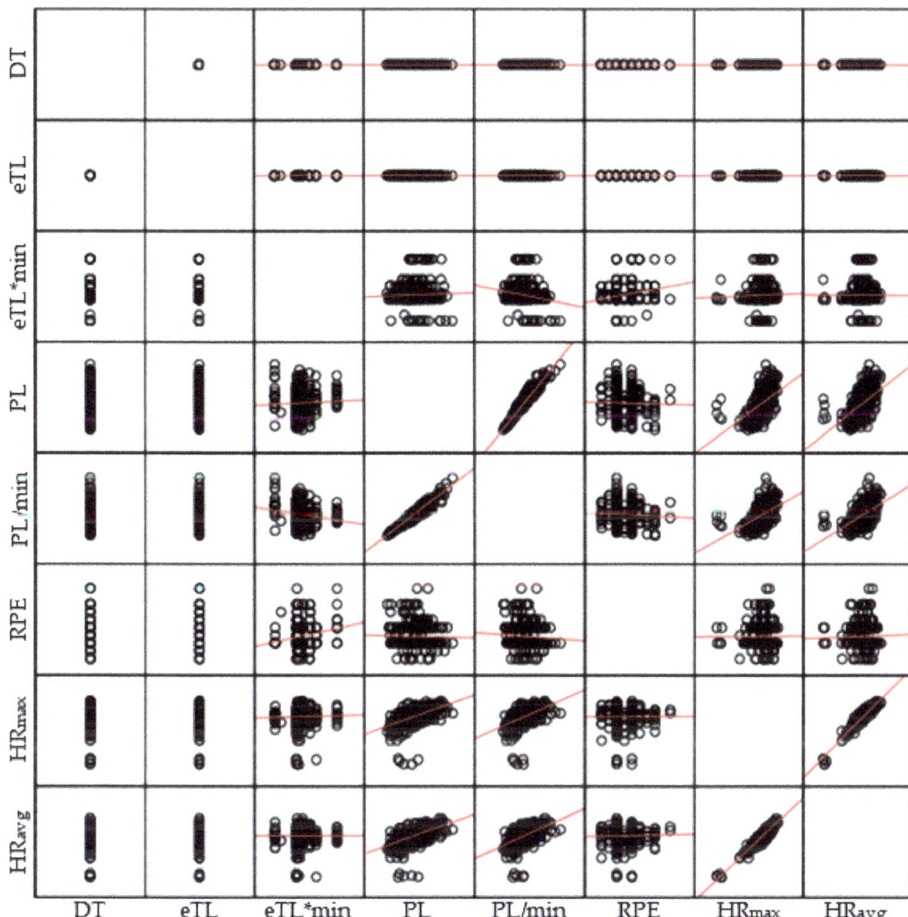

Figure 4. Matrix scatter plots between the intensities studied (in pairs) in the 3 vs. 3 matches. Note: eTL = External Intensity; TD = Task Density, PL – Player Load, min – Minute, ITL = Internal Intensity; HR = Heart Rate; avg = Average; max = Maximum.

Table 5 shows the correlation analysis between all of the objective and subjective intensities studied. The 3 vs. 3 session design prior to the application implies the same subjective eTL ($M \pm SD$, 28.00 ± 0.00). Although the correlation coefficient associated with the RPE is low, the most salient results indicate significant positive correlations between the RPE and objective variables of eTL (PL and PL/min) and iTL (HR_{avg}). Thus, high values of the RPE also indicate high values of the objective eTL and iTL variables mentioned, and vice versa. PL continues to show moderate to strong correlation with HR.

Finally, Table 6 shows an equivalence of the iTL variables compared to Buceta's values [36]. Thus, students should be trained before using the RPE pictorial scale [23] because they tend to show lower values than the real ones.

Table 4. Correlation analysis between the intensities recorded when applying the programs.

Intensity	Spearman's Rho	Objective iTL		Objective eTL		Subjective eTL	
		HR_{avg}	HR_{max}	PL/min	PL	eTL*min	eTL
TD	r	0.26 **	0.31 **	0.23 **	0.30 **	0.78 **	0.87 **
	p	0.00	0.00	0.00	0.00	0.00	0.00
	n	3205	3175	3273	3273	1088	1480
eTL	r	0.26 **	0.33 **	0.24 **	0.36 **	0.89 **	
	p	0.00	0.00	0.00	0.00	0.00	
	n	3205	3175	3273	3273	1088	
eTL*min	r	0.12 **	0.22 **	0.09 *	0.27 **		
	p	0.00	0.00	0.02	0.00		
	n	1031	1031	1088	1088		
PL	r	0.47 **	0.53 **	0.66 **			
	p	0.00	0.00	0.00			
	n	3175	3175	3273			
PL/min	r	0.57 **	0.51 **				
	p	0.00	0.00				
	n	3175	3175				
HR_{max}	r	0.88 **					
	p	0.00					
	n	3175					

Note: n = Cases Analyzed; r = Spearman's Correlation; eTL = External Intensity; TD = Task Density; PL = Player Load; min = Minute; iTL = Internal Intensity; HR = Heart Rate; avg = Average; max = Maximum; ** Significant correlation at the 0.01 level; * Significant correlation at the 0.05 level.

Table 5. Correlation analysis between the intensities recorded when apply to the 3 vs. 3 matches.

Intensity	Spearman's Rho	Ob. iTL		Sub. iTL	Ob. eTL		Sub. eTL	
		HR_{avg}	HR_{max}	RPE	PL/min	PL	eTL/min	eTL
TD	r	0.15 **	0.11 **	-	0.04	0.06	0.60 **	1.0 **
	p	0.00	0.00	-	0.27	0.09	0.00	-
	n	711	711	674	722	722	400	728
eTL	r	0.15 **	0.11 **	-	0.04	0.06	0.60 **	
	p	0.00	0.00	-	0.27	0.09	0.00	
	n	711	711	674	722	722	400	
eTL*min	r	0.12 *	0.13 **	0.08	−0.19 **	−0.08		
	p	0.02	0.01	0.15	0.00	0.13		
	n	390	390	346	400	400		
PL	r	0.61 **	0.54 **	0.21 **	0.94 **			
	p	0.00	0.00	0.00	0.00			
	n	805	805	668	817			
PL/min	r	0.57 **	0.53 **	0.15 **				
	p	0.00	0.00	0.00				
	n	805	805	668				
RPE	r	0.14 **	0.08					
	p	0.00	0.05					
	n	658	658					
HR_{max}	r	0.91 **						
	p	0.00						
	n	805						

Note: n = Cases Analyzed; r = Spearman's Correlation; Ob. = Objective; Sub. = Subjective; eTL = External Intensity; TD = Task Density; PL = Player Load; min = Minute; iTL = Internal Intensity; HR = Heart Rate; avg = Average; max = Maximum; RPE = Ratings of Perceived Exertion; ** Significant correlation at the 0.01 level; * Significant correlation at the 0.05 level.

Table 6. Equivalence of the iTL variables compared to Buceta's values [36].

HR$_{avg}$	HR$_{max}$ M(SD)	RPE	Borg Scale Equivalence	Approximate in bpm	Degree of Stress Intensity (% of max. Capacity)
170.51 (22.18)	188.33 (21.37)	3.40 (1.86)	Fairly light	110–130	30

Note: M = Mean; SD = Standard Deviation; iTL = Internal Intensity; HR = Heart Rate; avg = Average; max = Maximum; RPE = Ratings of Perceived Exertion; bpm = Beats per Minute.

4. Discussion

In sports practices, more accessible and manageable devices would be necessary, to which not all teachers have access. The study aimed to investigate the correlations between both objective and subjective intensities (eTL and iTL) to analyze the reliability of subjective techniques with respect to objective techniques. For this purpose, it was necessary to quantify the intensities caused by different soccer and basketball tasks in primary education (5th and 6th grades). In general, despite the low levels obtained, the main findings indicated that the significant correlations recorded between the intensities were mostly positive/increasing. Thus, high intensity values recorded by objective techniques also mean high intensity values recorded by subjective techniques, and vice versa.

Study 1 (intensities refer to intervention programs). There were significant positive correlations, with an increasing trend, between the eTL variables recorded using objective and subjective techniques. In the sports training context, Gómez-Carmona et al. [13] found a very strong correlation between the objective (WIMU Pro^{TM} devices) and subjective (SIATE observation sheet) eTL variables when monitoring U–19 soccer players. Also, Reina et al. [12] indicated that there was a correlation between the values obtained by both eTL measurement techniques when monitoring players from a women's senior basketball team. Therefore, it could be stated that the subjective eTL variables, used when designing the learning tasks, influence the objective eTL variables recorded when these tasks are later applied in sports play.

Likewise, the objective iTL (HR) variables correlated positively with all eTL variables, both objective and subjective. In a previous correlational study [12], a direct correlation was confirmed between the intensities recorded by subjective categorization of basketball tasks and the PL and HR recorded using objective techniques.

Movement intensity and HR analysis are traditional techniques used by sports professionals to measure the physical [14] and physiological [17] demands in invasion sports, respectively. Thus, the SIATE observation sheet [21] is proposed as a subjective technique for monitoring eTL without the need for employing expensive and complex technologies to use. In turn, this subjective eTL technique is known to correlate with HR [12,13], and it has been employed by different researchers for the planning of soccer [5,27,37] and basketball [38–40] sessions in physical education. The highest correlations (moderate to strong) were recorded between the PL and HR variables. Thus, the higher the accelerometry, the higher the task intensity, and vice versa.

In conclusion, when learning tasks causing different subjective eTLs were employed, the hypotheses A and C were partially fulfilled (Table 4).

Study 2 (intensities refer to 3 vs. 3 sessions). The RPE (psycho-physiological demand) can provide an adequate estimate of real HR, allowing us to know the intensity during sports play [22]. In terms of iTL, despite the low values obtained, the main results indicated significant positive correlations between the RPE and HR$_{avg}$. In this regard, previous studies in the context of sports training have established a high correlation between both measurement techniques, and they indicate the use of RPE as a valid method to monitor exercise in the absence of other costly and sophisticated tools [41,42], e.g., HR monitors. Alexiou and Coutts [43] suggested that the RPE correlates better with the HR in less intermittent sessions working on aerobic endurance when analyzing elite women's soccer players. This may be because HR$_{max}$ and RPE are lower in aerobic sessions [33]. In this regard, the type of task (teaching means) has a direct effect on kinematic, neuromuscular, and internal demands [7,13] and, together with the measurement technique used, can condition the magnitude and uncertainty of the associations [44].

Furthermore, the RPE correlated positively with the objective variables of PL and PL/min. In line with the results described above, Lovell et al. [25] stated that a combination of objective eTL and iTL (HR) factors predicted the RPE in professional rugby players. Likewise, McLaren et al. [44] stated that RPE and HR showed positive associations with eTL derived from running and accelerometry in team sports. PL continues to show moderate to strong correlation with HR. Therefore, in this study, there was an increasing trend in the values recorded through objective and subjective techniques. It is necessary to previously train students in the use of the RPE pictorial scale [23] in physical education. This subjective iTL technique is very popular among sports professionals due to its easy accessibility, simplicity, and accuracy [45]. Despite this, the data obtained can be compromised by the interaction of multiple factors [24], e.g., the psychological perspective [33], or previous experience in its use. In addition, Borresen and Lambert [41] observed that the data are not as accurate when proportionally more time is spent training at low or high intensity, compared to HR.

Another relevant result was the negative correlation between eTL*min and PL/min variables. Thus, high values of eTL*min indicated low values of PL/min, and vice versa. These results are in line with those obtained in previous studies on school soccer [1,7]. The tasks involving playing situations, with the simultaneous participation of the players and in reduced game-spaces, invoke a medium-high subjective eTL [27] and high intensity [7], while the PL recorded is low because these tasks cause few accelerations as players cover longer distances at more constant speeds [46,47]. However, the tasks without opposition and with consecutive participation (organization in rows) of the players involve a medium-low subjective eTL [27], while the accelerations and PL recorded are higher than those obtained in playing situations [1]. This type of task is performed repeatedly and with short breaks. The players move a few meters, accelerating to the maximum from the beginning of the tasks to decelerating quickly [48]. For this reason, it is important to pay attention to the design of the learning tasks used in physical education because it subsequently influences the physical and physiological demands supported by the students.

In conclusion, when the same learning tasks involving equal subjective eTL were employed, hypotheses B and C were partially fulfilled (Table 5).

5. Limitations and Practical Applications

A study limitation was the sample size, as well as not having trained the students in the use of the RPE scale. Although no strong correlations were recorded, significant and positive correlations were found, indicating that the intensity values increased or decreased equally throughout the soccer and basketball sessions.

Subjective techniques (SIATE observation sheet and RPE scale) are a suitable alternative because they help teachers to know the impact of learning tasks on students and to improve their physical fitness, since high objective values imply high subjective values, and vice versa. In addition, subjective techniques can easily be used in physical education, and they have low economic costs. However, further studies are needed to confirm the homogeneity of the objective and subjective measures.

6. Conclusions

The design of the learning tasks and knowledge of their subjective external intensities provide feedback on the intensities that will subsequently occur when applying them: the neuromuscular and kinematic intensity variables, the heart rate, and the ratings of perceived exertion. In this way, the proposed fitness physical and health objectives indicated in the Spanish curriculum and scientific literature could be met, i.e., $\geq 50\%$ of class time students should spend in moderate to vigorous physical activity for adequate cardiovascular work and health improvement.

This study could confirm the usefulness of the SIATE observation sheet and RPE as subjective techniques for the prediction of objective intensities in physical education classes since subjective measures provide a significant (not strong) correlation when compared to an objective standard. Also, these subjective techniques can easily be used by teachers. Knowledge of subjective measures will allow teachers to optimize the teaching–learning process. It is recommended for further studies that students have previous experience in the use of the RPE scale in school. Furthermore, teachers must be trained in the use of the SIATE observation sheet.

Author Contributions: Conceptualization, J.M.G.-C., M.G.G., and S.F.; methodology, J.M.G.-C., M.G.G., and S.F.; formal analysis, J.M.G.-C.; writing—original draft preparation, J.M.G.-C. and M.G.G.; writing—review and editing, S.J.I. and S.F.; visualization, J.M.G.-C. and M.G.G.; supervision, S.J.I. and S.F. All authors have read and agreed to the published version of the manuscript.

Funding: This study has been partially subsidized by the Aid for Research Groups (GR21149) from the Regional Government of Extremadura (Department of Economy, Science and Digital Agenda), with a contribution from the European Union from the European Funds for Regional Development.

Institutional Review Board Statement: The study was conducted in accordance with the Declaration of Helsinki and approved by the ethics committee of the University of Extremadura (protocol code 247/2019; 18 December 2019).

Informed Consent Statement: Written informed consent was obtained from parents or legal guardians of the students who were involved in the study.

Data Availability Statement: Data will be available upon reasonable request to the corresponding author.

Acknowledgments: The authors thank the academic authorities of the two schools, the physical education teachers, and the students for their participation in the research.

Conflicts of Interest: The authors declare no conflict of interest.

References

1. García-Ceberino, J.M.; Antúnez, A.; Feu, S.; Ibáñez, S.J. Quantification of Internal and External Load in School Football According to Gender and Teaching Methodology. *Int. J. Environ. Res. Public Health* **2020**, *17*, 344. [CrossRef] [PubMed]
2. González-Espinosa, S.; Antúnez, A.; Feu, S.; Ibáñez, S.J. Monitoring the External and Internal Load under 2 Teaching Methodologies. *J. Strength Cond. Res.* **2020**, *34*, 2920–2928. [CrossRef] [PubMed]
3. García-Ceberino, J.M.; Gamero, M.G.; González-Espinosa, S.; García-Rubio, J.; Feu, S. Study of the external training load of tasks for the teaching of handball in pre-service teachers according to their genre. *E-Balonamo Com* **2018**, *14*, 45–54.
4. Sierra, R. Formación docente para el control de la carga en la clase de educación física. *Rev. Investig. Educ.* **2005**, *2*, 33–48.
5. García-Ceberino, J.M.; Feu, S.; Gamero, M.G.; Ibáñez, S.J. Pedagogical Variables and Motor Commitment in the Planning of Invasion Sports in Primary Education. *Sustainability* **2021**, *13*, 4529. [CrossRef]
6. Aznar, S.; Webster, T. *Actividad Física y Salud en la Infancia y la Adolescencia. Guía para Todas las Personas que Participan en su Educación*; Ministerio de Educacion y Cultura, Centro de Investigación y Documentación Educativa: Madrid, Spain, 2009.
7. García-Ceberino, J.M.; Feu, S.; Antúnez, A.; Ibáñez, S. Organization of Students and Total Task Time: External and Internal Load Recorded during Motor Activity. *Appl. Sci.* **2021**, *11*, 10940. [CrossRef]
8. World Organization Health. Physical Activity. Available online: https://www.who.int/news-room/fact-sheets/detail/physical-activity (accessed on 20 January 2022).
9. Oliveira, R.; Brito, J.P.; Moreno-Villanueva, A.; Nalha, M.; Rico González, M.; Clemente, F.M. Reference Values for External and Internal Training Intensity Monitoring in Young Male Soccer Players: A Systematic Review. *Healthcare* **2021**, *9*, 1567. [CrossRef]
10. Nobari, H.; Brito, J.P.; Pérez-Gómez, J.; Oliveira, R. Variability of External Intensity Comparisons between Official and Friendly Soccer Matches in Professional Male Players. *Healthcare* **2021**, *9*, 1708. [CrossRef]
11. Staunton, C.A.; Abt, G.; Weaving, D.; Wundersitz, D.W.T. Misuse of the term 'load' in sport and exercise science. *J. Sci. Med. Sport* **2021**, *in press*. [CrossRef]
12. Reina, M.; Mancha-Triguero, D.; García-Santos, D.; García-Rubio, J.; Ibáñez, S.J. Comparación de tres métodos de cuantificación de la carga de entrenamiento en baloncesto. *RICYDE Rev. Int. Cienc. Deporte* **2019**, *15*, 368–382. [CrossRef]
13. Gómez-Carmona, C.D.; Gamonales, J.M.; Feu, S.; Ibáñez, S.J. Estudio de la carga interna y externa a través de diferentes instrumentos. Un estudio de casos en fútbol formativo. *Sport. Sci. J.* **2019**, *5*, 444–468. [CrossRef]
14. Gómez-Carmona, C.D.; Bastida-Castillo, A.; Ibáñez, S.J.; Pino-Ortega, J. Accelerometry as a method for external workload monitoring in invasion team sports. A systematic review. *PLoS ONE* **2020**, *15*, e0236643. [CrossRef] [PubMed]

15. Buchheit, M.; Lacome, M.; Cholley, Y.; Simpson, B. Neuromuscular Responses to Conditioned Soccer Sessions Assessed via GPS-Embedded Accelerometers: Insights into Tactical Periodization. *Int. J. Sports Physiol. Perform.* **2018**, *13*, 577–583. [CrossRef] [PubMed]

16. Fox, J.L.; Stanton, R.; Sargent, C.; Wintour, S.A.; Scanlan, A.T. The Association between Training Load and Performance in Team Sports: A Systematic Review. *Sports Med.* **2018**, *48*, 2743–2774. [CrossRef]

17. Silva, P.; Dos Santos, E.; Grishin, M.; Rocha, J.M. Validity of Heart Rate-Based Indices to Measure Training Load and Intensity in Elite Football Players. *J. Strength Cond. Res.* **2018**, *32*, 2340–2347. [CrossRef]

18. Gabbett, T.J. The training—Injury prevention paradox: Should athletes be training smarter and harder? *Br. J. Sports Med.* **2016**, *50*, 273–280. [CrossRef]

19. Schelling, X.; Torres-Ronda, L. An Integrative Approach to Strength and Neuromuscular Power Training for Basketball. *Strength Cond. J.* **2016**, *38*, 72–80. [CrossRef]

20. Reche-Soto, P.; Cardona, D.; Díaz, A.; Gómez-Carmona, C.D.; Pino-Ortega, J. Acelt y player load: Dos Variables para la Cuantificación de la Carga Neuromuscular/acelt and player load: Two Variables to Quanty Neuromuscular Load. *Rev. Int. Med. Cienc. Act. Fís. Deporte* **2020**, *20*, 167–183.

21. Ibáñez, S.J.; Feu, S.; Cañadas, M. Sistema integral para el análisis de las tareas de entrenamiento, SIATE, en deportes de invasión. *E-Balonamo Com* **2016**, *12*, 3–30.

22. Borg, G. *Borg's Perceived Exertion and Pain Scales*; Human Kinetics: Champaign, IL, USA, 1998.

23. Eston, R.G.; Parfitt, C.G. Effort Perception. In *Paediatric Exercise Physiology*; Armstrong, N., Ed.; Elsevier: London, UK, 2007; pp. 275–297.

24. López, A.T. Propuesta de control de la carga de entrenamiento y la fatiga en equipos sin medios económicos. *Rev. Esp. Educ. Fís. Deportes* **2017**, *417*, 55–69.

25. Lovell, T.W.J.; Sirotic, A.C.; Impellizzeri, F.M.; Coutts, A.J. Factors Affecting Perception of Effort (Session Rating of Perceived Exertion) during Rugby League Training. *Int. J. Sports Physiol. Perform.* **2013**, *8*, 62–69. [CrossRef]

26. Thomas, J.R.; Nelson, J.K.; Silverman, S.J. *Research Methods in Physical Activity*, 7th ed.; Human Kinetics: Champaign, IL, USA, 2015.

27. García-Ceberino, J.M.; Feu, S.; Ibáñez, S.J. Comparative Study of Two Intervention Programmes for Teaching Soccer to School-Age Students. *Sports* **2019**, *7*, 74. [CrossRef] [PubMed]

28. García-Ceberino, J.M.; Antúnez, A.; Feu, S.; Ibáñez, S.J. Validación de dos programas de intervención para la enseñanza del fútbol escolar/Validation of Two Intervention Programs for Teaching School Soccer. *Rev. Int. Med. Cienc. Act. Fís. Deporte* **2020**, *20*, 257–274.

29. González-Espinosa, S.; Ibáñez, S.J.; Feu, S. Design of two basketball teaching programs in two different teaching methods. *E-Balonamo Com* **2017**, *13*, 131–152.

30. González-Espinosa, S.; Ibáñez, S.J.; Feu, S.; Galatti, L.R. Programas de intervención para la enseñanza deportiva en el contexto escolar, PETB y PEAB: Estudio preliminar. *Retos-Nuevas Tend. Educ. Fís. Deporte Recreación* **2017**, *31*, 107–113.

31. Molina-Carmona, I.; Gomez-Carmona, C.D.; Bastida-Castillo, A.; Pino-Ortega, J. Validez del dispositivo inercial WIMU PROTM para el registro de la frecuencia cardíaca en un test de campo. *Sport-TK Rev. Euroam. Cienc. Deporte* **2018**, *7*, 81–86. [CrossRef]

32. White, A.; Hills, S.P.; Cooke, C.B.; Batten, T.; Kilduff, L.P.; Cook, C.J.; Roberts, C.; Russell, M. Match-Play and Performance Test Responses of Soccer Goalkeepers: A Review of Current Literature. *Sports Med.* **2018**, *48*, 2497–2516. [CrossRef]

33. Murillo-Lorente, V.; Álvarez-Medina, J.; Manomelles-Marqueta, P. Control training loads through perceived exertion. Prediction of heart rate. *Retos-Nuevas Tend. Educ. Fís. Deporte Recreación* **2016**, *30*, 82–86.

34. Field, A. *Discovering Statistics Using SPSS Statistics*, 4th ed.; Sage Publications Ltd.: London, UK, 2013.

35. Martínez, R.M.; Tuya, L.C.; Martínez, M.; Pérez, A.; Cánovas, A.M. El coeficiente de correlación de los rangos de Spearman caracterizacion. *Rev. Habanera Cienc. Méd.* **2009**, *8*, 1–19.

36. Buceta, J.M. *Psicología del Entrenamiento Deportivo*; Dykinson: Madrid, Spain, 1998.

37. Gamero, M.G.; García-Ceberino, J.M.; Feu, S.; Antúnez, A. Estudio de las variables pedagógicas en tareas de enseñanza del fútbol en función de la parte de sesión. *Sport-TK Rev. Euroam. Cienc. Deporte* **2019**, *8*, 39–46. [CrossRef]

38. Gamero, M.G.; García-Ceberino, J.M.; Reina, M.; Feu, S.; Antúnez, A. Study of the pedagogical variables of basketball tasks by game phase. *Retos-Nuevas Tend. Educ. Fís. Deporte Recreación* **2020**, *37*, 552–558.

39. García-Ceberino, J.M.; Gamero, M.G.; Gómez-Carmona, C.D.; Antúnez, A.; Feu, S. Incidence of organizational parameters in the quantification of the external training load of the tasks designed for teaching of the school basketball. *Rev. Psicol. Deporte* **2019**, *28*, 35–41.

40. García-Ceberino, J.M.; Gamero, M.G.; Reina, M.; Feu, S.; Ibáñez, S.J. Study of external load in basketball tasks based on game phases. *Retos-Nuevas Tend. Educ. Fís. Deporte Recreación* **2020**, *37*, 536–541.

41. Borresen, J.; Ian Lambert, M. The Quantification of Training Load, the Training Response and the Effect on Performance. *Sports Med.* **2009**, *39*, 779–795. [CrossRef]

42. Fanchini, M.; Azzalin, A.; Castagna, C.; Schena, F.; Impellizzeri, F. Effect of bout duration on exercise intensity and technical performance of small-sided games in soccer. *J. Strength Cond. Res.* **2011**, *25*, 453–458. [CrossRef]

43. Alexiou, H.; Coutts, A.J. A comparison of methods used for quantifying internal training load in women soccer players. *Int. J. Sports Physiol. Perform.* **2008**, *3*, 320–330. [CrossRef]

44. McLaren, S.J.; Macpherson, T.W.; Coutts, A.J.; Hurst, C.; Spears, I.R.; Weston, M. The Relationships between Internal and External Measures of Training Load and Intensity in Team Sports: A Meta-Analysis. *Sports Med.* **2018**, *48*, 641–658. [CrossRef]
45. Abbott, W.; Brickley, G.; Smeeton, N.J. Positional Differences in GPS Outputs and Perceived Exertion during Soccer Training Games and Competition. *J. Strength Cond. Res.* **2018**, *32*, 3222–3231. [CrossRef]
46. Akenhead, R.; Hayes, P.R.; Thompson, K.G.; French, D. Diminutions of acceleration and deceleration output during professional football match play. *J. Sci. Med. Sport* **2013**, *16*, 556–561. [CrossRef]
47. Sampaio, J.; McGarry, T.; Calleja-González, J.; Sáiz, S.J.; Schelling, X.; Balciunas, M. Exploring game performance in the National Basketball Association using player tracking data. *PLoS ONE* **2015**, *10*, e0132894. [CrossRef]
48. Ballesta, A.S.; Abrunedo, J.; Caparros, T. Accelerometry in Basketball. Study of External Load during Training. *Apunts Educ. Fís. Deportes* **2019**, *135*, 100–117.

 *healthcare*

Article

Physical Fitness and Performance in Talented & Untalented Young Chinese Soccer Players

Alfredo Irurtia [1,2,*], Víctor M. Torres-Mestre [1], Álex Cebrián-Ponce [1,2], Marta Carrasco-Marginet [1,2], Albert Altarriba-Bartés [3], Marc Vives-Usón [1], Francesc Cos [1] and Jorge Castizo-Olier [4]

[1] INEFC-Barcelona Sports Sciences Research Group, National Institute of Physical Education of Catalonia (INEFC), University of Barcelona, 08038 Barcelona, Spain; vmtm3@yahoo.es (V.M.T.-M.); acebrian@gencat.cat (Á.C.-P.); mcarrascom@gencat.cat (M.C.-M.); mvivesu@gencat.cat (M.V.-U.); cosfrancesc@gmail.com (F.C.)

[2] Catalan School of Kinanthropometry, National Institute of Physical Education of Catalonia (INEFC), University of Barcelona (UB), 08038 Barcelona, Spain

[3] Sport and Physical Activity Studies Center (CEEAF), Sport Performance Analysis Research Group (SPARG), University of Vic-Central University of Catalonia, Vic, 08500 Barcelona, Spain; albert.altarriba@uvic.cat

[4] School of Health Sciences, TecnoCampus, Pompeu Fabra University, Mataró, 08302 Barcelona, Spain; jcastizo@tecnocampus.cat

* Correspondence: airurtia@gencat.cat

Abstract: Sports performance is a complex process that involves many factors, including ethnic and racial differences. China's youth soccer is in a process of constant development, although information about the characteristics of its players and their methodological systems is scarce. The aim of this retrospective study was to characterize the physical fitness and the competitive performance of 722 Chinese players of three sports categories (8.0–9.9, 10.0–11.9 and 12.0–13.9 years), who were classified by their coaches as talented ($n = 204$) or untalented ($n = 518$). Players were assessed for anthropometry (body height, body mass, body mass index), lung capacity (Forced Vital Capacity), jumping performance (Squat Jump, Countermovement Jump and Abalakov tests), sprinting performance (10 m and 30 m Sprint tests), agility performance (Repeated Side-Step test) and flexibility (Sit & Reach test). A descriptive, comparative, correlational and multivariate analysis was performed. Competitive ranking was created in order to act as dependent variable in multiple linear regression analysis. Results indicate that Chinese players classified as talented have better motor performance than untalented ones. However, these differences are neither related nor determine the competitive performance of one group or the other.

Keywords: training; long-term; talent identification; team sport; multivariate analysis

Citation: Irurtia, A.; Torres-Mestre, V.M.; Cebrián-Ponce, Á.; Carrasco-Marginet, M.; Altarriba-Bartés, A.; Vives-Usón, M.; Cos, F.; Castizo-Olier, J. Physical Fitness and Performance in Talented & Untalented Young Chinese Soccer Players. *Healthcare* **2022**, *10*, 98. https://doi.org/10.3390/healthcare10010098

Academic Editors: João Paulo Brito and Rafael Oliveira

Received: 28 November 2021
Accepted: 30 December 2021
Published: 4 January 2022

Publisher's Note: MDPI stays neutral with regard to jurisdictional claims in published maps and institutional affiliations.

1. Introduction

The People's Republic of China (PRC) has a policy interest in promoting and developing professional soccer [1,2]. However, despite being one of the world's major economic powers, PRC has had to establish pacts and alliances with European clubs in order to replicate its methodology and sports training systems [3,4]. Indeed, China currently lacks basic structures, and clubs are filled with professional soccer schools, which are generally advised by and/or linked to big European sports clubs, dedicated to developing and promoting possible future talented players [5]. Children share studies, social life, and sports training under increasingly well-equipped academic-sports structures. In return, European clubs reserve the right to select the best promising players and allow them to compete in their teams in the future [3,5].

Identifying sports talent is an exciting and complex area of sports science, due to its multidimensional nature [6–10]. Moreover, this complexity increases when competitive performance must be explained or predicted in a team sport such as soccer [11,12]. Cognitive constructs related to the greater or lesser success in the execution of certain tactical skills, performed by a player who always interacts with his teammates and opponents, are difficult to assess individually [13]. Thus, on many occasions, the criteria of expert coaches are applied to determine a general ranking of players based on a score [14,15]. Coach-based skill ratings have been adopted in previous research with demonstrated validity and reliability [6,16]. In relation to youth soccer, Hendry and colleagues [15] used a five-point scale (one = poor to five = excellent) to rate 102 elite youth players, according to their skill level. Another example is the contribution of Jukic and colleagues [14], who created a questionnaire for coaches with nine items that attempted to cover, multidimensionally, the quality they attributed to each player, according to a series of technical, tactical, physical, and psychological aspects. Each player was classified into four scores A = above average performance; B = average performance; C = low performance; D = below standards (the analysis model explained 58.49% of the variance).

If sports performance is taken as a continuous variable, multiple linear regression (MLR) is the most common statistical technique. Thus, this ranking determines the competitive performance and conforms the dependent or predicted variable to confront the different independent or predictive variables (e.g., anthropometric, physiological, technical skills, and/or physical fitness tests) when a multivariate statistical model is performed [17]. On the other hand, in connection with the independent variables, although no clear evidence has been found to support the use of physical skill or physiological tests in early stages of sports development to predict future success, its administration is an established process in all training programs, both in individual or team sports [18,19]. In this regard, a recent review on the identification of talented players in soccer highlighted the importance of evaluating different skills, scaling the children by age groups, and adding information on their anthropometric and physiological profiles [20]. However, although it is well characterized that young, successful soccer players have anthropometric and physical fitness characteristics that are differentiated from the rest of their peers [21,22], it is necessary to understand that it is unlikely that these differences provide a reliable source to predict success within an already talented group [23]. That is why it is recommended to use the collected data to guide the training programs, while considering its substantial limitations to predict future sport success [24,25].

Access to information on the methods applied for the identification of sports talent in the PRC and on the training processes of its young athletes is difficult. Language, culture, and the country's own socio-political reality limit its availability. In soccer, there are very few scientific contributions published in English about the anthropometric profile, motor skills, or physiological characteristics of children and/or adolescents [26–29]. Therefore, the primary aim of this study was to provide information about anthropometric characteristics and performance in some physical fitness tests commonly used in an important Chinese soccer academy. Likewise, to help their managers and all technical staff, the possible relationships between the results of each test and a ranking established by the number of competitive victories was analyzed according to the competitive category and performance level (talented and untalented teams). Finally, from a multidimensional perspective, a multivariate analysis was performed to find out if any of these tests, or a set of them, could explain the variance of the ranking.

2. Materials and Methods

2.1. Study Design

This was a cross-sectional and retrospective cohort study with a descriptive, comparative, correlational, and multivariate analysis strategy. A convenience sample was recruited and were stratified into 2 performance level groups (talented vs. untalented) for each chronological age group consistent with their respective sports category (8.0–9.9 years, 10.0–11.9 years, 12.0–13.9 years). Twelve independent variables were assessed: chronological age—AGE—(years), height—HT—(cm), body mass—BM—(kg), body mass index—BMI—(kg/m^2), squat jump—SJ—(cm), counter movement jump—CMJ—(cm), CMJ Abalakov—ABK—(cm), sprint 10 m—SP10—(s), sprint 30 m—SP30—(s), "repeated side-step"—RSS—(n), sit & reach test—SRT—(cm), forced vital capacity—FVC—(L). A performance ranking was established as the dependent variable for the correlation and multivariate analyses (multiple linear regression analysis).

2.2. Subjects

Data used in this study were collected in the 2013–14 midseason (late February to May 2014) and corresponded to 722 male Chinese children aged from 8.0 to 13.9 years, from 6 competitive categories and fifty-nine regional teams. Children pertained to the "Evergrande Football School" –EFS– (Qingyuan City, Guang Dong Province, China), a private school with a collaboration agreement with the Real Madrid Foundation (Real Madrid Football Club, Madrid, Spain), for the sports training and holistic education of soccer children. According to the expert Spanish coaches of EFS and following their technical and tactical criterion, based on the preview of a series of matches, 204 children were considered as talented (8.0–9.9 years: $n = 41$; 10.0–11.9 years: $n = 114$; 12.0–13.9 years: $n = 49$) while 518 as untalented (8.0–9.9 years: $n = 150$; 10.0–11.9 years: $n = 232$; 12.0–13.9 years: $n = 136$). This classification into 2 performance level groups for each competitive category conformed to a standard procedure commonly used in the EFS at the beginning of each sport season (the list of technical and tactical performance indicators can be consulted as Supplementary Information). Independently of this categorization, the authors of this study created a competitive ranking based on the previous league results of every team. The formula was: "final points achieved · 100/maximum possible points". Thus, each player, grouped according to their chronological age, obtained a scoring scale that was comparable to the rest of the sample. The eligibility criteria were as follows: (1) male EFS soccer players aged between 8.0 to 13.9 who performed all the tests with a minimum of 2 recorded repetitions; (2) to have competed in the regional soccer league during the past season; (3) to be free of previous injuries and/or illnesses that may have affected the test results. From a total of 757 players, 35 players were finally discarded for not meeting the inclusion criteria.

2.3. Ethical Issues

All the assessments corresponded to measures or tests commonly used during soccer training sessions and whose rights to data processing or assignment belonged to EFS. Consequently, the study had the signed approval of the person legally responsible for the EFS. The study was conducted following the Helsinki Declaration Statement [30]. Retrospective protocol and procedures were approved by the Ethics Committee for Clinical Sport Research of Catalonia (Ethical Approval Code: 19/CEICGC/2020).

2.4. Procedures

All tests were performed by a single investigator according to previously standardized protocols. He was responsible of the physical conditioning area of EFS in the age groups analyzed and is the second author of this study. Anthropometric measurements and physical fitness tests were performed over 14 weeks, corresponding to the middle of the 2013–14 season, after the Chinese New Year (from the end of February to the end of May 2014). Every morning from Monday to Friday, from 9:30 a.m. to approximately 12:30 p.m., a whole team (12–14 subjects) performed all the tests in the EFS gym. This facility had a controlled temperature, set at $24.0 \pm 1.0\,^{\circ}\text{C}$, with a relative humidity of $55.5 \pm 10.0\%$. Upon arriving at the gym, in an adjoining room, players were measured and weighed by the medical doctor responsible for the EFS, who also performed the spirometry test. Next, and even though all the players were accustomed with the tests, because they had previously performed them during training, they were reminded of protocols before the start. Then, all subjects performed a general warm-up (10–15 min at 60% heart rate) and a specific warm-up (10–15 min, including performing each of the tests as a pre-test). Except for the SRT, which was always performed as the last test, the execution order of the rest (SJ, CMJ, ABK, SP10, SP30, and SST) was randomized. Recovery time between tests was always complete

(>3 min) and for each evaluation 2 attempts were performed to ensure the reliability of the measurements. Once verified, the best attempt was the one finally registered.

Specific protocols and materials used for every assessment are detailed as follows: Anthropometric measurements were performed according to the standard criteria of the International Society for the Advancement of Kinanthropometry [31]. HT was assessed to the nearest 0.1 cm using a telescopic stadiometer (Seca 220®, Hamburg, Germany), and BM was measured to the nearest 0.05 kg using a calibrated weighing scale (Seca 710®, Hamburg, Germany). BMI (kg/m^2) was derived from BM/HT^2. FVC was registered with a battery-operated portable spirometer (FCS-10000®; Grows Instrument Ltd., Hong Kong, China), following the previous protocols of the European Respiratory Society [32] and the recommendations on forced exhalation time in young adolescents, which were set to 6 s [33]. Each physical fitness test included in the assessment fulfilled with previously published standards and validated measurement protocols for children and adolescents: SJ, CMJ, and ABK [34,35] were assessed with Chronojump Boscosystem®, composed of a contact platform and its corresponding software interface, "Chronojump v1.3.9" (Chronojump, Boscosystem®, Barcelona, Spain). SP10 and SP30 followed previous methodological recommendations [36] and validated protocols for assessing youth sprint ability [37]. Both tests were automatically measured with 2 photoelectric barriers (Chronojump, Boscosystem®, Barcelona, Spain). SRT was performed according to the methodology of Eurofit [38] once its criterion-related validity had been verified for estimating the hamstring flexibility in children and adolescents [39]. Finally, regarding the RSS, despite being a widely used test in Asian schools and sports clubs to assess agility [40], no studies have been found on its validity prior to 2014, corresponding to the data of the present investigation. Recently, a research group from the Shanghai Sports University has published its protocol obtaining a strong concurrent validity when comparing its results against other previously published agility and movement skill tests [41]. Basically, RSS requires participants to demonstrate as many as possible repeated sideward steps in 20 s between 2 lines located 1 m apart (50 cm from the center, which would correspond to the starting position). The outcome measure is the total precise steps.

2.5. Statistical Analysis

Data are expressed by means and standard deviations. Normality and equal variance of the distributions were confirmed by the Kolmogorov-Smirnov and Levene tests, respectively. Test–retest reliability was examined using the intraclass correlation coefficient (ICC) with a two-way mixed average measures model. The coefficient of variation (CV) was calculated for all tests to determine the stability of measurement among trials with the 95% confidence interval. The Student's unpaired t-test was performed to analyze possible differences between talented and untalented groups. For all those that were statistically significant, the effect size (ES) was calculated using Cohen's d test. In order to check the differences between age groups, a one way ANOVA was performed using the Tukey post-hoc test. The Pearson's correlation coefficient was calculated to assess the relationship of each independent variable with the performance level (competitive results ranking). Finally, a stepwise multiple linear regression analysis (MLR) was performed to assess the ability of independent variables to explain the variance of the performance level. Precise p values were reported and $p \leq 0.05$ was considered significant. All data were analyzed using SPSS 22® (IBM Inc., New York, NY, USA). Effect size was calculated with G*Power v3.1.9.2 package (University of Kiel, Kiel, Germany).

3. Results

Test–retest reliability results are shown in Table 1. Anthropometric outcomes (HT, BM, BMI) did not differ in practically all 722 cases between the first and second measurements, hence we report the maximum reliability results. Regarding to the FVC and all physical fitness tests, high test–retest reliability was obtained (ICC values ranging from 0.77 to 0.97, and CV from 1.5 to 11.0% in both groups).

Table 1. Test–retest reliability statistics for 722 children, talented and untalented Chinese soccer players.

Tests	Talented ($n = 204$)		CV	Untalented ($n = 518$)		CV
	ICC			ICC		
	r	CI 95%	%	r	CI 95%	%
HT (cm)	1.00	–	0.0	1.00	–	0.0
BM (kg)	1.00	–	0.0	1.00	–	0.0
BMI (kg/m^2)	1.00	–	0.0	1.00	–	0.0
FVC (L)	0.83	0.77–0.87	11.0	0.96	0.95–0.97	10.8
SJ (cm)	0.88	0.84–0.91	7.9	0.92	0.91–0.94	9.9
CMJ (cm)	0.89	0.86–0.92	6.5	0.94	0.93–0.95	7.3
ABK (cm)	0.89	0.85–0.91	5.6	0.94	0.93–0.95	7.3
SP10 (s)	0.91	0.88–0.93	1.6	0.84	0.80–0.86	5.2
SP30 (s)	0.95	0.94–0.97	1.5	0.94	0.92–0.95	4.2
RSS (n)	0.90	0.87–0.93	6.2	0.93	0.91–0.94	8.2
SRT (cm)	0.96	0.95–0.97	8.5	0.97	0.96–0.97	9.4

HT: height, BM: body mass, BMI: body mass index, SJ: squat jump, CMJ: counter movement jump, ABK: Abalakov, SP10: sprint 10 m, SP30: sprint 30 m, RSS: repeated side-step, SRT: sit & reach test, FVC: forced vital capacity. ICC: intraclass correlation coefficient, CV: coefficient of variation, CI 95%: confidence interval.

Results of the whole sample ($n = 722$) and statistical differences between talented ($n = 204$) and untalented players ($n = 518$), are shown in Table 2. No significant differences were recorded in AGE, FVC, or in any anthropometric variables (HT, BM, BMI). In contrast, both in the jumping tests (SJ, CMJ, ABK) and in the sprint tests (SP10, SP30), talented players showed significantly better performance than their counterparts ($p \leq 0.05$), although with a small effect size ($d \leq 0.25$). The largest difference was registered in the RSS test ($p = 0.001$), again in favor of the talented group, with a medium effect size magnitude ($d = 0.49$). Finally, SRT values were higher in the talented group, although these were not statistically significant ($p = 0.39$).

Table 2. Descriptive results and comparison between anthropometric and physical fitness tests for 722 talented and untalented Chinese children soccer players.

Tests	All ($n = 722$)	Talented ($n = 204$)	Untalented ($n = 518$)	t-Test (p)	ES (d-Cohen)
AGE (years)	11.0 ± 1.4	11.1 ± 1.4	10.9 ± 1.4	0.08	–
HT (cm)	146.9 ± 10.7	147.0 ± 11.1	146.8 ± 10.5	0.84	–
BM (kg)	37.6 ± 8.9	37.3 ± 8.7	37.6 ± 9.0	0.69	–
BMI (kg/m^2)	17.2 ± 2.2	17.0 ± 1.9	17.2 ± 2.4	0.21	–
FVC (L)	2.7 ± 0.7	2.7 ± 0.7	2.7 ± 0.7	0.52	–
SJ (cm)	27.0 ± 5.1	27.6 ± 5.1	26.8 ± 5.1	0.05 *	0.16
CMJ (cm)	27.6 ± 5.2	28.3 ± 5.1	27.3 ± 5.2	0.02 *	0.19
ABK (cm)	31.8 ± 5.9	32.5 ± 6.1	31.5 ± 5.8	0.04 *	0.17
SP10 (s)	2.2 ± 0.1	2.1 ± 0.1	2.2 ± 0.1	0.01 *	0.22
SP30 (s)	5.4 ± 0.4	5.3 ± 0.4	5.4 ± 0.4	0.001 *	0.25
RSS (n)	40.0 ± 4.6	41.6 ± 4.5	39.4 ± 4.5	0.001 *	0.49
SRT (cm)	8.9 ± 5.0	9.2 ± 4.9	8.8 ± 5.0	0.39	–

AGE: chronological age, HT: height, BM: body mass, BMI: body mass index, SJ: squat jump, CMJ: counter movement jump, ABK: Abalakov, SP10: sprint 10 m, SP30: sprint 30 m, RSS: repeated side-step, SRT: sit & reach test, FVC: forced vital capacity. t-test (talented vs. untalented): p: * statistical significance at $p \leq 0.05$. ES: effect size (d-Cohen).

Differences for each age group and performance level are shown in Table 3. In the youngest group (8.0–9.9 years) there were no significant differences in any of the assessments performed between talented and non-talented players. In the 10.0–11.9 age group, talented players were significantly lighter and, consequently, had a lower BMI. They reported better performance in the CMJ, ABK, SP10, SP30, and RSS ($p \leq 0.03$), although in all cases the effect size of these differences was small ($d = 0.01$–0.33), with the exception of

RSS, in which it was medium ($d = 0.51$). The oldest talented group (12.0–13.9 years) only registered significantly higher values than their untalented counterparts in FVC ($p = 0.05$; $d = 0.32$) and RSS ($p = 0.001$; $d = 0.83$). On the other hand, when analyzing the differences between the 3 age groups, ANOVA results show that, regardless of the performance group (talented or untalented), players older than the preceding age were significantly heavier and taller, and they had a better performance in all tests performed. However, in SRT, no significant differences were found between younger and intermediate age groups, or between 8 years and 12 year-old players.

Table 3. Comparative analysis applied to 722 Chinese children soccer players: One-way ANOVA between age groups (results are showed in rows) and Student's t tests (talented vs. untalented), with the effect size showed in columns for significant differences.

Tests		8.0–9.9 Years	10.0–11.9 Years	12.0–13.9 Years
AGE (years)	Talented	9.0 ± 0.6	11.1 ± 0.5 [*]	12.9 ± 0.7 [‡†]
	Untalented	9.2 ± 0.5	11.0 ± 0.6 [*]	12.7 ± 0.6 [‡†]
p t-test & (*d*-Cohen)		ns	ns	ns
HT (cm)	Talented	135.9 ± 6.6	146.3 ± 8.1 [*]	158.1 ± 9.9 [‡†]
	Untalented	137.4 ± 6.5	146.9 ± 7.6 [*]	157.1 ± 8.6 [‡†]
p t-test & (*d*-Cohen)		ns	ns	ns
BM (kg)	Talented	30.8 ± 6.3	35.9 ± 6.0 [*]	46.2 ± 9.0 [‡†]
	Untalented	31.4 ± 6.2	37.4 ± 7.4 [*]	44.8 ± 8.8 [‡†]
p t-test & (*d*-Cohen)		ns	0.05 (0.22)	ns
BMI (kg/m^2)	Talented	16.5 ± 2.1	16.7 ± 1.5	18.3 ± 1.9 [‡†]
	Untalented	16.5 ± 2.3	17.2 ± 2.3 [*]	18.0 ± 2.4 [‡†]
p t-test & (*d*-Cohen)		ns	0.02 (0.26)	ns
FVC (L)	Talented	2.2 ± 0.5	2.6 ± 0.5 [*]	3.5 ± 0.8 [‡†]
	Untalented	2.2 ± 0.4	2.7 ± 0.5 [*]	3.3 ± 0.7 [‡†]
p t-test & (*d*-Cohen)		ns	ns	0.05 (0.32)
SJ (cm)	Talented	24.4 ± 3.2	27.0 ± 4.4 [*]	31.6 ± 5.4 [‡†]
	Untalented	24.1 ± 3.8	26.2 ± 4.0 [*]	30.7 ± 5.7 [‡†]
p t-test & (*d*-Cohen)		ns	ns	ns
CMJ (cm)	Talented	24.5 ± 3.3	27.9 ± 4.5 [*]	32.2 ± 5.3 [‡†]
	Untalented	24.7 ± 3.8	26.8 ± 4.2 [*]	31.1 ± 6.0 [‡†]
p t-test & (*d*-Cohen)		ns	0.02 (0.25)	ns
ABK (cm)	Talented	27.5 ± 3.3	32.0 ± 5.0 [*]	37.7 ± 6.2 [‡†]
	Untalented	28.3 ± 4.2	30.8 ± 4.5 [*]	36.0 ± 6.6 [‡†]
p t-test & (*d*-Cohen)		ns	0.03 (0.25)	ns
SP10 (s)	Talented	2.3 ± 0.1	2.2 ± 0.2 [*]	2.1 ± 0.1 [‡†]
	Untalented	2.3 ± 0.1	2.2 ± 0.1 [*]	2.1 ± 0.1 [‡†]
p t-test & (*d*-Cohen)		ns	0.01 (0.01)	ns
SP30 (s)	Talented	5.7 ± 0.3	5.3 ± 0.3 [*]	4.9 ± 0.4 [‡†]
	Untalented	5.8 ± 0.4	5.4 ± 0.3 [*]	5.0 ± 0.3 [‡†]
p t-test & (*d*-Cohen)		ns	0.01 (0.33)	ns
RSS (n)	Talented	37.1 ± 3.2	41.2 ± 3.2 [*]	46.2 ± 3.8 [‡†]
	Untalented	35.9 ± 3.6	39.5 ± 3.5 [*]	43.0 ± 3.9 [‡†]
p t-test & (*d*-Cohen)		ns	0.001 (0.51)	0.001 (0.83)
SRT (cm)	Talented	8.2 ± 3.7	8.7 ± 5.1	11.2 ± 4.8 [‡†]
	Untalented	8.8 ± 4.2	8.4 ± 5.1	9.6 ± 5.5 [†]
p t-test & (*d*-Cohen)		ns	ns	ns

AGE: chronological age, HT: height, BM: body mass, BMI: body mass index, SJ: squat jump, CMJ: counter movement jump, ABK: Abalakov, SP10: sprint 10 m, SP30: sprint 30 m, RSS: repeated side-step, SRT: sit & reach test, FVC: forced vital capacity, ns: not statistically significant. ANOVA: age groups statistical differences set at $p \leq 0.01$: [*] (8.0–9.9 vs. 10.0–11.9); [‡] (8.0–9.9 vs. 12.0–13.9); [†] (10.0–11.9 vs. 12.0–13.9). *p t*-test: "Talented vs. Untalented" statistical differences at $p \leq 0.05$. Effect Size (*d*-Cohen): ≈ 0.20 small effect; ≈ 0.50 medium; ≈ 0.80 large.

Statistically significant correlations between the competitive ranking and each of the variables analyzed were scarce and registered low ($r = 0.3$ to 0.5 or -0.3 to -0.5) or negligible values ($r < 0.3$ or -0.3). In the youngest age group, competitive performance of talented players seemed to maintain a certain relationship with their height ($r = 0.36$; $p = 0.02$) and their ability to cover 30 m in the shortest time possible ($r = -0.31$; $p = 0.05$). While in this age group (8.0–9.9 years) no correlation was registered in the untalented group, in the following group (10.0–11.9 years), older children ($r = 0.18$; $p = 0.01$) with a greater stature ($r = 0.20$; $p = 0.003$), a lower BM ($r = -0.18$; $p = 0.01$), and a greater lung capacity ($r = 0.16$; $p = 0.01$) seemed to obtain better competitive performance. This did not occur in talented players in this intermediate age group, where only lung capacity seemed to show a positive relationship with their competitive performance ($r = 0.19$; $p = 0.05$). In the oldest talented group (12.0–13.9 years), chronological age ($r = 0.28$; $p = 0.05$) and better results in SP30 ($r = -0.31$; $p = 0.03$) and RSS ($r = 0.30$; $p = 0.04$) seemed to be related to their position in the competitive ranking. Also, the ability to sprint, but this time in the SP10 ($r = -0.19$; $p = 0.03$), and the agility manifested in the RSS ($r = 0.24$; $p = 0.01$), were slightly related to performance of the least talented players in this age group. Multiple linear regression analysis reinforces the results of these correlations, with a low explanation of the variance of competitive performance (R^2 ranged between 5.0 to 20.0%) in all age groups and independently of their classification as talented or untalented (Table 4).

Table 4. Stepwise multiple linear regression analysis by chronological age groups in talented and untalented Chinese children soccer players.

Age (Years)	Talented (T) Untalent. (U)	Explicative Equations	F	df_1	df_2	p	R^2 Exact	Adjust.	SEE
8.0–9.9	T ($n = 41$)	$-64.63 + (0.98\ HT) - (2.09\ BMI)$	5.97	1	38	0.01	0.24	0.20	9.33
	U ($n = 150$)	No variables selected in multivariate model	–	–	–	–	–	–	–
10.0–11.9	T ($n = 114$)	$29.35 + (0.01\ FVC) - (0.47\ BM)$	4.70	1	111	0.01	0.08	0.06	8.90
	U ($n = 232$)	$15.72 - (0.04\ HT) - (0.80\ SP30) - (0.15\ CMJ) + (0.11\ ABK)$	6.46	1	227	0.01	0.10	0.09	1.13
12.0–12.9	T ($n = 49$)	$156.29 - (17.89\ SP30) - (0.72\ ABK)$	4.83	1	46	0.01	0.17	0.14	8.89
	U ($n = 136$)	$2.48 + (0.07\ RSS)$	8.04	1	134	0.01	0.06	0.05	1.13

HT: height, BMI: body mass index, FVC: forced vital capacity, BM: body mass, SP30: sprint 30 m, CMJ: counter movement jump, ABK: Abalakov; RSS: repeated side-step test, ANOVA statistics: Fentry ($p \leq 0.05$) to Fexit ($p \geq 0.10$), R2: coefficient of determination, SEE: standard error of the estimate, p: level of significance ($p \leq 0.05$).

4. Discussion

The purpose of this study was to provide relevant information on the basic anthropometric characteristics and some physical fitness tests commonly used in sports training in a large sample of young Chinese soccer players. To our knowledge, this is the first time that a multivariate model of competitive performance in young Chinese soccer players has been applied. The major finding of this study has been to verify the null or scarce relationship between anthropometric measures and/or physical fitness tests that coaches normally use during training sessions, and the classification of each player in a ranking based on their competitive league results as members of their respective teams.

Sport-specific technical skills assessments have been demonstrated a great sensitivity to discriminate different competitive levels and predict future sports performance in the talent identification area [42]. The present multidimensional proposal is based on a MLR analysis and has been previously applied to predict or explain sports performance, both in individual sports, such as swimming [43], triathlons [44], or artistic gymnastics [45], and in team sports, such as field hockey [46], handball [17,47], volleyball [48], or soccer [49–51]. However, its practical application in team sports is more complex, due to the necessary configuration of a ranking which acts as a response or dependent variable in MLR analysis, and which in many individual sports is already established by its own rules and regulations (e.g., ATP ranking in tennis, FINA points in swimming, ITRA points in trail running, etc.). In soccer, as well as in other team sports, the individual ranking must be configured based on quantitative parameters of game analysis (scouting reports) or by the criteria of expert coaches [15]. In this research, we have applied an individual score to each player based on the competitive results that they obtained as members of their respective teams. For example, the players of the team that was first classified in their respective league all obtained the same score. Although this criterion has certain limitations, because it does not consider individual intra-team differences, the expert coaches of the EFS considered it adequate because it was based on the results of a whole sport season and at the end of this period, in a sample of 722 players, this ranking differentiated between highest and lowest competitive level players.

Before discussing findings related to the rest of motor tests, it is important to note that sports performance is a complex process that involves many factors, including ethnic and racial differences [52,53]. Talent identification studies must consider cross-cultural or country analysis when attempting to identify—or to compare with other studies—the sports performance attributes [54]. Regarding Asian soccer players, racial differences have been found in the anthropometric and physiological profile of sixteen young Chinese elite male soccer players (16.2 ± 0.6 years) that, when compared to European and African counterparts, registered shorter stature and lower CMJ and SP30 values, among other physical fitness differences [28]. Comparing our results with those of other studies conducted with Chinese soccer players of similar ages is difficult, due to the lack of scientific information published in English. Only one study has been found that describes and correlates, by playing positions, some anthropometric and physical fitness attributes of seventy Chinese soccer players under 14 years [26]. According to our results in the 12.0–13.9 year-old players, HT (talented: 158.1 ± 9.9 cm, untalented: 157.1 ± 8.6 cm; $p > 0.05$) and BM (talented: 46.2 ± 9.0 kg, untalented: 44.8 ± 8.8 kg; $p > 0.05$) are comparable to their forwards soccer players (HT: 1.56 ± 0.11 m, BM: 43.9 ± 9.5 kg), being shorter and lighter than the rest of the players from other playing positions. As for fitness tests, average CMJ results, achieved in a study by Wong et al., were always greater than 50 cm, which were remarkable results for boys under 14 years, and therefore far from the 32.2 ± 5.3 cm (talented) and 31.1 ± 6.0 cm (untalented) values attained in our study. SP30 values are comparable again, as they stand between 4.81 ± 0.36 s (defenders) and 4.96 ± 0.4 s (forwards), and our results oscillate between 4.9 ± 0.4 s (talented) and 5.0 ± 0.3 s (untalented). Lastly, the authors found significant relationships of SP30 with BM (r = −0.54), HT (r = −0.64), and BMI (r = −0.24). CMJ also correlated with HT (r = 0.36). They conclude that talent identification in young Chinese soccer players should take into account the anthropometric profile, since it is related to some physical fitness performances, although these should not be an absolute selection criteria, since long-term performance depends on many other factors not sufficiently investigated at present.

In our case, there are no anthropometric differences between players classified by coaches as talented or untalented. Furthermore, no anthropometric variable correlates or explains the variance of performance. Similarly, although talented players are significantly better in jumping and sprinting tests, these outcomes appear to have poor or no relation to their subsequent competitive performance. These results are consistent with the practice of not overvaluing anthropometric measures or physical skill tests as tools for sports talent identification [18]. Several studies conducted with young European elite soccer players also relativize the motor tests' importance for the selection processes [55–57], including, specifically, the sprint and jump performances [58]. Although it seems to be confirmed that covering larger distances at a high speed is a key performance factor in highly trained prepubertal soccer players [59], most current, related literature conclude that the best training strategy in these early stages is to let the players play, and that it is better to give them a broad spectrum of fundamental motor skills rather than prioritize the specific conditioning capacities in the training sessions to a large extent [14,15,60]. Related to this, positioning, and deciding when to play, seem to be key factors for talent development in soccer [61] within some technical characteristics related with the dribbling [62,63] and the ball kicking speed [64].

The early development process of young talented athletes must necessarily include not only the evaluation of physical, physiological, and technical skill components, but also the psychological and sociological factors that longitudinally affect sports performance [20,65,66].

This holistic, multidisciplinary approach to talent identification is currently recognized and accepted by both the scientific community and by a large part of sports coaches [54]. However, its practical implementation implies that both scientists and coaches must come together to perform holistic models to predict sports performance, in order to optimize training and obtain the maximum competitive achievements in the medium and long term [54,67]. The expert eyes or "gut instinct" of coaches and stakeholders are particularly important, but not everything can depend on this [68–70]. For example, in our study, the selection criteria between talented and untalented players was subjectively pre-established by the EFS coaches, based on the visualization of a series of matches. This suggests that they predictably focused on something more than in the physical skills of the young players. However, the final result of this classification was contradictory, because it was precisely only the physical differences that discriminated one group from the other, and not the competitive performance finally demonstrated by both, as reported by the results of the multivariate model.

The youth soccer Chinese training system is constantly growing and optimizing. To continue moving forward, next steps should focus on optimizing procedures for the detection, recruitment, and final selection of those players with the best aptitudes, skills, and competencies associated with future competitive success. Our research is consistent with currently related literature about talent identification: it is necessary to multidimensionally assume the complexity of sports performance, understanding it as a longitudinal, dynamic, and specific process dependent on each situation and context [71,72]. The opinion, criterion and experience of coaches should be complemented with scientific analysis based on psychologically and sociologically validated predictors, together with anthropometric, physiological, physical, and technical–tactical tests, among others [70]. For example, regarding to psychological aspects and specifically those referring to perceptual cognitive predictors [73], decision making during a soccer match seemed to be the most relevant variable to explain the competitive performance of 127 U12 and U15 soccer players ($d = 0.81$). Concerning to personality-related factors, some significant predictor variables were: hope for success, fear of failure, self-esteem, self-efficacy [74]. Finally, sociological variables have been studied to a lesser extent, although parental support, socio-economic background, education, coach–child interaction, hours in practice, and cultural background seem to explain the variance in the performance of young elite football players, as reported in a recent systematic review [75].

This is absolutely necessary in order to avoid the dissonance noted in this study between the opinion of the coaches when they identify talented or untalented players and the poor or null relationship with the subsequent tests performed to evaluate their present and/or future competitive performance.

Finally, this is a retrospective study with a convenience-recruited sample. This research design has some limitations, such as the possible bias derived from the quality of the data during its registration [76], that the sample is not representative of the population [77], or specifically in this study, two remarkable aspects: that it would have been desirable to introduce: an inter- and intra-rater reliability analysis, and a somatic biological maturity indicator (i.e., prediction of age at peak height velocity). In this context, some methodological considerations must be considered in order to define the contributions and limitations of the present study: (1) a large sample was recruited, a circumstance that minimizes possible rare outcomes and gives strength to the results; (2) the second author of this research was responsible for the execution and supervision of all the tests, that were performed twice under strict execution of the previously validated protocols. Only the RSS protocol was not validated in 2014, but it has recently been published [41]; (3) it is not considered indispensable to perform an inter- or intra-rater reliability analysis for three reasons: (a) assessed tests are commonly used in sports training and have validated protocols for young people, (b) operators responsible for the tests' evaluation had extensive experience in these and, (c) test–retest reliability reports very high consistency results; (4) lastly, although it would have been scientifically desirable to operate with a biological age indicator, the present study has classified the age groups according to the corresponding competitive categories, a circumstance that will facilitate the understanding and application of the results by soccer coaches of many Chinese sports clubs.

5. Conclusions

This retrospective study provides relevant information on some basic anthropometric characteristics and physical skills of a large sample of young Chinese soccer players between 8.0 and 13.9 years old. A multivariate model configured from real competitive results is proposed in a novel way. Some practical considerations for coaches, clubs and training soccer institutions should be noted: Chinese soccer players classified as talented by expert coaches have better physical skills than untalented ones. However, these differences are neither related nor determine the competitive performance of one group or the other. The Chinese institutions and soccer professionals should consider these results to relativize the importance of anthropometric measurements and physical fitness tests in talent identification and development areas.

Supplementary Materials: The following supporting information can be downloaded at: https://www.mdpi.com/article/10.3390/healthcare10010098/s1, Table S1: Technical and tactical aspects analyzed by the Evergrande Football School Spanish team of expert coaches.

Author Contributions: Conceptualization, A.I., F.C. and J.C.-O.; data curation, V.M.T.-M. and M.V.-U.; formal analysis, A.I. and A.A.-B.; investigation, A.A.-B., M.V.-U. and F.C.; methodology, A.I., V.M.T.-M. and M.C.-M.; project administration, A.I. and V.M.T.-M.; supervision, Á.C.-P. and J.C.-O.; writing—original draft, A.I.; writing—review and editing, Á.C.-P., J.C.-O. and M.C.-M. All authors have read and agreed to the published version of the manuscript.

Funding: This research received no external funding.

Institutional Review Board Statement: The study was conducted in accordance with the Declaration of Helsinki, and approved by the Ethics Committee for Clinical Sport Research of Catalonia (Ethical Approval Code: 19/CEICGC/2020).

Informed Consent Statement: Informed consent was obtained from all subjects involved in the study.

Data Availability Statement: All data generated analyzed during the current study are available from the corresponding author on reasonable request.

Acknowledgments: The authors thank the Evergrande Football School managers, coaches, and technical staff for making this study possible. A special thanks to Huang Hai, head of the EFS Physical Conditioning Area.

Conflicts of Interest: The authors declare no conflict of interest.

References

1. Ma, Y.; Kurscheidt, M. Governance of the Chinese Super League: A struggle between governmental control and market orientation. *Sport Bus. Manag. Int. J.* **2019**, *9*, 4–25. [CrossRef]
2. Peng, Q.; Skinner, J.; Houlihan, B. An analysis of the Chinese Football Reform of 2015: Why then and not earlier? *Int. J. Sport Policy Politics* **2019**, *11*, 1–18. [CrossRef]
3. Liang, Y. The development pattern and a clubs' perspective on football governance in China. *Soccer Soc.* **2014**, *15*, 430–448. [CrossRef]
4. Liu, D.; Zhang, J.; Desbordes, M. Sport business in China: Current state and prospect. *Int. J. Sports Mark. Spons.* **2017**, *18*, 2–10. [CrossRef]
5. Zhang, J. Problems and Prospects: A Study on the Development History of Chinese Football Industry. *Int. J. Hist. Sport* **2020**, *37*, 102–123. [CrossRef]
6. Unnithan, V.; White, J.; Georgiou, A.; Iga, J.; Drust, B. Talent identification in youth soccer. *J. Sports Sci.* **2012**, *30*, 1719–1726. [CrossRef]
7. Reilly, T.; Bangsbo, J.; Franks, A. Anthropometric and physiological predispositions for elite soccer. *J. Sports Sci.* **2000**, *18*, 669–683. [CrossRef]
8. Reilly, T.; Williams, A.M.; Nevill, A.; Franks, A. A multidisciplinary approach to talent identification in soccer. *J. Sports Sci.* **2000**, *18*, 695–702. [CrossRef]

9.   Williams, A.M. Talent identification and development in soccer: An update and contemporary perspectives. *J. Sports Sci.* **2020**, *38*, 1197–1198. [CrossRef]
10. Johnston, K.; Wattie, N.; Schorer, J.; Baker, J. Talent Identification in Sport: A Systematic Review. *Sports Med.* **2018**, *48*, 97–109. [CrossRef]
11. Vaeyens, R.; Malina, R.M.; Janssens, M.; Van Renterghem, B.; Bourgois, J.; Vrijens, J.; Philippaerts, R.M. A multidisciplinary selection model for youth soccer: The Ghent Youth Soccer Project. *Br. J. Sports Med.* **2006**, *40*, 928–934. [CrossRef]
12. Pruna, R.; Miñarro Tribaldos, L.; Bahdur, K. Player talent identification and development in football. *Apunt. Med. l'Esport* **2018**, *53*, 43–46. [CrossRef]
13. Kilger, M.; Blomberg, H. Governing Talent Selection through the Brain: Constructing Cognitive Executive Function as a Way of Predicting Sporting Success. *Sport Ethics Philos.* **2020**, *14*, 206–225. [CrossRef]
14. Jukic, I.; Prnjak, K.; Zoellner, A.; Tufano, J.; Sekulic, D.; Salaj, S. The Importance of Fundamental Motor Skills in Identifying Differences in Performance Levels of U10 Soccer Players. *Sports* **2019**, *7*, 178. [CrossRef]
15. Hendry, D.T.; Williams, A.M.; Hodges, N.J. Coach ratings of skills and their relations to practice, play and successful transitions from youth-elite to adult-professional status in soccer. *J. Sports Sci.* **2018**, *36*, 2009–2017. [CrossRef]
16. Ali, A. Measuring soccer skill performance: A review. *Scand. J. Med. Sci. Sports* **2011**, *21*, 170–183. [CrossRef]
17. Saavedra, J.M.; Halldórsson, K.; Porgeirsson, S.; Einarsson, I.; Guðmundsdóttir, M.L. Prediction of Handball Players' Performance on the Basis of Kinanthropometric Variables, Conditioning Abilities, and Handball Skills. *J. Hum. Kinet.* **2020**, *73*, 229–239. [CrossRef]
18. Lidor, R.; Côté, J.; Hackfort, D. ISSP position stand: To test or not to test? The use of physical skill tests in talent detection and in early phases of sport development. *Int. J. Sport Exerc. Psychol.* **2009**, *7*, 131–146. [CrossRef]
19. Pearson, D.T.; Naughton, G.A.; Torode, M. Predictability of physiological testing and the role of maturation in talent identification for adolescent team sports. *J. Sci. Med. Sport* **2006**, *9*, 277–287. [CrossRef]
20. Sarmento, H.; Anguera, M.T.; Pereira, A.; Araújo, D. Talent Identification and Development in Male Football: A Systematic Review. *Sports Med.* **2018**, *48*, 907–931. [CrossRef]
21. Hulse, M.A.; Morris, J.G.; Hawkins, R.D.; Hodson, A.; Nevill, A.M.; Nevill, M.E. A field-test battery for elite, young soccer players. *Int. J. Sports Med.* **2013**, *34*, 302–311. [CrossRef]
22. Dodd, K.D.; Newans, T.J. Talent identification for soccer: Physiological aspects. *J. Sci. Med. Sport* **2018**, *21*, 1073–1078. [CrossRef]
23. Craig, T.P.; Swinton, P. Anthropometric and physical performance profiling does not predict professional contracts awarded in an elite Scottish soccer academy over a 10-year period. *Eur. J. Sport Sci.* **2020**, *21*, 1101–1110. [CrossRef]
24. Bergeron, M.F.; Mountjoy, M.; Armstrong, N.; Chia, M.; Côté, J.; Emery, C.A.; Faigenbaum, A.; Hall, G.; Kriemler, S.; Léglise, M.; et al. International Olympic Committee consensus statement on youth athletic development. *Br. J. Sports Med.* **2015**, *49*, 843–851. [CrossRef]
25. Den Hartigh, R.J.R.; Niessen, A.S.M.; Frencken, W.G.P.; Meijer, R.R. Selection procedures in sports: Improving predictions of athletes' future performance. *Eur. J. Sport Sci.* **2018**, *18*, 1191–1198. [CrossRef]
26. Wong, P.-L.; Chamari, K.; Dellal, A.; Wisløff, U. Relationship between anthropometric and physiological characteristics in youth soccer players. *J. Strength Cond. Res.* **2009**, *23*, 1204–1210. [CrossRef]
27. Wong, D.P.; Chamari, K.; Chaouachi, A.; Luk, T.C.; Lau, P.W.C. Heart Rate Response and Match Repeated-Sprint Performance in Chinese Elite Youth Soccer Players. *Asian J. Phys. Educ. Recreat.* **2011**, *17*, 42–49. [CrossRef]
28. Wong, D.P.; Wong, S.H. Physiological of professional Asian athlete elite youth soccer players. *J. Strength Cond. Res.* **2009**, *23*, 1383–1390. [CrossRef]
29. Rowat, O.; Fenner, J.; Unnithan, V. Technical and physical determinants of soccer match-play performance in elite youth soccer players. *J. Sports Med. Phys. Fit.* **2017**, *57*, 369–379. [CrossRef]
30. World Medical Association (WMA). World Medical Association Declaration of Helsinki: Ethical principles for medical research involving human subjects. *JAMA* **2013**, *310*, 2191–2194. [CrossRef]
31. Stewart, A.; Marfell-Jones, M.; Olds, T.; de Ridder, H. *International Standards for Anthropometric Assessment*; International Society for the Advancement of Kinanthropometry: Murcia, Spain, 2011.
32. Moore, V.C. Spirometry: Step by step. *Breathe* **2012**, *8*, 232–240. [CrossRef]
33. Menezes, A.M.B.; Wehrmeister, F.C.; Muniz, L.C.; Perez-Padilla, R.; Noal, R.B.; Silva, M.C.; Gonçalves, H.; Hallal, P.C. Physical activity and lung function in adolescents: The 1993 Pelotas (Brazil) birth cohort study. *J. Adolesc. Health* **2012**, *51*, S27–S31. [CrossRef]
34. Bosco, C.; Luhtanen, P.; Komi, P.V. A simple method for measurement of mechanical power in jumping. *Eur. J. Appl. Physiol.* **1983**, *50*, 273–282. [CrossRef]
35. Martín Acero, R.; Fernández-del Olmo, M.; Andrés Sánchez, J.; Luis Otero, X.; Aguado, X.; Rodríguez, F.A. Reliability of squat and countermovement jump tests in children 6 to 8 years of age. *Pediatr. Exerc. Sci.* **2011**, *23*, 151–160. [CrossRef]
36. Rumpf, M.C.; Cronin, J.B.; Oliver, J.L.; Hughes, M. Assessing youth sprint ability-methodological issues, reliability and performance data. *Pediatr. Exerc. Sci.* **2011**, *23*, 442–467. [CrossRef]
37. Vicente-Rodríguez, G.; Rey-López, J.P.; Ruíz, J.R.; Jiménez-Pavón, D.; Bergman, P.; Ciarapica, D.; Heredia, J.M.; Molnar, D.; Gutierrez, A.; Moreno, L.A.; et al. Interrater reliability and time measurement validity of speed-agility field tests in adolescents. *J. Strength Cond. Res.* **2011**, *25*, 2059–2063. [CrossRef]

38. Council of Europe. *Handbook for the Eurofit Tests of Physical Fitness*; Italian National Olympic Committee, Central Direction for Sport's Technical Activities Documentation and Information Division: Rome, Italy, 1988.
39. Castro-Pinero, J.; Chillon, P.; Ortega, F.B.; Montesinos, J.L.; Sjostrom, M.; Ruiz, J.R. Criterion-related validity of sit-and-reach and modified sit-and-reach test for estimating hamstring flexibility in children and adolescents aged 6–17 years. *Int. J. Sports Med.* **2009**, *30*, 658–662. [CrossRef]
40. Sakamaki, T.; Kato, N.; Fukumitsu, N.; Hasebe, A.; Adachi, C.; Takemori, K.; Yunoki, H. Studies on the method of measurement repeated side steps. *Jpn. J. Phys. Fit. Sports Med.* **1974**, *23*, 77–84. [CrossRef]
41. Cao, Y.; Zhang, C.; Guo, R.; Zhang, D.; Wang, S. Performances of the Canadian Agility and Movement Skill Assessment (CAMSA), and validity of timing components in comparison with three commonly used agility tests in Chinese boys: An exploratory study. *PeerJ* **2020**, *8*, e8784. [CrossRef]
42. Koopmann, T.; Faber, I.; Baker, J.; Schorer, J. Assessing Technical Skills in Talented Youth Athletes: A Systematic Review. *Sports Med.* **2020**, *50*, 1593–1611. [CrossRef]
43. Saavedra, J.M.; Escalante, Y.; Rodríguez, F.A. A multivariate analysis of performance in young swimmers. *Pediatr. Exerc. Sci.* **2010**, *22*, 135–151. [CrossRef]
44. Puccinelli, P.J.; Lima, G.H.O.; Pesquero, J.B.; de Lira, C.A.B.; Vancini, R.L.; Nikolaids, P.T.; Knechtle, B.; Andrade, M.S. Predictors of performance in Amateur Olympic distance triathlon: Predictors in amateur triathlon. *Physiol. Behav.* **2020**, *225*, 113110. [CrossRef]
45. Kochanowicz, A.; Kochanowicz, K.; Niespodziúski, B.; Mieszkowski, J.; Aschenbrenner, P.; Bielec, G.; Szark-Eckardt, M. Maximal Power of the Lower Limbs of Youth Gymnasts and Biomechanical Indicators of the Forward Handspring Vault Versus the Sports Result. *J. Hum. Kinet.* **2016**, *53*, 33–40. [CrossRef]
46. Elferink-Gemser, M.T.; Visscher, C.; Lemmink, K.A.P.M.; Mulder, T.W. Relation between multidimensional performance characteristics and level of performance in talented youth field hockey players. *J. Sports Sci.* **2004**, *22*, 1053–1063. [CrossRef]
47. Fernández, J.; Vila, M.; Rodríguez, F. Model study of the structure through a conditional multivariate analysis focused on talent identification in handball players. *Eur. J. Hum. Mov.* **2004**, *12*, 175–191. [CrossRef]
48. Gabbett, T.; Georgieff, B. Physiological and anthropometric characteristics of Australian junior national, state, and novice volleyball players. *J. Strength Cond. Res.* **2007**, *21*, 902–908. [CrossRef]
49. Deprez, D.N.; Fransen, J.; Lenoir, M.; Philippaerts, R.M.; Vaeyens, R. A retrospective study on anthropometrical, physical fitness, and motor coordination characteristics that influence dropout, contract status, and first-team playing time in high-level soccer players aged eight to eighteen years. *J. Strength Cond. Res.* **2015**, *29*, 1692–1704. [CrossRef]
50. Güllich, A.; Kovar, P.; Zart, S.; Reimann, A. Sport activities differentiating match-play improvement in elite youth footballers—A 2-year longitudinal study. *J. Sports Sci.* **2017**, *35*, 207–215. [CrossRef]
51. Sullivan, C.; Kempton, T.; Ward, P.; Coutts, A.J. The efficacy of talent selection criteria in the Australian Football League. *J. Sports Sci.* **2020**, *38*, 773–779. [CrossRef]
52. Samson, J.; Yerlès, M. Racial differences in sports performance. *Can. J. Sport Sci.* **1988**, *13*, 109–116.
53. Suminski, R.R.; Mattern, C.O.; Devor, S.T. Influence of racial origin and skeletal muscle properties on disease prevalence and physical performance. *Sports Med.* **2002**, *32*, 667–673. [CrossRef]
54. Larkin, P.; O'Connor, D. Talent identification and recruitment in youth soccer: Recruiter's perceptions of the key attributes for player recruitment. *PLoS ONE* **2017**, *12*, e0175716. [CrossRef]
55. Höner, O.; Leyhr, D.; Kelava, A. The influence of speed abilities and technical skills in early adolescence on adult success in soccer: A long-term prospective analysis using ANOVA and SEM approaches. *PLoS ONE* **2017**, *12*, e0182211. [CrossRef]
56. Arcos, A.L.; Martins, J. Physical fitness performance of young professional soccer players does not change during several training seasons in a Spanish elite reserve team: Club study, 1996–2013. *J. Strength Cond. Res.* **2018**, *32*, 2577–2583. [CrossRef]
57. Arcos, A.L.; Martinez-Santos, R.; Castillo, D. Spanish Elite Soccer Reserve Team Configuration and the Impact of Physical Fitness Performance. *J. Hum. Kinet.* **2020**, *71*, 211–218. [CrossRef]
58. Martinez-Santos, R.; Castillo, D.; Los Arcos, A. Sprint and jump performances do not determine the promotion to professional elite soccer in Spain, 1994–2012. *J. Sports Sci.* **2016**, *34*, 2279–2285. [CrossRef]
59. Fenner, J.S.J.; Iga, J.; Unnithan, V. The evaluation of small-sided games as a talent identification tool in highly trained prepubertal soccer players. *J. Sports Sci.* **2016**, *34*, 1983–1990. [CrossRef]
60. Serra-Olivares, J.; García López, L.M.; Calderón, A. Learning and talent in soccer. *Apunt. Educ. Física Deportes* **2017**, *33*, 64–77. [CrossRef]
61. Kannekens, R.; Elferink-Gemser, M.T.; Visscher, C. Positioning and deciding: Key factors for talent development in soccer. *Scand. J. Med. Sci. Sports* **2011**, *21*, 846–852. [CrossRef]
62. Höner, O.; Votteler, A.; Schmid, M.; Schultz, F.; Roth, K. Psychometric properties of the motor diagnostics in the German football talent identification and development programme. *J. Sports Sci.* **2015**, *33*, 145–159. [CrossRef]
63. Huijgen, B.C.H.; Elferink-Gemser, M.T.; Lemmink, K.A.P.M.; Visscher, C. Multidimensional performance characteristics in selected and deselected talented soccer players. *Eur. J. Sport Sci.* **2014**, *14*, 2–10. [CrossRef]
64. Rada, A.; Kuvačić, G.; De Giorgio, A.; Sellami, M.; Ardigò, L.P.; Bragazzi, N.L.; Padulo, J. The ball kicking speed: A new, efficient performance indicator in youth soccer. *PLoS ONE* **2019**, *14*, e0217101. [CrossRef]

65. Davids, K.; Baker, J. Genes, environment and sport performance: Why the nature-nurture dualism is no longer relevant. *Sports Med.* **2007**, *37*, 961–980. [CrossRef]
66. Ziv, G.; Lidor, R. Anthropometrics, Physical Characteristics, Physiological Attributes, and Sport-Specific Skills in Under-14 Athletes Involved in Early Phases of Talent Development—A Review. *J. Athl. Enhanc.* **2014**, *3*. [CrossRef]
67. Zuber, C.; Zibung, M.; Conzelmann, A. Holistic patterns as an instrument for predicting the performance of promising young soccer players—A3-years longitudinal study. *Front. Psychol.* **2016**, *7*, 1088. [CrossRef]
68. Schorer, J.; Rienhoff, R.; Fischer, L.; Baker, J. Long-term prognostic validity of talent selections: Comparing national and regional coaches, laypersons and novices. *Front. Psychol.* **2017**, *8*, 1146. [CrossRef]
69. Roberts, A.H.; Greenwood, D.A.; Stanley, M.; Humberstone, C.; Iredale, F.; Raynor, A. Coach knowledge in talent identification: A systematic review and meta-synthesis. *J. Sci. Med. Sport* **2019**, *22*, 1163–1172. [CrossRef]
70. Sieghartsleitner, R.; Zuber, C.; Zibung, M.; Conzelmann, A. Science or coaches' eye?—Both! beneficial collaboration of multidimensional measurements and coach assessments for efficient talent selection in elite youth football. *J. Sports Sci. Med.* **2019**, *18*, 32–43. [CrossRef]
71. Suppiah, H.T.; Low, C.Y.; Chia, M. Detecting and developing youth athlete potential: Different strokes for different folks are warranted. *Br. J. Sports Med.* **2015**, *49*, 878–882. [CrossRef]
72. Pol, R.; Balagué, N.; Ric, A.; Torrents, C.; Kiely, J.; Hristovski, R. Training or Synergizing? Complex Systems Principles Change the Understanding of Sport Processes. *Sports Med. Open* **2020**, *6*, 28. [CrossRef]
73. O'Connor, D.; Larkin, P.; Mark Williams, A. Talent identification and selection in elite youth football: An Australian context. *Eur. J. Sport Sci.* **2016**, *16*, 837–844. [CrossRef]
74. Höner, O.; Feichtinger, P. Psychological talent predictors in early adolescence and their empirical relationship with current and future performance in soccer. *Psychol. Sport Exerc.* **2016**, *25*, 17–26. [CrossRef]
75. Reeves, M.J.; McRobert, A.P.; Littlewood, M.A.; Roberts, S.J. A scoping review of the potential sociological predictors of talent in junior-elite football: 2000–2016. *Soccer Soc.* **2018**, *19*, 1085–1105. [CrossRef]
76. Euser, A.M.; Zoccali, C.; Jager, K.J.; Dekker, F.W. Cohort studies: Prospective versus retrospective. *Nephron-Clin. Pract.* **2009**, *113*, c214–c217. [CrossRef]
77. Taherdoost, H. Sampling Methods in Research Methodology; How to Choose a Sampling Technique for Research. *SSRN Electron. J.* **2018**, *5*, 18–27. [CrossRef]

 *healthcare*

Article

Change of Direction Speed and Reactive Agility in Prediction of Injury in Football; Prospective Analysis over One Half-Season

Dragan Mijatovic [1], Dragan Krivokapic [2], Sime Versic [3,4], Goran Dimitric [5] and Natasa Zenic [3,*]

[1] Faculty of Health Sciences, University of Mostar, 88000 Mostar, Bosnia and Herzegovina; mijatovicdragan@ymail.com
[2] Faculty for Sport and Physical Education, University of Montenegro, 81400 Niksic, Montenegro; dr.agan@t-com.me
[3] Faculty of Kinesiology, University of Split, 21000 Split, Croatia; sime.versic@kifst.eu
[4] HNK Hajduk, 21000 Split, Croatia
[5] Faculty of Sport and Physical Education, University of Novi Sad, 21000 Novi Sad, Serbia; goran.dimitric@uns.ac.rs
[*] Correspondence: natasazenic@gmail.com

Abstract: Agility is an important factor in football (soccer), but studies have rarely examined the influences of different agility components on the likelihood of being injured in football. This study aimed to prospectively evaluate the possible influences of sporting factors, i.e., flexibility, reactive agility (RAG), and change of direction speed (CODS), on injury occurrence over one competitive half-season, in professional football players. Participants were 129 football professional players (all males, 24.4 ± 4.7 years), who underwent anthropometrics, flexibility, and RAG and CODS (both evaluated on non-dominant and dominant side) at the beginning of second half-season 2019/20 (predictors). Over the following half-season, occurrence of injury was registered (outcome). To identify the differences between groups based on injury occurrence, t-test was used. Univariate and multivariate logistic regressions were calculated to identify the associations between predictors and outcome. Results showed incidence of 1.3 injuries per 1000 h of training/game per player, with higher likelihood for injury occurrence during game than during training (Odds Ratio (OR) = 3.1, 95%CI: 1.63–5.88) Univariate logistic regression showed significant associations between players' age (OR = 1.65, 95%CI: 1.25–2.22), playing time (OR = 2.01, 95%CI: 1.560–2.58), and RAG (OR = 1.21, 95%CI: 1.09–1.35, and OR = 1.18, 95%CI: 1.04–1.33 for RAG on dominant- and non-dominant side, respectively), and injury occurrence. The multivariate logistic regression model identified higher risk for injury in those players with longer playing times (OR = 1.81, 95%CI: 1.55–2.11), and poorer results for RAG for the non-dominant side (OR = 1.15, 95%CI: 1.02–1.28). To target those players who are more at risk of injury, special attention should be paid to players who are more involved in games, and those who with poorer RAG. Development of RAG on the non-dominant side should be beneficial for reducing the risk of injury in this sport.

Keywords: agility; soccer; flexibility; predictors; outcome

Citation: Mijatovic, D.; Krivokapic, D.; Versic, S.; Dimitric, G.; Zenic, N. Change of Direction Speed and Reactive Agility in Prediction of Injury in Football; Prospective Analysis over One Half-Season. *Healthcare* **2022**, *10*, 440. https://doi.org/10.3390/healthcare10030440

Academic Editors: João Paulo Brito and Rafael Oliveira

Received: 14 January 2022
Accepted: 24 February 2022
Published: 25 February 2022

Publisher's Note: MDPI stays neutral with regard to jurisdictional claims in published maps and institutional affiliations.

1. Introduction

From the aspect of kinesiological analysis, football (soccer) is a contact sport of intermittent and poly-structural nature, with high technical-tactical and physical demands [1–3]. Each player's performance is directly influenced by the environment (i.e., teammates, opponents, and ball). To these stimuli, the player must constantly adapt and react in order to assure situational efficiency. Over the last decade, the intensity of the game has increased significantly, and studies have recorded an increase in sprinting distance (speeds above 25 km/h) by approximately 35% over a period of 7 years [1,4]. Additionally, the frequency of games is constantly increasing [5]. As a result of such increases in the psycho-physiological demands, and despite the evident progress of sports medicine, better sports equipment and training grounds and improved recovery methods and prevention programs, injuries are one of the most evident problems in football [6]. Non-contact muscular injuries represent a special problem, which, although being mostly preventable through primary and secondary prevention strategies, account for one third of all time-loss injuries in men's professional football [7–9].

Injuries in professional football represent multiple problems. Firstly, they can remove a player from training and competition for a while, thereby reducing the value of the team and its chances of success. Additionally, they represent a financial burden, either through direct treatment costs or through indirect costs for salaries of the players that are not performing [1,2,10]. In particular, at the elite European level, a one-month absence of a starting player costs the club in question approximately EUR 500,000 on average [11]. Studies have shown that players in the European leagues have an average of two injuries per year and spend around 37 days out of team training [12,13]. All of this has led scientists to study the predictors of injuries in football players to identify potential risks and try to reduce the number of injuries [14].

Recent findings suggest that causes of injuries should be analyzed in a multifactorial and dynamic etiological model with intrinsic (person-related) and extrinsic (environment-related) risk factors, along with training and match workloads [13,14]. Intrinsic factors include the individual biological and psychosocial characteristics of athletes. These, more specifically, include age and gender determinants, muscle strength and flexibility, functional instabilities, imbalances, previous injuries, and maladaptive rehabilitation processes [14].

Many studies have addressed the associations of different motor and functional abilities with injuries [9,14,15]. In general, insufficient muscular strength and endurance, poor cardiorespiratory endurance, and lack of flexibility, power, speed, and balance represent risk factors for musculoskeletal injuries [16,17]. More specifically, in terms of flexibility, higher ranges of motion for the hamstrings and ankles were associated with lower injury risk [9]. There is also moderate evidence that lower levels of body strength, power and balance are good predictors of future muscle injuries [9,15]. Finally, cardiorespiratory endurance showed an association with injury: strong evidence was found that poor performance on timed shuttle runs increased the risk of injury [18]. Interestingly, the authors of a recent comprehensive review, which overviewed the associations between variables of physical fitness and musculoskeletal injury risk, clearly noted that they did not find any study linking agility and injury occurrence in athletes [9].

Specifically, agility represents the ability to quickly and accurately change the direction of motion of the whole body in response to a stimulus [19]. It is generally accepted that agility in professional sports should be graded according to two relatively independent qualities: (i) change of direction speed (CODS) or pre-planned agility, and (ii) reactive agility or non-planned agility (RAG) [20]. From the perspective of injury occurrence, it is important to highlight the specifics of CODS and RAG. While CODS is a change in the speed and direction of movement according to a predetermined pattern, this facet of agility is predominantly influenced by speed, power and anthropometrics [21]. On the other hand, during RAG the athlete must change his direction of movement in response to an external stimulus (i.e., opponent, ball, or teammate), which is logically and additionally influenced by perceptual and cognitive capacities [22,23].

Both CODS and RAG have been repeatedly confirmed as important determinants of success in sports [24–26]. This is especially emphasized in team sports, such as football, where the movement of the athlete is directly determined by the movements of teammates, opponents, and the ball itself [27–29]. Therefore, there is no doubt that both RAG and CODS capacities are directly responsible for accurate, effective, and (most importantly) safe execution of directional changes, which are themselves known to be movement scenarios where a great deal of injuries occur both in training and in competition [30–32]. Surprisingly, there is a lack of studies considering agility as a factor influencing injury occurrence in sports, and to the best of our knowledge no study has reported this association in professional football.

Given the lack of knowledge on the association between agility and musculoskeletal injuries in football, the main aim of this research was to prospectively observe the influences of agility components (RAG and CODS) on injury occurrence in professional football players during one half-season. Additionally, we observed flexibility and sporting factors (playing position, age, experience in football, playing time) as factors of influence on injury occurrence. We hypothesized that RAG and CODS status at the beginning of the season would be inversely related to occurrence of injury, meaning lower risks of injury in players who performed better on football-specific tests of CODS and RAG.

2. Materials and Methods

2.1. Participants

In this study we observed 129 professional football players from Bosnia and Herzegovina (all males, 24.4 ± 4.7 years of age) All of them were members of teams competing at the highest competitive level in the country, including the national champions for the observed season.

The invitation for study participation was sent by the National Football Federation, and players were informed of potential benefits and risks. Involvement in the research was voluntary, and participants' personal data were protected with an identification code, known only to the main/head researcher. Prior to the study, participants signed consent forms for participation and the Ethical Board of University of Split, Faculty of Kinesiology approved the investigation (approval number: 2181-205-02-05-14-001). Based on (i) a previously evidenced injury occurrence of 20%, (ii) a population sample of 250 professional players competing in the first national division in Bosnia and Herzegovina (goalkeepers and players younger than 18 years were excluded from total population), (iii) a margin of error of 0.05, and (iv) a confidence level of 0.95, the required sample size for this investigation was calculated to be 125 participants (calculated by Statistica, Tibco Inc., Palo Alto, CA, USA).

Inclusion criteria were: (i) a minimum of 8 years of active football training, (ii) competing at the highest national level, (iii) no injury/illness 15 days prior to baseline testing. Exclusion criteria included: (i) being a goalkeeper, (ii) pain, illness, or injury which prevented the player from performing the tests used in the study, (iii) playing less than 15 min per game prior to occurrence of injury in the following half-season.

This prospective study included baseline testing and during a follow-up period (first competitive half-season in 2019/20) (Figure 1). At baseline (January 2020), we tested predictors, and during the follow-up period (continuously over a period of six months after baseline) we observed outcomes (injury status) (Figure 1).

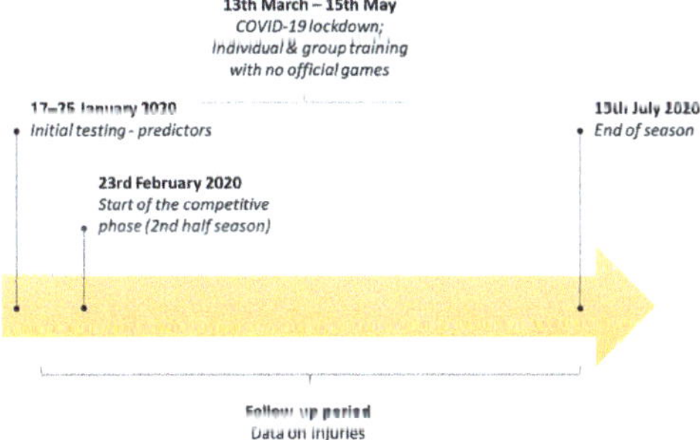

Figure 1. Study design.

2.2. Variables and Measurements

In this investigation we observed predictors and outcomes. Predictors included age (in years), age when players started to play football (later described as experience in football), anthropometric indices, flexibility, reactive agility, and change of direction speed. Injury status over the competitive half-season was the outcome.

Anthropometrics included (i) body height (in 0.5 cm) (Seca, Birmingham, UK), (ii) body mass (in 0.1 kg) (Tanita TBF-300, Tanita, Tokyo, Japan), both measured with standardized techniques and calibrated equipment, and (iii) calculated body mass index (BMI; in kg/m^2).

We used two tests of flexibility, the sit and reach test (SIT&REACH), and a test of maximal abduction (MAXABD). The SIT&REACH was commenced under standardized conditions. In brief, the participant was instructed to sit on the floor, with his bare feet placed vertically against a measuring box. He had to lean forward maximally while keeping his knees and arms fully extended, to reach maximally along the measuring tape on the box, and to maintain the final reaching position for approximately 3 s. The measurement was carried out over three trials, and after the reliability analysis showed high consistency of the measurements (intra class correlation (ICC) of 0.96), the highest score was kept as the final result for each participant [33].

The MAXABD was performed using standardized settings and equipment. Each participant sat on the floor, resting his back on the wall, and was instructed to perform maximal voluntary leg abduction while keeping his legs constantly in contact with the floor, and knees fully extended. The maximal distance between the inner malleoli of the right and left legs was measured. The test was performed over three trials and, following the reliability analysis which confirmed appropriate reliability (ICC = 0.93), the maximal score was recorded as the final result for each participant [34].

The CODS and RAG were tested using an originally developed hardware system based on an ATMEL micro-controller (model AT89C51RE2; ATMEL Corp., San Jose, CA, USA). This measurement equipment has been repeatedly used in studies involving athletes from various sports, including football [21,27]. For the measurement, a photoelectric infrared sensor (E18-D80NK) was used as the time triggering input, and LEDs placed in the 30-cm-high cones were used as controlled outputs. Distances used in RAG and CODS were identical (Figure 2), but execution of the tests differed with regard to testing scenarios (e.g., pre-planned and unplanned testing templates—see the following text for details).

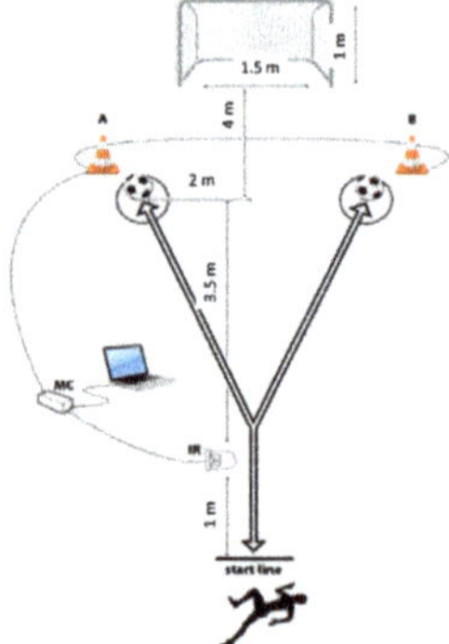

Figure 2. Testing of reactive agility and change of direction speed in football.

For both agility tests, the player ran at maximal intensity through the gate, and when he crossed the infrared signal timing began. At the same moment (no time delay), one cone (either "A" or "B") was lit. The player had to run at maximum speed in the designated direction, kick the ball with the inside of the foot toward the goal, and then turn and run back as quickly as possible to the starting gate. The timing stopped when the player passed the gate on the way back. For CODS testing, players were informed about which cone would be lit in each trial, and therefore the movement template could be pre-planned. Four trials were performed (A, B, A, B for trials 1–4, respectively), and there was a rest of 20 s between trials. For RAG testing, players did not have information on which cone would be lit up, so their movements were not pre-planned. Despite the fact the players did not know the template of RAG testing in advance, all players were tested using the same sequence throughout the five trials (trial 1: A cone, trial 2: B, trial 3: B, trial 4: A, trial 5: A), with 20 s of rest between trials.

In this study we separately observed performances on dominant and non-dominant sides for both CODS and RAG, as recently suggested [35]. In evidencing dominant and non-dominant side performance, we calculated the average of the first and third trial for CODS (e.g., average performance in the direction of cone A), and the average of the second and fourth trial (average for direction cone B). The better score marked the dominant side, while the poorer result was considered to mark the non-dominant side. Similarly, the average of the first and fourth trial was compared to the average of the second and third trial for RAG, and dominant vs. non-dominant sides were established accordingly. The reliability of the CODS was better than the reliability of the reactive agility (intra class correlations of 0.89 and 0.79 for CODS and reactive agility, respectively), but this was expected and has been explained in detail previously [23].

The Oslo Sports Trauma Research Center Overuse Injury Questionnaire (OSTRC) was used for injury recording [36]. Players were questioned with the OSTRC once a week by the team physician, who sent the results to the main investigator (first author). The outcomes of this study were the incidences of musculoskeletal injuries that occurred during the study in four body regions: ankle, knee, back, and shoulder. Each answer in the OSTRC corresponds to a score. For each question (body location), a score between 0 and 25 is given, and a theoretical score (sum) ranging from 0 to 100 is calculated for four body regions. Reported scores of >39 were classified as injuries. Together with injury reports, physicians reported the total number of hours of training/games for their team. Finally, injuries were reported in total, along with in number of injuries per 1000 h of training/game. Although we observed multiple injuries occurring, for the purposes of statistical analysis and identification of the relationships between predictors and outcome, we did not differentiate single injuries from multiple injuries [37].

2.3. Statistics

The Kolmogorov–Smirnov test was used to check for the normality of the distributions for all variables, and all variables but injury occurrence were found to be normally distributed. Consequently, means and standard deviations were reported for all normally distributed variables, while frequencies and percentages were reported for injury occurrence.

T-test for independent samples was performed to identify the differences between groups based on injury occurrence. Additionally, differences between injured and non-injured players were evaluated by magnitude-based Cohen's effect size (ES) statistics (including 95% Confidence Intervals—95%CI) with modified qualitative descriptors, using the following criteria: <0.02 = trivial; 0.2–0.6 = small; >0.6–1.2 = moderate; >1.2–2.0 = large; and >2.0 very large differences.

The associations between the studied predictors and outcomes were evaluated by a logistic regression calculation using binarized criteria on the categorized OSTRC scale (0 = absence of injury, 1 = injury). The odds ratios (OR) with the corresponding 95% CI are reported. In the first phase, predictors were univariately correlated with outcomes. In the next phase, all significant predictors were simultaneously included in the multivariate logistic regression calculation, in order to control the possible confounding effects of the predictors. The Hosmer–Lemeshow test was used to check the model fit (with a significant χ^2 indicating an inappropriate model fit).

Statistica version 13.5 (Tibco Inc., Palo Alto, CA, USA) was used for all analyses, and a significance level of $p < 0.05$ was applied.

3. Results

During the period of this research, 35 players (32%) suffered from injury, while 75 (68%) did not experience any injury as evidenced by the OSTRC questionnaire.

Over approximately 40,000 h of training/game, 51 injuries occurred in total, resulting in 1.3 injury per 1000 h of exposure (training/game). Of all injuries, 61% (32 injuries) occurred during training, while 39% (19 injuries) occurred during games. Since games accounted for 12% of overall players' exposure during study period, the risk for being injured during a game was three times higher than the risk for being injured during training (OR = 3.1, 95%CI: 1.63–5.88).

A total of 8% of players suffered from multiple injuries. On average, each player suffered from 0.46 injuries over the study period.

Injured players were older (*t*-test = 2.43, $p < 0.05$, moderate ES differences), more experienced in football (*t*-test = 2.02, $p = 0.02$, moderate ES differences), more involved in the game (*t*-test = 4.23, $p = 0.01$, moderate ES differences), and achieved poorer results in RAG-D (*t*-test = 2.03, $p < 0.05$, moderate ES differences) and RAG-ND (*t*-test = 2.47, $p = 0.02$, moderate ES differences) (Table 1).

Table 1. Descriptive statistics and differences in study variables between injured and non-insured players.

Variables	Injured (*n* = 35)		Non-Injured (*n* = 75)		*t*-test		Effect Size
	Mean	SD	Mean	SD	*t*-test	*p*	d (95%CI)
Age (years)	26.2	4.4	23.8	4.8	2.43	0.02	0.51 (0.11–0.92)
Experience in football (years)	16.97	2.93	14.79	2.87	2.02	0.03	0.75 (0.34–1.16)
Playing time (min/game)	67.11	7.03	58.98	8.98	4.23	0.01	0.96 (0.54–1.38)
Body height (cm)	187.87	6.78	185.25	7.21	0.25	0.79	0.37 (−0.03–0.77)
Body mass (kg)	78.99	7.51	79.00	6.25	0.01	0.99	0.01 (−0.4–0.40)
Body mass index (kg/m^2)	23.32	1.23	23.26	1.18	0.22	0.81	0.05 (−0.35–0.45)
CODS-ND (s)	2.87	0.25	2.89	0.31	0.24	0.78	0.06 (−0.33–0.46)
CODS-D (s)	2.74	0.2	2.76	0.39	0.3	0.74	0.06 (−0.34–0.46)
RAG-ND (s)	3.15	0.20	2.94	0.24	2.47	0.02	0.92 (0.50–1.34)
RAG-D (s)	3.08	0.19	2.94	0.25	2.03	0.04	0.66 (0.25–1.07)
SIT&REACH (cm)	30.26	7.51	29.69	7.39	0.34	0.72	0.07 (−0.32–0.47)
MAXABD (cm)	138.35	12.79	138.57	13.25	0.07	0.8	0.016 (−0.38–0.42)

Legend: CODS—change of direction speed, RAG—reactive agility, D—dominant side, ND—nondominant side, SIT&REACH—sit and reach flexibility test, MAXABD—maximal abduction flexibility test

Results of the univariate logistic regression calculations are presented in Figure 3. Univariate logistic analysis evidenced significant correlations between playing time (OR = 2.01, 95%CI: 1.560–2.58, $p < 0.001$), players' age (OR = 1.65, 95%CI: 1.25–2.22, $p < 0.01$), experience in football (OR = 1.51, 95%CI: 1.03–2.08, $p < 0.05$), RAG-D (OR = 1.18, 95%CI: 1.04–1.33, $p < 0.05$) and RAG-ND (OR = 1.21, 95%CI: 1.09–1.35, $p < 0.05$), and injury occurrence. Evidently, the risk of injury over the competitive half-season was higher for players who were more involved in active play, older players, those who had more experience in football, and those who performed worse in either reactive agility test.

When all significant predictors were simultaneously included in multivariate logistic regression, significant associations were retained for playing time (OR = 1.81, 95%CI: 1.55–2.11, $p < 0.001$) and RAG-ND (OR = 1.15, 95%CI: 1.02–1.28, $p < 0.05$), with proper model fit as indicated by Hosmer Lemeshow test ($p > 0.05$). Specifically, players who were more involved in the game and who performed poorly in the reactive agility task on the non-dominant side were more prone to injury over the study period (Figure 4).

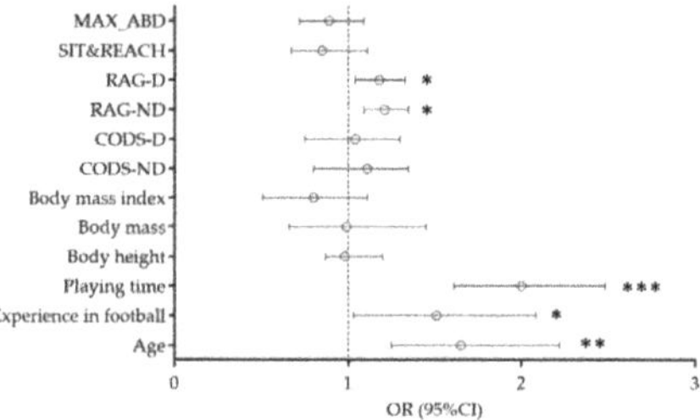

Figure 3. Results of the univariate logistic analysis for prediction of injury occurrence (*** $p < 0.001$, ** $p < 0.01$, * $p < 0.05$).

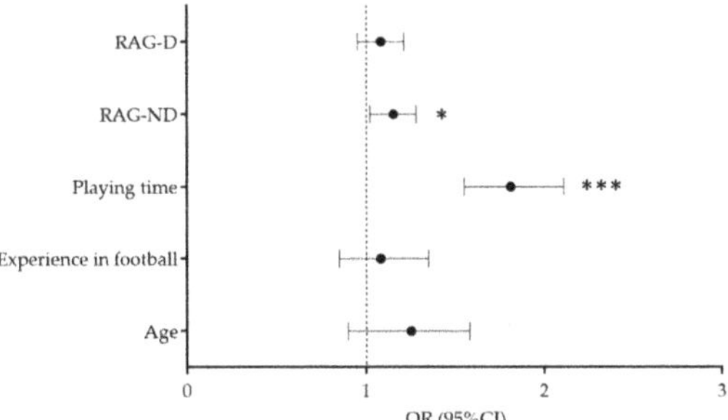

Figure 4. Results of the multivariate logistic analysis in prediction of injury occurrence (*** $p < 0.001$, * $p < 0.05$).

4. Discussion

The main aim of this study was to investigate the influences of RAG and CODS agility components on injury occurrence in professional football players, and we hypothesized lower risks of injury in players who performed better on the football-specific tests of CODS and RAG. Regarding this, there are several important findings of the study. First, at the univariate level, older age, longer playing time, and poor performances in both RAG tests (RAG-ND, and RAG-D) were identified as risk factors for injury occurrence. However, the multivariate model identified longer playing time and poorer RAG-ND as the only significant risk factors for injury occurrence. As a result, our initial hypothesis may be partially accepted.

4.1. Age and Playing Time as Predictors of Injury

Our results suggest that older football players are more prone to injuries, and this is generally supported by previous reports. Even studies carried out 30 years ago have shown that the incidence of injuries in football increases with the age of the players (0.0331 versus 4.021 injuries in youth and professional players, respectively) [38]. This difference can be seen even in younger age categories, where the incidence of injuries is five times lower among athletes aged 7 to 12 compared to those older than 12 [38]. This is not characteristic only for males, as age has been shown to be a significant risk factor in a prospective study of female football players also [39]. Although it is clear that injury incidence will be affected by the large difference in intensity between training and games (through it increases in both settings with age), older players will undoubtedly be exposed to higher training loads over the years, which itself increases the risk of injury [38].

A prospective cohort study of 23 elite European teams showed that over a 7-year period, a significantly higher incidence of injuries was observed in matches compared to training sessions (27.5 vs. 4.1, per 1000 h of exposure) [12]. Supportively, results of our study suggest that active involvement in the game (e.g., playing time) increases the risk of injury. However, it is important to note that in our multivariate logistic regression model (when all significant predictors were simultaneously included in the model), playing time actually diminished the influence of age on injury occurrence. In other words, older (i.e., more experienced) players spend more time playing in games and play more matches, which naturally increases their likelihood of being injured [12,10 12].

It is relatively well documented that that a larger amount of exposure to soccer directly correlates with the number of injuries [43,44]. A study of 41 teams in the four best divisions in Sweden stated that exposure to soccer and ankle sprain injuries per player were more frequent in higher divisions [44]. Similar findings were presented in a study of female soccer players, where multivariate logistic regression showed that higher exposure to soccer significantly increases the risk of traumatic leg injuries [43]. Additionally, a study of South African Premier Soccer League clubs showed that, besides other factors, exposure time to soccer has an impact of the risks and mechanisms of injuries [45]. Therefore, our results are generally supported by previous considerations regarding playing time as a risk factor for injuries in soccer.

4.2. Reactive Agility and Injury Occurence

One of the most important findings of this study was that players who achieved better results on RAG tests had lower incidences of injury. At the same time, results on the CODS tests were not significantly related to the frequency of injuries. In analyzing these findings, the differences between determinants of CODS and determinants of RAG should be considered. In brief, CODS is primarily influenced by motor capacities (e.g., speed, power, and coordination) and anthropometrics. On the other hand, the importance of these factors is not so pronounced in RAG, where perceptual and decision-making skills (e.g., visual scanning, anticipation, pattern recognition and knowledge of situations) are highly important [19]. It is therefore reasonable to conclude that players who achieved better results on RAG generally have better perception of the environment during games and react better to new situations.

The movements of players during a football game are constantly influenced by the movement of teammates, opposing players and the ball. As a result, faster and better decision making is expected to be a protective factor against dangerous and risky situations. More specifically, a player with better RAG will be able to recognize a potentially risky situation beforehand, and to avoid unnecessary duels or collisions with the opponent. Additionally, a player who has better RAG can react in a timely manner and prevent additional high-intensity stress to his muscles, such as sprinting, sudden breaking, and similar movements that risk injuries [7].

A recent systematic review of the associations between physical fitness and muscle injuries has found that certain levels of flexibility, power, speed, and balance present risk factors for injury occurrence, but the authors stated that no study has analyzed the relations between agility and muscle injuries in football [9]. That said, we did find one study where the authors examined the influence of agility on injury occurrence in Australian rugby. In this study, rugby players with poorer RAG performances, specifically longer decision times, had a lower risk of injury [46]. These results were surprising, even for the authors themselves, who originally expected that RAG would be protective against injuries. However, their explanation was that players with poorer reaction times can inadvertently avoid severe collisions which lead to injuries and have just partial contact with opponents which does not result in full force tackles [46].

Although the finding from the previously cited Australian rugby study disagrees with our findings (we found a lower injury risk for players with better RAG), several important facts could explain the disagreement. First of all, rugby is a contact sport where frequent hitting and tackling occurs, and in the aforementioned study, contact injuries were analyzed [46]. On the other hand, in this research, both contact- and non-contact injuries were analyzed. Furthermore, in the study on rugby players, the playing time was not analyzed as a predictor of injuries. Therefore, it was possible that those players who spent more time in the game had better RAG, and because of that were simply more exposed to high-risk situations. This is particularly important if we consider that better RAG in rugby is closely related to the qualitative level of players, and therefore it can be assumed that these players will spend more time in game play [47,48]. Finally, in the rugby study, a significant predictor of injuries was "decision time during RAG" (and not RAG itself). Although an important determinant of RAG, "decision time" is actually only one part of RAG performance. On the other hand, herein we used a sport-specific RAG test that includes a specific ball manipulation task, which entails a much more complex movement pattern, all the more so considering the dominant and non-dominant-side testing.

4.3. Dominant vs. Non-Dominant Side Reactive Agility Performances and Injury Occurence

RAG-ND stands out as the more significant predictor of injury occurrence than RAG-D. More specifically, when we calculated multiple logistic results, the RAGND was a significant predictor of injury occurrence, together with playing time. The first reason for the fact that RAGD was not a significant predictor in the multivariate calculation can probably be found in the level of the players (professional players with more than 10 years of experience in systematic football training). Therefore, all participants were highly skilled, and their dominant-side performances (irrespective of the unplanned nature of the task for RAG) were highly efficient. On the other hand, execution of the RAG test on the non-dominant side actually presents the "real-world" unpredictable movement template and is logically more related to eventual discrepancies in reactive agility tasks on the field [49].

Specifically, RAG represents coping in emerging situations [50]. Since the RAG-D test for our players enabled them to use their highly refined skills, the RAG-ND was more likely to be a significant predictor of injury, since it was more likely to be representative of unexpected circumstances. Indeed, during the game, footballers have 1200–1400 changes of direction, which occur on average every 2 to 4 s [51]. These events include path changes, jumping, acceleration, and deceleration, and themselves represent risky movements in terms of injury [52,53]. Since most of these high-risk-movements take place under the influence of external stimuli (i e , reactively), the importance of RAG (especially when executed on the non-dominant side) in the prevention of injuries is clearly substantial.

4.4. Limitations and Strengths

The most important limitation comes from the fact that we observed players over one half-season only. Therefore, we lacked data on injuries later in the season. Second, in this study we used OSTRC as an overall index of injury status. There is no doubt that some of the injuries recorded by the OSTRC in our study could not be (even theoretically) connected to the studied motor capacities but are clearly related to collisions and contact (shoulder injuries, for example). However, we were of the opinion that the arbitrary elimination of some injuries from the record would have resulted in even greater error, so we included all recorded injuries. Additionally, our study involved professional football players from only one country, and during a specific period of time (a season which was interrupted by COVID-19 lockdown). Due to the great deal of contextual factors which are known to be related to specifics of the sport (even injury occurrence), the results are generalizable to samples similar to the one observed in our research.

This is one of the first studies where different agility components were related to injury occurrence in professional team sports, and to the best of our knowledge the first one where this was done for football. We observed professional football players competing at the highest level of national competition. The usage of the football-specific tests of CODS and RAG, and separate evaluations of the performances on the dominant and non-dominant sides, are important strengths of the study. Therefore, although this investigation is not the final word on a topic, we hope it will increase knowledge and initiate further studies.

5. Conclusions

This study highlighted the importance of RAG as a predictor of injury occurrence. Generally, professional players perform numerous changes in direction per game in pre-planned and unplanned scenarios, and these manoeuvres are often performed rapidly when they do not have full control over their bodies. Therefore, it is clear that scenarios requiring reactive agility present particularly risky conditions for dysfunctional locomotion, and consequently may result in injury. As a result, coaches working with football athletes should pay particular attention to mastering technical movement skills which will assure musculoskeletal safety in RAG tasks.

Special attention should be paid to RAG performance on the non-dominant side. Specifically, we studied professional players, and almost certainly their dominant-side RAG performances were highly refined. In other words, despite the differences in RAG-D performance, the technical quality of the RAG-D execution for most of the players was almost certainly proficient, and therefore safe. Meanwhile, RAG-ND was found to be an important determinant of injury occurrence, even when statistically controlled for the most significant factor of influence—playing time. This clearly points to the necessity of: (i) independent evaluation of RAG performances on dominant and non-dominant sides, and (ii) specific mastering of RAG on the non-dominant side in professional football players.

Playing time was found to be most important risk factor for injury occurrence in the studied players. While games present the most important psycho-physiological stress on players, resulting in large numbers of injury-risk situations, such results are expected and in line with previous reports. As a result, in order to decrease the risk of injury in professional football, special attention should be paid to players who are on the field during games for a long time, especially if they are older and lack RAG.

Author Contributions: Data curation, D.M.; Formal analysis, G.D.; Funding acquisition, D.K.; Methodology, D.M. and N.Z.; Project administration, D.M.; Resources, D.K.; Supervision, G.D.; Validation, S.V.; Writing—original draft, S.V., G.D. and N.Z. All authors have read and agreed to the published version of the manuscript.

Funding: This research was funded by the Croatian Science Foundation (grant number IP-2018-01-8330).

Institutional Review Board Statement: The study was conducted according to the guidelines of the Declaration of Helsinki and approved by the Institutional Ethics Committee of Faculty of Kinesiology, University of Split (protocol code No.: 2181-205-02-05-14-001, 8 January 2018).

Informed Consent Statement: Informed consent was obtained from all subjects involved in the study.

Data Availability Statement: Data will be provided to all interested parties upon reasonable request.

Conflicts of Interest: The authors declare no conflict of interest.

References

1. Enright, K.; Green, M.; Hay, G.; Malone, J.J. Workload and injury in professional soccer players: Role of injury tissue type and injury severity. *Int. J. Sports Med.* **2020**, *41*, 89–97. [CrossRef]
2. Owen, A.L.; Forsyth, J.J.; Wong, D.P.; Dellal, A.; Connelly, S.P.; Chamari, K. Heart rate–based training intensity and its impact on injury incidence among elite-level professional soccer players. *J. Strength Cond. Res.* **2015**, *29*, 1705–1712. [CrossRef]
3. Fang, B.; Kim, Y.; Choi, M. Effect of Cycle-Based High-Intensity Interval Training and Moderate to Moderate-Intensity Continuous Training in Adolescent Soccer Players. *Healthcare* **2021**, *9*, 1628. [CrossRef] [PubMed]
4. Barnes, C.; Archer, D.; Hogg, B.; Bush, M.; Bradley, P. The evolution of physical and technical performance parameters in the English Premier League. *Int. J. Sports Med.* **2014**, *35*, 1095–1100. [CrossRef] [PubMed]
5. Dupont, G.; Nedelec, M.; McCall, A.; McCormack, D.; Berthoin, S.; Wisløff, U. Effect of 2 soccer matches in a week on physical performance and injury rate. *Am. J. Sports Med.* **2010**, *38*, 1752–1758. [CrossRef]
6. Ekstrand, J.; Lundqvist, D.; Davison, M.; D'Hooghe, M.; Pensgaard, A.M. Communication quality between the medical team and the head coach/manager is associated with injury burden and player availability in elite football clubs. *Br. J. Sports Med.* **2019**, *53*, 304–308. [CrossRef] [PubMed]
7. Ekstrand, J.; Hägglund, M.; Waldén, M. Epidemiology of muscle injuries in professional football (soccer). *Am. J. Sports Med.* **2011**, *39*, 1226–1232. [CrossRef]
8. Jaspers, A.; Brink, M.S.; Probst, S.G.; Frencken, W.G.; Helsen, W.F. Relationships between training load indicators and training outcomes in professional soccer. *Sports Med.* **2017**, *47*, 533–544. [CrossRef]
9. Sarah, J.; Lisman, P.; Gribbin, T.C.; Murphy, K.; Deuster, P.A. Systematic review of the association between physical fitness and musculoskeletal injury risk: Part 3—flexibility, power, speed, balance, and agility. *J. Strength Cond. Res.* **2019**, *33*, 1723–1735.
10. Daneshjoo, A.; Nobari, H.; Kalantari, A.; Amiri-Khorasani, M.; Abbasi, H.; Rodal, M.; Pérez-Gómez, J.; Ardigò, L.P. Comparison of Knee and Hip Kinematics during Landing and Cutting between Elite Male Football and Futsal Players. *Healthcare* **2021**, *9*, 606. [CrossRef]
11. Ekstrand, J. Keeping your top players on the pitch: The key to football medicine at a professional level. *Br. J. Sports Med.* **2013**, *47*, 723–724. [CrossRef]
12. Ekstrand, J.; Hägglund, M.; Waldén, M. Injury incidence and injury patterns in professional football: The UEFA injury study. *Br. J. Sports Med.* **2011**, *45*, 553–558. [CrossRef]
13. Jaspers, A.; Kuyvenhoven, J.P.; Staes, F.; Frencken, W.G.; Helsen, W.F.; Brink, M.S. Examination of the external and internal load indicators' association with overuse injuries in professional soccer players. *J. Sports Sci. Med.* **2018**, *21*, 579–585. [CrossRef] [PubMed]
14. Dvorak, J.; Junge, A.; Chomiak, J.; Graf-Baumann, T.; Peterson, L.; Rosch, D.; Hodgson, R. Risk factor analysis for injuries in football players. *Am. J. Sports Med.* **2016**. [CrossRef]

15. de la Motte, S.J.; Gribbin, T.C.; Lisman, P.; Murphy, K.; Deuster, P.A. Systematic Review of the Association Between Physical Fitness and Musculoskeletal Injury Risk: Part 2—Muscular Endurance and Muscular Strength. *J. Strength Cond. Res.* **2017**, *31*, 3218–3234. [CrossRef] [PubMed]
16. Ursej, E.; Sekulic, D.; Prus, D.; Gabrilo, G.; Zaletel, P. Investigating the Prevalence and Predictors of Injury Occurrence in Competitive Hip Hop Dancers: Prospective Analysis. *Int. J. Environ. Res. Public Health* **2019**, *16*, 3214. [CrossRef] [PubMed]
17. Bennett, H.; Chalmers, S.; Milanese, S.; Fuller, J. The association between Y-balance test scores, injury, and physical performance in elite adolescent Australian footballers. *J. Sci. Med. Sport* **2021**. [CrossRef]
18. Lisman, P.J.; de la Motte, S.J.; Gribbin, T.C.; Jaffin, D.P.; Murphy, K.; Deuster, P.A. A Systematic Review of the Association Between Physical Fitness and Musculoskeletal Injury Risk: Part 1—Cardiorespiratory Endurance. *J. Strength Cond. Res.* **2017**, *31*, 1744–1757. [CrossRef]
19. Sheppard, J.M.; Young, W.B. Agility literature review: Classifications, training and testing. *J. Sports Sci. Med.* **2006**, *24*, 919–932. [CrossRef]
20. Sekulic, D.; Spasic, M.; Esco, M.R. Predicting agility performance with other performance variables in pubescent boys: A multiple-regression approach. *Percept. Mot. Skills* **2014**, *118*, 447–461. [CrossRef]
21. Pehar, M.; Sisic, N.; Sekulic, D.; Coh, M.; Uljevic, O.; Spasic, M.; Krolo, A.; Idrizovic, K. Analyzing the relationship between anthropometric and motor indices with basketball specific pre-planned and non-planned agility performances. *J. Sports Med. Phys. Fitness* **2018**, *58*, 1037–1044. [CrossRef] [PubMed]
22. Sekulic, D.; Pehar, M.; Krolo, A.; Spasic, M.; Uljevic, O.; Calleja-González, J.; Sattler, T. Evaluation of Basketball-Specific Agility: Applicability of Preplanned and Nonplanned Agility Performances for Differentiating Playing Positions and Playing Levels. *J. Strength Cond. Res.* **2017**, *31*, 2278–2288. [CrossRef]
23. Sekulic, D.; Krolo, A.; Spasic, M.; Uljevic, O.; Peric, M. The development of a New Stop'n'go reactive-agility test. *J. Strength Cond. Res.* **2014**, *28*, 3306–3312. [CrossRef]
24. Gabbett, T.J.; Kelly, J.N.; Sheppard, J.M. Speed, change of direction speed, and reactive agility of rugby league players. *J. Strength Cond. Res.* **2008**, *22*, 174–181. [CrossRef]
25. Coh, M.; Vodicar, J.; Žvan, M.; Šimenko, J.; Stodolka, J.; Rauter, S.; Mackala, K. Are Change-of-Direction Speed and Reactive Agility Independent Skills Even When Using the Same Movement Pattern? *J. Strength Cond. Res.* **2018**, *32*, 1929–1936. [CrossRef] [PubMed]
26. Ribeiro, J.; Afonso, J.; Camões, M.; Sarmento, H.; Sá, M.; Lima, R.; Oliveira, R.; Clemente, F.M. Methodological Characteristics, Physiological and Physical Effects, and Future Directions for Combined Training in Soccer: A Systematic Review. *Healthcare* **2021**, *9*, 1075. [CrossRef] [PubMed]
27. Pojskic, H.; Åslin, E.; Krolo, A.; Jukic, I.; Uljevic, O.; Spasic, M.; Sekulic, D. Importance of reactive agility and change of direction speed in differentiating performance levels in junior soccer players: Reliability and validity of newly developed soccer-specific tests. *Front. Physiol.* **2018**, *9*, 506. [CrossRef] [PubMed]
28. Jovanovic, M.; Sporis, G.; Omrcen, D.; Fiorentini, F. Effects of speed, agility, quickness training method on power performance in elite soccer players. *J. Strength Cond. Res.* **2011**, *25*, 1285–1292. [CrossRef]
29. Trecroci, A.; Longo, S.; Perri, E.; Iaia, F.M.; Alberti, G. Field-based physical performance of elite and sub-elite middle-adolescent soccer players. *Res. Sports Med.* **2019**, *27*, 60–71. [CrossRef]
30. Purnell, M.; Shirley, D.; Nicholson, L.; Adams, R. Acrobatic gymnastics injury: Occurrence, site and training risk factors. *Phys. Ther. Sport* **2010**, *11*, 40–46. [CrossRef]
31. Faude, O.; Junge, A.; Kindermann, W.; Dvorak, J. Injuries in female soccer players: A prospective study in the German national league. *Am. J. Sports Med.* **2005**, *33*, 1694–1700. [CrossRef] [PubMed]
32. Sanz, A.; Pablos, C.; Ballester, R.; Sanchez-Alarcos, J.V.; Huertas, F. Range of Motion and Injury Occurrence in Elite Spanish Soccer Academies. Not Only a Hamstring Shortening-Related Problem. *J. Strength Cond. Res.* **2020**, *34*, 1924–1932. [CrossRef]
33. Lemmink, K.A.; Kemper, H.C.; Greet, M.H.; Rispens, P.; Stevens, M. The validity of the sit-and-reach test and the modified sit-and-reach test in middle-aged to older men and women. *Res. Q. Exerc. Sport* **2003**, *74*, 331–336. [CrossRef]
34. Malliaras, P.; Hogan, A.; Nawrocki, A.; Crossley, K.; Schache, A. Hip flexibility and strength measures: Reliability and association with athletic groin pain. *Br. J. Sports Med.* **2009**, *43*, 739–744. [CrossRef] [PubMed]
35. Versic, S.; Pehar, M.; Modric, T.; Pavlinovic, V.; Spasic, M.; Uljevic, O.; Corluka, M.; Sattler, T.; Sekulic, D. Bilateral Symmetry of Jumping and Agility in Professional Basketball Players: Differentiating Performance Levels and Playing Positions. *Symmetry* **2021**, *13*, 1316. [CrossRef]
36. Clarsen, B.; Ronsen, O.; Myklebust, G.; Florenes, T.W.; Bahr, R. The Oslo Sports Trauma Research Center questionnaire on health problems: A new approach to prospective monitoring of illness and injury in elite athletes. *Br. J. Sports Med.* **2014**, *48*, 754–760. [CrossRef]
37. Sekulic, D.; Prus, D.; Zevrnja, A.; Peric, M.; Zaletel, P. Predicting Injury Status in Adolescent Dancers Involved in Different Dance Styles: A Prospective Study. *Children* **2020**, *7*, 297. [CrossRef]
38. Keller, C.S.; Noyes, F.R.; Buncher, C.R. The medical aspects of soccer injury epidemiology. *Am. J. Sports Med.* **1988**, *16*, S-105–S-112. [CrossRef]
39. Östenberg, A.; Roos, H. Injury risk factors in female European football. A prospective study of 123 players during one season. *Scand. J. Med. Sci. Sports* **2000**, *10*, 279–285. [CrossRef]

40. Kalén, A.; Rey, E.; de Rellán-Guerra, A.S.; Lago-Peñas, C. Are Soccer Players Older Now Than Before? Aging Trends and Market Value in the Last Three Decades of the UEFA Champions League. *Front. Psychol.* **2019**, *10*, 76. [CrossRef]
41. Bowen, L.; Gross, A.S.; Gimpel, M.; Li, F.-X. Accumulated workloads and the acute: Chronic workload ratio relate to injury risk in elite youth football players. *Br. J. Sports Med.* **2017**, *51*, 452–459. [CrossRef] [PubMed]
42. Lundblad, M.; Waldén, M.; Magnusson, H.; Karlsson, J.; Ekstrand, J. The UEFA injury study: 11-year data concerning 346 MCL injuries and time to return to play. *Br. J. Sports Med.* **2013**, *47*, 759–762. [CrossRef] [PubMed]
43. Söderman, K.; Alfredson, H.; Pietilä, T.; Werner, S. Risk factors for leg injuries in female soccer players: A prospective investigation during one out-door season. *Knee Surg. Sports Traumatol. Arthrosc.* **2001**, *9*, 313–321. [CrossRef] [PubMed]
44. Ekstrand, J.; Tropp, H. The incidence of ankle sprains in soccer. *Foot Ankle Int.* **1990**, *11*, 41–44. [CrossRef]
45. Calligeris, T.; Burgess, T. The incidence of injuries and exposure time of professional football club players in the Premier Soccer League during football season. *S. Afr. J. Sports Med.* **2015**, *27*, 16–19. [CrossRef]
46. Gabbett, T.J.; Ullah, S.; Jenkins, D.; Abernethy, B. Skill qualities as risk factors for contact injury in professional rugby league players. *J. Sports Sci.* **2012**, *30*, 1421–1427. [CrossRef]
47. Gabbett, T.; Kelly, J.; Ralph, S.; Driscoll, D. Physiological and anthropometric characteristics of junior elite and sub-elite rugby league players, with special reference to starters and non-starters. *J. Sci. Med. Sport* **2009**, *12*, 215–222. [CrossRef]
48. Gabbett, T.; Wake, M.; Abernethy, B. Use of dual-task methodology for skill assessment and development: Examples from rugby league. *J. Sports Sci.* **2011**, *29*, 7–18. [CrossRef]
49. McNeil, D.G.; Spittle, M.; Mesagno, C. Imagery training for reactive agility: Performance improvements for decision time but not overall reactive agility. *Int. J. Sport Exerc. Psycho.* **2021**, *19*, 429–445. [CrossRef]
50. Jeffreys, I. A task-based approach to developing context-specific agility. *Strength Cond. J.* **2011**, *33*, 52–59. [CrossRef]
51. Sporis, G.; Jukic, I.; Ostojic, S.M.; Milanovic, D. Fitness profiling in soccer: Physical and physiologic characteristics of elite players. *J. Strength Cond. Res.* **2009**, *23*, 1947–1953. [CrossRef] [PubMed]
52. Robinson, G.; O'Donoghue, P. A movement classification for the investigation of agility demands and injury risk in sport. *Int. J. Perform. Anal. Sport* **2008**, *8*, 127–144. [CrossRef]
53. Bloomfield, J.; Polman, R.; O'Donoghue, P. Turning movements performed during FA Premier League soccer matches. *J. Sports Sci. Med.* **2007**, *6*, 9–10.

healthcare

Article

What Does Provide Better Effects on Balance, Strength, and Lower Extremity Muscle Function in Professional Male Soccer Players with Chronic Ankle Instability? Hopping or a Balance Plus Strength Intervention? A Randomized Control Study

Hadi Mohammadi Nia Samakosh [1,*], João Paulo Brito [2,3,4], Seyed Sadredin Shojaedin [1], Malihe Hadadnezhad [1] and Rafael Oliveira [2,3,4]

[1] Department of Biomechanics and Corrective Exercises and Sports Injuries, University of Kharazmi, Tehran 15719-14911, Iran
[2] Sports Science School of Rio Maior–Polytechnic Institute of Santarém, 2040-413 Rio Maior, Portugal
[3] Life Quality Research Centre, 2040-413 Rio Maior, Portugal
[4] Research Center in Sport Sciences, Health Sciences and Human Development, 5001-801 Vila Real, Portugal
* Correspondence: shahramsamakosh92@yahoo.com

Citation: Mohammadi Nia Samakosh, H.; Brito, J.P.; Shojaedin, S.S.; Hadadnezhad, M.; Oliveira, R. What Does Provide Better Effects on Balance, Strength, and Lower Extremity Muscle Function in Professional Male Soccer Players with Chronic Ankle Instability? Hopping or a Balance Plus Strength Intervention? A Randomized Control Study. *Healthcare* **2022**, *10*, 1822. https://doi.org/10.3390/healthcare10101822

Academic Editor: Maria Chiara Gallotta

Received: 31 July 2022
Accepted: 19 September 2022
Published: 21 September 2022

Publisher's Note: MDPI stays neutral with regard to jurisdictional claims in published maps and institutional affiliations.

Abstract: Chronic ankle instability (CAI) has a higher frequency in soccer due to the rapid changes in body movement. Thus, this study compared the effects of eight weeks of a hopping protocol and a combined protocol of balance plus strength in a within-between group analysis. Thirty-six male professional soccer players participated in this study and were randomly allocated in three groups: control group (CG, n = 12), hopping group (HG, n = 12), and balance plus strength group (BSG, n = 12). Strength, static and dynamic balance, and function were assessed at baseline and eight weeks post intervention. First, Foot and Ankle Ability Measure (FAAM) and FAAM sport scales were applied. Then, a dynamometer was used to measure strength of the muscles around the hip, knee, and ankle joints. The Bass stick measured static balance and the Y balance test measured dynamic balance. Additionally, functional tests were carried out by Triple Hop, the Figure 8 hop, and vertical jump. A repeated measures ANOVA [(3 groups) × 2 moments] was used to compare the within and between group differences. In general, all tests improved after eight weeks of training with both protocols. Specifically, the BSG improved with large ES for all tests, while the HG improved all test with small to large effect sizes (ES). Furthermore, HG showed higher values for vertical jump (p < 0.01, ES = 1.88) and FAAMSPORT (p < 0.05, ES = 0.15) than BSG. BSG showed higher values for hip abduction (p < 0.05, ES = 2.77), hip adduction (p < 0.05, ES = 0.87), and ankle inversion (p < 0.001, ES = 1.50) strength tests, while HG showed higher values for knee flexion [ES = 0.86, (0.02, 1.69)] and ankle plantarflexion [ES = 0.52, (−0.29, 1.33)]. Balance plus strength protocol showed more positive effects than the hopping protocol alone for soccer players with CAI.

Keywords: dynamic balance; isometric; jump; stability; static balance

1. Introduction

Chronic ankle instability (CAI) is characterized by a wide spectrum of long-term sensorimotor and mechanical deficits that cause recurrent damage to the lateral ankle-ligament complex, and consequently, a subjective feeling of ankle-joint instability during functional or dynamic movements [1]. CAI can be caused by lateral ankle sprain, which is one of the most common injuries during exercise and daily living activities [2–4]. This type of injury accounts for 15 to 30 percent of sports injuries and has a return rate to training of ~75 percent [3]. Generally, injuries such CAI occur more frequently in soccer, futsal, basketball, volleyball, and sports that require a rapid change in direction [5]. Consequently, it is known that one of the most important lower extremity joints that helps maintain one's center of mass and body control stability is the ankle [6]. Postural control has both static and dynamic dimensions. The effort to maintain a high level of support with the least amount of movement is called static posture control, and the attempt to maintain a high level of support while performing several movements is called dynamic posture control [7,8]. As one of the controversial concepts of the sensory-motor system, the topic of postural control explores the complex and Interrelated relationship between sensory inputs and the motor responses required to maintain or modify the posture [5].

An ankle injury could be related to various mechanical receptors, such as the articular capsule, ligaments, and tendons of the ankle joint [2]. When afferent inputs change after injury, muscle contractions change, and therefore, mechanical receptor injuries can lead to functional and balance issues, deficits in strength, and chronic instability [2,5]. However, several exercise training programs, such as balance, Pilates, core stability, aquatic therapy, and proprioception exercise, have been shown to improve static and dynamic balance of people with CAI [2,5]. Despite evidence on improving muscle strength, there has been conflicting research concerning effects on proprioception of balance [9]. Another training approach that recently showed improvements in athletes with CAI is the use of hopping protocols [10,11]. Ardakani et al. [11] showed that a six-week hopping protocol improved Foot and Ankle Ability Measures (FAAM), Cumberland Ankle Instability Tool (CAIT), Foot and Ankle Outcome Score, Peak Ground Reaction Forces, Time to Peak Ground Reaction, and several movements of the hip, knee, and ankle in male university basketball players with CAI [11]. Minoonejad et al. [10] also showed improvements in FAAM, CAIT, and Foot and Ankle Outcome Scores, as well as preparatory muscle activation, reactive muscle activation, and muscle onset time across the lower extremity with six weeks hopping protocol. However, none of the previous studies analyzed static and dynamic balance [10,12]. The literature is scarce on the effects of isolated plyometric training and those that combine balance plus strength training on neuromuscular adaptation on chronic ankle instability in soccer players. On the one hand, ankle sprain is common among soccer players; on the other hand, dynamic and static balance are essential to the soccer game. Therefore, the main aim of the present study was to compare the effects of two exercise training programs (a hopping protocol and a combined protocol of balance plus strength) in professional soccer players. The primary hypothesis was that both training protocols can improve function, balance, and strength of lower extremity muscles in soccer players with CAI. The secondary hypothesis was that balance plus strength would provide better effects than a hopping protocol.

2. Materials and Methods

2.1. Design

The present study is a randomized clinical trial with an IRCT code of IRCT20190830044643N1, with control group, parallel groups, and two blinding testers (certified strength and conditioning professionals).

2.2. Participants

This study was conducted with men professional soccer players suffering from CAI in Iran. Thirty-nine players were selected from two first-division professional clubs (Azadegan league) across the country.

A questionnaire was used to collect information about exercise history, and CAI was diagnosed by the team physician, who has a sports medicine degree, following the International Ankle Consortium, which means that for an injury to be considered as CAI, the following inclusion criteria were adopted [13]: at least one severe ankle sprain in the past; the first sprain must have happened at least 12 months before enrolling in the trial; the initial sprain has to be accompanied by inflammatory signs (i.e., pain, edema, and so on); at least one day of targeted physical activity must have been disrupted by the first injury; the most recent injury must have occurred at least three months before enrolling in the study; previous ankle joint injury that gives way, recurrent sprain, or feelings of instability; in the six months leading up to study enrolment, participants must report at least two occurrences of their ankle giving way; experiencing a recurring sprain, which is defined as two or more sprains on the same ankle; instability in the ankle joint; Cumberland Ankle Instability Tool < 27, which confirmed self-reported ankle instability [14]. Additionally, the following exclusion criteria were applied: history of past musculoskeletal surgery in either lower extremity limb; history of a fracture in one of the lower extremity limbs that necessitated a realignment; acute damage to the musculoskeletal structures of other joints in the lower extremity in the past three months that resulted in at least one day of missed physical activity; ankle anterior drawer test was positive; only one training session was missed.

The selection of subjects with CAI was conducted through a clinical anterior drawer test of ankle, and scores were obtained using the CAI questionnaire (i.e., Cumberland Ankle Instability Tool) [13,14]. The questionnaire, which consists of nine multiple-choice questions, provides information on the extent and onset of pain and symptoms associated with functional ankle instability. A lower score indicates greater severity of the injury as experienced by the person answering. The validity and reliability of this questionnaire in assessing the severity of ankle sprain injury have been confirmed by previous studies [15]. The functional stability score of the ankle is between 0 and 30; a score range of 27 to 30 represents good health of the ankle, while a score between 0 and 27 indicates more severe ankle instability [16]. The following eligibility criteria to participate in the training protocols were applied. Inclusion criteria: (1) players had unilateral ankle instability (dominant leg); (2) no musculoskeletal abnormalities of the lower extremity and no lower back pain; (3) players had to complete a minimum of 85% of training sessions; (4) no injuries in the last three months. Exclusion criteria: (1) absence in three training sessions and/or tests; (2) stop following the intended exercise protocol.

Before staring the study, players and their parents signed the informed consent form to be examined according to the ethical principles contained in the Declaration of Helsinki (2013), and they declared their voluntary participation. Ethical permission was given by the Human Ethics Committee at the University of Tarbiat Modares Iran code of IR.MODARES.REC.1397.043.

2.3. Sample Size

In order to achieve sample power, a priori F-test family, with the statistical test of ANOVA: repeated measures, within-between interaction ($\alpha = 0.05$, $1 - \beta = 0.95$ and d: 0.5), through G*Power software (RRID:SCR_013726) was used [17]. A minimum of 21 subjects should be included in the analysis to achieve 97% of actual power. Thus, a total of 39 players were selected to this study. Three players did not complete the assessments. All players were committed to finish the training session unless they were injured. In the present study, three players withdrew from the intervention due to injuries, such as an injury in a within-team match in the second week (two players in control group). In addition, one player from BSG did not participate in the post test. Therefore, 36 players were considered for further analysis into the three groups: hopping group (HG), BSC, and control group (CG).

2.4. Randomization

A blinded person generated the allocation sequence, and two other researchers as trainers enrolled players and assigned them to interventions. The allocation sequence was random. Additionally, the allocation sequence was concealed until players were enrolled and assigned to interventions.

Players were assigned to one of three groups with an equal randomization (ratio of 1:1:1) (Figure 1). Randomization was performed by a blinded person who did not know the aims or design of the study. Randomization was performed with each person-specific naming code (up to 39) previously sealed: the blind person was asked to place 13 cards in three balls [18].

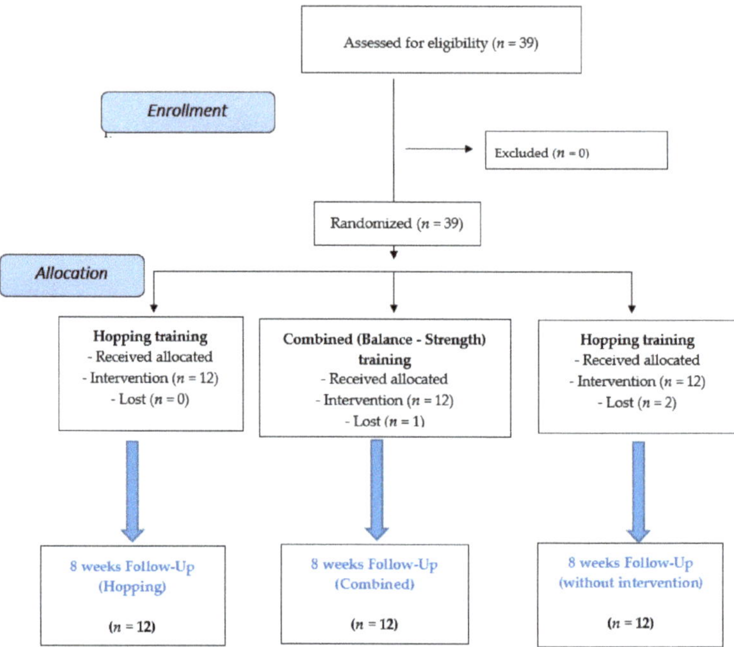

Figure 1. Transparent reporting of trials.

2.5. Procedures

All players performed a routine of exercises planned by the coaches and certified strength and conditioning professionals. Prior to the tests, the participants executed a standardized 5-min warm-up protocol consisting of a series of double leg squats (2 × 8 repetitions) and double leg maximum jumps (2 × 5 repetitions), followed by dynamic calf-stretching with a straight and bent knee. The intervention groups performed three additional training sessions 1–2 h before the start of their routine training per week (45–60 min, each session) accordingly to the group attributed, while the CG only performed the regular soccer training sessions. The players who were in intervention groups rested between 30–60 min until starting the regular soccer sessions. These sessions were held on Mondays, Wednesdays, and Fridays and were composed of speed endurance sessions (e.g., long sprints, repeated sprints), aerobic high-intensity sessions (e.g., interval training, medium-to-large sized games), and ball-possession games or team/opponent tactics sessions, respectively. On Tuesdays, Thursdays, and Saturdays, all players rested. On Sundays, they played one within-team match per week (intra-team friendly match).

The intervention occurred in the pre-season and lasted eight weeks. The intervention groups were assessed one week before the start of training and one week after, while the CG was assessed one week before and two weeks after the training programs finished.

Before intervention protocols, demographic data and exercise history were collected. Static and dynamic balance were assessed using Bass-Stick and Y-balance tests (Move 2 Perform, Evansville, IL, USA), respectively [19,20]. Muscle strength was assessed by a handheld manual dynamometer (MMT) (North Coast Made in USA) [21]. Lower extremity function was assessed through the Functional Indices of Triple Hop test, Figure 8 Hop and vertical jump tests. Additionally, FAAM and FAAMSPORT disability questionnaires were used [22]; it should be noted that all evaluations were performed with the dominant leg, which was specified using the ball kicking test [23]. The test-retest reliability was clinically acceptable for FAAM (intraclass correlation, ICC = 0.87) and FAAMSPORT (ICC = 0.87) [22]. Both questionnaires included several questions regarding ankle pain and function, including the FAAM Activities of Daily Living (ADL) and Sport scales. The FAAM comprises two scales: the ADL subscale and the Sport subscale. The ADL subscale consists of 21 questions that pertain to various functional activities one would encounter in normal daily activity. The Sport subscale contains eight questions pertaining to different activities that are related to sport participation. Participants were asked to rate the activity as no difficulty at all (4 points), slight difficulty (3 points), moderate difficulty (2 points), extreme difficulty (1 point), unable to do (0 points), or N/A (not applicable). The FAAM scores were recorded as a percentage of 84 points [22].

2.5.1. Assessment of Dynamic Balance

To begin the Y-balance test, the actual lower limb length from the anterior superior iliac spine to the medial was measured to normalize the data and compare subjects. To measure leg length, the subject was first asked to lie on the back of the table, then the distance between the anterior superior iliac spine to the distal portion of the medial ankle malleolus was repeated twice for each subject and each leg. The mean is then calculated as the leg length. The subject stood in the center of the room, then rested on one leg and then on the other leg, returned to normal position on two legs, and remained in this position for 10 to 15 s before the next attempt [20]. Three efforts in one direction had to be completed before moving in the other direction and had to be performed in a sequential clockwise or counterclockwise manner. The subject's toe touched the highest possible point in the specified directions, while the distance between the contact point center and touched point was measured in centimeters. The soccer player cannot touch their foot down on the floor before returning to the starting position. Any loss of balance will result in a failed attempt. However, once they have returned to the starting position, they are permitted to place their foot down behind the stance foot. The soccer player cannot place their foot on top of the reach indicator to gain support during the reach—they must push the reach indicator using the red target area. The soccer player must keep their foot in contact with the target indicator until the reach is finished. They cannot flick or kick the reach indicator to achieve a better performance [20]. Each participant completed 2–3 trials to get familiarized. The rest time was 30 s between three efforts, and the rest time was 60 s when changing to the other direction.

The following formula is used to determine the mean balance scores (Y-balance test) in each direction: total score = Y balance–Anterior (cm) + Y balance–Posteromedial (cm) + Y balance–Posterolateral (cm)/3. The score was expressed in centimeters [20].

2.5.2. Isometric Tests
Assessment of Isometric Hip Abductor Strength

Hip abductors were measured in a position at the examination table, with a strap at the desired angle (the hip abducted 10 degrees to maintain normal position). The subject's trunk was fixed using a strap attached to the iliac crest superior and around the table. In the tested leg, the center of pressure of the dynamometer was placed at a point 5 cm proximal to the lateral line of the knee joint. After zeroing the isometric dynamometer (North Coast Medical Inc, CA, USA) [21], the subject was requested to perform a maximal isometric contracture of the hip and to hold it for 5 s three times with an interval of 15 s [24]. The rest time was 60 s between three trials, and the rest time was 120 s between each of the isometric tests.

Assessment of Isometric Hip Adductor Strength

The hip adductor strength was measured when the subject was lying sideways on the examination table, while the hip joint was tested in extension and while it was flexed to $90°$. The subject with the upper hand grasped the edge of the table, and the lower side was in a comfortable position under the subject. When the dynamometer was positioned 5 cm above the femur's medial condyle, the subject was required to perform a maximal contracture and to hold it for 5 s, three times, with an interval of 15 s between each test [24].

Assessment of Isometric Knee Flexor Strength

Knee flexor strength was measured in a sitting position on the table with the knee positioned at 90° relative to the femur, while the femur was fixed with two bands. The position of the head dynamometer is considered to measure the flexor force on the knee, the surface of the distal of the posterior leg. After zeroing the dynamometer, the subject was requested to perform a maximal isometric contracture of the knee flexion and to hold it for 5 s. This test was performed three times with an interval of 15 s between each test [24].

Assessment of Isometric Knee Extensor Strength

Knee extensor strength measured in a sitting position on a table with the knee positioned at 90° relative to the hip while the femur was fixed with two bands. The position of the head dynamometer is considered to measure the extensor force on the knee and the surface of the distal of the anterior leg. After zeroing the dynamometer, the subject was requested to perform a maximal isometric contracture of the knee extension and to hold it for 5 s. For this test, three trials with an interval of 15 s between each test were performed [24].

Assessment of Isometric Ankle Plantarflexion and Dorsiflexion Strength

The subject was in a sitting position with the knee joint extending to the ankle at 0°. To measure plantar strength, a dynamometer was positioned on the proximal metatarsophalangeal surface on the palmar surface while lowering the ankle, and to measure dorsiflexor strength. After zeroing the dynamometer, the subject was requested to perform a maximal isometric contraction of the ankle plantarflexion and dorsiflexion and to hold it for 5 s. For this test, three trials with an interval of 15 s between each test were performed [24].

Assessment of Isometric Ankle Inversion and Eversion Strength

Inversion and eversion ankle strength was measured in a lying-back position with the legs extended outward and forward. After zeroing the dynamometer, the subject was requested to perform a maximal isometric contraction of the ankle inversion and eversion and to hold it for 5 s. For this test, three trials with an interval of 15 s between each test were performed [24].

For all isometric tests, the average of every three isometric contractions performed was recorded in kg.

2.5.3. Assessment of Triple Hop

This test required a narrow measuring tape that was 6 m long, securely positioned on the ground (Figure 2). The test involved performing three consecutive hops by traveling the maximum distance possible and landing on the same foot in each hop, and ultimately maintaining the landing mode for at least 3 s. Hand movements can be used to maintain balance. After performing two or three attempts, the subject does a triple hop two times for the dominant leg, and the total distance traveled is recorded [25]. The rest time was 60 s between three trials, and the rest time was 120 s between each of the hop/jump test.

Figure 2. Triple hop test.

2.5.4. Assessment of Figure 8 Hop

The Figure 8 Hop test was used to measure power, speed, and balance of the lower extremity, emphasizing one leg control. The path is 5 m long and 1 m wide, with seven obstacles (three obstacles at the top, three obstacles at the bottom, and one obstacle in the center). The subject stood behind the starting line with his dominant leg, while his other leg was slightly bent from the knee and hip joints. The subject was then instructed to hop at maximum speed to travel the specified path twice, and the elapsed time was recorded with a precision of 0.01 s. The subject was requested to hold his hands on the iliac crest during the test to avoid oscillating hand movements. It is worth mentioning that the subject performed this trial with shoes. The subject performed one to three experimental attempts, with a rest interval of 30 s. If the subjects lost balance during the test or something went wrong, the test was considered an error, and the test was repeated [25].

2.5.5. Assessment of Vertical Jump

The vertical jump test was performed so that the subject would first stand on the dominant leg wall (the location of the foot on the ground was predetermined). Then, with a stretched-out hand, the wall was marked to indicate the initial height. Then, the subject jumped to their maximum, and a sign was put on the wall. This jump was performed three times. Then, the distance between the base mark and the final mark in each jump was measured and recorded as the subject's vertical jump height [26].

2.6. Interventions

2.6.1. Balance plus Strength Protocol

The balance plus strength training protocol consisted of balance and strength types of exercises (examples in Figure 3) that were performed over eight weeks (three sessions per week), 45–60 min per session. Additionally, the intensity of exercise increased, with an increasing number of sets and repeated movements, every two weeks. Specifically, the intensity of exercise increased (sets and repetitions) by observing the principle of overload for each person. Rest between sets lasted 30 s and rest between exercises lasted 1 min (Table 1) [7]. Before and during the balance and exercises, the players were instructed to focus their strength to hold proper alignment of their body, to maintain a visual point on the front floor, to ensure the full alignment of the body, to not bend forward from the waist while undertaking a single leg squat and squat, to not drop the pelvis on the "bridge to plank", and to raise your knees as much as possible in the tuck jump.

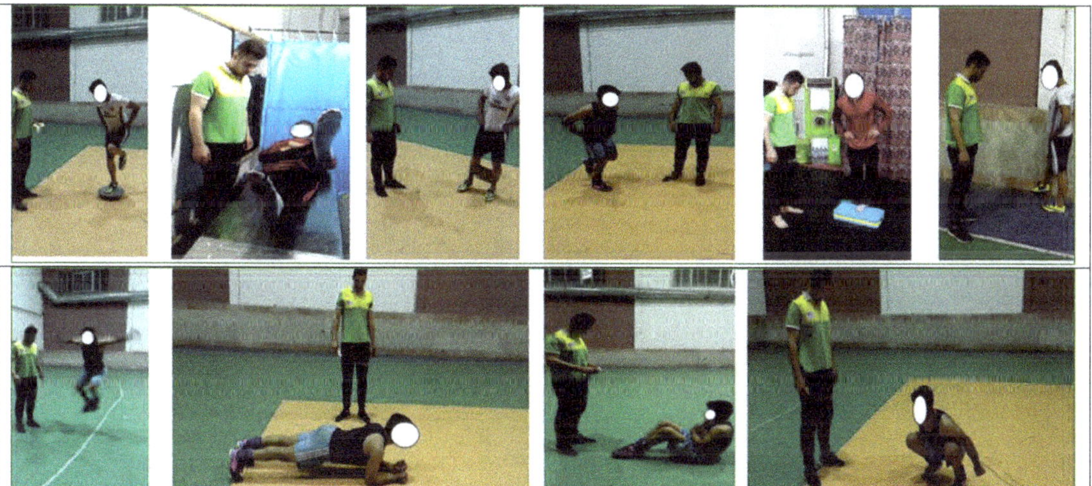

Figure 3. Examples of the balance plus strength exercises.

Table 1. Balance plus strength protocol.

Exercise	Period of Training at the Week							
	1–2 Weeks		3–4 Weeks		5–6 Weeks		7–8 Weeks	
	Set	R/S	Set	R/S	Set	R/S	Set	R/S
Bridge to plank (s)	3	20	3	25	4	30	5	30
60-degree trunk flexion (s)	3	20	3	25	4	30	5	30
Scissor movement (R)	3	10	3	15	4	15	5	15
Working with resistance band [a] (R)	3	10	3	15	4	15	5	15
Movement by wall [b] (R)	3	10	3	15	4	15	5	15
Single leg squat	3	10	3	15	4	15	5	15
Standing on one foot (s)	3	8	3	12	4	15	5	15
Working with a wobble board [c] (S)	3	8	3	12	4	15	5	15
Squat Jump (R)	3	8	3	12	4	15	5	15
Tuck jump (R)	3	8	3	12	4	15	5	15
Longitudinal jump (R)	3	8	3	12	4	15	5	15
Lateral jump (R)	3	8	3	12	4	15	5	15

S = Second, R = Repetition; [a] flexion, extension, abduction, adduction hip; [b] abduction and adduction hips, inversion ankle; [c] balance training (standing).

2.6.2. Hopping Protocol

The protocol consisted of six types of hopping exercises (examples in Figure 4) performed for eight weeks, three sessions per week, and 30 min per session. The intensity of the exercises was progressively increased by adding exercises and sets. Table 2 present the Hopping training protocol [27]. Rest between sets lasted 30 s and rest between exercises lasted 1 min. During training, soccer players were given feedback on postural control to avoid injury. They were also instructed on the position of the arms (hands-free, hands behind, or hands-on chest), "keep the focus", and "feel the foot contact with the ground, avoiding a valgus overall lower limb alignment on landing, and reducing ground reaction forces".

Table 2. Hopping protocol.

Weeks	Number of Hops	Type of Exercise	Set $\times$ Repetition
1	70	Side hop with two legs (hands-free)	3×10
		Forward and backward hop (hands-free)	2×10
		Forward hop with two legs (hands-free)	2×10
2	90	Side hop with two legs (hands-on chest)	2×15
		Forward and backward hop (hands-free)	2×10
		Forward hop with two legs (hands-free)	2×10
		Side hop with one leg (hands-free)	5×4
3	100	Side hop with one leg (hands-on chest)	3×10
		Forward and backward hop with one leg (hands-free)	2×10
		Forward hop with two legs (hands-on chest)	2×10
		Zigzag hop with two legs (hands-free)	2×10

Table 2. *Cont.*

Weeks	Number of Hops	Type of Exercise	Set × Repetition
4	110	Side hop with one leg (hands behind)	2 × 10
		Forward and backward hop with one leg (hands-on chest)	2 × 10
		Forward hop with on leg (hands-free)	2 × 10
		Zigzag hop with one leg (hands-free)	2 × 10
		Square hop with two legs (hands-free)	2 × 10
5–6	120	Side hop with one leg (hands behind)	2 × 10
		Forward hop with on leg (hands behind)	2 × 10
		Forward and backward hop with one leg (hands behind)	2 × 10
		Zigzag hop with one leg (hands-on chest)	2 × 10
		Square hop with one leg (hands-free)	2 × 10
		Figure 8 hop with two legs (hands-free)	2 × 10
7–8	130	Side hop with one leg (hands behind)	3 × 10
		Forward hop with on leg (hands behind)	2 × 10
		Forward and backward hop with one leg (hands behind)	2 × 10
		Zigzag hop with one leg (hands behind)	2 × 10
		Square hop with one leg (hands-on chest)	2 × 10
		Figure 8 hop with one leg (hands-free)	2 × 10

Figure 4. Examples of hopping exercises.

2.6.3. Statistical Analysis

Data were analyzed using Statistical Package IBM SPSS Statistics for Windows, version 22.0 (IBM Corp., Armonk, NY, USA). A Shapiro–Wilk test was used to assess the normal distribution of data. Then, Levene's test evaluated the equality of variances to verify homogeneity. Data were presented as means and standard deviations (SDs). Repeated measures ANOVA analysis with partial eta squared (ηp^2) and the Bonferroni adjustment post hoc test were used to compare pre-post interventions and groups [(3 groups) $\times$ 2 moments]. The significance level was set at $p \leq 0.05$. In addition, relative changes (%) and Cohen's D effect size (ES) with confidence intervals (CI, 95%) were calculated to measure the magnitude effects. Cohen's D method suggested that d = 0.2 is considered a 'small' effect size, d = 0.5 is considered a 'medium' effect size, and d = 0.8 is considered a 'large' effect size [28].

3. Results

Characteristics of the study are reported in Table 3 for each group at baseline. There were no significant differences between the groups, all $p > 0.05$.

Table 3. Characteristics of the study groups at baseline (mean $\pm$ SD).

Variables	BSG (*n* = 12)	HG (*n* = 12)	CG (*n* = 12)	*p*-Value
Age [years]	21.08 ± 1.78	20.83 ± 1.80	20.58 ± 1.37	0.76
Body weight [kg]	76.41 ± 9.21	75.41 ± 9.62	71.83 ± 7.60	0.42
Body height [cm]	177.58 ± 7.15	176.58 ± 5.59	178.58 ± 2.88	0.60
BMI [kg·m^{-2}]	24.20 ± 0.71	24.18 ± 0.71	21.13 ± 0.13	0.15
Dominant leg length [cm]	100.41 ± 5.93	101.16 ± 3.95	98.50 ± 3.20	0.34
Activity history [years]	6.16 ± 2.62	6.08 ± 2.67	6.25 ± 2.37	0.98

BMI: body mass index; BSG: Balance-Strength training; HG: Hopping training; CG: control group.

The outcomes of the two-way ANOVA and the Bonferroni post hoc test regarding balance and function tests are presented in Table 4. The ANOVA analysis group x time provided the following significant results: static balance Bass-stick test ($F_{2.33} = 53.06$, $\eta p^2 = 0.82$, $p < 0.001$), Y balance–Anterior ($F_{2.33} = 69.70$, $\eta p^2 = 0.84$, $p < 0.001$), Y-balance–posteromedial ($F_{2.33} = 30.39$, $\eta p^2 = 0.73$, $p < 0.001$), posterolateral ($F_{2.33} = 6.84$, $\eta p^2 = 0.38$, $p = 0.005$), total scores ($F_{2.33} = 60.80$, $\eta p^2 = 0.84$, $p < 0.001$), triple single leg hop test ($F_{2.33} = 4.45$, $\eta p^2 = 0.28$, $p = 0.02$), Figure 8 hop ($F_{2.33} = 4.07$, $\eta p^2 = 0.27$, $p = 0.03$), vertical jump ($F_{2.33} = 71.90$, $\eta p^2 = 0.86$, $p < 0.001$), FAAM ($F_{2.33} = 5.32$, $\eta p^2 = 0.32$, $p = 0.01$), and FAAMSPORT ($F_{2.33} = 69.87$, $\eta p^2 = 0.86$, $p < 0.001$).

Table 4. Training effects (with CI, 95%) for the balance and function variables between groups.

Groups	Pre Test (Mean ± SD)	Post Test (Mean ± SD)	Pre vs. Post Performance Change (%)	Pre vs. Post ES (CI, 95%)
		Static balance Bass–stick (s)		
BSG	9.33 ± 1.85	15.18 ± 2.64 [c]	↑ 62.70	2.56 (1.50 to 3.63) ***
HG	10.07 ± 2.24	11.22 ± 2.13	↑ 11.42	0.52 (−0.24 to 1.39) *
CG	8.85 ± 1.57	8.97 ± 1.73	↑1.35	0.07 (−0.74 to 0.86)
		Y balance–Anterior (cm)		
BSG	82.35 ± 6.18	88.56 ± 5.21	↑ 7.54	1.08 (0.70 to 2.53) ***
HG	82.41 ± 6.30	83.76 ± 5.64	↑ 1.63	0.22 (−0.57 to 1.03) *
CG	78.20 ± 6.23	78.97 ± 3.11	↑ 0.98	0.09 (−0.65 to 0.96)
		Y balance–Posteromedial (cm)		
BSG	87.42 ± 6.63	95.71 ± 4.21 [c]	↑ 9.48	1.49 (0.65 to 2.47) ***
HG	86.59 ± 4.37	92.27 ± 4.67 [a]	↑ 6.55	1.25 (0.15 to 1.84) ***
CG	87.81 ± 2.78	88.63 ± 2.81	↑ 0.93	0.29 (−0.58 to 1.03) *
		Y balance–Posterolateral (cm)		
BSG	82.54 ± 4.33	89.69 ± 4.01 [a]	↑ 8.66	1.71 (1.09 to 3.05) ***
HG	80.69 ± 8.09	85.16 ± 6.20 [a]	↑ 5.53	0.62 (−0.19 to 1.45) **
CG	78.19 ± 3.79	79.75 ± 5.46	↑ 1.99	0.33 (−0.49 to 1.12) *

Table 4. *Cont.*

Groups	Pre Test (Mean ± SD)	Post Test (Mean ± SD)	Pre vs. Post Performance Change (%)	Pre vs. Post ES (CI, 95%)
		Y balance–total (cm)		
BSG	84.10 ± 3.19	91.32 ± 2.33 [c]	↑ 8.58	2.58 (1.28 to 3.31) ***
HG	83.23 ± 4.90	87.07 ± 4.23 [a]	↑ 4.61	0.83 (0.18 to 1.87) **
CG	81.40 ± 1.33	82.45 ± 1.85	↑ 1.28	0.65 (−0.52 to 1.09) **
		Triple single leg hop (cm)		
BSG	414.99 ± 8.40	434.98 ± 12.67 [b]	↑ 1.28	1.85 (1.22 to 3.22) ***
HG	419.63 ± 14.49	442.22 ± 13.56 [b,d]	↑ 5.38	1.60 (1.03 to 2.97) ***
CG	409.44 ± 15.31	418.15 ± 18.95	↑ 2.12	0.50 (−0.33 to 1.33) *
		Figure 8 hop (s)		
BSG	9.00 ± 0.89	7.64 ± 0.34 [b,d]	↓ 15.11	2.01 (−0.93 to 0.68) ***
HG	9.11 ± 1.11	7.03 ± 0.61 [c,f]	↓ 22.83	2.32 (7−3.82 to−1.63) ***
CG	9.76 ± 0.70	8.88 ± 1.02	↓ 9.01	1 (−1.66 to−0.79) ***
		Vertical jump (cm)		
BSG	21.55 ± 2.21	25.46 ± 1.93 [b,d,h]	↑ 18.14	1.88 (1.61 to 3.78) ***
HG	22.57 ± 1.40	28.89 ± 1.71 [c,f]	↑ 28.00	4.04 (2.01 to 4.38) ***
CG	21.04 ± 3.22	22.19 ± 3.20	↑ 5.46	0.35 (−0.34 to 1.28) *
		FAAM		
BSG	76.16 ± 3.35	82.33 ± 3.89 [a,d]	↑ 8.10	1.69 (0.80 to 2.66) ***
HG	76.25 ± 5.04	83.16 ± 3.88 [b]	↑ 9.06	1.53 (0.95 to 2.86) ***
CG	75.41 ± 4.05	77.16 ± 3.48	↑ 2.32	0.46 (−0.40 to 1.21) *
		FAAMSPORT		
BSG	64.16 ± 1.52	70.00 ± 16.51 [c,d,i]	↑ 9.10	0.49 (−0.33 to 1.30) *
HG	65.25 ± 2.34	72.00 ± 8.59 [c]	↑ 10.34	1.07 (0.24 to 1.95) ***
CG	65.58 ± 2.06	67.00 ± 4.60	↑ 2.16	0.39 (−0.42 to 1.20) *

SD: standard deviation; CI Confidence Interval; BSG: Balance-Strength training; HG: Hopping training; CG: control group; * Small effect size; ** Medium effect size; *** Large effect size; ↓ Decrease; ↑ Increase; [a] Denotes significant difference from pre to post training ($p < 0.05$). [b] Denotes significant difference from pre to post training ($p < 0.01$). [c] Denotes significant difference from pre to post training ($p < 0.001$). [d] Denotes significant difference with the CG post training ($p < 0.05$). [f] Denotes significant difference with the CG post training ($p < 0.001$). [h] Denotes significant difference with the HG ($p < 0.01$). [i] Denotes significant difference with the HG post training ($p < 0.001$).

Regarding pre to post comparisons, BSG showed significant improvements in static balance Bass-stick test, Y balance–Anterior, Y-balance–posteromedial, posterolateral, and total scores, triple single leg hop test, Figure 8 hop, vertical jump, FAAM and FAAM-SPORT, while HG showed significant results in Y-balance–posteromedial, posterolateral, and total scores, Figure 8 hop, vertical jump, FAAM and FAAMSPORT. CG did not present significant changes.

Regarding comparisons between groups, BSG presented higher values for Figure 8 hop, vertical jump, FAAM and FAAMSPORT than CG. HG presented higher values for triple single leg hop and Figure 8 hop than CG. Finally, when comparing intervention groups, HG showed higher values for vertical jump [ES = 1.88, (0.93, 2.93)] and FAAMSPORT [ES = 0.15, (−0.65, 0.95)] than BSG. No other significant differences were noted.

Figure 5 also highlights some of the main results for Y balance–total, triple single leg hop, vertical jump and FAAMSport.

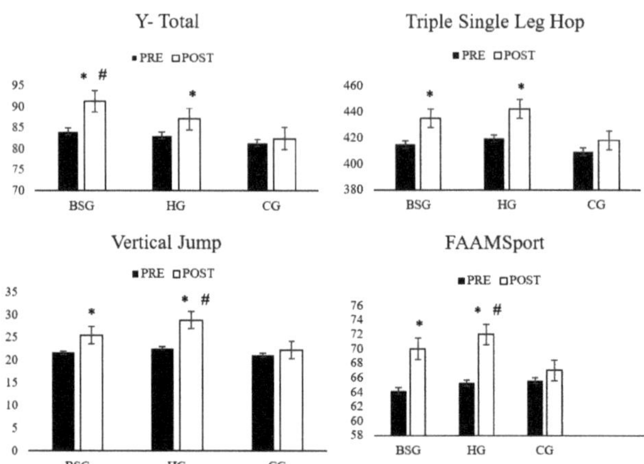

Figure 5. Pre to post test of Y- Total, Triple single leg hop, Vertical Jump and FAAMsport. * denotes difference between pre to post test ($p < 0.05$). # denotes difference between BSG versus HG ($p < 0.05$).

The outcomes of the two-way ANOVA and the Bonferroni post hoc test regarding strength tests are presented in Table 5. The ANOVA analysis group × time provided the following significant results: hip abduction ($F_{2.33} = 26.84$, $\eta p^2 = 0.71$, $p < 0.001$), hip adduction ($F_{2.33} = 16.15$, $\eta p^2 = 0.59$, $p < 0.001$), knee flexion ($F_{2.33} = 25.56$, $\eta p^2 = 0.70$, $p < 0.001$), knee extension ($F_{2.33} = 14.67$, $\eta p^2 = 0.57$, $p < 0.001$), ankle plantarflexion ($F_{2.33} = 9.12$, $\eta p^2 = 0.45$, $p = 0.001$), ankle dorsiflexion ($F_{2.33} = 16.91$, $\eta p^2 = 0.60$, $p < 0.001$), ankle eversion ($F_{2.33} = 15.29$, $\eta p^2 = 0.58$, $p < 0.001$) and ankle inversion ($F_{2.33} = 36.82$, $\eta p^2 = 0.77$, $p < 0.001$).

Regarding pre to post comparisons, BSG showed significant increases in all strength tests, while HG showed significant improvements in hip abduction, hip adduction, knee flexion, knee extension, ankle plantarflexion, ankle dorsiflexion, ankle eversion, and ankle inversion.

Regarding comparisons between groups, BSG presented higher values for hip abduction, hip adduction, knee flexion, knee extension, ankle plantarflexion, ankle dorsiflexion ankle dorsiflexion, and ankle inversion than CG. HG presented higher values for knee extension, knee flexion, ankle plantarflexion, ankle dorsiflexion, and ankle inversion than CG. Finally, when comparing intervention groups, BSG showed higher values for hip abduction [ES = 2.77, (1.67, 3.87)], hip adduction [ES = 0.87, (0.04, 1.71)], and ankle inversion [ES = 1.50, (0.48, 2.52)], while HG showed higher values for knee flexion [ES = 0.86, (0.02, 1.69)] and ankle plantarflexion [ES = 0.52, (−0.29, 1.33)].

Table 5. Training effects for the strength variables between pre-post intervention and groups.

Groups	Pre Test (Mean ± SD)	Post Test (Mean ± SD)	Pre vs. Post Performance Change (%)	Pre vs. Post ES (95% CI)
		Hip abduction (Kg)		
BSG	11.75 ± 1.74	14.79 ± 0.82 [c,j,f]	↑ 25.87	2.23 (0.98 to 2.90) ***
HG	12.57 ± 1.87	13.58 ± 1.99	↑ 8.03	0.52 (−0.27 to 1.35) *
CG	11.73 ± 1.03	11.91 ± 0.97	↑ 1.53	0.17 (−0.68 to 0.92)
		Hip adduction (Kg)		
BSG	10.51 ± 1.25	13.19 ± 1.36 [c,j,f]	↑ 25.49	2.05 (1.22 to 3.23) ***
HG	10.46 ± 1.66	12.14 ± 1.02 [b]	↑ 16.06	1.21 (0.58 to 2.37) ***
CG	10.67 ± 0.93	10.83 ± 0.88	↑ 1.49	0.17 (−0.68 to 0.92)

Table 5. *Cont.*

Groups	Pre Test (Mean ± SD)	Post Test (Mean ± SD)	Pre vs. Post Performance Change (%)	Pre vs. Post ES (95% CI)
		Knee flexion (Kg)		
BSG	11.49 ± 0.72	13.56 ± 0.82 [c,j,f]	↑ 18.01	2.68 (1.32 to 3.37) ***
HG	11.79 ± 1.27	13.61 ± 1.43 [c,f]	↑ 15.43	1.34 (0.69 to 2.52) ***
CG	11.36 ± 0.77	11.58 ± 0.76	↑ 1.93	0.28 (−0.59 to 1.01) *
		Knee extension (Kg)		
BSG	11.82 ± 1.56	12.56 ± 1.42 [d]	↑ 6.26	2.50 (−0.17 to 1.47) ***
HG	11.45 ± 1.55	13.58 ± 0.91 [c,f]	↑ 21.57	1.67 (0.75 to 2.59) ***
CG	11.17 ± 0.79	11.38 ± 0.38	↑ 1.88	0.33 (−0.62 to 0.99) *
		Ankle plantarflexion (Kg)		
BSG	8.93 ± 1.07	11.30 ± 1.07 [c,d,j]	↑ 26.53	2.21 (1.46 to 3.58) ***
HG	9.24 ± 1.23	11.90 ± 1.23 [c,f]	↑ 27.13	2.16 (1.29 to 3.33) ***
CG	9.36 ± 0.79	9.66 ± 0.71	↑ 3.20	0.33 (−0.51 to 1.10) *
		Ankle dorsiflexion (Kg)		
BSG	8.20 ± 0.73	10.91 ± 1.18 [c,d]	↑ 33.04	2.76 (10.91 to 8.2) ***
HG	8.71 ± 0.73	11.41 ± 1.28 [c,f]	↑ 24.69	2.59 (11.41 to 8.71) ***
CG	9.15 ± 0.68	9.42 ± 0.66	↑ 2.95	0.40 (9.42 to 9.15) *
		Ankle eversion (Kg)		
BSG	4.69 ± 0.47	6.09 ± 0.72 [c,f]	↑ 29.85	2.30 (0.91 to 2.80) ***
HG	4.95 ± 0.52	5.48 ± 0.71	↑ 10.70	0.85 (0.04 to 1.72) **
CG	4.70 ± 0.48	4.87 ± 0.51	↑ 3.61	0.34 (−0.53 to 1.08) *
		Ankle inversion (Kg)		
BSG	5.53 ± 0.54	7.45 ± 0.51 [c,f,h]	↑ 34.71	3.65 (2.55 to 5.20) ***
HG	5.32 ± 0.74	6.49 ± 0.75 [c,d]	↑ 21.99	1.57 (0.85 to 2.73) ***
CG	5.52 ± 0.63	5.69 ± 0.72	↑ 3.07	0.25 (−0.57 to 1.04) *

SD: standard deviation; CI Confidence Interval; BSG: Balance-Strength training; HG: Hopping training; CG: control group; ↓ Decrease; ↑ Increase; * Small. ** Medium. *** Large. [b] Denotes significant difference from pre to post training ($p < 0.01$). [c] Denotes significant difference from pre to post training ($p < 0.001$). [d] Denotes significant difference with the CG post training ($p < 0.05$). [f] Denotes significant difference with the CG post training ($p < 0.001$). [j] Denotes significant difference with the HG ($p < 0.05$). [h] Denotes significant difference with the HG ($p < 0.01$).

4. Discussion

In the present study, we analyzed two types of hopping and balance plus strength protocols on balance, strength, and lower extremity strength muscle function in soccer players with CAI. The overall results of the present study showed positive effects on those variables which confirmed our first study hypothesis. Specifically, BSG improved with large ES for all tests, while HG improved all test with small to large ES, which confirmed our second study hypothesis. The results showed higher values for static and dynamic balance, as well as strength in the group that performed BSG protocol, which could be associated when more emphasis is given to the neuromuscular training, which consequently contribute to a greater effectiveness of this type of training protocol. Another justification could be the isolated focus on proprioception and muscle strength components in BSG exercises compared to HG exercises.

Research has shown that muscle weakness and subsequent increased ankle joint loosening and motor-sensory deficits resulting from the sprain are associated with balance deficits and postural control [3]. Moreover, it has been shown that muscle weakness and subsequent increased ankle joint laxity and motor-sensory deficits resulting from the sprain are associated with balance deficits and postural control [3].

In this sense, two training protocols (combined training of hopping and core stability and hopping training alone) provided more desirable outcomes than core stability training [29]. However, the same authors did not find any significant difference between the HG that involved the ankle area and the combined protocol (core stability plus hopping) that included the trunk area in addition to the ankle section. The results of the present study corroborated the previous findings considering the positive results and the differences in the protocols.

A research paper examined the effect of 6 weeks of balance training and combined (balance and plyometric) training using the time to stabilization (TTS) test on the force plate and balance by the one-leg stand test. The results showed that combined training (Balance-Plyometric) more effectively reduced postural oscillations in static and dynamic conditions compared to balance training alone [30].

To justify the greater effectiveness of BSG, it seems that a combination of many types of exercises together improved the balance and strength of soccer players with CAI. The study of Hall et al. [30] examined the effects of two types of BSG on balance, strength, and lower extremity function in people with CAI and showed that both kinds of training improved those variables [30], which is also supported by the results of the present study.

Exercise interventions on hip and lower extremity strength among players should be considered because lack of control in these areas will lead to knee valgus dynamic [31]. In this regard, it is stated that the strength of the hip joint is essential for proper walking mechanics and foot posture when heel contact impact changes postural stability and muscle strength absorption patterns in the hip and ankle after injury, which can be effective in reinjury in the future [32]. As stated by the authors, there are training protocols that improve the strength and endurance of the core muscles that can explain the effect of enhancing balance so that the core muscles create a solid cylinder, and subsequently, produce inertia against body perturbation, providing the body with a movement-stable surface. The abdominal muscles, including the transverse abdominis, rectus abdominis, medial oblique, and lateral oblique, are all integrated to provide spinal stability and, thus, a stronger support surface for lower extremity movement [32].

In some cases, the importance of proper activation and stability of the trunk while maintaining static stance control has also been emphasized [33]. Kibler et al. [34] described that the activation of the central muscular structure of the body, in patterns associated with limb movement, contributes to the development of balance and subsequent function. According to the same authors, the body triggering the core muscles used to yield the necessary rotational torque around it and to produce limb movement [34]. The second part of the BSG, balance and proprioception training, can have a greater impact on training than hopping training. As stated, injury can have a profound effect on proprioception and neuromuscular control. Munn et al. [35], in a review article on individuals suffering CAI, reported that secondary postural defect was due to control, neuromuscular, and proprioception deficits [35]. The outcomes of the present study showed that BSG protocol was significantly more effective in improving Y-balance test in all directions than the results of other studies that examined the effect of balance training. Hall et al. also applied a balance training protocol in the anterior, posteromedial, and posterolateral direction and reported significant improvements [2].

Researchers have claimed that hopping training creates a link between strength and coordination and directly enhances competitive performance [36]. The present study showed positive effects of the hopping protocol on the static and dynamic balance. This is in line with the outcomes reported by Myer et al. [37] through the application of a plyometric training program [37]. Therein, Kramer et al. [38] pointed out the importance of using this kind of training to maintain muscle strength and stated that using plyometric training may prevent the neuromuscular dysfunction of people with low activity and people with impaired posture due to prolonged sitting. In addition, balance and muscle strength are essential parts of hopping protocols [38]. In that regard, Ulrich et al. [39]

pointed to the greater impact of short-term and explosive training compared to traditional functional training on activities such as soccer [39].

In terms of comparing the strength of the muscles of the lower extremity, positive effects were found for both groups with a higher effect for BSG than HG. For instance, the Figure 8 hop and single-leg triple hop tests, which require consecutive jumps and landing movements, were improved after eight weeks of training.

Nonetheless, HG showed higher values for vertical jump than BSG. A possible justification may be related to many hops and jumps in the hopping training, thus presenting a greater impact on this factor. The hopping training has been included as a sport-specific exercise and as a drill designed to improve speed, agility, and aerobic conditioning [40,41] and is appreciated by players and coaches increasing player compliance and participation [42–44]. The available evidence suggests that hopping exercises can elicit change in the stiffness of various elastic components of the muscle-tendon complex of plantar flexors [45]. The hopping training improves the stretch-shortening cycle of muscle function and induces numerous positive changes in the neural and musculoskeletal systems, and in athletic performance [46]. Additionally, the hopping exercise has been shown to enhance neuromuscular fitness and lower extremity muscle strength, which Kramer et al. [38] pointed out in their research. One of the possible reasons for advancing the functional characteristics after participating in HG is to improve neuromuscular adaptation and increase the strength of the muscles involved in lateral hop, Figure 8 hop, and single-leg triple hop protocols [39]. Another possible reason might be the strength of the joints and their construction muscles to stabilize the lower extremity joints with deep receptor activity and neuromuscular control to maintain balance when jumping in different directions. BSG training that also includes strength training with improving the ankle joint forces improves ankle strength scores by increasing plantarflexion and dorsiflexion performance as well as strengthening the ligaments around the ankle, which reduced the pressure applied to the ankle joint and improved the scores obtained in the ankle and foot ability questionnaire in individuals with CAI.

Consistent with the previous results, McKeon et al. [47] showed that performing four weeks of dynamic balance training can improve self-reported function measured by the FAAM and FAAMSPORTS of non-athletes suffering CAI [47]. Webster and Gribble also systematically reviewed functional rehabilitation interventions for CAI and stated that using closed chain rehabilitation training for four to six weeks and three to four sessions per week significantly improved the self-reported function of people with CAI [48]. One of the reasons for the improvement in the self-reported function of the participants in the neuromuscular training is due to the reduction of the limitations on the sensorimotor system as a result of these protocols. Following the results of previous studies that applied hopping protocols [10,11], the present study showed improvements on FAAM and FAAMSPORT. However, our results also revealed that participants from both groups still report symptoms of CAI [49]. Thus, more studies are needed to confirm the present findings.

It is relevant to highlight that both training protocols were applied in same days of the regular soccer training sessions. Regular soccer training included endurance sessions with light and long-term activities (such as jogging and long-distance running on the grass). Gradually, interval training was implemented during the eight-week period. Nonetheless, the present study did not address the intensity nor the exercises performed during the study period by both teams from where participants were recruited, which is a limitation that should be considered in future studies. For example, future research can employ body composition and internal and external load variables to address soccer-specific training, as well as consider other key factors, such as playing position and player's starting status [50,51].

Furthermore, other limitations of the present research must be acknowledged. Although we present a sample power calculation, our sample is small, and it originates from two different teams. For that reason, the results must be carefully interpreted. In addition, players with bilateral ankle instability were not excluded, and the testing was simply performed on the ankle joint with the lower CAIT score, which reinforces that future studies should assess both limbs instead of only the dominant limb. The number of sets and repetitions for load progression were balanced between protocols; even so, and considering the soccer training skills, we believe a comparison of both training protocols is helpful for coaches and their staff. In addition, it was not possible to access the CG in the same week of the experimental groups to not change training routines of the players, which could affect some results, although it is possible to confirm that their physical routines were kept. Moreover, we could not confirm the reliability of the tests applied, which should be considered in future studies. Finally, the protocols only have a duration of eight weeks during the pre-season. It would be beneficial for future studies to test longitudinal approaches to access the long-term effects of the interventions and to test them during in-season periods.

Beyond the previous suggestion, future studies can use different tests and training monitoring practices to provide more knowledge on the training protocols applied. For instance, the traditional squat, counter movement, or drop jumps can be easily added. A change of direction and cardiorespiratory tests would also be beneficial to amplify the results. Lastly, the combination of body composition with internal training load and global positioning system variables may also consider other key factors, such as playing position, player's starting status, or type of week [52].

From a practical point of view, the training protocols of this study lasted 45–60 min for BSG and 30 min for HG. Balance plus strength training seems to be very long; however, there were previous studies with such durations [21,53]. Depending on the time available for training, coaches and/or strength and conditioning professionals can choose the protocol that better suits their context.

5. Conclusions

This study showed that both BSG and HG improved the balance, strength, and function of soccer players with CAI. However, the BSG protocol seems to be more effective because it emphasizes each of these factors separately. It can be more effective than hopping training (which has strength and balance in nature) and has a greater impact on faster recovery for soccer players with CAI.

Author Contributions: Conceptualization, H.M.N.S. and M.H.; methodology, H.M.N.S., J.P.B. and R.O.; software, H.M.N.S.; validation, H.M.N.S., M.H. and S.S.S.; formal analysis, H.M.N.S.; investigation, H.M.N.S.; resources, H.M.N.S.; data curation, H.M.N.S.; writing—original draft preparation H.M.N.S. and R.O.; writing—review and editing, H.M.N.S., J.P.B. and R.O.; visualization, H.M.N.S.; supervision, S.S.S., J.P.B. and R.O.; project administration, H.M.N.S.; funding acquisition, J.P.B. and R.O. All authors have read and agreed to the published version of the manuscript.

Funding: This research was funded by the Portuguese Foundation for Science and Technology, I.P., Grant/Award Number UIDP/04748/2020.

Institutional Review Board Statement: Ethical permission from the Human Ethics Committee at the University of Tarbiat Modares Iran code of IR.MODARES.REC.1397.043. Informed consent was obtained from all subjects involved in the study.

Informed Consent Statement: Written informed consent was obtained from the participants to publish this paper.

Data Availability Statement: The data presented in this study are available on request from the corresponding author.

Acknowledgments: We thank all participants of the study.

Conflicts of Interest: The authors declare no conflict of interest. The funders had no role in the design of the study; in the collection, analyses, or interpretation of data; in the writing of the manuscript, or in the decision to publish the results.

References

1. Hertel, J.; Corbett, R.O. An updated model of chronic ankle instability. *J. Athl. Train.* **2019**, *54*, 572–588. [CrossRef] [PubMed]
2. Hale, S.A.; Fergus, A.; Axmacher, R.; Kiser, K. Bilateral improvements in lower extremity function after unilateral balance training in individuals with chronic ankle instability. *J. Athl. Train.* **2014**, *49*, 181–191. [CrossRef] [PubMed]
3. Hubbard, T.J.; Wikstrom, E.A. Ankle sprain: Pathophysiology, predisposing factors, and management strategies. *Open Access J. Sports Med.* **2010**, *1*, 115. [CrossRef] [PubMed]
4. Silva, D.C.F.; Santos, R.; Vilas-Boas, J.P.; Macedo, R.; Montes, A.M.; Sousa, A.S.P. Different cleat models do not influence side hop test performance of soccer players with and without chronic ankle instability. *J. Hum. Kinet.* **2019**, *70*, 156–164. [CrossRef] [PubMed]
5. Schiftan, G.S.; Ross, L.A.; Hahne, A.J. The effectiveness of proprioceptive training in preventing ankle sprains in sporting populations: A systematic review and meta-analysis. *J. Sci. Med. Sport* **2015**, *18*, 238–244. [CrossRef]
6. Lee, A.J.; Lin, W.-H. The influence of gender and somatotype on single-leg upright standing postural stability in children. *J. Appl. Biomech.* **2007**, *23*, 173–179. [CrossRef]
7. Kachouri, H.; Borji, R.; Baccouch, R.; Laatar, R.; Rebai, H.; Sahli, S. The effect of a combined strength and proprioceptive training on muscle strength and postural balance in boys with intellectual disability: An exploratory study. *Res. Dev. Disabil.* **2016**, *53*, 367–376. [CrossRef]
8. Gribble, P.; Hertel, J.; Denegar, C.R.; Buckley, W.E. The effects of fatigue and chronic ankle instability on dynamic postural control. *J. Athl. Train.* **2004**, *39*, 321–329.
9. Smith, B.I.; Docherty, C.L.; Simon, J.; Klossner, J.; Schrader, J. Ankle strength and force sense after a progressive, 6-week strength-training program in people with functional ankle instability. *J. Athl. Train.* **2012**, *47*, 282–288. [CrossRef]
10. Minoonejad, H.; Ardakani, M.K.; Rajabi, R.; Wikstrom, E.A.; Sharifnezhad, A. Hop stabilization training improves neuromuscular control in college basketball players with chronic ankle instability: A randomized controlled trial. *J. Sport Rehabil.* **2019**, *28*, 576–583. [CrossRef]
11. Ardakani, M.K.; Wikstrom, E.A.; Minoonejad, H.; Rajabi, R.; Sharifnezhad, A. Hop-stabilization training and landing biomechanics in athletes with chronic ankle instability: A randomized controlled trial. *J. Athl. Train.* **2019**, *54*, 1296–1303. [CrossRef] [PubMed]
12. Taube, W.; Kullmann, N.; Leukel, C.; Kurz, O.; Amtage, F.; Gollhofer, A. Differential reflex adaptations following sensorimotor and strength training in young elite athletes. *Endoscopy* **2007**, *28*, 999–1005. [CrossRef] [PubMed]
13. Hertel, J. Functional anatomy, pathomechanics, and pathophysiology of lateral ankle instability. *J. Athl. Train.* **2002**, *37*, 364–375. [PubMed]
14. Martin, R.L.; Irrgang, J.J.; Burdett, R.G.; Conti, S.F.; Van Swearingen, J.M. Evidence of validity for the foot and ankle ability measure (FAAM). *Foot Ankle Int.* **2005**, *26*, 968–983. [CrossRef] [PubMed]
15. Hiller, C.E.; Refshauge, K.M.; Bundy, A.C.; Herbert, R.D.; Kilbreath, S.L. The cumberland ankle instability tool: A report of validity and reliability testing. *Arch. Phys. Med. Rehabil.* **2006**, *87*, 1235–1241. [CrossRef]
16. Hiller, C.E.; Kilbreath, S.L.; Refshauge, K.M. Chronic ankle instability: Evolution of the model. *J. Athl. Train.* **2011**, *46*, 133–141. [CrossRef] [PubMed]
17. Prajapati, B.; Dunne, M.; Armstrong, R. Sample size estimation and statistical power analyses. *Optom. Today* **2010**, *16*, 10–18.
18. Gorji, S.M.; Samakosh, H.M.N.; Watt, P.; Marchetti, P.H.; Oliveira, R. Pain neuroscience education and motor control exercises versus core stability exercises on pain, disability, and balance in women with chronic low back pain. *Int. J. Environ. Res. Public Health* **2022**, *19*, 2694. [CrossRef]
19. Chan, K.W.; Ding, B.C.; Mroczek, K.J. Acute and chronic lateral ankle instability in the athlete. *Bull. NYU Hosp. Jt. Dis.* **2011**, *69*, 17.
20. Bulow, A.; Anderson, J.E.; Leiter, J.R.; MacDonald, P.B.; Peeler, J. The modified star excursion balance and y-balance test results differ when assessing physically active healthy adolescent females. *Int. J. Sports Phys. Ther.* **2019**, *14*, 192–203. [CrossRef]
21. Rad, Z.M.; Norasteh, A.A. The effect of a six-week core stability exercises on balance, strength, and endurance in female students with trunk defects. *Phys. Treat. Specif. Phys. Ther. J.* **2020**, *10*, 231–238. [CrossRef]
22. Burcal, C.J.; Chung, S.; Johnston, M.L.; Rosen, A.B. Does the method of administration affect reliability of the foot and ankle ability measure? *J. Sport Rehabil.* **2020**, *29*, 1038–1041. [CrossRef]
23. McBeth, J.M.; Earl-Boehm, J.E.; Cobb, S.C.; Huddleston, W.E. Hip muscle activity during 3 side lying hip strengthening exercises in distance runners. *J. Athl. Train.* **2012**, *47*, 15–23. [CrossRef] [PubMed]
24. Daloia, L.M.T.; Figueiredo, M.M.L.; Martinez, E.Z.; Mattiello-Sverzut, A.C. Isometric muscle strength in children and adolescents using Handheld dynamometry: Reliability and normative data for the Brazilian population. *Braz. J. Phys. Ther.* **2018**, *22*, 474–483. [CrossRef] [PubMed]
25. Swearingen, J.; Lawrence, E.; Stevens, J.; Jackson, C.; Waggy, C.; Davis, D.S. Correlation of single leg vertical jump, single leg hop for distance, and single leg hop for time. *Phys. Ther. Sport* **2011**, *12*, 194–198. [CrossRef] [PubMed]
26. Tveter, A.T.; Holm, I. Influence of thigh muscle strength and balance on hop length in one-legged hopping in children aged 7–12 years. *Gait Posture* **2010**, *32*, 259–262. [CrossRef]
27. Caffrey, E.; Docherty, C.L.; Schrader, J.; Klossnner, J. The ability of 4 single-limb hopping tests to detect functional performance deficits in individuals with functional ankle instability. *J. Orthop. Sports Phys. Ther.* **2009**, *39*, 799–806. [CrossRef]
28. Cohen, J. A power primer. *Psychol. Bull.* **1992**, *112*, 155. [CrossRef]

29. Huang, P.-Y.; Jankaew, A.; Lin, C.-F. Effects of plyometric and balance training on neuromuscular control of recreational athletes with functional ankle instability: A randomized controlled laboratory study. *Int. J. Environ. Res. Public Health* **2021**, *18*, 5269. [CrossRef]
30. Hall, E.A.; Chomistek, A.K.; Kingma, J.J.; Docherty, C.L. Balance-and strength-training protocols to improve chronic ankle instability deficits, part I: Assessing clinical outcome measures. *J. Athl. Train* **2018**, *53*, 568–577. [CrossRef]
31. Ford, K.; Nguyen, A.-D.; Dischiavi, S.; Hegedus, E.; Zuk, E.; Taylor, J. An evidence-based review of hip-focused neuromuscular exercise interventions to address dynamic lower extremity valgus. *Open Access J. Sports Med.* **2015**, *6*, 291–303. [CrossRef] [PubMed]
32. Mok, N.W.; Yeung, E.W.; Cho, J.C.; Hui, S.C.; Liu, K.C.; Pang, C.H. Core muscle activity during suspension exercises. *J. Sci. Med. Sport* **2015**, *18*, 189–194. [CrossRef]
33. Gordon, A.T.; Ambegaonkar, J.P.; Caswell, S.V. Relationships between core strength, hip external rotator muscle strength, and star excursion balance test performance in female lacrosse players. *Int. J. Sports Phys. Ther.* **2013**, *8*, 97–104. [PubMed]
34. Ben Kibler, W.; Press, J.; Sciascia, A. The role of core stability in athletic function. *Sports Med.* **2006**, *36*, 189–198. [CrossRef] [PubMed]
35. Munn, J.; Sullivan, S.J.; Schneiders, A. Evidence of sensorimotor deficits in functional ankle instability: A systematic review with meta-analysis. *J. Sci. Med. Sport* **2010**, *13*, 2–12. [CrossRef] [PubMed]
36. Holm, I.; Tveter, A.T.; Fredriksen, P.M.; Vøllestad, N. A normative sample of gait and hopping on one leg parameters in children 7–12 years of age. *Gait Posture* **2009**, *29*, 317–321. [CrossRef]
37. Myer, G.D.; Ford, K.R.; Brent, J.L.; Hewett, T.E. The effects of plyometric vs. dynamic stabilization and balance training on power, balance, and landing force in female athletes. *J. Stren. Cond. Res.* **2006**, *20*, 345.
38. Kramer, A.; Kümmel, J.; Gollhofer, A.; Armbrecht, G.; Ritzmann, R.; Belavy, D.; Felsenberg, D.; Gruber, M. Plyometrics can preserve peak power during 2 months of physical inactivity: An RCT including a one-year follow-up. *Front. Physiol.* **2018**, *9*, 633. [CrossRef]
39. Ullrich, B.; Pelzer, T.; Pfeiffer, M. Neuromuscular effects to 6 weeks of loaded countermovement jumping with traditional and daily undulating periodization. *J. Strength Cond. Res.* **2018**, *32*, 660–674. [CrossRef]
40. Noyes, F.R.; Barber-Westin, S.D.; Smith, S.T.T.; Campbell, T. A training program to improve neuromuscular and performance indices in female high school soccer players. *J. Strength Cond. Res.* **2013**, *27*, 340–351. [CrossRef]
41. Hewett, T.E.; Stroupe, A.L.; Nance, T.A.; Noyes, F.R. Plyometric training in female athletes: Decreased impact forces and increased hamstring torques. *Am. J. Sports Med.* **1996**, *24*, 765–773. [CrossRef] [PubMed]
42. Noyes, F.R.; Barber-Westin, S.D.; Smith, S.T.; Campbell, T.; Garrison, T.T. A Training program to improve neuromuscular and performance indices in female high school basketball players. *J. Strength Cond. Res.* **2012**, *26*, 709–719. [CrossRef] [PubMed]
43. Barber-Westin, S.D.; Hermeto, A.A.; Noyes, F.R. A six-week neuromuscular training program for competitive junior tennis players. *J. Strength Cond. Res.* **2010**, *24*, 2372–2382. [CrossRef] [PubMed]
44. Noyes, F.R.; Barber-Westin, S.D.; Smith, S.T.; Campbell, T. A Training program to improve neuromuscular indices in female high school volleyball players. *J. Strength Cond. Res.* **2011**, *25*, 2151–2160. [CrossRef] [PubMed]
45. Markovic, G.; Mikulic, P. Neuro-musculoskeletal and performance adaptations to lower-extremity plyometric training. *Sports Med.* **2010**, *40*, 859–895. [CrossRef] [PubMed]
46. Butler, R.J.; Crowell III, H.P.; Davis, I.M. Lower extremity stiffness: Implications for performance and injury. *Clin. Biomech.* **2003**, *18*, 511–517. [CrossRef]
47. McKeon, P.O.; Hertel, J. Systematic review of postural control and lateral ankle instability, part I: Can deficits be detected with instrumented testing? *J. Athl. Train.* **2008**, *43*, 293–304. [CrossRef]
48. Webster, K.A.; Gribble, P.A. Functional rehabilitation interventions for chronic ankle instability: A systematic review. *J. Sport Rehabil.* **2010**, *19*, 98–114. [CrossRef]
49. Carcia, C.R.; Martin, R.L.; Drouin, J.M. Validity of the foot and ankle ability measure in athletes with chronic ankle instability. *J. Athl. Train.* **2008**, *43*, 179–183. [CrossRef]
50. Teixeira, J.; Forte, P.; Ferraz, R.; Leal, M.; Ribeiro, J.; Silva, A.; Barbosa, T.; Monteiro, A. Monitoring accumulated training and match load in football: A systematic review. *Int. J. Environ. Res. Public Health* **2021**, *18*, 3906. [CrossRef]
51. Miguel, M.; Oliveira, R.; Loureiro, N.; García-Rubio, J.; Ibáñez, S. Load measures in training/match monitoring in soccer: A systematic review. *Int. J. Environ. Res. Public Health* **2021**, *18*, 2721. [CrossRef] [PubMed]
52. Oliveira, R.; Francisco, R.; Fernandes, R.; Martins, A.; Nobari, H.; Clemente, F.M.; Brito, J.P. In-season body composition effects in professional women soccer players. *Int. J. Environ. Res. Public Health* **2021**, *18*, 12023. [CrossRef] [PubMed]
53. Kotzamanidis, C.; Chatzopoulos, D.; Michailidis, C.; Papaiakovou, G.; Patikas, D. The effect of a combined high-intensity strength and speed training program on the running and jumping ability of soccer players. *J. Strength Cond. Res.* **2005**, *19*, 369–375. [CrossRef] [PubMed]

Article

Does Inspiratory Muscle Training Affect Static Balance in Soccer Players? A Pilot Randomized Controlled Clinical Trial

Silvana Loana de Oliveira-Sousa [1], Martha Cecilia León-Garzón [2], Mariano Gacto-Sánchez [1,*], Alfonso Javier Ibáñez-Vera [3], Luis Espejo-Antúnez [4] and Felipe León-Morillas [2]

[1]   Department of Physiotherapy, University of Murcia, 30003 Murcia, Spain
[2]   Department of Physiotherapy, Jerónimos Campus, 135. Catholic University of Murcia UCAM, Guadalupe, 30107 Murcia, Spain
[3]   Department of Health Sciences, Campus de las Lagunillas, University of Jaén, 23071 Jaén, Spain
[4]   Department of Medical-Surgical Therapeutics, Faculty of Medicine and Health Sciences, University of Extremadura, 06006 Badajoz, Spain
*   Correspondence: marianogacto@um.es

Abstract: Inspiratory muscle training (IMT) is effective in improving postural stability and balance in different clinical populations. However, there is no evidence of these effects in soccer players. A single-blind, two-arm (1.1), randomized, placebo-controlled pilot study on 14 soccer players was performed with the main aim of assessing the effect of IMT on static balance, and secondarily, of examining changes in the respiratory muscle function. The experimental group (EG) received an IMT program with progressive intensity, from 20% to 80%, of the maximal inspiratory pressure (MIP). The sham group (SG) performed the same program with a fixed load of 20% of the MIP. Static balance and respiratory muscle function variables were assessed. A two-factor analysis of variance for repeated measures was used to assess differences after training. Statistical significance was set at $p < 0.05$. Significant increases were observed in the EG on length of sway under eyes open (from 2904.8 ± 640.0 to 3522.4 ± 509.0 mm, $p = 0.012$) and eyes closed (from 3166.2 ± 641.3 to 4173.3 ± 390.8 mm, $p = 0.004$). A significant increase in the maximal voluntary ventilation was observed for both groups (EG $p = 0.005$; SG $p = 0.000$). No significant differences existed between the groups. IMT did not improve the static balance in a sample of soccer players. Conducting a high-scale study is feasible and could refine the results and conclusions stemming from the current pilot study.

Keywords: inspiratory muscle training; static balance; center of pressure; soccer players

Citation: de Oliveira-Sousa, S.L.; León-Garzón, M.C.; Gacto-Sánchez, M.; Ibáñez-Vera, A.J.; Espejo-Antúnez, L.; León-Morillas, F. Does Inspiratory Muscle Training Affect Static Balance in Soccer Players? A Pilot Randomized Controlled Clinical Trial. *Healthcare* 2023, *11*, 262. https://doi.org/10.3390/healthcare11020262

Academic Editors: João Paulo Brito and Rafael Oliveira

Received: 7 December 2022
Revised: 10 January 2023
Accepted: 12 January 2023
Published: 14 January 2023

1. Introduction

Postural stability is an essential motor skill in soccer, and postural talent might be considered an indicator of performance [1]. Soccer is a sport with high physical demands. Some of them involve the respiratory muscles, which can significantly contribute to the limitation of exercise, due to muscle fatigue and the effects on blood flow in the limbs that perform physical work. In addition, postural stability can be altered by respiratory movements [2–4] since these interfere with the biomechanics of the torso and postural sway, reducing these aspects under apnea conditions [5] and increasing them with hyperventilation [6]. During a soccer match, players must use their motor skills and control their posture, while also gathering visual information about other team members as well as opponents. Due to the conditions of the game itself, it is of paramount importance for players to have a good balance to control, pass or shoot the ball [7,8].

Inspiratory muscle training (IMT) has been shown to be effective in improving variables related to postural stability in a wide variety of clinical and healthy populations [9–14]. These effects are supported by empirical data on the assistance of the respiratory muscles in trunk stabilization: the diaphragm coordinates with the abdominal muscles, therefore generating a hydraulic effect in the abdominal cavity which, in turn, assists the spinal stabilization by stiffening the lumbar spine through increased intra-abdominal pressure [15–18]. This action assists in maintaining postural stability in situations where external forces (i.e., rapid movement of the upper limb) destabilize the spine, and during reactive and dynamic tasks [16]. Inspiratory muscle training increases the efficiency of diaphragmatic phasic contractions, and the ability to increase intraabdominal pressure, improving balance abilities [14].

Under normal conditions, the respiratory and postural functions of the diaphragm can be coordinated when the stability of the trunk is challenged by repeated rapid movements of the limbs [16]. However, this coordination can be affected by several factors, such as increased respiratory demand due to exercise or weakness of the respiratory muscles [19]. During a soccer game, the respiratory demand is increased, due to high-intensity intermittent activity that requires performers to undertake regular repeated sprints across a 90-min game, and where the sustained level of effort approaches the anaerobic threshold (75% of maximal effort).

In a recent case–control study, we found that greater values of inspiratory muscle strength are associated with shorter path length and less lateral displacement in the closed eyes condition in a sample of soccer players [20]. This association may lead to the hypothesis that a specific IMT program in soccer players could improve their postural stability. Data supporting or rejecting this hypothesis are, however, scarce in these populations. In previous literature reviews encompassing different types of players and specifically involving soccer players, none of the included randomized controlled trials (RCTs) examined variables related to balance [21,22]. Thus, the primary objective of this study was to assess the effect of IMT on static balance, and secondarily, to examine changes in the respiratory muscle function.

2. Materials and Methods

2.1. Study Design

A single-blind (evaluators) pilot randomized controlled clinical trial was carried out in a sample of soccer players from the U23 soccer team of the Catholic University of Murcia (UCAM). The study was developed from February to May 2018. This study has followed the ethical principles of the Declaration of Helsinki, was approved by the corresponding Ethics Committee and has been registered in ClinicalTrials.gov under the code NCT03383900. The CONSORT statements for conducting and reporting this randomized controlled trial were also followed. All subjects agreed to take part in the study and signed the corresponding informed consent.

2.2. Participants

Soccer players from the UCAM U23 men's soccer team during the season 2017–2018 were included in this study. Inclusion criteria included attending training sessions regularly, alongside participating in over 80% of the games from the start of the season until the first measurement. Exclusion criteria included severe musculoskeletal injuries and lung-based pathologies. Informed consent forms were signed by every player included in the study.

All the athletes were requested to continue their current training regime, with no changes in volume or training intensity.

2.3. Randomization and Masking

After signing an informed consent form, participants were randomly assigned (1:1) to an experimental group (EG) or sham group (SG), using a computer-generated random number table. The randomization code was performed by a physiotherapist who did not participate in the measurements nor in the intervention. Until the end of training, these codes were known only by the physiotherapist responsible for randomization. Assessors were also blinded to the participants' intervention assignment.

2.4. Interventions

2.4.1. Experimental Group (EG)

The EG players performed IMT with a Power Breathe device model Heavy Resistance— HR plus (Power Breathe International Ltd., Southam, Warwickshire; England, UK). The IMT consisted of daily sessions (3 sets of 15 repetitions—inspirations), 6 days a week, for a period of 8 weeks. The IMT was performed prior to routine training in the changing rooms of the soccer field and was supervised by a physiotherapist. Whenever players did not train on the field and/or when a player did not attend the training session, they had to send a video performing the IMT at home to a WhatsApp group in which the physiotherapist supervised the activity. The EG players started by breathing at a resistance equal to 20% of their maximum inspiratory pressure (MIP) for one week. The load was then increased incrementally, 5–10% each session, to reach 80% of their MIP at the end of the first month. After week 4, the MIP value was reevaluated and a resistance corresponding to 80% of this new MIP value was used throughout the second month, therefore achieving a total of 8 weeks of training [23–26].

2.4.2. Sham Group (SG)

Players from the SG performed IMT with Power Breathe. The training program was the same as that of the EG in terms of frequency and duration, whereas the intensity was 20% of the MIP throughout the training period.

2.5. Measurements

Data were collected by two physiotherapists (FLM, SLOS) experienced in the field of spirometry. Data collection took place at the same time each day, in the morning timeframe, in the same facility (Physiotherapy practice room from the UCAM), and under similar temperature conditions (24–25 °C).

2.5.1. Static Balance

Static balance was measured through the analysis of stabilometric variables on a balance platform (Free Step platform, Rome, Italy), with an active surface of 400 × 400 mm and 8 mm thickness, and Free-Step v.1.0.3 software (Rome, Italy). Four stabilometric variables were examined in the bipodal eyes open (BOA) and bipodal eyes closed (BOC) conditions: length of sway (LS), surface of the ellipse (SE), lateral axis (DX) and antero-posterior axis (DY). The measurement procedure was performed according to a previously published protocol [20,27]. The players had to stand on the platform with bipodal support (45°) for 90 s with eyes open and 90 s with eyes closed. During the measurement under the open-eyes condition, the players had to keep their gaze on a fixed point that coincided with the height of their eyes and was about 2.5 m away from the platform.

2.5.2. Respiratory Muscle Function

To assess respiratory muscle function, both the inspiratory muscle strength and respiratory endurance were examined. Inspiratory muscle strength was measured indirectly, through the maximal inspiratory pressure (MIP), obtaining values of both the absolute value of MIP and the predictive value in % of MIP (values of normality in Caucasian population) [28]. A maximum inspiratory mouth pressure monitor Datospir Touch (Sibelmed, Barcelona, Spain) was used. With the individuals in a sitting position and wearing a nose clip, they were asked to perform a maximal expiration (close to the residual volume), followed by a maximal inspiration (close to the total lung capacity). The measurement was carried out by a previously trained physiotherapist, performing a maximum of eight measurements, until reaching three measurements that obtained a difference of ≤10%, thus selecting the best result according to the protocol established by the Spanish Society of Pneumology and Thoracic Surgery (SEPAR) [29].

Respiratory muscular endurance was obtained by measuring the maximal voluntary ventilation (MVV): it consists in mobilizing the maximum amount of air in a short period of time (15 s) with a respiratory rate beyond 80 breaths per minute, according to the criteria established by the SEPAR [29]. Subjects in a sitting position and wearing a nose clip were asked to breathe as fast and deep as possible for 15s. Three maneuvers were carried out with 1-min rest between maneuvers, and the best value among the three maneuvers was chosen. Data on both MVV in liters (MVV l, MVV% predict) and number of breaths per minute (MVV bpm) were recorded. These variables were measured with a digital spirometer (Datospir Touch, Sibelmed, Barcelona, Spain). In addition, data on age, weight and height were collected.

2.6. Data Analysis

Considering the required sample size, a calculation through G* Power software (ver. 3.1.9.7; Heinrich-Heine-Universität Düsseldorf, Düsseldorf, Germany; http://www.gpower.hhu.de/ -accessed on 26 November 2022-) was performed using the following parameters: one tail, large effect size d = 0.8, alpha = 0.05, statistical power = 0.80 and a 1:1 allocation ratio. The total sample size would consist of 42 subjects (i.e., 21 individuals per group); therefore, our study corresponds to a pilot study, since a total of 18 subjects were initially recruited and 14 completed the study (EG = 7; SG = 7). The perception that pilot trials are simply a casual prelude to a larger trial may somehow threaten the rigor with which they are implemented, and such an approach to pilot trial design and implementation runs the risk of providing misleading results, as stated by Arnold et al. [30]. For this reason, the current study followed the recommendations developed by Thabane et al. [31] for reporting the results of pilot studies.

All results are presented as mean (SD) when applicable. Data were tested for normality using the Shapiro–Wilk test (considering the final size of 14 subjects). A comparison of means for independent samples was carried out, using Student's *t*-test to determine possible differences between the experimental and sham groups at baseline. Additionally, for inspiratory muscle strength at baseline, a descriptive analysis was performed, based on the predictive values of normality. A two-factor analysis of variance for repeated measures was used to observe differences after training, both within the same group at different times (pre-intervention and post-intervention), and between groups (experimental versus sham) for each one of the variables of interest. Mean differences between groups were also calculated for the variables approached. Statistical significance was set at $p < 0.05$. The effect size was calculated using partial eta-squared and interpreted as small ($\eta^2 > 0.01$), medium ($\eta^2 > 0.06$), or large ($\eta^2 > 0.14$) [32]. All analyses were performed using SPSS software version 22.0 (IBM, Armonk, NY, USA).

3. Results

3.1. Selection Process

The UCAM U23 soccer team was composed of a total of twenty-two players. The initial sample consisted of eighteen players, since four of them did not meet the inclusion criteria. The random assignment performed led to a distribution of nine players for the EG and nine for the SG. During the study, there were sample losses. Finally, fourteen players completed the study (EG = 7; SG = 7) (Figure 1).

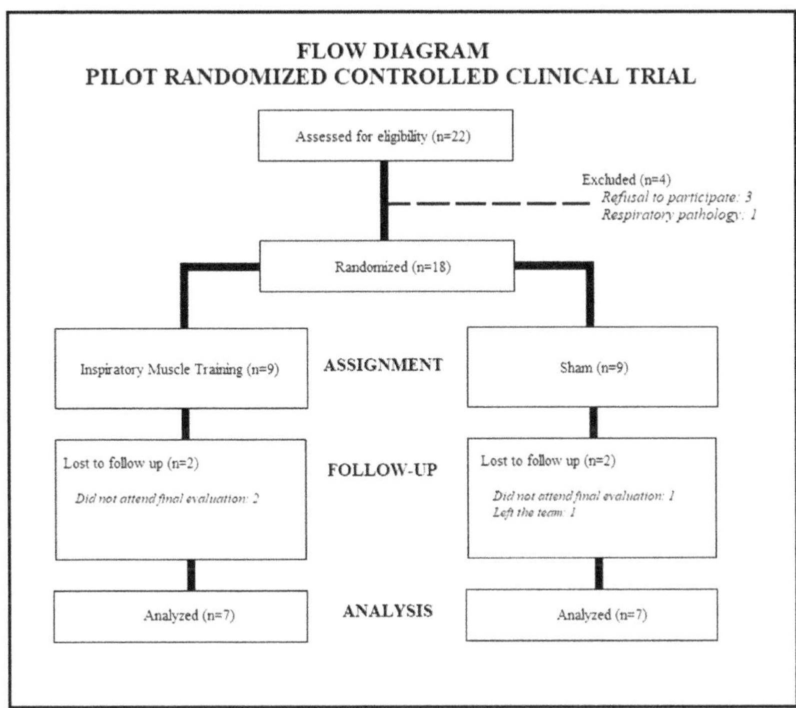

Figure 1. Participant selection and randomization process.

3.2. Characteristics of the Participants

Table 1 shows the sociodemographic and anthropometric characteristics alongside the values corresponding to the respiratory muscle function and the static balance in both groups. Groups were homogeneous at baseline, with no statistically significant differences across variables.

Table 1. General characteristics of the participants (n = 14) at baseline.

Variables	IMT (n = 7)	Sham (n = 7)	p
Sociodemographic			
Age (years)	20.00 ± 0.81	20.00 ± 0.57	1.000
Anthropometric			
Size (cm)	177.57 ± 4.23	180.42 ± 5.79	0.313
Weight (kg)	71.88 ± 3.79	77.45 ± 8.70	0.147
Respiratory muscle function			
MIP (cm H_2O)	161.57 ± 31.11	175.42 ± 30.27	0.415
MIP (% pred)	104.80 ± 21.4	111.50 ± 20.80	0.169
MVV (liters)	188.74 ± 30.06	201.78 ± 23.15	0.381
MVV (bpm)	89.89 ± 15.30	86.04 ± 21.58	0.707
MVV (% pred)	98.43 ± 14.00	103.39 ± 12.31	0.495

Table 1. *Cont.*

Variables	IMT ($n = 7$)	Sham ($n = 7$)	p
Static balance			
BOA_LS (mm)	2904.81 ± 640.03	4358.13 ± 1863.14	0.075
BOA_SE (mm^2)	70.44 ± 42.04	123.14 ± 56.10	0.070
BOA_DX (mm)	10.06 ± 3.16	12.05 ± 4.00	0.322
BOA_DY (mm)	10.91 ± 3.47	14.75 ± 3.92	0.077
BOC_LS (mm)	3166.22 ± 641.30	4290.61 ± 1453.5	0.086
BOC_SE (mm^2)	78.37 ± 63.52	127.21 ± 74.39	0.211
BOC_DX (mm)	10.01 ± 5.72	12.60 ± 2.85	0.305
BOC_DY (mm)	11.19 ± 4.41	17.43 ± 8.75	0.118

Data are expressed as mean ± standard deviation. MIP = maximum inspiratory pressure; % pred = % predictive; MVV = maximum voluntary ventilation; bpm = breaths per minute; BOA = bipodal eyes open; LS = length of sway; SE = surface of the ellipse; mm^2 = square millimeters; DX = lateral axis; DY = antero-posterior axis; BOC = bipodal eyes closed.

3.3. Static Balance

Table 2 shows the results of the analysis of stabilometric variables under the condition of eyes open and eyes closed with bipodal support. Significant changes were found in the EG on the variable length of sway under the condition of eyes open (BOA LS) with an increase from an initial value of 2904.8 ± 640.0 to 3522.4 ± 509.0 mm ($p = 0.012$). No significant changes were found in the other stabilometric variables under the open-eyes condition analyzed. Regarding the closed-eyes condition, the length of sway (BOC LS) also showed significant changes in the EG with a baseline value of 3166.2 ± 641.3 mm and a post-intervention value of 4173.3 ± 390.8 mm ($p = 0.004$). For the rest of the variables under the aforementioned condition, there were no significant changes in any of the groups.

In relation to the differences between the groups, no significant changes were found.

Table 2. Differences in static balance variables between groups pre–post-intervention.

	IMT ($n = 7$)				Sham ($n = 7$)				Differences within Interventions (Post-I–Pre-I)		Differences between Interventions (Post-I–Pre-I)
	Pre-IN	Post-IN	p	η^2	Pre-IN	Post-IN	p	η^2	IMT	Sham	IMT-Sham
BOA_LS (mm)	2904.8 ± 640.0	3522.4 ± 509.0	0.012	0.675	4358.1 ± 1863.1	3643.3 ± 596.7	0.286	0.186	617.6 ± 817.7	−714.8 ± 1956.2	1332.4 (−787.5, 3452.3)
BOA_SE (mm^2)	70.4 ± 42.0	177.5 ± 171.3	0.175	0.282	123.1 ± 56.1	162.8 ± 115.9	0.375	0.133	107.1 ± 176.3	39.7 ± 128.7	67.4 (−150.6, 285.6)
BOA_DX (mm)	10.0 ± 3.1	12.8 ± 5.3	0.220	0.238	12.0 ± 4.0	15.1 ± 7.7	0.329	0.159	2.8 ± 6.1	3.1 ± 8.6	−0.3 (−10.9, 10.3)
BOA_DY (mm)	10.9 ± 3.4	12.3 ± 3.4	0.090	0.405	14.7 ± 3.9	18.4 ± 8.1	0.223	0.235	1.4 ± 4.8	3.7 ± 8.9	−2.3 (−12.4, 7.8)
BOC_LS (mm)	3166.2 ± 6413	4173.3 ± 390.8	0.004	0.779	4290.6 ± 1453.5	4006.5 ± 787.1	0.554	0.062	1007.1 ± 750.7	−284.0 ± 1650.4	1291.2 (−521.9, 3104.3)
BOC_SE (mm^2)	78.3 ± 63.5	78.4 ± 51.2	0.998	0.000	127.2 ± 74.3	120.8 ± 127.7	0.888	0.004	0.1 ± 81.6	−6.4 ± 147.7	6.5 (−162.2, 175.2)
BOC_DX (mm)	10.0 ± 5.7	12.8 ± 5.3	0.338	0.153	12.6 ± 2.8	15.1 ± 7.7	0.341	0.152	2.7 ± 7.8	2.5 ± 8.2	0.3 (−11.0, 11.6)
BOC_DY (mm)	11.1 ± 4.4	12.3 ± 3.4	0.370	0.135	17.4 ± 8.7	18.4 ± 8.1	0.808	0.011	1.2 ± 5.5	1.0 ± 11.9	−2.2 (−15.4, 11.0)

Data are expressed as mean ± standard deviation or mean (95% confidence interval). Pre-IN: pre-intervention; Post-IN: post-intervention; η^2 = effect size. BOA = bilateral eyes open; LS = length of sway; SE = surface of the ellipse; mm^2 = squared millimeters; DX = lateral axis; DY = antero-posterior axis; BOC = bilateral eyes closed.

3.4. Respiratory Muscle Function

Regarding the inspiratory muscle strength (MIP% predictive) at baseline, the mean predictive percentage of the fourteen players (total sample) was 108.1% (96.30–120.0; 95% CI). A total of 25% of the players presented average values below their normality, that is, below 100% of the predictive one. On the other hand, 50% showed predictive values on the borderline value of normality, with an average of 102.4% (Figure 2).

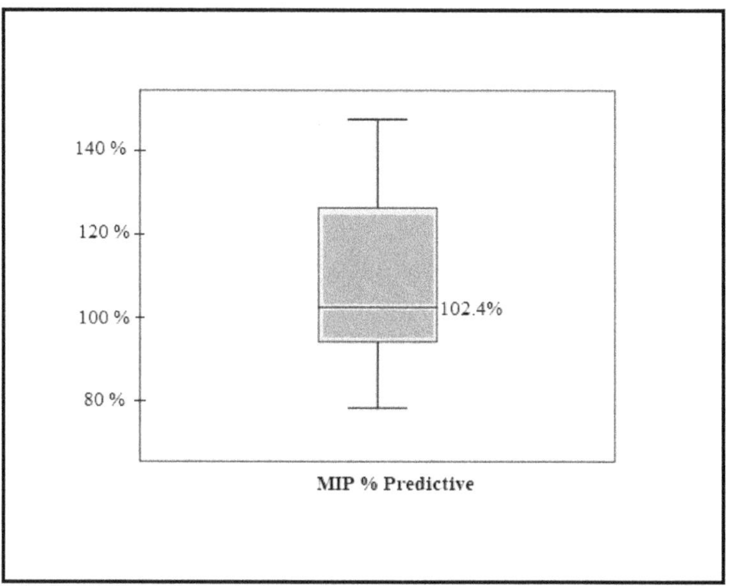

Figure 2. Baseline inspiratory muscle strength (MIP% predictive).

After the training period (8 weeks), the EG showed an increase from 161.5 ± 31.1 to 184.4 ± 21.5 cm H_2O with a significance level of $p = 0.076$. Regarding the SG, it ranged from 175.4 ± 30.2 to 176.0 ± 16.4 cm H_2O ($p = 0.951$) (Table 2). In terms of predictive values, the EG improved from 104.8 ± 21.4 to 119.5 ± 14.5 cmH2O, while the change in the SG was negligible, as well as the effect size (η^2), which was very large in the EG compared to that of the SG (0.432 vs. 0.000, respectively).

Regarding MVV (Table 3), increases in the number of breaths per minute with statistically significant changes were observed in both groups. The EG obtained an initial value of 89.8 ± 15.3 bpm and post-intervention of 143.2 ± 21.9 bpm with $p = 0.005$, while the SG obtained a pre-intervention result of 86.0 ± 21.5 bpm and post-intervention of 147.8 ± 33.4 bpm with $p = 0.000$. Differences between groups were non-significant. Regarding MVV in liters and in % of the predictive value, none of the groups presented significant increases and no significant differences between groups were found either.

Table 3. Differences in respiratory function variables between groups pre–post-intervention.

	IMT ($n = 7$)				Sham ($n = 7$)				Differences within Interventions (Post-I–Pre-I)		Differences between Interventions (Post-I–Pre-I)
	Pre-I	Post-I	p	η^2	Pre-I	Post-I	p	η^2	IMT	Sham	IMT-Sham
MIP (cm H_2O)	161.5 ± 31.1	184.4 ± 21.5	0.076	0.434	175.4 ± 30.2	176.0 ± 16.4	0.951	0.001	22.9 ± 37.8	0.6 ± 34.3	22.3 (-28.74, 73.34)
MIP (% pred)	104.8 ± 21.4	119.5 ± 14.5	0.076	0.432	111.5 ± 20.8	111.6 ± 10.5	0.981	0.000	14.7 ± 25.8	0.1 ± 23.3	14.6 (-20.1, 49.3)
MVV (liters)	188.7 ± 30.0	169.5 ± 37.9	0.180	0.278	201.7 ± 23.1	197.2 ± 26.6	0.579	0.054	-19.2 ± 48.3	-4.5 ± 35.2	-14.7 (-74.4, 45.0)
MVV (bpm)	89.8 ± 15.3	143.2 ± 21.9	0.005	0.756	86.0 ± 21.5	147.8 ± 33.4	0.000	0.890	53.4 ± 26.7	61.8 ± 39.7	-8.4 (-56.2, 39.4)
MVV (% pred)	98.4 ± 14.0	$88.4 = 19.0$	0.187	0.270	103.3 ± 12.3	101.2 ± 15.6	0.611	0.046	-10.0 ± 23.6	-2.1 ± 19.8	-7.9 (-38.7, 22.9)

Data are expressed as mean $\pm$ standard deviation or mean (95% confidence interval). Pre-I = pre-intervention; Post-I = post-intervention; MIP = maximum inspiratory pressure; % pred = % predictive; MVV = maximum voluntary ventilation; bpm = breaths per minute; $\eta2$ = effect size.

4. Discussion

The main objective of this study was to determine whether an 8-week inspiratory muscle-specific training program can improve static balance in soccer players. The results revealed no improvement in any of the static balance variables studied. Despite the fact that significant changes in the variable length of sway were stated concerning the experimental group, these changes corresponded to an increase after the training period, unexpectedly and counterintuitively. Secondarily, we observed significant increases in the maximum voluntary ventilation (breaths per minute) in both groups and an important increase, bordering statistical significance, in the inspiratory muscle strength in the IMT group. However, no significant differences on any of the variables examined were found between the groups.

Across all the stabilometric variables examined, significant changes were found only on the length of sway for the IMT group. Surprisingly, this group showed an increase in the length of the postural oscillations, both in the eyes open and eyes closed conditions (although not reaching statistically significant differences when compared to the placebo group). These results are contradictory to previous studies conducted in other healthy populations [13,14]. Rodrigues et al. [13] investigated the effect of a 4-week IMT program on postural balance responses during orthostatic stress in healthy elderly women and observed a reduction in the CoP distance and velocity in the IMT group compared to the sham group.

Unfortunately, the scarcity of data on soccer players hinders a possible comparison with other RCTs carried out on soccer players. In our recent case–control study, we observed that the inspiratory muscle strength was negatively correlated with sway oscillations [20]. This fact led us to the hypothesis that an IMT program could reduce the oscillations of the center of pressure in those players. Two possible reasons could explain the contradictory response observed among our sample of players: On the one hand, the postural strategies adopted by athletes may differ considerably from those of patients; therefore, the response to an IMT may be subsequently different. On the other hand, the increase in the length of sway can perhaps be understood from an ecological-approach perspective, since some authors affirm that the postural oscillation based on the support of asymptomatic young subjects, generated by postural fluctuations, can provide sensory exploratory information about how the body itself interacts with the environment. These fluctuations generate shifts in the center of pressures and depend on the postural patterns of each individual. Therefore, from this perspective, the observed results could be interpreted as an increase in the oscillation to coordinate the postural fluctuations and to have a greater adaptability to the environment [27,33,34].

In spite of the growing interest concerning the role played by IMT in postural control and balance, most of the underlying mechanisms that may explain this relationship remain uncertain. The improvement in the inspiratory muscle strength (MIP) does not seem to entail a better postural control per se. The enhancement of MIP can have an impact on other variables which, in turn, can alter postural control. For example, improvement in the diaphragmatic strength can affect the functionality of the core as a whole, which, in turn, improves trunk stability and balance.

Regarding the changes on variables concerning the respiratory muscle function, the players in the IMT group showed an important increase (14% vs. 0.1% in SG) in the inspiratory muscle strength. However, this increase fell short of being statistically significant. Furthermore, both groups increased their number of breaths per minute in the maximal voluntary ventilation test, but there was no improvement in the number of liters of air mobilized per minute. The characteristics of the training protocol and the device used may have affected the results. In our study, we performed a semi-supervised intervention protocol, of one daily session, while most authors used two-session-per-day protocols to train athletes [23–26,34,35]. The frequency of one session per day could, therefore, be insufficient to achieve important changes across healthy populations. On another note, an inspiratory threshold load device was used, which may have little effect on the amount of air mobilized and, therefore, on the maximum minute ventilation when compared to devices for voluntary isocapnic hyperpnea [36]. Finally, the results of the evaluation of the respiratory muscle strength at baseline confirm that these players present a respiratory condition not very different from other non-athletic subjects [37]. Due to the high demand that these players present and the subsequent probability of fatigue of untrained respiratory muscles, we believe that monitoring and training these muscles is of paramount importance.

From a methodological perspective, the aim of the current pilot study was to analyze the feasibility of a larger-scale clinical trial through the analysis of its validity and feasibility. Thus, the randomization and blinding, the acceptability of the intervention and the selection of the primary outcome revealed a high level of feasibility of the study protocol [38]. Only the recruitment and consent aspect's results were somehow unsatisfactory since we obtained a 78% retention rate, a percentage below the standards expounded by Harris et al. [39] and Walters et al. [40], with a pooled percentage of 80% and a total percentage of 89%, respectively. Therefore, the retention rate should be specifically taken into consideration by means of developing specific strategies to enhance the retention rates, as stated elsewhere [41].

Even though the application of the results stemming from a pilot study for effect size and sample size estimations should be cautiously considered [38], based on the values of the Cohen's d (d = 1.17) obtained on the variable length of sway under the open-eyes condition for the experimental group (pre- to post-assessment), and considering a two-tailed hypothesis, the estimated sample size required would correspond to 26 subjects overall (13 subjects per study arm), which is somehow in line with the a priori size calculated (42 subjects).

5. Limitations

Although we have followed a rigorous methodological design, this study should be interpreted in light of its limitations. On the one hand, by including only one soccer team, the sample size was small, and we experienced a 22% patient loss rate after randomization. We could have increased the sample by including players from other teams, but we dismissed the idea after considering that different training routines would certainly affect the results. Therefore, the study corresponds to a pilot randomized controlled study and further research should focus on broader samples meeting the required minimal size. Moreover, our study was carried out with a sample of players from a club in a second division category; therefore, the results are not necessarily transferable to elite players or higher sport levels.

6. Conclusions

The results stemming from this pilot RCT have shown that an 8-week semi-supervised IMT program, performed with a threshold loading device, was not able to improve the static balance in a sample of soccer players. Moreover, no between-group differences were found regarding respiratory muscle function. Conducting a high-scale study is feasible and could refine the results and conclusions stemming from the current pilot study.

Author Contributions: Conceptualization, F.L.-M., M.C.L.-G. and S.L.d.O.-S.; methodology F.L.-M., M.C.L.-G., L.E.-A., A.J.I.-V., M.G.-S. and S.L.d.O.-S.; software, validation, resources and data curation, F.L.-M., L.E.-A., A.J.I.-V., M.G.-S. and S.L.d.O.-S.; formal analysis, S.L.d.O.-S.; investigation and writing—original draft preparation, F.L.-M. and S.L.d.O.-S.; writing—review and editing, F.L.-M., M.C.L.-G., L.E.-A., A.J.I.-V., M.G.-S. and S.L.d.O.-S.; visualization and supervision, L.E.-A., A.J.I.-V.; project administration, M.G.-S.; funding acquisition F.L.-M., M.C.L.-G., L.E.-A., A.J.I.-V., M.G.-S. and S.L.d.O.-S. All authors have read and agreed to the published version of the manuscript.

Funding: The APC was funded by Catholic University of Murcia. PMAFI project 12/16.

Institutional Review Board Statement: This study has followed the ethical principles of the Declaration of Helsinki and was approved by the UCAM Ethics Committee. The CONSORT statements for conducting and reporting this randomized controlled trial were also followed. All subjects agreed to take part in the study and signed the corresponding informed consent form.

Informed Consent Statement: Informed consent was obtained from all subjects involved in the study.

Data Availability Statement: The study was registered on ClinicalTrials.gov (accessed on 26 November 2022) under the following ID: NCT03383900.

Acknowledgments: All the authors thank the players of the soccer team and the Catholic University of Murcia for their funding and for their participation in this study.

Conflicts of Interest: The authors declare no conflict of interest.

References

1. Paillard, T.; Noe, F.; Riviere, T.; Marion, V.; Montoya, R.; Dupui, P. Postural performance and strategy in the unipedal stance of soccer players at different levels of competition. *J. Athl. Train.* **2006**, *41*, 172–176. [PubMed]
2. Hamaoui, A.; Gonneau, E.; Le Bozec, S. Respiratory disturbance to posture varies according to the respiratory mode. *Neurosci. Lett.* **2010**, *475*, 141–144. [CrossRef]
3. Hamaoui, A.; Hudson, A.L.; Laviolette, L.; Nierat, M.C.; Do, M.C.; Similowski, T. Postural disturbances resulting from unilateral and bilateral diaphragm contractions: A phrenic nerve stimulation study. *J. Appl. Physiol.* **2014**, *117*, 825–832. [CrossRef]
4. Stephens, R.J.; Haas, M.; Moore, W.L.; Emmil, J.R.; Sipress, J.A.; Williams, A. Effects of Diaphragmatic Breathing Patterns on Balance: A Preliminary Clinical Trial. *J. Manip. Physiol. Ther.* **2017**, *40*, 169–175. [CrossRef]
5. Caron, O.; Fontanari, P.; Cremieux, J.; Joulia, F. Effects of ventilation on body sway during human standing. *Neurosci. Lett.* **2004**, *366*, 6–9. [CrossRef]
6. Malakhov, M.; Makarenkova, E.; Melnikov, A. The Influence of Different Modes of Ventilation on Standing Balance of Athletes. *Asian J. Sports Med.* **2014**, *5*, e22767. [CrossRef] [PubMed]
7. Orchard, J. Is there a relationship between ground and climatic conditions and injuries in football? *Sports Med.* **2002**, *32*, 419–432. [CrossRef] [PubMed]
8. Gerbino, G.P.; Griffin, E.D.; Zurakowski, D. Comparison of standing balance between female collegiate dancers and soccer players. *Gait Posture* **2007**, *26*, 501–507. [CrossRef]
9. Lee, K.; Park, D.; Lee, G. Progressive Respiratory Muscle Training for Improving Trunk Stability in Chronic Stroke Survivors: A Pilot Randomized Controlled Trial. *J. Stroke Cerebrovasc. Dis.* **2019**, *28*, 1200–1211. [CrossRef]
10. Janssens, L.; McConnell, A.K.; Pijnenburg, M.; Claeys, K.; Goossens, N.; Lysens, R.; Troosters, T.; Brumagne, S. Inspiratory muscle training affects proprioceptive use and low back pain. *Med. Sci. Sports Exerc.* **2015**, *47*, 12–19. [CrossRef]
11. Tounsi, B.; Acheche, A.; Lelard, T.; Tabka, Z.; Trabelsi, Y.; Ahmaidi, S. Effects of specific inspiratory muscle training combined with whole-body endurance training program on balance in COPD patients: Randomized controlled trial. *PLoS ONE* **2021**, *16*, e0257595. [CrossRef]
12. Aydoğan-Arslan, S.; Uğurlu, K.; Sakizli-Erdal, E.; Keskin, E.D.; Demirgüç, A. Effects of Inspiratory Muscle Training on Respiratory Muscle Strength, Trunk Control, Balance and Functional Capacity in Stroke Patients: A single-blinded randomized controlled study. *Top Stroke Rehabil.* **2022**, *29*, 40–48. [CrossRef]
13. Rodrigues, G.D.; Gurgel, J.L.; Galdino, I.D.S.; da Nóbrega, A.C.L.; Soares, P.P.D.S. Inspiratory muscle training improves cerebrovascular and postural control responses during orthostatic stress in older women. *Eur. J. Appl. Physiol.* **2020**, *120*, 2171–2181. [CrossRef] [PubMed]
14. Ferraro, F.V.; Gavin, J.P.; Wainwright, T.; McConnell, A. The effects of 8 weeks of inspiratory muscle training on the balance of healthy older adults: A randomized, double-blind, placebo-controlled study. *Physiol. Rep.* **2019**, *7*, e14076. [CrossRef]
15. Hodges, P.W.; Butler, J.E.; McKenzie, D.K.; Gandevia, S.C. Contraction of the human diaphragm during rapid postural adjustments. *J. Physiol.* **1997**, *505*, 539–548. [CrossRef] [PubMed]
16. Hodges, P.W.; Gandevia, S.C. Activation of the human diaphragm during a repetitive postural task. *J. Physiol.* **2000**, *522*, 165–175. [CrossRef] [PubMed]
17. Kocjan, J.; Adamek, M.; Gzik-Zroska, B.; Czyżewski, D.; Rydel, M. Network of breathing. Multifunctional role of the diaphragm: A review. *Adv. Respir. Med.* **2017**, *85*, 224–232. [CrossRef] [PubMed]
18. Bordoni, B.; Zanier, E. Anatomic connections of the diaphragm: Influence of respiration on the body system. *J. Multidiscip. Healthc.* **2013**, *6*, 281–291. [CrossRef]
19. Hodges, P.W.; Heijnen, I.; Gandevia, S.C. Postural activity of the diaphragm is reduced in humans when respiratory demand increases. *J. Physiol.* **2001**, *537*, 999–1008. [CrossRef]
20. León-Morillas, F.; Lozano-Quijada, C.; Lérida-Ortega, M.A.; León-Garzón, M.C.; Ibáñez-Vera, A.J.; Oliveira-Sousa, S.L. Relationship between Respiratory Muscle Function and Postural Stability in Male Soccer Players: A Case-Control Study. *Healthcare* **2021**, *9*, 644. [CrossRef] [PubMed]

21. HajGhanbari, B.; Yamabayashi, C.; Buna, T.R.; Coelho, J.D.; Freedman, K.D.; Morton, T.A.; Palmer, S.A.; Toy, M.A.; Walsh, C.; Sheel, A.W.; et al. Effects of respiratory muscle training on performance in athletes: A systematic review with meta-analyses. *J. Strength Cond. Res.* **2013**, *27*, 1643–1663. [CrossRef] [PubMed]
22. Karsten, M.; Ribeiro, G.S.; Esquivel, M.S.; Matte, D.L. The effects of inspiratory muscle training with linear workload devices on the sports performance and cardiopulmonary function of athletes: A systematic review and meta-analysis. *Phys. Ther. Sport* **2018**, *34*, 92–104. [CrossRef] [PubMed]
23. Guy, J.H.; Edwards, A.M.; Deakin, G.B. Inspiratory muscle training improves exercise tolerance in recreational soccer players without concomitant gain in soccer-specific fitness. *J. Strength Cond. Res.* **2014**, *28*, 483–491. [CrossRef] [PubMed]
24. Najafi, A.; Ebrahim, K.; Ahmadizad, S.; Jahani Ghaeh Ghashlagh, G.R.; Javidi, M.; Hackett, D. Improvements in soccer-specific fitness and exercise tolerance following 8 weeks of inspiratory muscle training in adolescent males. *J. Sports Med. Phys. Fitness* **2019**, *59*, 1975–1984. [CrossRef]
25. Da Silva, H.P.; De Moura, T.S.; Dos Santos Silveira, F. Efeitos do treinamento muscular inspiratório em atletas de Futebol. *Rev. Bras. Prescrição E. Fisiol. Exerc.* **2018**, *12*, 616–623.
26. Nicks, C.R.; Morgan, D.W.; Fuller, D.K.; Caputo, J.L. The influence of respiratory muscle training upon intermittent exercise performance. *Int. J. Sports Med.* **2009**, *30*, 16–21. [CrossRef]
27. Lozano-Quijada, C.; Poveda-Pagán, E.J.; Segura-Heras, J.V.; Hernández-Sánchez, S.; Prieto-Castelló, M.J. Changes in Postural Sway after a Single Global Postural Reeducation Session in University Students: A Randomized Controlled Trial. *J. Manipulative Physiol. Ther.* **2017**, *40*, 467–476. [CrossRef]
28. Morales, P.; Sanchis, J.; Cordero, P.J.; Díez, J.L. Maximum static respiratory pressures in adults. The reference values for a Mediterranean Caucasian population. *Arch. Bronconeumol.* **1997**, *33*, 213–219. [CrossRef]
29. Giner, J.; Casan, P.; Berrojalbiz, M.A.; Burgos, F.; Macian, V.; Sanchis, J. Cumplimiento de las "recomendaciones SEPAR" sobre la espirometría. *Arch. Bronconeumol.* **1996**, *32*, 516–522. [CrossRef]
30. Arnold, D.M.; Burns, K.E.; Adhikari, N.K.; Kho, M.E.; Meade, M.O.; Cook, D.J.; McMaster Critical Care Interest Group. The design and interpretation of pilot trials in clinical research in critical care. *Crit. Care Med.* **2009**, *37*, 69–74. [CrossRef]
31. Thabane, L.; Ma, J.; Chu, R.; Cheng, J.; Ismaila, A.; Rios, L.P.; Robson, R.; Thabane, M.; Giangregorio, L.; Goldsmith, C.H. A tutorial on pilot studies: The what, why and how. *BMC Med. Res. Methodol.* **2010**, *10*, 1. [CrossRef]
32. López-Martín, O.; Segura Fragoso, A.; Rodríguez Hernández, M.; Dimbwadyo Terrer, I.; Polonio-López, B. Efectividad de un programa de juego basado en realidad virtual para la mejora cognitiva en la esquizofrenia. *Gac. Sanit.* **2016**, *30*, 133–136. [CrossRef] [PubMed]
33. Van Emmerik, R.E.A.; Van Wegen, E.E.H. On the functional aspects of variability in postural control. *Exerc. Sport Sci. Rev.* **2002**, *30*, 177–183. [CrossRef] [PubMed]
34. Haddad, J.M.; Ryu, J.H.; Seaman, J.M.; Ponto, K.C. Time-to-contact measures capture modulations in posture based on the precision demands of a manual task. *Gait Posture* **2010**, *32*, 592–596. [CrossRef] [PubMed]
35. Archiza, B.; Andaku, D.K.; Caruso, F.C.R.; Bonjorno, J.C.; De Oliveira, C.R.; Ricci, P.A.; Do Amaral, A.C.; Mattiello, S.M.; Libardi, C.A.; Phillips, S.A.; et al. Effects of inspiratory muscle training in professional women football players: A randomized sham-controlled trial. *J. Sports Sci.* **2018**, *36*, 771–780. [CrossRef]
36. Ozmen, T.; Gunes, G.Y.; Ucar, I.; Dogan, H.; Gafuroglu, T.U. Effect of respiratory muscle training on pulmonary function and aerobic endurance in soccer players. *J. Sports Med. Phys. Fitness* **2017**, *57*, 507–513. [CrossRef]
37. Mackała, K.; Kurzaj, M.; Okrzymowska, P.; Stodółka, J.; Coh, M.; Rożek-Piechura, K. The Effect of Respiratory Muscle Training on the Pulmonary Function, Lung Ventilation, and Endurance Performance of Young Soccer Players. *Int. J. Environ. Res. Public Health* **2019**, *17*, 234. [CrossRef]
38. In, J. Introduction of a pilot study. *Korean J. Anesthesiol.* **2017**, *70*, 601–605. [CrossRef]
39. Harris, L.K.; Skou, S.T.; Juhl, C.B.; Jäger, M.; Bricca, A. Recruitment and retention rates in randomised controlled trials of exercise therapy in people with multimorbidity: A systematic review and meta-analysis. *Trials* **2021**, *22*, 396. [CrossRef]
40. Walters, S.J.; Bonacho Dos Anjos Henriques-Cadby, I.; Bortolami, O.; Flight, L.; Hind, D.; Jacques, R.M.; Knox, C.; Nadin, B.; Rothwell, J.; Surtees, M.; et al. Recruitment and retention of participants in randomised controlled trials: A review of trials funded and published by the United Kingdom Health Technology Assessment Programme. *BMJ Open* **2017**, *7*, e015276. [CrossRef]
41. Mezquita Raya, P.; Reyes García, R.; de Torres Sánchez, A. Patients' retention strategies in clinical trials. *Endocrinol. Nutr.* **2015**, *62*, 475–477. [CrossRef] [PubMed]

Article

Twelve-Week Lower Trapezius-Centred Muscular Training Regimen in University Archers

Chien-Nan Liao [1,†], Chun-Hao Fan [2,†], Wei-Hsiu Hsu [2,3,4,*], Chia-Fang Chang [2], Pei-An Yu [2,3], Liang-Tseng Kuo [2,3], Bo-Ling Lu [2] and Robert Wen-Wei Hsu [2,3]

[1] Department of Athletic Sports, National Chung Cheng University, Chia Yi 621, Taiwan; liao.cody@gmail.com
[2] Sports Medicine Center, Chang Gung Memorial Hospital, Chia Yi 621, Taiwan; fun711009@cgmh.org.tw (C.-H.F.); kojia882@yahoo.com.tw (C.-F.C.); b9002065@cgmh.org.tw (P.-A.Y.); light71829@gmail.com (L.-T.K.); bolin271@yahoo.com.tw (B.-L.L.); wwh@cgmh.org.tw (R.W.-W.H.)
[3] Department of Orthopedic Surgery, Chang Gung Memorial Hospital, Chia Yi 621, Taiwan
[4] School of Medicine, Chang Gung University, Tao-Yuan 333, Taiwan
* Correspondence: 7572@cgmh.org.tw; Tel.: +886-5-36231000 (ext. 2855)
† These authors contributed equally to this work.

Abstract: Archery is a fine-motor-skill sport, in which success results from multiple factors including a fine neuromuscular tuning. The present study hypothesised that lower trapezius specific training can improve archers' performance with concomitant changes in muscle activity and shoulder kinematics. We conducted a prospective study in a university archery team. Athletes were classified into exercise and control groups. A supervised lower trapezius muscle training program was performed for 12 weeks in the exercise group. The exercise program focused on a lower trapezius-centred muscular training. Performance in a simulated game was recorded as the primary outcome, and shoulder muscle strength, kinematics, and surface electromyography were measured and analysed. In the exercise group, the average score of the simulation game increased from 628 to 639 after the training regimens (maximum score was 720), while there were no such increases in the control group. The lower trapezius muscle strength increased from 8 to 9 kgf after training regimens and shoulder horizontal abductor also increased from 81 to 93 body weight% for the exercise group. The upper/lower trapezius ratio decreased from 2.2 to 1.1 after training. The lower trapezius exercise training regimen could effectively improve the performance of an archer with a simultaneous increase in shoulder horizontal abductor and lower trapezius muscle strength.

Keywords: electromyography; lower trapezius; muscle strength; shoulder kinematics

Citation: Liao, C.-N.; Fan, C.-H.; Hsu, W.-H.; Chang, C.-F.; Yu, P.-A.; Kuo, L.-T.; Lu, B.-L.; Hsu, R.W.-W. Twelve-Week Lower Trapezius-Centred Muscular Training Regimen in University Archers. *Healthcare* **2022**, *10*, 171. https://doi.org/10.3390/healthcare10010171

Academic Editors: João Paulo Brito and Rafael Oliveira

Received: 8 December 2021
Accepted: 13 January 2022
Published: 17 January 2022
Corrected: 17 February 2022

Publisher's Note: MDPI stays neutral with regard to jurisdictional claims in published maps and institutional affiliations.

1. Introduction

Archery is a fine-motor-skill sport, in which success is defined by the ability to shoot a target repeatedly with high precision and accuracy [1–3]. Recurve archery is an Olympic sport that requires concentration, extreme precision, upper body strength, and endurance [4]. An archery shot has three phases: the stance phase; the arming phase, during which the archer pulls the bowstring and pushes the bow; and the sighting phase, which involves the final stretching of the bow while focusing on the target [2]. Because of the high precision required, minimising movements during aiming allows for greater repeatability and consistency of this closed-loop skill [5]. Therefore, specific neuromuscular training is necessary for precise aiming during the sighting phase [6].

In the sighting phase, archers are required to maintain certain angles of shoulder abduction, horizontal extension, and elbow joint flexion for shorter duration [7]. Therefore, a series of movements are necessary during a shot, which require strong activity of the muscles attached to the shoulder girdle and upper extremity [8]. Archer performance is affected by the muscle activity of the shoulder girdle and upper extremity, especially scapular stability and lower trapezius (LT) muscle activity, on the drawing side [8–12]. Similarly, kinematic studies have revealed that decreased scapular elevation and abduction and increased scapular extension are closely related to better performance [8,10,12]. Therefore, LT strengthening can enhance an archer's performance [8]. However, few studies have analysed the effect of LT-centred muscular training on overhead athletes' performance [13,14]. Accordingly, the present study investigated the effect of LT-centred muscular training exercises on the muscle strength, kinematics, and performance of university archers. The study hypothesised that LT-centred muscular training can improve university archers' performance with concomitant changes in muscle activity and shoulder kinematics.

2. Materials and Methods

2.1. Participants

This prospective study recruited participants from a university archery team, who attended a supervised LT-centred muscular training programme for 12 weeks from June 2020 to August 2020. All participants involved in the present study were recruited from a university archer team, which recruited students from high school because of their good performance. Performance in a simulated game was recorded as the primary outcome, and shoulder muscle strength, kinematics, and surface electromyography (EMG) were measured and analysed.

In previous archery surveys, most athletes were divided into different groups with the number of people in each group ranging from 6 to 10 [8,10,15–17]. Therefore, this study was expected to recruit 20 archery athletes, and finally, a total of fourteen archers (all of whom used recurve bows) participated in this study. All participants provided written informed consent before the study. They were all right handed; they used their right hand for the drawing movement during archery. The individuals were excluded if they (1) had a history of surgery to either shoulder, (2) had a shoulder muscle disease, or (3) had shoulder impingement syndrome. This study was approved by the Chang Gung Medical Foundation Institutional Review Board (IRB: 201800990B0). All participants provided written informed consent.

2.2. Study Design

According to the wishes of athletes, all participants were divided into exercise (3 male: 4 female) and control groups (2 male: 5 female). All participants underwent the routine exercise protocol (Table S1). The exercise group received additional exercise. The following anthropometric parameters were measured: height, body weight, shoulder muscle strength, and scores in the simulated championship. Shoulder kinematics and electromyography were simultaneously recorded during archery shots. All measurements were performed before and after a 12-week intervention.

2.3. Simulation Game

Moreover, the composition of participants or environmental factors, such as the location of and weather at the championship, may affect their competitive ranking or athletic performance [18,19]. Therefore, to better isolate the effect of exercise training on competition scores, we calculated the participants' scores on a simulation game before and after the 12-week training programme. The simulation game requires the player to perform 72 arrows shot at a distance of 70 m, and the highest possible score was 720. All simulation games were held at standard archery fields under the same weather conditions.

2.4. Upper Limb Joint Angle Motion

Arrow shooting data were collected in an indoor biomechanics laboratory. After a regular warm-up, each participant shot six arrows towards a target placed at a 5 m distance away from them. A mark of 1 cm in diameter was stuck on the target, and the archers aimed at this mark each time. This study used one optical camera (Bonita 480 m; Oxford Metrics, Oxford, UK) at 100 Hz to capture the moment of the thrown arrows and record the shoulder joint angle (shoulder abduction and horizontal extension) at that moment (Figure S1). The moment when the arrow was thrown was defined as the moment when the fingers on the string (the drawing side) were straightened. Shooting motion data were captured with an eight-camera system (T20; Oxford Metrics) at 100 Hz. The evaluation was performed using marker sets for modified Plug-In Gait full-body modelling, with the markers attached to the skin with adhesive surgical tape [20]. The data were processed using the Nexus motion analysis system v2.5 (Vicon; Oxford Metrics). The shooting of six arrows was recorded consecutively and the average of the six arrows was taken for analysis.

2.5. LT Strength

Participants were provided instructions regarding the test procedure. They lay in the prone position, with the upper extremity diagonally overhead, in line with the fibres of the LT muscle [21]. The participants were then instructed to maintain the arm in the test position while the examiner provided resistance. The handheld dynamometer (MicroFet 2; Hogan Health Industries, West Jordan, UT, USA) force sensor was applied to the distal one-third of the participant's radial forearm, and a downward force was applied by the examiner until the participant's maximal muscular effort was overcome. The maximum force on the handheld dynamometer was recorded. Three trials were recorded consecutively on each upper extremity, with a 30 s intertrial rest, and the average for each side was used for analysis.

2.6. Shoulder Horizontal Abductor/Adductor Muscle Strength

The shoulder horizontal abductors/adductors muscle strength of the upper limbs was tested with the HUMAC NORM system (CSMi, New York, NY, USA) with the mode of concentric at the $60°/s$ for horizontal abductor/adductor. All tests were repeated five times, and the maximum strength recorded each test. The participants lay prone and received verbal encouragement during peak torque exertion. Muscle strength was normalised to body weight [22].

2.7. Surface Electromyography

Prior to electrode application, the skin was cleaned by being scrubbed with alcohol to reduce skin impedance. Ag/AgCl electrodes (Medi-Trace 200, Covidien/Kendall, USA) with a centre-to-centre distance of 2 cm were placed longitudinally on the muscle belly along the bow and drawing sides upper trapezius (UT), LT, as well as drawing side deltoid middle (DM), deltoid posterior (DP), biceps brachii (BB), and triceps brachii (TB) [8,12]. The reference electrode was placed on the olecranon process. EMG data were amplified (BioNomadix, BIOPAC systems, Goleta, CA, USA) and subjected to analogue/digital (A/D) conversion at 1000 Hz. The data were processed using a Nexus motion analysis system v2.5 (Vicon; Oxford Metrics). We analysed the ratio of LT to UT activity (the UT/LT ratio) as an indicator of elite athletes, which represented a biomechanical factor involving shoulder function [8,12].

EMG activities of all muscles were quantified. EMG data from the six shots were full-wave rectified. The data were then analysed using an integral calculus level of 1 s before the arrows were thrown. Prior to the shots, the maximum voluntary contractions (MVC) of all six muscles were determined. Shot six arrows were recorded consecutively and take the average of the six arrows for analysis.

Manual muscle testing was used to calculate the isolated maximum muscle strength [8,12,23]. Participants performed 3 s maximum voluntary isometric contractions against manual resistance. UT strength was tested in the standing position, and resistance was applied to the scapular elevation. LT strength was tested in the prone position, with the shoulder at $140°$ flexion and $135°$ abduction, and resistance was applied to the point between the acromion and scapular spine. DM strength was tested in the standing position with $90°$ shoulder abduction, and resistance was applied to humeral abduction. DP strength was tested in the prone position with $90°$ shoulder abduction, and resistance was applied to the horizontal abduction of the humerus. BB strength was tested in the standing position with 0° shoulder flexion and 90° elbow flexion, and resistance was applied to elbow joint flexion. TB strength was tested in the prone position with the shoulder at 0° flexion and 90° abduction and 90° elbow flexion, and resistance was applied to elbow joint flexion. The MVC were performed and measured three times, and the mean values were considered for analysis. After full-wave rectification, EMG amplitudes were normalised to MVC.

2.8. Additional Exercise Protocol

Elastic resistance bands can be a viable option to conventional resistance-training equipment during single-joint resistance exercises [24,25]. Elastic band resistance training produces adaptations similar to those with weight resistance training in the early phases of strength training [26]. However, none of the participants had received the LT exercise program. To avoid the possible compensatory effects of using resistance equipment [27], which can reduce the efficiency of LT exercise program, elastic bands (Thera-Band Black, USA) were used for exercise program at the beginning (first 4 weeks), and cable stack machines (Matrix G3-MS50; Johnson Health Tech, Taichung, Taiwan) were used subsequently (Table S2). The LT, latissimus dorsi, rhomboid, middle trapezius, and serratus anterior would be trained in the training program.

In the first 4 weeks, elastic bands were used during the following exercises, which were performed 3 days a week: reverse fly, straight-arm pull-down, reverse straight-arm pull-down, and straight-arm seated row. In addition, the participants performed floor movements, including locust pose and superman pose. Each set included 15 s isometric contractions; each trial had five sets, with a 15 s break between sets.

In weeks 5–12, cable stack machines were used during the following exercises, which were performed 3 days a week: reverse cable fly, straight-arm pull-down, reverse straight-arm pull-down, straight-arm seated row, and Y shape lat pull-down. In addition, alternative isometric contractions in pull-up and pull-down positions were performed for 15 s per set, with five sets per trial and a 15 s rest between sets.

Ten minutes of warm-up and cool-down were performed before and after the exercise program. At least one trained physical coach supervised the motion and physical conditions of the participants during the exercise program.

2.9. Statistical Analysis

All data analyses were performed using SPSS for Windows (version 20.0; SPSS, Chicago, IL, USA). All continuous data were presented in terms of mean $\pm$ standard deviation. The 2×2 (group $\times$ time) repeated measures ANOVA was used to analyse intergroup and the between pre-and post-exercise training differences. Use Fisher's exact test to compare whether there was a difference in the proportion of athletes who have increased or decreased the score of the simulation game between the groups. No abnormal distributions were noticed in pre-tests according to Shapiro–Wilk results for both groups. The relative effect size (ES) for the performance data was calculated using Cohen's d. It is defined as the difference between two means divided by a standard deviation for the data. Additionally, the categorical variables' ES would use Cohen's w. Intraclass correlation coefficient (ICC), typical error (TE), and percentual coefficient of variation (CV) were calculated to access intertest correlation (pre-and post-exercise training). Significance was set at the level of $p \leq 0.05$. The power was determined using G*power software version 3.1.9.7. (Heinrich Heine University, Dusseldorf, Germany). The input parameters used for the Γ test were alpha = 0.05 and the primary outcome (simulated game) of effect size, 0.56. The power was calculated as 0.97.

3. Results

The 14 athletes (five male and nine female university archers) were divided into exercise and control groups. The mean age of the enrolled athletes was 19 years. No significant intergroup differences were found in demographic characteristics, including age, height, weight, body mass index, and years in archery ($p > 0.05$; Table S3). Furthermore, it was shown that the age, years of training, and score was similar between groups. It means that all athletes' performance and skill levels were similar [8].

In the exercise group, the average score of the simulation game increased from 628 ± 19 to 639 ± 20 after the training regimens ($p = 0.045$, Cohen's $d = 0.56$, ICC = 0.84, CV = 3%, TE = 4; Figure 1). The effect size was 0.56 which belongs to moderate Cohen's d [28,29], the ICC was 0.84 which belongs to good and the CV was 3% [30]. Additionally, a CV of 10% or less was set as the level at which a measure was considered reliable [31]. No such increases were noted in the control group (617 ± 29 to 601 ± 44, $p = 0.125$, Cohen's $d = 0.30$). In the final simulation game performance, the score of the exercise group was not significantly higher than that of the control group ($p = 0.062$, Cohen's $d = 1.11$). However, all athletes in the exercise group have improved their scores in the simulated game after the exercise program. The scores of two athletes in the control group increased. The number of simulated game score increases in the exercise group was significantly higher than that in the control group ($p = 0.026$, Cohen's $w = 0.043$).

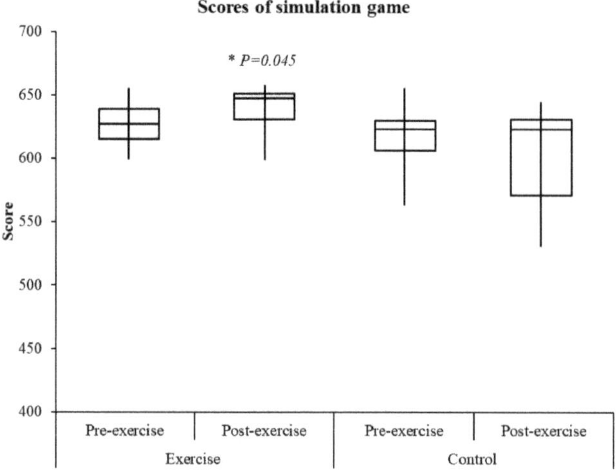

Figure 1. Scores of simulation game. * $p \leq 0.05$ between pre- and post-exercise.

Regarding muscle strength, LT strength increased from 8 ± 3 to 9 ± 3 kg-force (kgf) after the participants in the exercise group underwent the training regimen ($p = 0.045$, Cohen's $d = 0.33$, ICC = 0.93, CV = 35%, TE = 2; Figure 2A). Horizontal abductions strength also increased from 81 ± 28 to 93 ± 31 body weight% ($p = 0.048$, Cohen's $d = 0.45$, ICC = 0.87, CV = 30%, TE = 5; Figure 2B). No such increases were noted in the control group (Table S4). Regarding shoulder kinematics, horizontal extensions and shoulder abduction did not differ between both groups ($p > 0.05$; Table 1). On the drawing side, the activities of UT, BB, and TB decreased after training in the exercise group ($p \leq 0.05$, Cohen's $d \geq 0.58$, ICC ≥ 0.47, CV $\leq 55%$, TE ≤ 6; Table 2). No differences in LT were observed. The UT/LT ratio decreased from 2.2 ± 0.8 to 1.1 ± 0.9 after training ($p = 0.014$, Cohen's $d = 1.29$, ICC = 0.42, CV = 60%, TE = 1; Figure 3). In the control group, only differences in TB were detected on the drawing side.

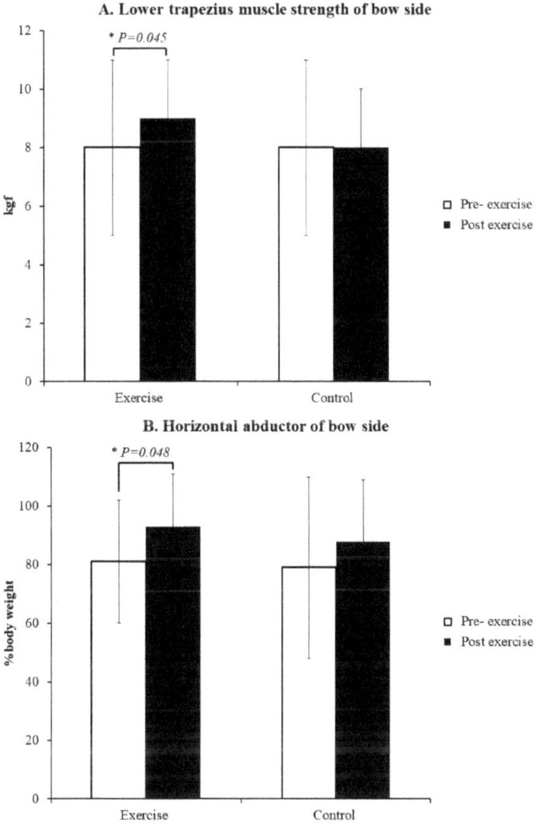

Figure 2. (**A**). Lower trapezius muscle strength of bow side; (**B**). Horizontal abductor of bow side. * $p \leq 0.05$ between pre- and post-exercise.

Table 1. Shoulder angle (degree).

			Pre-Exercise	Post-Exercise	ES	ICC	CV(%)	TE
			Mean ± SD	Mean ± SD				
Exercise	Bow side	Shoulder abduction	94 ± 6	91 ± 9	0.39	0.69	8	3
		Horizontal extension	141 ± 20	140 ± 28	0.04	0.91	17	5
	Drawing side	Shoulder abduction	120 ± 4	120 ± 4	0.01	0.76	3	2
		Horizontal extension	147 ± 5	146 ± 5	0.20	0.45	3	2
Control	Bow side	Shoulder abduction	95 ± 5	95 ± 5	0.01	0.61	5	2
		Horizontal extension	130 ± 33	137 ± 21	0.25	0.68	20	5
	Drawing side	Shoulder abduction	123 ± 6	120 ± 4	0.59	0.76	4	2
		Horizontal extension	144 ± 9	143 ± 9	0.11	0.94	6	3

CV: coefficient of variation; ES: effect size; ICC: intraclass correlation coefficient; SD: standard deviation; TE: typical error.

Table 2. Before releasing EMG %MVC.

			Pre-Exercise Mean ± SD	Post-Exercise Mean ± SD	ES	ICC	CV(%)	TE
Exercise	Bow side	Lower trapezius	49 ± 24	42 ± 17	0.33	0.88	44	4
		Upper trapezius	95 ± 35	69 ± 36 *	0.73	0.80	44	6
	Drawing side	Lower trapezius	38 ± 16	40 ± 10	0.15	0.84	33	4
		Upper trapezius	78 ± 22	44 ± 36 *	1.14	0.47	55	6
		Deltoid posterior	85 ± 13	74 ± 15	0.78	0.71	18	4
		Deltoid middle	59 ± 12	59 ± 14	0.01	0.69	21	4
		Triceps brachii	26 ± 10	19 ± 14 *	0.58	0.88	55	4
		Biceps brachii	42 ± 15	29 ± 12 *§	0.96	0.57	41	4
Control	Bow side	Lower trapezius	63 ± 25	52 ± 17	0.51	0.74	37	5
		Upper trapezius	95 ± 17	95 ± 21	0.01	0.48	19	4
	Drawing side	Lower trapezius	54 ± 30	50 ± 24	0.15	0.88	50	5
		Upper trapezius	77 ± 21	65 ± 30	0.46	0.37	36	5
		Deltoid posterior	93 ± 10	88 ± 4	0.66	0.35	8	3
		Deltoid middle	69 ± 16	56 ± 18	0.76	0.32	28	4
		Triceps brachii	34 ± 13	23 ± 12 *	0.88	0.77	46	4
		Biceps brachii	54 ± 32	51 ± 24	0.11	0.97	51	5

CV: coefficient of variation; ES: effect size; ICC: intraclass correlation coefficient; MVC: maximum voluntary contractions; SD: standard deviation; TE: typical error. * $p \leq 0.05$ between pre- and post-exercise. § $p \leq 0.05$ between exercise and control group.

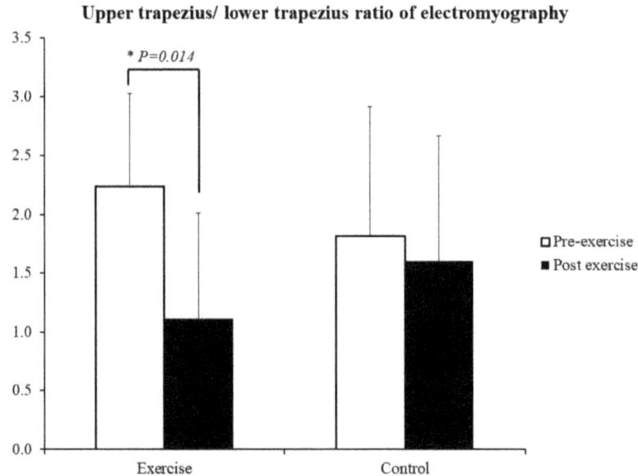

Figure 3. Upper trapezius/lower trapezius ratio as indicated through electromyography. * $p \leq 0.05$ between pre- and post-exercise.

4. Discussion

Notably, the current study revealed that an exercise protocol focusing on LT and other muscles, such as the rhomboid muscles, was associated with an increase in the simulation game score, in conjunction with an increase in LT and horizontal abductor strength. These increases were associated with decreases in the activities of UT, BB, and TB muscles.

Archery requires a high level of stability, which is achieved through the fine coordination of the shoulder girdle muscles. In general, muscle strengthening exercises of the upper extremities usually involve large muscle groups, such as the UT, pectoralis major, anterior deltoid, DP, and latissimus dorsi, to produce gains in strength and hypertrophy. The stabilising muscles, such as LT and serratus anterior, are often overlooked [13]. In addition, hyperactivity and early activation of the UT can lead to excessive scapular elevation when the arm is raised, which impairs performance [32–34]. Greater UT activity increases the higher risk of shoulder impingement [12,35–37]. Exercises to strengthen the middle trapezius, LT, and serratus anterior muscles can restore muscle balance and improve scapular stability [38,39]. In the present study, we developed a training exercise regimen focusing on the LT-centred muscular that improved the strength of LT and horizontal abductors and, simultaneously, dramatically decreased the muscle activities of UT, BB, and TB in the presence of persistent LT activities. This decrease in muscle activities may enable archers to enhance their stability through being able to meticulous balance themselves.

In the present study, shoulder kinematics did not differ after training. Indeed, scapular protraction and elevation usually result from increased DM and UT activity, along with a reduced activity of the middle trapezius and LT during shoulder abduction [37]. We did not observe decreases in abduction or increases in horizontal extensions. It was possible that the participant's posture was not modified in this short study period; further instruction to improve their posture should be given. The increased LT strength can aid the fine tuning of the shoulder muscles, such as the deltoid, UT, BB, and TB.

A study advocated focusing on the scapular stabilisers—in particular, the middle trapezius, LT, and serratus anterior—to maintain balance in the scapular force couple [39]. Excessive UT activation with a decreased activation of LT and serratus anterior leads to abnormal scapular movement [32,33,35,36]. The training programme in this study could reduce UT activity during archery and avoid excessive shrug-induced elevation of the shoulder girdle.

The general functions of the trapezius include the following: for the UT, scapular upward rotation and elevation; for the middle trapezius, retraction; and for the LT, upward rotation and depression [40]. In the literature on archery, LT activity is significantly higher among elite players than among pre-elite and beginner players, implying that the LT muscle of the draw arm is actively involved in scapular fixation during shooting [8].

During shoulder abduction, scapular protraction and elevation lead to increased activity of the DM and UT and reduced activity in the middle trapezius and LT [37]. In addition, the inferomedial directed fibres of the LT may also contribute to posterior tilt and the external rotation of the scapula during humeral elevation [41]. However, after training, archery players' UT activity decreases during archery. This increases the horizontal extension in the aiming stage, whereas shoulder abduction remains unchanged.

In fact, the differences in scores of simulation games could be multifactorial that biomechanics played a partial role. Although the lower trapezius-centred exercise was shown to decrease UT/LT ratio which was favoured biomechanically in the aiming phase, there were still other factors such as mental status/autonomous nervous functions, lower extremities function, and core stability that could contribute to the performance. We further analysed the score results with individual responses, i.e., the number of participants in the exercise and control group had higher scores for post testing, to eliminate the person-related bias. It was found that the number of simulated game score increases in the exercise group was significantly higher than that in the control group. Based on the present study, we suggested that lower trapezius-centred exercise is beneficial biomechanically in the aiming phase. Further interventions such as mindfulness, core muscle training was suggested to further enhance performance

This study only investigated university archers with small sample size and lack of randomization. In addition, this study uses statistical methods (2 × 2 repeated measures ANOVA) to confirm the baseline data in muscle strength, kinematics, EMG, and game scores were all compared and shown no differences between groups. Additionally, ES, ICC, CV, and TE were used to confirm the reliability of the pre-and post-exercise. In the present study, the additional lower trapezius-centred excise resulted in extra training time that could be partly responsible for the differences between groups. Although this lower trapezius-centred excise was low to medium intensity, further studies in the precise amount of training were suggested. Moreover, the study provided a training model specifically for archers to enhance their performance. Future studies are expected to expand the scope of the investigation and improve the technology involved in archery.

5. Conclusions

In conclusion, among archers, the biomechanics of the drawing side affects the performance and thus competition results. The main influencing factor was the strength of the shoulder horizontal abductor at the drawing side. In addition, the designed exercise training regimen can effectively improve competition ranking by increasing shoulder horizontal abductor and LT strength and reducing UT activity.

Supplementary Materials: The following are available online at https://www.mdpi.com/article/10.3390/healthcare10010171/s1, Table S1: Routine exercise protocol, Table S2: Additional to routine exercise protocol, Table S3: Participants' demographic characteristics, Table S4: Muscle strength; Figure S1: Shoulder angle of drawing side.

Author Contributions: Conceptualization, C.-N.L., C.-H.F. and W.-H.H.; methodology, C.-N.L., C.-H.F., C.-F.C. and W.-H.H.; software, C.-H.F., P.-A.Y. and L.-T.K.; validation, C.-N.L., C.-H.F. and W.-H.H.; formal analysis, C.-H.F., B.-L.L. and L.-T.K. and W.-H.H.; investigation, C.-N.L., C.-H.F. and C.-F.C.; resources, W.-H.H., P.-A.Y. and L.-T.K.; data curation, C.-N.L., C.-H.F. and W.-H.H.; writing—original draft preparation, C.-H.F. and W.-H.H.; writing—review and editing, C.-N.L.; supervision, R.W.-W.H. All authors have read and agreed to the published version of the manuscript.

Funding: Subsidies from the Taiwan Ministry of Education Sports Development Fund EMRPG6K0011, and Chang Gung Memorial Hospital Grant CMRPG6K0141-2 is appreciated.

Institutional Review Board Statement: This study was approved by the Chang Gung Medical Foundation Institutional Review Board (IRB: 201800990B0) that approved consent was written. All participants provided written informed consent.

Informed Consent Statement: Informed consent was obtained from the parents or guardians of players under the age of 20 before the examination. Written informed consent has been obtained from the players to publish this paper.

Data Availability Statement: Remaining data are available from the corresponding author upon reasonable request.

Conflicts of Interest: The authors declare no conflict of interest.

References

1. Taha, Z.; Musa, R.M.; Majeed, A.P.A.; Alim, M.M.; Abdullah, M.R. The identification of high potential archers based on fitness and motor ability variables: A support vector machine approach. *Hum. Mov. Sci.* **2018**, *57*, 184–193. [CrossRef]
2. Leroyer, P.; van Hoecke, J.; Helal, J.N. Biomechanical study of the final push-pull in archery. *J. Sports Sci.* **1993**, *11*, 63–69. [CrossRef] [PubMed]
3. Martin, P.E.; Siler, W.L.; Hoffman, D. Electromyographic analysis of bow string release in highly skilled archers. *J. Sports Sci.* **1990**, *8*, 215–221. [CrossRef]
4. Mann, D.L.; Littke, N. Shoulder injuries in archery. *Can. J. Sport Sci.* **1989**, *14*, 85–92. [PubMed]
5. Era, P.; Konttinen, N.; Mehto, P.; Saarela, P.; Lyytinen, H. Postural stability and skilled performance—A study on top-level and naive rifle shooters. *J. Biomech.* **1996**, *29*, 301–306. [CrossRef]
6. Hung, T.C.; Liao, Y.H.; Tsai, Y.S.; Ferguson-Stegall, L.; Kuo, C.H.; Chen, C.Y. Hot water bathing impairs training adaptation in elite teen archers. *Chin. J. Physiol.* **2018**, *61*, 118–123. [CrossRef]
7. Hideaki, T.; Yoko, K.; Masanobu, A. Characteristics of shooting time of the world's top level male archery athletes. *NSSU J. Sport Sci.* **2012**, *1*, 8–12.
8. Shinohara, H.; Urabe, Y. Analysis of muscular activity in archery: A comparison of skill level. *J. Sports Med. Phys. Fitness* **2018**, *58*, 1752–1758. [CrossRef]
9. Clarys, J.P.; Cabri, J.; Bollens, E.; Sleeckx, R.; Taeymans, J.; Vermeiren, M.; van Reeth, G.; Voss, G. Muscular activity of different shooting distances, different release techniques, and different performance levels, with and without stabilizers, in target archery. *J. Sports Sci.* **1990**, *8*, 235–257. [CrossRef]
10. Lin, J.J.; Hung, C.J.; Yang, C.C.; Chen, H.Y.; Chou, F.C.; Lu, T.W. Activation and tremor of the shoulder muscles to the demands of an archery task. *J. Sports Sci.* **2010**, *28*, 415–421. [CrossRef]
11. Spratford, W.; Campbell, R. Postural stability, clicker reaction time and bow draw force predict performance in elite recurve archery. *Eur. J. Sport Sci.* **2017**, *17*, 539–545. [CrossRef]
12. Shinohara, H.; Urabe, Y.; Maeda, N.; Xie, D.; Sasadai, J.; Fujii, E. Does shoulder impingement syndrome affect the shoulder kinematics and associated muscle activity in archers? *J. Sports Med. Phys. Fitness* **2014**, *54*, 772–779.

13. De Mey, K.; Danneels, L.; Cagnie, B.; Borms, D.; T'Jonck, Z.; van Damme, E.; Cools, A.M. Shoulder muscle activation levels during four closed kinetic chain exercises with and without redcord slings. *J. Strength Cond. Res.* **2014**, *28*, 1626–1635. [CrossRef]
14. Cools, A.M.; Johansson, F.R.; Cambier, D.C.; Velde, A.V.; Palmans, T.; Witvrouw, E.E. Descriptive profile of scapulothoracic position, strength and flexibility variables in adolescent elite tennis players. *Br. J. Sports Med.* **2010**, *44*, 678–684. [CrossRef]
15. Sarro, K.J.; Viana, T.C.; de Barros, R.M.L. Relationship between bow stability and postural control in recurve archery. *Eur. J. Sport Sci.* **2021**, *21*, 515–520. [CrossRef]
16. Ertan, H. Muscular activation patterns of the bow arm in recurve archery. *J. Sci. Med. Sport* **2009**, *12*, 357–360. [CrossRef]
17. Ertan, H.; Soylu, A.R.; Korkusuz, F. Quantification the relationship between fita scores and emg skill indexes in archery. *J. Electromyogr. Kinesiol.* **2005**, *15*, 222–227. [CrossRef]
18. Zhou, C.; Hopkins, W.G.; Mao, W.; Calvo, A.L.; Liu, H. Match performance of soccer teams in the chinese super league-effects of situational and environmental factors. *Int. J. Environ. Res. Public Health* **2019**, *16*, 4238. [CrossRef]
19. Williams, C. Environmental factors affecting elite young athletes. *Med. Sport Sci.* **2011**, *56*, 150–170. [CrossRef]
20. Kadaba, M.P.; Ramakrishnan, H.K.; Wootten, M.E. Measurement of lower extremity kinematics during level walking. *J. Orthop. Res.* **1990**, *8*, 383–392. [CrossRef]
21. Petersen, S.M.; Wyatt, S.N. Lower trapezius muscle strength in individuals with unilateral neck pain. *J. Orthop. Sports Phys. Ther.* **2011**, *41*, 260–265. [CrossRef]
22. Hu, G.; Jiang, Q.; Lee, J.Y.; Kim, Y.H.; Ko, D.H. Isokinetic strength and functional scores after rehabilitation in jiu-jitsu fighter with repair surgery of pectoralis major muscle rupture: A case report. *Healthcare* **2021**, *9*, 527. [CrossRef]
23. Celik, D.; Sirmen, B.; Demirhan, M. The relationship of muscle strength and pain in subacromial impingement syndrome. *Acta Orthop. Traumatol. Turc.* **2011**, *45*, 79–84. [CrossRef]
24. Iversen, V.M.; Mork, P.J.; Vasseljen, O.; Bergquist, R.; Fimland, M.S. Multiple-joint exercises using elastic resistance bands vs. Conventional resistance-training equipment: A cross-over study. *Eur. J. Sport Sci.* **2017**, *17*, 973–982. [CrossRef]
25. Bergquist, R.; Iversen, V.M.; Mork, P.J.; Fimland, M.S. Muscle activity in upper-body single-joint resistance exercises with elastic resistance bands vs. Free weights. *J. Hum. Kinet.* **2018**, *61*, 5–13. [CrossRef]
26. Colado, J.C.; Triplett, N.T. Effects of a short-term resistance program using elastic bands versus weight machines for sedentary middle-aged women. *J. Strength Cond. Res.* **2008**, *22*, 1441–1448. [CrossRef]
27. Schmidt, R.A.; Lee, T.D. *Motor Control and Learning: A Behavioral*, 5th ed.; Human Kinetics: Champaign, IL, USA, 2011.
28. Cohen, J. *Statistical Power Analysis for the Behavioral Sciences*, 2nd ed.; Routledge: New York, NY, USA, 1988; p. 567.
29. Sawilowsky, S. New effect size rules of thumb. *J. Mod. Appl. Stat. Methods* **2009**, *8*, 467–474. [CrossRef]
30. Koo, T.K.; Li, M.Y. A guideline of selecting and reporting intraclass correlation coefficients for reliability research. *J. Chiropr. Med.* **2016**, *15*, 155–163. [CrossRef] [PubMed]
31. Opar, D.A.; Piatkowski, T.; Williams, M.D.; Shield, A.J. A novel device using the nordic hamstring exercise to assess eccentric knee flexor strength: A reliability and retrospective injury study. *J. Orthop. Sports Phys. Ther.* **2013**, *43*, 636–640. [CrossRef]
32. Cools, A.M.; Witvrouw, E.E.; Declercq, G.A.; Vanderstraeten, G.G.; Cambier, D.C. Evaluation of isokinetic force production and associated muscle activity in the scapular rotators during a protraction-retraction movement in overhead athletes with impingement symptoms. *Br. J. Sports Med.* **2004**, *38*, 64–68. [CrossRef]
33. Cools, A.M.; Witvrouw, E.E.; Declercq, G.A.; Danneels, L.A.; Cambier, D.C. Scapular muscle recruitment patterns: Trapezius muscle latency with and without impingement symptoms. *Am. J. Sports Med.* **2003**, *31*, 542–549. [CrossRef]
34. Cools, A.M.; Borms, D.; Castelein, B.; Vanderstukken, F.; Johansson, F.R. Evidence-based rehabilitation of athletes with glenohumeral instability. *Knee Surg. Sports Traumatol. Arthrosc.* **2016**, *24*, 382–389. [CrossRef] [PubMed]
35. Ludewig, P.M.; Cook, T.M. Alterations in shoulder kinematics and associated muscle activity in people with symptoms of shoulder impingement. *Phys. Ther.* **2000**, *80*, 276–291. [CrossRef]
36. Cools, A.M.; Declercq, G.A.; Cambier, D.C.; Mahieu, N.N.; Witvrouw, E.E. Trapezius activity and intramuscular balance during isokinetic exercise in overhead athletes with impingement symptoms. *Scand. J. Med. Sci. Sports* **2007**, *17*, 25–33. [CrossRef] [PubMed]
37. Contemori, S.; Panichi, R.; Biscarini, A. Effects of scapular retraction/protraction position and scapular elevation on shoulder girdle muscle activity during glenohumeral abduction. *Hum. Mov. Sci.* **2019**, *64*, 55–66. [CrossRef]
38. Van de Velde, A.; de Mey, K.; Maenhout, A.; Calders, P.; Cools, A.M. Scapular-muscle performance: Two training programs in adolescent swimmers. *J. Athl. Train.* **2011**, *46*, 160–167, discussion 68–69. [CrossRef]
39. Cools, A.M.; Palmans, T.; Johansson, F.R. Age-related, sport-specific adaptions of the shoulder girdle in elite adolescent tennis players. *J. Athl. Train.* **2014**, *49*, 647–653. [CrossRef] [PubMed]
40. Escamilla, R.F.; Yamashiro, K.; Paulos, L.; Andrews, J.R. Shoulder muscle activity and function in common shoulder rehabilitation exercises. *Sports Med.* **2009**, *39*, 663–685. [CrossRef]
41. Ludewig, P.M.; Cook, T.M.; Nawoczenski, D.A. Three-dimensional scapular orientation and muscle activity at selected positions of humeral elevation. *J. Orthop. Sports Phys. Ther.* **1996**, *24*, 57–65. [CrossRef]

Correction

Correction: Liao et al. Twelve-Week Lower Trapezius-Centred Muscular Training Regimen in University Archers. *Healthcare* 2022, *10*, 171

Chien-Nan Liao [1], Chun-Hao Fan [2], Wei-Hsiu Hsu [2,3,4,*], Chia-Fang Chang [2], Pei-An Yu [2,3], Liang-Tseng Kuo [2,3], Bo-Ling Lu [2] and Robert Wen-Wei Hsu [2,3]

[1] Department of Athletic Sports, National Chung Cheng University, Chia Yi 621, Taiwan; liao.cody@gmail.com
[2] Sports Medicine Center, Chang Gung Memorial Hospital, Chia Yi 621, Taiwan; fun711009@cgmh.org.tw (C.-H.F.); kojia882@yahoo.com.tw (C.-F.C.); b9002065@cgmh.org.tw (P.-A.Y.); light71829@gmail.com (L.-T.K.); bolin271@yahoo.com.tw (B.-L.L.); wwh@cgmh.org.tw (R.W.-W.H.)
[3] Department of Orthopedic Surgery, Chang Gung Memorial Hospital, Chia Yi 621, Taiwan
[4] School of Medicine, Chang Gung University, Tao-Yuan 333, Taiwan
* Correspondence: 7572@cgmh.org.tw; Tel.: +886-5-36231000 (ext. 2855)

The authors would like to make the following corrections to the published paper [1]. Replacing the Institutional Review Board number 2019800990B0 with 201800990B0.

Reference

1. Liao, C.-N.; Fan, C.-H.; Hsu, W.-H.; Chang, C.-F.; Yu, P.-A.; Kuo, L.-T.; Lu, B.-L.; Hsu, R.W.-W. Twelve-Week Lower Trapezius-Centred Muscular Training Regimen in University Archers. *Healthcare* **2022**, *10*, 171. [CrossRef] [PubMed]

Citation: Liao, C.-N.; Fan, C.-H.; Hsu, W.-H.; Chang, C.-F.; Yu, P.-A.; Kuo, L.-T.; Lu, B.-L.; Hsu, R.W.-W. Correction: Liao et al. Twelve-Week Lower Trapezius-Centred Muscular Training Regimen in University Archers. *Healthcare* 2022, *10*, 171. *Healthcare* **2022**, *10*, 378. https://doi.org/10.3390/healthcare10020378

Received: 28 January 2022
Accepted: 30 January 2022
Published: 17 February 2022

Publisher's Note: MDPI stays neutral with regard to jurisdictional claims in published maps and institutional affiliations.

 *healthcare*

Article

Agility Skills, Speed, Balance and CMJ Performance in Soccer: A Comparison of Players with and without a Hearing Impairment

Hakan Yapici [1], Yusuf Soylu [2], Mehmet Gulu [1,*], Mehmet Kutlu [3], Sinan Ayan [1], Nuray Bayar Muluk [4], Monira I. Aldhahi [5] and Sameer Badri AL-Mhanna [6,*]

[1] Department of Coaching Education, Faculty of Sport Sciences, Kirikkale University, Kirikkale 71450, Turkey
[2] Department of Physical Education and Sports, Faculty of Sport Sciences, Tokat Gaziosmanpasa University, Tokat 60250, Turkey
[3] Department of Coaching Education, Faculty of Sport Sciences, Hitit University, Corum 19000, Turkey
[4] ENT Department, Faculty of Medicine, Kirikkale University, Kirikkale 71450, Turkey
[5] Department of Rehabilitation Sciences, College of Health and Rehabilitation Sciences, Princess Nourah bint Abdulrahman University, P.O. Box 84428, Riyadh 11671, Saudi Arabia
[6] Department of Physiology, School of Medical Sciences, Universiti Sains Malaysia, Kubang Kerian 16150, Kelantan, Malaysia
* Correspondence: mehmetgulu@kku.edu.tr (M.G.); sameerbadri9@gmail.com (S.B.A.-M.)

Abstract: This study investigates the differences in agility, speed, jump and balance performance and shooting skills between elite hearing-impaired national team soccer players (HISP) and without-hearing-impairment elite soccer players (woHISP). Players were divided into two groups, the HISP group (n = 13; 23.5 ± 3.1 years) and the woHISP group (n = 16; 20.6 ± 1.4 years), and were tested in three sessions, seven apart, for metrics including anthropometrics, speed (10 m, 20 m and 30 m), countermovement jump (CMJ), agility (Illinois, 505, zigzag), T test (agility and shooting skills), and balance. The results showed that 30 m, 20 m and 10 m sprint scores, agility / skills (sec), shooting skills (goals), zigzag, Illinois, and 505 agility skills, and countermovement jump scores were significantly lower among players with hearing impairments ($p < 0.05$). There were no significant T test differences between HISP and woHISP ($p > 0.05$). The HISP showed right posterolateral and posteromedial, and left posterolateral and posteromedial scores that were lower than the woHISP group ($p < 0.05$). Anterior scores were not significantly different between each leg ($p > 0.05$). In conclusion, the HISP group showed higher performance scores for speed (10 m, 20 m and 30 m), CMJ, agility (Illinois, 505, zigzag) and T test (sec and goals), but not balance. Hearing-impaired soccer players are determined by their skill, training, and strategy, not their hearing ability.

Keywords: soccer; hearing-impaired; athletic performance; skill; physical fitness

Citation: Yapici, H.; Soylu, Y.; Gulu, M.; Kutlu, M.; Ayan, S.; Muluk, N.B.; Aldhahi, M.I.; AL-Mhanna, S.B. Agility Skills, Speed, Balance and CMJ Performance in Soccer: A Comparison of Players with and without a Hearing Impairment. *Healthcare* **2023**, *11*, 247. https://doi.org/10.3390/healthcare11020247

Academic Editors: João Paulo Brito and Rafael Oliveira

Received: 6 November 2022
Revised: 8 January 2023
Accepted: 9 January 2023
Published: 13 January 2023

1. Introduction

World Health Organization data reveal that there are about 466 million people with hearing impairments worldwide [1]. As stated by the International Committee of Sports for the Deaf, deafness is the capacity to hear sound only at a frequency of 55 dB or higher in the better ear [2]. Hearing-impaired soccer players, just like all soccer players, are expected to have good speed, agility and balance. These skills enable soccer players to move quickly and efficiently on the field, change direction quickly, and maintain their balance when receiving or passing the ball or making a shot at a goal [3,4].

Soccer game characteristics involve intermittent high-intensity physical and technical tactical demands such as rapid changes of direction, jumps and dribbling [5,6]. Because of increasing game requirements, players must improve their physical skills based on different soccer match demands [7,8]. As a result, during the period of the season, players can usually maintain or increase their overall aerobic and anaerobic fitness [9]. Nevertheless, monitoring football's internal and external loads, mainly acute workload, differs between game periods [10]. Soccer-specific abilities refer to the specific physical, technical and tactical skills required to perform well. Training programs have been designed to improve motor abilities such as agility, sprinting and balance to enhance overall physical performance in soccer [11,12]. However, game performance has been affected by various obstacles, including physical, mental or psychological challenges [5,13].

Physical obstacles such as hearing loss, rather than a physical disability such as poor performance or injury, may affect athletes' performance compared to those without a hearing impairment [14]. Long-term hearing loss may also significantly alter sensory processing and lead to motor deficiencies [15]. From this point, various physical and physiological factors also affect how well hearing-impaired soccer players perform during training and games. However, hearing-impaired soccer players have been less able to perform due to poor body composition, and less aerobic endurance, strength, balance and agility compared to without-hearing-impairment players [16,17]. Therefore, developing anthropometrics and physical and physiological characteristics of hearing-impaired footballers is also necessary to increase performance to respond to the developments in soccer.

Even though previous studies have shown that hearing-impaired soccer players have differences in some physical performance parameters, there is no evidence that these characteristics are directly related to hearing ability, except for balance related to a damaged vestibular system [18]. Limited studies on hearing-impaired soccer players have focused on acute physical performance response [16] or compared hearing-impaired and non-hearing-impaired soccer players in laboratory fields [18], and female soccer players' biomechanical characteristics [19]. Therefore, more studies are needed comparing elite soccer groups without hearing impairments and similar hearing-impaired groups' results for performance outside the laboratory. When the performance responses of hearing-impaired soccer players were analyzed, we realized that more studies in this field are needed. We hypothesized that elite hearing-impaired national team soccer players (HISP) might show higher performance based on technical abilities and some physical parameters, except balance ability, than without-hearing-impairment soccer players (woHISP).

2. Materials and Methods

2.1. Participants

This study was conducted with a cross-sectional design. An initial power analysis (G*Power, University Duesseldorf, Duesseldorf, Germany) was performed to determine the sample size [20]. A total of 29 players were separated into two groups, both of which comprised elite soccer players: HISP (n = 13, age: 23.5 ± 3.1, height: 177.8 ± 5.1, weight: 72.8 ± 3.7, body mass index: 23.1 ± 1.9) and woHISP (n = 16, age: 20.6 ± 1.4, height: 178.4 ± 6.1, weight: 72.5 ± 8.7, body mass index: 23.2 ± 2.9). The inclusion criteria were as follows: (i) all physical evaluations conducted before the start of the season; (ii) no injuries at the time of the evaluations; and (iii) no injuries in the month prior to the evaluations (Figure 1). The HISP included the ICSD criteria to compete, a minimum hearing loss of 55 dB in the better ear or both ears, and being active athletes of the Turkish Deaf Sports Federation, competing at the national level at the time of the study. However, the HISP group had European, World and Olympic champions and medal-winners.

Before the testing, the volunteers were asked to provide information about their training experience, education level and hearing loss level. The sign language interpreter informed the HISP groups. This study was conducted at Kirikkale University, Sports Sciences Faculty, according to the principles outlined in the Declaration of Helsinki. Athletes were fully informed about the procedures of this study, and all volunteers signed a written consent form before study participation. This study was approved by the Kirikkale University Non-Invasive Researches Ethics Committee (Date: 10 December 2020, number: 7 December 2020).

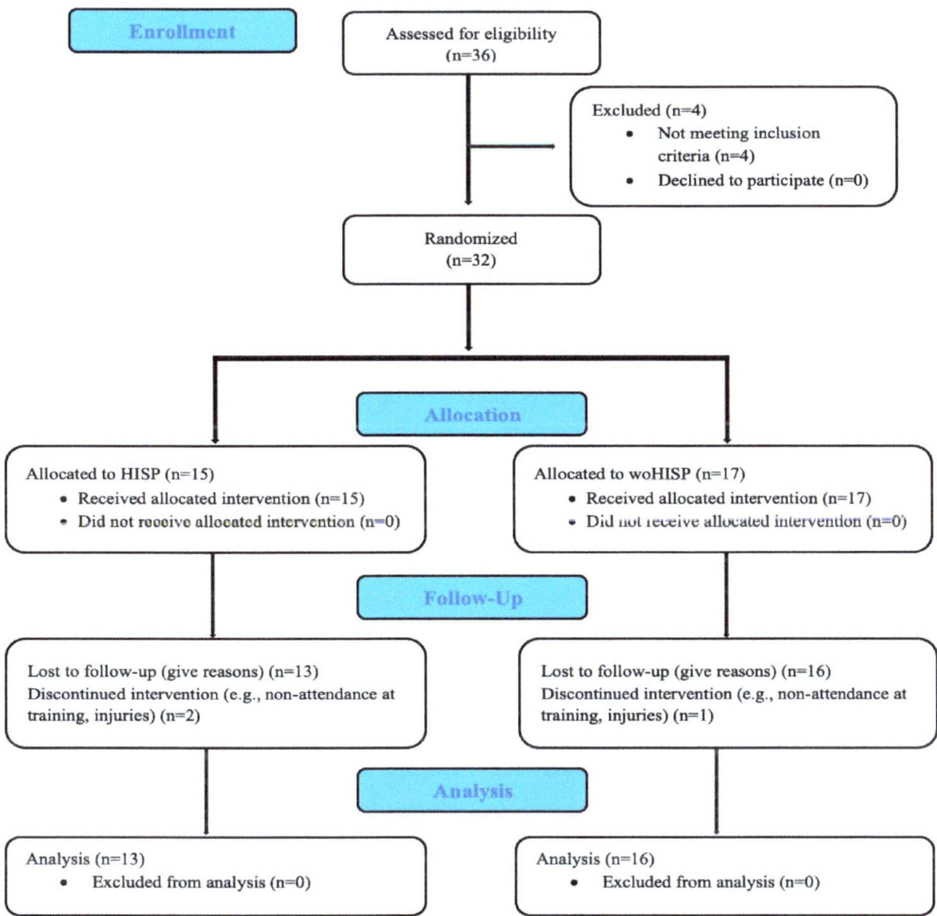

Figure 1. CONSORT flow diagram.

2.2. Measurements

2.2.1. Anthropometric Measurements

Body height (BH), body mass (BM) and composition measurements were performed using a bioelectrical impedance analysis (BIA) and body composition analyzer (Tanita Body Composition Analyzer BC 418 professional model, Japan). The subjects' heights were measured with an anthropometrics rod set on the day before performance tests, to the nearest 1 cm. We preferred using BIA testing in estimating body fat percentages (BFP), based on some research studies showing validity. Related literature stated that BIA is a suitable alternative for estimating body fat percentages when subjects are within a "normal" body fat range. However, there is a tendency for BIA to overestimate body fat in lean subjects and underestimate body fat in obese individuals.

2.2.2. Y Balance Test

The Y balance test was performed to evaluate postural control. Participants were allowed a maximum of six trials., with three successful attempts for each access direction. After three successful attempts in each direction, the rater recorded the maximum and average distance. In addition, the reach distance was recorded to the nearest 0.5 cm [21].

2.2.3. 10, 20 and 30 m Sprint Test

A Brower Timing System (USA) was placed 10, 20 and 30 m away from the starting point with a photocell device (0.01-s precision). When the athlete felt ready for the sprint test, they were asked to start with their preferred foot in front. The times between the start and finish gates at 10 m (first electronic timing gate) and 20 m (second electronic timing gate) were recorded. After a 10 min rest interval, the measurement was repeated three times, and the best performance was recorded for further analysis [22].

2.2.4. Illinois Test

The participant began the test by lying face down on the floor behind the starting line with arms at their sides and head to the side or forward. With the "start" command, the participant stood up and ran the test track, which consisted of three cones lined up in a straight line at intervals of 5 m in width, 10 m in length and 3.3 m in the middle section. Two trials were run, a rest period of 3 min was given between trials, and the best score was recorded. If the participant did not run the course by the instructions, did not reach the finish lines, did not complete the course or moved any cone, they were disqualified [23].

2.2.5. 505 Agility Test

After a 10 m approach run at the start of a 15 m track, the participants were asked to run 5 m forward and backwards as quickly as possible. During testing, the time between two passes was recorded. The best result from repeated measurements was recorded in seconds [24].

2.2.6. Zigzag Test

The test for agility, acceleration, deceleration and balance consisted of 4×5 m sections set at $100°$ angles. The test started with the exit of the participants from the photocell door (point A). They ran to point B, returned from point B, reached the middle point C again and, finally, after passing through the photocell at point D, the test was terminated. Each participant repeated the test three times, and the best score was recorded [22].

2.2.7. Countermovement Jump (cm)

Participants were asked to assume a full squat position with their hands on their waist and were told to jump as high as possible with maximum force, without making any springing movements from the knees. The participants were asked to have the same positions during jumping and landing on the platform again. During the test, athletes were told not to move forward, backward or sideways, and keep their hands on their waist. This test was conducted on each athlete three times, and the best score was used for analysis [24,25].

2.2.8. Agility and Shooting Skill Test

This test was applied to measure and determine the agility and mobility status of the athletes as well as their ability to score a goal. A Brower Timing System photocell device with the same start and end point was placed at the test site. After the necessary information was given to the subjects, the time began when they passed through the starting gate. The athletes kicked ball number 1 in the middle with their right foot, then ran to ball number 2, where they kicked the ball with their left foot. The subjects ran from here to ball number 3 on the right, kicked this with their right foot, ran to number 4, kicked the ball with their left foot, then ran to the starting point with their back turned, and time stopped when they passed through the finish gate. Participants were asked to score goals by hitting the four balls into the goal. The subjects hit four balls. The athletes' times were recorded in seconds. The raw time score was re-evaluated as the total skill and agility score. If the subjects managed four goals, three goals, two goals or one goal, 1.00s, 0.75 s, 0.50 s or 0.25 s, respectively, were subtracted from the completion time [26].

2.3. Procedures

The testing sessions were performed each day for seven days at the university's Exercise Physiology Laboratory and gymnasium. The first session of tests included measurements of body composition before breakfast. Sprint tests were determined after a 15 min standardized soccer warm-up. Agility tests were carried out during the second test session. Lastly, balance tests consisted of three different movement directions in the third test session: anterior, posteromedial and posterolateral. Participants were instructed to adhere to the following guidelines before all tests: dress in sports apparel, avoid vigorous exercise and alcohol consumption 24 h before testing and be ready to test correctly hydrated. All participants were instructed to perform each test with their maximum effort via verbal encouragement throughout each trial. All participants were tested in a specific order to standardize the testing process: weight, height, body composition, balance, sprint, and agility. Standardized procedures were followed for each assessment test and are published in the Procedures section below.

2.4. Statistical Analysis

Data were represented as mean ± standard deviation (SD). The data were evaluated to see whether they had normal distribution or not. The Shapiro–Wilk test was not verified for the normality assumption [27]. Comparison of physical and sports characteristics, speed, agility and balance results were analyzed using the Mann–Whitney U test. To analyze the relationship between independent variables, Spearman's correlation rho efficient test was used. The following scale was used to assess the magnitude of the Spearman's correlations and their differences: 0.00 to 0.10, trivial; 0.11 to 0.29, small; 0.30 to 0.49, moderate; and 0.5 to, large [28]. Cohen's (d) standardized effect size was used to determine the magnitude of differences, using the following thresholds [26]: 0.0 to 0.2, trivial; 0.2 to 0.6, small; 0.6 to 1.2, moderate; 1.2 to 2.0, large; >2.0, very large. A value of $p < 0.05$ was considered statistically significant. Variables were analyzed using IBM SPSS for Windows Version 21.0. (IBM Corp., Armonk, NY, USA). The American Psychological Association (APA) 6.0 style was used to report statistical differences [29].

3. Results

In terms of educational level, a total of three (23%) hearing-impaired subjects (HISP group) had bachelor's degrees and 10 (77%) had high and secondary school degrees. Measurement results in HISP and woHISP groups, and physical and training characteristics are shown in Table 1.

3.1. Physical and Sports Experience Characteristics

There were no significant differences between body height ($p = 0.774$; d = 0.11, trivial ES) and body mass ($p = 0.554$, d = trivial ES) of the HISP and woHISP groups. The sports experience of HISP was significantly greater than woHISP ($p = 0.000$; d = 2.48, very large ES). The BFP (%) of the HISP group (9.9 ± 1.2%) was significantly lower than the woHISP group ($p = 0.000$; d = 1.89, large ES) (Table 1).

Table 1. Measurement results in soccer HISP groups and woHISP groups.

| | | Group 1 (HISP) ($n = 13$) | | Group 2 (woHISP) ($n = 16$) | | CI 95% | | | | |
		Mean	SD	Mean	SD	Lower	Upper	Cohen's d	Descriptor	p *
Physical and sports experience characteristics										
Body height (cm)		177.84	5.17	178.43	6.19	−5.0	3.8	0.11	Trivial	0.774
Body mass (kg)		72.86	3.72	72.54	8.77	−5.0	5.7	0.05	Trivial	0.554
Sports experience (year)		12.30	3.30	6.25	1.00	−2.9	1.3	2.48	Very Large	0.000 *
BFP (%)		9.76	1.17	11.86	1.05	−1.1	−0.5	1.89	Large	0.000 *
Sprints and agility/skill variables										
30 m (Sec.)		4.91	0.21	5.70	0.52	−0.5	−0.3	2.06	Very Large	0.000 *
20 m (Sec.)		2.45	0.10	2.84	0.25	−0.3	−0.1	2.05	Very Large	0.000 *
10 m (Sec.)		1.75	0.10	1.91	0.20	−0.3	−0.1	1.01	Moderate	0.017 *
T Test		10.33	0.14	10.45	0.38	−1.1	−0.3	0.42	Small	0.895
Agility/Skill (Sec.)		11.45	0.30	12.18	0.60	−1.1	−0.5	1.54	Large	0.002 *
Agility and Shooting Skill Tests (goals)		10.59	0.22	11.38	0.51	−1.1	−0.5	2.01	Very Large	0.000 *
Zigzag (Sec.)		6.10	0.23	7.60	0.79	−1.9	−1.0	2.58	Very Large	0.000 *
Illinois (Sec.)		16.79	0.43	18.71	0.57	−2.3	−1.5	3.80	Very Large	0.000 *
505 Agility (Sec.)		2.40	0.17	2.86	0.37	−0.7	−0.3	1.60	Large	0.002 *
CMJ (cm)		51.10	1.77	48.10	3.07	1.0	4.9	1.20	Large	0.002 *
Y-balance test										
Right foot	Anterior	73.69	12.13	76.68	6.07	−10.1	4.1	0.31	Small	0.272
	Posterolateral	66.00	13.08	110.00	6.16	−51.6	−36.4	4.30	Very Large	0.000 *
	Posteromedial	72.07	14.33	99.87	9.38	−36.9	−18.7	2.30	Very Large	0.000 *
Left foot	Anterior	71.38	8.65	74.06	9.49	−9.7	4.3	0.30	Small	0.392
	Posterolateral	70.61	7.83	112.25	4.95	−46.6	−36.7	6.36	Very Large	0.000 *
	Posteromedial	66.69	13.09	101.62	5.42	−42.3	−27.6	3.49	Very Large	0.000 *

* = $p < 0.005$.

3.2. Sprints and Agility/Skill Variables

In the HISP group, sprint scores were 4.91 ± 0.21 s, 2.45 ± 0.10 s and 1.75 ± 0.10 s, respectively. In the woHISP group, sprint scores were 5.70 ± 0.52 s, 2.84 ± 0.25 s and 1.91 ± 0.20 s, respectively. The 30 m ($p = 0.000$; d = 2.06, very large ES), 20 m ($p = 0.000$; d = 2.05, very large ES) and 10 m ($p = 0.017$; d = 1.01, moderate ES) sprint scores of the HISP group were significantly lower than those of the woHISP group (Table 1).

The *T* test scores were not different between the HISP and the woHISP groups ($p = 0.895$; d = 0.42, small ES). However, the agility/skill (sec) of the HISP group was significantly lower than the woHISP group ($p = 0.002$; d = 1.54, large ES). Moreover, the shooting skill (goals) scores of the HISP group were significantly lower than the woHISP group ($p = 0.000$; d = 2.01, very large ES). The zigzag ($p = 0.000$; d = 2.58, very large ES), Illinois ($p = 0.000$; d = 3.80, very large ES) and 505 agility skill ($p = 0.002$; d = 1.60, large ES) scores of the HISP group were significantly lower than those of the woHISP group. The jumping scores of the HISP group were significantly higher ($p = 0.002$; d = 1.20, large ES) than the woHISP group (Table 1).

3.3. Y Balance Test Results

There were no significant differences between anterior Y balance scores of the HISP and woHISP groups for the right ($p = 0.272$; d = 0.531, small ES) and left ($p = 0.392$; d = 0.30, small ES). However, the right posterolateral ($p = 0.000$; d = 4.30, very large ES) and right posteromedial ($p = 0.000$; d = 2.30, very large ES), and left posterolateral ($p = 0.000$; d = 636, very large ES) and left posteromedial ($p = 0.000$; d = 3.49, very large ES) Y balance scores of the HISP group were significantly lower than those of the woHISP group (Table 1).

3.4. Correlation Test Results in the HISP Group Are Shown in Table 2

There were positive correlations between the 30 m and 20 m sprint scores ($p < 0.01$; r = 0.995, large), zigzag and 30 m ($p < 0.01$; r = 0.730, large), 20 m ($p < 0.01$; r = 0.690, large), agility skill and 10 m ($p < 0.05$; r = 0.584, large) and Illinois and 10 m ($p < 0.05$; r = 0.592, large). As 505 agility skills scores increased, jumping scores decreased ($p < 0.05$; r = −0.633, large). As shown in the results in Table 2, there were positive correlations between body fat and right posteromedial Y balance scores ($p < 0.05$; r = 0.567, large), left posterolateral ($p < 0.05$; r = 0.567, large), Illinois and left posterolateral Y balance ($p < 0.05$; r = 0.594, large) and right posterolateral and zigzag ($p < 0.05$; r = 0.631, large).

Table 2. Correlation test results in soccer HISP groups.

Variables (r)	% Fat	30 m	20 m	10 m	Zigzag	T Test	Illinois	505	CMJ	Y-Balance Test–Right			Y-Balance Test–Left		
										Right Ant **	Right PL **	Right PM **	Left Ant **	Left PL **	Left PM **
% Fat	1														
30 m.	−0.109	1													
20 m.	−0.095	0.995 **	1												
10 m.	0.350	0.264	0.248	1											
Zigzag	0.149	0.730 **	0.690 **	0.229	1										
T Test	0.062	0.466	0.482	0.584 *	0.216	1									
Illinois	0.203	0.387	0.370	0.592 *	0.374	0.054	1								
505	−0.337	0.332	0.311	−0.316	0.515	0.203	−0.262	1							
CMJ	0.243	−0.172	−0.164	0.171	−0.125	−0.070	0.080	−0.633 *	1						
Right Ant **	−0.085	0.069	0.078	−0.290	0.002	0.098	−0.394	0.070	0.226	1					
Right PL **	0.236	0.396	0.385	0.127	0.631 *	0.066	0.211	0.438	−0.294	0.297	1				
Right PM **	0.567 *	0.262	0.285	0.391	0.141	0.463	0.191	−0.154	0.184	0.479	0.514	1			
Left Ant **	0.089	0.127	0.107	−0.068	0.106	0.113	−0.037	−0.090	0.320	0.82 **	0.169	0.533	1		
Left PL **	0.584 *	−0.003	−0.042	0.330	0.083	−0.266	0.594 *	−0.480	0.120	−0.177	0.128	0.356	0.248	1	
Left PM **	0.401	0.222	0.218	0.234	0.487	0.230	0.003	0.126	0.206	0.545	0.78 **	0.67 *	0.447	0.060	1

** $p < 001$, * $p < 0.005$, Ant: anterior, PL: posterolateral, PM: posteromedial.

4. Discussion

This study aimed to determine sprint, agility, balance, jumping and skill-based performance response differences between HISP and woHISP groups. The main findings of this study revealed that the HISP group's sports experience, BFP, 10 m, 20 m, 30 m, agility/skill, agility and shooting skill tests, zigzag Illinois, 505, jumping, right and left posterolateral and posteromedial scores were significantly better than woHISP.

The current study results indicated that the HISP group's BFP (%) was largely higher than the woHISP group. The deaf Czech national soccer team have been shown to have higher BF% but lower free fat mass than without-hearing-impairment players [17]. Comparing hearing female players and hearing-impaired soccer players, hearing players have significantly different waist- and calf-circumferences, and waist–hip ratios [19]. Some studies have found changes in body composition throughout the season, demonstrating how these changes may be significantly influenced by factors such as training intensity, exposure to match time, or diet [30,31]. Players who are lean and have a high percentage of muscle mass may have an advantage in positions that require speed and agility. On the other hand, players who are heavier and have a higher percentage of body fat may have an advantage in positions that require strength and power. It is important to note that soccer players should aim to have a body composition that allows them to perform at their best and meet the demands of their position or role on the field. This may involve maintaining a healthy weight, sufficient muscle mass and balanced body fat distribution. Players should also be mindful of their overall health and wellness and aim to maintain a healthy diet and exercise routine that helps them meet their physical goals.

Regarding physical fitness results, our study showed that while the HISP group had better 30 m, 20 m and 10 m sprints, agility, agility skills and jumping performance than woHISP, balance performance was reduced. During soccer games, high-intensity movements can be classified as actions requiring quick acceleration (10 m sprint), actions demanding maximum speed (30 m sprint), or actions requiring agility [22]. Additionally, it is well known that professional athletes perform better during sprints (10, 20, 30 and 40 m) in games than their less-skilled counterparts [32].However, these sprints typically last between two and four seconds and cover distances of less than 20 m [33]. Sprint ability is essential for soccer players, as it allows them to accelerate quickly and reach high speeds when running with the ball or chasing down opponents. This study has important findings, and the HISP group performance revealed significantly better agility abilities and skills with the ball to woHISP. While the classic definition of agility is the simple capacity for rapid direction changes, agility is a crucial performance variable comprising perceptual and decision-making characteristics such as visual scanning and anticipation in soccer [34]. Considering the effect of the reaction time on athletic performance, hearing-impaired athletes scored better in visual reaction time [35]. Visual reaction time is an essential factor in athletic performance, as it can affect an athlete's ability to react to changes in the game or competition. The visual reaction time or technical skills of deaf athletes will depend on their abilities and training, and are not necessarily related to their hearing status. Our main finding demonstrated that CMJ performance was higher in the HISP group than the woHISP group. Neuls et al. [17] reported that Czech national team soccer players had lower scores than the first league, but these were not significant. In contrast to our results, Soslu et al. [14] performed a study on repeated countermovement jump tests for volleyball and basketball players who were hearing-impaired and deaf. The results showed that deaf players' scores, in particular, were reduced for jumps and jump heights, the force produced, the acceleration at the time of the jump and the jump velocity compared to non-deaf players. There need to be more studies in the hearing-impaired soccer player context to compare different soccer players.

In this study, balance tests scores recorded during the Y balance test showed a significant decrease in the HISP group compared to the woHISP group. Additionally, a significant difference was found between the posterolateral and posteromedial in both legs, with a very large effect. These findings align with a previous study on national deaf basketball players, who demonstrated lower balance performance than hearing-impaired players [36]. As a result of the vestibular, visual and proprioceptive systems being injured, the balance may be affected negatively. Without a visual component, the balance will likely be impaired in deaf individuals who already have damage to this region [37]. Balance is crucial in soccer, as it allows players to maintain their footing and control the ball while moving on the field. Factors affecting balance performance in soccer include strength, coordination and proprioception.

The second aim of this study was to analyze the correlation of agility with speed, jumping and balance. Our results revealed large correlations between agility and speed. Our results are similar to previous studies reporting strong correlation between speed, agility and agility/skills [22,38,39]. Compared to non-elite soccer players, elite players have higher levels of physical fitness and talent [40]. The relationship between speed and agility in soccer is complex, as both abilities are important for success on the field. However, agility is often more important in soccer, as it is more relevant to the movements and actions players need to perform during a game. Both speed and agility can be improved through training, and many soccer players incorporate specific drills and exercises into their training programs to improve these abilities.

Another finding of this study was the negative correlation between jumping and 505 agility performance. Our results are similar to the previous study of Köklü et al. [38], which reported a significant negative relationship ($r = -0.769$, $p = 0.01$) between jumping and agility (angle $100°$) without the ball. Strong correlations between COD and CMJ would be expected, given that CMJ is an example of a long stretch-shortening cycle action (500 ms) that requires time to create a force for propulsion [41,42]. In the acceleration phase, where much force must be produced quickly to overcome the inertia in both vertical and horizontal orientation, lower limb power (represented by CMJ and cross hop) can, therefore, be considered a determinant of change in direction [43]. Our study revealed strong correlations between balance and agility. A previous study supports our results; dynamic balancing performance was recently shown to be significantly correlated with COD performance [44], and another study revealed that moderate correlations indicated that dynamic balance is an influential factor in agility, but is not the main limiting factor [41]. It is essential to recognize that hearing impairment is just one factor that can impact an athlete's performance, and it is not fair to generalize the abilities of hearing-impaired athletes. Each athlete is unique and has their strengths and challenges. The most important thing is to provide support and resources to help hearing-impaired athletes reach their full potential, regardless of their performance relative to non-deaf athletes. Considering the sample group of this study, the fact that they are Olympic-level athletes causes them to show better results than the results in other studies.

The current study has some limitations. A comparison has been made between the hearing-impaired group of athletes and others who were not similar in their sports experience and age. The soccer players without a hearing impairment had lower ages and sports experience. For this reason, further research is necessary to better match the larger sample size of elite adult soccer players in the same class, age category and sports experiences for the with- and without-hearing-impairment groups.

5. Conclusions

In conclusion, the current study highlighted that the HISP group had higher speed (10 m, 20 m, and 30 m), agility (zigzag, Illinois, 505), CMJ and agility skills (sec and goals). Balance performance was lower in the HISP group than the woHISP group, including the posterolateral and posteromedial of the right and left legs. These results show that hearing-impaired athletes have the potential to perform at the same level and even better than non-hearing-impaired athletes with appropriate training methods and planning. Additionally, hearing-impaired athletes are a special population with very different anthropometric and physical characteristics. Therefore, coaches must provide adequate anthropometrics and physical performance profiles for players with different characteristics and training stimuli during seasons. It seems to be worth investigating possible factors influencing players' workload.

Author Contributions: Conceptualization, Y.S., H.Y. and M.K.; methodology, H.Y., M.G. and N.B.M.; formal analysis, H.Y. and N.B.M.; investigation, Y.S., M.G. and H.Y.; data curation, H.Y., S.A., Y.S. and M.I.A.; writing—original draft preparation, Y.S., M.G., S.B.A.-M., M.I.A. and N.B.M.; writing—review and editing, Y.S., S.B.A.-M., M.I.A. and M.G.; visualization, H.Y.; supervision, M.K.; Funding acquisition, M.I.A. All authors have read and agreed to the published version of the manuscript.

Funding: This research was funded by Princess Nourah bint Abdulrahman University Researchers Supporting Project number (PNURSP2023R 286), Princess Nourah bint Abdulrahman University, Riyadh, Saudi Arabia.

Institutional Review Board Statement: This study was conducted at Kirikkale University, Sports Sciences Faculty, according to the principles outlined in the Declaration of Helsinki. Ethics committee approval was obtained from the Kirikkale University Non-Invasive Researches Ethics Committee (Date: 10 December 2020, number: 7 December 2020). The purpose of the study was explained to all athletes in advance, and all volunteers signed a written consent form before study participation.

Informed Consent Statement: The purpose of the study was explained to all athletes in advance, and all volunteers signed a written consent form before study participation.

Data Availability Statement: The data presented in this study are available on request from the corresponding author. The data are not publicly available due to restrictions on privacy.

Acknowledgments: We would like to thank all hearing-impaired soccer players, university soccer team athletes and their coaches and managers for their contribution to, interest in and support of our research study and sports sciences. We would like to thank the Princess Nourah bint Abdulrahman University for supporting this project through the Princess Nourah bint Abdulrahman University Researchers Supporting Project number (PNURSP2023R 286), Princess Nourah bint Abdulrahman University, Riyadh, Saudi Arabia.

Conflicts of Interest: The authors declare no conflict of interest.

References

1. *Addressing the Rising Prevalence of Hearing Loss*; World Health Organization: Geneva, Switzerland, 2018.
2. Kurková, P.; Válková, H.; Scheetz, N. Factors impacting participation of European elite deaf athletes in sport. *J. Sport. Sci.* **2011**, *29*, 607–618. [CrossRef]
3. Forsman, H.; Blomqvist, M.; Davids, K.; Liukkonen, J.; Konttinen, N. Identifying technical, physiological, tactical and psychological characteristics that contribute to career progression in soccer. *Int. J. Sport. Sci. Coach.* **2016**, *11*, 505–513. [CrossRef]
4. Zago, M.; Giuriola, M.; Sforza, C. Effects of a combined technique and agility program on youth soccer players' skills. *Int. J. Sport. Sci. Coach.* **2016**, *11*, 710–720. [CrossRef]
5. Clemente, F.M.; Soylu, Y.; Arslan, E.; Kilit, B.; Garrett, J.; van den Hoek, D.; Badicu, G.; Filipa Silva, A. Can high-intensity interval training and small-sided games be effective for improving physical fitness after detraining? A parallel study design in youth male soccer players. *PeerJ* **2022**, *10*, e13514. [CrossRef] [PubMed]
6. Fang, B.; Kim, Y.; Choi, M. Effect of Cycle-Based High-Intensity Interval Training and Moderate to Moderate-Intensity Continuous Training in Adolescent Soccer Players. *Healthcare* **2021**, *9*, 1628. [CrossRef]
7. Drust, B.; Atkinson, G.; Reilly, T. Future perspectives in the evaluation of the physiological demands of soccer. *Sport. Med.* **2007**, *37*, 783–805. [CrossRef]
8. Reilly, T. An ergonomics model of the soccer training process. *J. Sport. Sci.* **2005**, *23*, 561–572. [CrossRef]
9. Hammami, M.A.; Ben Abderrahmane, A.; Nebigh, A.; Le Moal, E.; Ben Ounis, O.; Tabka, Z.; Zouhal, H. Effects of a soccer season on anthropometric characteristics and physical fitness in elite young soccer players. *J. Sport. Sci.* **2013**, *31*, 589–596. [CrossRef]
10. Sioud, R.; Hammami, R.; Gene-Morales, J.; Juesas, A.; Colado, J.C.; van den Tillaar, R. Effects of Game Weekly Frequency on Subjective Training Load, Wellness, and Injury Rate in Male Elite Soccer Players. *Int. J. Environ. Res. Public Health* **2023**, *20*, 579. [CrossRef]
11. Arslan, E.; Soylu, Y.; Clemente, F.M.; Hazir, T.; Kin Isler, A.; Kilit, B. Short-term effects of on-field combined core strength and small-sided games training on physical performance in young soccer players. *Biol. Sport* **2021**, *38*, 609–616. [CrossRef]
12. Stølen, T.; Chamari, K.; Castagna, C.; Wisløff, U. Physiology of soccer: An update. *Sport. Med.* **2005**, *35*, 501–536. [CrossRef]
13. Soylu, Y.; Ramazanoglu, F.; Arslan, E.; Clemente, F.M. Effects of mental fatigue on the psychophysiological responses, kinematic profiles, and technical performance in different small-sided soccer games. *Biol. Sport* **2022**, *39*, 965–972. [CrossRef] [PubMed]

14. Soslu, R.; Özer, Ö.; Uysal, A.; Pamuk, Ö. Deaf and non-deaf basketball and volleyball players' multi-faceted difference on repeated counter movement jump performances: Height, force and acceleration. *Front. Sport. Act. Living* **2022**, *4*, 941629. [CrossRef] [PubMed]

15. Lévesque, J.; Théoret, H.; Champoux, F. Reduced procedural motor learning in deaf individuals. *Front. Hum. Neurosci.* **2014**, *8*, 343. [CrossRef] [PubMed]

16. Güzel, N.A.; Medeni, Ö.Ç.; Mahmut, A.; Savaş, S. Postural control of the elite deaf football players. *J. Athl. Perform. Nutr.* **2016**, *3*, 1–8. Available online: https://dergipark.org.tr/en/pub/japn/issue/18606/196359 (accessed on 3 January 2023).

17. Neuls, F.; Botek, M.; Krejci, J.; Panska, S.; Vyhnanek, J.; McKune, A. Performance-associated parameters of players from the deaf Czech Republic national soccer team: A comparison with hearing first league players. *Sport Sci. Health* **2019**, *15*, 527–533. [CrossRef]

18. Chilosi, A.M.; Comparini, A.; Scusa, M.F.; Berrettini, S.; Forli, F.; Battini, R.; Cipriani, P.; Cioni, G. Neurodevelopmental disorders in children with severe to profound sensorineural hearing loss: A clinical study. *Dev. Med. Child Neurol.* **2010**, *52*, 856–862. [CrossRef]

19. Szulc, A.M.; Buśko, K.; Sandurska, E.; Kołodziejczyk, M. The biomechanical characteristics of elite deaf and hearing female soccer players: Comparative analysis. *Acta Bioeng. Biomech.* **2017**, *19*, 127–133. [CrossRef]

20. Faul, F.; Erdfelder, E.; Lang, A.-G.; Buchner, A. G* Power 3: A flexible statistical power analysis program for the social, behavioral, and biomedical sciences. *Behav. Res. Methods* **2007**, *39*, 175–191. [CrossRef]

21. Linek, P.; Sikora, D.; Wolny, T.; Saulicz, E. Reliability and number of trials of Y Balance Test in adolescent athletes. *Musculoskelet. Sci. Pract.* **2017**, *31*, 72–75. [CrossRef]

22. Little, T.; Williams, A.G. Specificity of acceleration, maximum speed, and agility in professional soccer players. *J. Strength Cond. Res.* **2005**, *19*, 76–78. [CrossRef]

23. Negra, Y.; Chaabene, H.; Amara, S.; Jaric, S.; Hammami, M.; Hachana, Y. Evaluation of the Illinois Change of Direction Test in Youth Elite Soccer Players of Different Age. *J. Hum. Kinet.* **2017**, *58*, 215–224. [CrossRef]

24. Dugdale, J.H.; Arthur, C.A.; Sanders, D.; Hunter, A.M. Reliability and validity of field-based fitness tests in youth soccer players. *Eur. J. Sport Sci.* **2019**, *19*, 745–756. [CrossRef]

25. Gülü, M.; Akalan, C. A new peak-power estimation equations in 12 to 14 years-old soccer players. *Medicine* **2021**, *100*, e27383. [CrossRef] [PubMed]

26. Kutlu, M.; Yapıcı, H.; Yoncalık, O.; Celik, S. Comparison of a new test for agility and skill in soccer with other agility tests. *J. Hum. Kinet.* **2012**, *33*, 143–150. [CrossRef] [PubMed]

27. YAĞIN, F.H.; GÜLDOĞAN, E.; Colak, C. A web-based software for the calculation of theoretical probability distributions. *J. Cogn. Syst.* **2021**, *6*, 44–50. [CrossRef]

28. Cohen, J. *Statistical Power Analysis for the Social Sciences*; Lawrence Erlbaum Associates: New York, NY, USA, 1988.

29. Yağin, B.; Yağin, F.H.; Gözükara, H.; Colak, C. A web-based software for reporting guidelines of scientific researches. *J. Cogn. Syst.* **2021**, *6*, 39–43. [CrossRef]

30. Milanese, C.; Cavedon, V.; Corradini, G.; De Vita, F.; Zancanaro, C. Seasonal DXA-measured body composition changes in professional male soccer players. *J. Sport. Sci.* **2015**, *33*, 1219–1228. [CrossRef] [PubMed]

31. Caldwell, B.P.; Peters, D.M. Seasonal variation in physiological fitness of a semiprofessional soccer team. *J. Strength Cond. Res.* **2009**, *23*, 1370–1377. [CrossRef]

32. Bradley, P.S.; Di Mascio, M.; Peart, D.; Olsen, P.; Sheldon, B. High-intensity activity profiles of elite soccer players at different performance levels. *J. Strength Cond. Res.* **2010**, *24*, 2343–2351. [CrossRef]

33. Barnes, C.; Archer, D.; Hogg, B.; Bush, M.; Bradley, P. The evolution of physical and technical performance parameters in the English Premier League. *Int. J. Sport. Med.* **2014**, *35*, 1095–1100. [CrossRef] [PubMed]

34. Zouhal, H.; Abderrahman, A.B.; Dupont, G.; Truptin, P.; Le Bris, R.; Le Postec, E.; Sghaeir, Z.; Brughelli, M.; Granacher, U.; Bideau, B. Effects of Neuromuscular Training on Agility Performance in Elite Soccer Players. *Front. Physiol.* **2019**, *10*, 947. [CrossRef] [PubMed]

35. Soto-Rey, J.; Pérez-Tejero, J.; Rojo-González, J.J.; Reina, R. Study of reaction time to visual stimuli in athletes with and without a hearing impairment. *Percept. Mot. Ski.* **2014**, *119*, 123–132. [CrossRef] [PubMed]

36. Cobanoglu, G.; Suner-Keklik, S.; Gokdogan, C.; Kafa, N.; Savas, S.; Guzel, N.A. Static balance and proprioception evaluation in deaf national basketball players. *Balt. J. Health Phys. Act.* **2021**, *13*, 9–15. [CrossRef]

37. Akınoğlu, B.; Kocahan, T. The effect of deafness on the physical fitness parameters of elite athletes. *J. Exerc. Rehabil.* **2019**, *15*, 430–438. [CrossRef]

38. Köklü, Y.; Alemdaroğlu, U.; Özkan, A.; Koz, M.; Ersöz, G. The relationship between sprint ability, agility and vertical jump performance in young soccer players. *Sci. Sport.* **2015**, *30*, e1–e5. [CrossRef]

39. Jovanovic, M.; Sporis, G.; Omrcen, D.; Fiorentini, F. Effects of speed, agility, quickness training method on power performance in elite soccer players. *J. Strength Cond. Res.* **2011**, *25*, 1285–1292. [CrossRef]

40. Altmann, S.; Neumann, R.; Härtel, S.; Kurz, G.; Stein, T.; Woll, A. Agility testing in amateur soccer: A pilot study of selected physical and perceptual-cognitive contributions. *PLoS ONE* **2021**, *16*, e0253819. [CrossRef]

41. Falces-Prieto, M.; González-Fernández, F.T.; García-Delgado, G.; Silva, R.; Nobari, H.; Clemente, F.M. Relationship between sprint, jump, dynamic balance with the change of direction on young soccer players' performance. *Sci. Rep.* **2022**, *12*, 12272. [CrossRef]

42. Vescovi, J.D.; Mcguigan, M.R. Relationships between sprinting, agility, and jump ability in female athletes. *J. Sport. Sci.* **2008**, *26*, 97–107. [CrossRef]
43. Freitas, T.T.; Alcaraz, P.E.; Bishop, C.; Calleja-González, J.; Arruda, A.F.S.; Guerriero, A.; Reis, V.P.; Pereira, L.A.; Loturco, I. Change of Direction Deficit in National Team Rugby Union Players: Is There an Influence of Playing Position? *Sports* **2018**, *7*, 2. [CrossRef] [PubMed]
44. Sekulic, D.; Spasic, M.; Mirkov, D.; Cavar, M.; Sattler, T. Gender-specific influences of balance, speed, and power on agility performance. *J. Strength Cond. Res.* **2013**, *27*, 802–811. [CrossRef] [PubMed]

healthcare

Article

Lower Limb Unilateral and Bilateral Strength Asymmetry in High-Level Male Senior and Professional Football Players

Mário C. Espada [1,2,3,*], Marco Jardim [4], Rafael Assunção [4], Alexandre Estaca [5], Cátia C. Ferreira [1,6], Dalton M. Pessôa Filho [7,8], Carlos E. L. Verardi [7,9], José M. Gamonales [6,10] and Fernando J. Santos [1,2,11]

1   Instituto Politécnico de Setúbal, Escola Superior de Educação, 2914-504 Setúbal, Portugal; catia.ferreira@ese.ips.pt (C.C.F.)
2   Life Quality Research Centre, Complexo Andaluz, Avenida Dr. Mário Soares 110, 2040-413 Rio Maior, Portugal
3   CIPER, Faculdade de Motricidade Humana, Universidade de Lisboa, 1499-002 Lisboa, Portugal
4   Instituto Politécnico de Setúbal, Escola Superior de Saúde, 2914-503 Setúbal, Portugal; marco.jardim@ess.ips.pt (M.J.); rafael.assuncao@ess.ips.pt (R.A.)
5   Casa Pia Atlético Clube, Estádio Pina Manique, Parque de Monsanto, 1500-462 Lisboa, Portugal; alexandre.estaca@gmail.com
6   Research Group in Optimization of Training and Performance Sports, Faculty of Sport Science, University of Extremadura, 10005 Cáceres, Spain; martingamonales@unex.es
7   Department of Physical Education, São Paulo State University (UNESP), Bauru 17033-360, Brazil; dalton.pessoa-filho@unesp.br (D.M.P.F.); carlos.verardi@unesp.br (C.E.L.V.)
8   Graduate Programme in Human Development and Technology, São Paulo State University (UNESP), Rio Claro 13506-900, Brazil
9   Graduate Programme in Developmental Psychology and Learning, Faculty of Science, São Paulo State University (UNESP), Bauru 17033-360, Brazil
10   Faculty of Health Sciences, University of Francisco de Vitoria, 28223 Madrid, Spain
11   Faculdade de Motricidade Humana, Universidade de Lisboa, 1499-002 Lisboa, Portugal
*   Correspondence: mario.espada@ese.ips.pt; Tel.: +351-265-710-800

Citation: Espada, M.C.; Jardim, M.; Assunção, R.; Estaca, A.; Ferreira, C.C.; Pessôa Filho, D.M.; Verardi, C.E.L.; Gamonales, J.M.; Santos, F.J. Lower Limb Unilateral and Bilateral Strength Asymmetry in High-Level Male Senior and Professional Football Players. *Healthcare* **2023**, *11*, 1579. https://doi.org/10.3390/healthcare11111579

Academic Editor: Giovanni Morone

Received: 13 March 2023
Revised: 23 May 2023
Accepted: 25 May 2023
Published: 28 May 2023

Abstract: This study sought to assess the relationship between different jumping asymmetries and associated performance variables in high-level male senior and professional football players. Nineteen football players with at least 12 years of training experience (23.2 ± 3.1 years of age; 75.2 ± 4.8 kg of body mass and 181 ± 0.06 cm of height) participated in this study performing countermovement jump (CMJ), squat jump (SJ), single-leg CMJ and drop jump (DJ), associated performance variable eccentric utilization ratio (EUR), stretch-shortening cycle (SSC), bilateral deficit (BLD), and limb symmetry index (LSI) were determined. High correlations were observed between different methodologies of jump tests and associated performance indicators (SSC, BLD, EUR), except LSI. Moreover, CMJ and SJ results were different ($p < 0.05$), but no differences were found between interlimb in CMJ ($p = 0.19$) and DJ ($p = 0.14$). Between the same limbs and different jumps differences were detected in CMJ and DJ ($p < 0.01$), and it has also been found that the laterality effect size on strength was small in CMJ ($ES = 0.30$) and DJ ($ES = 0.35$). LSI between CMJ and DJ was not different despite higher mean values in CMJ, and although mean BLD was positive (>100%), the results highlight the need for individual evaluation since eight players scored negatively. An in-depth and accurate analysis of performance in preseason screening jump tests should be considered, aiming to detect injury risk, specifically evaluating different jumping test methodologies, and determining jumping associated performance variables for each test, namely EUR, SSC, BLD, and LSI. Specific muscle-strengthening exercises could be implemented based on this study results and outcomes, aiming to reduce injury risks and lower extremity asymmetries and to enhance individual football performance in high-level male senior and professional football players. Sports institutions should pay special attention regarding potential health problems in athletes exposed to daily high training loads.

Keywords: soccer; strength; interlimb; symmetry; injury risk; performance

1. Introduction

Football is one of the most popular sports worldwide [1], characterized by a variety of unpredictable movements involving rapid speed changes [2]. According to previous studies, this sport is associated with a higher injury rate compared to other competitive sports since the incidence of injury is estimated to be 15 to 35 injuries per 1000 h of training/matches [3,4], and specifically, muscle injuries (>90%) occur in noncontact situations [5,6], indications that are associated with a special care in training evaluation and prescription and monitoring of the competitive events.

Interlimb asymmetry (or bilateral difference, bilateral asymmetry), defined as differences between the dominant and non-dominant limb function or performance [7], may be related to long-term training in the same sport [8] and has been a popular source of investigation in recent years. Playing football demands fast and powerful movements, implying lower-limb muscles in maximal and rapid actions [9]. In football, players are exposed to an increasing number of competitive events and high volumes of regular training, associated with a preferred limb dominance [10,11]. Functional asymmetric adaptations may therefore relate to the asymmetric motor demands [12] which characterize particular sports [13].

The repetition of execution of unilateral movements or skills is one of the particularities of football in training and competitive events (e.g., cutting, changing direction, and kicking) [14]; consequently, the increased interlimb asymmetry may be important considering injury risks under the fatigued state [15]. A minimum symmetry target of 10% has been previously suggested for evaluation and rehabilitation protocols [16], being equally considered for clearance to return to practice after anterior cruciate ligament reconstruction (ACLR) [17]. Lower limb injuries are a serious issue in professional athletes because they may be associated with long-term absence from match and health-related problems and related to decreased performance and financial costs [18]. On average, for example, typical hamstring strain injuries causes athletes to miss 2 weeks of training or competitive play [19], with this timeline potentially representing up to 6 fixtures being missed, with professional football teams often experiencing 5–6 hamstring injuries per season [5].

Strength is a very important physical quality in football that underpins successful performance in the complex and varied motor skill tasks in this sport [20,21]. Jump tests are a viable method of quantifying interlimb asymmetries and have successfully been used to prospectively identify the risk of injury of athletes [22]. The countermovement jump (CMJ) and squat jump (SJ) are among the most regularly used vertical jumps for evaluation purposes [23,24], and drop jump (DJ) is frequently used as a plyometric exercise to develop and evaluate jumping performance [25]. The performance in CMJ (height) is on average greater than the height of the SJ [24,26], and the difference between the jumps, often indicated as the eccentric utilization ratio (EUR), has been suggested as an indicator of performance [26]. Lower values of EUR denote that the athlete should improve elasticity storage, which is usually developed with explosive exercises; in contrast, athletes with higher EUR are normally directed to training basic strength.

Associated with CMJ and SJ is also the stretch-shortening cycle (SSC), previously indicated as a stretching of the muscle–tendon unit prior to a shortening [27] and linked to increased force, mechanical work, torque, and power during the shortening phase of the SSC [28,29]. More specifically, during CMJ, bilateral deficit (BLD) has been discussed, although the elements that influence the BLD are not well defined in the literature [30]. Recently, Pleša et al. [31] stressed that higher BLD could reflect improved neuromuscular capacity in unilateral tasks compared with the bilateral and found that larger a BLD is associated with superior linear sprint and approach jump performance in volleyball players, suggesting BLD as a tool to assist practitioners in decision making in athletic training. A score below 100% is indicative of a bilateral strength deficit, while a bilateral index higher than 100% is indicative of bilateral facilitation (BLF), where the force of both legs during a bilateral exercise exceeds the sum of the unilateral movements.

Football demands explosive movement performance such as change of direction speed, heading, and shooting [32], and all these are often associated with single-leg movements at some point. The limb symmetry index (LSI) is associated with single-leg tests and considered, for example, in return-to-sport decision making for individuals after ACLR [17,33]. It was previously reported by Krych et al. [34] that 6 months after traditional ACLR, younger age was a key factor associated with exceptional strength and functional recovery (defined as strength LSI > 85%), and previous research highlighted that muscle function tests are strong determinants for between-limb asymmetry predictions in ACLR [35].

Even though some level of asymmetry is probable in football players, few investigators have examined the prevalence of asymmetry during commonly used screening tests in high-level football players, mainly because they are exposed to intense training and competitive schedules. It was previously indicated that asymmetries measured during common screening protocols may be associated with reductions in jumping performance [14] and that it is important to identify reliable methods of assessing the direction and magnitude of asymmetries to reduce the potential risk for injury and re-injury in all populations [19]. To the authors' knowledge, to date, no study with high-level male senior and professional football players has compared different jumping methods, considering unilateral and bilateral perspectives and the respective associated performance variables (EUR, SSC, BLD, and LSI), which may provide relevant information regarding evaluation purposes and specific intervention procedures aiming injury prevention and performance enhancement. Hence, the aim of the present study was to assess the relationship between different jumping asymmetries and associated performance variables in high-level male senior and professional football players.

2. Materials and Methods

2.1. Subjects

Nineteen football players with at least 12 years of training experience (23.2 ± 3.1 years of age; 75.2 ± 4.8 kg of body mass and 181 ± 0.06 cm of height) were involved in this research. All subjects, integrated in a Portuguese first division football team, trained for at least 36 weeks per year with regular 5 training sessions per week (3–4 h per day), plus 1 or 2 competitive events a week, in the preceding 12 months before participation in the research. The criteria for not participating in the study were injury associated with absence from training more than 4 days in the 3 months before the research and injury related to surgical intervention in the 12 months before data collection.

International ethical standards for sport and exercise science research [36] were followed and the study also considered the Declaration of Helsinki. The Ethical Committee of the Polytechnic Institute of Leiria approved the research (CE/IPLEIRIA/22/2021) and club authorization was obtained before contact with the athletes, with all of these providing written informed consent. No economic incentives were provided, and goalkeepers were excluded from this study because of their different training and physical characteristics. Participants and the clubs' medical staffs were informed about the potential risks of the current procedures and provided written informed consent.

2.2. Study Design

Subjects were tested on two occasions at the same time of day to avoid circadian rhythm effects, each separated by 24 h. All footballers arrived at the club's training facility wearing appropriate footwear and comfortable athletic clothing and refrained from intense training at least 48 h prior to testing.

Previous to data collection, no specific familiarization procedures were necessary because athletes were all experienced players, many of them with national team international participation in their national countries, with regular involvement in strength training and evaluation events. ChronoJump photoelectric cells (Bosco System, Barcelona, Spain) was used to evaluate the jump´s performance and a 30 cm step platform (Reebok®) during the DJ. The height of the athletes was evaluated on a measurement platform (Seca 274, Milan, Italy) and body mass (kg) determined on a calibrated physician scale (Seca 786 Culta, Milan, Italy).

2.3. Procedures

A dynamic 20 min warm-up was performed before data collection, and afterward all subjects practiced each jump test as many times as they wanted, although all players were instructed to practice each test a minimum of three times.

A 3 min rest period was prescribed between the end of warm-up and the first test. After that, athletes performed the tests, simultaneously assessed by two experienced observers with doctoral degrees in sports sciences, accompanied in the data collection by 3 members of the club's clinical staff. A total of three attempts were completed in all jumps, and observers were blinded to each other's scores. The higher value was considered for the final analysis [30], and a 1 min rest was granted between each jump to minimize the effect of fatigue [31].

During testing, no motivational communication took place because communication is a relevant element in the coach–athlete relationship [37] and could have influence the performance and score.

2.4. Testing Description

2.4.1. Squat Jump (SJ)

Subjects begin by standing on the designated testing leg with their hands on hips. Subjects were then instructed to perform with knees bent in a 90° flexion position and afterward, after a short pause moment, and jump as far forward as possible, landing on both legs. Upon landing, the players were required to 'hold and stick' their position for 2 s. Failure to stick the landing resulted in a void trial and the jump being retaken after a 60 s rest. This was a criterion for all jump tests.

2.4.2. Countermovement Jump (CMJ)

Participants begin by standing with their hands on hips and were instructed to start from the standing position and use an explosive countermovement (until the 90° knee flexion position), jumping as high as possible.

2.4.3. Single-Leg Countermovement Jump (SL-CMJ)

Subjects paused in an upright position with hands on hips and feet positioned hip-width apart. To start the test, one leg was lifted off the floor to approximately mid-shin-height of the standing leg. Subjects afterward performed a countermovement to 90° flexion position, followed by a quick upward vertical movement, triple extending at the ankle, knee, and hip, with the objective of jumping as high as possible. The jumping leg had to remain fully extended, and hands fixed to hips; any variation from this required the trial be re taken after a 60 s rest period [22].

2.4.4. Single Leg Drop Jump (SL-DJ)

Football players were asked to perform single-leg drop jump (SL-DJ) trials on both legs after hopping from a box of 30 cm height positioned 5 cm posterior to the ChronoJump photoelectric cells. It was requested that the participants take off standing on a single-leg, land on the same leg, and stabilize as quickly as possible to afterward perform a quick upward vertical movement, as described by [38].

2.5. Measurments

2.5.1. Limb Asymmetry Index (LSI)

The asymmetry index was calculated using the equation (dominant limb–non-dominant limb)/dominant limb) × 100 [39].

2.5.2. Bilateral Deficit (BLD)

The BLD was computed according to the following formula [40]:

$$\text{BLD (\%)} = [100 \times (\text{Bilateral/Right unilateral} + \text{Left unilateral})] - 100$$

2.5.3. Eccentric Utilization Ratio (EUR)

The EUR was calculated as the ratio between CMJ and SJ heights and considered an indicator of lower extremity SSC performance in athletes. An ideal EUR of ~1.1 was previously suggested, in which the CMJ score should be 1.1× (10%) that of the SJ [26].

2.5.4. Stretch-Shortening Cycle (SSC)

The theory behind SSC is that the muscles and tendons are able to store elastic energy in the pre-stretch phase of the movement. One example of this is a CMJ, which tendentially produces more force than an SJ. For SSC measurement, the difference between the two types of vertical jumps is used (CMJ-SJ) [41].

2.6. Statistical Analysis

Previous to data collection, the sample size required was determined (GPower, v.3.1.9, University of Kiel, Kiel, Germany). The sample was deemed as adequate based on a statistical power of 75%, effect size (5% change), and α error probability of 0.05 associated with 18 subjects. Data was initially organized as means and standard deviations (M $\pm$ SD) in Microsoft Excel™; all additional analyses were made in Statistical Package for Social Sciences v27.0 (SPSS, IBM Corp, Armonk, NY, USA). Reliability of tests measures was estimated using intraclass correlation coefficient (ICC) with absolute agreement and 95% confidence intervals. Interpretation considered previous research [42] specifically values > 0.9 = excellent, 0.75–0.9 = good, 0.5–0.75 = moderate, and <0.5 = poor.

The linear regression models between different test results were computed, the trend-line equations and determination coefficients (R^2) were calculated. Cohen's d was used for effect size (*ES*) calculation and considered extremally large (>4), very large (2–4), large (1.2–2), moderate (0.6–1.2), small (0.2–0.6), and trivial (0–0.2) [43].

Pearson's r correlation coefficients between assessment methods were determined, with the absolute value of the correlation demarcated as follows [44]: negligible correlation (r^2 < 30), weak correlation (r^2 = 0.30–0.50), moderate correlation (r^2 = 0.50–0.70), strong correlation (r^2 = 0.70–0.90), and very strong correlation (r^2 > 90). Paired-sample t-test was used to compare the studied variables and data obtained with both lower limbs (left and right). Statistical significance was accepted at $p \leq 0.05$.

3. Results

High levels of reliability (ICC > 0.87 and <0.94) were observed in all the test results and studied variables, with the lower levels associated with single-leg evaluations. Descriptive statistics of all tests results are shown in Table 1.

Table 1. Mean $\pm$ standard deviation (M $\pm$ SD), minimum and maximum of performance in all tests.

Variables	Mean	SD	Minimum	Maximum
SJ (cm)	36.38 *	6.98	26.30	50.40
CMJ (cm)	38.40 *	5.92	29.60	50.10
EUR (A.U.)	1.06	0.06	0.97	1.17
SSC (%)	6.27	5.50	−3.37	17.14
SL-CMJ (L) (cm)	20.68 #	4.73	12.70	30.90
SL-CMJ (R) (cm)	21.29 **	5.10	14.80	35.00
LSI-CMJ (A.U.)	92.57	4.98	83.13	99.44
BLD-CMJ (A.U)	11.71	27.68	−33.32	70.22
SL-DJ (L) (cm)	23.41 #	4.89	16.90	31.50
SL-DJ (R) (cm)	24.64 **	5.36	17.80	33.80
LSI-DJ (A.U.)	89.06	7.79	70.03	97.78

M, Mean; SD, Standard deviation; SJ, Squat jump; CMJ, Countermovement jump; EUR, Eccentric utilization ratio; SSC, Stretch-shortening cycle; SL-CMJ (L), Single-leg countermovement jump (left); SL-CMJ (R), Single-leg countermovement jump (right); LSI-CMJ, Limb symmetry index--countermovement jump; BLD-CMJ, Bilateral deficit countermovement jump; SL-DJ (L), Single-leg drop jump (left); SL-DJ (R), Single-leg-drop jump (right); LSI-DJ, Limb symmetry index--drop jump; * and # significant difference ($p < 0.01$); ** significant difference ($p < 0.05$).

Results associated with CMJ and DJ were different ($p < 0.01$). No statistically significant differences were determined between the right and left leg in the CMJ ($p = 0.19$) and DJ ($p = 0.14$) contrary to the same single-leg performance in both CMJ and SJ. It has also been found that the laterality effect size on strength was small in CMJ ($ES = 0.30$) and DJ ($ES = 0.35$). Moreover, despite the mean value of SSC being positive, two players exhibited negative values, -3.37 and -3.13, which represented as consequence in these two specific cases a EUR value of 0.97, the minimum observed.

Considering LSI and the cut-off threshold of 85%, it should be highlighted that it was detected in five players an LSI-DJ below 85%, while regarding LSI-CMJ, only two presented values below this boundary threshold. Regarding BLD, eight players scored lower values than 100%, indicative of bilateral strength deficit, while the other eleven cohorts, exhibited BLF (BLD > 100%), since the force of both legs during a bilateral exercise exceeds the sum of the unilateral movements. Table 2 shows the correlations between the studied variables.

Table 2. Correlations between the studied variables associated with jumps.

	CMJ	EUR	SSC	SL-CMJ (L)	SL-CMJ (R)	BDL-CMJ	SL-DJ (L)	SL-DJ (R)
SJ	0.97 **	−0.74 **	−0.73 **	0.71**	0.66 **	-	0.59 **	0.60 **
CMJ		−057 *	−0.56	0.73 **	0.65 **	-	0.59 **	0.59 **
EUR			0.99 **	-	-		-	-
SSC			-	-	-	-	-	-
SL-CMJ (L)				-	0.92 **	−0.67 **	0.79 **	0.79 **
SL-CMJ (R)					-	−0.57 *	0.71 **	0.71 **
BDL-CMJ						-	−0.55 *	−0.55 *
SL-DJ (L)							0.77 **	0.99 **
SL-DJ (R)								1

SJ, Squat jump; CMJ, Countermovement jump; EUR, Eccentric utilization ratio; SSC, Stretch-shortening cycle; SL-CMJ (L), Single-leg countermovement jump (left); SL-CMJ (R), Single-leg-countermovement jump (right); BLD-CMJ, Bilateral deficit countermovement jump; SL-DJ (L), Single-leg drop jump (left); SL-DJ (R), Single-leg drop jump (right). * ($p < 0.05$), ** ($p < 0.01$).

Several correlations were visible between performance in jumping tests and associated performance variables (EUR, SSC, BLD), although no correlations were observed between LSI-CMJ, LSI-DJ, and all the other studied variables. Figure 1 depicts the linear regression of SJ and CMJ on EUR.

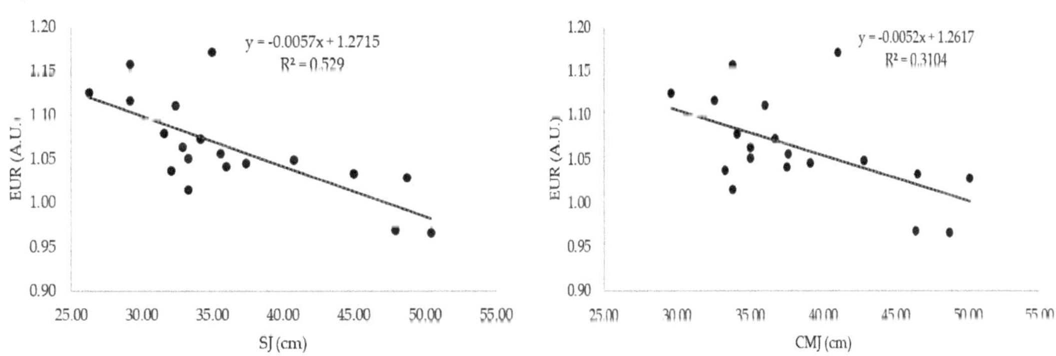

Figure 1. Linear regression of squat jump and countermovement jump on eccentric utilization ratio.

4. Discussion

The aim of the present study was to assess the relationship between different jumping asymmetries and associated performance variables in high-level male senior and professional football players. The main findings were: (1) high correlations were observed between different methodologies of jump tests (the execution dynamic of CMJ is different compared to SJ) and associated performance indicators (SSC, BLD, EUR), except LSI; (2) CMJ and SJ results were different and should not be used solely and interchangeably as screening tests performed during preseason to evaluate players physical condition; (3) no differences were found between interlimb in CMJ and DJ, although between the same limbs and different jumps differences were detected; (4) LSI between CMJ and DJ were not different, despite higher mean values in CMJ and; (5) although mean BLD was positive (>100%), the results highlight the need for individual evaluation in screening tests performed during preseason in high-level male senior and professional football since eight players scored negatively, which is also underlined by the maximum, minimum and standard deviation of the results in the present study.

Asymmetry was indicated as a potential risk factor for injury in football players [45] that may be the result of additional stress placed on the soft tissue structures of the non-dominant leg [46]. Previously, musculoskeletal imbalances of 10% on at least one jump test were observed in male youth football players [11]. Later, Lockie et al. [47] reported larger interlimb differences for the SL-CMJ (10.4%) in male team sport athletes aged 23 years, and more recently, asymmetries of 15% during landing tasks were shown in uninjured 10–18-year-old elite male football players [45]. Additionally, Bishop et al. [48] conducted a study with elite youth female football players and reported significantly larger asymmetries for SL-CMJ (12.5%). Similar observations were previously shown, reporting larger asymmetry for SL-CMJ (11.2%) in adolescent male handball players aged 16 years [49].

In the present study, CMJ and SJ results were different, and no differences were observed between interlimb in CMJ and DJ, more specifically, CMJ (p = 0.19) and DJ (p = 0.14). Although different jump differences were found between the same limbs and the laterality effect size on strength was small in CMJ (ES = 0.30) and DJ (ES = 0.35), these results lead us to suggest that CMJ and SJ should not be used solely and interchangeably as screening tests performed during preseason to evaluate physical condition of high-level male senior and professional footballers. Moreover, biceps femoris and soleus may impact jump performance because an early activation of the biceps femoris has been found to negatively influence the joint power transfer [50], consequently reducing the effect of the SSC, a key factor related to performance in vertical jumps [51], which allows us to speculate that CMJ interlimb asymmetry may be particularly related with lateral asymmetry of these muscles.

The associated mechanisms associated with SSC are highly debated in the literature as none of the mechanisms can completely explain this SSC-effect [52,53]. The preactivation of the muscle [54], the stretch-reflex [55], and the release of stored passive-elastic energy in the tendinous tissue [56,57] are mechanisms attributed to this effect. In this study, despite the mean value of SSC being positive, two players exhibited negative values, −3.37 and −3.13, which represented as a consequence in these two specific cases an EUR value of 0.97. This evidence highlights the importance of analyzing not only different methodology and performance of jump tests but also associated performance such as SCC, and very importantly, bearing in mind an individual perspective since football team mean values do not represent individual potential injury risk.

The EUR is used as an indicator of elastic storage in CMJ in strength and conditioning practice [58]. Larger EUR can be associated with superior CMJ performance but may also lower SJ performance. Moreover, poor SJ performance could be related to high levels of muscle slack [24] or to poor ability to develop force rapidly [59]. This is supported by the evidence that individuals with stiffer tendons, a trait that is beneficial for rapid force development, exhibit lower EUR [60], and it should also be noted that EUR was larger in track and field athletes compared to gymnasts and parkour practitioners, while the jumping performance (SJ and CMJ) was better in the latter cases compared to that in track and field athletes [61]. In the present study, we found no correlations between EUR and other jump tests associated performance variables, namely BDL and LSI, only correlations with SJ (r = −0.74, p < 0.01) and CMJ (r = −0.57, p < 0.05); however, r-squared of the regression analysis between EUR and both jump methodologies was not very strong, confirming the evidence that the utilization of stored elastic energy plays a short role in CMJ, except when executed with low amplitude [62].

The BLD phenomenon is related to the evidence that force production during a maximal bilateral contraction normally is lower compared to the sum of total force produced by the left and right limbs combined [63]. For this reason, true examination of the BLD is only possible when comparable bilateral and unilateral tasks have been completed (e.g., bilateral and unilateral CMJ). Previously, Bračič et al. [64] observed a BLD of 19.1% in 12 elite sprinters. and a jump height BLD of 8.5% was reported in trained volleyball players [65]. More recently, it was shown that the DJ test showed BLF contrarily to the CMJ, the authors emphasized that this may offer further support showing the advanced technical nature of reactive strength tasks [66]. Considering BLD and the results in the present study, although mean BLD was positive (>100%), we found that eight high-level male senior and professional football players scored lower than 100%, which is indicative of bilateral strength deficit, while the other eleven cohorts exhibited BLF (BLD > 100%), since the force of both legs during a bilateral exercise exceeds the sum of the unilateral movements. This evidence underlines that when left and right limbs were summed and still could not exceed the bilateral score, this demonstrates how challenging unilateral fast SSC activities are, which was previously supported in [67], research which showed that both endurance and power-based athletes showed a BLF of −3.8 to −13.8% in power during the DJ test, emphasizing that the wide variation in BLD scores in our and other studies led Bishop et al. [66] to suggest that this phenomenon should be investigated with multiple metrics for each test.

Nevertheless, absolute strength is an important factor influencing bilateral strength asymmetry [68], and the effects of strength asymmetry and/or limb preference on dynamic athletic movements such as running, jumping and kicking [69], some of the most common or routine movements in football during training and competition, were verified. In addition, it has been demonstrated that strength asymmetry of >15% caused a reduced jump height [68]; therefore, the cut-off threshold of 85% appears to be strengthened by the latter findings. In another study, Kotsifaki et al. [70] indicated that measuring only jump distance, even using the healthy leg as a reference, is not sufficient to fully assess knee function after ACLR, and Wang et al. ([71] suggested that natural mechanics could be responsible for asymmetry in able-bodied walking instead of neurophysiological mechanisms, for example, leg dominance, something that should be considered in football because the repetition in quadriceps activity can have as a consequence muscular imbalances between limbs [72].

In the present study, we found that although LSI between CMJ and DJ was not different despite higher mean values in CMJ, in five high-level male senior and professional football players an LSI-DJ below 85% was observed, while regarding LSI-CMJ, only two presented values were below this threshold boundary. Moreover, no correlations were noted between LSI-CMJ, LSI-DJ, and all the jumping associated performance variables. Considering this context, it must be mentioned that when the dominant leg of football players is associated with dynamic movements and powerful contractions, such as shooting the ball, the contralateral non-dominant leg is solicited in a different way by ensuring body stability related to isometric contractions. This motor mechanism repeatedly performed during training and competition could explain some of the differences between both limbs in asymmetrical sports. Noteworthy, the maximum, minimum, and standard deviation results found in our study suggest the need for individualized analysis in screening tests performed during preseason and also individualized training prescription and recovery methods in high-level male senior and professional football players.

Thus, from a practical applications perspective, it seems prudent to suggest that strength training should be considered in high-level male senior and professional football players, as recently suggested for football academy players [73,74]. Finally, football club structures (in concrete, some of those associated with decision making regarding daily training conditions) should assume that collaborative work developed by professionals of different areas is a plus for athlete's global condition and may represent an important support in the definition of evaluation procedures, specifically protocols aiming to improve injury prevention and, at the same time, to improve sports performance [18].

It is important to note a few limitations within this study: (a) The possibility that the results cannot be directly considered and applied to other, age categories, gender or level football players, because in this study participated male high-level male senior and professional football players, with experiences and training routines that do not characterize the vast majority of football players. (b) The study was conducted during the preseason (August); consequently, the protocols and results should not be considered and applicable in other moments such during high intensity training load (for example in training camps without competition events and double or triple training sessions) or weeks with several competitive events (weekends and mid-week).

Future studies should consider intervention strategies to effectively focus the interlimb asymmetry concern. It is also advisable to increase the sample size and expand data collection in more teams of the same football league with different training methodologies or even other-level teams of different leagues. We also suggest the replication of this study with other age-categories, gender, and in different events of the season or even longitudinally between seasons, including the possibility of injury occurrence analysis.

5. Conclusions

The findings of the present study emphasize that high-level male senior and professional football players are characterized by fairly good mean values of lower limb symmetry. However, it should be noted that an in-depth and accurate analysis of jumping performance in preseason screening tests should be considered aiming to detect injury risk and specifically evaluating different jumping test methodologies and determining jumping associated performance variables for each test, namely EUR, SSC, BLD, and LSI.

The correlations between most of the strength tests and jumping-associated performance variables underline the relationship between these, which should be considered for physical evaluation, in a longitudinal tracking performance during football season, and in a pluriannual perspective, and injury prevention and performance enhancement in football programs. From a training perspective, it should be noted that bilateral training should in no way be ignored, although unilateral strength exercises will aid in the development of unilateral movement competency.

The results in this study highlight that screening for muscle strength and asymmetry could be of particular importance for football injury prevention, and sports institutions should pay special attention regarding potential health problems in athletes exposed to daily high training loads. Specific muscle-strengthening exercises could be implemented based on this study's results and outcomes, aiming to reduce injury risks and lower extremity asymmetries and to enhance individual football performance in high-level male senior and professional football players.

Author Contributions: Conceptualization, M.C.E., R.A., A.E., and F.J.S.; methodology, M.C.E., R.A., A.E., and F.J.S.; formal analysis, M.C.E., R.A., A.E., C.C.F., D.M.P.F., C.E.L.V., J.M.G., and F.J.S.; investigation, M.C.E., R.A., A.E., and F.J.S.; supervision, M.C.E., R.A., A.E., and F.J.S.; data curation, M.C.E., C.C.F., D.M.P.F., C.E.L.V., J.M.G., and F.J.S.; writing—original draft preparation, M.C.E., C.C.F., D.M.P.F., J.M.G., and F.J.S.; writing—review and editing, M.C.E., M.J., C.C.F., D.M.P.F., C.E.L.V., J.M.G., and F.J.S.; visualization, M.C.E., M.J., R.A., A.E., C.C.F., D.M.P.F., C.E.L.V., J.M.G., and F.J.S.; funding acquisition, M.C.E., M.J., D.M.P.F., C.E.L.V., J.M.G., and F.J.S. All authors have read and agreed to the published version of the manuscript.

Funding: This research was funded by Foundation for Science and Technology, I.P., Grant/Award Number UIDB/04748/2020 and Instituto Politécnico de Setúbal. It was also partially subsidised by the Aid for Research Groups (GR21149) from the Regional Government of Extremadura (Department of Economy, Science and Digital Agenda), with a contribution from the European Union from the European Funds for Regional Development. Furthermore, the author José M. Gamonales is a beneficiary of a grant from the Spanish University System Upgrading Programme, Field of Knowledge: Biomedical (Grant Ref.: MS-18).

Institutional Review Board Statement: The study considered the Declaration of Helsinki guidelines and was submitted to the Ethical Committee of the Polytechnic Institute of Leiria (CE/IPLEIRIA/22/2021).

Informed Consent Statement: Informed consent was obtained from all participants in the study.

Data Availability Statement: The data that support the findings of this study are available from the corresponding and first authors (mario.espada@ese.ips.pt and fernando.santos@ese.ips.pt) upon reasonable request.

Conflicts of Interest: The authors declare no conflict of interest.

References

1. Ibrahim, A.; Murrell, G.A.; Knapman, P. Adductor Strain and Hip Range of Movement in Male Professional Soccer Players. *J. Orthop. Surg.* **2007**, *15*, 46–49. [CrossRef] [PubMed]
2. Kalinowski, P.; Bojkowski, Ł.; Śliwowski, R. Motor and Psychological Predispositions for Playing Football. *TRENDS Sport Sci.* **2019**, *2*, 51–56. [CrossRef]
3. Wong, P.; Hong, Y. Soccer Injury in the Lower Extremities. *Br. J. Sports Med.* **2005**, *39*, 473–482. [CrossRef] [PubMed]
4. Herrero, H.; Salinero, J.J.; Del Coso, J. Injuries Among Spanish Male Amateur Soccer Players: A Retrospective Population Study. *Am. J. Sports Med.* **2014**, *42*, 78–85. [CrossRef]
5. Ekstrand, J.; Hagglund, M.; Walden, M. Epidemiology of Muscle Injuries in Professional Football (soccer). *Am. J. Sports Med.* **2011**, *39*, 1226–1232. [CrossRef] [PubMed]
6. Hägglund, M.; Waldén, M.; Ekstrand, J. Risk Factors for Lower Extremity Muscle Injury in Professional Soccer: The UEFA Injury Study. *Am. J. Sports Med.* **2013**, *41*, 327–335. [CrossRef] [PubMed]
7. Hodges, S.J.; Patrick, R.J.; Reiser, R.F. Effects of Fatigue on Bilateral Ground Reaction Force Asymmetries During the Squat Exercise. *J. Strength Condition. Res.* **2011**, *25*, 3107–3117. [CrossRef]
8. Hart, N.H.; Nimphius, S.; Weber, J.; Spiteri, T.; Rantalainen, T.; Dobbin, M.; Newton, R.U. Musculoskeletal Asymmetry in Football Athletes: A Product of Limb Function Over Time. *Med. Sci. Sports Exerc.* **2016**, *48*, 1379–1387. [CrossRef]
9. Rodriguez-Rosell, D.; Mora-Custodio, R.; Franco-Márquez, F.; Yáñez-García, J.M.; González-Badillo, J.J. Traditional vs. Sport-Specific Vertical Jump Tests: Reliability, Validity and Relationship with the Legs Strength and Sprint Performance in Adult and Teen Soccer and Basketball Players. *J. Strength Cond. Res.* **2017**, *31*, 196–206. [CrossRef]
10. Fousekis, K.; Tsepis, E.; Vagenas, G. Multivariate Isokinetic Strength Asymmetries of the Knee and Ankle in Professional Soccer Players. *J. Sports Med. Phys. Fitness.* **2010**, *50*, 465–474.
11. Daneshjoo, A.; Rahnama, N.; Mokhtar, A.H.; Yusof, A. Bilateral and Unilateral Asymmetries of Isokinetic Strength and Flexibility in Male Young Professional Soccer Players. *J. Human Kinetics.* **2013**, *36*, 45–53. [CrossRef] [PubMed]
12. Bayliss, A.J.; Weatherholt, A.M.; Crandall, T.T.; Farmer, D.L.; McConnell, J.C.; Crossley, K.M.; Warden, S.J. Achilles Tendon Material Properties are Greater in the Jump Leg of Jumping Athletes. *J. Musculoskelet. Neuronal. Interact.* **2016**, *16*, 105–112.
13. Maloney, S.J. The Relationship Between Asymmetry and Athletic Performance: A Critical Review. *J. Strength Cond. Res.* **2019**, *33*, 2579–2593. [CrossRef] [PubMed]
14. Read, P.J.; McAuliffe, S.; Bishop, C.; Oliver, J.L.; Graham-Smith, P.; Farooq, M.A. Asymmetry Thresholds for Common Screening Tests and Their Effects on Jump Performance in Professional Soccer Players. *J. Athl. Train.* **2021**, *56*, 46–53. [CrossRef]
15. Bishop, C.; McAuley, W.; Read, P.; Gonzalo-Skok, O.; Lake, J.; Turner, A. Acute Effect of Repeated Sprints on Interlimb Asymmetries During Unilateral Jumping. *J. Strength Cond. Res.* **2021**, *35*, 2127–2132. [CrossRef]
16. Fousekis, K.; Tsepis, E.; Poulmedis, P.; Athanasopoulos, S.; Vagenas, G. Intrinsic Risk Factors of Non-contact Quadriceps and Hamstring Strains in Soccer: A Prospective Study of 100 Professional Players. *Br. J. Sports Med.* **2011**, *45*, 709–714. [CrossRef]
17. Kyritsis, P.; Bahr, R.; Landreau, P.; Miladi, R.; Witvrouw, E. Likelihood of ACL Graft Rupture: Not Meeting Six Clinical Discharge Criteria Before Return to Sport is Associated with a Four Times Greater Risk of Rupture. *Br. J. Sports Med.* **2016**, *50*, 946–951. [CrossRef] [PubMed]
18. Santos, F.J.; Valido, A.J.; Malcata, I.S.; Ferreira, C.C.; Pessôa Filho, D.M.; Verardi, C.E.L.; Espada, M.C. The Relationship between Preseason Common Screening Tests to Identify Inter-Limb Asymmetries in High-Level Senior and Professional Soccer Players. *Symmetry* **2021**, *13*, 1805. [CrossRef]
19. Cuthbert, M.; Comfort, P.; Ripley, N.; McMahon, J.J.; Evans, M.; Bishop, C. Unilateral vs. Bilateral Hamstring Strength Assessments: Comparing Reliability and Inter-limb Asymmetries in Female Soccer Players. *J. Sports Sci.* **2021**, *39*, 1481–1488. [CrossRef]
20. Morin, J.-B.; Gimenez, P.; Edouard, P.; Arnal, P.; Jiménez-Reyes, P.; Samozino, P.; Brughelli, M.; Mendiguchia, J. Sprint Acceleration Mechanics: The Major Role of Hamstrings in Horizontal Force Production. *Front. Physiol.* **2015**, *6*, 404. [CrossRef]
21. Chaabene, H.; Prieske, O.; Negra, Y.; Granacher, U. Change of Direction Speed: Toward a Strength Training Approach with Accentuated Eccentric Muscle Actions. *Sports Med.* **2018**, *48*, 1773–1779. [CrossRef] [PubMed]
22. Impellizzeri, F.M.; Rampinini, E.; Maffiuletti, N.; Marcora, S.M. A Vertical Jump Force Test for Assessing Bilateral Strength Asymmetry in Athletes. *Med. Sci. Sports Exerc.* **2007**, *39*, 2044–2050. [CrossRef]

23. Petrigna, L.; Karsten, B.; Marcolin, G.; Paoli, A.; D'Antona, G.; Palma, A.; Bianco, A. A Review of Countermovement and Squat Jump Testing Methods in the Context of Public Health Examination in Adolescence: Reliability and Feasibility of Current Testing Procedures. *Front. Physiol.* **2019**, *10*, 1384. [CrossRef] [PubMed]
24. Van Hooren, B.; Bosch, F. Influence of Muscle Slack on High-intensity Sport Performance: A Review. *Strength Cond. J.* **2016**, *38*, 75–87. [CrossRef]
25. Matic, M.S.; Pazin, N.R.; Mrdakovic, V.D.; Jankovic, N.N.; Ilic, D.B.; Stefanovic, D.L. Optimum Drop Height for Maximizing Power Output in Drop Jump: The Effect of Maximal Muscle Strength. *J. Strength Cond. Res.* **2015**, *29*, 3300–3310. [CrossRef] [PubMed]
26. McGuigan, M.R.; Doyle, T.L.; Newton, M.; Edwards, D.J.; Nimphius, S.; Newton, R.U. Eccentric Utilization Ratio: Effect of Sport and Phase of Training. *J. Strength Cond. Res.* **2006**, *20*, 992–995. [CrossRef]
27. Nicol, C.; Avela, J.; Komi, P.V. The Stretch-shortening Cycle: A Model to Study Naturally Occurring Neuromuscular Fatigue. *Sports Med.* **2006**, *36*, 977–999. [CrossRef]
28. Komi, P.V.; Gollhofer, A. Stretch Reflexes can have an Important Role in Force Enhancement During SSC Exercise. *J. Appl. Biomech.* **1997**, *13*, 451–460. [CrossRef]
29. Komi, P.V. Stretch-shortening Cycle: A Powerful Model to Study Normal and Fatigued Muscle. *J. Biomech.* **2000**, *33*, 1197–1206. [CrossRef]
30. Ascenzi, G.; Ruscello, B.; Filetti, C.; Bonanno, D.; Di Salvo, V.; Nuñez, F.J.; Mendez-Villanueva, A.; Suarez-Arrones, L. Bilateral Deficit and Bilateral Performance: Relationship with Sprinting and Change of Direction in Elite Youth Soccer Players. *Sports* **2020**, *8*, 82. [CrossRef]
31. Pleša, J.; Kozinc, Ž.; Šarabon, N. Bilateral Deficit in Countermovement Jump and Its Influence on Linear Sprinting, Jumping, and Change of Direction Ability in Volleyball Players. *Front. Physiol.* **2022**, *13*, 768906. [CrossRef] [PubMed]
32. Stølen, T.; Chamari, K.; Castagna, C.; Wisløff, U. Physiology of Soccer: An Update. *Sports Med.* **2005**, *35*, 501–536. [CrossRef] [PubMed]
33. Grindem, H.; Snyder-Mackler, L.; Moksnes, H.; Engebretsen, L.; Risberg, M.A. Simple Decision Rules can Reduce Reinjury Risk by 84% After ACL Reconstruction: The Delaware-Oslo ACL Cohort Study. *Br. J. Sports Med.* **2016**, *50*, 804–808. [CrossRef]
34. Krych, A.J.; Woodcock, J.A.; Morgan, J.A.; Levy, B.A.; Stuart, M.J.; Dahm, D.L. Factors Associated with Excellent 6-month Functional and Isokinetic Test Results Following ACL Reconstruction. *Knee Surg. Sport Traumatol. Arthrosc.* **2015**, *23*, 1053–1059. [CrossRef] [PubMed]
35. Nagai, T.; Schilaty, N.D.; Laskowski, E.R.; Hewett, T.E. Hop Tests can Result in Higher Limb Symmetry Index Values than Isokinetic Strength and Leg Press Tests in Patients Following ACL Reconstruction. *Knee Surg. Sports Traumatol. Arthrosc.* **2020**, *28*, 816–822. [CrossRef]
36. Harriss, D.; MacSween, A.; Atkinson, G. Ethical Standards in Sport and Exercise Science Research: 2020 Update. *Int. J. Sports Med.* **2019**, *40*, 813–817. [CrossRef]
37. Dos Santos, F.J.L.; Louro, H.G.; Espada, M.; Figueiredo, T.; Lopes, H.; Rodrigues, J. Relation of Coaches' Expectations with Instruction and Behavior of Athletes. *Cuad. De Psicol. Del Deporte* **2019**, *19*, 62–78. [CrossRef]
38. Hewett, T.E.; Myer, G.D.; Ford, K.R.; Heidt, R.S., Jr.; Colosimo, A.J.; McLean, S.G.; van den Bogert, A.J.; Paterno, M.V.; Succop, P. Biomechanical Measures of Neuromuscular Control and Valgus Loading of the Knee Predict Anterior Cruciate Ligament Injury Risk in Female Athletes: A Prospective Study. *Am. J. Sport Med.* **2005**, *33*, 492–501. [CrossRef]
39. Bishop, C.; Read, P.; Lake, J.; Chavda, S.; Turner, A. Interlimb Asymmetries: Understanding How to Calculate Differences from Bilateral and Unilateral Tests. *Strength Cond. J.* **2018**, *40*, 1–6. [CrossRef]
40. Howard, J.D.; Enoka, R.M. Maximum Bilateral Contractions are Modified by Neurally Mediated Interlimb Effects. *J. Appl. Physiol.* **1991**, *70*, 306–316. [CrossRef]
41. Komi, P.V.; Bosco, C. Utilization of Stored Elastic Energy in Leg Extensor Muscles by Men and Women. *Med. Sci. Sports.* **1978**, *10*, 261–265.
42. Koo, T.K.; Li, M.Y. A Guideline of Selecting and Reporting Intraclass Correlation Coefficients for Reliability Research. *J. Chiropr. Med.* **2016**, *15*, 155–163. [CrossRef]
43. Hopkins, W.; Marshall, S.; Batterham, A.; Hanin, J. Progressive Statistics for Studies in Sports Medicine and Exercise Science. *Med. Sci. Sports Exerc.* **2009**, *41*, 3–13. [CrossRef] [PubMed]
44. Mukaka, M.M. A Guide to Appropriate use of Correlation Coefficient in Medical Research. *Malawi Med. J.* **2012**, *24*, 69–71.
45. Read, P.J.; Oliver, J.L.; De Ste Croix, M.B.A.; Myer, G.D.; Lloyd, R.S. A Prospective Investigation to Evaluate Risk Factors for Lower Extremity Injury Risk in Male Youth Soccer Players. *Scand. J. Med. Sci. Sport.* **2018**, *28*, 1244–1251. [CrossRef]
46. Hewit, J.; Cronin, J.; Hume, P. Multidirectional Leg Asymmetry Assessment in Sport. *Strength Cond. J.* **2012**, *34*, 82–86. [CrossRef]
47. Lockie, R.G.; Callaghan, S.J.; Berry, S.P.; Cooke, E.R.; Jordan, C.A.; Luczo, T.M.; Jeffriess, M.D. Relationship Between Unilateral Jumping Ability and Asymmetry on Multidirectional Speed in Team-sport Athletes. *J. Strength Cond. Res.* **2014**, *28*, 3557–3566. [CrossRef] [PubMed]
48. Bishop, C.; Read, P.; McCubbine, J.; Turner, A. Vertical and Horizontal Asymmetries Are Related to Slower Sprinting and Jump Performance in Elite Youth Female Soccer Players. *J. Strength Cond. Res.* **2021**, *35*, 56–63. [CrossRef]

49. Madruga-Parera, M.; Bishop, C.; Read, P.; Lake, J.; Brazier, J.; Romero-Rodriguez, D. Jumping-based Asymmetries are Negatively Associated with Jump, Change of Direction, and Repeated Sprint Performance, but not Linear Speed, in Adolescent Handball Athletes. *J. Hum. Kinet.* **2020**, *71*, 47–58. [CrossRef] [PubMed]

50. Fukashiro, S.; Besier, T.F.; Barrett, R.; Cochrane, J.; Nagano, A.; Lloyd, D.G. Direction Control in Standing Horizontal and Vertical Jumps. *Int. J. Sport. Health Sci.* **2005**, *3*, 272–279. [CrossRef]

51. Falch, H.N.; Rædergård, H.G.; Van Den Tillaar, R. Relationship of Performance Measures and Muscle Activity between a 180_ Change of Direction Task and Different Countermovement Jumps. *Sports.* **2020**, *8*, 47. [CrossRef]

52. van Schenau, G.J.v.; Bobbert, M.F.; de Haan, A. Does Elastic Energy Enhance Work and Efficiency in the Stretch-shortening Cycle? *J. Appl. Biomech.* **1997**, *13*, 389–415. [CrossRef]

53. Tomalka, A.; Weidner, S.; Hahn, D.; Seiberl, W.; Siebert, T. Cross-bridges and Sarcomeric Non-cross-bridge Structures Contribute to Increased Work in Stretch-shortening Cycles. *Front. Physiol.* **2020**, *11*, 921. [CrossRef] [PubMed]

54. Bobbert, M.F.; Casius, L.J.R. Is the Effect of a Countermovement on Jump Height Due to Active State Development? *Med. Sci. Sports Exerc.* **2005**, *37*, 440–446. [CrossRef] [PubMed]

55. Dietz, V.; Schmidtbleicher, D.; Noth, J. Neuronal Mechanisms of Human Locomotion. *J. Neurophysiol.* **1979**, *42*, 1212–1222. [CrossRef]

56. Kawakami, Y.; Muraoka, T.; Ito, S.; Kanehisa, H.; Fukunaga, T. In Vivo Muscle Fibre Behaviour During Counter-movement Exercise in Humans Reveals a Significant Role for Tendon Elasticity. *J. Physiol.* **2002**, *540*, 635–646. [CrossRef] [PubMed]

57. Finni, T.; Ikegawa, S.; Komi, P.V. Concentric Force Enhancement During Human Movement. *Acta Physiol. Scand.* **2001**, *173*, 369–377. [CrossRef]

58. Flanagan, E.P.; Comyns, T.M. The Use of Contact Time and the Reactive Strength Index to Optimize Fast Stretch-shortening Cycle Training. *Strength Cond. J.* **2008**, *30*, 32–38. [CrossRef]

59. Mclellan, C.P.; Lovell, D.I.; Gass, G.C. The Role of Rate of Force Development on Vertical Jump Performance. *J. Strength Cond. Res.* **2011**, *25*, 379–385. [CrossRef]

60. Kubo, K.; Kawakami, Y.; Fukunaga, T. Influence of Elastic Properties of Tendon Structures on Jump Performance in Humans. *J. Appl. Physiol.* **1999**, *87*, 2090–2096. [CrossRef]

61. Grospretre, S.; Lepers, R. Performance Characteristics of Parkour Practitioners: Who are the Traceurs? *Eur. J. Sport Sci.* **2016**, *16*, 526–535. [CrossRef]

62. Kopper, B.; Csende, Z.; Trzaskoma, L.; Tihanyi, J. Stretch-shortening Cycle Characteristics During Vertical Jumps Carried Out with Small and Large Range of Motion. *J. Electromyogr. Kinesiol.* **2014**, *24*, 233–239. [CrossRef]

63. Janzen, C.; Chilibeck, P.; Davison, K. The Effect of Unilateral and Bilateral Strength Training on the Bilateral Deficit and Lean Tissue Mass in Post-menopausal Women. *Eur. J. Appl. Physiol.* **2006**, *97*, 253–260. [CrossRef]

64. Bračič, M.; Supej, M.; Peharec, S.; Bačić, P.; Čoh, M. An Investigation of the Influence of Bilateral Deficit on the Counter-movement Jump Performance in Elite Sprinters. *Kinesiol.* **2010**, *42*, 73–81.

65. van Soest, A.J.; Roebroeck, M.E.; Bobbert, M.F.; Huijing, P.A.; van Ingen Schenau, G.J. Comparison of One-legged and Two-legged Countermovement Jumps. *Med. Sci. Sports Exerc.* **1985**, *17*, 635–639. [CrossRef] [PubMed]

66. Bishop, C.; Berney, J.; Lake, J.; Loturco, I.; Blagrove, R.; Turner, A.; Read, P. Bilateral Deficit During Jumping Tasks: Relationship with Speed and Change of Direction Speed Performance. *J. Strength Cond. Res.* **2021**, *35*, 1833–1840. [CrossRef] [PubMed]

67. Pain, M. Considerations for Single and Double Leg Drop Jumps: Bilateral Deficit, Standardizing Drop Height, and Equalizing Training Load. *J. Appl. Biomech.* **2014**, *30*, 722–727. [CrossRef]

68. Bailey, C. Force Production Symmetry During Static, Isometric, and Dynamic Tasks. 2014. Available online: http://dc.etsu.edu/etd/2388/ (accessed on 5 December 2022).

69. Bailey, C.A.; Sato, K.; Burnett, A.; Stone, M.H. Force-production Asymmetry in Male and Female Athletes of Differing Strength Levels. *Int. J. Sports Physiol. Perform.* **2015**, *10*, 504–508. [CrossRef]

70. Kotsifaki, A.; Korakakis, V.; Whiteley, R.; Van Rossom, S.; Jonkers, I. Measuring only Hop Distance During Single Leg Hop Testing is Insufficient to Detect Deficits in Knee Function after ACL Reconstruction: A Systematic Review and Meta-analysis. *Br. J. Sports Med.* **2020**, *54*, 139–153. [CrossRef]

71. Wang, H.; Fleischli, J.E.; Nigel Zheng, N. Effect of Lower Limb Dominance on Knee Joint Kinematics after Anterior Cruciate Ligament Reconstruction. *Clin. Biomech.* **2012**, *27*, 170–175. [CrossRef]

72. Holcomb, W.R.; Rubley, M.D.; Lee, H.J.; Guadagnoli, M.A. Effect of Hamstring-emphasized Resistance Training on Hamstring: Quadriceps Strength Ratios. *J. Strength Cond. Res.* **2007**, *21*, 41–47. [CrossRef] [PubMed]

73. Beato, M.; Young, D.; Stiff, A.; Coratella, G. Lower-Limb Muscle Strength, Anterior-Posterior and Inter-Limb Asymmetry in Professional, Elite Academy and Amateur Soccer Players. *J. Hum. Kinet.* **2021**, *77*, 135–146. [CrossRef] [PubMed]

74. Raya-González, J.; Castillo, D.; de Keijzer, K.L.; Beato, M. The Effect of a Weekly Flywheel Resistance Training Session on Elite U-16 Soccer Players' Physical Performance During the Competitive Season. A Randomized Controlled Trial. *Res. Sports Med.* **2021**, *29*, 571–585. [CrossRef] [PubMed]

 *healthcare*

Systematic Review

How to Improve the Reactive Strength Index among Male Athletes? A Systematic Review with Meta-Analysis

André Rebelo [1,2,3], João R. Pereira [1,2,3,4,*], Diogo V. Martinho [1,2,4,5,6], João P. Duarte [1,2,4,7], Manuel J. Coelho-e-Silva [4] and João Valente-dos-Santos [1,2,3,4]

1   LX Applied Sport Sciences Research Group, 1000-289 Lisbon, Portugal; andre94rebelo@hotmail.com (A.R.); dvmartinho92@hotmail.com (D.V.M.); joaopedromarquesduarte@gmail.com (J.P.D.); j.valente-dos-santos@hotmail.com (J.V.-d.-S.)
2   CIDEFES, Centro de Investigação em Desporto, Educação Física e Exercício e Saúde, Universidade Lusófona, 1749-024 Lisboa, Portugal
3   COD, Center of Sports Optimization, Sporting Clube de Portugal, 1600-464 Lisbon, Portugal
4   Research Unity in Sport and Physical Activity (CIDAF, UID/DTP/04213/2020), Faculty of Sport Sciences and Physical Education, University of Coimbra, 3040-248 Coimbra, Portugal; mjcesilva@hotmail.com
5   Polytechnic of Coimbra, Coimbra Health School, Dietetics and Nutrition, 3045-093 Coimbra, Portugal
6   Laboratory for Applied Health Research (LabinSaúde), 3045-093 Coimbra, Portugal
7   Porto Biomechanics Laboratory (LABIOMEP-UP), University of Porto, 4200-450 Porto, Portugal
*   Correspondence: p5916@ulusofona.pt

Citation: Rebelo, A.; Pereira, J.R.; Martinho, D.V.; Duarte, J.P.; Coelho-e-Silva, M.J.; Valente-dos-Santos, J. How to Improve the Reactive Strength Index among Male Athletes? A Systematic Review with Meta-Analysis. *Healthcare* **2022**, *10*, 593. https://doi.org/10.3390/healthcare10040593

Academic Editors: João Paulo Brito and Rafael Oliveira

Received: 1 March 2022
Accepted: 14 March 2022
Published: 22 March 2022

Publisher's Note: MDPI stays neutral with regard to jurisdictional claims in published maps and institutional affiliations.

Abstract: The reactive strength index (RSI) describes the individual's capability to quickly change from an eccentric muscular contraction to a concentric one and can be used to monitor, assess, and reduce the risk of athlete's injury. The purpose of this review is to compare the effectiveness of different training programs on RSI. Electronic searches were conducted in MEDLINE, PubMed, Scopus, SPORTDiscus, and Web of Science from database inception to 11 February 2022. This meta-analysis was conducted in accordance with the recommendations of the preferred reporting items for systematic reviews and meta-analyses (PRISMA). The search returned 5890 records, in which 39 studies were included in the systematic review and 30 studies were included in the meta-analysis. Results from the randomized studies with the control group revealed that plyometric training improved RSI in adult athletes (0.84, 95% CI 0.37 to 1.32) and youth athletes (0.30, 95% CI 0.13 to 0.47). Evidence withdrawn from randomized studies without a control group revealed that resistance training also improved the RSI (0.44, 95% CI 0.08 to 0.79) in youth athletes but not in adults. Interventions with plyometric training routines have a relatively large, statistically significant overall effect in both adult and youth athletes. This supports the implementation of this type of interventions in early ages to better cope with the physical demands of the various sports. The impact of resistance training is very low in adult athletes, as these should seek to have a more power-type training to see improvements on the RSI. More interventions with sprint and combined training are needed.

Keywords: strength; power; reactive strength; players; plyometric training; resistance training

1. Introduction

The reactive strength index (RSI) describes the individual's capability to quickly change from an eccentric muscular contraction to a concentric one [1]. In other words, the RSI was created to assess the athlete's reactive strength, and it was originally measured with the drop jump (DJ) test [1]. For this test, the athletes must perform a vertical jump as soon as they land on the ground from a specific height [1]. The hands can stay on the athletes' hips throughout the test or not, as both methods have shown good levels of reliability [2]. This test should incorporate various drop heights to assess at what height the athlete can elevate more his centre of gravity, and it was already proven to be a reliable and valid test to measure the RSI [3]. The RSI can be calculated by dividing the jump height by the ground contact time, providing valuable information for coaches, regarding plyometric performance (i.e., jump height) and how each jump is performed (i.e., ground contact time) [4]. Jump height can be measured directly or can be derived from flight time with the following mathematical formula [5]: jump height (m) = (gravity × (flight time)^2)/8, where gravity = 9.81 m/s and flight time is in seconds.

Most recently, with advancements in technology, more tests have been developed to measure the RSI, such as the countermovement (CMJ), tuck jump, squat jump, weighted CMJ, single-leg jump [6], 10/5 [7], single rebound jump [8], vertical rebound for 5 repetitions [9], vertical rebound for 15 repetitions [10], and vertical rebound for 10 s tests [11]. In those cases where there is no drop or rebound jump, the RSI is designated explicitly by reactive strength index modified (RSI_{mod}), since it is calculated by dividing the jump height by the time to take-off (time to produce force from the beginning of the eccentric muscular phase until the moment the athlete leaves the ground) [6].

To obtain these variables mentioned before (i.e., jump height, flight time, ground contact time, and time to take-off), three different methods can be used: (a) the flight time [12]; (b) the difference between the height of two marks during the jump [13]; and (c) the mathematical integration of the ground reaction force [14]. The first one requires the use of contact mats [12,15,16], photocell mats [13,15], or accelerometers [16,17]. The second method uses different devices to calculate displacement (i.e., linear position transducers) [18]. The third method is considered the best one, as its accuracy is extremely high if adequate sampling frequency methods are chosen and requires the use of one or two force plates [13,19].

The use of RSI is vital for high-performance sports professionals, as it can be used as a motivational tool, in a way that coaches can instantly deliver feedback to their athletes, according to their RSI value, in order to improve their physical performance [5]. Furthermore, both RSI and RSI_{mod} can be used as variables to potentially monitor athlete's neuromuscular readiness [20]. Moreover, the RSI has been shown to have a strong relationship with change of direction speed, acceleration speed [21], and agility [22]. Additionally, maximal strength, especially relative to body mass, appears to have a very strong relationship with RSI_{mod}, indicating that stronger athletes tend to have better reactive strength [23].

However, there is no clear information on which type of training would produce better improvements on RSI. Besides that, to the best of our knowledge, the scientific literature does not review this topic. Therefore, the aim of this study was to analyse the strategies that can improve the RSI of male athletes, through a systematic review of experimental research and meta-analysis.

2. Methods

2.1. Protocol and Registration

This meta-analysis was conducted in accordance with the recommendations of the preferred reporting items for systematic reviews and meta-analyses (PRISMA) [24]. The study protocol was registered with PROSPERO (CRD42020176616).

2.2. Information Sources

The literature search on five electronic databases (i.e., MEDLINE, PubMed, Scopus, SPORTDiscus, and Web of Science) started on 5 July 2020 and was conducted from database inception to 11 February 2022.

2.3. Search Strategy

All retrieved papers were exported to CADIMA software, a tool designed to increase the efficiency of the evidence synthesis process and facilitate reporting of all activities to maximize methodological rigor [25]. Duplicates were automatically removed. Titles and abstracts of potentially relevant papers were screened by two reviewers (A.R. and J.R.P.). Disagreements between authors were solved through discussion and, when necessary, three other authors (D.M., J.P.D., and J.V.-d.-S.) were involved. Full-text copies were acquired for all papers that met title and abstract screening criteria. Full-text screening was performed by two reviewers (A.R. and J.R.P.). Again, any discrepancies were discussed, until the authors reached an agreement and consulted the three other authors, when required.

The comprehensive search strategy is available in the Supplementary File (Supplemental Material S1).

2.4. Inclusion Criteria

Scientific peer-reviewed published papers written in English, Portuguese, French, and Spanish were eligible for the present systematic review. The review sought to identify all studies reporting exercise interventions to improve the RSI in both male adult and youth athletes. Therefore, studies were eligible if: (1) subjects were male athletes; (2) subjects had between 11 and 45 years old; (3) the study included at least two moments of evaluation, with a baseline RSI measurement and post-intervention RSI measurement; (4) the study included a training program that aimed to improve the RSI.

2.5. Exclusion Criteria

Studies that do not describe a protocol to induce effects on RSI or that used RSI during a recovery program were excluded from the present study.

2.6. Categorisation of Studies

We identified six categories of exercise interventions, through the process of reviewing the included studies. The definitions of these exercise interventions are provided in Table 1.

Table 1. Definition of types of interventions and comparators.

Type	Definition
	Intervention
Plyometrics	Exercises that are designed to enhance neuromuscular performance on the lower limbs. This involves application of jump, hopping, and bounding training.
Resistance training	Training program that aims to improve strength, power, or hypertrophy with resistances (e.g., elastic bands, barbells, dumbbells, kettlebells, or body weight).
Sprint training	Acceleration or maximal velocity training either resisted or unloaded.
Change of direction (COD) or sprint or plyometric or a combination of those	COD: Any exercise that enforces the participant to accelerate, decelerate and do a COD. This type of intervention is defined by a combination of one or more of sprint training, COD training, or plyometric training.
Sports-specific training	Sports-specific exercises training (e.g., small-sided games in soccer).
	Control
Maintained training routines	Sport training routines

2.7. Data Extraction

Pre-established data extraction criteria were created with seven items: (a) general information (authors name and year of the study), (b) sample characteristics (size and age), (c) sport, (d) training program, (e) measurement equipment, (f) methodology, and (g) results.

2.8. Risk of Bias Assessment

The revised Cochrane risk of bias tool for randomized trials (RoB 2.0) scale was used to quantify the risk of bias in eligible, individually randomized, parallel-group trials and provide information on the general methodological quality of studies. The RoB 2.0 scale rates internal study validity and the presence of replicable statistical information on a scale from low risk of bias to high risk of bias [26]. The risk of bias in non-randomized studies of interventions (ROBINS-I) tool was used to quantify the risk of bias in eligible non-randomized trials and provide information on the general methodological quality of the studies. The ROBINS-I scale rates internal study validity and the presence of replicable statistical information on a scale from 1 (low risk of bias) to 4 (high risk of bias) [27]. Inter-rater agreement was calculated using Cohen's kappa coefficient (k). Using different tools to assess the risk of bias on randomized and non-randomized studies was supported elsewhere [28].

2.9. Statistical Models

Meta-analyses were conducted to estimate the overall effects of the intervention programs to improve RSI for all available data, as well as for studies that only included randomized samples. Within the previous categorization, meta-analyses were also divided by studies with (intervention vs. control) and without a control group (pre-intervention vs. post-intervention). Additionally, subgroup meta-analyses were performed to identify the effects of each training program ("plyometrics", "resistance training", "sprint training", "sprint or plyometric or a combination of those", and "sports-specific training") on adults ($\geq$18 years old) and youth (<18 years old) athletes. Additionally, pre- and post-intervention results from all the included studies (randomized and non-randomized) were ranked on two different meta-analyses to understand the effectiveness of each training program on adult and youth athletes.

Meta-analysis, and the respective forest plots, were calculated when the mean, standard deviation, and sample sizes were introduced on the Cochrane collaboration's review manager computer program (RevMan version 5.4.1, Oxford, UK). When this data was impossible to retrieve from the manuscript [29–38], authors were contacted to provide the missing information. Most of the authors replied to the request [29–34]; therefore, the data was included on the meta-analyses. In addition, when the same study reported different RSI measurements (i.e., from different heights and/or tests), the method that resulted in the highest positive performance change was selected for the meta-analysis.

The pooled data for each outcome were reported as standardized mean differences (SMD), with a 95% confidence interval (CI). Each meta-analysis was performed using the random-effects model, and heterogeneity was assessed using I^2 statistic and chi-square (Q) tests. A Q value with a significance of $p \leq 0.05$ was considered significant heterogeneity, while, for the I^2 value, 25% was considered low, 50% was considered moderate, and 75% was considered high heterogeneity [39].

3. Results

3.1. Study Selection

The search strategy returned 5890 records, and the PRISMA flow diagram [40] is shown in Figure 1. Records were excluded based on the included participants (not male athletes), intervention or comparator (not a training program), post intervention data (not reporting RSI measurements), or testing on surfaces other than the floor (i.e., force sledge). In addition, for the quantitative synthesis (meta-analysis), ineligible studies were excluded for reporting RSI data only in figures and/or percentage [35–38] or for being unique, in terms of physical training method [41,42] and, therefore, not being able to pair it with other studies to conduct a meta-analysis. In total, thirty-nine studies were included in the systematic review and thirty-three were included in the meta-analysis.

3.2. Risk of Bias Assessment

Inter-rater agreement for the risk of bias assessment, using the RoB 2.0, was $\kappa = 0.933$; for the ROBINS-I, it was $\kappa = 1.0$. Thus, overall, the risk of bias within individual studies assessed using the RoB 2.0 scale ranged between low and high risk of bias (Supplemental Material S2), whereas the ROBINS-I scale ranged from a serious to critical risk of bias (Supplemental Material S3).

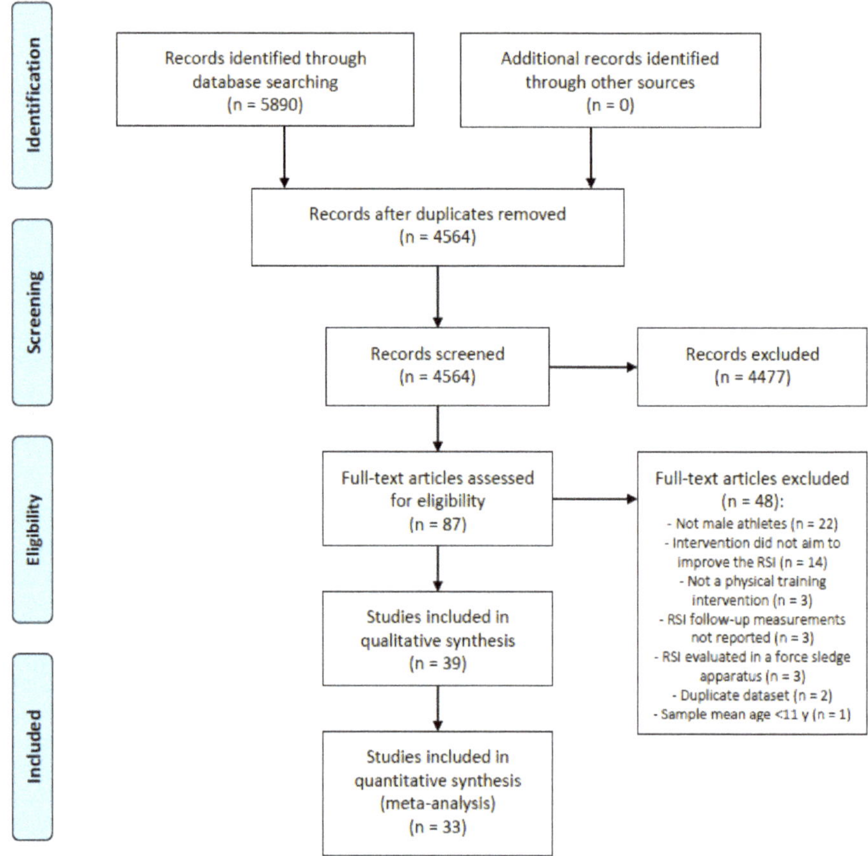

Figure 1. PRISMA statement flow chart. RSI, reactive strength index.

Of the thirty-three randomized studies assessed with the RoB 2.0, twelve (36%) [35,43–53] had a high risk of bias. Randomisation process (18%) and selection of reported results (15%) were the most common sources of high risk of bias.

Of the six non-randomized studies assessed with the ROBINS-I, one (17%) [54] was of critical risk of bias assessment, due to their departures from the intended interventions, participants being excluded because of missing data, and selection of participants based on their characteristics observed after the start of the intervention. The remaining five studies (83%) [36,42,55–57] were of serious risk of bias assessment.

3.3. Study Characteristics

Study characteristics (including training program characteristics) for all 39 included studies are presented in Supplemental Material S4. Overall, the most common tests used to quantify the RSI are DJs (79%) [29–36,38,41–43,45,46,49–54,56–66], vertical hops (21%) [44,47,48,67–71], and CMJs (8%) [37,55,72]. The most popular materials used to quantify the RSI are contact mats (46%) [29–34,36,38,45,46,49,53,61,62,68,69,71,72], force plates (36%) [35,37,41–43,48,50,52,54–56,60,65,66], and photoelectric systems (15%) [44,47,57,58,67,70]. Soccer is the most common sport studied (56%) [29–34,36,38,44–48,50,52,53,61,62,67–70], followed by rugby (18%) [35,43,50,52,58,60,65] and basketball (8%) [50,53,72]. Intervention duration ranged from four weeks [35,42,65] to two years [36].

3.4. Meta-Analysis

3.4.1. Studies with Control Group

Fourteen randomized studies (four with adult athletes [50,51,53,60] and ten with youth athletes [29–34,45,61,62,68]), with the control group, reporting plyometric training programs were included in the meta-analysis (Figure 2a). The overall standardized mean difference in adult athletes was 0.84 (95% CI 0.37 to 1.32), with a significant overall intervention effect. The overall standardized mean difference in youth athletes between groups was lower than adult athletes, at 0.30 (95% CI 0.13 to 0.47), with the intervention effect almost the same. Examination of heterogeneity statistics revealed a non-significant heterogeneity for both adult and youth athletes RSI results (χ^2 = 13.60 (p = 0.06), I^2 = 49% and χ^2 = 23.86 (p = 0.16), I^2 = 25%, respectively).

The meta-analysis included two randomized studies (one with adult athletes [52] and another with youth athletes [68]) reporting a combination of change of direction, plyometric, and/or sprint training programs (Figure 2b). The overall standardized mean difference in adult athletes was −0.01 (95% CI −0.83 to 0.82), with a non-significant overall intervention effect. The overall standardized mean difference in youth athletes between groups was higher than adult athletes, at 0.65 (95% CI −0.02 to 1.33), with a non-significant overall intervention effect.

Regarding the non-randomized studies, in Figure S1c, at Supplemental Material S5, three non-randomized studies with adult athletes [43,54,56] and one non-randomized studies with youth athletes [55] reporting resistance training programs can be seen. The overall standardized mean difference in adult athletes was 0.49 (95% CI 0.02 to 0.96), with a significant overall intervention effect. The overall standardized mean difference in youth athletes between groups was higher than adult athletes, at 0.65 (95% CI 0.15 to 1.16), also with a significant overall intervention effect. Examination of heterogeneity statistics revealed a non-significant heterogeneity for both adult and youth athletes RSI results (χ^2 = 0.49 (p = 0.92), I^2 = 0% and χ^2 = 0.12 (p = 0.94), I^2 = 0%, respectively). Non-randomized studies reporting plyometric interventions or a combination of change of direction, plyometric, and/or sprint training programs were not found. Therefore, no meta-analyses were performed for these interventions with non-randomized studies.

3.4.2. Studies without Control Group

Nine randomized studies (two with adult athletes [49,66] and seven with youth athletes [44,46–48,67,69,70]), without a control group, reporting plyometric training programs were included in the meta-analysis (Figure 3a). The studies with adults reported a significant intervention effect, with a standardized mean difference of 0.94 (95% CI 0.29 to 1.60). The overall standardized mean difference in youth athletes was 0.78 (95% CI 0.31 to 1.24), with a significant overall intervention effect. Examination of heterogeneity statistics revealed a high heterogeneity for youth athletes RSI results (χ^2 = 58.49 (p < 0.00001), I^2 = 76%).

Only one randomized study with adults [49] examined the impact of sprint training methods in RSI (Figure 3b). The study's standardized mean difference was 0.56 (95% CI −0.11 to 1.23) with a non-significant intervention effect. No studies were found for youth athletes.

Two randomized studies with adult athletes [49,72] and three randomized studies with youth athletes [58,65,72] reporting resistance training programs were included in the meta-analysis (Figure 3c). The overall standardized mean difference in adult athletes was 0.19 (95% CI −0.33 to 0.72), with a non-significant overall intervention effect. The overall standardized mean difference in youth athletes between groups was higher than adult athletes, at 0.44 (95% CI 0.12 to 0.75), with a significant overall intervention effect. Examination of heterogeneity statistics revealed a non-significant heterogeneity for both adult and youth athletes RSI results (χ^2 = 0.42 (p = 0.81), I^2 = 0% and χ^2 = 1.54 (p = 0.98), I^2 = 0%, respectively). One additional non-randomized study [57] reporting resistance training was found in youth athletes. However, the mentioned study did not change the significant effect previously calculated only for the randomized studies (Figure S2c at Supplemental Material S5).

(**a**)

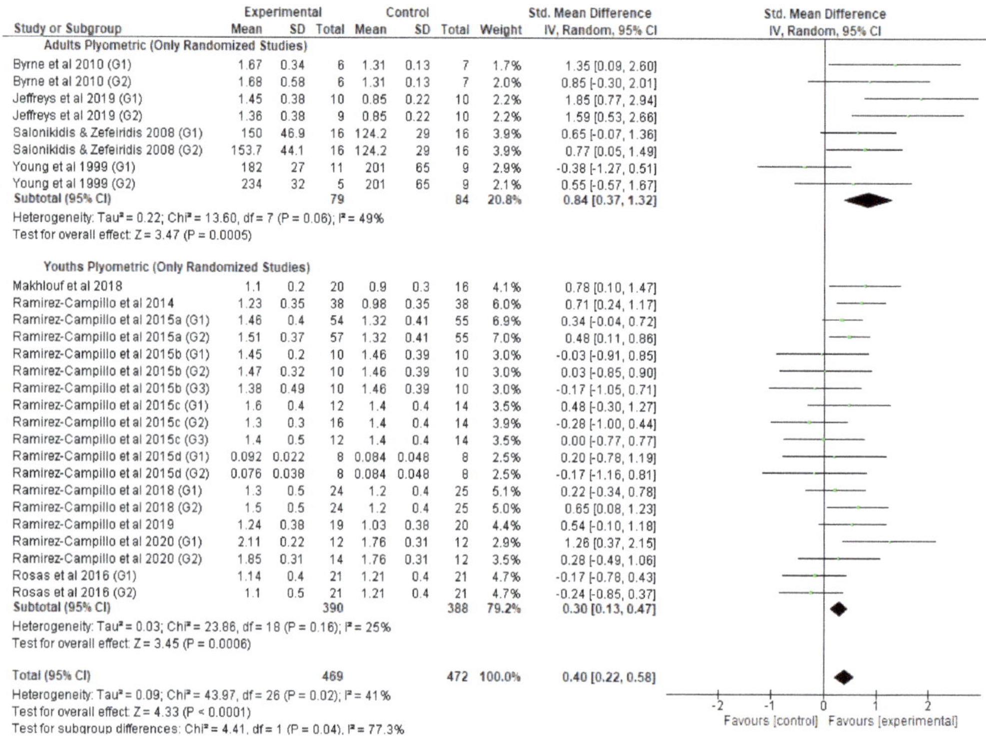

(**b**)

Figure 2. Forest plot of the overall standardized mean difference [95% CI] for each randomized study (the size of the green dot corresponds to the weight of the study within the meta-analysis) with control group with (**a**) a plyometric training intervention, and (**b**) combination of change of direction, sprint, or plyometric training intervention.

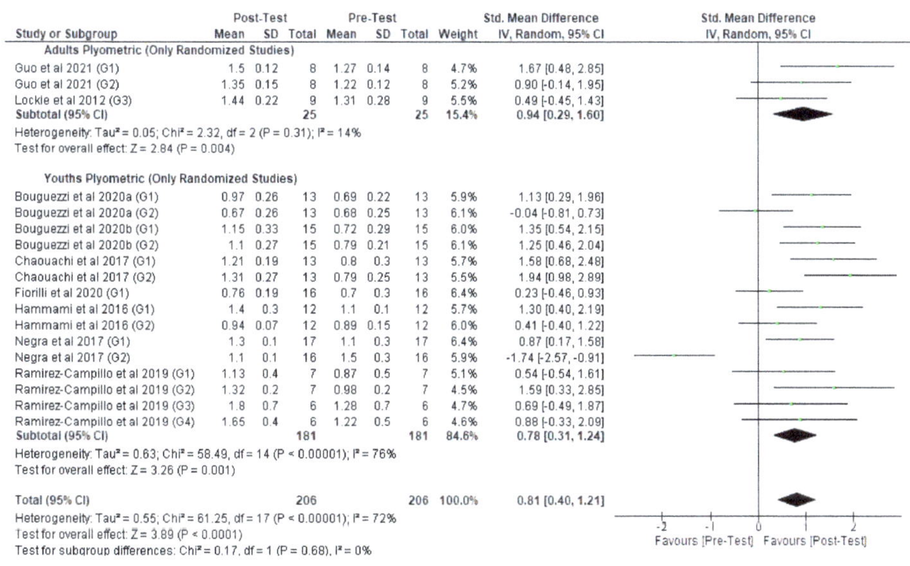

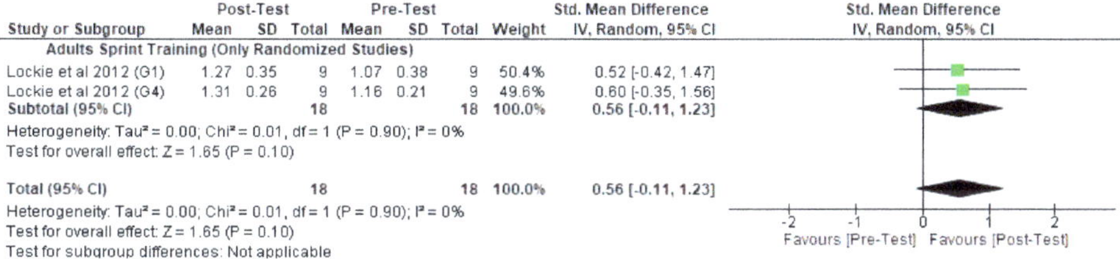

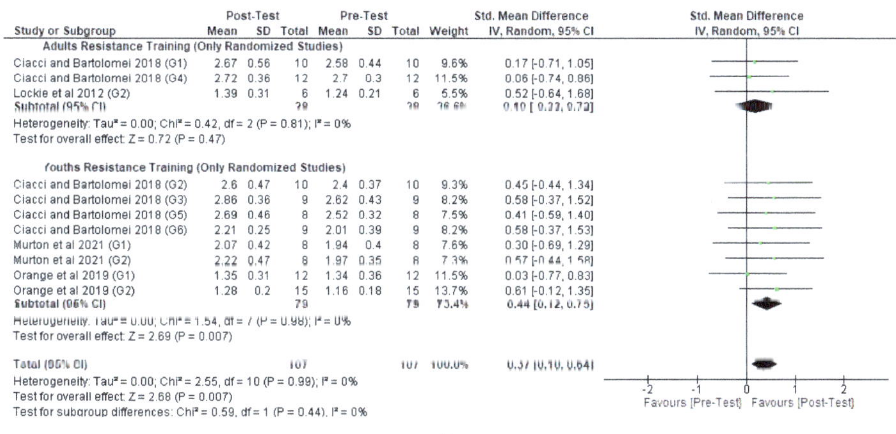

Figure 3. Forest plot of the overall standardized mean difference [95% CI] for each randomized study (the size of the green dot corresponds to the weight of the study within the meta-analysis), without a control group, with a (**a**) plyometric training intervention, (**b**) sprint training intervention, and (**c**) resistance training intervention included in the meta-analysis.

3.4.3. Studies Ranked by Intervention Effects

A ranked forest plot of the intervention effects on adult athletes demonstrates that, compared with pre-test, several interventions demonstrated similar differences: combined training, with an overall standardized mean difference of 0.58 (95% CI −0.06 to 1.21); sprint training, with an overall standardized mean difference of 0.56 (95% CI −0.11 to 1.23); and plyometric training, with an overall standardized mean difference of 0.54 (95% CI 0.26 to 0.82) (Figure 4a). Of those interventions, only plyometrics were considered significant. In youth athletes, combined training had the largest standardized mean difference (1.15, 95% CI 0.48 to 1.83), followed by plyometric training (0.61, 95% CI 0.42 to 0.80) (Figure 4b).

3.5. Qualitative Analysis

As previously reported, six studies were not included in the quantitative analysis. Two [37,38] had plyometric training interventions and, like those who were part of the meta-analysis, it was seen that this training method improves the RSI. Like the results from the meta-analyses on plyometric training, participants of those two studies also improved their RSI. Additionally, another study [41] with a plyometric training method inside water also improved the RSI performance of their participants. With respect to resistance training, one study [35] reported two different resistance training interventions in adult athletes, and the findings suggested that some individuals had performance decrements over the four week training period. However, one study [36] with youth athletes showed RSI improvements with a resistance training intervention. Lastly, one study [42], with adult athletes, had a badminton specific/change of direction intervention, and it was seen that the RSI improved, mainly due to ground contact times enhancements, rather than jumping height. The characteristics of all these mentioned studies can be seen in Supplemental Material S4.

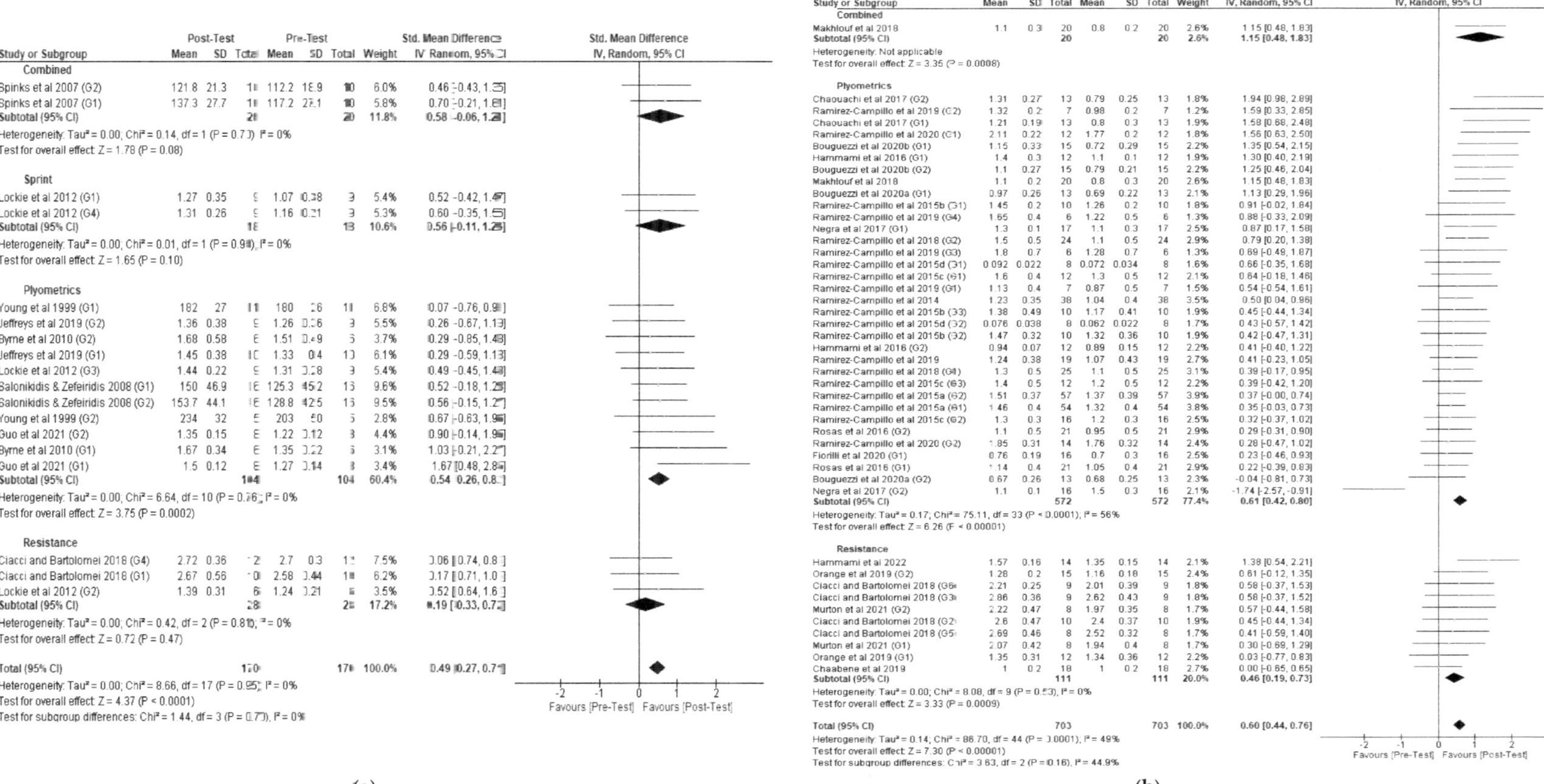

(a)

(b)

Figure 4. Forest plot, ranked by treatment effectiveness (the size of the green dot corresponds to the weight of the study within the meta-analysis), regarding (a) adult and (b) youth athletes.

4. Discussion

This is the first study using meta-analyses to investigate the comparative effectiveness of different exercise interventions to improve the RSI in both male adult and youth athletes. Given the importance that RSI may have in athletes' performance, in particular, acceleration, change of direction, [21], and agility [22], it is vital to investigate effective strategies to improve it. Randomized studies with a control group reported significant overall intervention effects, regarding plyometric and resistance training programs, whereas combined training interventions did not have a significant overall effect. Randomized studies without a control group reported similar tendencies in youth athletes; however, plyometric, sprint, and resistance training programs showed a non-significant overall intervention effect in adult athletes. Compared to randomized studies alone, results from a combination of randomized and non-randomized did not differ. The results ranked by treatment effect indicate that resistance training is inferior, compared to both sprint and plyometric training methods, in enhancing the RSI in adult and youth athletes.

Plyometric training is characterized by a pre-activation (stretch) of the extensors' muscles (e.g., quadriceps during a jump), followed by a shortening phase of these same extensors' muscles, which represents the stretch-shortening cycle (SSC) [73]. The duration of the SSC, usually measured by the ground contact time, can be categorized into slow (>250 milliseconds; CMJ, changes of direction) or fast (<250 milliseconds; DJ, sprints) [74]. Studies have shown that low correlations exist between these two types of SSC [75–77], and reactive strength training is commonly referred as "plyometrics".

The effectiveness of plyometric training methods to improve jumping height ability have been shown [75]; since the RSI can be enhanced by improving this variable (by the formula: RSI = jump height/ground contact time), it is not surprising that plyometrics were successful in increasing the RSI in both adult and youth athletes. Nevertheless, one of the plyometric studies [70] reported that the intervention group declined in the RSI after the training intervention. However, it should be noted that this intervention group performed the plyometric training on unstable surfaces, which may be done with longer ground contact times, resulting in worse RSI values.

According to a motor learning perspective, exercises performed in a similar way to the target task produce an enhanced performance, since they generate a greater transfer, due to their specificity [78]. With this under consideration, both plyometric and sprint exercises are usually performed with a maximum acceleration during the triple-extension phase, and the same applies to the various tests used to quantify the RSI (i.e., vertical hops, CMJ, and DJ) [79]. Although some resistance training exercises might also have this triple-extension phase, the movement will always be slower because of the higher loads used, compared to plyometric and sprint training methods [80]. Additionally, sprint training methods enhance the SSC muscle function, by decreasing ground contact time and increasing flight time [81], because of the importance of the eccentric phase during sprinting to maximize the power output during the concentric phase [82]. Moreover, during the maximal velocity phase of a sprint, the ground contact time is minimal (~80–120 ms); therefore, there is no time to produce muscle force [83]. Thus, the force is only produced by the tendons, being that the tendon stiffness is a particularly important property to generate high forces in a truly short time (fast SSC) [84]. The ability to sprint and generate high forces in short periods of time is somehow related to the RSI, as it is calculated by the jump height, divided by the contact time. Consequently, sprints might reduce the denominator of the RSI formula (i.e., contact time), and subsequently, increase the RSI value. Furthermore, during the acceleration phase of a sprint, an athlete needs to produce more horizontal force to propel himself, while, on the maximal velocity phase, an athlete needs to produce more vertical force [83]. Thus, the force vector direction is the same during a vertical jump (used to access the RSI) as it is on the maximal velocity phase of a sprint. Consequently, sprint training (in particular, maximal velocity), might have some transfer to vertical jumps and, therefore, to better RSI performances, since this type of training may also enhance motor unit firing frequency, ultimately benefitting strength−power characteristics [85]. Nevertheless, only few studies that used sprint training to improve the RSI were found and more research is needed to corroborate this tendency.

Although resistance training induced improvements on RSI, youth athletes seem to take more advantage of this type of training, compared to adult athletes. In general, adult athletes are more skilled and have higher levels of strength, compared to youth athletes. Thus, after achieving specific strength standards, to improve their performance, adult athletes must shift towards a power-type training, while maintaining their strength levels [86]. In youth athletes, on the other hand, resistance training enhances motor control and coordination, which are the base for bigger future improvements in other physical qualities, such as velocity and power [86]. These neural changes that youth athletes experience during resistance training might explain why this training method produces better RSI improvements, compared to adult athletes.

The plyometric intervention was the training program most reported in the present review. Nevertheless, due to different designs (i.e., randomized study vs. non-randomized study or presence vs. absence of a control group) on the individual studies, it was not possible to band together with the different plyometric training programs into different categories (e.g., fast SSC, slow SSC, extensive plyometrics, intensive plyometrics, unilateral, bilateral, vertical emphasis, and horizontal emphasis). The resistance training program is also a broad category. Once again, due to different study designs, it was not possible to group these studies into more specific categories (e.g., high volume, strength training, power training, and accentuated eccentric training). As it was previously mentioned, reactive strength describes the individual's capability to quickly change from an eccentric muscular contraction to a concentric one [1]. Consequently, it is plausible to consider that an athlete that is able to produce higher rates of force development during the eccentric phase will be able to express his/her concentric potential quicker than an athlete with lower levels of eccentric rates of force development. Therefore, it is expected that resistance training focusing on producing more force (eccentrically) in less time will also improve the RSI. Three studies [43,44,65] that were included in the present review reported this type of training. Whereas, in the studies from Murton and colleagues [65] and Douglas and colleagues [43], the adult rugby players in the intervention group improved their performance on RSI after 4 and 12 weeks, respectively, of accentuated eccentric resistance training. In the study from Fiorilli and colleagues [44], the youth soccer players from the intervention group did not improve their RSI after 6 weeks of flywheel eccentric overload training (pre-test: 0.84 ± 0.18; post-test: 0.83 ± 0.18). The flywheel device is an isoinertial equipment that, in a given movement, returns during the eccentric phase, the exact same force produced on the concentric phase. Therefore, using this device does not guarantee that the eccentric overload has been reached per se. As the methodology used by Fiorilli and colleagues [44] was not fully detailed, this may justify the results previously mentioned. Consequently, future research should aim to understand not only the effect of the resistance or plyometric training on RSI but, essentially, the effects of different types/categories of resistance and plyometric training on RSI.

Furthermore, it was noted that only 36% of the studies included used the gold standard method of measurement (i.e., force plates) [13]. Despite the fact that contact mats are cheaper and easier to use than force plates [13], investigators should be warned that the flight times derived from the contact mats are not always consistent, when compared with the flight times derived from the force plates [87]. More than half of the studies included in the review (i.e., 56%) had a sample of soccer athletes. It would be important that future research dedicate itself to study other sport modalities.

This review highlighted the best training intervention to improve the RSI and indicated possible directions for future research in this topic. Nevertheless, a limitation of this review is that some studies could not be included in the meta-analysis, due to not reporting RSI values in a continuous way. Therefore, fewer studies were included in the meta-analysis than in the systematic review; it is possible that the inclusion of these studies could modify the results observed. Additionally, 36% of the randomized studies included had a high risk of bias. Although, it is crucial to notice that most of the high risk of bias was due to the randomisation process.

5. Conclusions

Results from this systematic review and meta-analysis suggest that interventions with plyometric training routines have a relatively large, statistically significant overall effect in both adult and youth athletes. Thus, evidence supports implementing these types of interventions at an early age, in order to better cope with the physical demands of the various sports. Resistance training seems to have less impact on trained adult athletes; therefore, trained adult athletes should seek to have a more power-type training to improve the RSI. More research on specific resistance (e.g., strength vs. power type vs. eccentric overload) and plyometric (e.g., fast vs. short SSC) interventions are needed. Likewise, more research with sprint and combined training interventions is needed to understand the effects of these methods on the RSI.

Supplementary Materials: The following supporting information can be downloaded at: https://www.mdpi.com/article/10.3390/healthcare10040593/s1, Supplemental Material S1: Comprehensive search strategy; Supplemental Material S2: Risk of bias assessment using the Revised Cochrane risk-of-bias tool for randomized trials (RoB 2.0) scale; Supplemental Material S3: Risk of bias assessment using the Risk of Bias in Non-randomized Studies—of Interventions (ROBINS-I) tool; Supplemental Material S4: Characteristics of the included studies, interventions, and outcomes; Supplemental Material S5: Meta-analyses of all study designs.

Author Contributions: All authors reviewed the manuscript. A.R., J.R.P., D.V.M., J.P.D., M.J.C.-e.-S. and J.V.-d.-S. generated the hypotheses; A.R. did the literature search, analysed the data, and wrote the first draft of the manuscript. All authors revised the manuscript critically for important content. All authors had full access to all the data (including statistical reports and tables) in the study and can take responsibility for the integrity of the data and the accuracy of the data analysis. A.R. and J.R.P. extracted the data. A.R., J.R.P., D.V.M. and J.P.D. assessed bias. A.R., J.R.P., D.V.M., J.P.D., M.J.C.-e.-S. and J.V.-d.-S. evaluated the studies for inclusion. All authors have read and agreed to the published version of the manuscript.

Funding: This research received no external funding.

Institutional Review Board Statement: Not applicable.

Informed Consent Statement: Not applicable.

Conflicts of Interest: The authors declare no conflict of interest for this article.

References

1. Young, W. Laboratory strength assessment of athletes. *New Study Athl.* **1995**, *10*, 88–96.
2. Flanagan, E.P.; Ebben, W.P.; Jensen, R.L. Reliability of the reactive strength index and time to stabilization during depth jumps. *J. Strength Cond. Res* **2008**, *22*, 1677–1682. [CrossRef] [PubMed]
3. Markwick, W.J.; Bird, S.P.; Tufano, J.J.; Seitz, L.B.; Haff, G.G. The intraday reliability of the Reactive Strength Index calculated from a drop jump in professional men's basketball. *Int. J. Sports Physiol. Perform.* **2015**, *10*, 482–488. [CrossRef] [PubMed]
4. McClymont, D.; Hore, A. Use of the reactive strength index (RSI) as an indicator of plyometric training conditions. In Proceedings of the Science and Football V: The Proceedings of the Fifth World Congress on Sports Science and Football, Lisbon, Portugal, 11–15 April 2003; pp. 408–416.
5. Flanagan, E.P.; Comyns, T.M. The Use of Contact Time and the Reactive Strength Index to Optimize Fast Stretch-Shortening Cycle Training. *Strength Cond. J.* **2008**, *30*, 32–38. [CrossRef]
6. Ebben, W.P.; Petushek, E.J. Using the Reactive Strength Index Modified to Evaluate Plyometric Performance. *J. Strength Cond. Res.* **2010**, *24*. [CrossRef] [PubMed]
7. Harper, D.; Hobbs, S.; Moore, J. The Ten to Five Repeated Jump Test. A New Test for Evaluation of Reactive Strength. In Proceedings of the BASES Student Conference, Chester, UK, 12–13 April 2011.
8. Flanagan, E. An Examination of the Slow and Fast Stretch Shortening Cycle in Cross Country Runners and Skiers. In Proceedings of the 25th International Society of Biomechanics in Sports, Ouro Preto, Brazil, 23–27 August 2007.
9. Lloyd, R.S.; Oliver, J.L.; Hughes, M.G.; Williams, C.A. Reliability and validity of field-based measures of leg stiffness and reactive strength index in youths. *J. Sports Sci.* **2009**, *27*, 1565–1573. [CrossRef]

10. Hobara, H.; Inoue, K.; Omuro, K.; Muraoka, T.; Kanosue, K. Determinant of leg stiffness during hopping is frequency-dependent. *Eur. J. Appl. Physiol.* **2011**, *111*, 2195–2201. [CrossRef]
11. Chelly, S.M.; Denis, C. Leg power and hopping stiffness: Relationship with sprint running performance. *Med. Sci. Sports Exerc.* **2001**, *33*, 326–333. [CrossRef]
12. García-López, J.; Peleteiro, J.; Rodríguez-Marroyo, J.A.; Morante, J.C.; Herrero, J.A.; Villa, J.G. The validation of a new method that measures contact and flight times during vertical jump. *Int. J. Sports Med.* **2005**, *26*, 294–302. [CrossRef]
13. Glatthorn, J.F.; Gouge, S.; Nussbaumer, S.; Stauffacher, S.; Impellizzeri, F.M.; Maffiuletti, N.A. Validity and reliability of Optojump photoelectric cells for estimating vertical jump height. *J. Strength Cond. Res.* **2011**, *25*, 556–560. [CrossRef]
14. Kibele, A. Possibilities and Limitations in the Biomechanical Analysis of Countermovement Jumps: A Methodological Study. *J. Appl. Biomech.* **1998**, *14*, 105. [CrossRef]
15. Impellizzeri, F.M.; Rampinini, E.; Castagna, C.; Martino, F.; Fiorini, S.; Wisloff, U. Effect of plyometric training on sand versus grass on muscle soreness and jumping and sprinting ability in soccer players. *Br. J. Sports Med.* **2008**, *42*, 42–46. [CrossRef] [PubMed]
16. Casartelli, N.; Müller, R.; Maffiuletti, N.A. Validity and reliability of the Myotest accelerometric system for the assessment of vertical jump height. *J. Strength Cond. Res.* **2010**, *24*, 3186–3193. [CrossRef] [PubMed]
17. Nuzzo, J.L.; Anning, J.H.; Scharfenberg, J.M. The reliability of three devices used for measuring vertical jump height. *J. Strength Cond. Res.* **2011**, *25*, 2580–2590. [CrossRef]
18. García-López, J.; Morante, J.C.; Ogueta-Alday, A.; Rodríguez-Marroyo, J.A. The type of mat (Contact vs. Photocell) affects vertical jump height estimated from flight time. *J. Strength Cond. Res.* **2013**, *27*, 1162–1167. [CrossRef]
19. Baca, A. A comparison of methods for analyzing drop jump performance. *Med. Sci. Sports Exerc.* **1999**, *31*, 437–442. [CrossRef]
20. Ronglan, L.; Raastad, T.; Børgesen, A. Neuromuscular fatigue and recovery in elite female handball players. *Scand. J. Med. Sci. Sports* **2006**, *16*, 267–273. [CrossRef]
21. Young, W.B.; Miller, I.R.; Talpey, S.W. Physical qualities predict change-of-direction speed but not defensive agility in Australian rules football. *J. Strength Cond. Res.* **2015**, *29*, 206–212. [CrossRef]
22. Young, W.B.; Murray, M.P. Reliability of a Field Test of Defending and Attacking Agility in Australian Football and Relationships to Reactive Strength. *J. Strength Cond. Res.* **2017**, *31*, 509–516. [CrossRef]
23. Beckham, G.; Suchomel, T.; Bailey, C.; Sole, C.; Grazer, J. The Relationship of the Reactive Strength Index-Modified and Measures of Force Development in the Isometric Mid-Thigh Pull. In Proceedings of the 32nd International Conference of Biomechanics in Sports, Johnson City, TN, USA, July 2014.
24. Liberati, A.; Altman, D.G.; Tetzlaff, J.; Mulrow, C.; Gøtzsche, P.C.; Ioannidis, J.P.A.; Clarke, M.; Devereaux, P.J.; Kleijnen, J.; Moher, D. The PRISMA statement for reporting systematic reviews and meta-analyses of studies that evaluate healthcare interventions: Explanation and elaboration. *BMJ* **2009**, *339*, b2700. [CrossRef]
25. Kohl, C.; McIntosh, E.J.; Unger, S.; Haddaway, N.R.; Kecke, S.; Schiemann, J.; Wilhelm, R. Online tools supporting the conduct and reporting of systematic reviews and systematic maps: A case study on CADIMA and review of existing tools. *Environ. Evid.* **2018**, *7*, 8. [CrossRef]
26. Sterne, J.A.C.; Savović, J.; Page, M.J.; Elbers, R.G.; Blencowe, N.S.; Boutron, I.; Cates, C.J.; Cheng, H.-Y.; Corbett, M.S.; Eldridge, S.M.; et al. RoB 2: A revised tool for assessing risk of bias in randomised trials. *BMJ* **2019**, *366*, l4898. [CrossRef]
27. Sterne, J.A.; Hernán, M.A.; Reeves, B.C.; Savović, J.; Berkman, N.D.; Viswanathan, M.; Henry, D.; Altman, D.G.; Ansari, M.T.; Boutron, I.; et al. ROBINS-I: A tool for assessing risk of bias in non-randomised studies of interventions. *BMJ* **2016**, *355*, i4919. [CrossRef]
28. Buttner, F.; Winters, M.; Delahunt, E.; Elbers, R.; Lura, C.B.; Khan, K.M.; Weir, A.; Ardern, C.L. Identifying the 'incredible'! Part 2: Spot the difference—A rigorous risk of bias assessment can alter the main findings of a systematic review. *Br. J. Sports Med.* **2020**, *54*, 801–808. [CrossRef] [PubMed]
29. Ramírez-Campillo, R.; Meylan, C.M.; Álvarez-Lepín, C.; Henríquez-Olguín, C.; Martinez, C.; Andrade, D.C.; Castro-Sepúlveda, M.; Burgos, C.; Baez, E.I.; Izquierdo, M. The effects of interday rest on adaptation to 6 weeks of plyometric training in young soccer players. *J. Strength Cond. Res.* **2015**, *29*, 972–979. [CrossRef] [PubMed]
30. Ramírez Campillo, R.; Gallardo, F.; Henríquez-Olguín, C.; Meylan, C.M.; Martínez, C.; Álvarez, C.; Caniuqueo, A.; Cadore, E.L.; Izquierdo, M. Effect of Vertical, Horizontal, and Combined Plyometric Training on Explosive, Balance, and Endurance Performance of Young Soccer Players. *J. Strength Cond. Res.* **2015**, *29*, 1784–1795. [CrossRef] [PubMed]
31. Ramírez-Campillo, R.; Burgos, C.H.; Henríquez-Olguín, C.; Andrade, D.C.; Martínez, C.; Álvarez, C.; Castro-Sepúlveda, M.; Marques, M.C.; Izquierdo, M. Effect of unilateral, bilateral, and combined plyometric training on explosive and endurance performance of young soccer players. *J. Strength Cond. Res.* **2015**, *29*, 1317–1328. [CrossRef] [PubMed]
32. Ramírez-Campillo, R.; Henríquez-Olguín, C.; Burgos, C.; Andrade, D.C.; Zapata, D.; Martínez, C.; Álvarez, C.; Baez, E.I.; Castro-Sepúlveda, M.; Peñailillo, L.; et al. Effect of Progressive Volume-Based Overload During Plyometric Training on Explosive and Endurance Performance in Young Soccer Players. *J. Strength Cond. Res.* **2015**, *29*, 1884–1893. [CrossRef]
33. Ramírez-Campillo, R.; Meylan, C.; Alvarez, C.; Henríquez-Olguín, C.; Martínez, C.; Cañas-Jamett, R.; Andrade, D.C.; Izquierdo, M. Effects of in-season low-volume high-intensity plyometric training on explosive actions and endurance of young soccer players. *J. Strength Cond. Res.* **2014**, *28*, 1335–1342. [CrossRef]

34. Rosas, F.; Ramirez-Campillo, R.; Diaz, D.; Abad-Colil, F.; Martinez-Salazar, C.; Caniuqueo, A.; Cañas-Jamet, R.; Loturco, I.; Nakamura, F.Y.; McKenzie, C.; et al. Jump Training in Youth Soccer Players: Effects of Haltere Type Handheld Loading. *Int. J. Sports Med.* **2016**, *37*, 1060–1065. [CrossRef]

35. Argus, C.K.; Gill, N.D.; Keogh, J.W.; McGuigan, M.R.; Hopkins, W.G. Effects of two contrast training programs on jump performance in rugby union players during a competition phase. *Int. J. Sports Physiol. Perform.* **2012**, *7*, 68–75. [CrossRef] [PubMed]

36. Keiner, M.; Sander, A.; Hartmann, H.; Mickel, C.; Wirth, K. Do long-term strength training and age affect the performance of drop jump in adolescents? *J. Aust. Strength Cond.* **2018**, *26*, 24–38.

37. Dello Iacono, A.; Martone, D.; Milic, M.; Padulo, J. Vertical- vs. Horizontal-Oriented Drop Jump Training: Chronic Effects on Explosive Performances of Elite Handball Players. *J. Strength Cond. Res.* **2017**, *31*, 921–931. [CrossRef] [PubMed]

38. Ramirez-Campillo, R.; Alvarez, C.; Gentil, P.; Moran, J.; García-Pinillos, F.; Alonso-Martínez, A.M.; Izquierdo, M. Inter-individual Variability in Responses to 7 Weeks of Plyometric Jump Training in Male Youth Soccer Players. *Front. Physiol.* **2018**, *9*, 1156. [CrossRef] [PubMed]

39. Higgins, J.P.T.; Thompson, S.G.; Deeks, J.J.; Altman, D.G. Measuring inconsistency in meta-analyses. *BMJ* **2003**, *327*, 557–560. [CrossRef]

40. Moher, D.; Liberati, A.; Tetzlaff, J.; Altman, D.G.; The, P.G. Preferred Reporting Items for Systematic Reviews and Meta-Analyses: The PRISMA Statement. *PLoS Med.* **2009**, *6*, e1000097. [CrossRef]

41. Sporri, D.; Ditroilo, M.; Pickering Rodriguez, E.C.; Johnston, R.J.; Sheehan, W.B.; Watsford, M.L. The effect of water-based plyometric training on vertical stiffness and athletic performance. *PLoS ONE* **2018**, *13*, e0208439. [CrossRef]

42. Wee, E.H.; Low, J.Y.; Chan, K.Q.; Ler, H.Y. Effects of Specific Badminton Training on Aerobic and Anaerobic Capacity, Leg Strength Qualities and Agility Among College Players. In *Sport Science Research and Technology Support*; Springer: Cham, Switzerland, 2016; pp. 192–203.

43. Douglas, J.; Pearson, S.; Ross, A.; McGuigan, M. Effects of Accentuated Eccentric Loading on Muscle Properties, Strength, Power, and Speed in Resistance-Trained Rugby Players. *J. Strength Cond. Res.* **2018**, *32*, 2750–2761. [CrossRef]

44. Fiorilli, G.; Mariano, I.; Iuliano, E.; Giombini, A.; Ciccarelli, A.; Buonsenso, A.; Calcagno, G.; di Cagno, A. Isoinertial Eccentric-Overload Training in Young Soccer Players: Effects on Strength, Sprint, Change of Direction, Agility and Soccer Shooting Precision. *J. Sports Sci. Med.* **2020**, *19*, 213–223.

45. Ramirez-Campillo, R.; Alvarez, C.; García-Pinillos, F.; Gentil, P.; Moran, J.; Pereira, L.A.; Loturco, I. Effects of Plyometric Training on Physical Performance of Young Male Soccer Players: Potential Effects of Different Drop Jump Heights. *Pediatr. Exerc. Sci.* **2019**, *31*, 306–313. [CrossRef]

46. Ramirez-Campillo, R.; Alvarez, C.; Sanchez-Sanchez, J.; Slimani, M.; Gentil, P.; Chelly, M.S.; Shephard, R.J. Effects of plyometric jump training on the physical fitness of young male soccer players: Modulation of response by inter-set recovery interval and maturation status. *J. Sports Sci.* **2019**, *37*, 2645–2652. [CrossRef] [PubMed]

47. Bouguezzi, R.; Chaabene, H.; Negra, Y.; Ramirez-Campillo, R.; Jlalia, Z.; Mkaouer, B.; Hachana, Y. Effects of Different Plyometric Training Frequencies on Measures of Athletic Performance in Prepuberal Male Soccer Players. *J. Strength Cond. Res.* **2020**, *34*, 1609–1617. [CrossRef] [PubMed]

48. Hammami, R.; Granacher, U.; Makhlouf, I.; Behm, D.G.; Chaouachi, A. Sequencing Effects of Balance and Plyometric Training on Physical Performance in Youth Soccer Athletes. *J. Strength Cond. Res.* **2016**, *30*, 3278–3289. [CrossRef]

49. Lockie, R.G.; Murphy, A.J.; Schultz, A.B.; Knight, T.J.; Janse de Jonge, X.A. The effects of different speed training protocols on sprint acceleration kinematics and muscle strength and power in field sport athletes. *J. Strength Cond. Res.* **2012**, *26*, 1539–1550. [CrossRef] [PubMed]

50. Byrne, P.J.; Moran, K.; Rankin, P.; Kinsella, S. A comparison of methods used to identify 'optimal' drop height for early phase adaptations in depth jump training. *J. Strength Cond. Res.* **2010**, *24*, 2050–2055. [CrossRef] [PubMed]

51. Salonikidis, K.; Zafeiridis, A. The effects of plyometric, tennis-drills, and combined training on reaction, lateral and linear speed, power, and strength in novice tennis players. *J. Strength Cond. Res.* **2008**, *22*, 182–191. [CrossRef] [PubMed]

52. Spinks, C.D.; Murphy, A.J.; Spinks, W.L.; Lockie, R.G. The effects of resisted sprint training on acceleration performance and kinematics in soccer, rugby union, and Australian football players. *J. Strength Cond. Res.* **2007**, *21*, 77–85. [CrossRef]

53. Young, W.B.; Wilson, G.J.; Byrne, C. A comparison of drop jump training methods: Effects on leg extensor strength qualities and jumping performance. *Int. J. Sports Med.* **1999**, *20*, 295–303. [CrossRef]

54. Beattie, K.; Carson, B.P.; Lyons, M.; Rossiter, A.; Kenny, I.C. The Effect of Strength Training on Performance Indicators in Distance Runners. *J. Strength Cond. Res.* **2017**, *31*, 9–23. [CrossRef]

55. Dobbs, I.J.; Oliver, J.L.; Wong, M.A.; Moore, I.S.; Lloyd, R.S. Effects of a 12-Week Training Program on Isometric and Dynamic Force-Time Characteristics in Pre- and Post-Peak Height Velocity Male Athletes. *J. Strength Cond. Res.* **2020**, *34*, 653–662. [CrossRef]

56. Li, F.; Wang, R.; Newton, R.U.; Sutton, D.; Shi, Y.; Ding, H. Effects of complex training versus heavy resistance training on neuromuscular adaptation, running economy and 5-km performance in well-trained distance runners. *PeerJ* **2019**, *7*, e6787. [CrossRef] [PubMed]

57. Chaabene, H.; Prieske, O.; Lesinski, M.; Sandau, I.; Granacher, U. Short-Term Seasonal Development of Anthropometry, Body Composition, Physical Fitness, and Sport-Specific Performance in Young Olympic Weightlifters. *Sports* **2019**, *7*, 242. [CrossRef] [PubMed]

58. Orange, S.T.; Metcalfe, J.W.; Robinson, A.; Applegarth, M.J.; Liefeith, A. Effects of In-Season Velocity- Versus Percentage-Based Training in Academy Rugby League Players. *Int. J. Sports Physiol. Perform.* **2019**, *15*, 554–561. [CrossRef] [PubMed]
59. Doma, K.; Leicht, A.S.; Boullosa, D.; Woods, C.T. Lunge exercises with blood-flow restriction induces post-activation potentiation and improves vertical jump performance. *Eur. J. Appl. Physiol.* **2020**, *120*, 687–695. [CrossRef]
60. Jeffreys, M.A.; De Ste Croix, M.B.A.; Lloyd, R.S.; Oliver, J.L.; Hughes, J.D. The Effect of Varying Plyometric Volume on Stretch-Shortening Cycle Capability in Collegiate Male Rugby Players. *J. Strength Cond. Res.* **2019**, *33*, 139–145. [CrossRef]
61. Ramirez-Campillo, R.; Alvarez, C.; Gentil, P.; Loturco, I.; Sanchez-Sanchez, J.; Izquierdo, M.; Moran, J.; Nakamura, F.Y.; Chaabene, H.; Granacher, U. Sequencing Effects of Plyometric Training Applied Before or After Regular Soccer Training on Measures of Physical Fitness in Young Players. *J. Strength Cond. Res.* **2020**, *34*, 1959–1966. [CrossRef]
62. Ramirez-Campillo, R.; Alvarez, C.; García-Pinillos, F.; Sanchez-Sanchez, J.; Yanci, J.; Castillo, D.; Loturco, I.; Chaabene, H.; Moran, J.; Izquierdo, M. Optimal Reactive Strength Index: Is It an Accurate Variable to Optimize Plyometric Training Effects on Measures of Physical Fitness in Young Soccer Players? *J. Strength Cond. Res.* **2018**, *32*, 885–893. [CrossRef]
63. Cloak, R.; Nevill, A.; Smith, J.; Wyon, M. The acute effects of vibration stimulus following FIFA 11+ on agility and reactive strength in collegiate soccer players. *J. Sport Health Sci.* **2014**, *3*, 293–298. [CrossRef]
64. Tsoukos, A.; Bogdanis, G.C.; Terzis, G.; Veligekas, P. Acute Improvement of Vertical Jump Performance After Isometric Squats Depends on Knee Angle and Vertical Jumping Ability. *J. Strength Cond. Res.* **2016**, *30*, 2250–2257. [CrossRef]
65. Murton, J.; Eager, R.; Drury, B. Comparison of flywheel versus traditional resistance training in elite academy male Rugby union players. *Res. Sports Med.* **2021**. *Online ahead of print.* [CrossRef]
66. Guo, Z.; Huang, Y.; Zhou, Z.; Leng, B.; Gong, W.; Cui, Y.; Bao, D. The Effect of 6-Week Combined Balance and Plyometric Training on Change of Direction Performance of Elite Badminton Players. *Front. Psychol.* **2021**, *12*, 684964. [CrossRef]
67. Bouguezzi, R.; Chaabene, H.; Negra, Y.; Moran, J.; Sammoud, S.; Ramirez-Campillo, R.; Granacher, U.; Hachana, Y. Effects of jump exercises with and without stretch-shortening cycle actions on components of physical fitness in prepubertal male soccer players. *Sport Sci. Health* **2020**, *16*, 297–304. [CrossRef]
68. Makhlouf, I.; Chaouachi, A.; Chaouachi, M.; Ben Othman, A.; Granacher, U.; Behm, D.G. Combination of Agility and Plyometric Training Provides Similar Training Benefits as Combined Balance and Plyometric Training in Young Soccer Players. *Front. Physiol.* **2018**, *9*, 1611. [CrossRef] [PubMed]
69. Chaouachi, M.; Granacher, U.; Makhlouf, I.; Hammami, R.; Behm, D.G.; Chaouachi, A. Within Session Sequence of Balance and Plyometric Exercises Does Not Affect Training Adaptations with Youth Soccer Athletes. *J. Sports Sci. Med.* **2017**, *16*, 125–136.
70. Negra, Y.; Chaabene, H.; Sammoud, S.; Bouguezzi, R.; Mkaouer, B.; Hachana, Y.; Granacher, U. Effects of Plyometric Training on Components of Physical Fitness in Prepuberal Male Soccer Athletes: The Role of Surface Instability. *J. Strength Cond. Res.* **2017**, *31*, 3295–3304. [CrossRef] [PubMed]
71. Hammami, R.; Gene-Morales, J.; Abed, F.; Amin Selmi, M.; Moran, J.; Colado, J.C.; Hem Rebai, H. An eight-weeks resistance training programme with elastic band increases some performance-related parameters in pubertal male volleyball players. *Biol. Sport* **2022**, *39*, 219–226. [CrossRef]
72. Ciacci, S.; Bartolomei, S. The effects of two different explosive strength training programs on vertical jump performance in basketball. *J. Sports Med. Phys. Fit.* **2018**, *58*, 1375–1382. [CrossRef]
73. Komi, P.V. Physiological and biomechanical correlates of muscle function: Effects of muscle structure and stretch-shortening cycle on force and speed. *Exerc. Sport Sci. Rev.* **1984**, *12*, 81–121. [CrossRef]
74. Schmidtbleicher, D. Training for power events. In *The Encyclopedia of Sports Medicine: Strength and Power in Sport*; Komi, P., Ed.; Blackwell: Oxford, UK, 1992; Volume 3, pp. 169–179.
75. Markovic, G. Does plyometric training improve vertical jump height? A meta-analytical review. *Br. J. Sports Med.* **2007**, *41*, 349–355. [CrossRef]
76. Hennessy, L.; Kilty, J. Relationship of the stretch-shortening cycle to sprint performance in trained female athletes. *J. Strength Cond. Res.* **2001**, *15*, 326–331.
77. Sáez-Sáez de Villarreal, E.; Requena, B.; Newton, R.U. Does plyometric training improve strength performance? A meta-analysis. *J. Sci. Med. Sport* **2010**, *13*, 513–522. [CrossRef] [PubMed]
78. Stone, M.; Plisk, S.; Collins, D. Strength and conditioning. *Sports Biomech.* **2002**, *1*, 79–103. [CrossRef] [PubMed]
79. Cormie, P.; McGuigan, M.R.; Newton, R.U. Developing maximal neuromuscular power: Part 2—Training considerations for improving maximal power production. *Sports Med.* **2011**, *41*, 125–146. [CrossRef]
80. Cormie, P.; Mccaulley, G.O.; Triplett, N.T.; Mcbride, J.M. Optimal Loading for Maximal Power Output during Lower-Body Resistance Exercises. *Med. Sci. Sports Exerc.* **2007**, *39*, 340–349. [CrossRef] [PubMed]
81. Markovic, G.; Jukic, I.; Milanovic, D.; Metikos, D. Effects of sprint and plyometric training on muscle function and athletic performance. *J. Strength Cond. Res.* **2007**, *21*, 543–549. [CrossRef]
82. Mero, A.; Komi, P.V.; Gregor, R.J. Biomechanics of sprint running. A review. *Sports Med.* **1992**, *13*, 376–392. [CrossRef]
83. Nagahara, R.; Mizutani, M.; Matsuo, A.; Kanehisa, H.; Fukunaga, T. Association of Sprint Performance With Ground Reaction Forces During Acceleration and Maximal Speed Phases in a Single Sprint. *J. Appl. Biomech.* **2018**, *34*, 104–110. [CrossRef]
84. Douglas, J.; Pearson, S.; Ross, A.; McGuigan, M. Reactive and eccentric strength contribute to stiffness regulation during maximum velocity sprinting in team sport athletes and highly trained sprinters. *J. Sports Sci.* **2020**, *38*, 29–37. [CrossRef]

85. Suchomel, T.J.; Nimphius, S.; Bellon, C.R.; Stone, M.H. The Importance of Muscular Strength: Training Considerations. *Sports Med.* **2018**, *48*, 765–785. [CrossRef]
86. Suchomel, T.J.; Nimphius, S.; Stone, M.H. The Importance of Muscular Strength in Athletic Performance. *Sports Med.* **2016**, *46*, 1419–1449. [CrossRef]
87. Whitmer, T.D.; Fry, A.C.; Forsythe, C.M.; Andre, M.J.; Lane, M.T.; Hudy, A.; Honnold, D.E. Accuracy of a vertical jump contact mat for determining jump height and flight time. *J. Strength Cond. Res.* **2015**, *29*, 877–881. [CrossRef] [PubMed]

healthcare

Review

Isotonic and Isometric Exercise Interventions Improve the Hamstring Muscles' Strength and Flexibility: A Narrative Review

Akhmad Fajri Widodo [1], Cheng-Wen Tien [2], Chien-Wei Chen [1,3,*] and Shih-Chiung Lai [3]

[1] International Sport Science Master's Program, College of Human Development and Health, National Taipei University of Nursing and Health Sciences, Taipei 112303, Taiwan; fajriwidodo@gmail.com
[2] Physical Education Office, General Education Centre, National Taipei University of Nursing and Health Sciences, Taipei 112303, Taiwan; chengwen@ntunhs.edu.tw
[3] Department of Exercise and Health Science, College of Human Development and Health, National Taipei University of Nursing and Health Sciences, Taipei 112303, Taiwan; shihchiung@ntunhs.edu.tw
* Correspondence: gogozipper130@gmail.com

Abstract: Background: Hamstring weakness has been associated with an increased risk of hamstring strain, a common sports injury that occurs when athletes perform actions such as quick sprints. The hamstring complex comprises three distinct muscles: the long and short heads of the bicep femoris, the semimembranosus, and the semitendinosus. Methods: The researchers collected the data from different electronic databases: PubMed, Google Scholar, and the Web of Science. Results: Many studies have been conducted on the numerous benefits of hamstring strength, in terms of athletic performance and injury prevention. Isotonic and isometric exercises are commonly used to improve hamstring strength, with each exercise type having a unique effect on the hamstring muscles. Isotonic exercise improves the muscles' strength, increasing their ability to resist any force, while isometric training increases strength and the muscles' ability to produce power by changing the muscle length. Conclusions: These exercises, when performed at low intensity, but with high repetition, can be used by the healthy general population to prepare for training and daily exercise. This can improve hamstring muscle strength and flexibility, leading to enhanced performance and reduced injury risk.

Keywords: athlete; exercise; injury; performance

Citation: Widodo, A.F.; Tien, C.-W.; Chen, C.-W.; Lai, S.-C. Isotonic and Isometric Exercise Interventions Improve the Hamstring Muscles' Strength and Flexibility: A Narrative Review. *Healthcare* **2022**, *10*, 811. https://doi.org/10.3390/ healthcare10050811

Academic Editors: João Paulo Brito and Rafael Oliveira

Received: 22 March 2022
Accepted: 24 April 2022
Published: 27 April 2022

Publisher's Note: MDPI stays neutral with regard to jurisdictional claims in published maps and institutional affiliations.

1. Introduction

Hamstring strains are frequently observed in various athletes, from recreational to elite, and they typically occur during sports that require rapid sprinting and concurrent excessive muscle stretches, such as athletics and ball games [1–4]. Hamstring injuries are often slow to heal, and it is not uncommon for them to recur [5]. Nearly one-third of individuals who have a hamstring injury will reinjure themselves within a year of resuming their sport [6], and hamstring weakness has also been linked to an increased risk of hamstring strain [7].

Hamstring injuries are intricate and multifactorial in nature [8]. The proposed general risk factors for a hamstring injury can be classified into intrinsic (player-related) components and extrinsic (environment-related) factors [9]. Intrinsic risk factors for a hamstring injury include muscle weakness, instability, weariness, poor flexibility, poor technique, and psychosocial problems. Extrinsic risk factors include an insufficient warm-up period, training conditions, and playing surfaces [10]. Hamstring strength is also critical for knee joint stabilization, which directly affects the characteristics of agility manoeuvres, such as accelerations, decelerations, direction changes, and cutting [11].

Isotonic [12] and isometric exercises [13,14] are popular methods for developing hamstring strength. These interventions are frequently used as part of the physical preparation for a wide range of people, from injured members of the public to elite strength and power athletes. Changes that are frequently recorded following hamstring isotonic and isometric exercises include improvements in muscle mass, tendon quality, strength, power, range of motion, muscle fatigue delay, and voluntary activation [15].

2. Materials and Methods

The researchers collected the data from three different electronic databases: PubMed, Google Scholar, and the Web of Science. The data collected were limited to English language studies. Boolean logic (OR, AND) was used with keywords and other free terms to combine searches: ('exercise' OR 'training' OR 'isotonic exercise' OR 'isometric exercise') AND ('muscle' OR 'hamstring' OR 'injury' OR 'biceps femoris' OR 'semitendinosus' OR 'semimembranosus'). The data collected in different databases followed their specific syntax rules. We also collected articles present in the reference lists of the resulting articles that were within the scope of this review.

3. Results

We selected 7 main articles out of 879 after deletion of the non-English literature. The articles are original studies that focus more on isotonic and isometric exercises (Table 1).

Table 1. Summary of related studies.

Authors	Year	Result
Rich A., Cook J.L., Hahne A.J., et al. [16]	2021	Isotonic (eccentric) exercise helps to increase hamstring muscle strength.
Rio E., Kidgell D., Purdam C., Gaida J., Moseley G.L., Pearce A.J., Cook J. [17]	2015	Quadriceps strength on maximal voluntary isometric contraction (MVIC) increased significantly following the isometric condition.
Seagrave, et al. [18]	2014	Nordic hamstring exercises may decrease the incidence or acute hamstring injuries by increasing the strength.
Nelson, R.T., Bandy, W.D. [19]	2004	Eccentric exercise helps to improve hamstring muscle flexibility.
Delahunt E., McGroarty M., De Vito G., Ditroilo M. [20]	2016	Improvement in hamstring muscle control during eccentric contractions.
Marusic J., Vatovec R., Markovic G., Sarabon N. [21]	2020	Short-term eccentric hamstring strengthening at long muscle length can have significant favorable effects on various architectural and functional characteristics of the hamstrings.
Van der Horst N., Smits D.W., Petersen J., Goedhart E.A., Backx F.J. [17]	2015	Nordic hamstring exercise protocol in regular amateur training significantly reduces hamstring injury incidence.

4. Discussion

4.1. Hamstring Structure and Function

Numerous studies have demonstrated a strong correlation between hamstring strength and athletic abilities, such as running and jumping [22,23]. The human hamstring muscle complex comprises three separate muscles: the semitendinosus, semimembranosus, and biceps femoris. This complex is involved in a wide variety of activities, from standing to explosive movements, such as sprinting and jumping [24]. Two of the muscles of the hamstring complex originate in the pelvis and run posteriorly along the length of the femur, via the femoroacetabular and tibiofemoral joints. The short head of the biceps femoris, in contrast, originates on the lateral lip of the femoral linea aspera, distal to the femoroacetabular joint [24]. As a result, some argue that it is not a true hamstring muscle. Unlike the biceps femoris, the other two hamstring muscles originate from the ischial tuberosity. The proximal long heads of the biceps femoris and semitendinosus muscles are connected by the aponeurosis. It extends roughly 7 cm from the ischial tuberosity. The distal hamstrings define the superolateral (biceps femoris) and superomedial (semimembranosus and semitendinosus) margins of the popliteal fossa. The gastrocnemius is primarily important for establishing the inferior border of the popliteal fossa [24].

The hamstring muscle complex is responsible for hip extension and knee flexion. It activates during the final 25% of the swing phase of the gait cycle, providing hip extension force and resisting knee extension. Additionally, the complex acts as a dynamic stabilizer of the knee joint. In conjunction with the anterior cruciate ligament (ACL), the muscles prevent the tibia from undergoing anterior translation relative to the femur during heel strike [25].

4.2. Muscle Types

The hamstring muscles are the three major muscles of the posterior aspect of the thigh. The semimembranosus is the medial-most muscle, the long and short heads of the biceps femoris are the lateral-most, and the semitendinosus is between them. This group of muscles is clinically significant, as it is highly susceptible to injury, especially in athletes [26].

Woods et al. described the anatomical distribution of hamstring injuries, observing that 53% of cases involved the biceps femoris, 16% the semitendinosus, and 13% the semimembranosus [27]. The biceps femoris is the strongest muscle in the hamstring complex, and it is responsible for knee flexion, external rotation, and posterolateral stability [28].

The biceps femoris muscle has both a long and short head, although the short head may be missing in some common variations. The long head originates from the medial ischial tuberosity and the inferior sacrotuberous ligament. The short head originates from the lateral lip of the femur's linea aspera, the proximal two-thirds of the supracondylar line, and the lateral intermuscular septum. The two heads converge to produce the biceps femoris tendon, which is attached to the fibular head, crural fascia, and proximal tibia [28].

The semitendinosus muscle has been shown to exhibit stronger neuronal, metabolic, and stiffness responses than the other hamstring muscles during a knee flexion test, implying that the way in which the load is shared between these muscles results in a greater workload on the semitendinosus. However, because the hamstring heads differ physiologically and mechanically [29], the reactions to fatigue may also differ, in terms of active stiffness between the muscle components [30].

The semitendinosus is a solitary muscle, but it is more appropriately considered a digastric muscle physiologically, because of the presence of an intervening raphe into which the proximal fibres insert. These fibres arise from the upper region of the ischial tuberosity's inferomedial impression via a tendon that connects to the long head of the biceps femoris. Of the three hamstring muscles, the semitendinosus has the shortest proximal tendon and the smallest physiological cross-sectional area (8.08 cm^2) [31].

Caudal to the ischial tuberosity, the semitendinosus muscle becomes bulbous, with the semimembranosus tendon lying anterior to it. The semimembranosus muscle is frequently confused with the semitendinosus muscle, since the latter's proximal tendon is not always a separate tissue. Further distally, the semitendinosus muscle produces a long tendon. This elongated distal tendon may exacerbate the muscle's proneness to rupture [32].

Although the semimembranosus muscle may contribute to creating this conjoint tendon, this structure could be considered an anatomic variation [27]. The semimembranosus muscle originates on the superolateral part of the ischial tuberosity, beneath the proximal half of the semitendinosus muscle. The semimembranosus tendon runs between the medial and anterior sides of the other hamstring tendons. The proximal tendon is a lengthy structure that connects to the adductor magnus tendon and serves as the origin of the long head of the biceps femoris [25].

On the other hand, the semimembranosus, with its shorter fibre length, larger cross-sectional area, and shorter tendon, may contribute more to knee stabilization by stiffening the medial joint and resisting rotational loads. This selectivity of muscle activation within adjacent muscles has been demonstrated in studies of the lower and upper limbs using electromyography [33]. The semimembranosus is the largest muscle in the leg, with the longest proximal tendon. A broad aponeurosis joins the ischial tuberosity to its proximal insertion. The distal segment comprises tendinous branches joined to the popliteal fascia, the oblique popliteal ligament, and the posterior portion of the medial tibial condyle [34].

4.3. Injury

The majority of hamstring strains occur during intense sports, such as sprinting, which cause the muscles to overextend in response to abrupt changes in speed or direction. The most frequently damaged hamstring muscle starts from the biceps femoris, the semimembranosus to the semitendinosus [35]. Hamstring injuries typically cause posterior thigh pain, which is aggravated by knee flexion and hip extension. Patients may occasionally report hearing a "popping" sound with serious injuries. When clinicians evaluate a patient with a suspected hamstring injury, they must rule out other probable diagnoses, such as lumbosacral radiculopathy, adductor strain, or femoral stress fracture [2].

A grade 1 hamstring strain is defined as slight discomfort or swelling, little or minimal loss of function, absence of pain and functional impairment, and the presence of minor disruption of the hamstring myofibrils. A grade 2 hamstring strain is defined as recognizable partial tissue disruption with considerable pain, oedema, and function loss due to partial-thickness tears of the musculotendinous fibres. A grade 3 hamstring strain is defined as musculotendinous unit disruption or tear with severe discomfort, oedema, and loss of function [36].

Protection, rest, ice, compression, and elevation should be used to limit inflammation and swelling in the acute phase of hamstring injuries [37]. The patient's pain threshold should regulate their range of motion, since excessive hamstring stretching may result in scar tissue formation [38]. The role of nonsteroidal anti-inflammatory drugs (NSAIDs) in treating hamstring injuries is debatable, with some studies indicating little benefit, while others indicate possible adverse effects [39]. However, short courses of NSAIDs (5–7 days) have been shown to not adversely affect recovery, but should be used purely as analgesics. Alternative pharmacologic therapies, such as platelet-rich plasma, have been found to assist recovery in athletes, although there are currently insufficient data to support the use of platelet-rich plasma to treat muscle strain injuries [40].

Fast and slow myosins could be markers of skeletal muscle injury [41]. The biceps femoris long head myosin heavy chain composition was not found to be 'fast', and, therefore, the composition does not appear to explain the high incidence of hamstring strain injury [42]. Furthermore, in vitro studies demonstrate that type II fibres (MHC-II) have a higher rate of force development [43] and greater maximum force at high velocities of shortening, yet the influence of MHC composition on in vivo hamstring function remains unknown.

Exercise programmes focused on eccentric contraction have been shown to significantly reduce the recovery time in patients who have healed sufficiently to resume therapeutic activities [24]. These regimens can be tailored to the patient's recovery level and can contribute to reducing the injury recurrence rate [44]. Although hamstring stretching is widely recommended to reduce the risk of reinjury, there is no evidence that hamstring flexibility training reduces the rate of hamstring reinjury. Additionally, research has highlighted the crucial role of neuromuscular control in the lumbopelvic region. According to a 2004 prospective randomized research study, patients with acute hamstring strain injuries, who received rehabilitation using a progressive agility and trunk stabilization programme, had a reduced recurrence rate compared to those who received more typical progressive stretching and strengthening [45].

4.4. Isotonic Exercise

Isotonic exercise is a form of resistance training, and can be performed using free weights to strengthen a specific muscle area. Isotonic workouts emphasize comprehensive fitness, so they are popular with those looking to lose weight, improve muscle tone, increase physical power, or build their muscles. Examples of isotonic exercises include dumbbell curls and bench presses. However, isotonic exercise may result in delayed-onset muscular soreness [16].

In one study, a three-week regimen of isotonic hamstring exercise resulted in no improvement in pain or function [46]. Additionally, tendon pain may impair a patient's ability to engage in sports, employment, and other daily activities. The effect of a hamstring injury on work and social activities is especially significant, since pain during sitting is often the predominant complaint [47,48]. Isotonic exercise has been well described in the literature, and it is clear from recent reviews that the greatest improvements in force and power production are observed at movement velocities similar to those used in training [49].

A typical isotonic exercise regimen is a standard gym programme, but with the weights moving at varying speeds. Throughout the workout, the approach emphasizes isotonic muscular contractions. A static weight is used to maintain muscle tension during the action. This affects the muscle, expanding the muscle cells and increasing the amount of weight that the muscle can lift. Alternatively, an isotonic exercise regimen is a form of muscular exercise intended to increase muscle strength and endurance. Since hamstring strains frequently occur during eccentric muscle contraction, stressing these muscles with eccentric exercise may help to avoid further hamstring strains [50].

The terms "concentric" and "eccentric" originate from basic physiology and refer to muscle contraction. Two types of muscular contractions have been described: isometric (no change in muscle length during contraction) and isotonic (constant tension in the contraction as the muscle length changes) [51]. Isotonic contractions are classified into two types: concentric and eccentric. Concentric contractions cause the muscular tension to increase in response to resistance, and then stabilize as the muscle shortens. Eccentric contraction causes the muscle to lengthen as the resistance exceeds the force produced by the muscle [51].

When the force applied to the muscle exceeds the instantaneous force produced by the muscle, eccentric (lengthening) muscular contraction occurs, resulting in the forced lengthening of the muscle–tendon system while the muscle contracts [52]. In comparison to concentric or isometric (constant-length) muscular contractions, eccentric muscle actions exhibit numerous distinguishing characteristics that may account for their unique adaptations [53]. Although they are not usually evident, eccentric muscular contractions are a necessary component of most motions performed during daily or athletic activities. Skeletal muscles contract eccentrically to support the body's weight against gravity, absorb shock, and store elastic recoil energy for concentric (or accelerating) contractions [54].

The importance of such contractions in decelerating is revealed in downhill jogging or walking downstairs, which emphasizes the eccentric effort of the knee extensor muscles [55]. The favourable adaptive responses in hamstring strength and the architecture of the biceps femoris long head to eccentric hamstring exercise training are well documented [56].

Eccentric training has been found to elicit neuromuscular changes that are largely specific to each kind of muscle movement [57]. The eccentric role of muscles is critical in most sports. Eccentric muscular contractions enhance performance during the concentric phase of stretch-shortening cycles, which is critical for activities such as sprinting, jumping, and throwing. Muscles stimulated during lengthening motions can also operate as shock absorbers, assisting the athlete in decelerating during landing activities or accurately responding to strong external loading in sports such as alpine skiing [58]. Shorter muscle length has been shown to be a strong predictor of 100 m times in well-trained sprint runners [59], and increases in fascicle length have been observed in both a group of athletes who eliminated heavy resistance training and increased their high-speed training [60] and a group of resistance trainers who performed the concentric phase of their lifts at maximum velocity [60,61].

Isotonic exercise is effective for the development of hamstring muscle strength. Additionally, eccentric strength training has been shown to have a dramatic effect on overall hamstring muscle strength, with 29% and 19% improvement from eccentric and concentric exercise, respectively [62]. Both eccentric and concentric exercises have been shown to result in significant improvement in isotonic strength in the knee extensor muscles, which was related to hamstring strength, after a 12-week intervention [63]. In another study, hamstring stiffness increased significantly with isometric training (15.7%), suggesting that such training may be an important addition to ACL injury prevention programmes [13]. Eccentric hamstring exercises also help to improve some physical fitness components, such as the ability to sprint and jump over 5, 10, and 20 m and to countermovement jump [12].

Isotonic training consists of both eccentric and concentric exercises; thus, it positively affects the hamstring muscles, enabling them to produce more force and power, supporting performance. During eccentric and concentric muscle contraction, the muscle length changes, producing power. This training also affects the flexibility and stiffness of the muscles. Performing isotonic training may, thus, help to enhance the capacity of the hamstring muscles, to improve performance and reduce the risk of injury.

4.5. Isometric Exercise

Resistance training is used by a wide variety of people, from injured members of the general public to elite strength and power athletes [64]. Isometric training is gentle on the joints, while maintaining and increasing strength, making it ideal for individuals who require low-impact exercise due to an injury or arthritic condition. It also provides an excellent workout that is achievable without having to leave the house or go to the gym. Isometric exercises have recently been recommended to help alleviate and manage tendon pain [17].

Isometric exercises build muscle mass, strength, and bone density, while reducing cholesterol and improving digestion. Isometric contractions may, indeed, be more useful for improving muscles than eccentric contractions and flexibility [65]. Isometric exercise is, thus, a popular method of developing hamstring strength [12]. This exercise is a self-controlled bodyweight resistance variation of the traditional leg curl exercise. It does not require any equipment and provides an eccentric stimulus using bodyweight. Isometric training has been shown to alter physiological characteristics, such as muscle architecture, tendon stiffness and health [66], joint angle-specific torque [67], and metabolic activity [68]. Isometric contractions have been shown to carry a lower metabolic cost than concentric dynamic contractions. This leads to the hypothesis that performing maximal isometric contractions during warm-up may enhance subsequent explosive performance, while limiting the detrimental effects of accumulated fatigue [69]. Maximal isometric contractions during warm-up may also increase hamstring stiffness [13] because such stiffness is significantly correlated with isotonic muscle performance [70].

Numerous benefits have been attributed to isometric contraction training. Such training enables the application of force in rehabilitative settings to be firmly controlled within pain-free joint angles [71]. Isometric squat exercises that improve hamstring muscle contraction help to improve strength performance and kayaking sprints, relative to traditional exercises [72]. One study showed that isometric training resulted in improved hamstring isometric peak torque (9%), increased fascicle length (22%), and reduced pennation angle (−17%) in the biceps femoris long head [73]. Isometric training also enables the induction of force overload, since the maximal isometric force is larger than the force produced by concentric contractions [15]. Further, a practitioner who is familiar with the physical demands of a sport may be able to use isometric training to target specific weak areas in the range of motion, which can have a beneficial effect on performance [74] and injury prevention [9]. Finally, isometric contractions can also be used to provide analgesic relief and enable painless dynamic loading [17].

The previous study of isometric exercise suggested that a TheraBand (Hygenic Corp., Akron, OH, USA) was wrapped around the heel and the subject held the ends of the TheraBand in each hand [19]. The subject was instructed to keep the right knee locked in full extension and the hip in neutral internal and external rotation throughout the entire activity. The subject was then instructed to bring the right hip into full hip flexion by pulling on the TheraBand, attached to the foot, with both arms, making sure the knee remained locked in full extension at all times.

By performing a 70% maximal voluntary isometric contraction, study results show that a greater improvement in hamstring musculotendinous stiffness [13] and increasing the loading strategy with a heavy slow resistance program was helpful to the proximal hamstring tendinopathy [75].

Isometric exercises, thus, train the muscle to maintain its resistance ability against any force or under any conditions. They provide a simple and effective approach that can help to improve the hamstring muscles, and increase their resistance and power during activity. Most injuries to hamstrings occur because the muscles lack the strength to maintain the force or produce the amount of energy required.

5. Conclusions

The hamstring muscle group comprises the biceps femoris short and long heads, semitendinosus, and semimembranosus. Hamstring injuries are common in athletes of all levels, from beginners to professionals, and as is also explained, a myosin heavy chain does not appear to explain the high incidence of hamstring strain injury. Increasing hamstring muscle strength has traditionally been used as a preventative measure to reduce the risk of injury. Resistance training, consisting of both isotonic and isometric exercise, is widely used to treat hamstring muscle injuries. Isotonic exercise comprises both eccentric and concentric muscular contractions, and is defined as exercise that changes the muscle length, while isometric exercise results in no change in the muscle length. Isotonic training has been well described as the most effective method for increasing force and power generation, and isometric exercise improves hamstring muscle stiffness. While each of these exercise types has its own beneficial effects on hamstring muscle strength, judiciously combining the two may result in the greatest improvement in hamstring muscle strength for performance.

Applying both isotonic and isometric exercise may have different effects on the hamstring muscle. These exercises should be performed at low intensity and with high repetition by beginners, for 6–12 weeks. Once the muscle has adjusted to the exercise and requires a greater load to improve further, the load can be increased, and the volume or repetition of the exercise can be decreased. Most athletes use this type of exercise during the general preparation phase of their training, but the general population can use this exercise as well, depending on their needs. Muscle training is important for improving hamstring muscle capacity and reducing injury risk. Isometric exercise will train the muscles to maintain their ability to resist any force or under any conditions. Isometric exercise is a simple and effective approach to improve the resistance and power of the hamstring muscles during activity. Most hamstring injuries occur because the muscles lack the strength to produce the force or energy required.

Author Contributions: Conceptualization, all authors; methodology, A.F.W. and C. W.T.; software, A.F.W.; validation, all authors; investigation, all authors; data curation, A.F.W.; writing—original draft preparation, A.F.W. and C.-W.T., writing—review and editing, C.-W.C. and S.-C.L.; supervision, C.-W.C. and S.-C.L.; project administration, C.-W.C.; funding acquisition, C.-W.C. All authors have read and agreed to the published version of the manuscript.

Funding: This work was financially supported through research grants from the Ministry of Science and Technology, Taipei, Taiwan (MOST 110-2320-B-227-001-).

Conflicts of Interest: The authors declare no conflict of interest.

References

1. Ayuob, A.; Kayani, B.; Haddad, F.S. Acute Surgical Repair of Complete, Nonavulsion Proximal Semimembranosus Injuries in Professional Athletes. *Am. J. Sports Med.* **2020**, *48*, 2170–2177. [CrossRef] [PubMed]
2. Chu, S.K.; Rho, M.E. Hamstring Injuries in the Athlete: Diagnosis, Treatment, and Return to Play. *Curr. Sports Med. Rep.* **2016**, *15*, 184–190. [CrossRef] [PubMed]
3. Martin, R.L.; Cibulka, M.T.; Bolgla, L.A.; Koc, T.A., Jr.; Loudon, J.K.; Manske, R.C.; Weiss, L.; Christoforetti, J.J.; Heiderscheit, B.C. Hamstring Strain Injury in Athletes. *J. Orthop. Sports Phys. Ther.* **2022**, *52*, CPG1–CPG44. [CrossRef] [PubMed]
4. Gronwald, T.; Klein, C.; Hoenig, T.; Pietzonka, M.; Bloch, H.; Edouard, P.; Hollander, K. Hamstring injury patterns in professional male football (soccer): A systematic video analysis of 52 cases. *Br. J. Sports Med.* **2022**, *56*, 165–171. [CrossRef]
5. Lee, J.W.Y.; Mok, K.M.; Chan, H.C.K.; Yung, P.S.H.; Chan, K.M. Eccentric hamstring strength deficit and poor hamstring-to-quadriceps ratio are risk factors for hamstring strain injury in football: A prospective study of 146 professional players. *J. Sci. Med. Sport* **2018**, *21*, 789–793. [CrossRef]
6. Heiderscheit, B.C.; Sherry, M.A.; Silder, A.; Chumanov, E.S.; Thelen, D.G. Hamstring strain injuries: Recommendations for diagnosis, rehabilitation, and injury prevention. *J. Orthop. Sports Phys. Ther.* **2010**, *40*, 67–81. [CrossRef] [PubMed]
7. Danielsson, A.; Horvath, A.; Senorski, C.; Alentorn-Geli, E.; Garrett, W.E.; Cugat, R.; Samuelsson, K.; Hamrin Senorski, E. The mechanism of hamstring injuries—A systematic review. *BMC Musculoskelet. Disord.* **2020**, *21*, 641. [CrossRef]
8. Mendiguchia, J.; Alentorn-Geli, E.; Brughelli, M. Hamstring strain injuries: Are we heading in the right direction? *Br. J. Sports Med.* **2012**, *46*, 81–85. [CrossRef]
9. Van Beijsterveldt, A.M.; van de Port, I.G.; Vereijken, A.J.; Backx, F.J. Risk factors for hamstring injuries in male soccer players: A systematic review of prospective studies. *Scand J. Med. Sci. Sports* **2013**, *23*, 253–262. [CrossRef]
10. Goldman, E.F.; Jones, D.E. Interventions for preventing hamstring injuries: A systematic review. *Physiotherapy* **2011**, *97*, 91–99. [CrossRef]
11. Jones, P.A.; Thomas, C.; Dos'Santos, T.; McMahon, J.J.; Graham-Smith, P. The Role of Eccentric Strength in 180 degrees Turns in Female Soccer Players. *Sports* **2017**, *5*, 42. [CrossRef] [PubMed]
12. Chaabene, H.; Negra, Y.; Moran, J.; Prieske, O.; Sammoud, S.; Ramirez-Campillo, R.; Granacher, U. Effects of an Eccentric Hamstrings Training on Components of Physical Performance in Young Female Handball Players. *Int. J. Sports Physiol. Perform.* **2019**, *15*, 91–97. [CrossRef] [PubMed]
13. Blackburn, J.T.; Norcross, M.F. The effects of isometric and isotonic training on hamstring stiffness and anterior cruciate ligament loading mechanisms. *J. Electromyogr. Kinesiol.* **2014**, *24*, 98–103. [CrossRef]
14. Burgess, K.E.; Connick, M.J.; Graham-Smith, P.; Pearson, S.J. Plyometric vs. isometric training influences on tendon properties and muscle output. *J. Strength Cond. Res.* **2007**, *21*, 986–989. [CrossRef] [PubMed]
15. Oranchuk, D.J.; Storey, A.G.; Nelson, A.R.; Cronin, J.B. Isometric training and long-term adaptations: Effects of muscle length, intensity, and intent: A systematic review. *Scand J. Med. Sci. Sports* **2019**, *29*, 484–503. [CrossRef] [PubMed]
16. Rich, A.; Cook, J.L.; Hahne, A.J.; Rio, E.K.; Ford, J. Randomised, cross-over trial on the effect of isotonic and isometric exercise on pain and strength in proximal hamstring tendinopathy: Trial protocol. *BMJ Open Sport Exerc. Med.* **2021**, *7*, e000954. [CrossRef]
17. Rio, E.; Kidgell, D.; Purdam, C.; Gaida, J.; Moseley, G.L.; Pearce, A.J.; Cook, J. Isometric exercise induces analgesia and reduces inhibition in patellar tendinopathy. *Br. J. Sports Med.* **2015**, *49*, 1277–1283. [CrossRef]
18. Seagrave, R.A., 3rd; Perez, L.; McQueeney, S.; Toby, E.B.; Key, V.; Nelson, J.D. Preventive Effects of Eccentric Training on Acute Hamstring Muscle Injury in Professional Baseball. *Orthop. J. Sports Med.* **2014**, *2*, 2325967114535351. [CrossRef]
19. Nelson, R.T.; Bandy, W.D. Eccentric Training and Static Stretching Improve Hamstring Flexibility of High School Males. *J. Athl. Train.* **2004**, *39*, 254–258.
20. Delahunt, E.; McGroarty, M.; De Vito, G.; Ditroilo, M. Nordic hamstring exercise training alters knee joint kinematics and hamstring activation patterns in young men. *Eur. J. Appl. Physiol.* **2016**, *116*, 663–672. [CrossRef]
21. Marusic, J.; Vatovec, R.; Markovic, G.; Sarabon, N. Effects of eccentric training at long-muscle length on architectural and functional characteristics of the hamstrings. *Scand J. Med. Sci. Sports* **2020**, *30*, 2130–2142. [CrossRef] [PubMed]
22. Vaczi, M.; Fazekas, G.; Pilissy, T.; Cselko, A.; Trzaskoma, L.; Sebesi, B.; Tihanyi, J. The effects of eccentric hamstring exercise training in young female handball players. *Eur. J. Appl. Physiol.* **2022**, *122*, 955–964. [CrossRef] [PubMed]
23. Markovic, G.; Sarabon, N.; Boban, F.; Zoric, I.; Jelcic, M.; Sos, K.; Scappaticci, M. Nordic Hamstring Strength of Highly Trained Youth Football Players and Its Relation to Sprint Performance. *J. Strength Cond. Res.* **2020**, *34*, 800–807. [CrossRef] [PubMed]
24. Rodgers, C.D.; Raja, A. *Anatomy, Bony Pelvis and Lower Limb, Hamstring Muscle*; StatPearls: Tampa, FL, USA, 2022.
25. Koulouris, G.; Connell, D. Hamstring muscle complex: An imaging review. *Radiographics* **2005**, *25*, 571–586. [CrossRef]
26. Davis, K.W. Imaging of the hamstrings. *Semin. Musculoskelet. Radiol.* **2008**, *12*, 28–41. [CrossRef]
27. Farfan, E.; Rojas, S.; Olive-Vilas, R.; Rodriguez-Baeza, A. Morphological study on the origin of the semitendinosus muscle in the long head of biceps femoris. *Scand J. Med. Sci. Sports* **2021**, *31*, 2282–2290. [CrossRef]
28. Hsu, D.; Anand, P.; Mabrouk, A.; Chang, K.V. *Biceps Tendon Rupture of the Lower Limb*; StatPearls: Tampa, FL, USA, 2022.
29. Kellis, E. Intra- and Inter-Muscular Variations in Hamstring Architecture and Mechanics and Their Implications for Injury: A Narrative Review. *Sports Med.* **2018**, *48*, 2271–2283. [CrossRef]

30. Mendes, B.; Firmino, T.; Oliveira, R.; Neto, T.; Cruz-Montecinos, C.; Cerda, M.; Correia, J.P.; Vaz, J.R.; Freitas, S.R. Effects of knee flexor submaximal isometric contraction until exhaustion on semitendinosus and biceps femoris long head shear modulus in healthy individuals. *Sci. Rep.* **2020**, *10*, 16433. [CrossRef]
31. Woodley, S.J.; Mercer, S.R. Hamstring muscles: Architecture and innervation. *Cells Tissues Organs* **2005**, *179*, 125–141. [CrossRef]
32. Koulouris, G.; Connell, D. Evaluation of the hamstring muscle complex following acute injury. *Skeletal. Radiol.* **2003**, *32*, 582–589. [CrossRef]
33. Schiaffino, S.; Reggiani, C. Fiber types in mammalian skeletal muscles. *Physiol. Rev.* **2011**, *91*, 1447–1531. [CrossRef] [PubMed]
34. Lempainen, L.; Kosola, J.; Pruna, R.; Sinikumpu, J.J.; Valle, X.; Heinonen, O.; Orava, S.; Maffulli, N. Tears of biceps femoris, semimembranosus, and semitendinosus are not equal-a new individual muscle-tendon concept in athletes. *Scand J. Surg.* **2021**, *110*, 483–491. [CrossRef] [PubMed]
35. Opar, D.A.; Williams, M.D.; Shield, A.J. Hamstring strain injuries: Factors that lead to injury and re-injury. *Sports Med.* **2012**, *42*, 209–226. [CrossRef] [PubMed]
36. Paoletta, M.; Moretti, A.; Liguori, S.; Snichelotto, F.; Menditto, I.; Toro, G.; Gimigliano, F.; Iolascon, G. Ultrasound Imaging in Sport-Related Muscle Injuries: Pitfalls and Opportunities. *Medicina* **2021**, *57*, 1040. [CrossRef] [PubMed]
37. Brukner, P. Hamstring injuries: Prevention and treatment-an update. *Br. J. Sports Med.* **2015**, *49*, 1241–1244. [CrossRef]
38. Jarvinen, M.J.; Lehto, M.U. The effects of early mobilisation and immobilisation on the healing process following muscle injuries. *Sports Med.* **1993**, *15*, 78–89. [CrossRef]
39. Reynolds, J.F.; Noakes, T.D.; Schwellnus, M.P.; Windt, A.; Bowerbank, P. Non-steroidal anti-inflammatory drugs fail to enhance healing of acute hamstring injuries treated with physiotherapy. *S. Afr. Med. J.* **1995**, *85*, 517–522.
40. Engebretsen, L.; Steffen, K.; Alsousou, J.; Anitua, E.; Bachl, N.; Devilee, R.; Everts, P.; Hamilton, B.; Huard, J.; Jenoure, P.; et al. IOC consensus paper on the use of platelet-rich plasma in sports medicine. *Br. J. Sports Med.* **2010**, *44*, 1072–1081. [CrossRef]
41. Guerrero, M.; Guiu-Comadevall, M.; Cadefau, J.A.; Parra, J.; Balius, R.; Estruch, A.; Rodas, G.; Bedini, J.L.; Cusso, R. Fast and slow myosins as markers of muscle injury. *Br. J. Sports Med.* **2008**, *42*, 581–584, discussion 584. [CrossRef]
42. Evangelidis, P.E.; Massey, G.J.; Ferguson, R.A.; Wheeler, P.C.; Pain, M.T.G.; Folland, J.P. The functional significance of hamstrings composition: Is it really a "fast" muscle group? *Scand J. Med. Sci. Sports* **2017**, *27*, 1181–1189. [CrossRef]
43. Metzger, J.M.; Moss, R.L. Calcium-sensitive cross-bridge transitions in mammalian fast and slow skeletal muscle fibers. *Science* **1990**, *247*, 1088–1090. [CrossRef] [PubMed]
44. Brooks, J.H.; Fuller, C.W.; Kemp, S.P.; Reddin, D.B. Incidence, risk, and prevention of hamstring muscle injuries in professional rugby union. *Am. J. Sports Med.* **2006**, *34*, 1297–1306. [CrossRef] [PubMed]
45. Sherry, M.A.; Best, T.M. A comparison of 2 rehabilitation programs in the treatment of acute hamstring strains. *J. Orthop. Sports Phys. Ther.* **2004**, *34*, 116–125. [CrossRef] [PubMed]
46. Cacchio, A.; Rompe, J.D.; Furia, J.P.; Susi, P.; Santilli, V.; De Paulis, F. Shockwave therapy for the treatment of chronic proximal hamstring tendinopathy in professional athletes. *Am. J. Sports Med.* **2011**, *39*, 146–153. [CrossRef] [PubMed]
47. Goom, T.S.; Malliaras, P.; Reiman, M.P.; Purdam, C.R. Proximal Hamstring Tendinopathy: Clinical Aspects of Assessment and Management. *J. Orthop. Sports Phys. Ther.* **2016**, *46*, 483–493. [CrossRef]
48. Davenport, K.L.; Campos, J.S.; Nguyen, J.; Saboeiro, G.; Adler, R.S.; Moley, P.J. Ultrasound-Guided Intratendinous Injections With Platelet-Rich Plasma or Autologous Whole Blood for Treatment of Proximal Hamstring Tendinopathy: A Double-Blind Randomized Controlled Trial. *J. Ultrasound Med.* **2015**, *34*, 1455–1463. [CrossRef]
49. Golik-Peric, D.; Drapsin, M.; Obradovic, B.; Drid, P. Short-term isokinetic training versus isotonic training: Effects on asymmetry in strength of thigh muscles. *J. Hum. Kinet.* **2011**, *30*, 29–35. [CrossRef]
50. Brockett, C.L.; Morgan, D.L.; Proske, U. Human hamstring muscles adapt to eccentric exercise by changing optimum length. *Med. Sci. Sports Exerc.* **2001**, *33*, 783–790. [CrossRef]
51. Padulo, J.; Laffaye, G.; Chamari, K.; Concu, A. Concentric and eccentric: Muscle contraction or exercise? *Sports Health* **2013**, *5*, 306. [CrossRef]
52. Hody, S.; Croisier, J.L.; Bury, T.; Rogister, B.; Leprince, P. Eccentric Muscle Contractions: Risks and Benefits. *Front. Physiol.* **2019**, *10*, 536. [CrossRef]
53. Duchateau, J.; Baudry, S. Insights into the neural control of eccentric contractions. *J. Appl. Physiol.* **2014**, *116*, 1418–1425. [CrossRef] [PubMed]
54. LaStayo, P.C.; Woolf, J.M.; Lewek, M.D.; Snyder Mackler, L.; Reich, T.; Lindstedt, S.L. Eccentric muscle contractions. Their contribution to injury, prevention, rehabilitation, and sport. *J. Orthop. Sports Phys. Ther.* **2003**, *33*, 557–571. [CrossRef] [PubMed]
55. Gault, M.L.; Willems, M.E. Aging, functional capacity and eccentric exercise training. *Aging Dis.* **2013**, *4*, 351–363. [CrossRef] [PubMed]
56. Gerard, R.; Gojon, L.; Decleve, P.; Van Cant, J. The Effects of Eccentric Training on Biceps Femoris Architecture and Strength: A Systematic Review With Meta-Analysis. *J. Athl. Train.* **2020**, *55*, 501–514. [CrossRef]
57. Blazevich, A.J.; Cannavan, D.; Coleman, D.R.; Horne, S. Influence of concentric and eccentric resistance training on architectural adaptation in human quadriceps muscles. *J. Appl. Physiol.* **2007**, *103*, 1565–1575. [CrossRef]
58. Vogt, M.; Hoppeler, H.H. Eccentric exercise: Mechanisms and effects when used as training regime or training adjunct. *J. Appl. Physiol.* **2014**, *116*, 1446–1454. [CrossRef]

59. Kumagai, K.; Abe, T.; Brechue, W.F.; Ryushi, T.; Takano, S.; Mizuno, M. Sprint performance is related to muscle fascicle length in male 100-m sprinters. *J. Appl. Physiol.* **2000**, *88*, 811–816. [CrossRef]

60. Blazevich, A.J.; Gill, N.D.; Bronks, R.; Newton, R.U. Training-specific muscle architecture adaptation after 5-wk training in athletes. *Med. Sci. Sports Exerc.* **2003**, *35*, 2013–2022. [CrossRef]

61. Alegre, L.M.; Jimenez, F.; Gonzalo-Orden, J.M.; Martin-Acero, R.; Aguado, X. Effects of dynamic resistance training on fascicle length and isometric strength. *J. Sports Sci.* **2006**, *24*, 501–508. [CrossRef]

62. Kaminski, T.W.; Wabbersen, C.V.; Murphy, R.M. Concentric versus enhanced eccentric hamstring strength training: Clinical implications. *J. Athl. Train.* **1998**, *33*, 216–221.

63. Unlu, G.; Cevikol, C.; Melekoglu, T. Comparison of the Effects of Eccentric, Concentric, and Eccentric-Concentric Isotonic Resistance Training at Two Velocities on Strength and Muscle Hypertrophy. *J. Strength Cond. Res.* **2020**, *34*, 337–344. [CrossRef] [PubMed]

64. Kraemer, W.J.; Ratamess, N.A.; French, D.N. Resistance training for health and performance. *Curr. Sports Med. Rep.* **2002**, *1*, 165–171. [CrossRef] [PubMed]

65. Stasinopoulos, D.; Stasinopoulos, I. Comparison of effects of eccentric training, eccentric-concentric training, and eccentric-concentric training combined with isometric contraction in the treatment of lateral elbow tendinopathy. *J. Hand Ther.* **2017**, *30*, 13–19. [CrossRef]

66. Kubo, K.; Ohgo, K.; Takeishi, R.; Yoshinaga, K.; Tsunoda, N.; Kanehisa, H.; Fukunaga, T. Effects of isometric training at different knee angles on the muscle-tendon complex in vivo. *Scand J. Med. Sci. Sports* **2006**, *16*, 159–167. [CrossRef] [PubMed]

67. Noorkoiv, M.; Nosaka, K.; Blazevich, A.J. Effects of isometric quadriceps strength training at different muscle lengths on dynamic torque production. *J. Sports Sci.* **2015**, *33*, 1952–1961. [CrossRef]

68. Schott, J.; McCully, K.; Rutherford, O.M. The role of metabolites in strength training. II. Short versus long isometric contractions. *Eur. J. Appl. Physiol. Occup. Physiol.* **1995**, *71*, 337–341. [CrossRef] [PubMed]

69. Vargas-Molina, S.; Salgado-Ramirez, U.; Chulvi-Medrano, I.; Carbone, L.; Maroto-Izquierdo, S.; Benitez-Porres, J. Comparison of post-activation performance enhancement (PAPE) after isometric and isotonic exercise on vertical jump performance. *PLoS ONE* **2021**, *16*, e0260866. [CrossRef]

70. Langan, S.P.; Murphy, T.; Johnson, W.M.; Carreker, J.D.; Riemann, B.L. The Influence of Active Hamstring Stiffness on Markers of Isotonic Muscle Performance. *Sports* **2021**, *9*, 70. [CrossRef]

71. Krebs, D.E.; Staples, W.H.; Cuttita, D.; Zickel, R.E. Knee joint angle: Its relationship to quadriceps femoris activity in normal and postarthrotomy limbs. *Arch. Phys. Med. Rehabil.* **1983**, *64*, 441–447.

72. Lum, D.; Barbosa, T.M.; Balasekaran, G. Sprint Kayaking Performance Enhancement by Isometric Strength Training Inclusion: A Randomized Controlled Trial. *Sports* **2021**, *9*, 16. [CrossRef]

73. Ribeiro-Alvares, J.B.; Marques, V.B.; Vaz, M.A.; Baroni, B.M. Four Weeks of Nordic Hamstring Exercise Reduce Muscle Injury Risk Factors in Young Adults. *J. Strength Cond. Res.* **2018**, *32*, 1254–1262. [CrossRef] [PubMed]

74. Tsoukos, A.; Bogdanis, G.C.; Terzis, G.; Veligekas, P. Acute Improvement of Vertical Jump Performance After Isometric Squats Depends on Knee Angle and Vertical Jumping Ability. *J. Strength Cond. Res.* **2016**, *30*, 2250–2257. [CrossRef] [PubMed]

75. Krueger, K.; Washmuth, N.B.; Williams, T.D. The Management of Proximal Hamstring Tendinopathy in a Competitive Powerlifter with Heavy Slow Resistance Training—A Case Report. *Int. J. Sports Phys. Ther.* **2020**, *15*, 814–822. [CrossRef] [PubMed]

Review

The Effect of Velocity Loss on Strength Development and Related Training Efficiency: A Dose–Response Meta–Analysis

Xing Zhang [1], Siyuan Feng [2] and Hansen Li [1,*]

[1] Key Laboratory of Physical Fitness Evaluation and Motor Function Monitoring, Institute of Sports Science, College of Physical Education, Southwest University, Chongqing 400715, China
[2] Laboratory of Genetics, University of Wisconsin-Madison, Madison, WI 53706, USA
* Correspondence: hanson-swu@foxmail.com; Tel.: +86-18980730787

Abstract: The velocity loss method is often used in velocity–based training (VBT) to dynamically regulate training loads. However, the effects of velocity loss on maximum strength development and training efficiency are still unclear. Therefore, we conducted a dose–response meta–analysis aiming to fill this research gap. A systematic literature search was performed to identify studies on VBT with the velocity loss method via PubMed, Web of Science, Embase, EBSCO, and Cochrane. Controlled trials that compared the effects of different velocity losses on maximum strength were considered. One–repetition maximum (1RM) gain and 1RM gain per repetition were the selected outcomes to indicate the maximum strength development and its training efficiency. Eventually, nine studies with a total of 336 trained males (training experience/history $\geq$ 1 year) were included for analysis. We found a non–linear dose–response relationship (reverse U–shaped) between velocity loss and 1RM gain ($p_{dose-response\ relationship} < 0.05$, $p_{non-linear\ relationship} < 0.05$). Additionally, a negative linear dose–response relationship was observed between velocity loss and 1RM gain per repetition ($p_{dose-response\ relationship} < 0.05$, $p_{non-linear\ relationship} = 0.23$). Based on our findings, a velocity loss between 20 and 30% may be beneficial for maximum strength development, and a lower velocity loss may be more efficient for developing and maintaining maximum strength. Future research is warranted to focus on female athletes and the interaction of other parameters.

Keywords: velocity–based training; resistance training; one–repetition maximum; maximum strength; training efficiency

Citation: Zhang, X.; Feng, S.; Li, H. The Effect of Velocity Loss on Strength Development and Related Training Efficiency: A Dose–Response Meta–Analysis. *Healthcare* **2023**, *11*, 337. https://doi.org/10.3390/healthcare11030337

Academic Editors: João Paulo Brito and Rafael Oliveira

Received: 31 December 2022
Revised: 21 January 2023
Accepted: 22 January 2023
Published: 23 January 2023

1. Introduction

Maximum strength development is critical for improving force–time characteristics and thereby, various athletic performances [1], including but not limited to jumping [2,3], sprinting [4], changing direction [5], and even specific sports skills [1]. Furthermore, greater strength is beneficial for the structural strength of ligaments, tendons, tendons/ligaments to bone junctions, joint cartilage, and connective tissue sheaths within the muscle, thereby reducing the risks in sports [6]. For these reasons, many efforts have been made to find effective training methods for developing maximum strength.

In the past decades, traditional resistance training (RT) methods have been viewed as a "gold criterion" in prescribing training loads for maximizing muscular strength [7]. However, traditional RT loads are usually designed based on individuals' 1RM (one–repetition maximum) before starting an RT session [8]. Such predesigned training loads rarely consider athletes' daily fluctuation in training state or performance [9,10], which may lead to inappropriate training loads, lower training benefits, and even degeneration or injuries [10]. In this context, a series of regulable and flexible RT methods, known as autoregulation methods, were invented to avoid this limitation of traditional RT.

Velocity loss (VL) is a critical index/parameter of velocity–based training (VBT) and is often applied to determine the number of repetitions in a training set [9]. Specifically, a training set will be called to stop when the velocity loss exceeds a target value. To date, some scholars have focused on the role of velocity loss methods in developing athletic performances [11–14], but the optimal velocity loss for maximizing strength adaption is still inconclusive.

In practice, competitive athletes usually have complex training arrangements, especially in competitive seasons. They may have fewer RT sessions and, in turn, a lower maximum strength [15,16]. Therefore, how to maintain and enhance maximum strength with a lower RT volume becomes a valuable research question. Several previous studies on VBT have implied that a lower velocity loss and its higher counterpart may be or similarly effective in maximum strength development, but the lower velocity loss may result in a lower training volume under the same relative intensity [17,18]. In other words, one repetition in a training intervention may lead to a greater gain of maximum strength on average under a lower velocity setting. This topic of training efficiency, however, has only been studied concerning traditional RT [19], while the relationship between velocity loss and training efficiency in VBT is understudied yet. Therefore, we conducted this dose–response meta–analysis aiming to:

1. Examine the effect of velocity loss on maximum strength development;
2. Examine the effect of velocity loss on the efficiency of maximum strength development.

2. Materials and Methods

This review was conducted following the recommendations for the preferred reporting items for systematic reviews and meta–analyses (PRISMA) [20]. However, due to the nature of the research question, outcome, and paper organization, we could not register our review in PROSPERO or other alternative databases [21].

2.1. Eligibility Criteria

Aiming to serve athletic training, studies on sportspeople who have at least a year of RT experience/history were considered. Studies that selected squat, bench press, and deadlift as the major training event were considered [22]. To investigate the long–term effect, interventions that lasted for at least four weeks were considered [23]. Guided by previous reviews, we used the one repetition maximum (1RM) for indicating maximum strength [24].

Our eligibility criteria obeyed the PICOS principle as follows [25]:

–P (population): people who have RT experience/history for at least one year;

–I (intervention): at least four weeks of VBT intervention that selected squat, bench press, and deadlift as their training events;

–C (comparison): compared the effect of VBT with different velocity losses on maximum strength;

–O (outcomes): 1RM was measured in the training events;

–S (study design): controlled trials based on pre–post designs that evaluated the effect of different velocity loss methods on maximum strength.

2.2. Information Sources and Search Strategy

A systematic literature search was conducted through the following English electronic databases: PubMed, Web of Science, Embase, EBSCO, and Cochrane. The time range was between the inception of each database and January 12, 2023. The detailed search strategy is shown in Table 1.

Table 1. Searching strategy for study inclusion.

Steps	Searching Command	Field
#1	Velocity–based training OR VBT OR velocity–based resistance training OR VBRT OR velocity loss OR VL	Title or abstract
#2	Maximum strength OR one–repetition maximum OR 1RM OR strength performance	Title or abstract
#3	#1 AND #2	

2.3. Study Selection

Two authors (XZ and HSL) independently selected relevant studies based on their titles, abstracts, and full texts. Any discrepancies were resolved by discussion or judgments from another author (SYF) [26].

2.4. Data Collection Process and Data Items

Two authors (XZ and HSL) independently extracted data from the included articles, including title, publication year, author name, study design, participant profile, sample size, intervention, measurement, and outcomes. Any discrepancies were resolved by discussion or judgments from a third author (SYF). Since 1RM is the most commonly used measure for evaluating maximum strength performance [8,27], it was selected as the outcome in the current study.

2.5. Risk of Bias Assessment

The risk of bias in the included studies was assessed using the PEDro scale (physiotherapy evidence–based database). According to the previous study, the PEDro scale was evaluated to have a high reliability and validity [28,29]. Items 2 to 11 were used to calculate the PEDro score. The methodology criteria were scored as: "Yes" (one point), "No" (zero points), or "Don't know" (zero points). The total PEDro score indicates the overall quality of each study (9–10 = excellent; 6–8 good; 4–5 = fair; and <4 = poor). Two authors (XZ and HSL) independently assessed the potential risk of bias. Any discrepancies were resolved by discussion or judgments from another author (SYF).

2.6. Statistical Analysis

2.6.1. Variable Calculation

The maximum strength development was represented by the 1RM gain throughout the intervention, which was calculated as the mean change from the baseline to the final test (post–pre interventions). The change in SD of the 1RM was calculated according to the Cochrane Handbook for Systematic Reviews of Interventions (Section 6.5.2.8) [30].

As the authors of the included studies did not provide correlations upon our request, we followed the method of others and assumed a conservative correlation coefficient of 0.5 for calculating [31]. This formula is as follows:

$$SD_{change} = \sqrt{\left(SD_{baseline}^2 + SD_{final}^2 - \left(2 \times 0.5 \times SD_{baseline} \times SD_{final}\right)\right)}$$

where SD indicates the standard deviation, and baseline and final indicate the outcome measured before and after an intervention.

The 1RM gain per repetition was used to assess the efficiency of different velocity loss methods and was calculated as the mean change (post–pre) in the maximum strength results divided by the total repetition throughout the intervention.

2.6.2. Dose–Response Meta–Analysis

The dose–response meta–analysis was carried out to examine the "velocity loss–1RM gain" and the "velocity loss–1RM gain per repetition" relationships. The robust error meta–regression (REMR) model was used for analyzing dose–response relationships [32]. This is a one–stage method that considers all included studies as a whole and treats each study as a cluster in order to validate the fitting of the dose–response curve. In addition, in order to consider both linearity and non–linearity in one model, we used the restricted cubic spline (RCS) function to fit the dose–response trend. The random knots were used in the RCS function [33]. Three random knots divided the data into four pieces, and the dose–response curve was fitted within each piece and was further smoothed at the knots. The first and last pieces were restricted as linear, while for the second and third pieces, a cubic spline was fitted. When the cubic and quadratic terms in the function are equal to zero, the RCS automatically degrades into a simple linear function. The Wald test was used to test the probability that the cubic and quadratic terms equal zero [34]. Egger's test was conducted to evaluate the publication bias [35]. All analyses were performed using STATA 14.0 (StataCorp LLC, Texas, US), and a p–value < 0.05 was considered statistically significant [36,37].

3. Results

3.1. Study Selection

The systematic literature search and study selection processes are outlined in Figure 1. A total of 3438 studies were identified in the search. After that, 3423 were excluded due to duplication, title, and abstract. Six studies were excluded for inappropriate controls or missing outcomes. Eventually, nine studies were included in the current study.

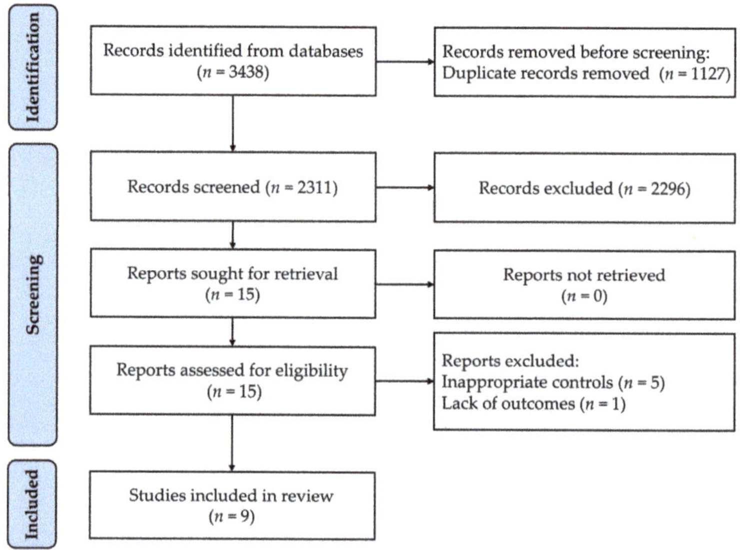

Figure 1. Flow diagram of the screening and selection of studies.

3.2. Study Characteristics

Nine studies and 336 subjects were included in the current study (Table 2). The studies were published between 2017 to 2022. All subjects were trained males (at least one year of RT history or from professional clubs). Most studies ($n = 6$) deployed an 8–week VBT intervention, and three other studies respectively performed a 5–week, a 6–week, and a 7–week VBT intervention, respectively. Most studies ($n = 6$) selected squat as the training events, and three studies used bench presses. All studies tested the 1RM as the primary outcome. All the included studies were controlled for relative intensity, the number of sets, and recovery time/interval. The range of velocity was between 0% and 50%.

Table 2. The characteristics of the included studies.

Authors	Velocity Loss (Sample)	Gender	Age (Years)	Experience (Years)	Intervention (Weeks)	Training Design	Total (Repetitions)	Outcome
Galiano, et al. [38]	VL5 (n = 15) VL20 (n = 13)	Male	23.0 ± 3.2	At least 1.5 years	7 weeks	Event: squats Intensity: 50% 1RM Set: 3 sets Recover time: 3 min Frequency: twice a week	VL5 (156.9) VL20 (480.5)	1RM
Rodríguez-Rosell, et al. [39]	VL10 (n = 11) VL30 (n = 11) VL45 (n = 11)	Male	22.8 ± 3.9	At least 1 year	8 weeks	Event: squats Intensity: 55–70% 1RM Set: 3 sets Recover time: 4 min Frequency: twice a week	VL10 (180.8) VL30 (347.9) VL45 (501.1)	1RM
Rodríguez-Rosell, et al. [40]	VL10 (n = 12) VL30 (n = 13)	Male	≈22.49	At least 1 year	8 weeks	Event: squats Intensity: 70–85% 1RM Set: 3 sets Recover time: 4 min Frequency: twice a week	VL10 (109.6) VL30 (228.0)	1RM
Rodiles-Guerrero, et al. [41]	VL10 (n = 15) VL30 (n = 15) VL50 (n = 15)	Male	23.0 + 2.0	At least 1 year	5 weeks	Event: bench presses Intensity: 65–85% 1RM Set: 4 sets Recover time: 3 min Frequency: three times a week	VL10 (211.1) VL30 (398.1) VL50 (444.4)	1RM
Rodiles-Guerrero, et al. [42]	VL0 (n = 12) VL15 (n = 13) VL25 (n = 13) VL50 (n = 12)	Male	23.3 + 3.3	At least 1.5 years	8 weeks	Event: bench presses Intensity: 55–70% 1RM Set: 3 sets Recover time: 4 min Frequency: twice a week	VL0 (48) VL15 (189.4) VL25 (310.2) VL50 (490.9)	1RM
Pareja-Blanco, et al. [43]	VL15 (n = 8) VL30 (n = 8)	Male	23.8 ± 3.4	Professional soccer club	6 weeks	Event: squats Intensity: 50–70% 1RM Set: 2–3 sets Recover time: 4 min Frequency: three times a week	VL15 (251.2) VL30 (414.6)	1RM
Pareja-Blanco, et al. [44]	VL20 (n = 12) VL40 (n = 10)	Male	22.7 ± 1.9	At least 1.5 years	8 weeks	Event: squats Intensity: 69–85% 1RM Set: 3 sets Recover time: 4 min Frequency: twice a week	VL20 (185.9) VL40 (310.5)	1RM
Pareja-Blanco, et al. [18]	VL0 (n = 14) VL10 (n = 14) VL20 (n = 13) VL40 (n = 14)	Male	24.1 ± 4.3	At least 1.5 years	8 weeks	Event: squats Intensity: 70–85% 1RM Set: 3 sets Recover time: 4 min Frequency: twice a week	VL0 (48.0) VL10 (143.6) VL20 (168.5) VL40 (305.6)	1RM
Pareja-Blanco, et al. [17]	VL0 (n = 15) VL15 (n = 16) VL25 (n = 15) VL50 (n = 16)	Male	24.1 ± 4.3	At least 1.5 years	8 weeks	Event: bench presses Intensity: 70–85% 1RM Set: 3 sets Recover time: 4 min Frequency: twice a week	VL0 (48.0) VL15 (136.6) VL25 (191.1) VL50 (316.4)	1RM

Note: VL, velocity loss; 1RM, one–repetition maximum; The same raining variables, if the main training variable is the same within the study, including the relative intensity, number of sets, and recover time.

3.3. Risk of Bias

Seven studies were assessed as good and two as fair quality according to the PEDio scale (Table 3). All studies were short in blinding, including subject, therapist, and assessor blinding.

Table 3. The risk of bias assessment for the included studies.

Studies	PEDro Item											Assessment
	1	2	3	4	5	6	7	8	9	10	11	
Galiano, et al. [38]	Yes	1	–	1	–	–	–	1	1	1	1	good
Rodríguez-Rosell, et al. [39]	Yes	1	–	1	–	–	–	1	1	1	1	good
Rodríguez-Rosell, et al. [40]	Yes	1	–	1	–	–	–	1	1	1	1	good
Rodiles-Guerrero, et al. [41]	Yes	1	–		–	–	–	1	1	1	1	fair
Rodiles-Guerrero, et al. [42]	Yes	1	–	1	–	–	–	1	1	1	1	good
Pareja-Blanco, et al. [43]	Yes	1	–	1	–	–	–		1	1	1	fair
Pareja-Blanco, et al. [44]	Yes	1	–	1	–	–	–	1	1	1	1	good
Pareja-Blanco, et al. [18]	Yes	1	–	1	–	–	–	1	1	1	1	good
Pareja-Blanco, et al. [17]	Yes	1	–	1	–	–	–	1	1	1	1	good

Note: Item 1. Eligibility criteria were specified. 2. Subjects were randomly allocated to groups (in a crossover study, subjects were randomly allocated an order in which treatments were received). 3. Allocation was concealed. 4. The groups were similar at baseline regarding the most important prognostic indicators. 5. There was blinding of all subjects. 6. There was blinding of all therapists who administered the therapy. 7. There was blinding of all assessors who measured at least one key outcome. 8. Measures of at least one key outcome were obtained from more than 85% of the subjects initially allocated to groups. 9. All subjects for whom outcome measures were available received the treatment or control condition as allocated, or where this was not the case, data for at least one key outcome was analyzed by "intention to treat." 10. The results of between–group statistical comparisons were reported for at least one key outcome. 11. This study provides both point measures and measures of variability for at least one key outcome.

3.4. Dose–Response Meta–Analysis

3.4.1. Maximum Strength Gain

Nine studies were included in the dose–response meta–analysis concerning the effect of velocity loss on maximum strength development. Figure 2 and Table 4 demonstrate a non–linear dose–response relationship (reverse U–shaped) between velocity loss and 1RM gain ($p_{dose-response\ relationship} < 0.05$, $p_{non-linear\ relationship} < 0.05$). The 1RM gain increased and then decreased with the velocity loss, and the greatest values were observed between VL20% and 30%.

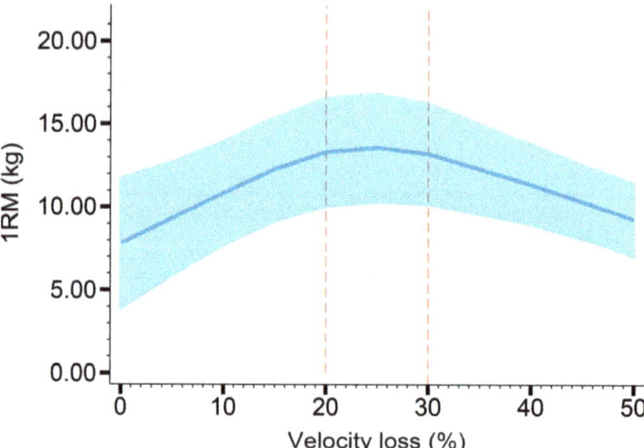

Figure 2. The dose–response curve of 1RM gain.

Table 4. The dose–specific effects of velocity loss on maximum strength gain and the model fit.

Velocity	ES	95%CI
VL0	7.82	3.84 to 11.80
VL5	9.34	5.88 to 12.79
VL10	10.86	7.68 to 14.03
VL15	12.30	9.12 to 15.49
VL20	13.32	10.00 to 16.64
VL25	13.59	10.28 to 16.89
VL30	13.21	10.12 to 16.30
VL40	11.38	8.95 to 13.82
VL45	10.33	8.07 to 12.58
VL50	9.27	7.02 to 11.53

Wald test		
Dose–response relationship		$p < 0.05$
Non–linear relationship		$p < 0.01$

Note: ES, estimate; CI, confidence interval.

3.4.2. Maximum Strength Gain Per Repetition

Nine studies were included in the dose–response meta–analysis concerning the effect of velocity loss on the efficiency of maximum strength development. Figure 3 and Table 5 demonstrate a linear dose–response relationship between velocity loss and 1RM gain per repetition ($p_{dose–response\ relationship} < 0.05$, $p_{non–linear\ relationship} = 0.23$). The efficiency consistently decreased with velocity loss, and the highest efficiency was observed at VL0%.

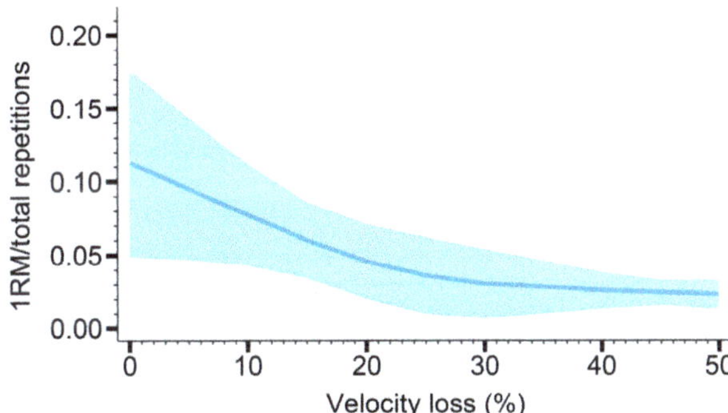

Figure 3. The dose–response curve of the maximum strength gain per repetition.

Table 5. The dose–specific effects of velocity loss on the maximum strength gain per repetition and the model fit.

Velocity	ES	95%CI
VL0	0.11	0.05 to 0.18
VL5	0.10	0.05 to 0.14
VL10	0.08	0.04 to 0.11
VL15	0.06	0.03 to 0.09
VL20	0.05	0.02 to 0.07
VL25	0.04	0.01 to 0.06
VL30	0.03	0.01 to 0.05
VL40	0.03	0.01 to 0.04
VL45	0.02	0.02 to 0.03
VL50	0.02	0.01 to 0.03

Wald test		
Dose–response relationship		$p < 0.05$
Non–linear relationship		$p = 0.23$

Note: ES, estimate; CI, confidence interval.

3.5. Publication Bias Assessment

The Egger's test revealed a significant publication bias in the results of 1RM gain ($p < 0.05$) and 1RM gain per repetition ($p < 0.05$), indicating a negative impact on our estimation.

4. Discussion

This study aimed to explore the effects of velocity loss of VBT on maximum strength development and training efficiency. The results generally showed an optimal range of velocity loss (20–30%) for maximum strength development. Moreover, we found that training efficiency broadly decreased with velocity loss, which means that a lower velocity loss may help to reduce the required training volume of RT sessions. These findings may offer some indications for the users and developers of VBT.

4.1. The Relationship between Velocity Loss on Maximum Strength Development

The optimal velocity loss range for maximizing strength adaptations has been widely discussed by strength and conditioning professionals [17,38,39,42]. For instance, a narrative review by Włodarczyk, et al. [45] demonstrated that a velocity loss between 10% and 20% could be helpful for maximum strength development among elite athletes. However, this finding is not supported by quantitative methods and may not address the existing controversies. A later study by Hernández-Belmonte and Pallarés [46] further provided a meta–analysis and compared the effects of the low–moderate (VL $\leq$ 25%) and the moderate–high velocity loss groups (VL > 25%) on maximum strength growth. Their observed differences were not statistically significant, which indicates a similar role of low–moderate and moderate–high velocity loss in strength development. However, the pairwise comparison that roughly divided the velocity range into two categories could not contribute to uncovering the relationship between velocity loss and training outcomes, particularly given the fact that some emerging studies have used other velocity loss settings. Our study, in comparison, included a wider velocity loss range and evaluated the dose–response relationship. We observed a reverse U–shaped relationship between velocity loss and 1RM gain. According to these results, the range between VL20% and VL30% might be most helpful for maximum strength development.

In physiological theories, changes in any athletic performance can be explained by the physiological adaptations induced by training stimulations [47,48]. Pareja-Blanco, et al. [44] found that VBT with a high velocity loss increased the cross–sectional area of slow–twitch fibers, suggesting a negative impact on maximum strength development. A recent review suggests that the velocity loss of VBT is negatively associated with the IIX (MHC–IIX) percentage and is positively associated with the myosin heavy chain I (MHC–I) percentage [9], which may explain the selective hypertrophy of skeletal muscles in VBT. In other words, a high velocity loss is beneficial for endurance–related performance instead of for maximum strength. These theories appear to go against our findings. One potential reason is that a too–low velocity loss may lead to an insufficient training volume. Specifically, our included studies controlled their training sets, so their velocity loss might largely influence their repetitions and therefore the training volume per session. For example, two included studies deployed six repetitions a week (1 rep $\times$ 3 sets $\times$ 2 sessions) for their VL0% groups. This "tiny" training volume may lead to insufficient stimulation for muscle fibers' growth and maximum strength development [49].

4.2. The Relationship between Velocity Loss and the Efficiency of Maximum Strength Development

Previous experimental studies and reviews have suggested the high efficiency of low velocity loss in promoting strength adaptions [17,18,46]. To our knowledge, no quantitative evidence has been found in relevant studies. This research gap has motivated this dose–response meta–analysis to examine the relationship between them. Our results revealed a consistently negative association between velocity loss and the 1RM gain per repetition. This finding indicates that a lower velocity loss may develop maximum strength with a lower training volume, which supports the efficiency of low velocity loss methods, although a low velocity loss is not the best option for maximizing strength adaption.

As mentioned in the Introduction, maintaining maximum strength is important for competitive athletes during a competitive season. Unfortunately, the truth is that the decline in strength is a highly common phenomenon in competitive seasons, especially among rugby [16], football [50], basketball [15], and baseball [51] players. A major reason for this phenomenon is the reduction in RT sessions resulting from frequent competitions [16,52,53]. Based on our findings, we recommend using VBT with a low velocity loss as an RT prescription during a competitive season because it may require a lower training volume and may therefore save energy and time for other training arrangements.

4.3. Limitations

Several limitations should be noted. First, only male samples were involved in the relevant interventions, so the findings of the current study may not be generalized to females. Second, samples were insufficient for assessing some velocity loss settings. For example, only one included study used VL5%, and no included study used VL35%. Thus, re–examination is needed when more relevant studies emerge. Third, we could not perform extra dose–response meta–analyses for squat or bench press due to insufficient samples. Hence, lower– and upper–body differences are unclear. Fourth, the current study only focused on maximum strength. Future research may check the effect of velocity loss on other indices, such as the maximal number of repetitions (MNR), countermovement jump (CMJ), and sprint time. Finally, and the most importantly, we only focused on the relationship between velocity loss and the outcomes of interest, while the impacts of other parameters (e.g., number of sets) were not controlled, and we could not reveal the potential interaction of different training parameters. These are the inherent limitations of such a research approach [54,55]. Thus, our findings must be applied to practice with caution.

5. Conclusions

In the current study, we examined the effects of velocity loss on maximum strength development and relevant training efficiency. Our results revealed a reverse U–shaped relationship between velocity loss and maximum strength gain, and the velocity loss range between 20 and 30% might most effectively contribute to maximum strength development among trained individuals. Meanwhile, a negative relationship was observed between velocity loss and the efficiency of maximum strength development, implying a direction to reduce total training volume and save energy for other training in some special scenarios. These findings may offer quantitative evidence to reinforce some previous studies and address some controversies, but they should also be carefully applied to practice given the several mentioned limitations.

Author Contributions: Conceptualization, H.L.; data curation, X.Z.; formal analysis, X.Z.; investigation, X.Z. and H.L.; methodology, X.Z.; project administration, S.F. and H.L.; supervision, S.F. and H.L.; visualization, X.Z.; writing—original draft, X.Z. and H.L.; writing—review and editing, X.Z. and H.L. All authors have read and agreed to the published version of the manuscript.

Funding: This research received no external funding.

Institutional Review Board Statement: Not applicable.

Informed Consent Statement: Not applicable.

Conflicts of Interest: The authors declare no conflict of interest.

References

1. Suchomel, T.J.; Nimphius, S.; Stone, M.H. The importance of muscular strength in athletic performance. *Sport. Med.* **2016**, *46*, 1419–1449. [CrossRef] [PubMed]
2. Kraska, J.M.; Ramsey, M.W.; Haff, G.G.; Fethke, N.; Sands, W.A.; Stone, M.E.; Stone, M.H. Relationship between strength characteristics and unweighted and weighted vertical jump height. *Int. J. Sports Physiol. Perform.* **2009**, *4*, 461–473. [CrossRef]
3. Wisløff, U.; Castagna, C.; Helgerud, J.; Jones, R.; Hoff, J. Strong correlation of maximal squat strength with sprint performance and vertical jump height in elite soccer players. *Br. J. Sport. Med.* **2004**, *38*, 285–288. [CrossRef]

4. Seitz, L.B.; Reyes, A.; Tran, T.T.; de Villarreal, E.S.; Haff, G.G. Increases in lower-body strength transfer positively to sprint performance: A systematic review with meta-analysis. *Sport. Med.* **2014**, *44*, 1693–1702. [CrossRef] [PubMed]

5. Spiteri, T.; Newton, R.U.; Binetti, M.; Hart, N.H.; Sheppard, J.M.; Nimphius, S. Mechanical determinants of faster change of direction and agility performance in female basketball athletes. *J. Strength Cond. Res.* **2015**, *29*, 2205–2214. [CrossRef]

6. Fleck, S.J.; Falkel, J.E. Value of resistance training for the reduction of sports injuries. *Sport. Med.* **1986**, *3*, 61–68. [CrossRef] [PubMed]

7. Mann, J.B.; Ivey, P.A.; Sayers, S.P. Velocity-based training in football. *Strength Cond. J.* **2015**, *37*, 52–57. [CrossRef]

8. Zhang, X.; Li, H.; Bi, S.; Luo, Y.; Cao, Y.; Zhang, G. Auto-Regulation Method vs. Fixed-Loading Method in Maximum Strength Training for Athletes: A Systematic Review and Meta-Analysis. *Front. Physiol.* **2021**, *12*, 651112. [CrossRef]

9. Weakley, J.; Mann, B.; Banyard, H.; McLaren, S.; Scott, T.; Garcia-Ramos, A. Velocity-based training: From theory to application. *Strength Cond. J.* **2021**, *43*, 31–49. [CrossRef]

10. Poliquin, C. Five steps to increasing the effectiveness of your strength training program. *NSCA J.* **1988**, *10*, 34–39. [CrossRef]

11. Montalvo-Pérez, A.; Alejo, L.B.; Valenzuela, P.L.; Gil-Cabrera, J.; Talavera, E.; Lucia, A.; Barranco-Gil, D. Traditional Versus Velocity-Based Resistance Training in Competitive Female Cyclists: A Randomized Controlled Trial. *Front. Physiol.* **2021**, *12*, 52. [CrossRef]

12. Held, S.; Hecksteden, A.; Meyer, T.; Donath, L. Improved Strength and Recovery after Velocity-Based Training: A Randomized Controlled Trial. *Int. J. Sport. Physiol. Perform.* **2021**, *16*, 1185–1193. [CrossRef] [PubMed]

13. Dorrell, H.F.; Moore, J.M.; Gee, T.I. Comparison of individual and group-based load-velocity profiling as a means to dictate training load over a 6-week strength and power intervention. *J. Sport. Sci.* **2020**, *38*, 2013–2020. [CrossRef] [PubMed]

14. Orange, S.T.; Hritz, A.; Pearson, L.; Jeffries, O.; Jones, T.W.; Steele, J. Comparison of the effects of velocity-based vs. traditional resistance training methods on adaptations in strength, power, and sprint speed: A systematic review, meta-analysis, and quality of evidence appraisal. *J. Sport. Sci.* **2022**, *40*, 1220–1234. [CrossRef]

15. Hoffman, J.R.; Fry, A.C.; Howard, R.; Maresh, C.M.; Kraemer, W.J. Strength, Speed and Endurance Changes During the Course of a Division I Basketball Season. *J. Strength Cond. Res.* **1991**, *5*, 144–149.

16. Argus, C.K.; Gill, N.D.; Keogh, J.W.L.; Hopkins, W.G.; Beaven, C.M. Changes in Strength, Power, and Steroid Hormones during a Professional Rugby Union Competition. *J. Strength Cond. Res.* **2009**, *23*, 1583–1592. [CrossRef]

17. Pareja-Blanco, F.; Alcazar, J.; Cornejo-Daza, P.J.; Sánchez-Valdepeñas, J.; Rodriguez-Lopez, C.; Hidalgo-de Mora, J.; Sánchez-Moreno, M.; Bachero-Mena, B.; Alegre, L.M.; Ortega-Becerra, M. Effects of velocity loss in the bench press exercise on strength gains, neuromuscular adaptations, and muscle hypertrophy. *Scand. J. Med. Sci. Sport.* **2020**, *30*, 2154–2166. [CrossRef]

18. Pareja-Blanco, F.; Alcazar, J.; Sánchez-Valdepeñas, J.; Cornejo-Daza, P.J.; Piqueras-Sanchiz, F.; Mora-Vela, R.; Sánchez-Moreno, M.; Bachero-Mena, B.; Ortega-Becerra, M.; Alegre, L.M. Velocity Loss as a Critical Variable Determining the Adaptations to Strength Training. *Med. Sci. Sport. Exerc.* **2020**, *52*, 1752–1762. [CrossRef]

19. Latella, C.; Owen, P.J.; Davies, T.; Spathis, J.; Mallard, A.; Van Den Hoek, D. Long-Term Adaptations in the Squat, Bench Press, and Deadlift: Assessing Strength Gain in Powerlifting Athletes. *Med. Sci. Sport. Exerc.* **2022**, *54*, 841–850. [CrossRef]

20. Moher, D.; Shamseer, L.; Clarke, M.; Ghersi, D.; Liberati, A.; Petticrew, M.; Shekelle, P.; Stewart, L.A. Preferred reporting items for systematic review and meta-analysis protocols (PRISMA-P) 2015 statement. *Syst. Rev.* **2015**, *4*, 1. [CrossRef]

21. Rico-González, M.; Pino-Ortega, J.; Clemente, F.; Los Arcos, A. Guidelines for performing systematic reviews in sports science. *Biol. Sport* **2022**, *39*, 463–471. [CrossRef] [PubMed]

22. Androulakis-Korakakis, P.; Fisher, J.P.; Steele, J. The minimum effective training dose required to increase 1RM strength in resistance-trained men: A systematic review and meta-analysis. *Sport. Med.* **2020**, *50*, 751–765. [CrossRef]

23. Grgic, J.; Schoenfeld, B.J.; Davies, T.B.; Lazinica, B.; Krieger, J.W.; Pedisic, Z. Effect of resistance training frequency on gains in muscular strength: A systematic review and meta-analysis. *Sport. Med.* **2018**, *48*, 1207–1220. [CrossRef] [PubMed]

24. Marshall, J.; Bishop, C.; Turner, A.; Haff, G.G. Optimal training sequences to develop lower body force, velocity, power, and jump height: A systematic review with meta-analysis. *Sport. Med.* **2021**, *51*, 1245–1271. [CrossRef] [PubMed]

25. Amir-Behghadami, M.; Janati, A. Population, Intervention, Comparison, Outcomes and Study (PICOS) design as a framework to formulate eligibility criteria in systematic reviews. *Emerg. Med. J.* **2020**, *37*, 387. [CrossRef] [PubMed]

26. Rahmati, M.; Gondin, J.; Malakoutinia, F. Effects of Neuromuscular Electrical Stimulation on Quadriceps Muscle Strength and Mass in Healthy Young and Older Adults: A Scoping Review. *Phys. Ther.* **2021**, *101*, pzab144. [CrossRef] [PubMed]

27. Zhang, X.; Feng, S.; Peng, R.; Li, H. The Role of Velocity-Based Training (VBT) in Enhancing Athletic Performance in Trained Individuals: A Meta-Analysis of Controlled Trials. *Int. J. Environ. Res. Public Health* **2022**, *19*, 9252. [CrossRef]

28. Maher, C.G.; Sherrington, C.; Herbert, R.D.; Moseley, A.M.; Elkins, M. Reliability of the PEDro scale for rating quality of randomized controlled trials. *Phys. Ther.* **2003**, *83*, 713–721. [CrossRef]

29. de Morton, N.A. The PEDro scale is a valid measure of the methodological quality of clinical trials: A demographic study. *Aust. J. Physiother.* **2009**, *55*, 129–133. [CrossRef]

30. Higgins, J.P.; Thomas, J.; Chandler, J.; Cumpston, M.; Li, T.; Page, M.J.; Welch, V.A. *Cochrane Handbook for Systematic Reviews of Interventions*; John Wiley & Sons: Hoboken, NJ, USA, 2019.

31. Chiu, S.; Sievenpiper, J.; De Souza, R.; Cozma, A.; Mirrahimi, A.; Carleton, A.; Ha, V.; Di Buono, M.; Jenkins, A.; Leiter, L. Effect of fructose on markers of non-alcoholic fatty liver disease (NAFLD): A systematic review and meta-analysis of controlled feeding trials. *Eur. J. Clin. Nutr.* **2014**, *68*, 416–423. [CrossRef]

32. Xu, C.; Doi, S.A. The robust error meta-regression method for dose–response meta-analysis. *JBI Evid. Implement.* **2018**, *16*, 138–144. [CrossRef]
33. Durrleman, S.; Simon, R. Flexible regression models with cubic splines. *Stat. Med.* **1989**, *8*, 551–561. [CrossRef]
34. Xu, C.; Liu, Y.; Jia, P.-L.; Li, L.; Liu, T.-Z.; Cheng, L.-L.; Deng, K.; Borhan, A.; Thabane, L.; Sun, X. The methodological quality of dose-response meta-analyses needed substantial improvement: A cross-sectional survey and proposed recommendations. *J. Clin. Epidemiol.* **2019**, *107*, 1–11. [CrossRef] [PubMed]
35. Begg, C.B.; Mazumdar, M. Operating characteristics of a rank correlation test for publication bias. *Biometrics* **1994**, *50*, 1088–1101. [CrossRef]
36. Zhao, Y.; Hu, F.; Feng, Y.; Yang, X.; Li, Y.; Guo, C.; Li, Q.; Tian, G.; Qie, R.; Han, M. Association of Cycling with Risk of All-Cause and Cardiovascular Disease Mortality: A Systematic Review and Dose–Response Meta-analysis of Prospective Cohort Studies. *Sport. Med.* **2021**, *51*, 1439–1448. [CrossRef]
37. Jayedi, A.; Gohari, A.; Shab-Bidar, S. Daily Step Count and All-Cause Mortality: A Dose–Response Meta-analysis of Prospective Cohort Studies. *Sport. Med.* **2022**, *52*, 89–99. [CrossRef]
38. Galiano, C.; Pareja-Blanco, F.; Hidalgo de Mora, J.; Sáez de Villarreal, E. Low-velocity loss induces similar strength gains to moderate-velocity loss during resistance training. *J. Strength Cond. Res.* **2022**, *36*, 340–345. [CrossRef]
39. Rodríguez-Rosell, D.; Yáñez-García, J.M.; Mora-Custodio, R.; Sánchez-Medina, L.; Ribas-Serna, J.; González-Badillo, J.J. Effect of velocity loss during squat training on neuromuscular performance. *Scand. J. Med. Sci. Sport.* **2021**, *31*, 1621–1635. [CrossRef]
40. Rodriguez-Rosell, D.; Yáñez-García, J.M.; Mora-Custodio, R.; Pareja-Blanco, F.; Ravelo-García, A.G.; Ribas-Serna, J.; González-Badillo, J.J. Velocity-based resistance training: Impact of velocity loss in the set on neuromuscular performance and hormonal response. *Appl. Physiol. Nutr. Metab.* **2020**, *45*, 817–828. [CrossRef]
41. Rodiles-Guerrero, L.; Pareja-Blanco, F.; León-Prados, J.A. Effect of velocity loss on strength performance in bench press using a weight stack machine. *Int. J. Sport. Med.* **2020**, *41*, 921–928. [CrossRef]
42. Rodiles-Guerrero, L.; Cornejo-Daza, P.J.; Sánchez-Valdepeñas, J.; Alcazar, J.; Rodriguez-López, C.; Sánchez-Moreno, M.; Alegre, L.M.; León-Prados, J.A.; Pareja-Blanco, F. Specific Adaptations to 0%, 15%, 25%, and 50% Velocity-Loss Thresholds during Bench Press Training. *Int. J. Sport. Physiol. Perform.* **2022**, *17*, 1–11. [CrossRef] [PubMed]
43. Pareja-Blanco, F.; Sánchez-Medina, L.; Suárez-Arrones, L.; González-Badillo, J.J. Effects of velocity loss during resistance training on performance in professional soccer players. *Int. J. Sport. Physiol. Perform.* **2017**, *12*, 512–519. [CrossRef] [PubMed]
44. Pareja-Blanco, F.; Rodríguez-Rosell, D.; Sánchez-Medina, L.; Sanchis-Moysi, J.; Dorado, C.; Mora-Custodio, R.; Yáñez-García, J.M.; Morales-Alamo, D.; Pérez-Suárez, I.; Calbet, J. Effects of velocity loss during resistance training on athletic performance, strength gains and muscle adaptations. *Scand. J. Med. Sci. Sport.* **2017**, *27*, 724–735. [CrossRef] [PubMed]
45. Włodarczyk, M.; Adamus, P.; Zieliński, J.; Kantanista, A. Effects of Velocity-Based Training on Strength and Power in Elite Athletes—A Systematic Review. *Int. J. Environ. Res. Public Health* **2021**, *18*, 5257. [CrossRef] [PubMed]
46. Hernández-Belmonte, A.; Pallarés, J.G. Effects of Velocity Loss Threshold during Resistance Training on Strength and Athletic Adaptations: A Systematic Review with Meta-Analysis. *Appl. Sci.* **2022**, *12*, 4425. [CrossRef]
47. Fry, A.C. The role of resistance exercise intensity on muscle fibre adaptations. *Sport. Med.* **2004**, *34*, 663–679. [CrossRef]
48. Rahmati, M.; McCarthy, J.J.; Malakoutinia, F. Myonuclear permanence in skeletal muscle memory: A systematic review and meta-analysis of human and animal studies. *J. Cachexia Sarcopenia Muscle* **2022**, *13*, 2276–2297. [CrossRef] [PubMed]
49. Schoenfeld, B.J.; Ogborn, D.; Krieger, J.W. Dose-response relationship between weekly resistance training volume and increases in muscle mass: A systematic review and meta-analysis. *J. Sport. Sci.* **2017**, *35*, 1073–1082. [CrossRef]
50. Hrysomallis, C.; Buttifant, D. Influence of training years on upper-body strength and power changes during the competitive season for professional Australian rules football players. *J. Sci. Med. Sport* **2012**, *15*, 374–378. [CrossRef]
51. Toby, B.; Glidewell, E.; Morris, B.; Key, V.H.; Nelson, J.D.; Schroeppel, J.P.; Mar, D.; Melugin, H.; Bradshaw, S.; McIff, T. Strength of dynamic stabilizers of the elbow in professional baseball pitchers decreases during baseball season. *Orthop. J. Sport. Med.* **2015**, *3*, 2325967115S2325900163. [CrossRef]
52. Rønnestad, B.R.; Nymark, B.S.; Raastad, T. Effects of in-season strength maintenance training frequency in professional soccer players. *J. Strength Cond. Res.* **2011**, *25*, 2653–2660. [CrossRef]
53. Gorostiaga, E.M.; Granados, C.; Ibañez, J.; Gonzalez-Badillo, J.J.; Izquierdo, M. Effects of an entire season on physical fitness changes in elite male handball players. *Med. Sci. Sport. Exerc.* **2006**, *38*, 357. [CrossRef]
54. Lesinski, M.; Prieske, O.; Granacher, U. Effects and dose–response relationships of resistance training on physical performance in youth athletes: A systematic review and meta-analysis. *Br. J. Sport. Med.* **2016**, *50*, 781–795. [CrossRef]
55. Hickmott, L.M.; Chilibeck, P.D.; Shaw, K.A.; Butcher, S.J. The effect of load and volume autoregulation on muscular strength and hypertrophy: A systematic review and meta-analysis. *Sport. Med. Open* **2022**, *8*, 9. [CrossRef]

Article

Effects of Aquatic Training and Bicycling Training on Leg Function and Range of Motion in Amateur Athletes with Meniscal Allograft Transplantation during Intermediate-Stage Rehabilitation

Yake Chen [1], Yonghwan Kim [2] and Moonyoung Choi [3,*]

[1] Department of Public Sports, Luoyang Normal University, Luoyang 471934, China; chenyake@lynu.edu.cn
[2] Department of Physical Education, Gangneung-Wonju National University, Gangneung 25457, Korea; yhkim@gwnu.ac.kr
[3] Department of Sports Science Convergence, Dongguk University, Seoul 04620, Korea
* Correspondence: dory0301@dongguk.edu; Tel.: +82-2-2260-8741; Fax: +82-2-2260-3741

Abstract: Meniscal allograft transplantation (MAT) is a treatment modality for restoring knee function in patients with irreversible meniscal injury. Strengthening programs to promote functional recovery are treated with caution during the intermediate rehabilitation phase following MAT. This study analyzed the effects of aquatic training (AQT) and bicycling training (BCT) during the intermediate stage of rehabilitation in amateur athletes that underwent MAT. Participants ($n = 60$) were divided into AQT ($n = 30$) and BCT ($n = 30$) groups. Both groups performed training three times per week from 6 to 24 weeks following surgery. International Knee Documentation Committee Subjective Knee Evaluation Form (IKDC) score, knee joint range of motion (ROM), isokinetic knee strength, and Y-balance test (YBT) performance were evaluated. All measured variables for the AQT and BCT groups improved significantly after training compared with pre-training values. The IKDC score and YBT were significantly higher for AQT than for BCT. The knee flexion ROM and isokinetic muscle strength were significantly improved in the BCT group compared to those in the AQT group. The AQT group exhibited greater improvement in dynamic balance, whereas BCT provided greater improvement in isokinetic muscle strength. AQT and BCT were effective in reducing discomfort and improving knee symptoms and functions during intermediate-stage rehabilitation following MAT in amateur athletes.

Keywords: aquatic training; bicycling training; meniscal allograft transplantation; rehabilitation; strength; dynamic balance; range of motion

Citation: Chen, Y.; Kim, Y.; Choi, M. Effects of Aquatic Training and Bicycling Training on Leg Function and Range of Motion in Amateur Athletes with Meniscal Allograft Transplantation during Intermediate-Stage Rehabilitation. *Healthcare* **2022**, *10*, 1090. https://doi.org/10.3390/healthcare10061090

Academic Editors: João Paulo Brito and Rafael Oliveira

Received: 15 May 2022
Accepted: 9 June 2022
Published: 11 June 2022

Publisher's Note: MDPI stays neutral with regard to jurisdictional claims in published maps and institutional affiliations.

1. Introduction

The meniscus is a fibrocartilaginous tissue that efficiently performs major knee functions such as load distribution, shock absorption, joint lubrication, and stability [1]. Damage or tears to the meniscus can cause articular cartilage degeneration and knee instability, ultimately leading to the early progression of osteoarthritis [2]. Therefore, the younger and more active the patient, the more appropriate is treatment aimed at preserving the tissue after meniscal injury [3]. Generally, meniscus repair as a surgical option to mend the damaged meniscus is preferred over meniscectomy to remove the meniscus [4]. However, suturing of the damaged meniscus is not possible in all situations. Depending on the type, extent, and location of the damage, repair of the horn is occasionally possible; in this case, meniscectomy is inevitable [5]. However, while removal of the meniscus improves pain and function in the short term, it may alter normal knee mechanics over the long term, thereby unbalancing the load on the tibiofemoral joint and increasing the risk of premature osteoarthritis [6]. Therefore, in young, active patients, meniscal allograft transplantation (MAT) may be an effective option for improving knee pain and function, improving biomechanics, and delaying osteoarthritis progression [7].

Regarding the rehabilitation protocol after MAT, previous studies have recommended performing range-of-motion (ROM) exercises as early as possible immediately after surgery [8,9]. However, considering the need for revascularization, fixation, and healing of the transplanted meniscus during the early post-surgical stage, partial weight-bearing is recommended rather than full weight-bearing for 4 to 6 weeks [10–12]. To minimize the posterior shear and rotational stress on the transplanted meniscus, full weight-bearing exercise is possible from 6 weeks after surgery, and it is stressed that patients should proceed slowly [8,13]. The closed kinetic chain movement that places a load on the meniscus is initially restricted; however, patients are allowed to perform light walking at 6 weeks and light running at 12 weeks after surgery [10,14].

Bicycling training (BCT) is recognized as a relatively safe load exercise during the early stages of rehabilitation for patients who underwent MAT and is commonly used to improve ROM and muscle function [14]. However, BCT does not reflect actual functional movements compared to dynamic running-based sports, and stimulation of the muscles is relatively limited [15]. Moreover, BCT is a beneficial aerobic exercise that can be performed during the early stage, but it requires more effort and time to reach the target exercise intensity because the load and fatigue are concentrated on the lower body [16]. Several previous studies have emphasized the importance of BCT in parallel with muscle strength and proprioception training for symptom improvement and functional recovery in the early rehabilitation stage following MAT [9,17].

Meanwhile, aquatic training (AQT) minimizes pain without interfering with the healing and fixation of the transplanted meniscus; AQT enables sport-specific movements, improves knee function by stimulating various muscles, and can be an effective intervention [18]. Furthermore, AQT can elicit an aerobic cardiorespiratory response similar to BCT with relatively little effort because hydrostatic pressure, the pressure exerted by water, provides load and fatigue stimuli to the entire body [19]. However, despite the positive effects of AQT, studies comparing AQT to BCT during the intermediate postoperative stage are rare.

Therefore, in this study, the subjective knee score, knee ROM, isokinetic knee strength, and dynamic balance were compared and analyzed between the AQT group conducted in water and the BCT group conducted on land during the intermediate rehabilitation stage in amateur athletes who underwent MAT. Through this study, the effectiveness of the two training methods was tested, and a better rehabilitation training method was discovered. Finally, the study was intended to contribute to the production of a safe and effective rehabilitation training program.

2. Materials and Methods

2.1. Experimental Design

This study complied with the Declaration of Helsinki and was approved by the Gangneung-Wonju National University Institutional Review Board (approval number: GWNU IRB 2021-13; approval date: 25 February 2021). Male and female amateur athletes who underwent MAT were recruited through a hospital bulletin board. Participation was voluntary, and only patients confirmed by a surgical specialist as capable of safely proceeding with the rehabilitation program were included. Recruited patients were divided into AQT and BCT groups, and the training intervention was performed. AQT or BCT was assigned through consultation and reflected the preferences of the participants. Subjective evaluations and ROM tests were performed at 6, 12, and 24 weeks after surgery. Due to safety considerations regarding the surgical site, muscle strength and dynamic balance were not tested at 6 weeks and were only measured at 12 and 24 weeks.

2.2. Participants

Sample sizes were calculated using G*power software (G*power 3.1, University of Düsseldorf, Düsseldorf, Germany). The main analysis method in our study was non-parametric comparison between the two groups. The setting conditions are as follows: Mann–Whitney test (two groups); effect size d = 0.5, α error probability = 0.05, and power (1-β error probability) = 0.80. The suggested sample size is 106 people.

Initially, 111 amateur athletes (88 male and 23 female, age 25–35 years) who underwent MAT were recruited. We use only male data for analysis; females are excluded. Five athletes dropped out during the study period, and the 18 women who completed training were far too few compared to men. Therefore, the female athletes were trained but were excluded from the analysis. The final analysis included 60 male athletes (AQT, n = 30; BCT, n = 30). The exclusion criteria were as follows: other lesions of the knee confirmed by radiological examination (n = 7), accompanying injuries such as anterior cruciate ligament rupture (n = 7), past knee injuries or surgery history (n = 6), dropouts, and patients who did not attend the final visit (n = 8). Athletes participated in soccer (n = 17), badminton (n = 7), tennis (n = 2), basketball (n = 6), martial arts (n = 5), baseball (n = 10), handball (n = 2), volleyball (n = 1), taekwondo (n = 4), wrestling (n = 1), judo (n = 2), and other sports (n = 3).

2.3. Subjective Knee Score

Knee scores related to patients' subjective symptoms and function were measured using the International Knee Documentation Committee Subjective Knee Evaluation Form (IKDC) [20]. The IKDC consists of 18 items related to sports participation, including symptoms, functions, and activities of daily living affected by knee injury or surgery. Questions related to knee symptoms evaluated pain, stiffness, swelling, locking/catching, and giving way. Questions related to knee function and sports participation rated the level of sporting activity, ascending and descending stairs, kneeling, squatting, flexing the knee, sitting, running, jumping, starting and stopping quickly, and subjective current knee function. A score was assigned to each question according to its importance, and after summing the scores for all questions, that sum was converted according to the calculation formula to obtain a total score. The highest possible total score is 100. A total score of 100 indicates that there are no knee symptoms or functional limitations and no limitations in sports or activities of daily living. The total score conversion method for all questions is as follows:

$$\text{Total score} = (\text{sum of items}/\text{maximum possible score}) \times 100 \tag{1}$$

2.4. Range of Motion

The ROM was measured using a manual goniometer. Each measurement was performed twice, and the average value of the measured ROM was used for analysis. The axis of joint movement was aligned with the lateral epicondyle of the femur. The stationary arm of the goniometer was aligned with the femur using the greater trochanter as a reference; the movement arm was aligned with the fibula using the lateral malleolus [21].

The knee flexion ROM was measured with the participant in the prone position. The torso was fixed, and the patient was instructed to bend the knee to the maximum, taking care not to cause any movement of the spine and pelvis. ROM recorded the endpoint as the maximum flexion angle. The knee extension ROM was measured with the participant in the supine position. A support was placed under the thigh such that the knee was fully extended without the patella touching the ground, and the foot was placed beyond the edge of the examination table. The endpoint of the ROM, at which the patient can extend the knee maximally, was recorded as the maximum extension angle.

2.5. Isokinetic Knee Strength

For the isokinetic knee strength test, the strengths of the extensor and flexor muscles of the knee joint were measured using an isokinetic dynamometer (Humac Norm; CSMi,

Stoughton, MA, USA). The isokinetic dynamometer measures the maximum resistance of the patient through muscle contraction against a mechanically controlled constant velocity [22]. The tests were performed at angular velocities of 60°/s and 180°/s. To maintain a consistent examination posture, the participants sat in the examination chair and aligned the axis of the dynamometer with the anatomical axis of the knee and the lateral epicondyle of the femur. In addition, to minimize compensatory movements, straps were fixed around the thighs, pelvis, and torso. Measurements were performed as uniaxial contractions for the continuous extension and flexion of the knee. To help the participants understand, the test method was adequately explained, and several prior exercises were conducted. The joint ROM for flexion and extension of the knee for the examination was set from 0° to 90°, and the maximum knee extension was set to 0°. The participant was prepared while waiting for the examiner's signal with the knee flexed at 90°.

Upon the examiner's start signal, extension was first measured with maximum muscle contraction, and flexion was subsequently measured with maximum muscle contraction using the following protocol: The patient repeated measurements four times at an angular velocity of 60°/s and four times at an angular velocity of 180°/s. For muscle strength, peak torque (Nm) was measured. The average muscle power (W) was also measured. The absolute values of the measured muscle strength and power were divided by the patient's body weight to obtain relative values, thereby removing differences based on body weight. Finally, to compare the muscle strength ratio of the involved and uninvolved knees between groups, the limb symmetry index (LSI) was calculated using the following formula:

$$LSI = (Involved\ limb/Uninvolved\ limb) \times 100 \qquad (2)$$

2.6. Y-Balance Test

Dynamic balance ability was measured using the Y-balance test (YBT) equipment (Y Balance Test™, Cerder Park, TX, USA) [23]. The examiner demonstrated the examination posture and sequence of movements and allowed participants adequate practice. The participants stood on one leg with the foot on the central stance plate for the examination. Then, while maintaining balance in a single-leg stance, a series of motions were performed to extend the opposite leg and push the reach indicator as far as possible with the tip of the toe in the anterior (ANT), posteromedial (PM), and posterolateral (PL) directions. Failure was considered to have occurred if the foot of the reaching limb touched the ground or when balance of the stance limb was lost. The test was conducted measuring the healthy leg first and then the operated leg. After measuring each of the three directions twice, the higher score was used in the analysis. The results were recorded in centimeters, and leg length was used to calculate the final score. Leg length was measured using a tape measure to determine the distance between the anterior superior iliac spine of the pelvis and the medial malleolus of the distal tibia. The total score for the three directions was calculated, and the LSI was compared between the groups in the same manner as muscle strength.

$$YBT\ total\ score = (sum\ of\ the\ three\ reach\ directions/3 \times limb\ length) \times 100 \qquad (3)$$

2.7. Training Program

The rehabilitation program was conducted as shown in Table 1. Early-stage training was performed identically without any difference between the AQT and BCT groups. According to MAT rehabilitation guidelines, ROM and partial weight-bearing exercises were performed in the initial stage [9,24]. All participants wore a brace that kept the knee fully extended for 6 weeks after surgery, and ambulation using crutches was performed with partial weight-bearing. Passive ROM exercises to restore knee motion were started immediately after surgery. The goal was to restore the knee ROM to 0–90° for the first 2 weeks, followed by 120° for 4 weeks and 135° for 6 weeks. Immediately after surgery, isometric quadriceps contraction, straight leg raise, and active knee extension were performed to strengthen the quadriceps muscles in an open kinetic chain. In the closed kinetic chain, the participant wore a brace and performed weight shifts and calf raises with the knee fully extended. Three weeks after surgery, wall squats with a limited ROM of 60° were allowed.

Table 1. Rehabilitation protocol.

	1–2 Weeks	3–4 Weeks	5–6 Weeks	7–8 Weeks	9–12 Weeks	3–6 Months
Brace	O	O	O			
Crutch	O	O	O			
Weight-bearing						
1/4 of body weight	O					
2/4 of body weight		O				
3/4 of body weight			O			
Full				O	O	O
Range of motion						
0–90°	O					
0–120°		O				
0–135°			O			
Stretching						
Hamstring, Quadriceps, GCM, ITB	O	O	O	O	O	O
Strengthening						
Quadriceps sets, Straight leg raise	O	O	O			
Active knee extension	O	O	O			
Active knee flexion				O	O	
Heel raises	O	O	O			
Wall squat			O			
Squat, Lunge, Step-ups				O	O	O
Leg press machine				O	O	O
Leg extension machine				O	O	O
Leg curl machine						O
Proprioception training						
Weight shift	O					
Tandem stance		O				
Single leg balance			O			
Single leg balance with leg swing				O	O	O
Single leg squat				O	O	O
Running, Jump-landing						O

GCM, gastrocnemius; ITB, iliotibial band.

Complete weight-bearing accompanied by knee flexion greater than 90° without the use of crutches and braces was allowed from 6 weeks after surgery. After 6 weeks, closed kinetic chain exercises, such as squats, lunges, and step-ups, were initiated under full weight-bearing without an orthosis, and single-leg balance was included to improve the muscular and nervous systems. At this stage, knee extension was allowed to increase the load by adding resistance, but knee flexion allowed only active motion without resistance for 12 weeks. Light running and jump-landing training were initiated 12 weeks after surgery. Light sports activities were allowed after six months, and vigorous contact sports were permitted after nine months.

2.7.1. Intervention Program: Aquatic Training

The AQT program was conducted 3 times per week by applying the continuous water aerobic routine (CWAR) described in the study by Kruel et al. [25]. The CWAR comprised eight routines, in which four water aerobic exercises (stationary running, cross-country skiing, jumping jacks, and frontal kicks) were each repeated twice. Each routine was performed continuously for 4 min without an interval. Therefore, the total training time was 32 min. All lower body movements were performed simultaneously with bilateral arm push–pulls for whole-body exercise. The training intensity of the AQT program was controlled using Borg's rating of perceived exertion (RPE) scale and an electronic heart-rate-monitoring device (Polar H10, Polar Electro, Bethpage, NY, USA). Verbal scales were used to express the effort level on a scale of 13 to 15 [26].

The examiner trained participants on the standard guidelines of the RPE scale to aid participants in verbally expressing their perceived level of effort as accurately as possible. In addition, adequate practice was performed before the actual training to familiarize participants with the feelings corresponding to minimum and maximum effort. Participants performed the routines as directed by the instructor at intensity levels corresponding to 'somewhat hard, 13' to 'hard, 15'. The suggested heart rate exercise intensity was 60–75% of the maximum heart rate.

2.7.2. Intervention Program: Bicycling Training

The BCT program was conducted 3 times per week, similar to the AQT, following the training intensity and duration of the continuous bicycling program proposed in a previous study [27]. A stationary friction-loaded cycle ergometer (Monark Model 864, Monark Crescent AB, Varberg, Sweden) was used for training. The saddle height of the stationary bicycle ergometer was individually adjusted based on the participant's body structure such that one leg was extended to a maximum of ~ 25° when the participant was sitting on the saddle [28] The BCT group performed continuous cycling for 32 min at an intensity of 60–75% of maximum heart rate while trying to maintain a pedaling speed of 60 RPM. To control the exercise intensity of the participants during training, the heart rate change was monitored in real time using an electronic heart-rate-monitoring device (Polar H10, Polar Electro, Bethpage, NY, USA), as was done in the AQT group.

2.8. Statistical Analysis

Data analysis was performed using SPSS Statistics (version 25.0; IBM Corp., Armonk, NY, USA). The normality test of the main variables was performed using Kolmogorov–Smirnov and Shapiro–Wilk tests. Since the variables did not exhibit a normal distribution, we performed a non-parametric analysis. Continuous variables are expressed as means and standard deviations, and non-continuous variables are expressed as numbers and percentages of patients. The Kruskal–Wallis test and the post hoc Bonferroni test were used for within-group tests of IKDC and ROM, which were tested three times. The Wilcoxon signed-rank test was used for comparison pre- and post-training within groups of twice-tested strength and YBT. Additionally, the Mann–Whitney U test was performed for between-group comparisons. The significance level was set at $p < 0.05$.

3. Results

3.1. General Characteristics of Participants

The participants were classified according to the intervention group, and their general characteristics are summarized in Table 2. When the AQT and BCT groups were compared, there were no statistically significant differences between the groups regarding age, height, weight, body mass index, injury site, or dominant side.

3.2. Subjective Knee Score

Table 3 shows the differences in IKDC scores analyzed by group and measurement week to evaluate the subjective knee score after MAT. Both the AQT and BCT groups exhibited significantly improved IKDC scores over time following surgery ($p < 0.001$ and $p < 0.001$, respectively). There was no significant difference at 6 and 24 weeks in the comparison between groups for each week, but at 12 weeks, the AQT group achieved significantly higher IKDC scores than the BCT group ($p = 0.033$).

Table 2. General characteristics of participants.

Variables	AQT (n = 30)	BCT (n = 30)	t or χ^2	p-Value
Age, years	28.7 ± 3.8	29.1 ± 4.0	−1.391	0.412
Height, cm	173.8 ± 2.9	174.1 ± 3.7	0.279	0.774
Weight, kg	68.7 ± 4.6	69.3 ± 5.8	0.573	0.615
BMI, kg/m^2	22.7 ± 1.5	22.9 ± 1.7	0.384	0.631
Involved side, n (%)				
Right	17 (56.7%)	16 (53.3%)	0.384	0.551
Left	13 (43.3%)	14 (46.7%)		
Dominant side, n (%)				
Right	25 (83.3%)	23 (76.7%)	0.207	0.415
Left	5 (16.7%)	7 (23.3%)		
Involved meniscus site, n (%)				
Medial	9 (30.0%)	11 (36.7%)	0.211	0.258
Lateral	21 (70.0%)	19 (63.3%)		

AQT, aquatic training; BCT, bicycling training; BMI, body mass index.

Table 3. Subjective knee score according to group and weeks.

Variables	Weeks	AQT (n = 30)	BCT (n = 30)	t	E.S	p-Value
IKDC score	6	65.5 ± 15.1	63.4 ± 14.4	2.631	0.142	0.512
	12	82.4 ± 17.9 [a]	72.8 ± 15.7 [a]	1.346	0.570	0.033
	24	93.1 ± 11.3 [b,c]	95.4 ± 13.2 [b,c]	−0.831	0.187	0.794
	p	<0.001 *	<0.001 *			

* $p < 0.05$; IKDC, International Knee Documentation Committee; [a]: 6 weeks vs. 12 weeks; [b]: 12 weeks vs. 24 weeks; [c]: 6 weeks vs. 24 weeks.

3.3. Knee Range of Motion

Table 4 shows the changes according to group and weeks post-surgery of knee ROM after MAT. Both the AQT and BCT groups showed significant improvement in flexion (p = 0.010 and p = 0.009, respectively) and extension ROM over time following surgery (p = 0.012 and p = 0.005, respectively). In the between-group comparison for each week, there was no significant difference in flexion ROM at 6 and 24 weeks, but at 12 weeks, the BCT group showed significantly greater ROM than the AQT group (p = 0.015). There was no significant between-group difference regarding extension at any number of weeks.

Table 4. Knee range of motion according to group and weeks.

Variables	Weeks	AQT (n = 30)	BCT (n = 30)	t	E.S	p-Value
Flexion (degree)	6	115.2 ± 8.1	113.3 ± 7.6	−3.379	0.241	0.651
	12	120.6 ± 5.4 [a]	132.0 ± 6.5 [a]	−1.080	1.907	0.015
	24	135.0 ± 3.9 [c]	134.9 ± 3.5 [c]	0.191	0.026	0.485
	* p	0.010	0.009			
Extension (degree)	6	10.4 ± 3.5	9.3 ± 3.1	−0.824	0.332	0.684
	12	2.1 ± 0.9 [a]	1.1 ± 0.9 [a]	−986	1.111	0.123
	24	−1.6 ± 0.7 [c]	−2.0 ± 0.8 [c]	−4.541	0.532	0.115
	* p	0.012	0.005			

* $p < 0.05$; AQT, aquatic training; BCT, bicycling training; E.S, effect size; [a]: 6 weeks vs. 12 weeks; [c]: 6 weeks vs. 24 weeks.

3.4. Isokinetic Knee Strength

Table 5 shows isokinetic knee strength according to group and weeks after MAT. At an angular velocity of 60°/s, muscle strength improved at 24 weeks compared to measured values at 12 weeks in both the AQT and BCT groups. Similarly, at an angular velocity of 180°/s, extension and flexion of the knee in the AQT and BCT groups improved at 24 weeks compared to the 12-week values. The BCT group exhibited significantly higher LSI than the AQT group in both extension and flexion strengths at week 12 at an angular velocity of $60°/s$. However, at an angular velocity of $180°/s$, there was no significant difference in LSI between the groups in either extension or flexion.

Table 5. Isokinetic knee strength according to group and weeks.

Variables	Weeks	AQT ($n = 30$)		LSI (%)	BCT ($n = 30$)		LSI (%)	Inter–Group LSI p-Value
		Uninvolved	Involved		Uninvolved	Involved		
60°/s extension	12	243.3 ± 45.7	105.3 ± 45.3	43.2 *	239.1 ± 38.0	127.1 ± 45.6	52.3 *	0.003
	24	258.4 ± 56.5	237.2 ± 49.2	91.9	260.3 ± 48.1	245.3 ± 51.9	94.2	0.350
	p	0.025	<0.001		0.021	<0.001		
60°/s flexion	12	136.2 ± 21.9	100.1 ± 24.7	73.5 *	130.0 ± 28.3	112.4 ± 30.8	86.2 *	0.005
	24	149.0 ± 22.1	136.3 ± 24.0	91.3	151.9 ± 20.8	140.4 ± 25.1	92.7	0.102
	p	0.014	<0.001		0.017	<0.001		
180°/s extension	12	145.3 ± 31.0	102.7 ± 39.7	70.3 *	142.6 ± 32.4	109.0 ± 35.1	76.8 *	0.213
	24	150.4 ± 29.7	142.4 ± 31.6	94.7	154.7 ± 30.6	145.9 ± 34.0	94.2	0.109
	p	0.210	<0.001		0.330	<0.001		
180°/s flexion	12	97.9 ± 19.9	90.4 ± 15.3	92.8	100.6 ± 11.4	98.5 ± 15.7	98.0	0.067
	24	110.1 ± 12.1	103.0 ± 10.8	93.6	115.8 ± 12.5	108.4 ± 16.4	93.9	0.153
	P	0.140	<0.001		0.417	<0.001		

$* p < 0.05$; AQT, aquatic training; BCT, bicycling training; LSI, limb symmetry index.

3.5. Y-Balance Test

Table 6 shows the changes in YBT after MAT. In the involved knee of the AQT and BCT groups, the YBT direction and total score were significantly improved at 24 weeks compared to the corresponding values at 12 weeks. In the intergroup comparison of LSI, the AQT group had significantly higher scores than the BCT group at 12 weeks in all directions, as well as higher total scores. However, there were no significant differences between the groups at 24 weeks.

Table 6. Y-balance test according to group and weeks.

Variables	Weeks	AQT ($n = 30$)		LSI (%)	BCT ($n = 30$)		LSI (%)	Inter–Group LSI p-Value
		Uninvolved	Involved		Uninvolved	Involved		
ANT	12	62.0 ± 9.0	40.3 ± 12.3	77.4 *	61.5 ± 9.9	30.4 ± 11.0	62.3 *	0.007
	24	65.5 ± 9.3	60.5 ± 13.4	92.3	63.4 ± 11.3	59.7 ± 12.9	93.7	0.651
	p	0.121	<0.001		0.231	<0.001		
PM	12	80.3 ± 12.5	60.4 ± 16.3	75.0 *	74.3 ± 13.8	49.3 ± 13.6	66.2 *	0.003
	24	81.6 ± 11.8	79.6 ± 15.3	97.5	80.3 ± 12.3	72.0 ± 14.1	90.0	0.591
	p	0.110	<0.001		0.210	<0.001		
PL	12	78.9 ± 13.8	63.6 ± 11.4	80.8 *	75.1 ± 12.9	51.3 ± 12.9	68.0 *	0.002
	24	82.3 ± 12.4	78.6 ± 14.3	95.1	81.4 ± 13.0	76.1 ± 13.4	93.8	0.794
	p	0.257	<0.001		0.098	<0.001		
Total	12	86.5 ± 11.2	67.6 ± 12.1	77.7 *	82.3 ± 11.6	54.0 ± 12.3	65.7 *	0.005
	24	89.9 ± 10.3	85.1 ± 11.6	95.2	87.9 ± 11.9	81.7 ± 13.1	92.7	0.510
	p	0.150	<0.001		0.254	<0.001		

$* p < 0.05$, AQT: aquatic training, BCT: bicycling training, LSI: limb symmetry index, ANT: anterior, PM: postero-medial, PL: posterolateral.

4. Discussion

AQT and BCT are partially weight-bearing and non-weight-bearing, respectively, thereby reducing the weight on the knee. AQT is often used in rehabilitation because of the hydrodynamic properties of water, including buoyancy and water pressure [29]. In addition, the stationary bicycling ergometer is unlikely to pose a risk to the recovering tissue, as it biomechanically provides controlled movement during flexion and extension [9]. In this study, a modified weight-bearing environment was provided, and the effects were compared during intermediate-stage rehabilitation of patients who had undergone MAT.

The subjective knee score evaluated by the IKDC improved significantly after training in both the AQT and BCT groups. The AQT group exhibited significantly greater improvement than the BCT group at week 12, which translates to relatively rapid recovery in the AQT group. In a similar study, the effect of early improvement was analyzed using minimal clinically important differences (MICDs). In the study by Liu et al. [30], the MICD of the IKDC score that can be applied to evaluate the outcome of patients who received MAT was proposed as 9.9. Compared to 6 weeks after surgery, at 12 weeks, the IKCD score of the AQT group increased by 16.9 points to achieve a significant MICD, whereas the BCT group showed an increase of 9.4, which was slightly insufficient for achieving an MICD.

The greater improvement exhibited by the AQT group at 12 weeks means that knee symptoms such as pain, stiffness, and swelling were significantly ameliorated by AQT compared to the results obtained by BCT. Water provides resistance such as turbulence and hydrostatic pressure while simultaneously reducing weight-bearing due to buoyancy, which may have had a combined effect [29]. In a study investigating the characteristics of water, AQT was reported to reduce pain and swelling, and promote recovery from fatigue by increasing blood circulation [31]. Therefore, the physical properties of AQT may have aided recovery after surgery, leading to physiological changes and subjective improvement of the knee condition.

Patients who have undergone MAT should be careful not to generate posterior shear and rotational forces for stable fixation and healing of implants during the early stage of rehabilitation. Therefore, the range of knee flexion is limited along with weight-bearing, and absolute immobilization is performed for at least three weeks [14]. After absolute immobilization, it is important to restore the ROM [32]. In a non-weight-bearing state, passive ROM has limitations in promoting active muscle contraction and mechanoreceptor activation; therefore, it is difficult to restore the functional movement patterns required in sports [33].

The results of this study showed that both the AQT and BCT groups displayed significantly improved knee flexion and extension ROM at 12 and 24 weeks, and the BCT group showed significantly greater knee flexion ROM than the AQT group at 12 weeks. This could be because the AQT program comprises various movements of the extremities, whereas the BCT program entails continuous pedaling involving repeated knee flexion and extension [34]. Training performed in the AQT environment is known to provide an advantage in ROM recovery, as there is less stress on the joints, owing to buoyancy resulting from the hydrodynamic properties of the aquatic medium [19].

Recovery of weakened knee muscle strength after MAT is an important factor in determining the success of postoperative outcomes and returning to sports [9]. In a study examining the long-term outcomes of patients who received MAT, decreased levels of function and activity were associated with decreased maximal quadriceps muscle strength [35]. In this study, the defect rate was evaluated using the LSI value, which is a reference value used to assess the involved side compared to the uninvolved side. In this study, both the AQT and BCT groups showed significantly improved muscle strength and achieved more than 90% on the LSI at 24 weeks. These results suggest that the inclusion of AQT and BCT in the rehabilitation process is an effective intervention for improving muscle function in patients with MAT. However, in the weekly comparison, the BCT group showed a significantly higher LSI than the AQT group at 12 weeks at an angular velocity of $60°/s$.

These results suggest that the intervention effect was faster for BCT than that for AQT because BCT was concentrated on the lower extremities.

The meniscus generates sensorimotor information necessary to control the stabilizing muscles surrounding the knee, as well as mechanical stability in the knee joint [36]. Free nerve endings (nociceptors) and three types of mechanoreceptors (Ruffini endings, Pacinian corpuscles, and Golgi tendon organs) exist in the anterior and posterior horns of the meniscus and the peripheral two-thirds of the body and are responsible for proprioception in the knee joint [37]. Proprioceptors transmit signals to the central nervous system (CNS) resulting from physical stimuli, such as tension or compression forces applied to the knee joint, thereby helping to regulate reflex responses and muscle coordination related to postural stability [38]. Thijs et al. [37] observed that patients with meniscal removal had significant deficits in proprioception of the knee joint and reported that MAT may contribute to restoring the reactivity of damaged knee proprioceptors. However, restoration of proprioceptive function alone cannot achieve complete improvement in postural stability, which can be further facilitated by training interventions performed during the rehabilitation stage [37]. In this study, the effects of AQT and BCT on the improvement of postural stability after MAT were evaluated using the YBT. The YBT is currently the most commonly used measurement tool for assessing the dynamic balance of the lower extremities [23]. Dynamic balance refers to the ability to maintain postural stability while moving the body or changing the position of a limb and is an important component of most daily life and sports activities [39]. AQT and BCT significantly improved the YTB results in this study. At 12 weeks, the AQT group showed significantly higher dynamic balance ability than the BCT group. Training in an AQT environment may induce instability that alters information in the somatosensory system under the influence of water turbulence, an intrinsic property of the environment [29]. In this study, unlike the BCT group, the AQT group was continuously affected by aquatic perturbation caused by water turbulence throughout the training session. This perturbation is believed to provide an additional balance stimulus for participants, such as further activation of the neuromuscular muscles of the ankle and knee joints, to restore balance. As a result, it is believed that the dynamic characteristics of AQT may have improved results earlier compared to the BCT group.

Based on the results of this study, realistic suggestions can be made for athletes who underwent MAT. If an underwater facility is available, we recommend a combination of AQT and BCT after MAT. In addition, if BCT is mainly performed without AQT, more careful observation is required to ensure that the recovery of dynamic balance is not delayed.

Based on the results of this study, it is recommended to use both AQT and BCT for athletes who have undergone MAT, but because this study was conducted without a control group, more research is needed to provide scientific evidence. This study had additional limitations. A control group was not established due to ethical considerations. Muscle strength, balance, and subjective satisfaction are possible parts of recovery over time after MAT. However, limiting rehabilitation training for research purposes to people visiting rehabilitation centers can be an ethical issue. In addition, assignment of AQT and BCT was not random because the training method chosen was based on the preference of the athlete. In particular, the AQT reflected individual preferences because it required the use of swimsuits. This study was conducted at a single rehabilitation center, there were relatively few participants, and the participants specialized in various sports; therefore, the influence of variable characteristics cannot be excluded. Analyzing only male data is one of the major limitations. In the future, it will be necessary to conduct a study by recruiting more female participants who underwent MAT. Although YBT was used for dynamic balance in this study, experiments with neuromuscular control through one-leg stabilometric measurement should be performed in future studies.

5. Conclusions

AQT and BCT after MAT improved subjective knee score, knee joint range of motion, isokinetic knee strength, and YBT at 24 weeks compared to pre-training values. Interim measurements performed 12 weeks after the intervention revealed that subjective knee score and YBT were higher in the AQT group, and flexion ROM and isokinetic knee strength were higher in the BCT group. Therefore, AQT and BCT with reduced weight-bearing could be effective training interventions to overcome challenges and improve symptoms and functions during the intermediate rehabilitation stage of MAT.

Author Contributions: Conceptualization, Y.C. and M.C.; methodology, Y.K.; formal analysis, Y.C. and M.C.; investigation, Y.K.; writing—original draft preparation, Y.C. and Y.K.; writing—review and editing, Y.C. and M.C.; supervision, Y.C. and M.C. All authors have read and agreed to the published version of the manuscript.

Funding: This research received no external funding.

Institutional Review Board Statement: The study was conducted according to the guidelines of the Declaration of Helsinki and approved by the Institutional Review Board of Gangeung-Wonju National University GWNUIRB-2021-13; approval date: 25 February 2021).

Informed Consent Statement: Informed consent was obtained from all participants involved in the study and written informed consent has been obtained from the participants to publish this paper.

Data Availability Statement: The data are not publicly available due to privacy or ethical reasons.

Conflicts of Interest: The authors declare no conflict of interest.

References

1. Rosso, F.; Bisicchia, S.; Bonasia, D.E.; Amendola, A. Meniscal allograft transplantation: A systematic review. *Am. J. Sports Med.* **2015**, *43*, 998–1007. [CrossRef] [PubMed]
2. Frank, R.M.; Cole, B.J. Meniscus transplantation. *Curr. Rev. Musculoskelet. Med.* **2015**, *8*, 443–450. [CrossRef] [PubMed]
3. Abrams, G.D.; Frank, R.M.; Gupta, A.K.; Harris, J.D.; McCormick, F.M.; Cole, B.J. Trends in meniscus repair and meniscectomy in the United States, 2005–2011. *Am. J. Sports Med.* **2013**, *41*, 2333–2339. [CrossRef] [PubMed]
4. Montgomery, S.R.; Zhang, A.; Ngo, S.S.; Wang, J.C.; Hame, S.L. Cross-sectional analysis of trends in meniscectomy and meniscus repair. *Orthopedics* **2013**, *36*, e1007–e1013. [CrossRef] [PubMed]
5. Smith, N.; Costa, M.; Spalding, T. Meniscal allograft transplantation: Rationale for treatment. *Bone Jt. J.* **2015**, *97*, 590–594. [CrossRef] [PubMed]
6. Bai, B.; Shun, H.; Yin, Z.X.; Liao, Z.-w.; Chen, N. Changes of contact pressure and area in patellofemoral joint after different meniscectomies. *Int. Orthop.* **2012**, *36*, 987–991. [CrossRef] [PubMed]
7. Hergan, D.; Thut, D.; Sherman, O.; Day, M.S. Meniscal allograft transplantation. *Arthrosc. J. Arthrosc. Relat. Surg.* **2011**, *27*, 101–112. [CrossRef]
8. Cole, B.J.; Dennis, M.G.; Lee, S.J.; Nho, S.J.; Kalsi, R.S.; Hayden, J.K.; Verma, N.N. Prospective evaluation of allograft meniscus transplantation: A minimum 2-year follow-up. *Am. J. Sports Med.* **2006**, *34*, 919–927. [CrossRef] [PubMed]
9. Noyes, F.R.; Heckmann, T.P.; Barber-Westin, S.D. Meniscus repair and transplantation: A comprehensive update. *J. Orthop. Sports Phys. Ther.* **2012**, *42*, 274–290. [CrossRef] [PubMed]
10. Lee, D.W.; Lee, J.H.; Kim, D.H.; Kim, J.G. Delayed rehabilitation after lateral meniscal allograft transplantation can reduce graft extrusion compared with standard rehabilitation. *Am. J. Sports Med.* **2018**, *46*, 2432–2440. [CrossRef] [PubMed]
11. Marcacci, M.; Marcheggiani Muccioli, G.M.; Grassi, A.; Ricci, M.; Tsapralis, K.; Nanni, G.; Bonanzinga, T.; Zaffagnini, S. Arthroscopic meniscus allograft transplantation in male professional soccer players: A 36-month follow-up study. *Am. J. Sports Med.* **2014**, *42*, 382–388. [CrossRef] [PubMed]
12. Zaffagnini, S.; Grassi, A.; Marcheggiani Muccioli, G.M.; Benzi, A.; Roberti di Sarsina, T.; Signorelli, C.; Raggi, F.; Marcacci, M. Is sport activity possible after arthroscopic meniscal allograft transplantation? Midterm results in active patients. *Am. J. Sports Med.* **2016**, *44*, 625–632. [CrossRef] [PubMed]
13. Kempshall, P.; Parkinson, B.; Thomas, M.; Robb, C.; Standell, H.; Getgood, A.; Spalding, T. Outcome of meniscal allograft transplantation related to articular cartilage status: Advanced chondral damage should not be a contraindication. *Knee Surg. Sports Traumatol. Arthrosc.* **2015**, *23*, 280–289. [CrossRef] [PubMed]
14. Rucinski, K.; Cook, J.L.; Crecelius, C.R.; Stucky, R.; Stannard, J.P. Effects of compliance with procedure-specific postoperative rehabilitation protocols on initial outcomes after osteochondral and meniscal allograft transplantation in the knee. *Orthop. J. Sports Med.* **2019**, *7*, 2325967119884291. [CrossRef] [PubMed]

15. Prosser, L.A.; Stanley, C.J.; Norman, T.L.; Park, H.S.; Damiano, D.L. Comparison of elliptical training, stationary cycling, treadmill walking and overground walking. Electromyographic patterns. *Gait Posture* **2011**, *33*, 244–250. [CrossRef] [PubMed]
16. Rønnestad, B.R.; Mujika, I. Optimizing strength training for running and cycling endurance performance: A review. *Scand. J. Med. Sci. Sports* **2014**, *24*, 603–612. [CrossRef]
17. Marcacci, M.; Zaffagnini, S.; Marcheggiani Muccioli, G.M.; Grassi, A.; Bonanzinga, T.; Nitri, M.; Bondi, A.; Molinari, M.; Rimondi, E. Meniscal allograft transplantation without bone plugs: A 3-year minimum follow-up study. *Am. J. Sports Med.* **2012**, *40*, 395–403. [CrossRef] [PubMed]
18. Villalta, E.M.; Peiris, C.L. Early aquatic physical therapy improves function and does not increase risk of wound-related adverse events for adults after orthopedic surgery: A systematic review and meta-analysis. *Arch. Phys. Med. Rehabil.* **2013**, *94*, 138–148. [CrossRef]
19. Torres-Ronda, L.; i del Alcázar, X.S. The properties of water and their applications for training. *J. Hum. Kinet.* **2014**, *44*, 237–248. [CrossRef]
20. Anderson, A.F.; Irrgang, J.J.; Kocher, M.S.; Mann, B.J.; Harrast, J.J.; Committee, I.K.D. The International Knee Documentation Committee subjective knee evaluation form: Normative data. *Am. J. Sports Med.* **2006**, *34*, 128–135. [CrossRef]
21. Norkin, C.C.; White, D.J. *Measurement of Joint Motion: A Guide to Goniometry*; FA Davis: Philadelphia, PA, USA, 2016.
22. Habets, B.; Staal, J.B.; Tijssen, M.; van Cingel, R. Intrarater reliability of the Humac NORM isokinetic dynamometer for strength measurements of the knee and shoulder muscles. *BMC Res. Notes* **2018**, *11*, 15. [CrossRef]
23. Coughlan, G.F.; Fullam, K.; Delahunt, E.; Gissane, C.; Caulfield, B.M. A comparison between performance on selected directions of the star excursion balance test and the Y balance test. *J. Athl. Train.* **2012**, *47*, 366–371. [CrossRef] [PubMed]
24. Heckmann, T.P.; Barber-Westin, S.D.; Noyes, F.R. Meniscal repair and transplantation: Indications, techniques, rehabilitation, and clinical outcome. *J. Orthop. Sports Phys. Ther.* **2006**, *36*, 795–814. [CrossRef] [PubMed]
25. Kruel, L.F.M.; Posser, M.S.; Alberton, C.L.; Pinto, S.S.; Oliveira, A.d.S. Comparison of energy expenditure between continuous and interval water aerobic routines. *Int. J. Aquat. Res. Educ.* **2009**, *3*, 186–196. [CrossRef]
26. Williams, N. The Borg rating of perceived exertion (RPE) scale. *Occup. Med.* **2017**, *67*, 404–405. [CrossRef]
27. Alansare, A.; Alford, K.; Lee, S.; Church, T.; Jung, H.C. The effects of high-intensity interval training vs. moderate-intensity continuous training on heart rate variability in physically inactive adults. *Int. J. Environ. Res. Public Health* **2018**, *15*, 1508. [CrossRef] [PubMed]
28. Quesada, J.I.P.; Jacques, T.C.; Bini, R.R.; Carpes, F.P. Importance of static adjustment of knee angle to determine saddle height in cycling. *J. Sci. Cycl.* **2016**, *5*, 26–31.
29. Becker, B.E. Aquatic therapy: Scientific foundations and clinical rehabilitation applications. *PmR* **2009**, *1*, 859–872. [CrossRef] [PubMed]
30. Liu, J.N.; Gowd, A.K.; Redondo, M.L.; Christian, D.R.; Cabarcas, B.C.; Yanke, A.B.; Cole, B.J. Establishing clinically significant outcomes after meniscal allograft transplantation. *Orthop. J. Sports Med.* **2019**, *7*, 2325967118818462. [CrossRef]
31. Holmberg, P.M.; Gorman, A.D.; Jenkins, D.G.; Kelly, V.G. Lower-Body Aquatic Training Prescription for Athletes. *J. Strength Cond. Res.* **2021**, *35*, 859–869. [CrossRef]
32. Afonso, J.; Ramirez-Campillo, R.; Moscão, J.; Rocha, T.; Zacca, R.; Martins, A.; Milheiro, A.A.; Ferreira, J.; Sarmento, H.; Clemente, F.M. Strength Training versus stretching for improving range of motion: A systematic review and meta-analysis. *Healthcare* **2021**, *9*, 427. [CrossRef]
33. Koch, M.; Memmel, C.; Zeman, F.; Pfeifer, C.G.; Zellner, J.; Angele, P.; Weber-Spickschen, S.; Alt, V.; Krutsch, W. Early functional rehabilitation after Meniscus surgery: Are currently used orthopedic rehabilitation standards up to date? *Rehabil. Res. Pract.* **2020**, *2020*, 3989535–3989542. [CrossRef]
34. Cavanaugh, J.T.; Powers, M. ACL rehabilitation progression: Where are we now? *Curr. Rev. Musculoskelet. Med.* **2017**, *10*, 289–296. [CrossRef] [PubMed]
35. Hurley, E.T.; Davey, M.S.; Jamal, M.S.; Manjunath, A.K.; Kingery, M.T.; Alaia, M.J.; Strauss, E.J. High rate of return-to-play following meniscal allograft transplantation. *Knee Surg. Sports Traumatol. Arthrosc.* **2020**, *28*, 3561–3568. [CrossRef] [PubMed]
36. Palm, H.; Laufer, C.; von Lübken, F.; Achatz, G.; Friemert, B. Do meniscus injuries affect postural stability? *Der Orthop.* **2010**, *39*, 486–494. [CrossRef] [PubMed]
37. Thijs, Y.; Witvrouw, E.; Evens, B.; Coorevits, P.; Almqvist, F.; Verdonk, R. A prospective study on knee proprioception after meniscal allograft transplantation. *Scand. J. Med. Sci. Sports* **2007**, *17*, 223–229. [CrossRef]
38. Al-Dadah, O.; Shepstone, L.; Donell, S. Proprioception following partial meniscectomy in stable knees. *Knee Surg. Sports Traumatol. Arthrosc.* **2011**, *19*, 207–213. [CrossRef]
39. Boyle, M.J.; Butler, R.J.; Queen, R.M. Functional movement competency and dynamic balance after anterior cruciate ligament reconstruction in adolescent patients. *J. Pediatric Orthop.* **2016**, *36*, 36–41. [CrossRef]

Article

It Is Not Just Stress: A Bayesian Approach to the Shape of the Negative Psychological Features Associated with Sport Injuries

Aurelio Olmedilla Zafra [1], Bruno Martins [2], F. Javier Ponseti-Verdaguer [3,*], Roberto Ruiz-Barquín [4] and Alejandro García-Mas [5]

[1] Department of Personality, Evaluation, and Psychological Treatment, Campus Regional Excellence Mare Nostrum, University of Murcia, 30100 Murcia, Spain; olmedilla@um.es
[2] GICAFE (Research Group of Sports Sciences—UIB), University of Lisbon, 1649-004 Lisbon, Portugal; bruno.pt@gmail.com
[3] GICAFE (Research Group of Sports Sciences), Department of Pedagogy, University of the Balearic Islands, 07122 Palma, Spain
[4] Department of Evolutive and Educational Psychology, Autonomous University of Madrid, 28049 Madrid, Spain; roberto.ruiz@uam.es
[5] GICAFE (Research Group of Sports Sciences), Department of Psychology, University of the Balearic Islands, 07122 Palma, Spain; alex.garcia@uib.es
* Correspondence: xponseti@uib.es; Tel.: +34-971173253

Abstract: The main objective of this study is to extend the stress and injury model of Andersen and Williams to other "negative" psychological variables, such as anxiety and depression, encompassed in the conceptual model of Olmedilla and García-Mas. The relationship is studied of this psychological macro-variable with two other variables related to sports injuries: the search for social support and the search for connections between risk and the environment of athletes. A combination of classic methods and probabilistic approaches through Bayesian networks is used. The study samples comprised 455 traditional and indoor football players (323 male and 132 female) of an average age of 21.66 years ($\pm$4.46). An ad hoc questionnaire was used for the corresponding sociodemographic data and data relating to injuries. The variables measured were the emotional states of: stress, depression and anxiety, the attitude towards risk-taking in different areas, and the evaluation of the perception of social support. The results indicate that the probabilistic analysis conducted gives a boost to the classic model focused on stress, as well as the conceptual planning derived from the Global Model of Sports Injuries (GMSI), supporting the possibility of extending the stress model to other variables, such as anxiety and depression ("negative" triad).

Keywords: sports injuries; psychological factors; football; Bayesian network

Citation: Zafra, A.O.; Martins, B.; Ponseti-Verdaguer, F.J.; Ruiz-Barquín, R.; García-Mas, A. It Is Not Just Stress: A Bayesian Approach to the Shape of the Negative Psychological Features Associated with Sport Injuries. *Healthcare* **2022**, *10*, 236. https://doi.org/10.3390/healthcare10020236

Academic Editors: João Paulo Brito and Rafael Oliveira

Received: 20 December 2021
Accepted: 25 January 2022
Published: 26 January 2022

Publisher's Note: MDPI stays neutral with regard to jurisdictional claims in published maps and institutional affiliations.

1. Introduction

Sports injuries have been the focus of many studies in different scientific disciplines, showing that they relate to a complex and multicausal phenomena [1–5]. In addition, their incidence has an important impact both on sports competitions and on the primary health care system, as well as on the athletes themselves. Epidemiological data show high levels of probability of suffering sports injuries, and on many occasions the clinics of sports federations cannot attend all injuries, referring them to the general health system, especially in competitions of young people in training stages [6–9]. On the other hand, the impact on the injured athlete often compromises his or her emotional and psychological state, in addition to other problems of a sporting, social, and even economic nature [10,11].

In fact, there are several factors related with the injuries' occurrence, such as biomedical, dispositional, nutritional, or postural ones [12], related to the field or pitch conditions, or with some psychosocial factors other than the practitioners themselves [13]. Another one of the aspects demonstrated in research conducted over recent decades is the role of psychological factors in the risk of athlete injuries [14–18]. The stress and injury model of Andersen and Williams [19,20] shows that stress is a key factor in the origin of injuries. Amid potentially stressful situations in the sporting field, athletes may have a stress response that increases muscle tension, which, in turn, hinders motor coordination and reduces flexibility. Furthermore, the stress response already mentioned also reduces the visual range of athletes, causing a significant loss of peripheral information and increasing distraction levels. The model also assumes that the response to stress is moderated by three main factors: the personality of the athlete, history of stressors, and coping resources. These psychosocial variables may alleviate or aggravate the stress response and may eventually affect the vulnerability of athletes to injury [3,21–25]. Although stress is the cornerstone of the model, other psychological and psychosocial variables also play a significant role (different aspects of personality, history of stressors—including the history of injuries—coping resources), as indicated in the Stress and Injury Model of Andersen and Williams [19]. Specifically, one of the aspects most emphasised is anxiety as a personal trait, which may determine the stress response of athletes [20], including recent expansions of the model, such as the ones linking the stress features with the Self Determination theory (frustration) [26].

Different studies have shown that high anxiety trait levels are linked to the vulnerability of athletes to injury [19,27–29]. In the review of Cagle et al. [30], it was found that four out of the six papers analysed identified anxiety as an injury predictor, while the other two did not. In any case, they ratified prior research conclusions [31] indicating that other psychosocial characteristics, including coping resources, worry, irritability, and stress, often seem to combine with anxiety in the prediction of sports injuries.

Furthermore, in addition to the history of stressors of athletes, the coping resources available may reduce the stress-injury link. One of these resources is the social support perceived. The scientific literature points to non-conclusive results when this variable has been studied [17,32–35]. The inconsistency of these results is probably due to the fact that the mediator role of social support is more complicated and unstable than initially thought [34].

In any case, it seems that a positive psychological disposition, which facilitates good management of competitive stress, may act to curb the risk of injury of athletes. Conversely, a negative psychological disposition may favour and increase vulnerability to injuries. Therefore, the Global Psychological Model of Sports Injuries (GPMSIs, or MGPsLD in Spanish) of Olmedilla and García-Mas [36] proposes theoretically delving deeper into the study of the specific psychological variables of greatest significance in the literature, those considered as such in the model of Andersen and Williams [19,20] and those present in the Conceptual Axis of the GPMSIs: psychosocial stress, coping resources (social support), and risk related behaviour. Furthermore, other negative psychological variables, such as depression, have been studied in research on injury rehabilitation resulting from serious injuries [37–39]. However, there are few studies that consider depression as one of the causal factors of injuries [40,41].

Bayesian networks (BNs) are beginning to be widely used in social sciences [42,43], and were recently presented as a useful methodology in sports psychology, given their ability to provide information on the probability of occurrence of events related to performance in sports or, for example, the likelihood of sports injuries. BNs, also referred to as causal networks or belief networks, are a form of statistical modelling that allows us to obtain a graphical network describing the dependencies and conditional independencies from empirical data. The graphical representation of BNs captures the compositional structure of the relations and the general aspects of all probability distributions that are factorised according to that structure. They have proven to be a promising tool for discovering relationships between negative features in sport [44], and in many other sport-related studies, such as cooperative teamwork, motivation and types of sporting cooperation among players in competing teams, motivational climate and competitive anxiety, psychological variables related to athlete injuries [45], the relative effect of age [46–49], and the relation between sport and educational performance [50]. In line with our study, a number of papers have recently been published that use a new approach, called dynamic BNs, which strives to predict and then mitigate the probability of injuries occurring in athletes [51]. Therefore, taking into consideration the importance of the incidence of sports injuries in the population of athletes and in the general health system, as well as their psychological impact on the athletes involve, the main objective of this study is to extend the stress and injury model of Andersen and Williams [19,20]—repeatedly verified—to other "negative" psychological variables, such as anxiety associated to competition and depression, encompassed in the conceptual model of Olmedilla and García-Mas [36]. Additionally, as a parallel objective, the relationship is sought to be studied of this psychological macro-variable with two other variables related to sports injuries: the search for social support, and the search for connections of risk with the environment surrounding the athletes. Finally, these variables and their relationships are studied with two specific characteristics: (1) on a wide sample of traditional and indoor football players, with a 50% combination (approximately) of injured and non-injured players; and (2) methodologically, a combination of classic methods and probabilistic approaches through Bayesian networks is used.

2. Materials and Methods

2.1. Design

The research conducted corresponds to an ex post facto study [52]. Furthermore, it is based on a descriptive and longitudinal correlational design. The independent and predictive variables used are: personal data, socio-demographic and injury records; stress levels; anxiety and depression levels; and social support perceived. As dependent variables for prediction purposes (variable criterion), a record of injuries sustained over the season is used.

2.2. Participants

The study sample is a non-random and non-probabilistic sample of an incidental and intentional nature. The number of participants assessed entails that the results to be obtained in the study do not exceed a sample error of 5% ($n > 400$; 95% reliability level) [52].

Regarding sporting data, in Table 1, the mean (X) number of years the participants have been federated for is XYF = 11.83 years (SD = 5.41), with 55.2% having spent at least 12 years as a federated athlete. In terms of training days per week, the mean is 3.66 days/week, SD = 1.18), with a broad range of training days, ranging from 2 to 7.

Table 1. Table of contingency of the sample considering the variables of sex and type of sport.

			Sport Played		Total		
			Football	Indoor Football		X Age	SD
Sex	Male	Number	300	23	323	23.66	4.36
		% of the total	65.9%	5.1%	71%		
	Female	Number	99	33	132	19.61	4.00
		% of the total	21.8%	7.3%	29%		
Total		Number	399	56	455	21.66	4.18
		% of the total	87.7%	12.3%	100.0%		

2.3. Instruments

To collect the personal information, an ad-hoc questionnaire on personal and socio demographic data was used. The questionnaire comprised nine questions referring to age, sex, sport played, current club, position played, competitive level, days of training per week, and the duration of daily training sessions.

To collect information on injuries sustained, an ad-hoc self-report on injuries was used. It was a self-reporting questionnaire comprising nine blocks of information to determine the injuries sustained during the previous season. The following questions were included: the number of injuries sustained during the previous season; when the injury happened (month); the time spent out before returning to the sport without discomfort; the specificity of the injury (type, place, detailed description of the injury); establishment of the main causal factors; degree of severity; degree of injury impact on subsequent performance; and casual attribution of the athlete to the injury.

The DASS-21 Questionnaire was used to assess depression, anxiety, and stress. The adapted Spanish version of the DASS-21 Questionnaire of Lovibond and Lovibond [53], undertaken by Román et al. [54], was used. The main objective of this scale is to assess the emotional states of stress, depression, and anxiety. The participants respond through a four-step Likert-like scale (0 = not applicable to me; 1 = slightly applicable to me or a small part of the time; 2 = largely applicable to me or a large part of the time; 3 = very applicable to me or most of the time). Each scale comprised seven items with a score ranging from 0 points (minimum) to 21 points (maximum). The reliability levels obtained by Lovibond and Lovibond [53], with a sample of 717 psychology students, were high or very high: Stress: $\alpha = 0.89$; Depression: $\alpha = 0.91$; Anxiety: $\alpha = 0.81$.

To assess the risk tendency, the DOSPERT-S Questionnaire was used. The adapted Spanish version of the scale created by Weber, Blais and Betz [55], translated by Rubio and Narváez [56] and validated by Lozano et al. [57], was used to assess attitude to risk-taking in different areas. The original and principal scale comprised 40 items. A total of eight additional items were added to the revised scale. However, after an exploratory factorial analysis, it was reduced to 30 items. Furthermore, the range of possible answer options was extended, ranging from five to seven (from 1: extremely unlikely; to 7: extremely likely). The questionnaire comprised five factors, each with six items. The following indicators were observed in this study: Social: $\alpha = 0.70$; Recreational: $\alpha = 0.80$; Finance: $\alpha = 0.77$; Health/Safety: $\alpha = 0.63$; and Ethics: $\alpha = 0.58$.

To assess social support, the Multidimensional Scale of Perceived Social Support of Zimet et al. [58], was used. The adapted Spanish version produced by Landeta and Calvete [59] was used in this study. The original scale [58] comprises 12 items that assess the perception of social support. The response system entails a four-step Likert scale ("almost never = 1"; "sometimes = 2"; "frequently = 3" and "almost always or always = 4"). These 12 items are distributed into three factors. The reliability levels found by Zimmet et al. [58] were: Family: $\alpha = 0.87$; friends: $\alpha = 0.85$ and significant others: $\alpha = 0.91$; in the Spanish version, a seven-step scale was used, which ranged from 1 = completely disagree up to 7 = completely agree, with 4 = neither agree nor disagree).

2.4. Procedure

The Regional Football Federation of Murcia (FFRM, Spain) was informed of the study, and both permission from and collaboration with it was requested. Traditional and indoor football teams that met the convenience sampling requirements were recruited. Subsequently, a meeting with the coach, players and parents of minors was scheduled. Finally, the players willing to participate in the research signed an informed consent form. In the case of minors, the form was signed by their parents or guardians. The tests were undertaken on an individual basis. Furthermore, this study was conducted in accordance with the recommendations of the Declaration of Helsinki, and it was approved by the Ethics Committee of the University of Murcia (ID: UM 1551/2017).

2.5. Data Analysis

All the standard statistical analyses were conducted using the IBM Corp. (released 2013) IBM SPSS Statistics for Windows, Version 22.0 (IBM Corp., Armonk, NY, USA). Subsequently, a Bayesian network analysis was conducted, making it necessary to determine the structure via a directed acyclic graph (DAG) and to assign conditional probabilities to each node of the DAG. Therefore, learning a BN involves the following two tasks: (i) structural learning, in other words, identifying the topology of the BN, and (ii) parametric learning or estimating the numerical parameters (conditional probabilities) given the topology of the network.

Structural learning was used to obtain the BN through the "BN learn package" [60] using R language [61]. To obtain the structure, the options were to use either a search and score algorithm [62], which assigns a score to each BN structure and selects the model structure with the highest score, or a constraint-based search algorithm [63], which establishes a conditional independence analysis on the data to generate an undirected graph and convert it into a BN using an additional independence test. The score-based algorithm Tabu [62] was used, which was a plausible model for our data, looking for the structure that best improves the score, e.g., using the highest one.

The final model was obtained by repeating structure learning several times (applying bootstrap resampling to our dataset); many network structures were explored (500 BNs) to reduce the impact of locally optimal (but globally suboptimal) networks on learning. The networks learned were averaged to obtain a more robust model. The averaged network structure was obtained using the arcs present in at least 85% of the networks, which gives a measure of the strength of each arc and establishes its significance when given a particular threshold (85%) [64].

The BN is used to make inferences, that is, to calculate new probabilities when new information is introduced [65]. Therefore, after building the BN, some instantiations were conducted (injection of hypothetical variables) to the bottom variables, as well as observation on how the node, bottom, and top variables change their probability values [66].

3. Results

With regard to the number of injuries, all the participants assessed ranged from having no injuries (0) to a maximum of five during the previous season, with a mean of 0.59 injuries/season (SD = 0.72), as can be seen in Table 2.

Table 2. Table of contingency of the sample considering the variables of number of injuries, sex and type of sport.

Sex	Sport		Number of Injuries						Total
			0	1	2	3	4	5	
Male	Football	Number	155	120	20	4	0	1	300
		% of the total	48.0%	37.2%	6.2%	1.2%	0.0%	0.3%	92.9%
	Indoor football	Number	16	6	0	0	1	0	23
		% of the total	5.0%	1.9%	0.0%	0.0%	0.3%	0.0%	7.1%
	Total	Number	171	126	20	4	1	1	323
		% of the total	52.9%	39.0%	6.2%	1.2%	0.3%	0.3%	100.0%
Female	Football	Number	46	42	10	1			99
		% of the total	34.8%	31.8%	7.6%	0.8%			75.0%
	Indoor football	Number	18	13	2	0			33
		% of the total	13.6%	9.8%	1.5%	0.0%			25.0%
	Total	Number	64	55	12	1			132
		% of the total	48.5%	41.7%	9.1%	0.8%			100.0%
Total	Football	Number	201	162	30	5	0	1	399
		% of the total	44.2%	35.6%	6.6%	1.1%	0.0%	0.2%	87.7%
	Indoor football	Number	34	19	2	0	1	0	56
		% of the total	7.5%	4.2%	0.4%	0.0%	0.2%	0.0%	12.3%
	Total	Number	235	181	32	5	1	1	455
		% of the total	51.6%	39.8%	7.0%	1.1%	0.2%	0.2%	100.0%

As regards the analyses of injury frequency, Table 2 shows how 51.6% of the sample ($n = 235$) stated they did not have an injury; 39.8% ($n = 181$) stated they had one injury; 7% ($n = 32$) stated they had two injuries; and 1.1% ($n = 5$) stated they had three injuries during the season. It was found that only one athlete had had four injuries, and another had had five injuries during the season (2%, respectively).

Regarding the athletes that stated they had sustained at least one injury ($n = 220$; 48.4% of the total), in the record of the first injury, referring to the month in which the injury took place, 83.6% ($n = 184$) responded. The frequency of injuries per month is as follows (from low to high incidence): October ($n = 27$; 14.7%); February ($n = 26$; 14.1%); November ($n = 23$; 12.5%); March ($n = 20$; 10.9%); January ($n = 19$; 10.3%); September ($n = 17$; 9.2%); May ($n = 14$; 7.6%); April ($n = 12$; 6.5%); August ($n= 5$; 2.7%); and July ($n = 1$; 0.5%). The frequency distribution corresponds almost exactly with the distribution of training sessions and games during the year and the active season.

With regard to time spent out injured before returning to the sport, 151 athletes responded, and the mean number of days without playing was 54.13 (SD = 65.29). The number of days spent out ranged from a minimum of two to a maximum of 420. The frequency analyses show that the number of inactive days in the greatest number of cases is 14 ($n = 32$; 7%), followed by 60 days ($n = 23$; 5.1%), and 30 days ($n = 22$; 4.8%), in that order. Almost half of the sample corresponded to a period under 29 days (45.7%), while the remaining 54.3% spent over 29 days out without playing.

As can be seen in Table 3, comparison between the number of injuries according to the sports category variable per sex, the prevalence rate is greater in the female subgroup (X = 0.62; SD – 0.683) than in the male subgroup (X – 0.58, SD = 0.737), which is slightly lower despite having a greater value dispersion. Analysis of mean differences using Student's *t*-test shows the absence of statistically significant differences between groups (Levene's test: F = 0.072; $p = 0.789$; t = −0.567; $p = 0.571$).

Table 3. Descriptive and mean difference analysis applying Student's *t*-statistic for two independent samples considering the variables: number of injuries, sport category per sex, type of sport and age of the players.

		N	*M*	*SD*	*F*	Sig.	*T*	Sig.	Cohen's d
Sex category	Male	323	0.58	0.737	0.072	0.789	−0.567	0.571	−0.056
	Female	132	0.62	0.683					
Sport	Football	399	0.61	0.718	0.218	0.641	1210	0.227	0.121
	Indoor football	56	0.48	0.738					
Age	>18 years old	75	0.45	0.664	0.819	0.366	−1821	0.069	0.244
	≤18 years old	378	0.62	0.730					

Regarding the type of sport, a higher prevalence of injuries was observed in the traditional football group (X = 0.61; SD = 0.718) than in the indoor football group (X = 0.48; SD = 0.738). Analysis of mean differences using Student's *t*-test shows the absence of statistically significant differences between groups (Levene's test: F =0.218; $p = 0.641$; t = 1.210; $p = 0.227$).

Considering the age variable, the athletes aged 18 years old or over show a higher prevalence (X = 0.62; SD = 0.730) than the group under 18 years old (X = 0.45, SD = 0.664). Despite not obtaining statistically significant differences, there is a trend towards statistical significance when comparing the two groups (Levene's test, F = 0.819; $p = 0.366$; t = −1.821; $p = 0.069$; $p < 0.10$). Despite not obtaining statistically significant results in this comparison, Cohen's d [67] was applied to calculate the effect size for this group. The results show a value of d = 0.244, which is a relatively small value.

Therefore, based on the studies carried out and the results shown in Tables 2 and 3, we can see that no statistically significant differences have been found (and that the trends have a small effect size) among the various variables in the study. As such, this justifies moving on to the second phase of the study, based on the search for probabilistic relationships through the analysis and generation of Bayesian Networks.

As for the study of correlations between the psychological variables and the epidemiological variables (those of the participants and those of the injury itself), the results obtained are in line with those found in the analysis of statistical differences: No significant correlations were found between the two types of variables.

3.1. Graph and Generated Bayesian Network

As can be seen in Figure 1, the resulting graph shows that the antecedent and trigger variable associated with the likelihood of having a sports injury is stress.

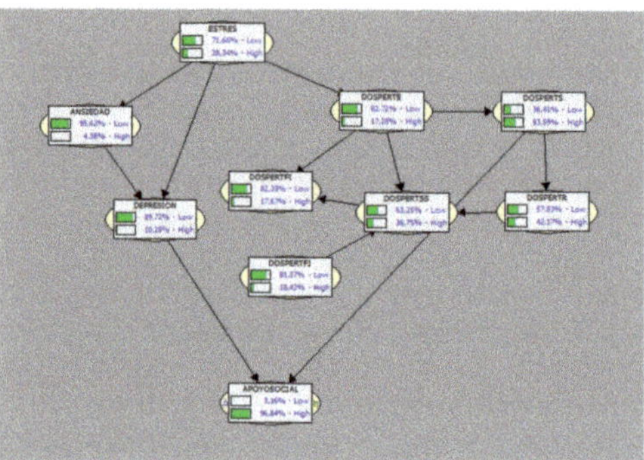

Figure 1. Graph and Bayesian Network discovered with values (high and low probability) of the study variables. Legend: Dospert: E (Ethical); FI (Financial/Business); FJ (Financial/Betting); SS (Health/Safety); S (Social); R (Recreational/Sport).

Two nodes are clearly formed, one around the psychological variables (stress, anxiety and depression –all three with low probabilities of occurrence in the population studied–the latter being twice as likely as anxiety), and the other with all the risk tendency factors, although they have different probabilities of occurrence associated with the sports injury. One of them, the FJ, acts autonomously (although from low probability values in our population), as an antecedent of this node, i.e., it would seem that the risk tendency in the financial section related to betting, even in the field of sports, would be a trigger for the other tendencies towards risk behaviors (all with low values of occurrence, except two: the risk factors associated with social habit, above all, and secondly, the sporting and/or recreational one.

As for the bottom (or descendant) variables, it is very clear in the BN conducted that Social Support, with a very high probability of occurrence (almost 100%), would not be a predictor factor, but rather would be an almost inevitable consequence of the athletes' perception given the injury.

Finally, it should be noted that the probability values found in the population studied are generally very low, which is in line with the low prevalence of injuries (around 50% of the sample did not suffer any injury, as can be seen in Table 2). Furthermore, the probabilistic weight of psychological factors is not very high, perhaps having a contributory role rather than a determining one—in terms of the probability of triggering the injury—in the anxiety-stress of the athletes.

3.2. BN Validation

The BN was validated using a 10-fold cross validation, taking the area under the curve (AUC), accuracy, sensitivity, and sensibility into consideration. Certain terms should first be defined to understand the validation used: true positive (TP), true negative (TN), false positive (FP), and false negative (FN). If an observation is labelled correctly within its class, it is considered a true positive. Conversely, if an observation is labeled correctly as not belonging to a specific class, it is a true negative. Both TP and TN suggest a consistent result in the classifier.

However, no classifier is perfect and if the model incorrectly labels an observation as belonging to a certain class, it is a false positive; and when incorrectly labelled as not belonging to a certain class, it is designated as a false negative [68]. Both FP and FN indicate that the results from the classifier are contrary to the actual label [59].

Sensitivity, specificity and accuracy are described in terms of these concepts: Sensitivity = TP / (TP + FN); Specificity = TN / (TN + FP); and Accuracy = (TN + TP) / (TN + TP + FN + FP).

The AUC shows that the probability of a randomly chosen positive datum being correctly ranked is much higher than for a randomly chosen negative datum [69]. The readings provide a complete overview of the performance of the BN. As Table 4 shows, the validation tables provided some average results, along with some medium values. These validation values should be considered when undertaking the next step in the Bayesian analysis process (the instantiations) and the final conclusions.

Table 4. Validation of the obtained BN through AUC indicators.

	AUC	Accuracy	Sensitivity	Specificity
STRESS	0.6	0.75	0.95	0.23
ANXIETY	0.5	0.96	1	0
DEPRESSION	0.6	0.92	0.99	0.18
DOSPERTE	0.63	0.83	0.94	0.32
DOSPERTFI	0.62	0.85	0.99	0.26
DOSPERTFJ	0.58	0.8	0.93	0.22
DOSPERTS	0.7	0.68	0.77	0.63
DOSPERTSS	0.64	0.72	0.92	0.36
DOSPERTR	0.66	0.67	0.72	0.6
SOCIAL SUPPORT	0.5	0.97	0	1

3.3. Instantiations

Based on the above results, it was decided to explore, as in previous studies [48,66,70], the changes in the probabilities of occurrence of antecedent variables when hypothetical values are "injected" into bottom or consequent variables. To be more precise, and also considering the meaning of the variables, the instantiations that have been carried out are as follows:

(a)   Social support, passing to 0% HIGH;
(b)   DOSPERT-S, Social/Safety, passing to 100% HIGH;
(c)   Depression, passing to 100% HIGH;
(d)   The union of the three above, simultaneously.

As can be seen in Table 5, removing the probability of occurrence of Social Support entails: increase in the likelihood of the depression, anxiety and—to a lesser degree—stress triad; decrease in the tendency of risk behavior, albeit small, and, above all, in ethical, social ("go it alone") and recreational/sporting behavior. In any case, the results of this first instantiation reaffirm those found in the original BN generated.

Table 5. Instantiation 1. Social Support, from 96.64% to 0% HIGH Likelihood. Instantiation 2. Depression, from 10.24% to 100% HIGH Likelihood. Instantiation 3. DESPORT SS, from 36.75% to 100% HIGH Likelihood. Instantiation 4. DESPORT SS, from 36.75% to 100% HIGH, Depression, from 10.24% to 100% HIGH, and Social Support, from 96.63% to 0% HIGH, likelihoods simultaneously.

	Social Sup.	Depression	Anxiety	Stress	DE	DS	DFI	DFJ	DR	DSS
Original BN										
HIGH	96.63	10.24	4.38	28.34	17.28	63.59	17.67	18.43	42.17	36.75
LOW	3.16	89.72	95.62	71.64	82.72	36.41	82.33	81.57	57.43	63.25
1. After Social Support = 0% High										
HIGH	0	44.41	14.09	39.38	16.21	50.15	17.22	18.43	36.46	35.09
LOW	100	55.59	85.91	60.62	83.79	49.85	82.78	81.57	63.54	64.91
2. After Depression = 100% HIGH, 0% LOW										
HIGH	86.91	100	30.39	55.97	20.02	64.27	18.44	18.43	35.09	37.92
LOW	13.09	0	69.61	44.03	79.98	35.73	81.56	81.57	64.91	62.08
3. After Dosport SS = 100% High, 0% LOW										
HIGH	96.98	10.66	4.61	30.73	34.23	72.8	30.2	24.58	57.94	100
LOW	3.02	89.34	95.39	69.27	65.77	27.2	69.8	75.42	42.06	0
4. After Dosport SS = 100% High, 0% LOW; Depression = 100% HIGH, 0% LOW, and Social Support = 0% HIGH, 100% LOW										
HIGH	0	100	30.07	58.17	41.34	87.3	32.152	3.46	64.18	100
LOW	100	0	69.93	41.87	58.66	12.7	67.857	6.54	32.82	0

4. Discussion

The results obtained in this study largely confirm the aims expressed in the initial objectives, both from the perspective of the quality of the explanations they allow for and the validity values found. No differences were found between the variables relating to players or the nature of the injuries. Along the same lines, no correlations were found between said variables, which occurred repeatedly when attempts were made to "connect" the objective characteristics of the injury or sport with the psychological variables considered in different studies [36,71].

The information collected in the ad hoc injury questionnaire applied in this research provides valuable information regarding injuries sustained during the season. However, once analyzed –beyond the limitations in its design that will subsequently be discussed–we observe that, once again, no clear pattern emerges that could justify a behavioral approach to the occurrence of sport injuries.

In terms of the probabilistic analysis carried out, consideration should be given, firstly, to providing support to the classical model centered on stress, as well as to the conceptual ordering derived from the Global Sports Injury Model [36], as they can be considered—among several other factors or variables, e.g., related with the characteristics of the practice, the pitch, or the participants' biomedical features as antecedents to the others. Furthermore, these results—obtained with a methodology based on probability analysis—support the possibility of extending the stress model to other variables (anxiety and depression, the so-called negative "triad"), while maintaining the same characteristics with respect to sport injuries.

Thus, the trigger variable and from which all the other "negative" psychological variables (gathered in a very clear node in the graph obtained) descend is the stress perceived by the athletes. The probability values found are low (it should be remembered that the incidence of injury in the sample studied is only about 50%), and here too stress is pre-eminent in terms of the probability of occurrence. The relatively low probability of depression (although it is the one with the highest value within the BN) and, secondly, of anxiety associated with competition, makes us reflect on whether it is possible to contemplate a component of "relief" from the injury—cease competing–in comparison with the low probability of anxiety found, perhaps related to an early return to competition and its demands. This result had already appeared—in a much smaller form—in a very early study by Liberal et al. [71]. At what point in the natural life of the sports injury does each of these variables carry the most weight? As the study is retrospective, further studies are needed to better understand whether this variable affects prevention (as it seems to do) rather than recovery. Were it to regard learned avoidance behavior, we would be entering more complex territory, which would require observation and a different theoretical approach to those usually carried out [72,73].

It is very interesting to have found that the probability values related to the various factors associated with risk-seeking tendencies (in different domains) constitute another node isolated from the previous node of the "negative triad". However, this result had already been partially reported in another study using the same methodology [74]. From the clear results found, an important issue emerges: can these be two different sources of probability of occurrence of a sports injury?

It should not be forgotten, on the other hand, that this characteristic of seeking out and undertaking risky behavior may have some bearing on the issue—which is critical on many levels today—of sports betting. This would seem to be confirmed by the fact that the highest values for the factors in this variable correspond to financial, social, and sport/leisure factors. This separation of the two nodes opens, in our opinion, an important collateral avenue for furthering our understanding of the psychological components of sport injuries.

Furthermore, the findings of this study can—as indicated, in part, above—fit very well into the proposed integrative model (the GMSI) [36] as they can be considered as antecedents of injury and, therefore, enter into the realm of possible injury prevention, rather than rehabilitation or recovery and the concept of 'returning to the sport'.

When we analyze the results of the instantiations on the bottom (or consequent) variables with hypothetically unachievable maximum and/or minimum values, the results found in the BN are clearly reaffirmed.

The results obtained can contribute to both theoretical and applied aspects of the scientific problem. On the one hand, from a theoretical perspective, the inclusion of "anxiety" and "depression" at the level of "stress" in the model of Andersen and Williams may allow a conceptually different approach, being able to speak of a "negative triad", and the need for empirical studies that can support this aspect in line with the proposals of Olmedilla and García-Mas in their GMSI [36]. On the other hand, the results seem to show that the presence of mental health indicators in sports practice is a fact of great importance, and not only from the perspective of injury [75]. In this sense, it seems necessary to implement psychology actions and programs for the promotion of mental health, and for the prevention of the basic indicators of the "negative triad", in line with Brenner et al. [76] and Henriksen et al. [77].

5. Conclusions

First of all, the existence of the "triad" node is consolidated, as the variables contained in it are affected jointly, with an almost "homogeneous" function (possibly, both in the occurrence and in the rehabilitation process of the injury) on modifying the stress values upwards. In addition, the separation of increase/decrease of the occurrence values that may hypothetically reach the probability of the risk tendency is maintained.

Likewise, we also see that a variable that is much discussed in many aspects and theoretical frameworks (see, for example, the corresponding scale of the updated sport engagement questionnaire, the SCQ-2 [78]), social support, is highly compromised both in the BN and, particularly, in the instantiations, as its probability of occurrence does not increase when the "negative" variables do. It is true that it has previously been pointed out within different areas that the line between support and social pressure is not entirely clear-cut and immovable [79,80], and this seems to be another example of it. As such, what seems to happen when the probability of occurrence of the SS decreases, is that it leads to an increase in the other factors (D, A and E) of the same variable, and a decrease in the probability values of risk-tendency factors. Perhaps the fact of not having a 'safety net' makes the players implement strategies and resources with greater implication ("...there is no family or anyone to protect me, I have to get on with it", "...either I do it or...I don't have anyone"), which may help to have greater focus, decide more quickly and stabilize behavior, along the lines of what is foreseen in the stress model.

And this immediately opens up another very relevant question, especially regarding youth sport: which psychosocial factor can be more, and which can be less, effective in preventing or aiding recovery from a sports injury? In this case, we fear that we will have to resort to methodologies of a more qualitative nature to complement the quantitative ones in order to obtain answers adapted to the specific characteristics of the athletes. As in our study, it has been shown that the family "umbrella" does not seem to have relevant probabilistic weight.

Finally, based on the results obtained and on this critical discussion, it can be stated that stress is not only found within the framework of psychological variables traditionally considered as "negative", and that these have different degrees of severity and complexity for the players. It can also be affirmed that intervention plans should separate the intervention on these psychological components from those related to the learning of avoidance and risk behavior.

As a last conclusion, we can state that stress is not the only factor in the classic and repeatedly proven model linking it to sports injuries.

5.1. Limitations and Future Developments

Obviously, there are some limitations in this study. Firstly, there is a transversal analysis, due to the rapid and dynamic composition of sports teams, which see changes regarding some of their players almost every season. This classic problem when working with sport competition teams is really difficult to address, hindering somewhat the longitudinal approach. Secondly, there is no work on the best way to better integrate the classic statistical methods with the Bayesian analysis (we are currently conducting them in a sequential way, gaining information on one over the other).

Thirdly, regarding the ad hoc questionnaire, its design does not provide enough specific information (is it a new injury, a relapse of an old one or another injury different from the one previously sustained?); the place of the injury (practice, competition or off the pitch); time the injury took place (opening the door to the study of the chronobiological factors in this field), and, in a more psychological fashion, its design must include the causal attribution made by the injured athlete about the cause of the injury.

Finally, the authors didn't have the opportunity to study the physical and body indications associated with the stress present in some players. This kind of evaluation may be a valuable complement to the paper and pencil data collection, when studying the anxiety and stress associated with the injury.

Regarding future developments, there are three lines of work in the medium term. Firstly, and given the results obtained, it would be interesting to compare the graphs and BNs of the two semi-samples of this study (injured vs. non-injured players) in order to observe the differences, if any, between the variables, their nodes and their probabilities of occurrence.

Secondly, once this extension of the stress model is "secured", it may also be relevant—in this case, following the GMSI conceptual model—whether depression is pre-injury (antecedent) or post-injury (consequent) and, in this case, we will have to work with a longitudinal follow-up or survival methodology and sample.

Thirdly, it would be good to improve our knowledge of the injuries and specific features, working with bigger samples including indoor football and female teams, regarding some particularities discovered in the kind of injuries sustained.

Further studies should also look into the differences between different sports and the injuries that occur, following, on the one hand, the classical differentiation between individual and team sports, as well as direct and indirect opposition sports. Furthermore, the analysis between more socially developed sports (such as biking and trekking), which currently take place in a very high percentage outside the competitive area and are done by amateurs, should be developed (this has not been systematically conducted so far). This type of analysis could—along the lines of the BN analyses—indicate different probabilities of sustaining an injury associated with doing different sports.

Finally, the role of competitive anxiety in relation to many sporting situations is very controversial indeed [46]. In this case, a line of study and development regarding anxiety (three-dimensionally considered) in relation to "returning to the sport" may be of great value from a theoretical and applied intervention perspective. This study should perhaps be carried out using a qualitative rather than quantitative methodology.

5.2. Practical Implications

Despite the limitations indicated above, this study has high practical implications for the professional practice of both sports coaches and other support professionals (sports doctors, sports psychologists, physical trainers, physiotherapists, etc.), given that traditional descriptive analyses indicate a trend in the results according to the sports age category (but not according to the sex category), thus we know that this is a variable that should be addressed in the pre- and post-injury intervention. Likewise, the analyses developed through BNs indicate the importance of generating adequate coping strategies in athletes (especially considering the specificity of possible social support at the sporting and extra-sporting level), with a view to preventing and coping with two possible consequences of a failure in the adaptation process of athletes: anxiety and depression.

Author Contributions: Conceptualization, A.O.Z. and A.G.-M.; methodology, R.R.-B., B.M., A.O.Z. and A.G.-M.; software, F.J.P.-V. and B.M.; formal analysis, R.R.-B. and F.J.P.-V.; investigation, A.O.Z. and A.G.-M.; resources, A.G.-M.; data curation, R.R.-B. and B.M.; writing—original draft preparation, F.J.P.-V.; project administration, A.O.Z. and A.G.-M.; funding acquisition, F.J.P.-V., A.O.Z. and A.G.-M. All authors have read and agreed to the published version of the manuscript.

Funding: This research was funded in part by the programme "Values in Sport", 2016/1584, from the Balearic Islands Government (Spain) and the Murcia (Spain) Regional Football Association's Football Project (Grant FFRMUMU-040092 321B 64502 14704).

Institutional Review Board Statement: Authorization to conduct the research and approval of the study was granted by the Ethical Committee of the University of Murcia (ID2303/2019). Furthermore, this study was conducted in accordance with the approved guidelines, the Declaration of Helsinki and with the informed consent of the participants.

Informed Consent Statement: Informed consent was obtained from all subjects involved in the study.

Data Availability Statement: Data supporting the reported results are found in Department of Personality, Evaluation, and Psychological Treatment, Faculty of Psychology, Murcia. Responsible is Aurelio Olmedilla Zafra.

Conflicts of Interest: The authors declare no conflict of interest.

References

1. Alahmad, T.A.; Kearney, P.; Cahalan, R. Injury in elite women's soccer: A systematic review. *Physician Sportsmed.* **2020**, *48*, 259–265. [CrossRef] [PubMed]
2. Bittencourt, N.F.; Meeuwisse, W.H.; Mendonça, L.D.; Nettel-Aguirre, A.; Ocarino, J.M.; Fonseca, S.T. Complex systems approach for sports injuries: Moving from risk factor identification to injury pattern recognition—narrative review and new concept. *Br. J. Sports Med.* **2016**, *50*, 1309–1314. [CrossRef] [PubMed]

3. Gabbett, T.J.; Nielsen, R.O.; Bertelsen, M.L.; Bittencourt, N.F.; Fonseca, S.T.; Malone, S.; Windt, J. In pursuit of the 'Unbreakable' Athlete: What is the role of moderating factors and circular causation? *Br. J. Sports Med.* **2019**, *53*, 394–395. [CrossRef] [PubMed]

4. Meeuwisse, W.H.; Tyreman, H.; Hagel, B.; Emery, C. A dynamic model of etiology in sport injury: The recursive nature of risk and causation. *Clin. J. Sports Med.* **2007**, *17*, 215–219. [CrossRef]

5. Meeuwisse, W.; Hagel, B. The multicausality of injury—current concepts. *Sports Inj. Res.* **2010**, *16*, 99.

6. Hootman, J.M.; Dick, R.; Agel, J. Epidemiology of collegiate injuries for 15 sports: Summary and recommendations for injury prevention initiatives. *J. Athl. Train.* **2007**, *42*, 311.

7. Prieto-González, P.; Martínez-Castillo, J.L.; Fernández-Galván, L.M.; Casado, A.; Soporki, S.; Sánchez-Infante, J. Epidemiology of sports-related injuries and associated risk factors in adolescent athletes: An injury surveillance. *Int. J. Environ. Res. Public Health* **2021**, *18*, 4857. [CrossRef]

8. Pujals, C.; Rubio, V.J.; Marquez, M.O.; Sánchez, I.; Ruiz-Barquin, R. Comparative sport injury epidemiological study on a Spanish sample of 25 different sports. *Rev. Psicol. Deporte* **2016**, *25*, 271–279.

9. Viljoen, C.T.; van Rensburg, D.C.J.; Verhagen, E.; Van Mechelen, W.; Tomás, R.; Schoeman, M.; Korkie, E. Epidemiology of injury and illness among trail runners: A systematic review. *Sports Med.* **2021**, *51*, 917–943. [CrossRef]

10. Von Rosen, P.; Kottorp, A.; Fridén, C.; Frohm, A.; Heijne, A. Young, talented and injured: Injury perceptions, experiences and consequences in adolescent elite athletes. *Eur. J. Sport Sci.* **2018**, *18*, 731–740. [CrossRef]

11. Wiese-Bjornstal, D.M. Psychological predictors and consequences of injuries in sport settings. In *APA Handbook of Sport and Exercise Psychology, Volume 1. Sport Psychology*; Anshel, M.H., Petrie, T.A., Steinfeldt, J.A., Eds.; American Psychological Association: Washington, DC, USA, 2019; pp. 699–725. [CrossRef]

12. Aicale, R.; Tarantino, D.; Mafulli, N. Oversuse injuries in sport: A comprehensive overview. *J. Orthop. Surg. Res.* **2018**, *13*, 309. [CrossRef]

13. Ahern, D.K.; Lohr, B.A. Psychosocial factors in Sport Injury Rehabilitation. *Clin. Sport Med.* **1997**, *16*, 755–768. [CrossRef]

14. Almeida, P.L.; Olmedilla, A.; Rubio, V.J.; Palou, P. Psychology in the realm of sport injury. *Rev. Psicol. Deporte* **2014**, *23*, 395–400.

15. Hausken-Sutter, S.E.; Pringle, R.; Schubring, A.; Grau, S.; Barker-Ruchti, N. Youth sport injury research: A narrative review and the potential of interdisciplinarity. *BMJ Open Sport Exerc. Med.* **2021**, *7*, e000933. [CrossRef]

16. Ivarsson, A.; Johnson, U.; Andersen, M.B.; Tranaeus, U.; Stenling, A.; Lindwall, M. Psychosocial factors and sport injuries: Meta-analyses for prediction and prevention. *Sports Med.* **2017**, *47*, 353–365. [CrossRef]

17. Johnson, U.; Tranaeus, U.; Ivarsson, A. Current status and future challenges in psychological research of sport injury prediction and prevention: A methodological perspective. *J. Sport Psychol.* **2014**, *23*, 401–409.

18. Slimani, M.; Bragazzi, N.L.; Znazen, H.; Paravlic, A.; Azaiez, F.; Tod, D. Psychosocial predictors and psychological prevention of soccer injuries: A systematic review and meta-analysis of the literature. *Phys. Ther. Sport* **2018**, *32*, 293–300. [CrossRef]

19. Andersen, M.B.; Williams, J.M. A model of stress and athletic injury: Prediction and prevention. *J. Sport Exerc. Psychol.* **1988**, *10*, 294–306. [CrossRef]

20. Williams, J.M.; Andersen, M.B. Psychosocial antecedents of sport injury: Review and critique of the stress and injury model'. *J. Appl. Sport Psychol.* **1998**, *10*, 5–25. [CrossRef]

21. Johnson, U.; Ivarsson, A. Psychosocial factors and sport injuries: Prediction, prevention and future research directions. *Curr. Opin. Psychol.* **2017**, *16*, 89–92. [CrossRef]

22. Olmedilla, A.; Rubio, V.J.; Ortega, E. Predicting and preventing sport injuries: The role of stress. In *Sports Injuries: Prevention, Management and Risk Factors*; Hopkins, G., Ed.; Nova Science Publishers: Hauppauge, NY, USA, 2015; pp. 87–104.

23. Petrie, T.; Perna, F. Psychology of injury: Theory, research, and practice. In *Sport Psychology: Theory, Applications, and Issues*; Morris, T., Summers, J., Eds.; Wiley: Westmeadows, Australia, 2004; pp. 547–571.

24. Rees, T.I.M.; Freeman, P. Social support moderates the relationship between stressors and task performance through self-efficacy. *J. Soc. Clin. Psychol.* **2009**, *28*, 244–263. [CrossRef]

25. Tranaeus, U.; Ivarsson, A.; Johnson, U. Evaluation of the effects of psychological prevention interventions on sport injuries: A meta-analysis. *Sci. Sports* **2015**, *30*, 305–313. [CrossRef]

26. Li, C.; Ivarsson, A.; Lam, L.T.; Sun, J. Basic psychological needs satisfaction and frustration, stress, and sports injury among university athletes: A four-wave prospective survey. *Front. Psychol.* **2019**, *10*, 665. [CrossRef]

27. Petrie, T.A. Coping skills, competitive anxiety, and playing status: Moderating effects on the life stress–injury relationship. *J. Sport Exerc. Psychol.* **1993**, *15*, 261–274. [CrossRef]

28. Lavallée, L.; Flint, F. The relationship of stress, competitive anxiety, mood state, and social support to athletic injury. *J. Athl. Train.* **1996**, *31*, 296–299.

29. Johnson, U.; Ivarsson, A. Psychological predictors of sport injuries among junior soccer players. *Scand. J. Med. Sci. Sports* **2011**, *21*, 129–136. [CrossRef] [PubMed]

30. Cagle, J.A.; Overcash, K.B.; Rowe, D.P.; Needle, A.R. Trait anxiety as a risk factor for musculoskeletal injury in athletes: A critically appraised topic. *Int. J. Athl. Ther. Train.* **2017**, *22*, 26–31. [CrossRef]

31. García-Mas, A.; Pujals, C.; Fuster-Parra, P.; Nuñez, A.; Rubio, V.J. Determination of the psychological and sportive variables related to sports injuries. *Rev. Psicol. Deporte* **2014**, *23*, 423–429.

32. Hardy, C.J.; Richman, J.M.; Rosenfeld, L.B. The role of social support in the life stress/injury relationship. *Sport Psychol.* **1991**, *5*, 128–139. [CrossRef]
33. Malinauskas, R. The associations among social support stress and life satisfaction as perceived by injured college athletes. *Soc. Behav. Pers.* **2010**, *38*, 741–752. [CrossRef]
34. Petrie, T.A.; Deiters, J.; Harmison, R.J. Mental toughness, social support, and athletic identity: Moderators of the life stress–injury relationship in collegiate football players. *Sport Exerc. Perform. Psychol.* **2014**, *3*, 13–27. [CrossRef]
35. Rees, T.; Mitchell, I.; Evans, L.; Hardy, L. Stressors, social support and psychological responses to sport injury in high- and low-performance standard participants. *Psychol. Sport Exerc.* **2010**, *11*, 505–512. [CrossRef]
36. Olmedilla, A.; García-Mas, A. The globlal psychological model of the sport injuries. *Acc. Psicol.* **2009**, *6*, 77–91.
37. Appaneal, R.N.; Levine, B.R.; Perna, F.M.; Roh, J.L. Measuring postinjury depression among male and female competitive athletes. *J. Sport Exerc. Psychol.* **2009**, *31*, 60–76. [CrossRef]
38. Guo, E.W.; Cross, A.G.; Hessburg, L.; Koolmees, D.; Bernstein, D.N.; Elhage, K.G.; Makhni, E.C. The Presence of Preoperative Depression Symptoms Does Not Hinder Recovery After Anterior Cruciate Ligament Reconstruction. *Orthop. J. Sports Med.* **2021**, *9*, 2325967120970219. [CrossRef]
39. Lichtenstein, M.B.; Gudex, C.; Andersen, K.; Bojesen, A.B.; Jørgensen, U. Do exercisers with musculoskeletal injuries report symptoms of depression and stress? *J. Sport Rehab.* **2019**, *28*, 46–51. [CrossRef]
40. Olmedilla, A.; Ortega, E.; Murcia, J.A.; García-Angulo, A. Relationship between levels of depression and sports injuries in football and indoor football players. *Sport-Tk* **2017**, *6*, 35–39.
41. Yang, J.; Cheng, G.; Zhang, Y.; Covassin, T.; Heiden, E.O.; Peek-Asa, C. Influence of symptoms of depression and anxiety on injury hazard among collegiate American football players. *Res. Sports Med.* **2014**, *22*, 147–160. [CrossRef]
42. Fuster-Parra, P.; Vidal-Conti, J.; Borràs, P.A.; Palou, P. Bayesian networks to identify statistical dependencies. A case study of Spanish University students' habits. *Inform. Health Soc. Care* **2017**, *42*, 166–179. [CrossRef]
43. Chen, W.; Huang, S. Evaluating flight crew performance by a bayesian network model. *Entropy* **2018**, *20*, 178. [CrossRef]
44. Fuster-Parra, P.; Garcia-Mas, A.; Ponseti, F.J.; Palou, P.; Cruz, J. A Bayesian network to discover relationships between negative features in sport. A case study of teen players. *Qual. Quant.* **2013**, *48*, 1473–1491. [CrossRef]
45. Núñez, A.; Ponseti, F.X.; Sesé, A.; García-Mas, A. Ansiedad y rendimiento percibido en atletas y músicos: Revisitando a Martens. *J. Sport Psychol.* **2020**, *29*, 21–28.
46. Ishigami, H. Relative age and birthplace effect in Japanese professional sports: A quantitative evaluation using a Bayesian hierarchical poisson model. *J. Sports Sci.* **2016**, *34*, 143–154. [CrossRef]
47. Ponseti, F.J.; Garcia-Mas, A.; Palou, P.; Cantallops, J.; Fuster-Parra, P. Self-determined motivation and types of sportive cooperation among players on competitive teams A Bayesian network analysis. *Int. J. Sport Psychol.* **2016**, *47*, 428–442. [CrossRef]
48. Ponseti, F.J.; Almeida, P.L.; Lameiras, J.; Martins, B.; Olmedilla, A.; López-Walle, J.; Garcia-Mas, A. Self-Determined Motivation and Competitive Anxiety in Athletes/Students: A Probabilistic Study Using Bayesian Networks. *Front. Psychol.* **2019**, *10*, 1947. [CrossRef]
49. Olmedilla, A.; Rubio, V.J.; Fuster-Parra, P.; Pujals, C.; García-Mas, A. A Bayesian Approach to Sport Injuries Likelihood: Does Player's Self-Efficacy and Environmental Factors Plays the Main Role? *Front. Psychol.* **2018**, *9*, 1174. [CrossRef]
50. Gavala-González, J.; Martins, B.; Ponseti, F.J.; Garcia-Mas, A. Studying Well and Performing Well: A Bayesian Analysis on Team and Individual Rowing Performance in Dual Career Athletes. *Front. Psychol.* **2020**, *11*, 583409. [CrossRef]
51. Peterson, K.D.; Evans, L.C. Decision support system for mitigating athletic injuries. *Int. J. Comput. Sci. Sport* **2019**, *18*, 45–63. [CrossRef]
52. León, O.G., y Montero, I. *Métodos de Investigación en Psicología y Educación. Las Tradiciones Cuantitativa y Cualitativa*; McGraw-Hill: New York, NY, USA, 2015.
53. Lovibond, P.; Lovibond, S. The structure of negative emotional states: Comparison of the depression anxiety stress scales (DASS) with the Beck depression and anxiety inventories. *Behav. Res. Ther.* **1995**, *33*, 335–343. [CrossRef]
54. Román, F.; Santibáñez, P.; Vinet, E. Uso de las Escalas de Depresión Ansiedad Estrés (DASS-21) como Instrumento de Tamizaje en Jóvenes con Problemas Clínicos. *Acta Investig. Psicol.* **2016**, *6*, 2325–2336. [CrossRef]
55. Weber, E.U.; Blais, A.R.; Betz, N.E. A Domain-specific Risk-attitude Scale: Measuring Risk Perceptions and Risk Behaviors. *J. Behav. Decis. Mak.* **2002**, *15*, 263–290. [CrossRef]
56. Rubio, V.; Narváez, M. *Escala de Evaluación de Tendencia al Riesgo por Dominios –Versión Española. Traducción de la Escala DOSPERT*; Autonomous Univerity of Madrid: Madrid, Spain, 2007.
57. Lozano, L.M.; Megías, A.; Catena, A.; Perales, J.C.; Baltruschat, S.; Cándido, A. Spanish validation of the domain-specific risk-taking (DOSPERT-30) scale. *Psicothema* **2017**, *29*, 111–118.
58. Zimet, G.D.; Dahlem, N.W.; Zimet, S.G.; Farley, G.K. The Multidimensional Scale of Perceived Social Support. *J. Pers. Assess.* **1988**, *52*, 30–41. [CrossRef]
59. Landeta, O.; Calvete, E. Adaptation and Validation of the Multidimensional Scale of Perceived Social Support. *Ansiedad Estrés* **2002**, *8*, 30–41.
60. Scutari, M. Learning Bayesian Networks with the bnlearnR Package. *J. Stat. Softw.* **2010**, *35*, 1–22. [CrossRef]
61. Rosselet, A. Language definition-based compiler development. *J. Syst. Softw.* **1987**, *7*, 145–161. [CrossRef]
62. Korb, K.B.; Nicholson, A.E. *Bayesian Artificial Intelligence*; CRC Press: Boca Raton, FL, USA, 2010.

63. Spirtes, P.; Glymour, C.; Scheines, J. *Causation, Prediction and Search*; Springer: New York, NY, USA, 1993.
64. Nagarajan, R.; Scutari, M.; Lèbre, S. *Bayesian Networks in R: With Applications in Systems Biology*; Springer: Berlin/Heidelberg, Germany, 2013.
65. Butz, C.J.; Hua, S.; Chen, J.; Yao, H. A simple graphical approach for understanding probabilistic inference in Bayesian networks. *Inf. Sci.* **2009**, *179*, 699–716. [CrossRef]
66. Fuster-Parra, P.; García-Mas, A.; Cantallops, J.; Ponseti, F.J.; Luo, Y. Ranking features on psychological dynamics of cooperative team work through Bayesian networks. *Symmetry* **2016**, *8*, 34. [CrossRef]
67. Cohen, J. *Statistical Power Analysis for the Behavioral Sciences*, 2nd ed.; Erlbaum: Hillsdale, NJ, USA, 1988. [CrossRef]
68. Lalkhen, A.G.; McCluskey, A. Clinical tests: Sensitivity and specificity. *Contin. Educ. Anaesth. Crit. Care Pain* **2008**, *8*, 221–223. [CrossRef]
69. Bradley, A.P. The use of the area under the ROC curve in the evaluation of machine learning algorithms. *Pattern Recognit.* **1997**, *30*, 1145–1159. [CrossRef]
70. Fuster-Parra, P.; García-Mas, A.; Ponseti, F.J.; Leo, F.M. Team performance and collective efficacy in the dynamic psychology of competitive team: A Bayesian network analysis. *Hum. Mov. Sci.* **2015**, *40*, 98–118. [CrossRef]
71. Liberal, R.; Escudero, J.T.; Cantallops, J.; Ponseti, J. Impacto psicológico de las lesiones deportivas en relación al bienestar psicológico y la ansiedad asociada a deportes de competición. *Rev. Psicol. Deporte* **2014**, *23*, 451–456.
72. Olmedilla, A.; Sánchez-Aldeguer, M.F.; Almansa, C.M.; Gómez-Espejo, V.; Ortega, E. Entrenamiento psicológico y mejora de aspectos psicológicos relevantes para el rendimiento deportivo en jugadoras de fútbol. *Rev. Psicol. Apl. Deporte Ejerc. Fís.* **2018**, *3*, 8. [CrossRef]
73. Moreno-Fernández, I.M.; Gómez-Espejo, V.; Olmedilla-Caballero, B.; Ramos-Pastrana, L.M.; Ortega-Toro, E.; Olmedilla, A. Eficacia de un programa de preparación psicológica en jugadores jóvenes de fútbol. *Rev. Psicol. Apl. Deporte Ejerc. Fís.* **2019**, *4*, 14. [CrossRef]
74. Rubio, V.J.; Pujals, C.; de la Vega, R.; Aguado, D.; Hernández, J.M. Autoeficacia y lesiones deportivas:¿ factor protector o de riesgo? *Rev. Psicol. Deporte* **2014**, *23*, 439–444.
75. Armino, N.; Gouttebarge, V.; Mellalieu, S.; Schlebusch, R.; van Wyk, J.P.; Hendricks, S. Anxiety and depression in athletes assessed using the 12-item General Health Questionnaire (GHQ-12)-a systematic scoping review. *S. Afr. J. Sports Med.* **2021**, *33*, 1–13. [CrossRef]
76. Brenner, J.S.; LaBotz, M.; Sugimoto, D.; Stracciolini, A. The psychosocial implications of sport specialization in pediatric athletes. *J. Athl. Train.* **2019**, *54*, 1021–1029. [CrossRef]
77. Henriksen, K.; Schinke, R.; Moesch, K.; McCann, S.; Parham, W.D.; Larsen, C.H.; Terry, P. Consensus statement on improving the mental health of high performance athletes. *Int. J. Sport Exerc. Psychol.* **2020**, *18*, 553–560. [CrossRef]
78. Sánchez-Miguel, P.A.; Chow, G.M.; Sousa, C.; Scanlan, T.K.; Ponseti, F.J.; Scanlan, L.; Garcia-Mas, A. Adapting the Sport Commitment Questionnaire-2 for Spanish Usage. *Percept. Mot. Ski.* **2019**, *126*, 267–285. [CrossRef]
79. Borrás, P.A.; Ponseti, F.J.; Pulido, D.; Sanchez-Romero, E.I. Preocupación deportiva, percepción de apoyo parental hacia el deporte y disposición al engaño y astucia en el deporte base. *J. Sport Health Res.* **2020**, *12*, 169–178.
80. Griffin, L.J.; Moll, T.; Williams, T.; Evans, L. Rehabilitation from Sport Injury: A Social Support Perspective. In *Essentials of Exercise and Sport Psychology: An Open Access Textbook*; Society for the Transparency, Openness, and Replication in Kinesiology: Utrecht, The Netherlands, 2021; pp. 734–758. [CrossRef]

 *healthcare*

Article

Influence of Mindfulness on Levels of Impulsiveness, Moods and Pre-Competition Anxiety in Athletes of Different Sports

Laura C. Sánchez-Sánchez [1,*], Clemente Franco [2], Alberto Amutio [3,4], Jaqueline García-Silva [1] and Juan González-Hernández [1]

1   Department of Personality, Evaluation and Psychological Treatment, Faculty de Psychology, University of Granada, Campus Cartuja, 180071 Granada, Spain
2   Department of Psychology, University of Almería, Carretera Sacramento, S/N, La Cañada de San Urbano, 04120 Almería, Spain
3   Department of Social Psychology, University of the Basque Country (UPV/EHU), Barrio Sarriena, S/N, 48940 Leioa, Spain
4   Facultad de Educación y Ciencias Sociales, Universidad Andres Bello, Santiago de Chile 7591538, Chile
*   Correspondence: lcsanchezsa@ugr.es; Tel.: +34-958-249-543

Abstract: Training in emotional regulation skills is one of the most important resources for the adaptation of athletes to contexts of sports pressure, especially during competitions. This study explored the effects of a mindfulness programme (Flow Meditation) on levels of impulsivity, mood and pre-competition anxiety-state in a sample of athletes (N = 41, 22.83 ± 5.62 years). Participants were randomly assigned to an intervention group (N = 21; 14 males and 7 females) which received the intervention over 10 weeks (a weekly session) and a control group (wait-list; N = 20; 13 males and 7 females). The variables under study were assessed through different questionnaires at pre- and post-test (T1–T2) in both groups. The mindfulness intervention was effective in reducing impulsivity (cognitive ($t = -4.48$, $p \leq 0.001$, Cohen's $d = 1.40$), both motor ($t = -4.03$, $p \leq 0.001$, Cohen's $d = 1.20$) and unplanned ($t = -5.32$, $p \leq 0.001$, Cohen's $d = 1.66$)), mood (tension ($t = -4.40$, $p \leq 0.001$, Cohen's $d = 1.37$), depression ($t = -4.56$, $p \leq 0.001$, Cohen's $d = 1.42$), anger ($t = -7.80$, $p \leq 0.001$, Cohen's $d = 2.47$), somatic anxiety ($t = -5.28$, $p \leq 0.001$, Cohen's $d = 1.65$), and cognitive anxiety ($t = -6.62$, $p \leq 0.001$, Cohen's $d = 2.07$) in the intervention group compared to the control group and with large to very large effect sizes. Mindfulness is a factor that enhances athletes' ability to cope with high sport pressure and the healthy management of competition (e.g., fear of failure), or with their daily life.

Keywords: mindfulness; awareness; sport; psychosocial variables

Citation: Sánchez-Sánchez, L.C.; Franco, C.; Amutio, A.; García-Silva, J.; González-Hernández, J. Influence of Mindfulness on Levels of Impulsiveness, Moods and Pre-Competition Anxiety in Athletes of Different Sports. *Healthcare* **2023**, *11*, 898. https://doi.org/10.3390/healthcare11060898

Academic Editors: João Paulo Brito and Rafael Oliveira

Received: 31 January 2023
Revised: 15 March 2023
Accepted: 16 March 2023
Published: 21 March 2023

1. Introduction

Physical exercise and sport have a positive effect on physical and mental health and on personal development [1]. However, in competitive sport, athletes must fulfil high expectations and be physically and mentally prepared to achieve peak performance. Today, more and more sport professionals recognise the importance of mental preparation to optimise performance and maintain mental health [2]. The relevance of psychological functioning to an athlete's performance is reflected both in sporting outcomes and in the athlete's well-being and mood. Within sports, there are a number of psychological traits or behaviours (e.g., impulsivity, anxiety, negative moods, or fatigue) that have been associated with deficits in emotional regulation skills and can negatively influence performance [3,4]. On the other hand, positive emotions and moods, such as joy and vigour, may contribute to increased self-confidence and sport performance [5,6].

From a cognitive–behavioural perspective, impulsivity could be defined as the tendency to behave without a sufficient process of the observation, analysis, and/or evaluation of consequences, in the form of tendencies towards risk-taking behaviour, rapid decision making, and lack of planning [7]. Impulsivity is generally considered a predisposition to perform quick and thoughtless actions in response to internal and/or external stimuli despite negative consequences for oneself and/or others. It is described in terms of the process involved: motor (i.e., action without thought), perceptual–attentional (i.e., lack of concentration on the task), and lack of planning or transcendence (i.e., prioritising when, how, and where the action occurs without giving value to the consequences for the future) [8–10]. In sport contexts, impulsive responding has been related both to the pursuit of greater and faster efforts to be perfect [11], fear of failure [12], and even self-destruction [13], and to negative emotions such as anger, aggressive behaviours, concentration difficulties, and poor performance [14,15].

As for anxiety, although scientific evidence has widely demonstrated that anxiety negatively affects psychosocial resources and individual conditions for sport performance [16,17], some athletes are able to perform under great pressure, while others drown under the combined weight of intense competition and high levels of physiological arousal and psychological stress. Anticipatory fear makes the individual more vulnerable [18], altering the functionality and adaptive capacities in terms of emotional [19,20], cognitive [21], and behavioural [22] responses linked to sport action, thus impacting on mood states and exacerbating activation processes [23,24]. Models such as the Multidimensional Theory of Competitive Anxiety identify the components of anxiety related to sport contexts [25], namely, cognitive anxiety (negative thoughts, expectations, and/or self-talk related to the competitive event), somatic anxiety (affective and physiological elements that directly affect the central nervous system), and self-confidence (related to the level of confidence and the perception of being prepared for the competition).

Understanding how the components of the impulsive and anxious response and mood states work also allows intervention on the psychological risks associated with sport situations and actions [26]. Early learning and/or training in psychological resources that promote emotional and behavioural regulation reduce discomfort in athletes (e.g., anger-anxiety, depression, fatigue, or withdrawal), mainly when they need their performance to adapt adequately and fluently to competitive situations or sporting demands [16,27], positively influencing sporting performance.

A growing number of studies have demonstrated the efficacy of psychological interventions, including mindfulness, on emotional and behavioural regulation in sports [28–31], hence the relevance of teaching regulatory skills, especially to young athletes participating in competitions [11,32,33], as they are subjected to greater sporting pressures (e.g., the obligation to win or to prove themselves to be the best).

In the 20 years since the incorporation of mindfulness in the context of sports, empirical findings have revealed effective outcomes associated with performance and personal well-being, as well as supporting the theorised mechanisms of change. Mindfulness is a form of meditation that involves intentional and non-judgmental awareness of the present moment, including physical sensations and experienced affective states [34,35]. The use of mindfulness facilitates directing one's attention towards coping with the pressure coming from technical, physical, biological, and even professional aspects. However, although the main objective of mindfulness is to learn to accept the internal events (i.e., thoughts, emotions, etc.) involved in any activity in the present, in studies with elite athletes, it has been observed that their coping with negative thoughts improved and their sports performance was also enhanced [36–38].

Currently, we have several mindfulness interventions or programmes applied to the field of sport, including the Mindfulness–Acceptance–Commitment Approach (MAC) [38], Mindful Sport Performance Enhancement (MSPE) [39], Mindful Performance Enhancement, Awareness, and Knowledge (mPEAK) [40], Mindfulness Meditation Training for Sport (MMTS) [41], Berlin Mindfulness-Based Training for Athletes (BAT) [42], the Mindfulness–Acceptance–Insight–Commitment program (MAIC) [43], among others [44]. In Spain, the Flow Meditation programme (Flow Meditation) [45] has been applied in those with different disorders and different populations, facilitating the reduction in impulsivity and aggressiveness in students and of anxiety and depression levels in older adults [46,47].

Expanding the repertoire of mindfulness strategies available to athletes and health professionals is a useful and relevant topic in this field. Furthermore, mindfulness interventions in the sport domain have mainly focused on variables such as coping, concentration levels, and relaxation [44]. Nowadays, trainers and sport professionals have no doubt that possessing good physical skills does not guarantee optimal sport performance, so they have started to take the psychological skills and personality characteristics of players seriously (i.e., anxiety and/or anger management, feelings of depression and fatigue, resilience, and even self-efficacy) [5,6].

The present study is the first to include a number of psychological variables that so far have barely been studied in the field of mindfulness applications to sport, but which are critical for improving well-being and sport performance. Thus, the aim of this study was to explore the efficacy of a mindfulness programme called Flow Meditation [45] on impulsivity, mood, and pre-competition anxiety in a group of athletes. On the basis of the scientific literature mentioned above, the starting hypothesis postulates that the intervention will improve impulsive response, mood (i.e., depression, tension, anger, fatigue, and vigour), and pre-competition anxiety in the intervention group compared to the control group.

2. Materials and Methods

2.1. Participants

The sample of the present study consisted of 41 federated athletes who participated in official competitions at the provincial, regional, or national level and who were studying different degrees at the University of Almeria (Spain). The ages of the participants ranged from 18 to 32 years (22.83 ± 5.62 years). In total, 6% of the participants were involved in athletics, 8% in tennis, 13% in swimming, 14% in basketball, 16% in handball, 21% in volleyball, and 22% in football. The intervention group consisted of 21 participants (14 males and 7 females), (21.74 ± 4.93 years), while the control group consisted of 20 athletes (13 males and 7 females) (23.92 ± 6.31 years). The sample was randomly assigned to one or the other group, controlling the variables of gender and sporting activity, so that there was a similar number of men and women and athletes from the different sporting disciplines in both groups, thus avoiding the interference of these variables in the results of the study. Informed consent was obtained from all subjects involved in the study.

2.2. Instruments

- Impulsivity. The Spanish adaptation of Oquendo et al. [48] of the Barratt Impulsivity Scale (BIS-11) [49,50] was applied. This questionnaire is composed of 30 items that are grouped into 3 subscales of impulsivity: (i) Cognitive impulsivity (8 items): tendency to make quick decisions; (ii) motor impulsivity (10 items): propensity to act solely on the stimulus of the moment, without thinking about the consequences; and (iii) impulsivity due to lack of planning (12 items): indicates the lack of planning of future actions. Each item consists of 4 Likert-type response options (never or almost never, sometimes, quite often, and always or almost always). The internal consistency of the different scales, obtained using Cronbach's alpha coefficient, showed values of 0.78 for the cognitive impulsivity subscale, 0.75 for motor impulsivity, and 0.66 for impulsivity due to non-planning.
- Mood. The original reduced version of the Profile of Mood States (POMS) [51], validated in Spanish by Fuentes et al. [52], was used. The abbreviated form of the Profile of Mood States consists of a list of 29 adjectives assessing five mood states: tension (6 items), depression (6 items), anger (6 items), vigour (6 items), and fatigue (5 items). The items are rated on a five-point Likert-type scale, from 0 = not at all; 1 = a little; 2 = moderately; 3 = quite a lot; to 4 = very much. The extent to which each of the moods described in each adjective has been experienced during the last week, including the day on which the questionnaire is answered, is assessed. Cronbach's alpha values for the reduced versions showed the following scores: tension (0.64), depression (0.69), anger (0.61), vigour (0.72), and fatigue (0.66).

- Competitive anxiety. The Spanish version of the Competitive Anxiety Inventory CSAI-2R by Andrade et al. [53] was administered in athletes. It consists of 18 items distributed in 3 dimensions: (i) somatic anxiety (8 items), which represents the perception of bodily indicators of anxiety, such as muscle tension, increased heart rate, sweating, and stomach discomfort; (ii) cognitive anxiety (5 items), which encompasses the negative feelings that the subject has about their performance and the consequences of the possible outcome; and (iii) self-confidence (5 items), which refers to the degree of confidence that the athlete believes they have about their chances of success. A Likert-type response format of four alternatives from 1 "not at all" to 4 "very much" was used. Internal reliability showed adequate consistency with Cronbach's alphas of 0.91 for the somatic anxiety dimension, 0.71 for the cognitive anxiety dimension, and 0.65 for the self-confidence dimension.

2.3. Design and Procedure

A quasi-experimental design was carried out with a wait-list control group, with intra- and inter-group comparisons.

Firstly, we proceeded to obtain the study sample by convenience sampling, for which we offered a course entitled *"The development of mindfulness in sport"*, aimed at athletes enrolled at the University of Almeria who were participating in official competitions at the provincial, regional, or national level. Once the study sample (n = 41) was constituted, we started to obtain the pre-test measurements for the dimensions of the variables of impulsivity, mood states, and anxiety-state, for which all the participants enrolled in the course were provided with questionnaires for the evaluation of these variables to be completed individually, on the weekend before the start of the intervention programme. Once this pre-test score was obtained, the participants were randomly assigned to the control and intervention groups, controlling the variables of gender and sporting discipline, so that they would not interfere with the results of the study. In the same week, the intervention programme began to be applied to the intervention group, with one session per week for 10 consecutive weeks. The subjects in the control group (wait-list; n = 20) were informed that for reasons of space they would receive the course in a second turn. During the weekend after the end of the course in the intervention group, the levels of the different dimensions of the impulsivity, mood and anxiety variables were re-evaluated in both groups under the same conditions as in the pre-test phase.

2.4. Intervention Programme

The programme was applied in the intervention group over ten sessions, with a weekly session lasting one and a half hours each. The sessions were held in the evening in a soundproofed university classroom with a good room temperature and the necessary equipment for the meditations (e.g., movable chairs, mats, blankets). This intervention programme consisted of learning and daily practice for 40 min of a meditation technique for the training and development of mindfulness called Flow Meditation [45,54]. Participants were given basic instructions on how to meditate at home: to always choose the same time and place (preferably a quiet place, with little light and good temperature); to allow at least two hours after eating before meditating and not to carry it out at a time when they had to do something urgent afterwards; to choose a comfortable posture, with the back upright; not to look for any special result, but to abandon themselves to whatever arises during meditation, flowing and without expecting big changes at the beginning, since meditating is a habit that takes time to be acquired and the effects take their time to develop and be noticed. The main objective is not to try to control thoughts, sensations, or feelings, nor to modify or change these for another but, on the contrary, to leave them free, accepting any private event that may appear or arise spontaneously in our consciousness. In other words, during practice, one becomes aware of the presence of thoughts without analysing their content or veracity, but rather developing the awareness that thoughts (as well as sensations) change every moment and that they are constantly flowing. Therefore, during the practice of this mindfulness technique, participants will understand experientially that thoughts arise and disappear continuously and that they are subject to a continuous flow, learning, in this way, to be present, open, and balanced before any phenomenon or mental or emotional process that occurs in their minds.

In each of the 10 sessions, in addition to the learning and practice of mindfulness, various metaphors and exercises from Acceptance and Commitment Therapy [55,56] as well as stories from the Zen tradition [57] and from Vipassana meditation [58] were employed. The aim of these stories and exercises was to reinforce the idea that when we try to control or eliminate the upsetting and unpleasant thoughts, emotions, and bodily sensations we experience, they become chronic, thus aggravating the psychological discomfort. So, the best option is therefore to become aware of them, accept them as they appear, and let them flow freely. Other components of the mindfulness Flow Meditation programme were body scan exercises [34]. After the research was completed, the mindfulness course was conducted with the control group, as indicated at the beginning of the study. The intervention programme was developed by an instructor with extensive experience in both practising and teaching mindfulness meditation techniques.

2.5. Data Analysis

Data were analysed with the statistical software SPSS software—IBM SPSS Statistics for Windows, Version 25.0. Armonk, NY, USA: IBM Corp.

First, Shapiro's normality test was conducted. Since the variables conformed to the normal distribution, parametric tests were then performed. Levenne's test was applied to corroborate the homogeneity of variances. Second, Student's t-tests for independent samples were performed to test whether there were significant differences between the intervention group and the control group before starting the intervention. Third, intra-group measures (t-tests for dependent samples) were performed between pre-test or time 1 (hereafter T1) and post-test or time 2 (hereafter T2). Fourth, post-treatment measures were compared between groups (t-tests for independent samples) to verify the extent to which these changes were statistically significant and on which variables. For variables where statistically significant differences were found, the effect size was calculated using Cohen's d. The interpretation is as follows: 0.2 is considered a small effect, 0.6 a moderate effect, 1.2 a high effect, and 2.0 a very high effect. Finally, a multiple linear regression analysis was performed with the variable self-confidence at T2 as the dependent variable, due to the divergence found between intra-group and inter-group results on this variable. In addition, descriptive analyses were performed in both groups for all variables at T1 and T2.

3. Results

3.1. Descriptive Data

The descriptive data for both groups for all variables at T1 and T2 are presented in Table 1. Student's t tests for independent samples found no significant differences between the main variables between the groups before the intervention ($p < 0.05$).

3.2. Intra-Group Comparison (T1–T2)

A comparison of the changes between T1 and T2 measures in both groups was made. As shown in Table 1, there were statistically significant changes in the intervention group in the following variables: the three subscales of the BIS-11, cognitive impulsivity, motor impulsivity and impulsivity due to non-planning; the POMS, in tension, depression, and anger; and the CSAI-2R, in somatic anxiety, cognitive anxiety, and self-confidence. In all of them, these statistically significant changes meant an improvement in the intervention group. In this group, the effect size was in all cases between high and very high. However, in the control group, statistically significant changes only occurred in the somatic anxiety and cognitive anxiety scales of the CSAI-2R, but these changes meant a worsening, i.e., an increase in anxiety levels. In this group, the effect size for those significant changes was moderate (see Table 1).

Table 1. Intra-group comparison of T1–T2 measures in the intervention and control groups with *t*-tests for dependent samples of T1–T2 measures in both groups and effect sizes.

Scales Subscales	Intervention Group					Control Group				
	Mean (SD) T1	Mean (SD) T2	*t*	*p*	Effect Size (Cohen's *d*)	Mean (SD) T1	Mean (SD) T2	*t*	*p*	Effect Size (Cohen's *d*)
BIS-11										
Cognitive Impulsivity	20.62 (3.88)	17.62 (2.13)	5.86	<0.001	1.81	21 (3.18)	21.35 (3.13)	−0.41	0.687	______
Motor Impulsivity	20.76 (3.66)	18 (2.12)	6.50	<0.001	2.0	21.1 (3.48)	21.75 (3.67)	−0.49	0.631	______
Unplanned Impulsivity	21.76 (3.63)	19.43 (2.6)	4.72	<0.001	1.46	23.45 (3.3)	23.95 (2.84)	−0.52	0.609	______
POMS										
Tension	15.86 (2.31)	12.86 (2.48)	9.49	<0.001	2.93	16.15 (2.03)	15.95 (1.99)	0.37	0.718	______
Depression	12.62 (2.66)	10.14 (1.82)	7.55	<0.001	2.33	12.15 (1.95)	12.85 (1.98)	−1.61	0.125	______
Anger	10.43 (1.63)	8.19 (1.17)	6.93	<0.001	2.14	11.1 (1.52)	11.7 (1.66)	−2.04	0.055	______
Vigour	18.95 (1.16)	19.1 (0.94)	−0.47	0642	______	18.6 (1.18)	18.85 (1.27)	−0.61	0.549	______
Fatigue	8.95 (1.24)	8.38 (1.07)	1.39	0180	______	8.45 (1.36)	8.7 (1.08))	−0.68	0.506	______
CSAI-2R										
Somatic Anxiety	21.1 (2.84)	17.48 (2.91)	9.83	<0.001	3.03	21.95 (3.93)	22.7 (3.42)	−2.52	0.021	0.80
Cognitive Anxiety	14.1 (1.92)	11.1 (1.61)	12.55	<0.001	3.87	14.00 (1.41)	14.55 (1.73)	−2.15	0.045	0.68
Self Confidence	14.1 (1.3)	14.81 (1.17)	−3.1	0.006	0.96	14.9 (1.83)	14.65 (1.88)	1.23	0.234	______

Degrees of freedom, intervention group = 20; control group = 19.

3.3. Inter-Group Comparisons

Differences between the two groups were also calculated on T2 measures, and statistically significant differences were found in favour of the intervention group on the following variables: on the three subscales of the BIS-11, cognitive impulsivity, motor impulsivity, and impulsivity due to non-planning; on the POMS, on tension, depression, and anger; and the CSAI-2R, on somatic anxiety and cognitive anxiety. In terms of effect size, all these results ranged from high (1.2) to very high (2.47) (see Table 2).

Table 2. Independent sample *t*-tests for T2 measures in the intervention and control groups and effect sizes.

Scale	Subscales	t	df	Effect Size (Cohen's *d*)	p
	Cognitive Impulsivity T2	−4.48	39	1.40	<0.001
	Motor Impulsivity T2	−4.03	39	1.2	<0.001
	Unplanned Impulsivity T2	−5.32	39	1.66	<0.001
	Tension T2	−4.4	39	1.37	<0.001
	Depression T2	−4.56	39	1.42	<0.001
	Anger T2	−7.8	33.98	2.47	<0.001
	Vigour T2	0.71	39	_____	0485
	Fatigue T2	−0.95	39	_____	0348
	Somatic Anxiety T2	−5.28	39	1.65	<0.001
	Cognitive Anxiety T2	−6.62	39	2.07	<0.001
	Self-confidence T2	0.33	31.57	_____	0.747

df (degrees of freedom).

3.4. Multiple Linear Regression Analysis

Given that in the intra-group analysis, statistically significant differences were found in the self-confidence variable at T2 in the intervention group, but no statistically significant differences were found in the inter-group analyses, a multiple linear regression analysis (Table 3) was performed with self-confidence as the dependent variable, in order to discern which variables were the best predictors of its values. The independent variables were group (intervention versus control group), gender, type of sport played, and the score on the same variable at T1. The results reveal that the two most predictive variables were, in first place, the score in self-confidence at T1, followed by the group to which they belonged.

Table 3. Linear regression of factor associated with self-confidence in T2.

Variables	β	SE	Standard β	t	p	95.0% Confidence Interval for β	
						Lower Limit	Upper Limit
Group	−0.801	0.307	−0.264	−2.612	0.013	−1.422	−0.179
Gender	−0.353	0.318	−0.111	−1.113	0.273	−0.998	0.291
Type of Sport	−0.002	0.079	−0.003	−0.028	0.978	−0.163	0.159
Self-confidence T1	0.805	0.102	0.847	7.865	0.000	0.597	1.012

4. Discussion

The results of the application of the Flow Meditation intervention programme in the sample of athletes studied reveal an improvement in the intervention group in cognitive impulsivity, motor impulsivity, and impulsivity due to lack of planning. There was also an improvement in tension, depression, and anger, as well as in somatic anxiety, cognitive anxiety, and self-confidence, confirming the hypotheses formulated. In contrast, the control group showed an increase in somatic and cognitive anxiety.

Along the same lines, other studies have observed improvements in burnout [59], decreased stress and anxiety [3,37], improved mood [17], as well as higher levels of flow and improved sports performance [39,60–62] following the application of a mindfulness programme in athletes. Recent research suggests a reduction in depressive symptoms and an increase in psychological well-being [59,63] and self-compassion [3].

Mindfulness intervention with Flow Meditation can be indirectly considered a type of psychological training to optimise performance. Moreover, it is an effective strategy to improve concentration and emotion regulation in athletes. According to the systematic review on mindfulness intervention programmes by Bühlmayer et al. [64], mindfulness practice affects cognitive processes and is considered increasingly significant within sport psychology training. The effects of increased mindfulness promote greater concentration, awareness, and acceptance, which are necessary to counteract the negative effects of stressful sport performance [65,66].

In addition to performance anxiety and negative thoughts, the negative effects of competition include fatigue, boredom, and pain [65]. Mindfulness interventions do not focus on directly altering or changing dysfunctional thoughts and emotions, but rather the athlete's relationship with their physiological and psychological states. However, collateral changes occur, such as improved cognitive skills [67] and even increased pain tolerance in injured athletes. Thus, Mohammed et al. [68] indicate a notable decrease in stress and an improvement in mental health after the application of the MBSR programme (Mindfulness-Based Stress Reduction). Similarly, other authors report effects on reducing the risk of injury and even facilitating recovery from serious injuries [59,69]. Other studies suggest that even brief mindfulness training can prevent deterioration in athletic performance and produce psychological benefits [3,60].

The results obtained indicate an increase in the variable of self-confidence in the intervention group. Recent research supports the findings of this study, suggesting that mindfulness is closely related to this variable, which is essential for promoting sports performance [70–73], thus preventing the onset of mental disorders and improving quality of life [38]. In short, it is possible to conclude that mindfulness training enhances cognitive and emotional resources for managing highly demanding situations and, at the same time, will contribute to correcting maladaptive behaviours of athletes in competition.

The mechanisms involved, which can act both directly and indirectly, include: neuro-biological changes, the increased emotional regulation and reduction in negative emotions, such as anxiety and anger [74,75], increased attentional control [67,76,77], flow [78,79], reduced rumination and irrational beliefs (e.g., perfectionism), reduced worry and self-judgment [3,44], decreased experiential avoidance [80], increased positive emotions and self-efficacy [6,81], and even (self-)compassion [3,24].

Flow Meditation constitutes a Second-Generation Mindfulness programme, and it includes ethical, existential, and spiritual aspects. Thus, the aim is not only to improve sports performance and individual well-being, but also to promote mindfulness skills for everyday life and values such as equanimity and transcendence [35,45]. This objective comes along with new lines of future research proposed by some authors [44,82]. In this way, Kee et al. [44] suggest that mindfulness training in athletes should incorporate the Taoist concept of *wu-wei*, whereby athletes learn to set aside and transcend ideas of winning or losing while playing and concentrate solely on the task, without judging their actions and paying attention only to those external stimuli relevant to the task. In Western terms, the more related word might be flow (i.e., to let go of effort, to flow, or to go with the flow). The flow state has been defined as an optimal state of awareness in which one is totally absorbed and connected with what one is doing, even to the exclusion of all thought or emotion, and in which mind and body work together harmoniously, pleasantly, almost automatically, and without conscious effort. Experiencing this state is associated with a positive impact on self-perceived sport competence and sports performance [79,83]. In this sense, optimal performance does not require the volitional control of internal states [38].

Athletes trained from a young age, for whom mindfulness is already part of their daily lifestyle, develop what has been called trait mindfulness, rather than state mindfulness, which is limited to specific situations or novice meditators [29]. Both describe mindfulness about the task in the present moment and, when they occur together, constitute consolidated processes of acceptance and engagement with the task, without value judgments, promoting achievement and the processes to accomplish it [84]. Moreover, the application of mindfulness-based therapies, in which present-moment awareness is enhanced, seems to have a more positive influence on psychological variables related to competition than other therapies involving an altered state of consciousness, such as hypnosis [85].

Finally, the application of this type of treatment in athletes could also function as a protective factor against other problems linked to competition, such as disorders related to body image or eating behaviour [86].

The study described here has several limitations that need to be noted: (i) the small number of participants belonging to each sport activity; (ii) the absence of an active control group; (iii) the exclusive use of self-report measures; (iv) the absence of a follow-up phase of the treatment effects.

Future studies should continue to evaluate the efficacy of mindfulness in controlled trials and in larger populations in order to generalise the findings and confirm the effects of the intervention on different psychological variables that affect the mental health and performance of athletes. Finally, it is of utmost relevance to assess the long-term effects on the population studied.

Author Contributions: Conceptualisation, A.A., C.F., L.C.S.-S., and J.G.-H.; methodology, C.F.; formal analysis, L.C.S.-S.; investigation, C.F., data curation, C.F. and L.C.S.-S.; writing—original draft preparation, A.A., C.F., L.C.S.-S., J.G.-H., and J.G.-S.; writing—review and editing, A.A., C.F., and L.C.S.-S.; visualisation, A.A. and L.C.S.-S.; supervision, A.A. and L.C.S.-S.; project administration, A.A., C.F. and L.C.S.-S. All authors have read and agreed to the published version of the manuscript.

Funding: This research received no external funding.

Institutional Review Board Statement: The study was approved by the Committee of Bioethics of the University of Almería, Spain (Registration number: UAL-BIO2019/001). The registered data for each of the instruments was alphanumerically coded, ensuring confidentiality and anonymity, in order to comply with the Personal Data Protection Act by the Ethics Committee for Research related to Human Beings (CEISH). International ethical guidelines for studies with human subjects described in the Nuremberg Code and in the Declaration of Helsinki were applied.

Informed Consent Statement: Informed consent was obtained from all subjects involved in the study.

Data Availability Statement: Data are available on request from the authors.

Conflicts of Interest: The authors declare no conflict of interest.

References

1. Liu, C. Research on the influence of college students' participation in sports activities on their sense of inferiority based on self-esteem and general self-efficacy. *Front. Psychol.* **2022**, *13*, 994209. [CrossRef] [PubMed]
2. Tóth, R.; Turner, M.J.; Kökény, T.; Tóth, L. "I must be perfect": The role of irrational beliefs and perfectionism on the competitive anxiety of Hungarian athletes. *Front. Psychol.* **2022**, *13*, 994126. [CrossRef] [PubMed]
3. Hut, M.; Glass, C.R.; Degnan, K.A.; Minkler, T.O. The effects of mindfulness training on mindfulness, anxiety, emotion dysregulation, and performance satisfaction among female student-athletes: The moderating role of age. *Asian J. Sport Exerc. Psychol.* **2021**, *1*, 75–82. [CrossRef]
4. Tamminen, K.; Wolf, S.A.; Dunn, R.; Bissett, J.E. A review of the interpersonal experience, expression, and regulation of emotions in sport. *Int. Rev. Sport Exerc. Psychol.* **2022**, 1–38. [CrossRef]
5. Gu, S.; Li, Y.; Jiang, Y.; Huang, J.H.; Wang, F. Mindfulness Training Improves Sport Performance via Inhibiting Uncertainty Induced Emotional Arousal and Anger. *J. Orthop. Sports Med.* **2022**, *4*, 296–304. [CrossRef]
6. Wang, Y.; Wu, C.H.; Chen, L.H. A longitudinal investigation of the role of perceived autonomy support from coaches in reducing athletes' experiential avoidance: The mediating role of subjective vitality. *Psychol. Sport Exerc.* **2023**, *64*, 102304. [CrossRef]
7. Bakhshani, N.M. Impulsivity: A predisposition toward risky behaviors. *Int. J. High Risk Behav. Addict.* **2014**, *3*, 204–228. [CrossRef]
8. Bari, A.; Robbins, T.W. Inhibition and impulsivity: Behavioral and neural basis of response control. *Prog. Neurobiol.* **2013**, *108*, 44–79. [CrossRef]
9. Reynolds, B.; Ortengren, A.; Richards, J.B.; De Wit, H. Dimensions of impulsive behavior: Personality and behavioral measures. *Personal. Individ. Differ.* **2006**, *40*, 305–315. [CrossRef]
10. Whiteside, S.P.; Lynam, D.R.; Miller, J.D.; Reynolds, S.K. Validation of the UPPS impulsive behaviour scale: A four-factor model of impulsivity. *Eur. J. Personal.* **2005**, *19*, 559–574. [CrossRef]

11. González-Hernández, J.; Capilla Díaz, C.; Gómez-López, M. Impulsiveness and cognitive patterns. Understanding the perfectionistic responses in Spanish competitive junior athletes. *Front. Psychol.* **2019**, *10*, 1605. [CrossRef] [PubMed]

12. Lage, G.M.; Gallo, L.G.; Cassiano, G.J.; Lobo, I.L.; Vieira, M.V.; Salgado, J.V. Correlations between impulsivity and technical performance in handball female athletes. *Psychology* **2011**, *2*, 721–726. [CrossRef]

13. Samuel, R.D.; Englert, C.; Zhang, Q.; Basevitch, I. Hi ref, are you in control? Self-control, ego-depletion, and performance in soccer referees. *Psychol. Sport Exerc.* **2018**, *38*, 167–175. [CrossRef]

14. Byrd, M.M.; Kontos, A.P.; Eagle, S.R.; Zizzi, S. Preliminary Evidence for a Relationship Between Anxiety, Anger, and Impulsivity in Collegiate Athletes with Sport-Related Concussion. *J. Clin. Sport Psychol.* **2021**, *1*, 1–20. [CrossRef]

15. Gabrys, K.; Wontorczyk, A. A Psychological Profile of Elite Polish Short Track Athletes: An Analysis of Temperamental Traits and Impulsiveness. *Int. J. Environ. Res. Public Health* **2022**, *19*, 3446. [CrossRef]

16. González-Hernández, J.; Gomariz-Gea, M.; Valero-Valenzuela, A.; Gómez-López, M. Resilient resources in youth athletes and their relationship with anxiety in different team sports. *Int. J. Environ. Res. Public Health* **2020**, *17*, 5569. [CrossRef]

17. Hoover, S.J.; Winner, R.K.; McCutchan, H.; Beaudoin, C.C.; Judge, L.W.; Jones, L.M.; Hoover, D.L. Mood and performance anxiety in high school basketball players: A pilot study. *Int. J. Exerc. Sci.* **2017**, *10*, 604.

18. Sahranavard, S.; Esmaeili, A.; Dastjerdi, R.; Salehiniya, H. The effectiveness of stress-management-based cognitive-behavioral treatments on anxiety sensitivity, positive and negative affect and hope. *BioMedicine* **2018**, *8*, 23. [CrossRef]

19. Doug, H.H.; Kim, J.H.; Lee, Y.S.; Bae, S.J.; Bae, S.J.; Kim, H.J.; Lyoo, I.K. Influence of temperament and anxiety on athletic performance. *J. Sports Sci. Med.* **2006**, *5*, 381.

20. Pons, J.; Ramis, Y.; García-Mas, A.; López de la Llave, A.; Pérez-Llantada, M. Percepción de la ansiedad competitiva en relación al Nivel de Cooperación y Compromiso Deportivo en Jugadores de Baloncesto de Formación. *Cuad. Psicol. Deporte* **2016**, *16*, 45–54.

21. García-Mas, A.; Martins, B.; Núñez, A.; Ponseti, F.J.; Trigueros, R.; Alias, A.; Aguilar-Parra, J.M. Can we speak of a negative psychological tetrad in sports? A probabilistic Bayesian study on competitive sailing. *PLoS ONE* **2022**, *17*, e0272550. [CrossRef] [PubMed]

22. Sutcliffe, J.H.; Greenberger, P.A. Identifying psychological difficulties in college athletes. *J. Allergy Clin. Immunol.* **2020**, *8*, 2216–2219. [CrossRef] [PubMed]

23. Bertollo, M.; Bortoli, L.; Gramaccioni, G.; Hanin, Y.; Comani, S.; Robazza, C. Behavioural and psychophysiological correlates of athletic performance: A test of the multi-action plan model. *Appl. Psychophysiol. Biofeedback* **2013**, *38*, 91–99. [CrossRef] [PubMed]

24. Ceccarelli, L.A.; Giuliano, R.J.; Glazebrook, C.M.; Strachan, S.M. Self-compassion and psycho-physiological recovery from recalled sport failure. *Front. Psychol.* **2019**, *10*, 1564. [CrossRef]

25. Martens, R.; Burton, D.; Vealey, R.S.; Bump, L.A.; Smith, D.E. Development and validation of the Competitive State Anxiety Inventory-2. In *Competitive Anxiety in Sport*; Martens, R., Vealey, R.S., Burton, D., Eds.; Human Kinetics: Champaign, IL, USA, 1990; pp. 117–190.

26. Swann, C.; Crust, L.; Jackman, P.; Vella, S.A.; Allen, M.S.; Keegan, R. Performing under pressure: Exploring the psychological state underlying clutch performance in sport. *J. Sports Sci.* **2017**, *35*, 2272–2280. [CrossRef]

27. Caprara, G.V.; Barbaranelli, C.; Pastorelli, C.; Cervone, D. The contribution of self-efficacy beliefs to psychosocial outcomes in adolescence: Predicting beyond global dispositional tendencies. *Personal. Individ. Differ.* **2004**, *37*, 751–763. [CrossRef]

28. Josefsson, T.; Ivarsson, A.; Gustafsson, H.; Stenling, A.; Lindwall, M.; Tornberg, R.; Böröy, J. Effects of mindfulness-acceptance-commitment (MAC) on sport-specific dispositional mindfulness, emotion regulation, and self-rated athletic performance in a multiple-sport population: An RCT study. *Mindfulness* **2019**, *10*, 1518–1529. [CrossRef]

29. Noetel, M.; Ciarocchi, J.; Van Zanden, B.; Lonsdale, C. Mindfulness and acceptance approaches to sporting performance enhancement: A systematic review. *Int. Rev. Sport Exerc. Psychol.* **2017**, *12*, 139–175. [CrossRef]

30. Terres-Barcala, L.; Albaladejo-Blázquez, N.; Aparicio-Ugarriza, R.; Ruiz-Robledillo, N.; Zaragoza-Martí, A.; Ferrer-Cascales, R. Effects of Impulsivity on Competitive Anxiety in Female Athletes: The Mediating Role of Mindfulness Trait. *Int. J. Environ Res. Public Health* **2022**, *19*, 3223. [CrossRef]

31. Zhang, C.-Q.; Si, G.; Duan, Y.; Lyu, Y.; Keatley, D.A.; Chan, D.K.C. The effects of mindfulness training on beginners' skill acquisition in dart throwing: A randomized controlled trial. *Psychol. Sport Exerc.* **2016**, *22*, 279–285. [CrossRef]

32. Alqallaf, A.S. *Examining the Interactive Effects of Mental Toughness, Self-Regulated Training Behaviors, and Personality in Swimming*; Bangor University: Bangor, UK, 2016.

33. Fawver, B.; Beatty, G.F.; Mann, D.T.; Janelle, C.M. Staying cool under pressure: Developing and maintaining emotional expertise in sport. In *Skill Acquisition in Sport*; Routledge: Oxfordshire, UK, 2019; pp. 271–290.

34. Kabat-Zinn, J. *Vivir con Plenitud las Crisis. Cómo Utilizar la Sabiduría del Cuerpo y la Mente Para Afrontar el Estrés, el Dolor y la Enfermedad*; Kairós: Barcelona, Spain, 2003.

35. Van Gordon, W.; Shonin, E. Second-generation mindfulness-based interventions: Toward more authentic mindfulness practice and teaching. *Mindfulness* **2020**, *11*, 1–4. [CrossRef]

36. Carraça, B.; Serpa, S.; Rosado, A.; Palmi, J. The mindfulness-based soccer program (MBSoccerP): Effects on elite athletes. *Cuad. Psicol. Deporte* **2018**, *18*, 62–85.

37. Vidic, Z.; Martin, M.S.; Oxhandler, R. Mindfulness Intervention with a US Women's NCAA Division I Basketball Team: Impact on Stress, Athletic Coping Skills and Perceptions of Intervention. *Sport Psychol.* **2017**, *31*, 147–159. [CrossRef]

38. Gardner, F.L.; Moore, Z.E. Mindfulness-based and acceptance-based interventions in sport and performance contexts. *Curr. Opin. Psychol.* **2017**, *16*, 180–184. [CrossRef]
39. Kaufman, K.A.; Glass, C.R.; Arnkoff, D.B. Evaluation of Mindful Sport Performance Enhancement (MSPE): A new approach to promote flow in athletes. *J. Clin. Sports Psychol.* **2009**, *4*, 334–356. [CrossRef]
40. Haase, L.; Kenttä, G.; Hickman, S.; Baltzell, A.; Paulus, M. Mindfulness training in elite athletes: MPEAK with BMX cyclists. In *Mindfulness and Performance*; Baltzell, A., Ed.; University Press Cambridge: Cambridge, UK, 2016; pp. 186–208. [CrossRef]
41. Baltzell, A.; Summers, J. *The Power of Mindfulness: Mindfulness Meditation Training in Sport (MMTS)*; Springer: Berlin/Heidelberg, Germany, 2018.
42. Jekauc, D.; Kittler, C.; Schlagheck, M. Effectiveness of a mindfulness- based intervention for athletes. *Psychology* **2016**, *8*, 1–13. [CrossRef]
43. Su, N.; Si, G.; Zhang, C.Q. Mindfulness and acceptance-based training for Chinese athletes: The mindfulness-acceptance-insight-commitment (MAIC) program. *J. Sport Psychol. Action* **2019**, *10*, 255–263. [CrossRef]
44. Kee, Y.H.; Li, C.; Zhang, C.Q.; Wang, J.C.K. The wu-wei alternative: Effortless action and non-striving in the context of mindfulness practice and performance in sport. *Asian J. Sport Exerc. Psychol.* **2021**, *1*, 122–132. [CrossRef]
45. Franco, C. *Meditación Fluir Para Serenar el Cuerpo y la Mente*; Bubok: Madrid, Spain, 2009.
46. Franco, C.; Amutio, A.; López-González, L.; Oriol, X.; Martínez-Taboada, C. Effect of a mindfulness training program on the impulsivity and aggression levels of adolescents with behavioural problems in the classroom. *Front. Psychol.* **2016**, *7*, 1385. [CrossRef]
47. Franco, C.; Amutio, A.; Mañas, I.; Pérez, M.C.; Gázquez, J.J. Reducing anxiety, geriatric depression and worry in a sample of older adults through a mindfulness training program. *Ter. Psicol.* **2017**, *35*, 71–79. [CrossRef]
48. Oquendo, M.A.; Baca-García, E.; Graver, R.; Morales, M.; Montalvan, V.; Mann, J.J. Spanish adaptation of the Barratt impulsiveness scale (BIS-11). *Eur. J. Psychiatry.* **2001**, *15*, 147–155.
49. Barratt, E.; Patton, J.H. Impulsivity: Cognitive, behavioral, and psychophysiological Correlates. In *The Biological Basis of Impulsivity and Sensation Seekin*; Zuckerman, M., Ed.; Lawrence Erlbaum Associates: Mahwah, NJ, USA, 1983.
50. Patton, J.H.; Stanford, M.S.; Barratt, E.S. Factor structure of the Barratt Impusiveness Scale. *J. Clin. Psychol.* **1995**, *51*, 768–774. [CrossRef] [PubMed]
51. McNair, D.M.; Lorr, M.; Droppelman, L.F. *Manual for the Profile of Mood States*; Educational and Industrial Testing Service: San Diego, CA, USA, 1992.
52. Fuentes, I.; Balaguer, I.; Meliá, J.L.; García-Merita, M. Forma Abreviada del Perfil de Estados de Ánimo (POMS). In Proceedings of the Comunicación Presentada en el V Congreso de Psicología de la Actividad Física y del Deporte, Valencia, Spain, 22–24 March 1995.
53. Andrade, E.M.; Lois, G.; Arce, C. Propiedades psicométricas de la versión española del Inventario de Ansiedad Competitiva CSAI-2R en deportistas. *Psicothema* **2007**, *19*, 150–155.
54. Franco, C.; Amutio, A.; Mañas, I.; Sánchez-Sánchez, L.C.; Mateos-Pérez, E. Improving psychosocial functioning in mastectomized women through a mindfulness-based program: Flow meditation. *Int. J. Stress Manag.* **2020**, *27*, 74–81. [CrossRef]
55. Hayes, S.C.; Stroshal, K.D.; Wilson, K.G. *Acceptance and Commitment Therapy*; The Guilford Press: New York, NY, USA, 1999.
56. Wilson, K.G.; Luciano, M.C. *Terapia de Aceptación y Compromiso. Un Tratamiento Conductual Orientado a los Valores*; Pirámide: Madrid, Spain, 2022.
57. Deshimaru, T. *La Práctica del Zen*; RBA: Barcelona, Spain, 2006.
58. Hart, W. La Vippasana. In *El Arte de la Meditación*; Luz de Oriente: Madrid, Spain, 1994.
59. Zhang, C.Q.; Li, X.; Chung, P.K.; Huang, Z.; Bu, D.; Wang, D.; Si, G. The effects of mindfulness on athlete burnout, subjective well-being, and flourishing among elite athletes: A test of multiple mediators. *Mindfulness* **2021**, *12*, 1899–1908. [CrossRef]
60. Perry, J.E.; Ross, M.; Weinstock, J.; Weaver, T. Efficacy of a Brief Mindfulness Intervention to Prevent Athletic Task Performance Deterioration: A Randomized Controlled Trial. *Sport Psychol.* **2017**, *31*, 410–421. [CrossRef]
61. Thompson, R.W.; Kauffman, K.A.; De Petrillo, L.; Glass, C.R.; Arnkoff, D.B. One year follow-up of mindful sport performance enhancement (MSPE) with archers, golfers and runners. *J. Clin. Sport Psychol.* **2011**, *5*, 99–116. [CrossRef]
62. Wong, R.S.K.; How, P.N.; Cheong, J.P.G. The Effectiveness of a Mindfulness Training Program on Selected Psychological Indices and Sports Performance of Sub-Elite Squash Athletes. *Front. Psychol.* **2022**, *13*, 906729. [CrossRef]
63. Norouzi, E.; Gerber, M.; Masrour, F.F.; Vaezmosavi, M.; Puhse, U.; Brand, S. Implementation of a mindfulness-based stress reduction (MBSR) program to reduce stress, anxiety, and depression and to improve psychological well-being among retired Iranian football players. *Psychol. Sport Exerc.* **2020**, *47*, 101636. [CrossRef]
64. Bühlmayer, L.; Birrer, D.; Röthlin, P.; Faude, O.; Donath, L. Effects of mindfulness practice on performance-relevant parameters and performance outcomes in sports: A meta-analytical review. *J. Sports Med.* **2017**, *47*, 2309–2321. [CrossRef]
65. Corbally, L.; Wilkinson, M.; Fothergill, M.A. Effects of mindfulness practice on performance and factors related to performance in long-distance running: A systematic review. *J. Clin. Sport Psychol.* **2020**, *14*, 376–398. [CrossRef]
66. Röthlin, P.; Birrer, D.; Horvath, S. Psychological skills training and a mindfulness-based intervention to enhance functional athletic performance: Design of a randomized controlled trial using ambulatory assessment. *BMC Psychol.* **2016**, *4*, 39. [CrossRef] [PubMed]

67. Zadkhosh, S.M.; Zandi, H.G.; Ghorbannejad, M. The effects of Mindfulness-Based Cognitive Therapy (MBCT) on cognitive skills in young soccer players. *Exerc. Qual. Life* **2019**, *11*, 5–11. [CrossRef]
68. Mohammed, W.A.; Pappous, A.; Sharma, D. Effect of mindfulness based stress reduction (MBSR) in increasing pain tolerance and improving the mental health of injured athletes. *Front. Psychol.* **2018**, *9*, 722. [CrossRef]
69. Anderson, S.A.; Haraldsdottir, K.; Watson, D. Mindfulness in Athletes. *Curr. Sports Med. Rep.* **2021**, *20*, 655–660. [CrossRef]
70. Amemiya, R.; Sakairi, Y. Relationship between mindfulness and cognitive anxiety-impaired performance: Based on performance evaluation discrepancies. *Asian J. Sport Exerc. Psychol.* **2021**, *1*, 67–74. [CrossRef]
71. Amir, D.; Shahir, V.A.; Ghorbani, S. The impact of Mindfulness and Mental Skills Protocols on Athletes' Competitive Anxiety. *Biomed. Hum. Kinet.* **2022**, *14*, 135–142. [CrossRef]
72. Mehrsafar, A.H.; Strahler, J.; Gazerani, P.; Khabiri, M.; Sánchez, J.C.J.; Moosakhani, A.; Zadeh, A.M. The effects of mindfulness training on competition-induced anxiety and salivary stress markers in elite Wushu athletes: A pilot study. *Physiol. Behav.* **2019**, *210*, 112655. [CrossRef]
73. Oguntuase, S.B.; Sun, Y. Effects of mindfulness training on resilience, self-confidence and emotion regulation of elite football players: The mediating role of locus of control. *Asian J. Sport Exerc. Psychol.* **2022**, *3*, 198–205. [CrossRef]
74. Mistretta, E.G.; Glass, C.R.; Spears, C.A.; Perskaudas, R.; Kaufman, K.A.; Hoyer, D. Collegiate Athletes' Expectations and Experiences with Mindful Sport Performance Enhancement. *J. Clin. Sport Psychol.* **2017**, *11*, 201–221. [CrossRef]
75. Stephens, A.N.; O'Hern, S.; Young, K.L.; Chambers, R.; Hassed, C.; Koppel, S. Self-reported mindfulness, cyclist anger and aggression. *Accid. Anal. Prev.* **2020**, *144*, 105625. [CrossRef]
76. Glass, C.R.; Spears, C.A.; Perskaudas, R.; Kaufman, K.A. Mindful Sport Performance Enhancement: Randomized Controlled Trial of a Mental Training Program with Collegiate Athletes. *J. Clin. Sport Psychol.* **2019**, *13*, 609–628. [CrossRef]
77. Kittler, C.; Arnold, M.; Jekauc, D. Effects of a Mindfulness-Based Intervention on Sustained and Selective Attention in Young Top-Level Athletes in a School Training Setting: A Randomized Control Trial Study. *J. Sport Psychol.* **2022**, *36*, 77–88. [CrossRef]
78. Amutio, A.; Telletxea, S.; Mateos-Pérez, E.; Padoan, S.; Basabe, N. Social climate in university classrooms: A mindfulness-based educational intervention. *PsyCh J.* **2022**, *11*, 114–122. [CrossRef] [PubMed]
79. Fernández, M.A.; Godoy-Izquierdo, D.; Jaenes, J.C.; Bohórquez, M.R.; Vélez, M. Flow and performance in marathon runners. *Rev. Psicol. Deporte* **2015**, *24*, 9–19.
80. Amutio, A.; Franco, C.; Gázquez, J.J.; Mañas, I. Aprendizaje y práctica de la conciencia plena en estudiantes de bachillerato para potenciar la relajación y la autoeficacia en el rendimiento escolar. *Univ. Psychol.* **2015**, *14*, 433–443. [CrossRef]
81. Roychowdhury, D.; Ronkainen, N.; Guinto, M.L. The transnational migration of mindfulness: A call for reflective pause in sport and exercise psychology. *Psychol. Sport Exerc.* **2021**, *56*, 101958. [CrossRef]
82. Zhang, C.Q.; Li, X.; Gangyan, S.; Chung, P.K.; Huang, Z.; Gucciardi, D.F. Examining the roles of experiential avoidance and cognitive fusion on the effects from mindfulness to athlete burnout: A longitudinal study. *Psychol. Sport Exerc.* **2023**, *64*, 102341. [CrossRef]
83. Jackson, S.; Csikszentmihalyi, M. *Flow in Sports: The Keys to Optimal Experiences and Performances*; Human Kinetics: Champaign, IL, USA, 1999.
84. Ullrich-French, S.; Cox, A.E. *Mindfulness in Exercise Psychology. Sport, Exercise and Performance Psychology. Research Directions to Advance the Field*; Oxford University Press: Oxford, UK, 2021.
85. Fernández-García, R.; Sánchez-Sánchez, L.C.; Zurita Ortega, F. Eficacia de la hipnosis en la modificación de variables psicológicas y fisiológicas en deportistas. *Univ. Psychol.* **2013**, *12*, 483–491. [CrossRef]
86. Petisco-Rodríguez, C.; Sánchez-Sánchez, L.C.; Fernández-García, R.; Sánchez-Sánchez, J.; García-Montes, J.M. Disordered Eating Attitudes, Anxiety, Self-Esteem and Perfectionism in Young Athletes and Non-Athletes. *Int. J. Environ. Res. Public Health* **2020**, *17*, 6754. [CrossRef]

Article

The Effect of Ramadan Fasting on the Coping Strategies Used by Male Footballers Affiliated with the Tunisian First Professional League

Jamel Hajji [1,2], Aiche Sabah [3], Musheer A. Aljaberi [4,*], Chung-Ying Lin [5] and Lin-Yi Huang [6,*]

1   Higher Institute of Sport and Physical Education of Gafsa, Gafsa University, Gafsa 2100, Tunisia
2   Faculty of Human and Social Sciences, Tunis University, Gafsa 2100, Tunisia
3   Faculty of Human and Social Sciences, Hassiba Benbouali University of Chlef, Chlef 02076, Algeria
4   Faculty of Medicine and Health Sciences, Taiz University, Taiz 6803, Yemen
5   Institute of Allied Health Sciences, College of Medicine, National Cheng Kung University, Tainan 701, Taiwan
6   Department of Pediatrics, E-Da Cancer Hospital, I-Shou University, Kaohsiung 824, Taiwan
*   Correspondence: musheer.jaberi@gmail.com (M.A.A.); pednb2773@gmail.com (L.-Y.H.)

Abstract: This study aimed to discover coping strategies among professional male Tunisian footballers during the Ramadan 2021 fast. One hundred and eighty footballers who belong to twelve Tunisian professional clubs (age: 25.54 ± 4.41 years, weight: 77.19 ± 5.99 kg; height: 180.54 ± 7.28 cm; BMI: 23.67 ± 0.58) were tested during three sessions: one week before Ramadan, during the last week of Ramadan, and one week after Ramadan 2021. The footballers completed the Arabic version of the Inventory of Coping Strategies for Competitive Sport (ICSCS) scale in each session. Responses were recorded retrospectively one hour after a competition. The analysis of variance revealed a significant effect of Ramadan fasting on the adaptation profile of footballers (F = 3.51; *p*-value = 0.0001). Before and after Ramadan fasting, active coping dominates the adaptation profile of Tunisian professional footballers. During Ramadan, footballers use an irregular and unbalanced coping profile. The lifestyle change induced by the Ramadan fast significantly and negatively affected the adaptation profile of Tunisian professional footballers. Under the effect of the month of Ramadan, footballers developed a different coping profile from that of normal months.

Keywords: fast day; adaptation; lifestyle; athletes; men; mental health

Citation: Hajji, J.; Sabah, A.; Aljaberi, M.A.; Lin, C.-Y.; Huang, L.-Y. The Effect of Ramadan Fasting on the Coping Strategies Used by Male Footballers Affiliated with the Tunisian First Professional League. *Healthcare* **2023**, *11*, 1053. https://doi.org/10.3390/healthcare11071053

Academic Editors: João Paulo Brito and Rafael Oliveira

Received: 18 February 2023
Revised: 30 March 2023
Accepted: 3 April 2023
Published: 6 April 2023

1. Introduction

Ramadan fasting is one of the religious rituals of Muslims. It is a form of intermittent fasting, which is practiced by more than 1.5 billion Muslims annually throughout the month of Ramadan. Ramadan is the ninth month of the Islamic calendar. In this month, worldwide Muslims must fast for 29–30 days. The daily fasting duration in Ramadan may vary between seasons, from 12 to 18 h, depending on the season and geographical area [1,2]. During fasting, individuals do not eat anything from sunrise to sunset. From sunset to brightness, Muslims can eat freely. Hence, the time of sleeping and eating may be affected by Ramadan, as the frequency and quantity of food, the duration of night sleep, and sports activities are reduced [3,4].

Muslim athletes often encounter significant metabolic, behavioral, and dietary disruption when they engage in Ramadan fasting [5–8]. Athletes who fast during Ramadan can potentially confer hypohydration [9], disturbances in body mass [10–13], variations in aerobic qualities [14,15], intense anaerobic qualities [16,17], metabolic and hormonal disturbances [18,19], disorders in physical activity profile [20], sleep disorders including sleep deficit [21,22], mood swings, and general impairment of physical and psychomotor performance [5,23]. Accordingly, many studies have shown a negative effect of fasting in Ramadan on several aspects of physical performance [24,25]. For example, athletes experience varying stress levels during Ramadan due to the disruption and alteration of their biological clock [26–28].

Aside from the physical effects, mood disturbances with mental fatigue have also increased during Ramadan [16,24,25]. Several studies have shown the harmful effects of fasting in Ramadan on the mental aspects of performance [29–33]. For example, one study identified the adverse effects of Ramadan fasting on sleep and the performance of footballers participating in the 2012 Olympics [30]. Another study showed a negative impact of Ramadan fasting on cognitive performance [34]. The results of these studies suggest that decision-making behaviors during training and competition conditions may be negatively affected during Ramadan. Therefore, understanding how to minimize the adverse effects of Ramadan fasting on Muslim athletes is a crucial issue.

Some studies have designed mental preparation sessions for Muslim athletes before the beginning of the Ramadan fast to assimilate proactive coping strategies [24,25,29]. Experienced Muslim athletes, especially those with many years of sports experience, usually have developed strategies to cope with the Ramadan fast [30,35]. For example, they acquired learned behavioral patterns that they found effective in meeting the fasting challenge. Some intermittent practice fasting throughout the year as strategies to prepare for the month of Ramadan, which is characterized by continuous fasting, changes in eating behavior, physical activity, sleep, and emotional and cognitive behavior, and lower energy intake resulting from the restriction of feeding to a limited time [2,15,36]. The athlete's nutritional status, degree of training, temperature, humidity, and daily fluid intake may reflect differences such as intermittent or continuous fasting [36–38]. Adaptation essentially involves the individual response of an athlete to cope with a stressful situation [39]. On the other hand, preventive coping aims to develop resources to reduce the effects of uncertain and stressful future events [25,40,41]. Proactive adaptation involves strategies for developing public resources to achieve personal goals.

Ideally, a male Muslim athlete immerses himself in competition to achieve the best performance. Although fasting during Ramadan represents an obstacle, he does everything to develop coping strategies to achieve his goals [24,25,29,30]. Some Muslim athletes adopt good coping strategies and have been successful in combating the negative influence of Ramadan fasting on their subsequent physical performance. In contrast, those who are less capable of coping with the disturbances induced by Ramadan fasting may not have peak performance [5,6,23].

In general, commonly used coping strategies include task-focused coping, in which the athlete directly confronts the threatening situation, and avoidance-focused coping, in which the athlete attempts to disengage or distract themselves from the stressful situation [42–44]. In addition, some studies consider fasting during Ramadan as an unfavorable condition for achieving the desired performance [5–7].

Therefore, it is important to understand the adaptation profile adopted by Tunisian professional footballers during the Ramadan fast. The present study aimed to examine the self-generated coping strategies adopted by our participants before, during, and after Ramadan. Specifically, Tunisian professional footballers, such as most Muslim athletes, fast every year during Ramadan. Nevertheless, they continue sports and physical activities and simultaneously cope with this stressful situation using self-developed coping strategies without prior mental preparation.

This study aims to examine the effect of Ramadan fasting on the coping profile of Tunisian professional footballers before, during, and after Ramadan. This study also aims to determine the coping strategies used by participants during the month of Ramadan.

Hence, the study hypothesized that; there is a significant effect of Ramadan fasting on the footballers' coping strategies. It is expected that the results of this study will be a reference that will help coaches and mental trainers to improve the performance of their athletes.

2. Materials and Methods

2.1. Study Design

The study was conducted in Tunisia in 2021. Ramadan began on Tuesday, 13 April 2021, and ended on Wednesday, 12 May 2021. The duration of each daytime fast was approximately 16 h starting at ~3:22 a.m. and ending at ~7:20 p.m. After the approval of officials and technical staff, the footballers were first invited six weeks before Ramadan to familiarize themselves with all the procedures involved in the study (FBR: Familiarization Before Ramadan). Then, the fasted footballers from the respective clubs were asked to respond to the Arabic version of the Inventory of Coping Strategies for Competitive Sport (ICSCS) [45], and measure their height and weight for body mass index (BMI) calculation in three sessions. The first session was carried out a week before Ramadan (MBR) from 3–7 April 2021, following the 22nd and 23rd day of the Tunisian professional championship. The second session occurred in the last week of Ramadan from 5–9 May 2021 (MDR), following the 24th and 25th day. The previous session was carried out a week after Ramadan (MAR) on 19 May 2021, following the last day of the championship (Figure 1).

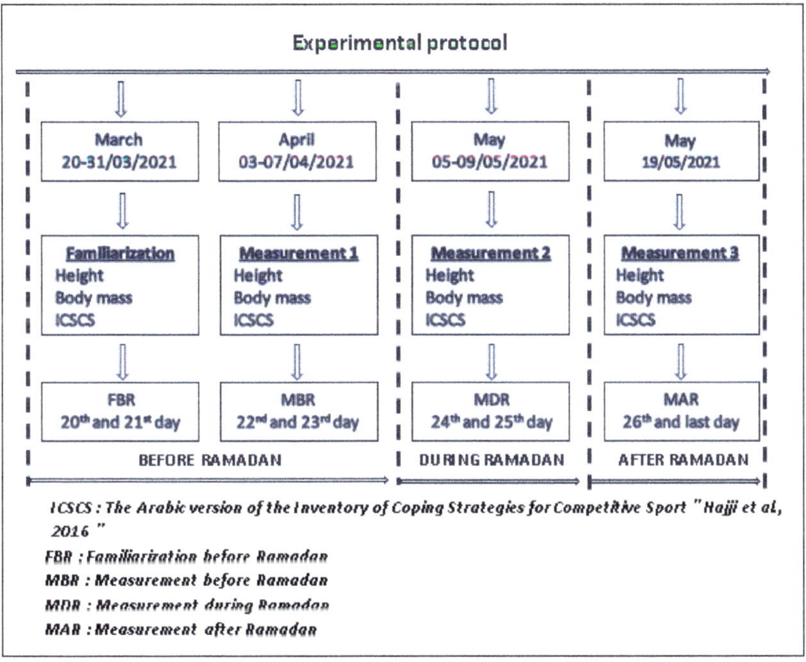

Figure 1. Experimental protocol adopted in the present study [45].

The study protocol was reviewed in-depth and fully approved by the "Ethics Committee for the Protection of Populations in the South" (C.P.P.SUD), Sfax, Tunisia (protocol reference C.P.P.SUD N 0031/2020). This study was based on the latest version of the Helsinki declaration and its subsequent amendments. The registration code for the trial is PACTR20211254567573.

2.2. Participants

After obtaining approval from club officials, coaches, and players, one hundred and eighty male footballers with a mean age of 25.54 years (SD = 4.41) volunteered to participate in this study. Fasting footballers were invited from 12 Tunisian clubs affiliated with Pro League 1 (CAB: Club Athlétique Bizertin; UST: Union Sportif de Tataouine; CA: Club African; ESS: Etoile Sportive du Sahel; ASS: Avenir Sportif de Soliman; CSS: Club Sportif

Sfaxien; ESM: Etoile Sportive de Métlaoui; USB: Union Sportive de Ben Guerdane; USMO: Union Sportive de Monastir; ASR: Avenir Sportif de Rejiche; EST: Espérance Sportive de Tunis; OB: Olympique de Béja). These 12 clubs, among 14 clubs, regularly participate in the Tunisian first professional league. The same activity continued during Ramadan; these players were recruited to observe Ramadan fasting. The characteristics of the participants are shown in Table 1.

Table 1. Mean and standard deviation of descriptive research variables (n = 15, N = 180).

Club		Club 1	Club 2	Club 3	Club 4	Club 5	Club 6	Club 7	Club 8	Club 9	Club 10	Club 11	Club 12
Age	Mean ± SD	24.0 ± 4.20	23.87 ± 3.9	24.60 ± 4.4	25.33 ± 4.5	24.80 ± 4.1	25.40 ± 4.0	25.67 ± 4.3	24.93 ± 4.3	25.87 ± 3.9	25.67 ± 4.3	26.87 ± 3.8	29.47 ± 5.5
Height (cm)	Mean ± SD	180.4 ± 7.1	182.2 ± 8.3	181.2 ± 7.1	182.13 ± 8.1	180.26 ± 7.9	179.6 ± 7.1	181 ± 7.6	178.46 ± 8.1	179.53 ± 6.7	181.47 ± 7.1	179.93 ± 7.6	180.86 ± 7.2
Weight (kg)	Mean ± SD	77.8 ± 5.9	78.67 ± 6.1	77.27 ± 6.5	78.13 ± 6.4	76.2 ± 6.2	75.73 ± 6.2	77.73 ± 6.1	76.06 ± 6.7	75.2 ± 4.5	78.2 ± 5.9	77 ± 6.1	78.33 ± 6.0
BMI	Mean ± SD	23.8 ± 0.5	23.83 ± 0.5	23.51 ± 0.6	23.53 ± 0.5	23.64 ± 0.8	23.43 ± 0.6	23.70 ± 0.5	23.84 ± 0.4	23.33 ± 0.6	23.72 ± 0.4	23.73 ± 0.8	23.92 ± 0.3

SD: standard deviation; BMI: weight divided by height squared, expressed in kg/m2; Club 1: CAB; Club 2: UST; Club 3: CA; Club 4: ESS; Club 5: ASS; Club 6: CSS; Club 7: ESM; Club 8: USB; Club 9: USMO; Club 10: ASR; Club 11: EST; Club 12: OB.

2.3. Applied Protocol and Measures

The Inventory of Competitive Sports Coping Strategies (ICSCS) (Supplementary Table S1) [43] was used to study the coping profile of Tunisian professional soccer players. The Gaudreau and Blondin [43] ICSCS is the most commonly used instrument in the sports context to assess coping strategies for competitive sports [44]. Due to the cultural specificity of the present sample, we used the Arabic version of ICSCS [45], which has been validated according to the methodology of Vallerand [46]. The Arabic version of the ICSCS has been validated for Tunisian athletes and has good psychometric properties [45]; the English version is attached as Supplementary Table S1. This tool makes it possible to measure ten coping strategies: mental imagery, thought control, expenditure of effort, search for support, relaxation, logical analysis, evacuation of unpleasant emotions, disengagement, social withdrawal, and mental distraction via 39 items rated on a 5-point Likert scale. Each strategy has four items [response scale ranging from 1 (never) to 5 (always)], except for the effort expenditure strategy, which has three items [response scale ranging from 1 (never) to 5 (always)]. The footballers were asked to answer all the items of the Arabic version of ICSCS one hour after their competition. The time to complete the ICSCS was about 15 min.

Measurements of body height were measured using a stadiometer, and weight was measured using a calibrated electronic scale (Tanita, Tokyo, Japan). BMI was calculated as body weight in kilograms divided by square height in meters. Weight status was defined according to age-specific BMI thresholds [47]. All measurements were taken by the same research group (4 investigators) directly in the stadium where the competition was taking place.

2.4. Statistical Analysis

The data obtained from the responses to the various items of the Arabic version of ICSCS were summarized in the form of means and standard deviation (SD) to illustrate the level of coping among footballers. Then, the scores obtained were analyzed by MANOVA on repeated measures [period (one week before Ramadan, during the last week of Ramadan, and one week after Ramadan 2021) × coping]. The Wilks Lambda test was centralized to compare the means and test the significant effect of Ramadan fasting on the coping profile in professional Tunisian footballers. All the observed differences were statistically significant when the p-value < 0.05 and effect sizes were calculated using eta square (η^2 = sum of squares in group effects/total sum of squares in the ANOVA). More specifically, a total sum of squares in the ANOVA included group effect, time effect, interaction effect, and error effect in the present study. The magnitude of eta squared is explained according to Cohen's suggestion: eta squared 0.01 as small effect; 0.06 as moderate effect; and 0.14 as large effect [48]. Data were analyzed using IBM SPSS statistics software (IBM SPSS software, France, version 21.0).

3. Results

The results of the MANOVA variance test demonstrated the effect of Ramadan fasting on the following coping strategies: effort expenditure [$F_{(11.17)}$ = 3.11; p-value = 0.01; η2 = 0.129], seeking support [$F_{(11.17)}$ = 1.99; p-value = 0.032; $η^2$ = 0.112], and venting of unpleasant emotions [$F_{(11.17)}$ = 2.31; p-value = 0.012; $η^2$ = 0.123] (Table 2).

Table 2. Inter-subject effects tests (N = 180).

Source	Dependent Variable		Before Ramadan					During Ramadan					After Ramadan			
			Mean ± SD	F	*p*-Value	$η^2$			Mean ± SD	F	*p*-Value	$η^2$		Mean ± SD	F	*p*-Value	$η^2$
Footballers	MI	Before Ramadan	3.83 ± 0.81	3.54	0.001	0.170	During Ramadan		3.72 ± 0.9	1.77	0.063	0.103	After Ramadan	3.74 ± 0.84	3.77	0.001	0.161
	TC		3.94 ± 0.82	4.75	0.001	0.185			3.51 ± 094	1.28	0.240	0.063		3.88 ± 0.88	4.06	0.001	0.211
	EE		4.19 ± 0.88	2.34	0.010	0.139			2.38 ± 1.04	3.11	0.001	0.129		3.33 ± 1.36	30.38	0.001	0.635
	SS		3.31 ± 0.89	2.08	0.024	0.119			3.69 ± 0.89	1.99	0.032	0.112		3.52 ± 0.89	2.46	0.007	0.166
	Rx		3.34 ± 0.82	1.79	0.061	0.094			3.82 ± 0.81	1.54	0.123	0.065		3.57 ± 0.87	3.46	0.001	0.176
	LA		3.64 ± 0.89	4.97	0.001	0.235			4.10 ± 0.78	1.22	0.279	0.077		3.90 ± 0.81	2.85	0.002	0.124
	VUE		3.15 ± 1.04	18.31	0.001	0.564			2.85 ± 0.97	2.31	0.012	0.123		3.23 ± 1.03	11.59	0.001	0.454
	Di		3.15 ± 1.15	26.52	0.001	0.617			3.71 ± 0.68	0.53	0.879	0.023		3.82 ± 0.76	5.21	0.001	0.279
	SW		2.84 ± 1.02	2.64	0.004	0.143			4.31 ± 0.78	0.93	0.515	0.047		3.48 ± 1.2	20.45	0.001	0.604
	MD		2.39 ± 1.04	5.79	0.001	0.282			3.53 ± 0.82	1.30	0.229	0.037		2.85 ± 1.12	10.92	0.001	0.408

MI: Mental imagery; TC: Thought control; EE: Effort expenditure; SS: Seeking support; Rx: Relaxation; LA: Logical analysis; VUE: Venting of unpleasant emotions; Di: Disengagement; SW: Social withdrawal; MD: Mental distraction. SD: standard deviation. F: Fisher test, η2; effect sizes.

One week before the Ramadan fast, Tunisian male professional soccer players were using task-oriented coping strategies [Effort expenditure: 4.19 ± 0.88]; [Thought control: 3.94 ± 0.82]; [Mental Imagery: 3.83 ± 0.81]; [Logic analysis: 3.64 ± 0.89]; [Rebound: 3.34 ± 0.82]; [Support seeking: 3.31 ± 0.89]. During the last week of Ramadan, Tunisian male professional footballers adopted the following coping strategies [Social withdrawal: 4.31 ± 0.68]; [Logic analysis: 4.10 ± 0.78]; [Rebound: 3.82 ± 0.81]; [Mental Imagery: 3.72 ± 0.90]; [Disengagement: 3.71 ± 0.68]. One week after Ramadan fasting, Tunisian male professional soccer players used the following coping strategies [Logical analysis: 3.90 ± 0.81]; [Thought control: 3.88 ± 0.88]; [Disengagement: 3.82 ± 0.76]; [Mental imagery: 3.74 ± 0.84] (Table 3).

A week before Ramadan, active coping dominates the coping profile in the majority of Tunisian professional clubs. During the last week of Ramadan, footballers from most Tunisian professional clubs use the strategy of social withdrawal. One week after Ramadan, effort expenditure and social withdrawal strategies dominate the adaptation profile among Tunisian professional clubs (Table 5). The results in Table 5 present the profile of coping among the different clubs of the Tunisian professional football league before, during, and after the month of Ramadan. This mass of information gives us an in-depth idea of the mental preparation system adopted by Tunisian clubs.

Table 3. Level of coping among professional footballers before, during, and after Ramadan (N = 180).

Coping Strategies	Mental Imagery	Thought Control	Effort Expenditure	Seeking Support	Relaxation	Logical Analysis	Venting of Unpleasant Emotions	Disengagement	Social Withdrawal	Mental Distraction	A	B	C
					LCBR								
Mean ± SD	3.83 ± 0.81	3.94 ± 0.82	4.19 ± 0.88	3.31 ± 0.89	3.34 ± 0.82	3.64 ± 0.89	3.15 ± 1.04	3.15 ± 1.15	2.84 ± 1.02	2.39 ± 1.04			
					LCDR						0.001	0.001	0.001
Mean ± SD	3.72 ± 0.9	3.51 ± 0.94	2.38 ± 1.04	3.69 ± 0.89	3.82 ± 0.81	4.10 ± 0.78	2.85 ± 0.97	3.71 ± 0.68	4.31 ± 0.78	3.53 ± 0.82			
					LCAR								
Mean ± SD	3.74 ± 0.84	3.88 ± 0.88	3.33 ± 1.36	3.52 ± 0.89	3.57 ± 0.87	3.90 ± 0.81	3.23 ± 1.03	3.82 ± 0.76	3.48 ± 1.20	2.85 ± 1.12			

SD: standard deviation; LCBR: level of coping before Ramadan; LCDR: level of coping during Ramadan; LCAR: level of coping after Ramadan. A: significant differences between LCBR and LCDR; B: significant differences between LCBR and LCAR. C: significant differences between LCDR and LCAR.

Table 4. Level of coping among Tunisian professional clubs before, during, and after Ramadan (n = 15; N = 180).

Club		Mental Imagery	Thought Control	Effort Expenditure	Seeking Support	Relaxation	Logical Analysis	Venting of Unpleasant Emotions	Disengagement	Social Withdrawal	Mental Distraction
					Level of coping before Ramadan among professional Tunisian clubs						
Club 1	Mean ± SD	3.23 ± 0.6	3.15 ± 0.7	3.62 ± 1.1	3.08 ± 0.7	3.28 ± 0.8	2.60 ± 0.6	2.41 ± 0.5	2.53 ± 0.5	3.40 ± 0.9	3.85 ± 1.0
Club 2	Mean ± SD	4.01 ± 1.1	3.48 ± 0.9	4.35 ± 0.4	3.36 ± 0.8	3.53 ± 0.7	3.91 ± 0.6	2.05 ± 0.7	2.51 ± 0.6	3.31 ± 1.4	2.51 ± 1.0
Club 3	Mean ± SD	3.96 ± 0.7	3.78 ± 0.6	3.86 ± 1.1	3.38 ± 0.9	3.55 ± 0.7	4.20 ± 0.8	2.32 ± 0.7	2.15 ± 0.7	3.23 ± 1.0	2.21 ± 1.1
Club 4	Mean ± SD	3.63 ± 0.7	4.08 ± 0.6	3.75 ± 1.1	3.30 ± 1.0	3.26 ± 0.6	3.16 ± 1.0	2.53 ± 0.9	2.63 ± 0.7	3.31 ± 0.7	2.75 ± 0.8
Club 5	Mean ± SD	3.86 ± 0.9	3.81 ± 0.8	4.15 ± 1.0	3.23 ± 0.9	3.60 ± 0.6	3.75 ± 0.9	2.80 ± 0.8	2.08 ± 0.8	2.98 ± 0.9	2.23 ± 0.8
Club 6	Mean ± SD	3.71 ± 0.6	4.03 ± 0.9	3.91 ± 1.1	3.25 ± 0.8	3.06 ± 1.0	3.66 ± 1.1	2.41 ± 0.7	2.23 ± 0.8	2.60 ± 1.2	2.02 ± 0.8
Club 7	Mean ± SD	3.46 ± 0.8	3.50 ± 0.9	4.29 ± 0.8	2.85 ± 0.9	2.85 ± 0.6	3.15 ± 0.8	3.35 ± 0.9	2.71 ± 1.1	2.66 ± 1.3	2.72 ± 1.0
Club 8	Mean ± SD	3.55 ± 0.8	3.91 ± 0.5	4.22 ± 0.8	2.85 ± 0.8	3.23 ± 0.7	3.60 ± 0.7	4.01 ± 0.5	3.80 ± 0.8	2.68 ± 1.0	2.53 ± 1.1
Club 9	Mean ± SD	3.68 ± 0.7	4.43 ± 0.7	4.49 ± 0.4	3.21 ± 0.8	3.60 ± 0.7	3.90 ± 0.6	4.08 ± 0.6	4.05 ± 0.6	2.33 ± 0.5	2.15 ± 0.7
Club 10	Mean ± SD	4.10 ± 0.6	4.40 ± 0.7	4.62 ± 0.5	3.71 ± 0.9	3.60 ± 0.9	3.85 ± 1.0	3.78 ± 0.8	4.30 ± 0.5	2.81 ± 0.8	1.65 ± 0.6
Club 11	Mean ± SD	4.62 ± 0.4	4.21 ± 0.7	4.44 ± 0.5	3.91 ± 0.9	3.55 ± 0.8	4.11 ± 0.4	4.11 ± 0.5	4.56 ± 0.6	2.50 ± 0.5	1.83 ± 0.9
Club 12	Mean ± SD	4.13 ± 0.8	4.45 ± 0.5	4.58 ± 0.5	3.61 ± 0.7	2.93 ± 1.1	3.81 ± 0.6	3.88 ± 0.7	4.23 ± 0.5	2.21 ± 0.7	2.16 ± 0.9

Table 5. Level of coping among Tunisian professional clubs before, during, and after Ramadan (n = 15; N = 180).

Club		Mental Imagery	Thought Control	Effort Expenditure	Seeking Support	Relaxation	Logical Analysis	Venting of Unpleasant Emotions	Disengagement	Social Withdrawal	Mental Distraction
		Level of coping before Ramadan among professional Tunisian clubs									
		Level of coping during Ramadan among professional Tunisian clubs									
Club 1	Mean ± SD	3.05 ± 0.9	3.32 ± 0.9	1.84 ± 0.5	3.06 ± 0.9	3.45 ± 0.7	3.81 ± 0.6	2.88 ± 0.9	3.56 ± 0.7	4.25 ± 0.8	3.71 ± 0.8
Club 2	Mean ± SD	3.82 ± 0.6	3.55 ± 0.7	2.02 ± 1.1	3.48 ± 1.0	3.65 ± 0.6	4.13 ± 0.9	2.43 ± 0.5	3.51 ± 0.5	4.42 ± 0.7	3.48 ± 0.4
Club 3	Mean ± SD	3.75 ± 0.7	3.56 ± 0.7	2.62 ± 1.0	3.43 ± 06	3.66 ± 05	4.00 ± 1.0	2.21 ± 0.8	3.83 ± 0.4	4.42 ± 0.5	3.18 ± 0.6
Club 4	Mean ± SD	3.56 ± 0.9	3.50 ± 0.8	2.51 ± 0.8	4.00 ± 0.6	3.86 ± 0.7	3.73 ± 0.6	3.21 ± 1.0	3.75 ± 0.7	4.42 ± 0.6	3.28 ± 0.8
Club 5	Mean ± SD	4.00 ± 1.0	3.40 ± 0.8	2.24 ± 1.0	3.95 ± 0.8	4.18 ± 0.5	3.91 ± 1.0	2.43 ± 0.7	3.85 ± 0.5	4.48 ± 0.5	3.40 ± 1.2
Club 6	Mean ± SD	3.66 ± 1.1	3.85 ± 1.0	3.13 ± 1.3	3.45 ± 1.0	3.62 ± 1.0	4.08 ± 0.5	2.81 ± 0.9	3.78 ± 0.8	4.32 ± 0.7	3.98 ± 0.9
Club 7	Mean ± SD	3.96 ± 0.7	3.48 ± 1.1	2.24 ± 1.0	4.00 ± 0.9	3.91 ± 1.1	4.35 ± 0.6	2.80 ± 1.0	3.46 ± 0.9	4.22 ± 1.1	3.35 ± 1.0
Club 8	Mean ± SD	4.01 ± 1.0	3.21 ± 1.1	2.28 ± 0.9	3.75 ± 0.9	3.91 ± 0.9	4.35 ± 0.9	3.43 ± 1.0	3.71 ± 0.6	4.40 ± 0.7	3.48 ± 0.7
Club 9	Mean ± SD	4.11 ± 0.7	4.05 ± 0.8	3.26 ± 1.2	3.93 ± 0.8	3.53 ± 0.8	4.21 ± 0.7	2.65 ± 1.2	3.78 ± 0.8	3.82 ± 0.5	3.46 ± 0.8
Club 10	Mean ± SD	3.65 ± 0.9	3.70 ± 0.9	2.20 ± 1.2	4.08 ± 0.8	4.31 ± 0.8	4.45 ± 0.5	3.06 ± 0.9	3.85 ± 0.6	4.50 ± 0.9	3.70 ± 0.8
Club 11	Mean ± SD	3.75 ± 0.9	3.03 ± 1.1	2.31 ± 0.8	3.73 ± 0.9	3.88 ± 0.9	4.12 ± 0.9	3.23 ± 1.0	3.70 ± 0.6	4.08 ± 0.9	3.40 ± 0.8
Club 12	Mean ± SD	3.35 ± 1.0	3.46 ± 0.8	1.84 ± 0.5	3.41 ± 0.8	3.83 ± 0.6	3.98 ± 0.6	3.01 ± 0.8	3.70 ± 0.7	4.36 ± 0.7	3.86 ± 0.7
		Level of coping after Ramadan among professional Tunisian clubs									
Club 1	Mean ± SD	3.33 ± 0.7	3.58 ± 1.0	3.93 ± 1.1	3.11 ± 0.8	2.86 ± 0.4	3.20 ± 0.9	3.40 ± 1.0	3.00 ± 1.1	2.73 ± 1.1	2.65 ± 0.9
Club 2	Mean ± SD	3.45 ± 0.9	3.88 ± 0.6	4.20 ± 0.7	2.76 ± 0.9	3.13 ± 0.9	3.61 ± 0.8	4.05 ± 0.6	3.61 ± 0.7	2.63 ± 0.9	2.88 ± 1.3
Club 3	Mean ± SD	3.45 ± 0.9	4.50 ± 0.6	4.31 ± 0.5	3.30 ± 0.8	3.41 ± 0.9	3.61 ± 0.8	3.90 ± 0.8	3.90 ± 0.6	2.75 ± 0.7	2.03 ± 0.6
Club 4	Mean ± SD	4.08 ± 0.7	4.41 ± 0.6	4.73 ± 0.2	3.63 ± 1.0	3.38 ± 0.8	3.78 ± 1.0	3.91 ± 0.8	4.38 ± 0.4	2.68 ± 1.0	1.81 ± 0.6
Club 5	Mean ± SD	4.50 ± 0.6	4.31 ± 0.6	4.60 ± 0.6	3.86 ± 0.9	3.88 ± 0.6	4.28 ± 0.3	3.95 ± 0.4	4.43 ± 0.5	2.46 ± 0.7	1.75 ± 0.7
Club 6	Mean ± SD	4.03 ± 0.7	4.38 ± 0.5	4.62 ± 0.5	3.70 ± 0.6	3.21 ± 1.1	3.88 ± 0.6	3.83 ± 0.6	4.25 ± 0.6	2.26 ± 0.8	2.13 ± 0.9
Club 7	Mean ± SD	3.03 ± 0.9	3.48 ± 0.8	2.04 ± 0.9	3.38 ± 1.1	3.75 ± 0.9	3.91 ± 0.6	3.10 ± 1.1	3.73 ± 0.7	4.16 ± 0.9	3.63 ± 0.9
Club 8	Mean ± SD	3.71 ± 0.6	3.51 ± 0.8	1.75 ± 0.8	3.70 ± 09	3.96 ± 0.7	4.18 ± 0.7	2.38 ± 0.5	3.76 ± 0.4	4.56 ± 0.6	3.70 ± 0.4
Club 9	Mean ± SD	3.78 ± 0.7	3.55 ± 0.8	2.57 ± 1.0	3.61 ± 0.7	3.81 ± 0.8	4.13 ± 1.1	2.20 ± 07	3.63 ± 0.4	4.53 ± 0.5	3.08 ± 0.6
Club 10	Mean ± SD	3.68 ± 0.9	3.65 ± 0.8	2.66 ± 0.9	3.80 ± 0.6	3.96 ± 0.5	3.68 ± 0.7	3.13 ± 1.1	3.80 ± 0.8	4.26 ± 0.8	3.38 ± 0.8
Club 11	Mean ± SD	3.83 ± 0.9	3.33 ± 1.0	2.13 ± 0.9	3.98 ± 0.7	4.10 ± 0.7	4.16 ± 1.0	2.58 ± 0.7	3.88 ± 0.5	4.48 ± 05	3.62 ± 1.2
Club 12	Mean ± SD	3.98 ± 06	3.91 ± 1.0	2.42 ± 1.0	3.36 ± 0.9	3.40 ± 0.9	4.33 ± 0.5	2.38 ± 1.0	3.48 ± 0.8	4.26 ± 0.7	3.52 ± 0.9

SD: standard deviation. Club 1: CA 3; Club 2: UST; Club 3: CA; Club 4: ESS; Club 5: ASS; Club 6: CSS; Club 7: ESM; Club 8: USB; Club 9: USMO; Club 10: ASR; Club 11: EST; Club 12: OB.

4. Discussion

This present study showed that Tunisian professional soccer players simultaneously use positive and negative coping strategies during the fasting month of Ramadan, whereas they use active coping strategies before and after Ramadan fasting. Furthermore, the results of the current study demonstrated that there is a significant effect of Ramadan fasting on the coping strategies of the footballers. Therefore, the study findings supported the proposed hypothesis. Understanding the associations between coping strategies and Ramadan fasting can help Muslim athletes obtain possible strategies for coping with Ramadan and further help them improve sports performance. Specifically, there are conflicting findings on the effects of Ramadan on performance in general and football [25]. Some studies provided evidence regarding how Ramadan fasting associated with several aspects of performance, as well as the cognitive element of adaptation, which agree with prior research's findings [15–17,23–25,34,49]. On the other hand, several other studies did not identify any effect of Ramadan fasting on the performance of Muslim athletes, especially for those who have developed coping strategies to cope with Ramadan fasting [7,44,50,51].

The present study revealed that task-oriented active coping was the most used to distinguish the coping pattern in most Tunisian professional soccer players before and after Ramadan fasting. This type of coping strategy, in which the athlete directly confronts the threatening situation, is positively associated with performance [42–44]. Although it comes from different contexts, the strategies of effort expenditure, thought control, mental imagery, logical analysis, relaxation, and seeking support remain the most used task-oriented coping strategies during Ramadan. The study of Farooq et al. [30] reported that elite footballers held negative beliefs and attitudes toward Ramadan fasting. Psychological stress in soccer players is a critical variable due to altered circadian clock and mood disturbances during Ramadan fasting [5–7,16]. On the other hand, the study of Zerguini et al. [32] concluded that biochemical, nutritional, subjective well-being, and performance variables were not negatively affected in young male national-level players who followed fasting of Ramadan in a controlled environment.

Although there is no special mental preparation aimed at helping Tunisian professional footballers during Ramadan, the self-adaptation that the Muslim athlete tries to use, in addition to the spiritual environment that surrounds him, all these factors help him to face the pressure imposed by the fasting system in the month of Ramadan. Studies show that social support around the athlete, along with the strength of spiritual beliefs [24], and choosing the best coping strategy may be a moderate variable in dealing with stressors that have arisen during Ramadan fasting [33]. According to Afacan [29], the adaptation strategies of Turkish professional footballers have been categorized into four dimensions: training, nutrition, lifestyle, and mental changes. It has also been determined that professional soccer players frequently use training and nutrition strategies. Meanwhile, other studies have recognized that mental preparation before fasting during Ramadan is necessary to acquire proactive coping skills [5,29,41,52]. The study of Fenneni et al. [53] confirms that experienced athletes, i.e., more trained athletes, have better-coping strategies than beginners. However, the coping methods of Ramadan fasting could be impacted by spiritual bias. Specifically, prior evidence shows that people engaging in Ramadan fasting may consider that physical discomfort was because of God's instruction [54]. In this regard, they may not be aware of using inappropriate coping methods and result in poor physical outcomes.

Limitations and Recomandations

In the present study, we examined the coping strategies used by footballers belonging to the Tunisian first professional league. Regarding the sample size of the current study, we focused on the main players who regularly participate in official matches. The experimental protocol adopted in the present study was carried out in three periods, before, during, and after the month of Ramadan. Therefore, submitting participants to four measures (familiarization, measures 1, 2, and 3), is considered among the limitations of this study. It was very difficult to pass the questionnaire, especially after a negative performance which can influence the credibility of the answers. In addition, measuring their height and weight for body mass index (BMI) calculation in three sessions within a short period is also one of the limitations of the current study.

This study attempts to shed light on the footballer's relationship with Ramadan fasting in Tunisia. But the Ramadan fast still requires studies and clarifications, particularly at the level of other categories of athletes (women and young athletes, etc.) as well as at the level of other sports disciplines (handball, athletics, combat sports...). Studies in this area can also be expanded to include sedentary people. Thus, it is necessary to carry out several other studies to determine whether fasting during the month of Ramadan has a positive or negative effect on sports performance and the extent of its impact on the diet and lifestyle of the athlete in particular and sedentary people in general.

Although revealing exciting results, this study has some limitations that should be mentioned. First, the most important is that all the study measurements were taken during the Coronavirus (Covid-19) period in 2021. Second, the use of the questionnaire via the self-report method is a limitation. According to Fortes [35], the use of Likert scale-based questionnaires in repeated measures surveys can make an educational impact. A measure of the effect of dietary change during Ramadan was not available. In addition, psychobiological signals such as cortisol and testosterone were not evaluated during testing as we were unable to collect blood and urine samples during this study. It was also more appropriate for us to measure the effect of Ramadan fasting on the coping strategies of Tunisian professional football players. Lastly, we did not use any external validity control to select our participants. Therefore, the present study might have some selection bias, and the present findings might have restricted external validity.

5. Conclusions

This study is one of the few attempts to examine the effects of Ramadan fasting on the coping profile of fasting Tunisian professional footballers. The results showed that the adaptation profile of footballers during Ramadan is unstable and irregular, which is significantly affected by fasting, which supported the current study hypothesis. In the period of Ramadan fasting, professional soccer players use both active task-oriented and passive coping strategies toward disengagement and distraction. In contrast, they use active coping strategies before and after Ramadan fasting. Thus, from a practical point of view, integrating a mental preparation program to develop coping strategies before the Ramadan fast seems to be an appropriate intervention strategy. However, further research is warranted due to the shortcomings mentioned above.

In summary, for Muslim athletes, fasting in the month of Ramadan is generally considered an essential religious and spiritual duty. Despite the difficulties caused by fasting in the month of Ramadan on the physical and mental levels, the Muslim athlete tries to resist these difficulties by using particular coping strategies which help him to face this stressful situation.

Supplementary Materials: The following supporting information can be downloaded at: https://www.mdpi.com/article/10.3390/healthcare11071053/s1, Table S1: Questionnaire for the assessment of coping strategies employed by athletes in competitive sport (Inventory of Competitive Sports Coping Strategies (ICSCS))

Author Contributions: Conceptualization, J.H., A.S. and M.A.A.; methodology, J.H., A.S. and M.A.A.; software, J.H. and C.-Y.L.; validation, J.H., A.S., C.-Y.L., L.-Y.H. and M.A.A.; formal analysis, J.H., A.S. and C.-Y.L.; investigation, J.H., A.S., C.-Y.L., L.-Y.H. and M.A.A.; resources, J.H., A.S., C.-Y.L., L.-Y.H. and M.A.A.; data curation, J.H. and A.S.; writing—original draft preparation, J.H., A.S., C.-Y.L., L.-Y.H. and M.A.A.; writing—review and editing, J.H., A.S., C.-Y.L., L.-Y.H. and M.A.A.; visualization, A.S., C.-Y.L. and M.A.A.; supervision, J.H. and A.S.; project administration, J.H. and A.S.; funding acquisition, L.-Y.H. All authors have read and agreed to the published version of the manuscript.

Funding: This study was funded by E-Da Hospital, Taiwan; (grant number: EDCHC112002).

Institutional Review Board Statement: The study protocol was reviewed in-depth and fully approved by the "Ethics Committee for the Protection of Populations in the South" (C.P.P.SUD), Sfax, Tunisia (protocol reference C.P.P.SUD N 0031/2020). This study was conducted in accordance with the Declaration of Helsinki and its subsequent amendments. The registration number for the trial is PACTR20211254567573.

Informed Consent Statement: Informed consent was obtained from all subjects involved in the study.

Data Availability Statement: The dataset that supports the findings of this study is not openly available and it will be available from the corresponding author upon reasonable request.

Acknowledgments: The authors warmly thank the club leaders, coaches, and professional footballers who participated in this study.

Conflicts of Interest: The authors declare no conflict of interest.

References

1. Al-Jafar, R.; Zografou Themeli, M.; Zaman, S.; Akbar, S.; Lhoste, V.; Khamliche, A.; Elliott, P.; Tsilidis, K.K.; Dehghan, A. Effect of Religious Fasting in Ramadan on Blood Pressure: Results From LORANS (London Ramadan Study) and a Meta-Analysis. *J. Am. Heart Assoc.* **2021**, *10*, e021560. [CrossRef] [PubMed]
2. Bandarian, F.; Namazi, N.; Atlasi, R.; Nasli-Esfahani, E.; Larijani, B. Research gaps in Ramadan fasting studies in health and disease. *Diabetes Metab. Syndr. Clin. Res. Rev.* **2021**, *15*, 831–835. [CrossRef]
3. Rouhani, M.H.; Azadbakht, L. Is Ramadan fasting related to health outcomes? A review on the related evidence. *J. Res. Med. Sci.* **2014**, *19*, 987–992.
4. Urooj, A.; Pai Kotebagilu, N.; Shivanna, L.M.; Anandan, S.; Thantry, A.N.; Siraj, S.F. Effect of Ramadan Fasting on Body Composition, Biochemical Profile, and Antioxidant Status in a Sample of Healthy Individuals. *Int. J. Endocrinol. Metab.* **2020**, *18*, e107641. [CrossRef] [PubMed]
5. Aziz, A.; Png, W. Practical tips to exercise training during the Ramadan fasting month. *ISN Bull.* **2008**, *1*, 13–19.
6. Anis, C.; Leiper, J.B.; Nizar, S.; Coutts, A.J.; Karim, C. Effects of Ramadan intermittent fasting on sports performance and training: A review. *Int. J. Sport. Physiol. Perform.* **2009**, *4*, 419–434. [CrossRef]
7. Damit, N.F.; Lim, V.T.W.; Muhamed, A.M.C.; Chaouachi, A.; Chamari, K.; Singh, R.; Chia, M.; Aziz, A.R. *Exercise Responses and Training during Daytime Fasting in the Month of Ramadan and Its Impact on Training-Induced Adaptations*; OMICS Group eBooks: Foster City, CA, USA, 2014.
8. Lotfi, S.; Madani, M.; Tazi, A.; Boumahmaza, M.; Talbi, M. [Variation of cognitive functions and glycemia during physical exercise in Ramadan fasting]. *Rev. Neurol.* **2010**, *166*, 721–726. [CrossRef]
9. Bouhlel, E.; Zaouali, M.; Miled, A.; Tabka, Z.; Bigard, X.; Shephard, R. Ramadan fasting and the GH/IGF-1 axis of trained men during submaximal exercise. *Ann. Nutr. Metab.* **2008**, *52*, 261–266. [CrossRef] [PubMed]
10. Bouhlel, E.; Salhi, Z.; Bouhlel, H.; Mdella, S.; Amamou, A.; Zaouali, M.; Mercier, J.; Bigard, X.; Tabka, Z.; Zbidi, A. Effect of Ramadan fasting on fuel oxidation during exercise in trained male rugby players. *Diabetes Metab.* **2006**, *32*, 617–624. [CrossRef]
11. Bouhlel, E.; Denguezli, M.; Zaouali, M.; Tabka, Z.; Shephard, R.J. Ramadan fasting's effect on plasma leptin, adiponectin concentrations, and body composition in trained young men. *Int. J. Sport Nutr. Exerc. Metab.* **2008**, *18*, 617–627. [CrossRef]
12. Güvenç, A. Effects of Ramadan fasting on body composition, aerobic performance and lactate, heart rate and perceptual responses in young soccer players. *J. Hum. Kinet.* **2011**, *29*, 79. [CrossRef] [PubMed]
13. Ziaee, V.; Razaei, M.; Ahmadinejad, Z.; Shaikh, H.; Yousefi, R.; Yarmohammadi, L.; Bozorgi, F.; Behjati, M.J. The changes of metabolic profile and weight during Ramadan fasting. *Singap. Med. J.* **2006**, *47*, 409.
14. Aziz, A.R.; Wahid, M.F.; Png, W.; Jesuvadian, C.V. Effects of Ramadan fasting on 60 min of endurance running performance in moderately trained men. *Br. J. Sport. Med.* **2010**, *44*, 516–521. [CrossRef] [PubMed]
15. Brisswalter, J.; Bouhlel, E.; Falola, J.M.; Abbiss, C.R.; Vallier, J.M.; Hauswirth, C. Effects of Ramadan intermittent fasting on middle-distance running performance in well-trained runners. *Clin. J. Sport Med.* **2011**, *21*, 422–427. [CrossRef] [PubMed]
16. Chtourou, H.; Hammouda, O.; Souissi, H.; Chamari, K.; Chaouachi, A.; Souissi, N. The effect of Ramadan fasting on physical performances, mood state and perceived exertion in young footballers. *Asian J. Sport. Med.* **2011**, *2*, 177. [CrossRef] [PubMed]
17. Memari, A.-H.; Kordi, R.; Panahi, N.; Nikookar, L.R.; Abdollahi, M.; Akbarnejad, A. Effect of Ramadan fasting on body composition and physical performance in female athletes. *Asian J. Sport. Med.* **2011**, *2*, 161. [CrossRef]
18. Chennaoui, M.; Desgorces, F.; Drogou, C.; Boudjemaa, B.; Tomaszewski, A.; Depiesse, F.; Burnat, P.; Chalabi, H.; Gomez-Merino, D. Effects of Ramadan fasting on physical performance and metabolic, hormonal, and inflammatory parameters in middle-distance runners. *Appl. Physiol. Nutr. Metab.* **2009**, *34*, 587–594. [CrossRef]
19. Maughan, R.J.; Bartagi, Z.; Dvorak, J.; Zerguini, Y. Dietary intake and body composition of football players during the holy month of Ramadan. *J. Sport. Sci.* **2008**, *26*, S29–S38. [CrossRef]
20. Aziz, A.R.; Che Muhamed, A.M.; Ooi, C.H.; Singh, R.; Chia, M.Y.H. Effects of Ramadan fasting on the physical activity profile of trained Muslim soccer players during a 90-minute match. *Sci. Med. Footb.* **2018**, *2*, 29–38. [CrossRef]

21. Faris, M.e.A.-I.E.; Jahrami, H.A.; Alhayki, F.A.; Alkhawaja, N.A.; Ali, A.M.; Aljeeb, S.H.; Abdulghani, I.H.; BaHammam, A.S. Effect of diurnal fasting on sleep during Ramadan: A systematic review and meta-analysis. *Sleep Breath.* **2020**, *24*, 771–782. [CrossRef]

22. Qasrawi, S.O.; Pandi-Perumal, S.R.; BaHammam, A.S. The effect of intermittent fasting during Ramadan on sleep, sleepiness, cognitive function, and circadian rhythm. *Sleep Breath.* **2017**, *21*, 577–586. [CrossRef]

23. Waterhouse, J. Effects of Ramadan on physical performance: Chronobiological considerations. *Br. J. Sport. Med.* **2010**, *44*, 509–515. [CrossRef]

24. Chamari, K.; Roussi, M.; Bragazzi, N.L.; Chaouachi, A.; Abdul, R.A. Optimizing training and competition during the month of Ramadan: Recommendations for a holistic and personalized approach for the fasting athletes. *Tunis Med.* **2019**, *97*, 1095–1103.

25. Kirkendall, D.T.; Chaouachi, A.; Aziz, A.R.; Chamari, K. Strategies for maintaining fitness and performance during Ramadan. *J. Sport. Sci.* **2012**, *30*, S103–S108. [CrossRef]

26. Almeneessier, A.S.; Pandi-Perumal, S.R.; BaHammam, A.S. Intermittent fasting, insufficient sleep, and circadian rhythm: Interaction and effects on the cardiometabolic system. *Curr. Sleep Med. Rep.* **2018**, *4*, 179–195. [CrossRef]

27. Almeneessier, A.S.; BaHammam, A.S. How does diurnal intermittent fasting impact sleep, daytime sleepiness, and markers of the biological clock? Current insights. *Nat. Sci. Sleep* **2018**, *10*, 439. [CrossRef]

28. Wehrens, S.M.; Christou, S.; Isherwood, C.; Middleton, B.; Gibbs, M.A.; Archer, S.N.; Skene, D.J.; Johnston, J.D. Meal timing regulates the human circadian system. *Curr. Biol.* **2017**, *27*, 1768–1775.e1763. [CrossRef]

29. Afacan, E. Coping Strategies That Football Players Use During Ramadan Fasting: Turkish Super League Sample. *Prog. Nutr.* **2021**, *23*, e2021251.

30. Farooq, A.; Herrera, C.P.; Zerguini, Y.; Almudahka, F.; Chamari, K. Knowledge, beliefs and attitudes of Muslim footballers towards Ramadan fasting during the London 2012 Olympics: A cross-sectional study. *BMJ Open* **2016**, *6*, e012848. [CrossRef] [PubMed]

31. DeLang, M.D.; Salamh, P.A.; Chtourou, H.; Saad, H.B.; Chamari, K. The effects of Ramadan intermittent fasting on football players and implications for domestic football Leagues over the next decade: A systematic review. *Sport. Med.* **2022**, *52*, 585–600. [CrossRef] [PubMed]

32. Zerguini, Y.; Ahmed, Q.A.; Dvorak, J. The Muslim football player and Ramadan: Current challenges. *J. Sport. Sci.* **2012**, *30*, S3–S7. [CrossRef] [PubMed]

33. Chaouachi, A.; Leiper, J.B.; Chtourou, H.; Aziz, A.R.; Chamari, K. The effects of Ramadan intermittent fasting on athletic performance: Recommendations for the maintenance of physical fitness. *J. Sport. Sci.* **2012**, *30*, S53–S73. [CrossRef] [PubMed]

34. Cherif, A.; Meeusen, R.; Farooq, A.; Briki, W.; Fenneni, M.A.; Chamari, K.; Roelands, B. Repeated sprints in fasted state impair reaction time performance. *J. Am. Coll. Nutr.* **2017**, *36*, 210–217. [CrossRef] [PubMed]

35. Fortes, L.d.S.; Almeida, S.S.; Ferreira, M.E.C. Influência da periodização do treinamento sobre os comportamentos de risco para transtornos alimentares em nadadoras. *Rev. Educ. Física/UEM* **2014**, *25*, 127–134. [CrossRef]

36. Bouhlel, H.; Latiri, I.; Zarrrouk, N.; Bigard, X.; Shephard, R.; Tabka, Z.; Bouhlel, E. Effect of Ramadan observance and maximal exercise on simple and choice reaction times in trained men. *Sci. Sport.* **2014**, *29*, 131–137. [CrossRef]

37. Maughan, R.; Fallah, J.; Coyle, E. The effects of fasting on metabolism and performance. *Br. J. Sport. Med.* **2010**, *44*, 490–494. [CrossRef] [PubMed]

38. Maughan, R.J.; Leiper, J.B.; Bartagi, Z.; Zrifi, R.; Zerguini, Y.; Dvorak, J. Effect of Ramadan fasting on some biochemical and haematological parameters in Tunisian youth soccer players undertaking their usual training and competition schedule. *J. Sport. Sci.* **2008**, *26* (Suppl. 3), S39–S46. [CrossRef]

39. Lazarus, R.S.; Folkman, S. *Stress, Appraisal, and Coping*; Springer Publishing Company: New York, NY, USA, 1984.

40. Djemai, H.; Hammad, R.; Al Qarra, S.; Dabayebeh, I.M. Self-Coping Strategies Among Jordanian Athletes During Ramadan Fasting: A Questionnaire Proposal. *Asian J. Sport. Med.* **2020**, *11*, e105569. [CrossRef]

41. Roy, J.; Hwa, O.C.; Singh, R.; Aziz, A.R.; Jin, C.W. Self-generated coping strategies among Muslim athletes during Ramadan fasting. *J. Sport. Sci. Med.* **2011**, *10*, 137.

42. Doron, J.; Stephan, Y.; Le Scanff, C. Les stratégies de coping: Une revue de la littérature dans les domaines du sport et de l'éducation. *Eur. Rev. Appl. Psychol.* **2013**, *63*, 303–313. [CrossRef]

43. Gaudreau, P.; Blondin, J.-P. Development of a questionnaire for the assessment of coping strategies employed by athletes in competitive sport settings. *Psychol. Sport Exerc.* **2002**, *3*, 1–34. [CrossRef]

44. Nicholls, A.R.; Levy, A.R.; Carson, F.; Thompson, M.A.; Perry, J.L. The applicability of self-regulation theories in sport: Goal adjustment capacities, stress appraisals, coping, and well-being among athletes. *Psychol. Sport Exerc.* **2016**, *27*, 47–55. [CrossRef]

45. Hajji, J.; Baaziz, M.; Mnedla, S.; Jannet, Z.B.; Elloumi, A. Validation of the Arabic Version of the Inventory of Coping Strategies of Competitive Sport (ISCCS). *Adv. Phys. Educ.* **2016**, *6*, 312–327. [CrossRef]

46. Vallerand, R.J. Vers une méthodologie de validation trans-culturelle de questionnaires psychologiques: Implications pour la recherche en langue française. *Can. Psychol. /Psychol. Can.* **1989**, *30*, 662. [CrossRef]

47. Cole, T.J.; Bellizzi, M.C.; Flegal, K.M.; Dietz, W.H. Establishing a standard definition for child overweight and obesity worldwide: International survey. *BMJ* **2000**, *320*, 1240. [CrossRef] [PubMed]

48. Lakens, D. Calculating and reporting effect sizes to facilitate cumulative science: A practical primer for t-tests and ANOVAs. *Front. Psychol* **2013**, *4*, 863. [CrossRef]

49. Aloui, A.; Chaouachi, A.; Chtourou, H.; Wong, D.P.; Haddad, M.; Chamari, K.; Souissi, N. Effects of Ramadan on the diurnal variations of repeated-sprint performance. *Int. J. Sport. Physiol. Perform.* **2013**, *8*, 254–263. [CrossRef]

50. Kordi, R.; Abdollahi, M.; Memari, A.-H.; Najafabadi, M.G. Investigating two different training time frames during Ramadan fasting. *Asian J. Sport. Med.* **2011**, *2*, 205. [CrossRef]

51. Shephard, R.J. The impact of Ramadan observance upon athletic performance. *Nutrients* **2012**, *4*, 491–505. [CrossRef]

52. Lim, V.; Dalmit, N.; Aziz, A. Recommendations for optimal competitive exercise performance and effective training-induced adaptations when Ramadan fasting. In *Effects of Ramadan Fasting on Health and Athletic Performance*; Chtourou, H., Ed.; OMICS Group eBooks: Foster City, CA, USA, 2015; pp. 204–221.

53. Fenneni, M.A.; Latiri, I.; Aloui, A.; Rouatbi, S.; Chamari, K.; Saad, H.B. Critical analysis of the published literature about the effects of Ramadan intermittent fasting on healthy children's physical capacities. *Libyan J. Med.* **2015**, *10*, 28351. [CrossRef] [PubMed]

54. Akkuş, Y.; Kiliç, S.P. Feelings, Difficulties and Attitudes in relation to Fasting: A Qualitative Study on Spiritual Coping Among Turkish Patients with Type 2 Diabetes. *J. Relig. Health* **2022**. [CrossRef] [PubMed]

Article

Coping Strategies and Perceiving Stress among Athletes during Different Waves of the COVID-19 Pandemic—Data from Poland, Romania, and Slovakia

Ryszard Makarowski [1], Radu Predoiu [2], Andrzej Piotrowski [3,*], Karol Görner [4], Alexandra Predoiu [2], Rafael Oliveira [5,6,7], Raluca Anca Pelin [8], Alina Daniela Moanță [2], Ole Boe [9,10], Samir Rawat [11] and Gayatri Ahuja [12]

[1] Faculty of Administration and Social Sciences, Academy of Applied Medical and Social Sciences in Elbląg, 82-300 Elbląg, Poland
[2] Faculty of Physical Education and Sport, National University of Physical Education and Sports, 060057 Bucharest, Romania
[3] Department of Personality Psychology and Forensic Psychology, University of Gdańsk, 80-309 Gdańsk, Poland
[4] Faculty of Sports, University of Presov, 08001 Presov, Slovakia
[5] Sports Science School of Rio Maior—Polytechnic Institute of Santarém, 2040-413 Rio Maior, Portugal
[6] Research Centre in Sport Sciences, Health Sciences and Human Development, 5001-801 Vila Real, Portugal
[7] Life Quality Research Centre, 2040-413 Rio Maior, Portugal
[8] Department of Physical Education and Sports-Kinetotherapy, Faculty of Medical Engineering, University Politehnica of Bucharest, 060042 Bucharest, Romania
[9] Department of Business, Strategy and Political Sciences, University of South-Eastern Norway, 3045 Drammen, Norway
[10] Institute of Psychology, Oslo New University College, 0456 Oslo, Norway
[11] Military MIND Academy, Pune 411060, India
[12] Department of Education, Ali Yavar Jung National Institute of Speech and Hearing Disabilities, Mumbai 400050, India
* Correspondence: andrzej.piotrowski@ug.edu.pl

Citation: Makarowski, R.; Predoiu, R.; Piotrowski, A.; Görner, K.; Predoiu, A.; Oliveira, R.; Pelin, R.A.; Moanță, A.D.; Boe, O.; Rawat, S.; et al. Coping Strategies and Perceiving Stress among Athletes during Different Waves of the COVID-19 Pandemic—Data from Poland, Romania, and Slovakia. *Healthcare* **2022**, *10*, 1770. https://doi.org/ 10.3390/healthcare10091770

Academic Editor: Herbert Löllgen

Received: 4 August 2022
Accepted: 13 September 2022
Published: 14 September 2022

Publisher's Note: MDPI stays neutral with regard to jurisdictional claims in published maps and institutional affiliations.

Abstract: Coronavirus disease (COVID-19), an infectious disease caused by the SARS-CoV-2 virus, has affected numerous aspects of human functioning. Social contacts, work, education, travel, and sports have drastically changed during the lockdown periods. The pandemic restrictions have severely limited professional athletes' ability to train and participate in competitions. For many who rely on sports as their main source of income, this represents a source of intense stress. To assess the dynamics of perceived stress as well as coping strategies during different waves of the COVID-19 pandemic, we carried out a longitudinal study using the Perception of Stress Questionnaire and the Brief COPE on a sample of 2020 professional athletes in Poland, Romania, and Slovakia. The results revealed that in all three countries, the highest intrapsychic stress levels were reported during the fourth wave (all, $p < 0.01$) and the highest external stress levels were reported before the pandemic ($p < 0.05$). To analyze the data, analyses of variance were carried out using Tukey's post hoc test and η2 for effect size. Further, emotional tension was the highest among Polish and Slovak athletes in the fourth wave, while the highest among Romanian athletes was in the pre-pandemic period. The coping strategies used by the athletes in the fourth wave were more dysfunctional than during the first wave (independent t test and Cohen's d were used). The dynamics of the coping strategies—emotion focused and problem focused—were also discussed among Polish, Romanian, and Slovak athletes. Coaches and sports psychologists can modify the athletes' perceived stress while simultaneously promoting effective coping strategies.

Keywords: sport; athlete; coping with stress; SARS-CoV-2; COVID-19; cross-cultural research

1. Introduction

SARS-CoV-2 (severe acute respiratory syndrome coronavirus 2) caused the coronavirus disease 2019 (COVID-19), which was classified as a pandemic by the World Health Organization in March of 2020, only a few months after the first case was detected in December of 2019 [1]. The dynamics of transmission, severe health consequences, and high mortality of infected patients were reported in many societies. Thus, many governments decided to introduce restrictions aimed at limiting the transmission of the virus [2]. Personal protective measures, such as face masks, frequent hand disinfecting, and social distancing, proved insufficient to stop the pandemic entirely. Many countries also introduced lockdowns, which forced people to stay home, thus, limiting physical and sports activity [3].

The lockdowns were intended to protect against infection but, unfortunately, they resulted in negative health consequences, such as weight anxiety and depression increases [4,5]. Individual countries tried to limit the spread of COVID-19 by means of various restrictions and staying indoors (people being confined to their homes), situations that impacted the general population's mental health (people did not have an adequate space in which to work or exercise) [6,7]. It was assumed that COVID-19's droplet transmission may be facilitated by strenuous physical activity, which results in deep lung ventilation [8]. Due to the lockdowns, sports halls and gyms were closed and outdoor physical activity was banned. Many people who were physically active were forced to suspend training almost overnight. Studies on athletes from over 140 countries showed that lockdowns led to lower intensity, frequency, and duration of training [9]. For professional athletes, lockdowns led to violations in long-term, rigorous training plans, inability to prepare for competitions, and canceling or severely limiting sports events [10]. For some of them, this may have caused an early end to their career due to their age and the inflexible schedules of global events [11]. The COVID-19 crisis decreased athletes' functional psychobiosocial states, e.g., cognitive, emotional, motivational, volitional, motor-behavioral, communicative, etc. (for psychobiosocial state examples, see [12]), while increasing their dysfunctional psychobiosocial states [13]. Even during the rebooting of sport activities, despite the resumption of sport activities, athletes experienced a detrimental situation, their mental health being still affected [14]. Further, increased stress and anxiety were common themes affecting student athletes' experiences when returning to sport amidst the COVID-19 pandemic [15].

Professional athletes who rely on sports as their main source of income found themselves in a difficult situation, as their physical activity was reduced and, thus, their income from broadcasting competitions, advertising, and awards dropped to almost zero. Restricting sports activities had a negative effect on athletes' health, but it was not effective in preventing the spread of COVID-19 in this population group [16]. Athletes' mental status naturally influences their functioning. A global sense of threat, social isolation, and uncertainty about the future may lead to anxiety, depression, and chronic stress [17].

Effective coping strategies reduced the stress experienced by athletes during the COVID-19 pandemic [15,18]. More specifically, positive reframing helps athletes to maintain a positive mood state and reduce distress, while self-blaming and behavioral disengagement are coping strategies that negatively influence athletes' mood. Regarding coping, it represents the conscious use of affective, cognitive, or behavioral efforts to effectively deal with demands, events that the individual perceives as potentially harmful or unpleasant [19]. The outcome of coping efforts is to reduce psychological distress, improve mental well-being and reduce physiological reactions that may impair performance [20,21]. The coping literature [22,23] discusses identifying the athlete's cognitive appraisal regarding the situation in which he/she is found (in training or competition). Larazus and Folkman [23] categorized appraisal as non-stressful (positive, harmless) or stressful, while stress appraisals were designated as threatening or challenging. Hoar et al. discussed (within another appraisal framework) the importance of perceived control. As authors mentioned, taking into account that perceived control can change over time, coping responses can become more or less effective (even well-learned coping strategies can be modified, as a consequence of environmental demands) [24]. Considering problem-focused and emotional approach coping, specialized literature underlines that men engage in more problem-focused coping, whereas women resort to more emotional-approach coping [25]. As stated, men who are using more emotion-focused coping strategies seemed to register higher levels of positive affect.

During the COVID-19 pandemic, physical activity can "generate and maintain resources combative of stress and protective of health" [26]. Significantly lower mental and physical health was found in individuals with the highest decrease in physical activity during the pandemic. Considering professional cyclists, those who followed a Sport Psychology Intervention (online) during the pandemic coped better with sport psychological stressors (no significant improvements were found, however, for sport social and for sport emotional well-being factors) [27]. In order to cope with negative psychological effects arising from the pandemic, researchers discussed the benefits of mental toughness training in the case of athletes [28], important aspects, since professional athletes obtained lower values for the agreeableness factor (during the pandemic crisis), compared to non-professionals [29]. Counselling athletes in an unprecedented situation, as in the COVID-19 pandemic, is very important for acquiring healthy behaviors. For example, mindful activities related to the body—the experience of one's body as trustworthy and safe—could reduce distress in athletes and increase positive stress [14]. Training regimens should be introduced as standard habits for well-being and health, especially for women and novice athletes, who registered higher levels of negative stress (distress) [13].

Each country is characterized by different dynamics of COVID-19 infections, resulting from, among others, the implemented prevention methods, the number of social contacts and foreign travels, national age average, healthcare quality, and economic conditions.

The purpose of the current research was to investigate the dynamics of stress perceived by Polish, Romanian, and Slovak athletes during the first four waves of the COVID-19 pandemic and to establish the changes in coping strategies between the first and fourth waves. The following research questions were put forward:

1. What were the dynamics of emotional, external, and intrapsychic stress before the pandemic and during different waves among athletes in Poland, Romania, and Slovakia (country split and wave split)?
2. What were the dynamics of perceived emotional tension, external stress, and intrapsychic stress in the total sample of athletes (regardless of country) throughout the research periods (until the fourth wave of the pandemic)?
3. What are the differences in the frequency of using strategies of coping with stress among athletes in the first and fourth waves of the pandemic?

2. Materials and Methods

2.1. Design

Data collection was carried out from November 2019 to January 2020 (the pre-pandemic period) as well as during or in the proximity of the first, second, third, and fourth waves of the COVID-19 pandemic (see Table 1). For example, in Romania, the peak of illnesses was recorded on 18 November 2020 (for the second wave) and on 25 March 2021 for the third wave. It is worth mentioning, also, that the first case of the SARS-CoV-2 coronavirus appeared in Poland on 4 March 2020, in Romania on 25 February, and in Slovakia on 6 March 2020. When completing the Perception of Stress Questionnaire, the instruction was as follows: "Please describe your thoughts, behaviors, fears and hopes as you have experienced them lately (in the last few weeks) and currently". Data collection was carried out in Poland, Romania, and Slovakia and was concluded at the beginning of 2022. It is important to emphasize that in November 2019 (when data collection began) nobody knew about COVID-19. The original research idea of analyzing the stress experienced by athletes from Poland, Romania, and Slovakia at time t_0 (the moment they completed the survey) and the coping strategies used were restructured to conduct a longitudinal study, in order to observe the dynamics of athletes' perceived stress (during different waves of the pandemic) and coping strategies used to deal with stress.

Table 1. Descriptive statistics and data collection timetable.

Country	Research Period	n	Men	Women	M_{age}	SD
Poland	Before the Pandemic 11.2019–1.2020	314	186	128	22.85	3.25
Romania		221	133	88	21.86	3.60
Slovakia		91	51	40	23.27	2.28
Poland	1st wave 4–6.2020	134	84	50	26.40	4.98
Romania		145	85	60	25.22	5.65
Slovakia		111	67	44	23.73	6.05
Poland	2nd wave 10–12.2020	63	31	32	24.25	4.77
Romania		171	112	59	20.82	5.77
Slovakia		99	48	51	23.32	9.10
Poland	3rd wave 4–6.2021	76	40	36	23.39	3.69
Romania		99	57	42	21.34	3.84
Slovakia		94	40	54	22.60	5.28
Poland	4th wave 10–12.2021	127	60	67	26.16	4.49
Romania		174	103	71	22.21	5.84
Slovakia		101	49	52	23.94	3.91

M, mean age in years; SD, standard deviation.

2.2. Participants

A total of 2020 professional athletes took part in the study, practicing various sports disciplines: handball, soccer, martial arts (kickboxing, judo, fencing, karate, MMA, taekwondo), rugby, basketball, athletics, aerobic and artistic gymnastics, volleyball, tennis, and swimming (a total of sixteen sports disciplines in each country). The inclusion criteria were a career of at least two years of training in a specific sport branch, under the supervision of a coach and a minimum age of 18 years (seniors). Athletes have been practicing the sports disciplines (in the entire sample) for an average of 8.3 years. About 82% of the participants achieved local/regional level performances, approximately 12% registered national performances (being national champions, vice-champions, or being part of the national teams in the branch of sport practiced), while about 6% obtained international results (at World or European level, only martial arts athletes). In each research period/wave of the pandemic and in each country, athletes having local/regional, national, and international performances were investigated. No missing values were identified due to the online survey/submission in which all items had to be rated. In the preliminary analysis of the data, using stem and leaf, eighteen cases (in the total sample) were recognized as outliers and excluded from further investigation. Thus, we retained 2020 athletes (from the total sample of 2038 eligible athletes). It is relevant, also, to highlight that approximately 70% of the athletes tested in each research period/in every wave of the pandemic (in each country) were tested also in the pre-pandemic period. Table 1 shows the athletes' descriptive statistics divided by gender, age, and data collection period.

The data collection carried out in the third wave of the pandemic (from April to the end of June 2021) presented a smaller number of people surveyed. To avoid a sample size reduction, 3rd wave data were not included in subsequent statistical analyses.

2.3. Instruments

Personal data were collected using an ad hoc questionnaire regarding personal and sociodemographic data. It comprised four items measuring the participants' age, gender, years of training, and sport type.

Stress was measured using the *Perception of Stress Questionnaire*, comprising 21 items which form three scales: *Emotional tension* (7 items, e.g., "I get nervous more often than I used to, and for no obvious reason"), *External stress* (7 items, e.g., "I feel drained by constantly having to prove I am right"), and *Intrapsychic stress* (7 items, e.g., "Thinking about my problems makes it hard for me to fall asleep") [30]. The generalized stress level (total score) is the sum of the following scales: *Emotional tension, External stress,* and *Intrapsychic stress.* Participants answer each item on a five-point Likert-type scale from 1 (definitely disagree) to 5 (definitely agree). The Cronbach's α reliability coefficient in the Polish sample was as follows: emotional tension: from 0.75 to 0.81; external stress: from 0.68 to 0.74; intrapsychic stress: from 0.77 to 0.80. The Cronbach's α reliability coefficient in the Romanian sample was as follows: emotional tension: from 0.59 to 0.79; external stress: from 0.65 to 0.82; intrapsychic stress: from 0.72 to 0.85. The Cronbach's α reliability coefficient in the Slovak sample was as follows: emotional tension: from 0.68 to 0.78; external stress: from 0.63 to 0.75; intrapsychic stress: from 0.72 to 0.80. The Perception of Stress Questionnaire has been used in studies of athletes [30] including Romanian and Slovak athletes. The translation of the questionnaire in Romanian and Slovak language was carried out with the consent of the author (Makarowski Ryszard) and respecting the author's recommendations. First, the original Polish version was translated into English and then translated into Polish by translators with psychological experience. The final Romanian and Slovak versions of the English version were created through retroversion, compared and used in the study (this procedure has been used in previous research [31]).

Using the Brief COPE questionnaire, we measured the strategies of coping with stress. It comprises 28 items covering 14 coping strategies: *self-distraction, active coping, denial, substance use, use of emotional support, use of instrumental support, behavioral disengagement, venting, positive reframing, planning, humor, acceptance, religion,* and *self-blame* (two items for each strategy) [32]. The participants indicate their frequency of using each coping strategy on a four-point Likert-type scale, from 1 (I have not been doing this at all) to 4 (I have been doing this a lot). In all data collection periods, the Cronbach's α reliability coefficients for each subscale in the Polish, Romanian, and Slovak versions ranged from 0.48 to 0.94.

The 14 coping strategies of the Brief COPE questionnaire can be grouped in several ways. In the current study, we decided to divide them into three groups: emotion-focused strategies (emotional support, positive reframing, acceptance, religion, humor), problem-focused strategies, (active coping, planning, use of informational support), and dysfunctional strategies (venting, denial, substance use, behavioral disengagement, self-distraction, self-blame), according to the model by Su et al. [33]. This model was also used in other studies on athletes and other samples in many countries [34–38].

2.4. Procedure

Participants were informed about the study aim and procedure. They were also informed about the anonymity of the collected data and the right to withdraw their participation at any time without having to provide a reason. Informed consent was obtained from all participants. Furthermore, this study was conducted in accordance with the recommendations of the Declaration of Helsinki, the Polish Psychological Association's Psychologist's Code of Ethics, the Slovak Psychological Association, and the Romanian Psychological Association and it was approved by the Ethics Committee of the National University of Physical Education and Sports in Bucharest, Romania (ID. 1185).

2.5. Data Analysis

All the standard statistical analyses were conducted using the Statistica v. 13 software. Data were presented by means and standard deviations. Analysis of variance was carried out using Tukey's T test for unequal sample sizes. Independent *t* test was also used. The statistical significance was set at a *p*-value of ≤ 0.05 and effect size (Cohen's d) was interpreted as follows: ≤ 0.2, trivial; >0.2, small; >0.6, moderate; >1.2, large; >2.0, very large; >4.0, nearly perfect [39]. Considering η2 the range intervals were: 0.01, small effect; 0.06, medium; 0.14, large effect [40]. All variables were normally distributed, with skewness coefficients in absolute value being less than 1 [41]. The assumption of the equality of variance was verified by Levene's test ($p > 0.05$, see Table 2).

Table 2. Stress levels in Polish, Romanian, and Slovak athletes before the COVID-19 pandemic and during the first, second, and fourth waves.

Country	Pandemic	n	Emotional Tension		External Stress		Intrapsychic Stress		Total Score	
			M	SD	M	SD	M	SD	M	SD
Poland	0	314	17.72	5.66	18.52	5.40	14.20	5.37	50.46	16.66
	1st wave	134	17.05	5.50	17.08	5.07	12.32	4.72	46.48	16.70
	2nd wave	63	15.73	6.30	17.49	5.99	13.31	5.33	46.54	16.61
	4th wave	127	18.19	6.65	17.60	5.58	15.83	5.67	51.64	16.31
	F		2.87 ($p = 0.035$)		2.61 ($p = 0.051$)		9.98 ($p = 0.001$)		3.97 ($p = 0.010$)	
	Levene's test		2.98 ($p = 0.531$)		1.34 ($p = 0.254$)		1.47 ($p = 0.211$)		73.63 ($p = 0.237$)	
	differences		0:2 *; 1:2 *; 4:(1.2) *		0:1 *		0:1 **; 0:4 *; 4:(1.2) **		0:(1.2) ***; 4:(1.2) ***; 0:4 *	
	Eta^2 (η2)		0.01		0.01		0.04		0.015	
Romania	0	221	18.41	5.89	19.68	5.53	15.50	4.82	53.60	14.69
	1st wave	145	15.92	6.12	16.50	5.42	13.48	4.97	45.91	15.23
	2nd wave	171	15.39	5.61	16.82	5.37	13.10	4.92	45.32	15.25
	4th wave	174	17.17	7.24	16.50	5.53	16.28	5.89	49.95	16.86
	F		8.79 ($p < 0.001$)		16.28 ($p < 0.001$)		14.95 ($p < 0.001$)		11.83 ($p = 0.002$)	
	Levene's test		7.68 ($p = 0.082$)		6.03 ($p = 0.092$)		4.15 ($p = 0.068$)		2.88 ($p = 0.544$)	
	differences		0:(1.2) **; 0:4 *; 4:(1.2) *		0:(1.2.4) ***		0:(1.2) **; 4:(1.2) **		0:(1.2.4) ***; 1:4 **; 2:4 **	
	Eta^2 (η2)		0.04		0.07		0.07		0.066	
Slovakia	0	91	17.97	4.82	20.25	4.46	15.75	4.12	53.98	13.06
	1st wave	111	17.44	5.00	18.90	4.11	15.09	4.15	51.43	12.17
	2nd wave	99	16.42	6.00	18.47	5.65	14.45	4.91	49.35	16.14
	4th wave	101	18.74	6.21	18.68	5.11	18.48	5.12	55.91	14.87
	F		3.11 ($p = 0.026$)		2.55 ($p = 0.054$)		14.99 ($p < 0.001$)		4.69 ($p = 0.003$)	
	Levene's test		6.41 ($p = 0.513$)		4.75 ($p = 0.083$)		2.85 ($p = 0.080$)		5.93 ($p = 0.062$)	
	differences		0:2 *; 2:4 **; 1:4 *		0:(1.2.4) *		0:(2.4) *; 4:(1.2) **		4:(1.2) ***	
	Eta^2 (η2)		0.02		0.02		0.11		0.031	

0—before the pandemic. * $p \leq 0.05$. ** $p \leq 0.01$. *** $p \leq 0.001$. η2 = 0.06 indicates a medium effect.

3. Results

Table 2 shows the analysis of variance results with the pandemic stage, that is, the pre-pandemic period and all four subsequent pandemic waves, as the grouping variable. The analyses were carried out separately for each national subsample (country split). Considering the significant differences observed between the research periods/waves of the pandemic (in each country and for each subscale of the Polish, Romanian, and Slovak versions), d value ranged from 0.22 to 0.80 (the smallest effect sizes were observed in each country for the total score).

The obtained results show that the highest overall stress levels among Polish and Slovak athletes were reported during the fourth wave of the pandemic. Romanian athletes reported the highest overall stress in the pre-pandemic period. In all three countries, the highest intrapsychic stress levels were reported during the fourth wave and the highest external stress levels were reported before the pandemic. Emotional tension was the highest among Polish and Slovak athletes in the fourth wave and, among Romanian athletes, before the pandemic. η2 (the overall effect size) indicates, generally, small or moderate to small differences between the examined waves of the pandemic (considering athletes' perceived stress). Only for intrapsychic stress were moderate to strong differences found (in Romanian and Slovak athletes).

The obtained data show that all stress dimensions significantly decreased during the first and second wave of the pandemic but increased significantly during the fourth wave (except external stress where no significant differences were found compared to the first two waves). The highest increase was observed for *intrapsychic stress.* Eta^2 values (the overall effect size) are: 0.02 (for emotional tension), 0.03 (for external stress), respectively,

0.06 (for intrapsychic stress), emphasizing moderate to small (respectively medium) differences between the research periods/waves of the pandemic, taking into account (in each investigated wave) the total sample of athletes (regardless of country).

Table 3 shows the differences in perceived stress (between countries, in each wave of the pandemic wave split) among the surveyed athletes.

Table 3. Stress levels in Polish, Romanian and Slovak athletes before the COVID-19 pandemic and during the first, second, and fourth waves (analysis of differences between countries).

Pandemic	Country	N	Emotional Tension		External Stress		Intrapsychic Stress		Total Stress	
			M	SD	M	SD	M	SD	M	SD
Before pandemic	Poland	314	17.72	5.66	18.52	5.41	14.20	5.37	50.46	16.66
	Romania	221	18.41	5.89	19.68	5.53	15.50	4.82	53.60	14.69
	Slovakia	91	17.79	4.82	20.25	4.46	15.75	4.12	53.98	13.06
	F		0.88		5.19		6.00		3.97	
	Levene's test		3.60 ($p = 0.058$)		4.78 ($p = 0.088$)		7.70 ($p = 0.051$)		7.57 ($p = 0.101$)	
	differences		NS		1:3 *		1:2 *		1:2 *; 1:3 *	
	Eta^2 (η2)		0.003		0.02		0.02		0.01	
1st wave	Poland	134	17.05	5.50	17.08	5.07	12.32	4.72	46.48	16.70
	Romania	145	15.92	6.12	16.50	5.42	13.48	4.97	45.91	15.23
	Slovakia	111	17.44	5.00	18.90	4.11	15.09	4.15	51.43	12.17
	F		2.61		7.71		10.64		6.99	
	Levene's test		10.39 ($p = 0.052$)		5.98 ($p = 0.070$)		2.62 ($p = 0.59$)		6.93 ($p = 0.065$)	
	differences		1:2 *; 2:3 *		1:3 *; 2:3 ***		1:2 *; 2:3 *		1:3 ***; 2:3 ***	
	Eta^2 (η2)		0.01		0.04		0.09		0.03	
2nd wave	Poland	63	15.73	6.30	17.49	5.99	13.31	5.33	46.54	16.61
	Romania	171	15.39	5.61	16.82	5.37	13.10	4.92	45.32	15.25
	Slovakia	99	16.42	6.00	18.47	5.65	14.45	4.91	49.35	16.14
	F		0.988		2.73		2.40		10.38	
	Levene's test		0.98 ($p = 0.37$)		2.76 ($p = 0.65$)		2.36 ($p = 0.095$)		2.25 ($p = 0.101$)	
	differences		2:3 *		2:3 *		1:3 *; 2:3 *		1:3 *; 2:3 **	
	Eta^2 (η2)		0.005		0.01		0.01		0.01	
4th wave	Poland	127	18.19	6.65	17.60	5.58	15.83	5.67	51.64	16.31
	Romania	174	17.17	7.24	16.50	5.53	16.28	5.89	49.95	16.86
	Slovakia	101	18.74	6.21	18.68	5.11	18.48	5.12	55.91	14.87
	F		1.90		5.27		7.01		4.36	
	Levene's test		2.54 ($p = 0.079$)		1.22 ($p = 0.295$)		1.85 ($p = 0.158$)		2.27 ($p = 0.103$)	
	differences		2:3 *		2:3 *		1:3 *; 2:3 *		1:2 *; 1:3 **; 2:3 ***	
	Eta^2 (η2)		0.01		0.02		0.03		0.02	

NS, not significant. 1-Poland, 2-Romania, 3-Slovakia. * $p \leq 0.05$. ** $p \leq 0.01$. *** $p \leq 0.001$. η2 = 0.06 indicates a medium effect.

The highest level of stress in individual waves of the pandemic occurred in Slovakia. The lowest level of general stress was recorded in athletes from Romania (except for the tests performed before the pandemic). The overall effect size (η2) shows, generally, small or moderate to small differences between the three countries when talking about athletes' perceived stress, in each examined wave of the pandemic.

To examine whether and how the frequency of using coping strategies by athletes changed between the first and fourth waves of the pandemic, independent t-test was carried out. The results are shown in Table 4.

Table 4. Strategies of coping with stress among Polish, Romanian, and Slovak athletes during the first and fourth waves of the pandemic.

Country	Coping Strategies	1 Wave		4 Wave		t	p	d
		M	SD	M	SD			
Poland	Active coping	4.48	1.32	4.27	1.16	1.391	0.165	0.17
	Planning	4.61	1.35	4.29	1.26	1.992	0.047	0.24
	Positive reframing	3.28	1.44	2.67	1.44	3.447	0.001	0.42
	Acceptance	4.14	1.37	4.11	1.24	0.177	0.860	0.02
	Humor	2.82	1.55	2.45	1.35	2.042	0.042	0.25
	Religion	2.38	2.12	1.68	1.87	2.835	0.005	0.35
	Emotional support	3.33	1.83	3.54	1.67	−0.988	0.324	0.12
	Use of informational support	3.20	1.81	3.17	1.43	0.154	0.878	0.02
	Self-distraction	3.16	1.45	3.29	1.46	−0.726	0.468	0.09
	Denial	1.60	1.46	1.70	1.37	−0.540	0.590	0.07
	Venting	2.51	1.39	2.84	1.23	−2.029	0.043	0.25
	Substance use	0.98	1.57	0.77	1.23	1.225	0.222	0.15
	Behavioral disengagement	1.00	1.13	1.43	1.18	−3.027	0.003	0.37
	Self-blame	2.45	1.66	2.65	1.64	−0.983	0.326	0.12
	Emotion-Focused strategies	15.94	5.30	14.45	4.82	2.390	0.018	0.29
	Problem-Focused strategies	12.29	3.68	11.73	2.62	1.428	0.154	0.18
	Dysfunctional strategies	11.70	4.94	12.66	4.78	−1.622	0.106	0.20
Romania	Active coping	4.85	1.19	4.55	1.39	2.022	0.044	0.23
	Planning	4.53	1.16	4.41	1.41	0.825	0.410	0.09
	Positive reframing	3.31	1.52	3.14	1.55	0.980	0.328	0.11
	Acceptance	4.07	1.35	4.05	1.51	0.141	0.888	0.02
	Humor	3.23	2.00	3.06	2.07	0.733	0.464	0.08
	Religion	2.63	1.98	2.39	1.87	1.113	0.267	0.12
	Emotional support	3.90	1.63	3.84	1.75	0.315	0.753	0.04
	Use of informational support	3.63	1.63	3.83	1.73	−1.093	0.275	0.12
	Self-distraction	3.24	1.48	3.47	1.70	−1.313	0.190	0.15
	Denial	1.41	1.65	1.75	1.72	−1.813	0.071	0.20
	Venting	2.96	1.81	3.29	1.85	−1.608	0.109	0.18
	Substance use	0.55	1.27	0.69	1.42	−0.902	0.368	0.10
	Behavioral disengagement	0.94	1.34	1.27	1.61	−1.991	0.047	0.22
	Self-blame	2.65	1.82	2.91	1.81	−1.280	0.202	0.14
	Emotion-Focused strategies	17.15	5.13	16.49	5.55	1.094	0.275	0.12
	Problem-Focused strategies	13.01	2.83	12.80	3.53	0.578	0.564	0.07
	Dysfunctional strategies	11.75	5.89	13.39	6.63	−2.318	0.021	0.26
Slovakia	Active coping	4.03	1.37	4.51	1.32	−2.388	0.018	0.36
	Planning	3.63	1.43	4.21	1.47	−2.666	0.008	0.40
	Positive reframing	3.04	1.41	2.87	1.40	0.852	0.395	0.13
	Acceptance	3.55	1.29	4.16	1.48	−2.849	0.005	0.44
	Humor	1.93	1.77	2.10	1.93	−0.611	0.542	0.09
	Religion	1.91	1.96	1.57	1.90	1.194	0.234	0.18
	Emotional support	3.82	1.46	3.81	1.56	0.033	0.974	0.00
	Use of informational support	3.24	1.44	3.70	1.66	−1.946	0.053	0.30
	Self-distraction	3.06	1.37	3.47	1.59	−1.802	0.073	0.28
	Denial	1.12	1.16	1.25	1.45	−0.658	0.511	0.10
	Venting	2.69	1.20	3.08	1.38	−2.003	0.047	0.31
	Substance use	0.70	1.28	0.69	1.33	0.076	0.940	0.01
	Behavioral disengagement	3.55	1.26	3.88	1.53	−1.520	0.130	0.23
	Self-blame	2.06	1.62	2.18	1.65	−0.486	0.628	0.07
	Emotion-Focused strategies	14.25	4.62	14.50	4.55	−0.360	0.719	0.05
	Problem-Focused strategies	10.90	2.90	12.42	3.07	−3.377	0.001	0.51
	Dysfunctional strategies	13.18	3.70	14.55	4.43	−2.184	0.030	0.34

M, mean. SD, standard deviation. d, Cohen's effect size.

Analyzing the dynamics of coping strategy use between the first and fourth waves of the COVID-19 pandemic, it can be observed that *emotion-focused strategies* became less frequent among Polish athletes. Regarding individual coping strategies, *behavioral disengagement* and *venting* became more frequent, while *planning, positive reframing,* and *humor* became less frequent. No significant changes in the frequency of the other individual coping strategies were observed.

1. Among Romanian athletes, *dysfunctional strategies* became more frequent. Regarding individual coping strategies, a decrease in the frequency of using *active coping* and an increase in using *behavioral disengagement* were observed. Neither of the other individual coping strategies was used significantly more frequently in the fourth wave.
2. In the Slovak athlete subsample, the frequency of using *problem-focused strategies* and *dysfunctional strategies* increased. Regarding individual coping strategies, *active coping, planning, acceptance,* and *venting* became more frequent. Neither of the other individual coping strategies was used significantly less or more frequently in the fourth wave.
3. It is worth noting that *dysfunctional strategies* became noticeably more frequent in each national subsample during the fourth wave (this difference was not statistically significant in the Polish athlete subsample).

4. Discussion

The emergence of the COVID-19 pandemic changed the structure and functioning of the world as we know it, permanently and suddenly. Such a significant threat has not been experienced by many European countries for a long time. The coronavirus disease has impacted nearly all aspects of human functioning. Numerous strains have increased the intensity of experienced stress and initiated the activization of coping strategies [42–44]. Social contacts, working, transport, spending free time, and engaging in physical activity have changed noticeably. For professional athletes, limiting the opportunities for training and canceling or delaying sports events represented a significant challenge [45]. These situations occurred due to the lockdown periods (as preventive measures for reducing COVID-19 spread), because of the infections (or the fear of infection) of athletes, coaches, sports managers, and organizers of sports competitions. In an attempt to identify most of the infected athletes worldwide before August 2020 (according to gender, age, symptoms, sport level, or location of the contraction of infection), researchers found 521 COVID-19-positive athletes [46]. It seems that most infected athletes practiced soccer and basketball (as authors asserted, the cases do not represent all of the infected athletes). Considering that globally, as of 5:44 p.m. CEST, 24 August 2022, there have been 595,219,966 confirmed cases of COVID-19 reported to the World Health Organization (WHO) [47], but it is very difficult to identify the number of COVID-19 infections among athletes (more so as there are professional athletes, amateur, college, junior, or senior athletes and some of them were asymptomatic).

Significantly restricted training opportunities, canceled or delayed events, and reduced income have compounded the universal concerns about one's own health and the health of one's family. Thus, the aim of the current study was to identify the stress level dynamics during the first four waves of the COVID-19 pandemic, in Polish, Romanian, and Slovak athletes, and to establish the changes in coping strategies between the first and fourth waves. The dependent variable was the level of stress under study, namely emotional tension, external stress, and intrapsychic stress. The independent variables (the variable playing the role of IVs) were three countries: Poland, Slovakia, and Romania; the research period: before the pandemic and the first, second, and fourth wave of the pandemic. An additional dependent variable was the way of coping with stress. The results revealed that in all three countries, the highest intrapsychic stress levels were reported during the fourth wave and the highest external stress levels were reported before the pandemic. Further, the coping strategies used by the athletes in the fourth wave were more dysfunctional than during the first wave.

Perceived stress levels among athletes differed depending on the country. The highest level of stress in individual waves of the pandemic was reported by Slovak athletes, while the lowest level of general stress was registered in athletes from Romania (except for the tests performed before the pandemic). Small or moderate to small differences were observed between the three investigated countries, when talking about athletes' experienced stress (in individual waves). The significant differences found (between countries) could be related to numerous variables: the intensity of the pandemic, the current economic and political situation in the country, and employment stability. The *Human Development Report*, which indicates the quality of life in a given country [48], is also important in this context.

In all countries, there was a noticeable trend in the overall pre-pandemic stress levels decreasing and remaining at a lower level throughout the first and second wave of the pandemic, before increasing during the fourth wave. Significant differences were observed in each investigated country (studied separately), between the waves of the pandemic, as well as small or moderate to small effect sizes (for emotional tension and external stress). In the case of intrapsychic stress, a moderate to strong effect size (Eta^2) was found in Romanian and Slovak athletes. Furthermore, when the three countries were studied together (the total sample), moderate to small (respectively, medium for intrapsychic stress) differences between the research periods/waves of the pandemic were highlighted. Similar results were observed for martial arts practitioners from Poland and Romania and also in athletes practicing various sports disciplines (non-martial-arts athletes) from Poland, Slovakia, and Romania, with stress levels decreasing during the height of the pandemic (during the lockdown and first wave), compared to the pre-pandemic period [30]. However, there are also studies that underline that about one month after the beginning of the lockdown (first wave), perceived stress increased in Italian athletes from individual and team sports [13] (martial arts athletes were not included in the sample).

Such differences (a significantly lower level of stress during the first three waves of the pandemic) can be explained when considering the psychological and social mechanisms behind the observed trend—it can be assumed that several phenomena co-occurred. First, habituation led to a lower intensity of reaction to a repeated stimulus. Due to a long-term presence of a constant stimulus, the stress reaction becomes reduced and, in time, it may become extinct [49–51]. Habituation has an adaptive function, as it allows for economical use of the individual's emotional and cognitive resources. Stress could have also decreased due to the cancellations or delays of upcoming sports events [52]. For many athletes, participation in competition involves intense psychological stress. In various sports disciplines (e.g., volleyball, tennis, track and field, cycling, boxing, soccer), higher perceived stress levels were observed upon resumption of competitions [14]. Further, for example, in swimmers, a significant release of stress hormones was observed, as a result of physical and mental stress associated with sports competition [53]. Cancelling or delaying such events may reduce psychological tension. Another reason for lowered stress levels could be the reduction in intensity of training and work, with the resulting rest and isolation allowing for a regeneration of psychological resources [54,55]. Due to training and competitions being limited, the perceived level of rivalry also decreased, as all athletes found themselves in similar circumstances [56,57]. It is worth underlining that the restrictions and possibilities in each time span (and in each investigated country) were relatively the same—during the lockdown period, coaches and athletes worked exclusively on different remote learning platforms (online) at home. When the conditions relaxed, athletes were able to practice outdoors, on sports fields or in parks, respecting the measures of social distancing. There are differences between sports branches. Some players, such as runners, were able to follow more easily the training plan during the COVID-19 pandemic (there was, almost, no break in the training). Sports competitions were also organized in Poland, Romania, and Slovakia (and televised) but without spectators (for months). Only coaches and athletes had access to the competition hall/area and they were previously tested against COVID-19. It is also important to mention that the vaccination campaign began (in the three countries) at the end of 2020—in Romania, on 27 December 2020. For example, during the third wave of the pandemic (on 7 April 2021), according to the National Institute of Public Health [58], only 1.288.487 Romanians (about 6.5% from the population) were vaccinated with both doses.

Sport is a stress-generating environment, as unpleasant remarks coming from supporters and noise from the stands (in many sports) increase athletes' stress levels [59,60]. The absence of spectators (during the first waves of the pandemic) can, therefore, reduce experienced stress in athletes. Here, also, we emphasize differences between sports disciplines, with team sports athletes feeling less negative stress [61] and reporting less anxiety and depression than in individual sports [62], while novice performers registered higher perceived stress than top athletes (potentially reflecting their less-adapted coping resources) [13]. Further, athletes with high athletic identity are less prone to higher levels of psychological distress compared to athletes with low athletic identity [61].

The pandemic has brought into focus the fundamental human issues of health and survival. Among stressors for athletes were the fear of COVID-19 infection (the fear of health deterioration), weight change, exercising at home, monthly income perception, and damaged performance in COVID-19 infection [63] (it is important to mention that athletes' trait anxiety values were below average). The risk of infection when rigorously following the hygiene and social isolation protocols is minimal and other life events are unable to cause intense stress reactions in the context of a global pandemic. When fundamental issues, such as health and survival, are threatened, people appraise everyday problems differently [64,65]. Moreover, the reduction in sports events resulted in a significant reduction in athletes' media exposure, which could have been a major source of pressure [66,67]. The increase in stress during the fourth wave of the pandemic could have been caused by the depletion of personal resources and poorer adaptation to the permanent conditions of the pandemic [68,69], as well as chronic fatigue syndrome [70]. All limitations and restrictions accumulated over time may cause higher levels of stress and greater health concerns. Further, athletes' worsening economic conditions and uncertain prospects for the future may be significant [71].

Emotional and external stress among athletes was lower during the first, second, and fourth wave than during the pre-pandemic period. However, a very high increase in intrapsychic stress was observed during the fourth wave. Increased intrapsychic stress is a consequence of prolonged negative events, with which the individual has not coped effectively (i.e., has used dysfunctional coping strategies) [30]. Past, present, as well as future anticipated events may be sources of stress [72]. In the context of a prolonged pandemic, this results in an accumulation of perceived stress.

The smaller number of people surveyed in the third wave of the pandemic resulted in the exclusion of this group of people due to the possibility of distorting the results of the research.

The long duration of the pandemic has also impacted strategies of coping with stress. On the one hand, athletes have learned which coping strategies are effective. On the other hand, permanent functioning during the pandemic may have modified the employed strategies. To identify changes in coping strategies among athletes, we compared them between the first and the fourth waves of the pandemic. In the Polish athlete subsample, we observed a decrease in the frequency of using *emotion-focused strategies*. It is assumed that emotion-focused strategies are ineffective in the long term, though they may negate the consequences of stress in some situations [73,74]. In the Romanian athlete subsample, the use of *dysfunctional strategies* increased in frequency. In the long term, using these coping strategies can be associated with a risk of depression, anxiety, and eating disorders [75,76]. Among Slovak athletes, the frequency of using *problem-focused strategies* and *dysfunctional strategies* increased simultaneously. Using *problem focused strategies* allows for coping with difficult situations in the most effective way [77].

Our study revealed a significant trend in coping strategy use among athletes. Comparing the frequency of using coping strategies, we observed an increase in using *dysfunctional strategies* in each country. Aggregated results for individual coping strategies show that, in each country, the frequency of using *dysfunctional strategies* during the fourth wave of the pandemic was higher than in the first wave. Long-term functioning in stressful situations may reduce personal resources. Thus, seeking easier solutions for a difficult situation, athletes more frequently used *dysfunctional strategies*. Long-term use of such strategies carried with it a risk of depression and worsened health [78]. However, these strategies can be modified with appropriate psychological intervention [79]. Considering elite athletes, as well as physical education students practicing sports most often, researchers highlight the important role of cognitive and behavioral strategies in coping with the stress generated by the COVID-19 pandemic [80]. It was found that "the sports level depended on the strategies of coping with the stress of the COVID-19 pandemic more strongly than gender".

Regarding individual coping strategies, in the Polish athlete subsample, the frequency of using *behavioral disengagement* and *venting* increased, while the frequency of using *planning, positive reframing*, and *humor* decreased. The first two of these are *dysfunctional strategies*. Increasing the frequency of using these strategies decreases the probability of effectively coping with stress. Even if in the short term, they may prove useful in reducing perceived stress, we cannot promote them, given their known long-term effects [30]. *Positive reframing* and *humor* are *emotion-focused strategies*. They are not effective in the long term for the subsequent pandemic waves. Decreasing the frequency of using these strategies decreases the probability of effectively coping with stress and minimizing its effect on wellbeing. The long duration of the pandemic was related to a decrease in the frequency of using *planning*, a *problem-focused strategy*. The unpredictability of the pandemic, together with a lack of control over many aspects of life, may have caused a decrease in using this strategy in the perspective of the pandemic's increasing duration.

In the Romanian athlete subsample, we noticed an increase in the frequency of using the individual strategy of *behavioral disengagement*, which is a *dysfunctional strategy*. It can be assumed that, similar to Polish athletes, the lack of control over the situation could have caused an increase in the frequency of using this strategy. However, a different pattern was observed among the Slovak athletes. They reported an increase in using *active coping, planning (problem-focused strategies)*, and acceptance (emotion-focused strategy), as well as a decrease in venting (dysfunctional strategy).

Permanent use of *dysfunctional strategies* is ineffective and is related to a risk for depression [81]. The noticeable increase in the frequency of using these coping strategies by athletes during the fourth wave of the pandemic, together with the increase in intrapsychic stress, should alert coaches and sport psychologists to the ways in which professional athletes modify their use of available coping strategies. Close cooperation with a sports psychologist and coach is essential in order to promote the most effective coping strategies for a given person [82]. Along with medical practitioners, members of the multidisciplinary team should work towards minimizing the strain experienced by athletes [83]. In order to reduce athletes' distress, specialists could use the so-called *internal* techniques (breathing and meditation, self-control techniques) [84], inner monologue (positive self-talk) increasing self-confidence, analytical relaxation, and autogenic training; self-monitoring of emotional reactions [85] could teach athletes positive conflict resolution strategies and guide them to get involved in motor and mental activities, which gives them great satisfaction [86]. Not least, specialists can use written emotional disclosure (WED) to support athletes during the COVID-19 pandemic and to promote their mental health [87] and the 4Ds for dealing with distress, an ultra-brief single session, which unifies strategies and exercises for problem-solving, emotion regulation, and for increasing resilience (restoring wellbeing) [88].

The present study has some limitations. The authors relied on self-report measures, supposing a possible recall bias and/or the issue of possible desirable answers (when talking about explicit evaluations), aspects being known [89] (however, the large number of athletes tested represents a strength of the study). Further, the results may be different if junior athletes were investigated, athletes from other countries, practicing a single sport discipline, if athletes were examined separately, according to the level of training, as well as according to their property status (these are relevant questions for future research). Moreover, the Cronbach's α reliability coefficients presented a range of low to very high level of reliability for the strategies of coping with stress, which should also be considered when interpreting the results of this study. Finally, even if there is a reciprocal relationship between stress and anxiety (the two dimensions having a mutual influence on each other), other investigation tools are recommended for anxiety, capturing the link between the anxiety of athletes (state anxiety and/or trait anxiety) and the size of a pandemic in a given country. This can be the subject of further research.

5. Conclusions

The conclusions of the current study, carried out in three countries, showed that the direct consequences of the pandemic are not related to an increase in perceived stress among athletes. Overall, stress levels during the fourth wave of the pandemic were not higher, in all countries, than during the pre-pandemic period. However, an increase in intrapsychic stress was noticeable between the first two waves and the fourth wave of the pandemic. The research is underlining the importance of athletes' experienced stress (which can influence, also, their anxiety level), capturing the dynamics of perceived emotional tension, external stress, and intrapsychic stress in athletes, before and throughout different waves of the COVID-19 pandemic.

Using constructive coping strategies allows for reducing the perceived stress. Using these strategies led to lower stress levels. Coaches and sport psychologists should continuously monitor stress levels among athletes, together with their coping efforts, in order to promote effective coping strategies. As the pandemic may have long-term consequences, it is particularly important to monitor athletes' psychological wellbeing also after its end, in a post-COVID-19 world.

Author Contributions: Conceptualization, R.M. and R.P.; methodology, R.M., A.P. (Andrzej Piotrowski) and K.G.; software, R.P.; validation, A.P. (Alexandra Predoiu), R.O., R.A.P. and A.D.M.; formal analysis, R.M. and R.P.; investigation, K.G. and A.P. (Alexandra Predoiu); resources, R.O., R.A.P. and A.D.M.; data curation, S.R. and G.A.; writing—original draft preparation, A.P. (Andrzej Piotrowski), O.B., S.R. and G.A.; writing—review and editing, A.P. (Andrzej Piotrowski), R.O., O.B., S.R. and G.A.; visualization, K.G. and A.P. (Alexandra Predoiu); supervision, A.P. (Andrzej Piotrowski), R.O. and O.B.; project administration, R.M.; funding acquisition, R.A.P. and A.D.M. All authors have read and agreed to the published version of the manuscript and equally contributed to the article.

Funding: This research received no external funding.

Institutional Review Board Statement: The study was conducted in accordance with the Declaration of Helsinki and approved by the Ethics Committee of the National University of Physical Education and Sports in Bucharest, Romania (ID: 1185).

Informed Consent Statement: Informed consent was obtained from all subjects involved in the study.

Data Availability Statement: Data are available upon request to the contact author

Conflicts of Interest: The authors declare no conflict of interest.

References

1. World Health Organization (WHO). Key Messages and Actions for COVID-19 Prevention and Control in Schools. 2020. Available online: https://www.who.int/docs/default-source/coronaviruse/key-messages-and-actions-for-covid-19-prevention-and-control-in-schools-march 2020.pdf?ofvran=baf81d52_4 (accessed on 14 May 2022).
2. Falvo, I.; Zufferey, M.C.; Albanese, E.; Fadda, M. Lived experiences of older adults during the first COVID-19 lockdown: A qualitative study. *PLoS ONE* **2021**, *16*, e0252101. [CrossRef] [PubMed]
3. Park, S.; Chang, Y.; Wolfe, R.R.; Kim, I.-Y. Prevention of Loss of Muscle Mass and Function in Older Adults during COVID-19 Lockdown: Potential Role of Dietary Essential Amino Acids. *Int. J. Environ. Res. Public Health* **2022**, *19*, 8090. [CrossRef] [PubMed]
4. Devoe, D.; Han, A.; Anderson, A.; Katzman, D.K.; Patten, S.B.; Soumbasis, A.; Flanagan, J.; Paslakis, G.; Vyver, E.; Marcoux, G.; et al. The impact of the COVID-19 pandemic on eating disorders: A systematic review. *Int. J. Eat. Disord* **2022**, *1*, 1–21. [CrossRef] [PubMed]
5. Albert, U.; Losurdo, P.; Leschiutta, A.; Macchi, S.; Samardzic, N.; Casaganda, B.; de Manzini, N.; Palmisano, S. Effect of SARS-CoV-2 (COVID-19) Pandemic and Lockdown on Body Weight, Maladaptive Eating Habits, Anxiety, and Depression in a Bariatric Surgery Waiting List Cohort. *Obes. Surg.* **2021**, *31*, 1905–1911. [CrossRef] [PubMed]

6. Amerio, A.; Brambilla, A.; Morganti, A.; Aguglia, A.; Bianchi, D.; Santi, F.; Costantini, L.; Odone, A.; Costanza, A.; Signorelli, C.; et al. COVID-19 Lockdown: Housing Built Environment's Effects on Mental Health. *Int. J. Environ. Res. Public Health* **2020**, *17*, 5973. [CrossRef]

7. Hansmann, R.; Fritz, L.; Pagani, A.; Clément, G.; Binder, C.R. Activities, Housing Situation and Other Factors Influencing Psychological Strain Experienced During the First COVID-19 Lockdown in Switzerland. *Front. Psychol.* **2021**, *12*, 735293. [CrossRef]

8. Sasser, P.; McGuine, T.; Haraldsdottir, K.; Biese, K.; Goodavish, L.; Stevens, B.; Watson, A.M. Reported COVID-19 Incidence in Wisconsin High School Athletes in Fall 2020. *J. Athl. Train.* **2021**, *57*, 59–64. [CrossRef]

9. Washif, J.A.; Farooq, A.; Krug, I.; Pyne, D.B.; Verhagen, E.; Taylor, L.; Wong, D.P.; Mujika, I.; Cortis, C.; Haddad, M.; et al. Training During the COVID-19 Lockdown: Knowledge, Beliefs, and Practices of 12,526 Athletes from 142 Countries and Six Continents. *Sports Med.* **2022**, *52*, 933–948. [CrossRef]

10. Håkansson, A.; Moesch, K.; Jönsson, C.; Kenttä, G. Potentially Prolonged Psychological Distress from Postponed Olympic and Paralympic Games during COVID-19—Career Uncertainty in Elite Athletes. *Int. J. Environ. Res. Public Health* **2021**, *18*, 2. [CrossRef]

11. Giusto, E.; Asplund, C.A. Persistent COVID and a Return to Sport. *Curr. Sports Med. Rep.* **2022**, *21*, 100–104. [CrossRef]

12. Robazza, C.; Bertollo, M.; Ruiz, M.C.; Bortoli, L. Measuring psychobiosocial states in sport: Initial validation of a trait measure. *PLoS ONE* **2016**, *11*, e0167448. [CrossRef] [PubMed]

13. di Fronso, S.; Costa, S.; Montesano, C.; Di Gruttola, F.; Ciofi, E.G.; Morgilli, L.; Robazza, C.; Bertollo, M. The effects of COVID-19 pandemic on perceived stress and psychobiosocial states in Italian athletes. *Int. J. Sport Exerc. Psychol.* **2020**, *20*, 79–91. [CrossRef]

14. di Fronso, S.; Montesano, C.; Costa, S.; Santi, G.; Robazza, C.; Bertollo, M. Rebooting in sport training and competitions: Athletes' perceived stress levels and the role of interoceptive awareness. *J. Sports Sci.* **2022**, *40*, 542–549. [CrossRef] [PubMed]

15. Levine, O.; Terry, M.; Tjong, V. The Collegiate Athlete Perspective on Return to Sport Amidst the COVID-19 Pandemic: A Qualitative Assessment of Confidence, Stress, and Coping Strategies. *Int. J. Environ. Res. Public Health* **2022**, *19*, 6885. [CrossRef]

16. Raimondi, S.; Cammarata, G.; Testa, G.; Bellerba, F.; Galli, F.; Gnagnarella, P.; Iannuzzo, M.L.; Ricci, D.; Sartorio, A.; Sasso, C.; et al. The Impact of Sport Activity Shut down during the COVID-19 Pandemic on Children, Adolescents, and Young Adults: Was It Worthwhile? *Int. J. Environ. Res. Public Health* **2022**, *19*, 7908. [CrossRef]

17. Haan, R.; Ali Alblooshi, M.E.; Syed, D.H.; Dougman, K.K.; Al Tunaiji, H.; Campos, L.A.; Baltatu, O.C. Health and Well-Being of Athletes during the Coronavirus Pandemic: A Scoping Review. *Front. Public Health* **2021**, *9*, 641392. [CrossRef]

18. Szczypińska, M.; Samełko, A.; Guszkowska, M. What Predicts the Mood of Athletes Involved in Preparations for Tokyo 2020/2021 Olympic Games During the Covid-19 Pandemic? The Role of Sense of Coherence, Hope for Success and Coping Strategies. *J. Sports Sci. Med.* **2021**, *20*, 421–430. [CrossRef]

19. Anshel, M.H.; Sutarso, T. Relationships between sources of acute stress and athletes' coping style in competitive sport as a function of gender. *Psych. Sport Exerc.* **2007**, *8*, 1–24. [CrossRef]

20. Nicholls, A.R.; Polman, R. Coping in sport: A systematic review. *J. Sports Sci.* **2007**, *25*, 11–31. [CrossRef]

21. Anshel, M.A.; Sutarso, T.; Sozen, D. Relationship between cognitive appraisal and coping style following acute stress among male and female Turkish athletes. *Int. J. Sport Exerc. Psychol.* **2012**, *10*, 290–304. [CrossRef]

22. Anshel, M.H. Qualitative validation of a model for coping with acute stress in sport. *J. Sport Behav.* **2001**, *24*, 223–246.

23. Lazarus, R.S.; Folkman, S. *Stress, Appraisal, and Coping*; Springer: New York, NY, USA, 1984.

24. Hoar, S.D.; Kowalski, K.C.; Gaudreau, P.; Crocker, P.R.E. A review of coping in sport. In *Literature Reviews in Sport Psychology*; Hanton, S., Mellalieu, S.D., Eds.; Nova Science: New York, NY, USA, 2006; pp. 47–90.

25. Baker, J.P.; Berenbaum, H. Emotional approach and problem focused coping: A comparison of potentially adaptive strategies. *Cogn. Emot.* **2007**, *21*, 95–118. [CrossRef]

26. Ruiz, M.C.; Devonport, T.J.; Chen-Wilson, C.H.; Nicholls, W.; Cagas, J.Y.; Fernandez-Montalvo, J.; Choi, Y.; Robazza, C. A Cross-Cultural Exploratory Study of Health Behaviors and Wellbeing During COVID-19. *Front. Psychol.* **2021**, *11*, 608216. [CrossRef]

27. Bertollo, M.; Forzini, F.; Biondi, S.; Di Liborio, M.; Vaccaro, M.G.; Georgiadis, E.; Conti, S. How Does a Sport Psychological Intervention Help Professional Cyclists to Cope With Their Mental Health During the COVID-19 Lockdown? *Front. Psychol.* **2021**, *12*, 607152. [CrossRef] [PubMed]

28. Dagnall, N.; Drinkwater, K.G.; Denovan, A.; Walsh, R.S. The Potential Benefits of Non-skills Training (Mental Toughness) for Elite Athletes: Coping with the Negative Psychological Effects of the COVID-19 Pandemic. *Front. Psychol.* **2021**, *3*, 581431. [CrossRef]

29. Clemente-Suárez, V.J.; Fuentes-García, J.P.; de la Vega Marcos, R.; Martínez Patiño, M.J. Modulators of the Personal and Professional Threat Perception of Olympic Athletes in the Actual COVID-19 Crisis. *Front. Psychol.* **2020**, *11*, 1985. [CrossRef]

30. Makarowski, R.; Piotrowski, A.; Predoiu, R.; Görner, K.; Predoiu, A.; Mitrache, G.; Malinauskas, R.; Bochaver, K.; Dovzhik, L.; Cherepov, E.; et al. Stress and coping during the COVID-19 pandemic among martial arts athletes—A crosscultural study. *Arch. Budo* **2020**, *16*, 161–171.

31. Makarowski, R.; Piotrowski, A.; Predoiu, R.; Görner, K.; Predoiu, A.; Mitrache, G.; Malinauskas, R.; Vicente-Salar, N.; Vazne, Z.; Cherepov, E.; et al. The English-speaking, Hungarian, Latvian, Lithuanian, Romanian, Russian, Slovak, and Spanish adaptations of Makarowski's Stimulating and Instrumental Risk Questionnaire for martial arts athletes. *Arch. Budo* **2021**, *17*, 1–33.

32. Carver, C.S. You want to measure coping but your protocol' too long: Consider the Brief Cope. *Int. J. Behav. Med.* **1997**, *4*, 92–100. [CrossRef]
33. Su, X.Y.; Lau, J.T.; Mak, W.W.; Choi, K.C.; Feng, T.J.; Chen, X.; Liu, C.L.; Liu, J.; Liu, D.; Chen, L.; et al. A preliminary validation of the Brief COPE instrument for assessing coping strategies among people living with HIV in China. *Infect. Dis. Poverty* **2015**, *4*, 41. [CrossRef]
34. Almeida, D.; Monteiro, D.; Rodrigues, F. Satisfaction with Life: Mediating Role in the Relationship between Depressive Symptoms and Coping Mechanisms. *Healthcare* **2021**, *9*, 787. [CrossRef] [PubMed]
35. Ștefenel, D.; Gonzalez, J.-M.; Rogobete, S.; Sassu, R. Coping Strategies and Life Satisfaction among Romanian Emerging Adults during the COVID-19 Pandemic. *Sustainability* **2022**, *14*, 2783. [CrossRef]
36. Piotrowski, A.; Sygit-Kowalkowska, E.; Hamzah, I. Work Engagement among Prison Officers. The Role of Individual and Organizational Factors in the Polish and Indonesian Penitentiary Systems. *Int. J. Environ. Res. Public Health* **2020**, *17*, 8206. [CrossRef] [PubMed]
37. Bragazzi, N.L.; Re, T.S.; Zerbetto, R. The relationship between nomophobia and maladaptive coping styles in a sample of Italian young adults: Insights and implications from a cross-sectional study. *JMIR Ment. Health* **2019**, *6*, e13154. [CrossRef]
38. Blakelock, D.; Chen, M.; Prescott, T. Coping and psychological distress in elite adolescent soccer players following professional academy deselection. *J. Sport Behav.* **2019**, *41*, 1–26.
39. Hopkins, W.; Marshall, S.; Batterham, A.; Hanin, J. Progressive statistics for studies in sports medicine and exercise science. *Med. Sci. Sports Exerc.* **2009**, *41*, 3–13. [CrossRef]
40. Larson-Hall, J. *A Guide to Doing Statistics in Second Language Research Using SPSS and R*, 2nd ed.; Routledge: New York, NY, USA, 2016.
41. Morgan, G.A.; Leech, N.L.; Gloeckner, G.W.; Barrett, K.C. *SPPS for Introductory Statistics: Use and Interpretation*; Lawrence Erlbaum Associates, Inc.: Mahwah, NJ, USA, 2004.
42. Kar, N.; Kar, B.; Kar, S. Stress and coping during COVID-19 pandemic: Result of an online survey. *Psychiatry Res.* **2021**, *295*, 113598. [CrossRef]
43. Polizzi, C.; Lynn, S.J.; Perry, A. Stress and coping in the time of COVID-19: Pathways to resilience and recovery. *Clin. Neuropsychiatry* **2020**, *17*, 59–62. [CrossRef]
44. Leguizamo, F.; Olmedilla, A.; Núñez, A.; Verdaguer, F.J.P.; Gómez-Espejo, V.; Ruiz-Barquín, R.; Garcia-Mas, A. Personality, coping strategies, and mental health in high-performance athletes during confinement derived from the COVID-19 pandemic. *Front. Public Health* **2021**, *8*, 561198. [CrossRef]
45. Lopes, L.R.; Miranda, V.A.R.; Goes, R.A.; Souza, G.G.A.; Souza, G.R.; Rocha, J.C.S.; Cossich, V.R.A.; Perini, J.A. Repercussions of the COVID-19 pandemic on athletes: A cross-sectional study. *Biol. Sport* **2021**, *38*, 703–711. [CrossRef]
46. You, M.; Liu, H.; Wu, Z. The spread of COVID-19 in athletes. *Sci. Sports* **2022**, *37*, 123–130. [CrossRef] [PubMed]
47. World Health Organization (WHO). WHO Coronavirus (COVID-19) Dashboard—Overview. 2020. Available online: https://covid19.who.int/ (accessed on 25 August 2022).
48. United Nations Development Programme. HUMAN DEVELOPMENT REPORT 2020: The Next Frontier-human Development and the Anthropocene. 2021. Available online: https://www.undp.org/serbia/publications/next-frontier-human-development-and-anthropocene (accessed on 18 July 2022).
49. Strahler, K.; Ehrlenspiel, F.; Heene, M.; Brand, R. Competitive anxiety and cortisol awakening response in the week leading up to a competition. *Psychol. Sport Exerc.* **2010**, *11*, 148–154. [CrossRef]
50. Hare, O.A.; Wetherell, M.A.; Smith, M.A. State anxiety and cortisol reactivity to skydiving in novice versus experienced skydivers. *Physiol. Behav.* **2013**, *118*, 40–44. [CrossRef] [PubMed]
51. Arvidson, E.; Dahlman, A.S.; Börjesson, M.; Gullstrand, L.; Jonsdottir, I.H. The effects of exercise training on hypothalamic-pituitary-adrenal axis reactivity and autonomic response to acute stress-a randomized controlled study. *Trials* **2020**, *21*, 888. [CrossRef] [PubMed]
52. Üngür, G.; Karagözoğlu, C. Do personality traits have an impact on anxiety levels of athletes during the COVID-19 pandemic? *Curr. Issues Pers. Psychol.* **2021**, *9*, 246–257. [CrossRef]
53. Carrasco Páez, L.; Martínez-Díaz, I.C. Training vs. Competition in Sport: State Anxiety and Response of Stress Hormones in Young Swimmers. *J. Hum. Kinet.* **2021**, *80*, 103–112. [CrossRef]
54. Abbas, N.; Nasir, A. Effect of Job Stress (Job, Role Management) Work Overload, Work Family Conflict, Job Embeddedness and Job Satisfaction on Job Performance of School Educators. *J. Educ. Psycho.* **2021**, *14*, 21–28.
55. Duxbury, L.; Bardoel, A.; Halinski, M. "Bringing the Badge home": Exploring the relationship between role overload, work family conflict, and stress in police officers. *Policing Soc.* **2021**, *31*, 997–1016. [CrossRef]
56. Mehrsafar, A.H.; Strahler, J.; Gazerani, P.; Khabiri, M.; Sánchez, J.; Moosakhani, A.; Zadeh, A.M. The effects of mindfulness training on competition-induced anxiety and salivary stress markers in elite Wushu athletes: A pilot study. *Physiol. Behav.* **2019**, *210*, 112655. [CrossRef]
57. Marshall, S.; McNeil, N.; Seal, E.L.; Nicholson, M. Elite sport hubs during COVID-19: The job demands and resources that exist for athletes. *PLoS ONE* **2022**, *17*, e0269817. [CrossRef]

58. National Institute of Public Health (INSP-CNSCBT). Actualizare Zilnică (07/04)—Evidența Persoanelor Vaccinate Împotriva COVID-19 [Daily Update (07/04)—List of People Vaccinated Against COVID-19]. Available online: https://vaccinare-covid.gov.ro/wp-content/uploads/2021/04/Tabel-situatie-vaccinari_07.04.2021.pdf (accessed on 24 August 2022).

59. Mellalieu, S.D.; Neil, R.; Hanton, S.; Fletcher, D. Competition stress in sport performers: Stressors experienced in the competition environment. *J. Sports Sci.* **2009**, *27*, 729–744. [CrossRef] [PubMed]

60. Gilbert, J.N.; Gilbert, W.; Morawski, C. Coaching strategies for helping adolescent athletes cope with stress: Reduce the stress about reducing stress in your athlete. *J. Phys. Educ. Recreat. Dance* **2007**, *78*, 13–24. [CrossRef]

61. Uroh, C.C.; Adewunmi, C.M. Psychological Impact of the COVID-19 Pandemic on Athletes. *Front. Psychol.* **2021**, *3*, 603415. [CrossRef] [PubMed]

62. Pluhar, E.; McCracken, C.; Griffith, K.L.; Christino, M.A.; Sugimoto, D.; Meehan, W.P., III. Team Sport Athletes May Be Less Likely to Suffer Anxiety or Depression than Individual Sport Athletes. *J. Sports Sci. Med.* **2019**, *18*, 490–496.

63. Yaman, M.S.; Kahveci, M.S.; Dönmez, A.; Hergüner, G.; Yaman, Ç.; Çakar, D.B.; Genç, H.I. Determining the anxiety of athletes during Covid-19 pandemic. *Pakistan J. Medical Health Sci.* **2021**, *15*, 1425–1432.

64. Christie, H.E.; Beetham, K.; Stratton, E.; Francois, M.E. "Worn-out but happy": Postpartum Women's Mental Health and Well-Being During COVID-19 Restrictions in Australia. *Front. Glob. Womens Health* **2022**, *2*, 793602. [CrossRef]

65. Corrêa, R.P.; Castro, H.C.; Quaresma, B.M.C.S.; Stephens, P.R.S.; Araujo-Jorge, T.C.; Ferreira, R.R. Perceptions and Feelings of Brazilian Health Care Professionals Regarding the Effects of COVID-19: Cross-sectional Web-Based Survey. *JMIR Form. Res.* **2021**, *5*, e28088. [CrossRef]

66. Pat-Horenczyk, R.; Bergman, Y.S.; Schiff, M.; Goldberg, A.; Cohen, A.; Leshem, B.; Jubran, H.; Worku- Mengisto, W.; Berkowitz, R.; Benbenishty, R. COVID-19 related difficulties and perceived coping among university and college students: The moderating role of media-related exposure and stress. *Eur. J. Psychotraumatol.* **2021**, *12*, 1929029. [CrossRef]

67. Smith, R.A.; Zhu, X.; Martin, M.A.; Myrick, J.G.; Lennon, R.P.; Small, M.L.; Van Scoy, L.J. Longitudinal study of an emerging COVID-19 stigma: Media exposure, danger appraisal, and stress. *Stigma Health* **2022**, *10*. [CrossRef]

68. Fernández-Valera, M.-M.; Soler-Sánchez, M.-I.; García-Izquierdo, M.; Meseguer de Pedro, M. Personal psychological resources, resilience and self-efficacy and their relationship with psychological distress in situations of unemployment. *Int. J. Soc. Psychol.* **2019**, *34*, 331–353. [CrossRef]

69. Zeidner, M.; Ben-Zur, H. Effects of an experimental social stressor on resources loss, negative affect, and coping strategies. *Anxiety Stress Coping* **2014**, *27*, 376–393. [CrossRef] [PubMed]

70. Moncorps, F.; Jouet, E.; Bayen, S.; Fornasieri, I.; Renet, S.; Las-Vergnas, O.; Messaadi, N. Specifics of chronic fatigue syndrome coping strategies identified in a French flash survey during the COVID-19 containment. *Health Soc. Care Community* **2022**, *30*, 1–10. [CrossRef] [PubMed]

71. Low, N.; Mounts, N.S. Economic stress, parenting, and adolescents' adjustment during the COVID-19 pandemic. *Fam. Relat.* **2022**, *71*, 90–107. [CrossRef]

72. Wong, P.T. Effective management of life stress: The resource-congruence model. *Stress Med.* **1993**, *9*, 51–60. [CrossRef]

73. Dumciene, A.; Pozeriene, J. The Emotions, Coping, and Psychological Well-Being in Time of COVID-19: Case of Master's Students. *Int. J. Environ. Res. Public Health* **2022**, *19*, 6014. [CrossRef] [PubMed]

74. Hannemann, J.; Abdalrahman, A.; Erim, Y.; Morawa, E.; Jerg-Bretzke, L.; Beschoner, P.; Geiser, F.; Hiebel, N.; Weidner, K.; Steudte-Schmiedgen, S.; et al. The impact of the COVID-19 pandemic on the mental health of medical staff considering the interplay of pandemic burden and psychosocial resources—A rapid systematic review. *PLoS ONE* **2022**, *17*, e0264290. [CrossRef]

75. Joshi, D.; Gonzalez, A.; Griffith, L.; Duncan, L.; MacMillan, H.; Kimber, M.; Vrkljan, B.; MacKillop, J.; Beauchamp, M.; Kates, N.; et al. The trajectories of depressive symptoms among working adults during the COVID-19 pandemic: A longitudinal analysis of the InHamilton COVID-19 study. *BMC Public Health* **2021**, *21*, 1895. [CrossRef]

76. Naeimijoo, P.; Arani, A.M.; Bakhtiari, M.; Farsani, G.M.; Yousefi, A. The relationship between Covid-related psychological distress and perceived stress with emotional eating in Iranian adolescents: The mediating role of emotion dysregulation. *J. Pract. Clin. Psychol.* **2021**, *9*, 329–338. [CrossRef]

77. LoPorto, J.; Spina, K. Risk Perception and Coping Strategies Among Direct Support Professionals in the Age of COVID-19. *J. Soc. Behav. Health Sci.* **2021**, *15*, 201–216. [CrossRef]

78. McIntyre, A.; Mehta, S.; Janzen, S.; Rice, D.; Harnett, A.; MacKenzie, H.M.; Vanderlaan, D.; Teasell, R. Coping strategies and personality traits among individuals with brain injury and depressive symptoms. *NeuroRehabilitation* **2020**, *47*, 25–34. [CrossRef]

79. Nielsen, M.B.; Knardahl, S. Coping strategies: A prospective study of patterns, stability, and relationships with psychological distress. *Scand. J. Psychol.* **2014**, *55*, 142–150. [CrossRef] [PubMed]

80. Szczypinska, M.; Samełko, A.; Guszkowska, M. Strategies for Coping With Stress in Athletes During the COVID-19 Pandemic and Their Predictors. *Front. Psychol.* **2021**, *12*, 624949. [CrossRef] [PubMed]

81. Cooper, C.; Katona, C.; Orrell, M.; Livingston, G. Coping strategies and anxiety in caregivers of people with Alzheimer's disease: The LASER-AD study. *J. Affect. Disord.* **2006**, *90*, 15–20. [CrossRef] [PubMed]

82. Tomczak, M.; Bręczewski, G.; Sokołowski, M.; Kaiser, A.; Czerniak, U. Personality traits and stress coping styles in the Polish National Cadet Wrestling Team. *Arch. Budo* **2013**, *9*, 161–168.

83. Ivarsson, A.; Johnson, U.; Podlog, L. Psychological Predictors of Injury Occurrence: A Prospective Investigation of Professional Swedish Soccer Players. *J. Sport. Rehabil.* **2013**, *22*, 19–26. [CrossRef]

84. Hernandez, J.; Anderson, K.B. Internal martial arts training and the reduction of hostility and aggression in martial arts students. *Psi. Chi. J. Psychol. Res.* **2015**, *20*, 169–176. [CrossRef]
85. Gould, D.; Dieffenbach, K.; Moffet, A. Psychological characteristics and their development in Olympic champions. *J. Appl. Sport. Psychol.* **2002**, *14*, 172–204. [CrossRef]
86. Lyubomirsky, S.; King, L.A.; Diener, E. The benefits of frequent positive affect: Does happiness lead to success? *Psychol. Bull.* **2005**, *131*, 803–855. [CrossRef]
87. Davis, P.A.; Gustafsson, H.; Callow, N.; Woodman, T. Written Emotional Disclosure Can Promote Athletes' Mental Health and Performance Readiness During the COVID-19 Pandemic. *Front. Psychol.* **2020**, *11*, 599925. [CrossRef]
88. Mansell, W.; Urmson, R.; Mansell, L. The 4Ds of Dealing With Distress—Distract, Dilute, Develop, and Discover: An Ultra-Brief Intervention for Occupational and Academic Stress. *Front. Psychol.* **2020**, *11*, 611156. [CrossRef]
89. Predoiu, R.; Makarowski, R.; Görner, K.; Predoiu, A.; Boe, O.; Ciolacu, M.V.; Grigoroiu, C.; Piotrowski, A. Aggression in martial arts coaches and sports performance with the COVID-19 pandemic in the background—A dual processing analysis. *Arch. Budo* **2022**, *18*, 23–36.

Article

The Personality and Resilience of Competitive Athletes as BMW Drivers—Data from India, Latvia, Lithuania, Poland, Romania, Slovakia, and Spain

Samir Rawat [1,†], Abhijit P. Deshpande [2,†], Radu Predoiu [3,†], Andrzej Piotrowski [4,*,†], Romualdas Malinauskas [5,†], Alexandra Predoiu [3,†], Zermena Vazne [6,†], Rafael Oliveira [7,8,9,†], Ryszard Makarowski [10,†], Karol Görner [11,†], Camelia Branet [12,†], Mihai Lucian Ciuntea [13,†], Doru Vasile Marineanu [14,†], Néstor Vicente-Salar [15,16,†] and Davide de Gennaro [17,†]

[1] Faculty of Management Sciences, Symbiosis International (Deemed University), Lavale, Pune 412115, India
[2] MBA (Sports Management) Program, Symbiosis School of Sports Sciences and Board of University Development, Symbiosis International (Deemed University), Lavale, Taluka Mulshi, Pune 412115, India
[3] Faculty of Physical Education and Sport, National University of Physical Education and Sports, 060057 Bucharest, Romania
[4] Institute of Psychology, University of Gdańsk, Jana Bażyńskiego 4 St., 80-309 Gdańsk, Poland
[5] Department of Physical and Social Education, Lithuanian Sports University, Sporto g. 6, 44221 Kaunas, Lithuania
[6] Department of Sport and Training Theory, Latvian Academy of Sport Education, Brivibas gatve 333, LV-1006 Riga, Latvia
[7] Sports Science School of Rio Maior—Polytechnic Institute of Santarém, 2040-413 Rio Maior, Portugal
[8] Research Centre in Sport Sciences, Health Sciences and Human Development, 5001-801 Vila Real, Portugal
[9] Life Quality Research Centre, 2040-413 Rio Maior, Portugal
[10] Faculty of Administration and Social Sciences, Academy of Applied Medical and Social Sciences in Elbląg, 82-300 Elblag, Poland
[11] Department of Sports Education and Humanistics, Faculty of Sports, University of Presov, St. 17. novembra n., 08001 Presov, Slovakia
[12] Faculty of Medical Engineering, Politehnica University of Bucharest, Splaiul Independentei 313, 060042 Bucharest, Romania
[13] Faculty of Movement, Sport and Health Sciences, Vasile Alecsandri University of Bacău, Calea Mărășești 157, 600115 Bacău, Romania
[14] Department of Psychology and Cognitive Sciences, Faculty of Psychology and Educational Sciences, University of Bucharest, 050663 Bucharest, Romania
[15] Department of Applied Biology—Nutrition, Institute of Bioengineering, Miguel Hernández University (UMH), 03202 Elche, Spain
[16] Institute for Health and Biomedical Research (ISABIAL), 03010 Alicante, Spain
[17] Department of Management and Innovation Systems, University of Salerno, Via Giovanni Paolo II 132, 84084 Fisciano, Italy
* Correspondence: andrzej.piotrowski@ug.edu.pl
† All authors have contributed equally to the work.

Citation: Rawat, S.; Deshpande, A.P.; Predoiu, R.; Piotrowski, A.; Malinauskas, R.; Predoiu, A.; Vazne, Z.; Oliveira, R.; Makarowski, R.; Görner, K.; et al. The Personality and Resilience of Competitive Athletes as BMW Drivers—Data from India, Latvia, Lithuania, Poland, Romania, Slovakia, and Spain. *Healthcare* **2023**, *11*, 811. https://doi.org/10.3390/healthcare11060811

Academic Editor: Manoj Sharma

Received: 9 February 2023
Revised: 3 March 2023
Accepted: 7 March 2023
Published: 9 March 2023

Abstract: Background: Individual differences in personality and resilience are related to a variety of social behaviors. The current study sought to answer the question of whether BMW drivers exhibit different personality profiles and resilience levels compared with drivers of other car brands. Participants and procedure: An international study was carried out in India, Latvia, Lithuania, Poland, Romania, Slovakia, and Spain on a sample of 448 athletes using the 20-item Mini-IPIP and the Resilience Scale. The results of BMW drivers ($n = 91$) were compared with the results of drivers of other German car brands ($n = 357$). Results: BMW drivers were characterized by higher neuroticism compared with drivers of other German car brands. They also showed higher resiliency, both in terms of total score and scores on the subscales of: personal coping competences and tolerance of negative emotions, tolerance of failures and perceiving life as a challenge, and optimistic attitude towards life and capacity for self-mobilization in difficult situations. The greatest difference was observed for the factor of tolerance of failures and perceiving life as a challenge. Using the Dwass-Steel-Critchlow-Fligner (DSCF) pairwise comparison test, gender differences between athletes (as BMW drivers and drivers of other German car brands, respectively) were discussed. Additionally, the results of the main logistic regression analyses emphasized that neuroticism represents a better predictor of BMW preference in the case of athletes (as drivers) than the scores obtained for resilience. Conclusions: BMW drivers differed from drivers of other German car brands only with regard to neuroticism. A higher level of neuroticism can affect mental health and the overall quality of life in athletes; aggression and distress management are essential. Athletes (as BMW drivers) also showed differences in resiliency levels. Understanding the mechanisms of behavior among BMW drivers is possible through considering their personality and individual differences.

Keywords: personality trials; Big Five; resilience; athletes; car drivers

---

1. Introduction

The Big Five personality factors (neuroticism, extraversion, openness to experience, agreeableness, and conscientiousness) continue to attract researchers' attention. This is likely the result of the stability of this taxonomy, as well as of the model's usefulness and effectiveness. The Big Five model's taxonomy is consistently used in research of various occupational, language, and cultural groups worldwide [1]. The Big Five model is predictive of behaviors in professional [2] and personal settings [3]. Commonly accepted, the Big Five model allows researchers worldwide to systematically gather and compare data [4].

The traits comprising the Big Five model, which represent universal adaptive mechanisms, facilitate functioning in various social, occupational, physical, and cultural contexts [5], which is underscored by the evolutionary theory of personality [6]. Personality traits predispose individuals to certain behaviors and serve as adaptive mechanisms determining goal-oriented behaviors in various specific situations, including sport-related settings [7]. Individuals displaying traits which are desirable from the point of view of a given role will adapt and function more effectively in those roles and in their associated environments [8].

Personality allows for predicting both positive (e.g., health behaviors, [9] Doornenbal, 2021) as well as negative behaviors (e.g., bullying, [10]). Considering the sports domain, "personality profiles of karatekas specializing in kata or kumite are similar to the personality profiles of athletes from other sports. Namely: the personality profile of athletes is low neuroticism, high extraversion and conscientiousness" [11]. The specific sports branches practiced shape the personality traits in athletes (the personality features are dependent on experience); moreover, "personality and sport activities are interconnected in an ongoing process of two way conditioning" [12]. It was found that emotional stability, openness to experience, extraversion, and conscientiousness are positively correlated with sports performance, while in the case of agreeableness, a negative link was reported [13]. Other researchers also highlighted that emotional stability and extraversion are associated with high conscientiousness and low neuroticism, respectively, in karate masters [14,15].

It is commonly believed that personality factors are important in the context of driving behaviors and are related to traffic accidents. Research has shown that certain personality traits, such as conscientiousness, neuroticism, and agreeableness, are strongly correlated with aggressive driving and road rage [16]. Dahlen and White point out that select personality traits (openness to experience, emotional stability, and agreeableness) may be useful in predicting dangerous driving behaviors [17]. Wei, Lee, Luo, and Lu have confirmed that among the personality traits, neuroticism is the most strongly related to risky driving behaviors: exceeding the speed limit, abnormal staying, and hard accelerating/decelerating [18]. Li established that drivers with high neuroticism and conscientiousness drive in a riskier manner, while individuals with high agreeableness are low-risk drivers [19]. Similar results were reported by Shen, Qu, Ge, Sun, and Zhang, who noted that neuroticism is negatively correlated with positive driving behaviors [20].

Driving safety is influenced by neurotic personality, which is related to risky and dangerous behaviors, as well as to an increase in the risk of traffic accidents [21]. Neurotic personality predicts dangerous driving behaviors [22]. High neuroticism leads drivers to become distracted and ignore traffic signs, including nonstandard ones, more often [23]. Drivers with high neuroticism also react more impulsively to traffic light changes [21]. Noticing slower reactions in other drivers may lead drivers with high neuroticism to aggressive behaviors. Drivers with high neuroticism generally display stronger stress reactions [24]. Meta-analyses showed that neuroticism and extraversion are related to dangerous behaviors, accidents, and injuries (employees of various professions were investigated) [25].

Cognitive abilities and personality factors function as independent predictors of driving behaviors [26]. Sutin et al. showed that neuroticism is related to lower memory performance, psychomotor performance, attention, executive function, and visuospatial abilities [27]. Emotional stability contributes to some cognitive abilities which are important for driving [28]. However, other individual differences are also related to driving behaviors. For instance, listening to rock music during driving increases speed variability and the frequency of changing lanes among individuals with choleric and sanguine personality profiles [29]. Another variable with relevance for driving behavior is resilience. Individuals with high resiliency cope better with stressful situations and situations which demand adaptation. Numerous studies highlight the role of resiliency in coping with difficult situations [30,31]. Resiliency increases psychological resources and allows for flexible adjustment to new life situations. Resiliency facilitates effective functioning in difficult situations as well as increased tolerance of negative emotions [32]. Individuals with high resiliency are able to better tolerate risks and experience lower levels of anxiety [33].

Currently, around twenty-three thousand lethal traffic accidents are reported in the European Union (EU) [34]. According to Tsai et al., "globally, more than 1.27 million people die in motor vehicle crashes each year, and 20–50 million people sustain injuries in vehicle crashes", and the three major risk factors for fatal vehicle crashes are speeding, a history of benzodiazepine use, and involvement in motor vehicle crashes [35].

The majority of traffic accidents are avoidable contingent on the drivers adjusting their behaviors to traffic conditions and rules. Traffic accidents are chiefly the domain of men—women cause roughly one in four traffic accidents, and the risk of death for women drivers is over eight times lower than for men [36]. However, it should be noted that the number of men and women with drivers' licenses is roughly equal. Cultural differences in traffic safety can also be observed. Countries with the highest number of lethal traffic accidents include Romania, Bulgaria, and Poland, while those with the lowest numbers are Sweden, Denmark, and Iceland [37]. For example, the number of victims per million inhabitants in Romania is double when compared with the European Union average [38].

In this sense, many insurance companies also require higher payments from drivers under the age of 26 [39]. This is because younger drivers are more often involved in accidents and drive in a more aggressive manner. Younger people are also more prone to risk-taking [40]. Insurance company data also tracks the car brands most often involved in traffic accidents. Not all brands are equally popular in all countries, leading to an over-representation of certain brands in the total number of cars, and, consequently, the total number of traffic accidents. Some studies showed that drivers of certain car brands are characterized by a less safe driving style than others. It is indicated that drivers of German brands (BMW, Audi, Opel), and BMW drivers in particular, drive less safely [41]. BMW drivers are stereotypically perceived as aggressive [42], driving faster than others [43], wild, and masculine [44]. Research showed that BMW drivers are more aggressive compared with other drivers [45]. It has also been noted that drivers of more expensive cars are more likely to drive in a more egotistic manner [46]. In the present study, the rationale behind choosing BMW vs. other brands refers to the driving style of BMW drivers; Ghayad [47], who summarized several studies, concluded that BMW drivers are the worst in terms of stopping for pedestrians, knowledge of traffic rules, and being the most disliked drivers on the roads (at a big distance, compared with drivers of other car brands). Moreover, BMW drivers were chosen for an exclusive comparison with other German car drivers because only BMW drivers have a social belief "to act in an aberrant (i.e., in abnormal) way" [45] (p. 89). Researches have suggested that BMW drivers act differently compared with other German cars "because they view their own presence on the road as a BMW-with-a-driver as qualitatively different to their presence as a non-BMW-with-a-driver, and therefore behave differently in their interactions with other road users" [45] (p. 89). Stereotypes were seen that poor driving is specific to drivers, "especially those who drive BMWs (although we have nothing against BMW)" [42] (p. 46).

In the current study, athletes were chosen rather than regular non-athlete drivers because professional athletes are often 'macho-persons' [48], "with 'macho' or narcissistic self-images" [49]. Not all athletes are champions, i.e., not all athletes have low levels of neuroticism [50], but neuroticism is the most strongly related to risky driving behaviors: exceeding the speed limit, abnormal staying, and hard accelerating/decelerating [18]. In addition, "macho-persons (i.e., athletes) not only act more aggressively on the road, but also they give preference to high-performance (for instance, often BMW) and sport-type cars and are more sensitive to image related issues" [45] (p. 90). Not least, athletes as drivers have not been the subject of much research so far. In the literature, we can find only a few studies on race car drivers [51,52]. They indicated the insufficiency of the research conducted so far in this field. For example, psychophysiological dynamics of race car drivers [53], the heart rate (HR) response in young male pilots in different conditions [54], or the physiological and thermal challenges in race car drivers were investigated [55]. Additionally, researchers were concerned with creating a mathematical model which can predict drivers' performance (reaction time was measured) in a variety of driving conditions [56], and were interested in increasing the psychomotor qualities and, therefore, quality of life in senior drivers, through Visual-Motor Useful Field of View training [57].

Considering the previous information, the aim of the current study was to examine whether athletes as BMW drivers differ from drivers of other German car brands with regard to basic personality traits and resiliency. Because personality and resiliency are better predictors of behavior when considered together [58], both these variables were included in the study. To the best of our knowledge, the current study is the first to examine resiliency among athlete drivers.

Therefore, the following research questions were formulated:

1. What are the resiliency levels among athlete drivers of various car brands (BMW and other brands)?
2. What are the personality trait levels among athlete drivers of various car brands?
3. Are there significant differences between BMW athlete drivers and athletes driving other German car brands with regard to basic personality traits and resiliency?
4. Can we predict athletes' preference for a car brand (based on personality traits or resiliency)?

2. Materials and Methods

2.1. Participants

A total of 448 participants from the following seven countries took part in the study: India, Latvia, Lithuania, Poland, Romania, Slovakia, and Spain. The participants were competitive athletes practicing different sports disciplines: track-and-field, martial arts (boxing, kickboxing, judo, fencing, karate, taekwondo), handball, soccer, aerobic and artistic gymnastics, swimming, volleyball, and basketball (fourteen sports disciplines in total). The inclusion criteria were: more than 18 years (seniors) and minimum five years of training in a specific sport discipline. Approximately 79% of the participants registered local/regional performances, while about 21% registered international/national performances, being considered experts/elite (with respect to the athletes' standard of performance—Swann et al., 2015).

BMW athlete drivers (n = 91) were identified in the sample and were compared with athlete drivers of other German car brands (n = 357).

2.2. Measures

2.2.1. Personality

Personality was measured with the 20-item *Mini-International Personality Item Pool* (Mini-IPIP), adapted by Topolewska et al. [59], which is a shortened version of the IPIP-Big-Five Factor Markers-50 (IPIP-BFM-50), measuring the Big Five personality traits according to Goldberg's lexical model [60]. It consists of 20 items, arranged into 5 subscales (4 items for each subscale): extraversion (e.g., "I am the life of the party."), agreeableness (e.g., "Sympathize with others' feelings."), conscientiousness (e.g., "Get chores done right away."), neuroticism (e.g., "Have frequent mood swings."), and intellect/imagination (e.g., "Have a vivid imagination."). The participants indicated how accurate each phrase was for them, using a 5-point Likert-type scale (from "1"—very inaccurate to "5"—very accurate). In the current sample, reliability for the subscales, measured with Cronbach's alpha, ranged from: 0.73—intellect/imagination to 0.85—extraversion (India); 0.78—conscientiousness to 0.88—neuroticism (Latvia); 0.79—intellect/imagination to 0.88—extraversion (Lithuania); 0.76—intellect/imagination to 0.86—neuroticism (Poland); 0.77—conscientiousness to 0.88—extraversion (Romania); 0.80—agreeableness to 0.88—conscientiousness (Slovakia); and 0.75—intellect/imagination to 0.85—extraversion (Spain).

2.2.2. Resilience

To measure resilience, the *Scale for Measuring Resilience* (SPP-25) by Ogińska-Bulik and Juczyński was used [32]. It consists of 25 items, arranged into 5 subscales (5 items for each subscale): perseverance and determination in action (e.g., "If I have to do something, I usually do it right away."), openness to new experiences and sense of humor (e.g., "I can look at situations from many different points of view."), personal coping competences and tolerance of negative emotions (e.g., " In difficult situations, I can cope with unpleasant feelings."), tolerance of failures and perceiving life as a challenge (e.g., "I can draw conclusions for the future from my failures and mistakes."), and optimistic attitude towards life and capacity for self-mobilization in difficult situations (e.g., "I always have an optimistic approach to life, regardless of the situation."). The respondents indicated their agreement with each item on a 5-point Likert scale from "0"—I definitely disagree" to "4"—I definitely agree. The higher the partial scores on each scale or the total score, the higher the degree of resilience. In the current sample, reliability for the subscales, measured with Cronbach's alpha, ranged from: 0.67—openness to new experiences and sense of humor to 0.78—personal coping competences and tolerance of negative emotions (India); 0.69—tolerance of failures and perceiving life as a challenge to 0.88—personal coping competences and tolerance of negative emotions (Latvia); 0.68—openness to new experiences and sense of humor to 0.75—optimistic attitude towards life and capacity for self-mobilization in difficult situations (Lithuania); 0.67—tolerance of failures and perceiving life as a challenge to 0.86—personal coping competences and tolerance of negative emotions (Poland); 0.68—perseverance and determination in action to 0.76—optimistic attitude towards life and capacity for self-mobilization in difficult situations (Romania); 0.70—openness to new experiences and sense of humor to 0.78—personal coping competences and tolerance of negative emotions (Slovakia); and 0.70—optimistic attitude towards life and capacity for self-mobilization in difficult situations to 0.76—perseverance and determination in action (Spain).

2.3. Procedure

The research was carried out between November 2021 and May 2022, with data collected (for the two questionnaires) through Google forms (Google LLC, Mountain View, CA, USA), as in previous investigations [61,62].

The snowball sampling technique was used as a recruitment technique to examine the participants: BMW athlete drivers and athletes driving another German car brand.

Regarding the 20-item Mini-IPIP questionnaire and the scale for measuring resilience (SPP-25), the final versions, in each language, were created through retroversion—a procedure used, also, in earlier studies [63].

2.4. Statistical Analysis

Mann–Whitney's U test was used to see whether scores on the measured variables differed significantly. Mann–Whitney's U test was chosen because it works fine with unequally sized samples. Condition, that the statistical power will diminish as the group sizes are unequal, has been taken into account. The calculated statistical power for unequally sized samples of the current study (a 1:4 ratio; total sample size: 448) is 0.83. G*Power tool to compute statistical power analyses was used [64]. The level of statistical significance was set at $p < 0.05$. The Dwass-Steel-Critchlow-Fligner (DSCF) pairwise comparison test was then used to examine the differences between four groups (male and female athletes as BMW drivers and male and female athletes driving other German car brands). The binomial logistic regression was carried out to predict athletes' preference (based on neuroticism and resilience) for a car brand. IBM SPSS Statistics 27.0 software (IBM Corp, Released 2020, IBM SPSS Statistics for Windows, Version 27.0. Armonk, NY, USA) was used in the statistical analysis. Effect sizes for U-statistics were expressed as r (0.1 weak effect size, 0.3 moderate, 0.5 strong, ≥ 0.7 very strong [65]).

3. Results

Participants from the following countries participated in the study: India ($n = 38$; $M_{age} = 30.81$; SD = 12.84), Latvia ($n = 72$; $M_{age} = 25.95$; SD = 8.71), Lithuania ($n = 56$; $M_{age} = 25.20$; SD = 6.96), Poland ($n = 121$; $M_{age} = 32.79$; SD = 9.28), Romania, ($n = 103$; $M_{age} = 24.52$; SD = 5.43), Slovakia ($n = 32$; $M_{age} = 25.83$; SD = 8.83), and Spain ($n = 26$; $M_{age} = 34.81$; SD = 11.83). Considering the training experience in a specific sport discipline, the average in the entire sample was 11.6 years. Ninety-one BMW athlete drivers (50 males and 41 females) were identified in the sample and there were 357 athlete drivers of other German car brands (198 males and 159 females): Audi (36 males, 27 female athletes), Volkswagen (85 males, 68 female athletes), Opel (49 males, 35 female athletes), and Mercedes-Benz (28 males, 29 female athletes). The mean age of the BMW athlete drivers was 25.75 (SD = 8.32). The mean age of the athlete drivers of other German brands was 29.36 (SD = 11.19).

First, data were screened for missing values or outliers [66]. Using stem-and-leaf, no outliers were identified. Additionally, no missing values were detected (due to the online survey in which all questions had to be rated).

Table 1 shows the results of the median scores comparison and Mann–Whitney's U test for BMW athlete drivers and athlete drivers of other German car brands.

The analyses show that BMW athlete drivers are more neurotic than athlete drivers of other German car brands. Effect size is r = 0.13, meaning a moderate to weak difference. Other personality traits (in terms of the 20-item Mini-IPIP scales) did not differentiate between these two groups. BMW athlete drivers also achieved statistically significantly higher total resiliency scores (r − 0.10—a weak effect size), as well as scores on the particular resiliency subscales. The subscale of *openness to new experiences and sense of humor* was also higher among BMW athlete drivers, although this difference did not reach statistical significance.

The resiliency results were roughly above 75 points for BMW athlete drivers (mean = 75.9, median = 76) and 72 points for the athlete drivers of other German car brands (mean = 72.9, median = 74). According to the norms, the mean score was 72.27; therefore, both of the results in the current study (at group level) are within the sixth sten which means an average level of resiliency.

In the next step, using the Dwass-Steel-Critchlow-Fligner (DSCF) pairwise comparison test, the existing differences between the four groups formed (male athletes BMW drivers, female athletes BMW drivers, male athletes driving other German car brands, and female athletes driving other German car brands, respectively) were examined, starting from the eleven dependent variables investigated (personality factors and resilience dimensions). Table 2 contains only the significant differences observed.

Table 1. Mann–Whitney U test results for the effect of car brand used on investigated variables.

Variables	Drivers	Median	Mann–Whitney U	Z	p
Extraversion	BMW	3.50	14,541.50	−1.551	0.121
	Others	3.25			
Agreeableness	BMW	3.75	16,128.00	−0.106	0.916
	Others	3.75			
Conscientiousness	BMW	3.75	16,005.50	−0.217	0.828
	Others	3.75			
Neuroticism	BMW	3.00	13,272.00	−2.710	0.007
	Others	3.00			
Intellect/Imagination	BMW	3.75	16,177.00	−0.061	0.952
	Others	3.75			
Perseverance and determination in action	BMW	15.00	14,962.50	−1.170	0.242
	Others	15.00			
Openness to new experiences and sense of humor	BMW	16.00	15,704.00	−0.494	0.621
	Others	16.00			
Personal coping competences and tolerance of negative emotions	BMW	15.00	14,058.00	−1.998	0.046
	Others	15.00			
Tolerance of failures and perceiving life as a challenge	BMW	16.00	13,732.00	−2.297	0.022
	Others	15.00			
Optimistic attitude towards life and capacity for self-mobilization in difficult situations	BMW	15.00	14,064.50	−1.989	0.047
	Others	14.00			
Total Resilience score	BMW	76.00	14,086.50	−1.957	0.050
	Others	74.00			

Table 2. DSCF test results for pairwise comparisons with significant differences.

Group (Median)		W	p
		Agreeableness	
1 (3.50)	2 (4.25)	7.13	<0.001
1(3.50)	4 (4)	6.44	<0.001
2 (4.25)	3 (3.50)	−7.44	<0.001
3 (3.50)	4 (4)	8.42	<0.001
		Conscientiousness	
3 (3.50)	4 (3.75)	6.613	<0.001
		Neuroticism	
1 (3.50)	2 (2.75)	−5.70	<0.001
1 (3.50)	3 (3.25)	−3.54	0.059
1 (3.50)	4 (2.75)	−8.73	<0.001
2 (2.75)	3 (3.25)	4.79	0.004
3 (3.25)	4 (2.75)	−9.22	<0.001
		Personal coping competences and tolerance of negative emotions	
1 (16)	4 (15)	−4.68	0.005
3 (15.5)	4 (15)	−3.73	0.042
		Optimistic attitude towards life and capacity for self-mobilization in difficult situations	
1 (15)	4 (13)	−5.17	0.001
3 (14)	4 (13)	−4.56	0.007
		Total Resilience score	
1 (77.5)	4 (73)	−3.69	0.045

Note. 1—Male athletes BMW drivers, 2—female athletes BMW drivers, 3—male athletes driving other German car brands, 4—female athletes driving other German car brands, W—Wilcoxon value.

Statistical analysis of the data shows that male athletes as BMW drivers are significantly less agreeable than female athletes as both BMW or other German car brands drivers.

Additionally, female athletes as BMW drivers are significantly more agreeable than male athletes who are drivers of other German car brands. Regarding neuroticism, it was found that male athletes as BMW drivers are significantly more neurotic than participants in all other three groups (the difference was marginally significant when compared with male athletes who are drivers of other German car brands). Additionally, when talking about resilience, male athletes as BMW drivers, as well as male athletes driving other German car brands, obtained a significantly higher score for personal coping competences and tolerance of negative emotions, optimistic attitude towards life and capacity for self–mobilization, and for the total Resilience score (only male athletes as BMW drivers in the case of the total Resilience result), compared with female athletes who were drivers of other German car brands.

Knowing that neuroticism and resilience are specific to competitive athletes as BMW drivers, we examined to what extent the two variables predict athletes' preference for a car brand. To achieve this aim, binomial logistic regression was used, considering the defined categories—athletes as BMW drivers and athletes driving other German car brands. The data of the main logistic regression analysis are underlined in Table 3, predicting the likelihood of athletes preferring and driving a BMW car based on neuroticism and resilience.

Table 3. Results of the binomial logistic regressions analysis.

	Neuroticism	**Resilience**
Omnibus test—model	0.001	0.042
Hosmer and Lemeshow test (chi-square, *p*-value)	14.46 (0.054)	12.97 (0.073)
Nagelkerke R^2	0.053	0.018
Overall percentage (Predicted − Percentage correct)	79.700	68.9
Wald test	11.737	3.923
B	0.550	0.020
SE	0.160	0.010
Odds ratio values	1.733	1.076
Confidence interval for *Exp(B)*	1.265–2.373	1.021–1.118

The main logistic regression analysis (two separate regressions were performed) emphasized that both models are statistically significant (omnibus test model, $p < 0.05$). With respect to the Hosmer and Lemeshow goodness of fit test, $p = 0.054$ (for neuroticism) and $p = 0.073$ (resilience), pointing out that the models are not a poor fit. NR $- \chi^2(1) = 12.28$, $p = 0.001$, RES $- \chi^2(1) = 4.12$, and $p = 0.042$, meaning that the logistic regression models were significant. In the case of the competitive athletes, the scores for neuroticism represent a better predictor of BMW preference (as drivers) than those of the results obtained for resilience. The models correctly classified 79.7% (neuroticism) and 68.9% (resilience) of cases, respectively. Nagelkerke R^2 (effect size index) shows a moderate to weak relation between neuroticism and athletes' preference for BMW, and a weak relation between resilience and athletes' preference (for driving a BMW).

It was asserted that higher levels of neuroticism and resilience are associated with an increased likelihood of driving a BMW car in the case of athletes.

4. Discussion

The human factor is responsible for the majority of traffic accidents. Excessive speed, not respecting right of way, and not adjusting speed for the weather conditions are responsible for over 95% of all traffic accidents [67]. The car's condition, the weather, and the road infrastructure are responsible for a very small proportion of all traffic accidents, and it would be difficult to eliminate these factors entirely. Insurance company data and media reports show that traffic accidents are most often caused by young men. Moreover, there is a common belief that BMW drivers are aggressive. Thus, the current study involved a sample of competitive athletes and divided them into subsamples based on the brand of car they drive.

The aim of the current study was to estimate the personality profile and resiliency of BMW athlete drivers compared with athlete drivers of other car brands. Based on the literature [18], we expected that personality traits may be related to driving BMWs. BMW athlete drivers were characterized by higher neuroticism compared with athlete drivers of other German car brands. BMW athlete drivers' neuroticism levels significantly exceeded the mean reported in the original IPIP-BFM-20 study [60]. Individuals characterized by higher neuroticism are more prone to reacting with frustration and anger at everyday events [68]. Individuals with high neuroticism do not cope well with stressful situations and have a tendency towards interpreting regular situations as threatening. In everyday situations, they more often interpret random occurrences as directed personally against them. They also show significant difficulties in controlling their own behavior [69]. Individuals with high neuroticism also engage in more reckless risk-taking behaviors [70].

As drivers, individuals with high neuroticism are characterized by a dangerous driving style, including purposefully breaking the speed limit, being aggressive towards other drivers [71], and using cellphones while driving [72]. Neuroticism seems to significantly impact driving behaviors from the point of view of causing dangerous situations, both for the drivers themselves as well as for other people. Neuroticism also plays an important part in shaping loyalty towards brands perceived as exciting [73]. Driving BMW cars can be seen as a source of excitement, partially due to the technical capabilities of the cars themselves. However, it remains an open question whether individuals with high neuroticism choose BMW cars or whether owning BMW cars increases neuroticism.

Considering success in sports and taking earlier reports talking about the Big Five model [1], researchers emphasized that "probably lower neuroticism and higher extraversion, openness to experience, agreeableness and conscientiousness" are specific to sport masters when compared with less successful athletes [74]. In the current study, higher levels of neuroticism were highlighted among athletes who were BMW drivers when compared with athletes driving other German car brands; the groups consisted of both top athletes and athletes with more local experience.

Macho personality is also related to an aggressive driving style [75]. This personality type is composed of such traits as, among others, perceiving violence as masculine and danger as exciting, and the cult of hardiness leading to self-control [76]. Such behaviors are commonly ascribed to BMW drivers [42,45]. It is possible that the social perception of BMW drivers encourages the purchase and use of these cars by individuals whose personality profiles are close to the macho stereotype [75]. At the same time, it is interesting to examine whether BMW cars are a differentiating factor among drivers of sports cars. Brands, including car brands, may be a reflection of the consumers' personality and identity [77]. According to the self-congruity theory, individuals purchase cars of those brands which best reflect their actual or desired personality [78]. It seems interesting to examine which personality factors determine the choice of the BMW brand. High horsepower, rapid acceleration, and technical solutions increasing driving safety allow for a more dynamic, but also riskier, driving style. It has been shown that drivers of high-powered cars engage in dangerous driving behaviors more often [75]. High horsepower may characterize all cars in this brand, regardless of the model.

It has been widely documented in the literature that individuals with high resiliency are able to effectively cope with stressful situations [30]. To the best of our knowledge, this is the first study examining resiliency among athletes as drivers. The current results show that BMW athlete drivers are characterized by higher resiliency compared with athlete drivers of other German car brands. They achieved both higher total resiliency scores and higher scores on the subscales of *personal coping competences and tolerance of negative emotions, tolerance of failures and perceiving life as a challenge, optimistic attitude towards life, and capacity for self-mobilization in difficult situations.* The results seem to be in line with the observation that BMW drivers cope better with difficult traffic situations. However, BMW drivers are characterized by a more aggressive and less socially accepted driving style [45]. On the other hand, it has to be noted that both BMW athlete drivers and athlete drivers of other German car brands in the current study achieved average resiliency scores despite significant differences in some resiliency indicators.

The obtained results are important in the context of sports psychology. The human personality is visible in everyday behavior, consumer decisions, and the type of preferred music [8,79,80]. Sports coaches and psychologists need to pay attention to many aspects of athlete behavior. The brand of the selected car may indirectly indicate what personality traits a given athlete has. This is especially important in the context of Perepjolkina's [45] (2009) research, which indicated greater aggressiveness in BMW drivers. Generally, athletes are closely monitored by their coaches, who must evaluate their behavior as a whole. The present research shows how car brand preferences are related to the sowing of basic personality traits and resilience. Resilience and controlled instrumental aggression are desirable and applauded in performance sport if they are within the boundaries of the game [81], and personality plays an important role in achieving high sports results [11]. However, sometimes "the fierce struggle for winning in competition and the *win-at-all-costs philosophy*" [82] can generate violent behaviors, mostly in the case of young athletes [83]. If, in this context, it can be added to a higher score for neuroticism—athletes are less emotionally stable and experience more negative effects, including irritability, anger, or anxiety [63]—the negative impact of such situations on athletes' mental health and overall quality of life is ensured.

Thereafter, the present research can contribute to a better understanding of the functioning of athletes in various fields of their activity. The findings can be helpful for coaches and sports psychologists as they pay attention to the non-sport decisions and attitudes of athletes, which may affect their sports results. Consumer choices dictated by personality profile provide additional information to coaches and sports psychologists and allow them to better understand possible patterns of their behavior. The main reason why the authors believe that this study will be useful to the community is based on the belief that special attention should be paid to athletes who choose to drive a BMW car, given the lack of stress and aggression management. Based on the findings of this study, a recommendation can be made for those athletes who choose to drive a BMW car to implement aggression management programs.

In order to reduce distress and increase emotional balance in athletes (developing a better ability to deal with unexpected situations or mistakes) members of the multidisciplinary team, and especially sports psychologists, can use: cognitive and behavioral strategies, which have proven their effectiveness [84]; analytical relaxation and autogenic training; positive self-talk (inner monologue) for self-confidence; self-monitoring of emotional reactions [85]; the 4Ds for dealing with distress—which unifies strategies and exercise with the aim of restoring wellbeing and improving emotion regulation [86]; stress inoculation training (SIT)—seen as an educational program to improve self-control [87], or *internal* techniques, e.g., breathing and meditation, which reduce hostility and impulsivity in athletes [88]. To acquire healthy behaviors, athletes with a higher level of neuroticism (in collaboration with specialists) could attempt mindful activities the experience of the body as trustworthy and safe [89], could learn appropriate conflict resolution strategies, and improve communication skills, which are verbal, nonverbal, and paraverbal. Additionally, eustress (positive stress) can be induced in more neurotic athletes by guiding them to get involved in pleasant activities that offer them satisfaction [90]. Additionally, "an integrative model of intervention against psychological (chronical and traumatic) stress" can be used by sports psychologists (working to increase athletes' emotional stability, mental health, and overall quality of life) comprising the following domains: emotional, behavioral, and cognitive [91].

However, a higher score for resilience (in the case of BMW drivers) offers an advantage in performance sports as well as allowing athletes to have a better ability to bounce back from failure and overcome barriers on the road to great performance.

Taken together, the results on personality and resiliency show that from among the personality dimensions, only neuroticism is negatively related with resiliency—the higher the neuroticism, the lower the resiliency [58]. Similar results have been reported in the original study on the scale for measuring resilience [32]. In the case of athletes as BMW drivers, resiliency and neuroticism are both at higher levels when compared with the values obtained by athletes who are drivers of other German car brands.

The findings of the current research extended previous investigations and also addressed gaps in the specialized literature, examining a less approached sample and topic—athletes as BMW drivers in terms of resilience and the Big Five personality traits.

Limitations and Future Directions

The current study was carried out in only a few countries. Thus, it is unclear whether the results would be replicated in other countries and, therefore, replicating the present research is warranted in other countries. In addition, the current study also did not measure how long the athletes were BMW or other brand drivers, which may also impact the results. Other variables potentially related to car brand choice were also not analyzed, such as socioeconomic status or the social perception of BMW drivers in a given country. In future studies, the economic status of the people who use these car brands should be asked, taking into account that the personality traits and resilience of people with high and low incomes may differ. Additionally, only self-reported measures were used, and the obtained results were declarative, the aspects of possible desirable responses being known [82]. Nonetheless, it is worth mentioning that the large number of studied athletes represents a strength of the current research. In further studies, data could be gained by combining questionnaires with behavioral observations. Participants could also be asked about their traffic citations, speeding tickets, and so forth, in the last 1–2 years. Detailed questions could concern the cars' specific parameters, such as tuning or applying various characteristic decals or decorative elements. Moreover, it is unclear how different the results will be when athletes as BMW drivers are compared with athletes driving other car brands (not German car brands), as well as a potential comparison between BMW athlete drivers from different countries of origin. Then, different studies must be conducted considering non-athletes (having less/greater driving experience), only novices or only elite/ expert athletes (as BMW drivers), and only athletes from a specific sport branch. Finally, it is not ensured that the findings were not influenced by the participants' expertise and were solely because of driving a certain brand of car, because the present study has a cross-sectional and ex-post facto design.

It seems interesting to analyze the relationship between the brand of car owned and the attitudes towards traffic rules and safety. These attitudes could concern such rules as speed limits, driving under the influence of alcohol, using baby seats, complying with mandatory technical check-ups, and so forth. Results on drivers of various car brands may be valuable when analyzed in conjunction with the contents of these brands' advertisements.

Future studies should examine the possible dynamics of personality and individual traits during a period of owning a BMW car (or a certain model of BMW car). It is also worthwhile to answer the question of whether differences in other traits can be distinguished among drivers based on the brand of car they drive, such as sensation seeking, narcissism, self-efficacy, or sense of coherence [92], considering the instrumental risk, the stimulating risk [62], or when talking about different factors of aggression: foul play, go-ahead and assertiveness—Makarowski [93], verbal aggression, anger, hostility, and physical aggression [94]. It is possible that the social perception of BMW drivers attracts individuals with a specific personality profile. This question also requires further study.

5. Conclusions

The conclusions of the current research, carried out in seven countries, emphasized that athletes as BMW drivers manifest a higher neuroticism level compared with drivers of other German car brands. Specifically, male athletes as BMW drivers are significantly more neurotic than participants in the other three groups (female athletes as BMW drivers and male and female athletes driving other German car brands), and significantly less agreeable than female athletes as both BMW and other German car brands drivers.

Athletes (male and female) as BMW drivers also showed higher resiliency in terms of total score and results on the following subscales: personal coping competences and tolerance of negative emotions, tolerance of failures and perceiving life as a challenge, and optimistic attitude towards life and capacity for self-mobilization in difficult situations compared with athletes driving other German car brands. Additionally, in terms of the total Resilience score, only male athletes (as BMW drivers) registered a significantly higher value than female athletes who were drivers of other German car brands (after pairwise comparisons).

In the case of the competitive athletes, the scores for neuroticism represent a better predictor of BMW preference as drivers than those results obtained for resilience. Various techniques are presented to be performed by sports psychologists in collaboration with athletes as BMW drivers, especially with male athletes, in order to reduce distress, increase emotional balance, to help them acquire healthy behaviors, and increase overall quality of life.

Author Contributions: Conceptualization: S.R., A.P. (Andrzej Piotrowski), R.M. (Romualdas Malinauskas), A.P. (Alexandra Predoiu), Z.V. and N.V.-S.; design: A.P.D., R.P., R.O., R.M. (Ryszard Makarowski) and K.G.; data collection: S.R., A.P.D., R.P., A.P. (Andrzej Piotrowski), R.M.(Romualdas Malinauskas), A.P. (Alexandra Predoiu), Z.V., R.M. (Ryszard Makarowski), K.G., C.B., M.L.C., D.V.M. and N.V.-S.; data analysis: S.R., R.P., A.P., R.M. (Romualdas Malinauskas), Z.V., K.G. and D.d.G.; data interpretation: A.P. (Alexandra Predoiu), R.O., R.M. (Ryszard Makarowski) and D.d.G.; writing—original draft preparation: S.R., A.P.D., R.P., A.P. (Andrzej Piotrowski), R.M. (Romualdas Malinauskas), Z.V., R.O., R.M. (Ryszard Makarowski), K.G. and N.V.-S.; writing—review and editing: A.P. (Alexandra Predoiu), C.B., M.L.C., D.V.M. and D.d.G. All authors are accountable for all aspects of the work in ensuring that questions related to the accuracy or integrity of any part of the work are appropriately investigated and resolved. All authors have read and agreed to the published version of the manuscript.

Funding: This research received no external funding.

Institutional Review Board Statement: This research was conducted in accordance with the recommendations of the Helsinki Declaration and was approved by the Ethics Committee—National University of Physical Education and Sports, Bucharest, Romania (ID: 945).

Informed Consent Statement: Informed consent was obtained from all participants involved in the study.

Data Availability Statement: The data presented in this study are available on request from the corresponding author.

Conflicts of Interest: The authors declare no conflict of interest. This study did not receive any funding. The paper has no commercial purpose.

References

1. Costa, P.T.; McCrae, R.R. Four ways five factors are basic. *Pers. Individ. Differ.* **1992**, *13*, 653–665. [CrossRef]
2. Pletzer, J.L.; Bentvelzen, M.; Oostrom, J.K.; de Vries, R.E. A meta-analysis of the relations between personality and workplace deviance: Big Five versus HEXACO. *J. Vocat. Behav.* **2019**, *112*, 369–383. [CrossRef]
3. Azucar, D.; Marengo, D.; Settanni, M. Predicting the Big 5 personality traits from digital footprints on social media: A meta-analysis. *Pers. Individ. Differ.* **2018**, *124*, 150–159. [CrossRef]
4. Voronin, I.; Tikhomirova, T.; Ismatullina, V.; Malykh, S. Cross-cultural differences in relationship between "Big Five" personality traits and intelligence in adolescents. *Pers. Individ. Differ.* **2016**, *101*, 524. [CrossRef]
5. Buss, D.M. Evolutionary personality psychology. *Annu. Rev. Psychol.* **1991**, *42*, 459–491. [CrossRef]
6. Maranges, H.M.; Reynolds, T. Evolutionary Theories of Personality. In *The Wiley-Blackwell Encyclopedia of Personality and Individual Differences*; Carducci, B.J., Nave, C.S., Eds.; John Wiley & Sons: Hoboken, NJ, USA, 2020; pp. 185–197.
7. Kaiseler, M.; Levy, A.; Nicholls, A.R.; Madigan, D.J. The independent and interactive effects of the Big-Five personality dimensions upon dispositional coping and coping effectiveness in sport. *Int. J. Sport Exerc. Psychol.* **2019**, *17*, 410–426. [CrossRef]

8. Piotrowski, A.; Martinská, M.; Boe, O.; Rawat, S.; Deshpande, A. The relationship between personality, self-esteem, emotional intelligence, and cultural intelligence. A military context. *Curr. Issues Pers. Psychol.* **2020**, *8*, 185–196. [CrossRef]

9. Doornenbal, B. Big Five personality as a predictor of health: Shortening the questionnaire through the elastic net. *Curr. Issues Pers. Psychol.* **2021**, *9*, 159–164. [CrossRef]

10. Fernández-del-Río, E.; Ramos-Villagrasa, P.J.; Escartín, J. The incremental effect of Dark personality over the Big Five in workplace bullying: Evidence from perpetrators and targets. *Pers. Individ. Differ.* **2021**, *168*, 110291. [CrossRef]

11. Piepiora, P.; Petre, L.; Witkowski, K. Personality of karate competitors due to their sport specialization. *Archiv. Budo* **2021**, *17*, 51–58.

12. Predoiu, A. Key personality traits of former handball players which can facilitate a successful career. *Discobolul Phys. Educ. Sport Kinetothe. J.* **2017**, *48*, 24–29.

13. Khan, B.; Ahmed, A.; Abid, G. Using the 'Big-Five'-For Assessing Personality Traits of the Champions: An Insinuation for the Sports Industry. *Pak. J. Commer Soc. Sci.* **2016**, *10*, 175–191.

14. Macovei, S.; Lambu, I.; Lambu, E.; Predoiu, R. Manifestations of some personality traits in karate do. *Procedia Soc. Behav. Sci.* **2014**, *117*, 269–274.

15. Bojanić, Z.; Nedeljković, J.; Šakan, D.; Mitic, P.M.; Milonovic, I.; Drid, P. Personality Traits and Self-Esteem in Combat and Team Sports. *Front. Psychol.* **2019**, *10*, 2280. [CrossRef] [PubMed]

16. Jovanović, D.; Lipovac, K.; Stanojević, P.; Stanojević, D. The effects of personality traits on driving-related anger and aggressive behaviour in traffic among Serbian drivers. *Transp. Res. F Traffic Psychol. Behav.* **2011**, *14*, 43–53. [CrossRef]

17. Dahlen, E.R.; White, R.P. The Big Five factors, sensation seeking, and driving anger in the prediction of unsafe driving. *Pers. Individ. Differ.* **2006**, *41*, 903–915. [CrossRef]

18. Wei, C.-H.; Lee, Y.; Luo, Y.-W.; Lu, J.-J. Incorporating Personality Traits to Assess the Risk Level of Aberrant Driving Behaviors for Truck Drivers. *Int. J. Environ. Res. Public Health* **2021**, *18*, 4601. [CrossRef]

19. Li, P.Y. Constructing the Risk Level Models Based on Inter-City Bus Drivers' Personalities. Master's Thesis, Department of Transportation and Communication Management Science, National Cheng Kung University, Tainan City, Taiwan, 2017.

20. Shen, B.; Qu, W.; Ge, Y.; Sun, X.; Zhang, K. The relationship between personalities and self-report positive driving behavior in a Chinese sample. *PLoS ONE* **2018**, *13*, 1–16. [CrossRef]

21. Zhang, W.; Zeng, Y.; Yang, Z.; Kang, C.; Wu, C.; Shi, J.; Ma, S.; Li, H. Optimal Time Intervals in Two-Stage Takeover Warning Systems With Insight Into the Drivers' Neuroticism Personality. *Front. Psych.* **2021**, *12*, 601536. [CrossRef]

22. Darzi, A.; Gaweesh, S.M.; Ahmed, M.M.; Novak, D. Identifying the causes of drivers' hazardous states using driver characteristics, vehicle kinematics, and physiological measurements. *Front. Neuro.* **2018**, *12*, 568. [CrossRef]

23. Wontorczyk, A.; Gaca, S. Study on the Relationship between Drivers' Personal Characters and Non-Standard Traffic Signs Comprehensibility. *Int. J. Environ. Res. Public Health* **2021**, *18*, 2678. [CrossRef] [PubMed]

24. Matthews, G.; Desmond, P.A. Stress and driving performance: Implications for design and training. In *Human Factors in Transportation. Stress, Workload, and Fatigue*; Hancock, P.A., Desmond, P.A., Eds.; Lawrence Erlbaum Associates Publishers: Mahwah, NJ, USA, 2001; pp. 211–231. [CrossRef]

25. Beus, J.M.; Dhanani, L.Y.; McCord, M.A. A meta-analysis of personality and workplace safety: Addressing unanswered questions. *J. Appl. Psychol.* **2015**, *100*, 481–498. [CrossRef] [PubMed]

26. Zicat, E.; Bennett, J.M.; Chekaluk, E.; Batchelor, J. Cognitive function and young drivers: The relationship between driving, attitudes, personality and cognition. *Transp. Res. F Traffic Psychol. Behav.* **2018**, *55*, 341–352. [CrossRef]

27. Sutin, A.R.; Stephan, Y.; Luchetti, M.; Terracciano, A. Five-factor model personality traits and cognitive function in five domains in older adulthood. *BMC Geriat.* **2019**, *19*, 343. [CrossRef] [PubMed]

28. Nechtelberger, M.; Vlasak, T.; Senft, B.; Nechtelberger, A.; Barth, A. Assessing Psychological Fitness to Drive for Intoxicated Drivers: Relationships of Cognitive Abilities, Fluid Intelligence, and Personality Traits. *Fron. Psych.* **2020**, *11*, 1002. [CrossRef]

29. Wen, H.; Sze, N.N.; Zeng, Q.; Hu, S. Effect of music listening on physiological condition, mental workload, and driving performance with consideration of driver temperament. *Int. J. Environ. Res. Public Health* **2019**, *16*, 2766. [CrossRef]

30. Süloğlu, D.; Güler, Ç. Predicting perceived stress and resilience: The role of differentiation of self. *Curr. Issues Pers. Psychol.* **2021**, *9*, 289–298. [CrossRef]

31. Rizal, F.; Egan, H.; Cook, A.; Keyte, R.; Mantzios, M. Examining the impact of mindfulness and self-compassion on the relationship between mental health and resiliency. *Curr. Issues Pers. Psychol.* **2020**, *8*, 279–288. [CrossRef]

32. Ogińska-Bulik, N.; Juczyński, Z. Resiliency Assessment Scale—SPP-25). *Now. Psych.* **2008**, *3*, 39–56.

33. McCleskey, J.; Gruda, D. Risk-taking, resilience, and state anxiety during the COVID-19 pandemic: A coming of (old) age story. *Pers. Individ. Differ.* **2021**, *170*, 110485. [CrossRef]

34. European Commission. 2018 Road Safety Statistics: What Is Behind the Figures. Available online: https://ec.europa.eu/commission/presscorner/detail/en/qanda_20_1004 (accessed on 9 January 2023).

35. Tsai, J.-H.; Yang, Y.-H.; Ho, P.-S.; Wu, T.-N.; Guo, Y.L.; Chen, P.-C.; Chuang, H.-Y. Incidence and Risk of Fatal Vehicle Crashes Among Professional Drivers: A Population-Based Study in Taiwan. *Front. Public Health* **2022**, *10*, 849547. [CrossRef] [PubMed]

36. Lardelli-Claret, P.; Espigares-Rodríguez, E.; Amezcua-Prieto, C.; Jiménez-Moleón, J.J.; de Dios Luna-del-Castillo, J.; Bueno-Cavanillas, A. Association of age, sex and seat belt use with the risk of early death in drivers of passenger cars involved in traffic crashes. *Int. J. Epidemiol.* **2009**, *38*, 1128–1134. [CrossRef]

37. Olszewski, P.; Szagała, P.; Rabczenko, D.; Zielińska, A. Investigating safety of vulnerable road users in selected EU countries. *J. Safety. Res.* **2019**, *68*, 49–57. [CrossRef] [PubMed]

38. Road Deaths in European Countries 2019 & 2020. Available online: https://www.statista.com/statistics/323869/international-and-uk-road-deaths/ (accessed on 9 January 2023).

39. Chliaoutakis, J.E.; Demakakos, P.; Tzamalouka, G.; Bakou, V.; Koumaki, M.; Darviri, C. Aggressive behavior while driving as predictor of self-reported car crashes. *J. Safety Res.* **2002**, *33*, 431–443. [CrossRef] [PubMed]

40. Loke, W.C.; Mohd-Zaharim, N. Resilience, decision-making and risk behaviours among early adolescents in Penang, Malaysia. *Vulnerable Child. Youth Stud.* **2019**, *14*, 229–241. [CrossRef]

41. Dailimail. Achtung! The Worst Drivers Are behind the Wheels of German Cars like BMWs, Audis and Mercs (but Relax if You See Someone in a Peugeot 206), Study Reveals. Available online: https://www.dailymail.co.uk/news/article-6525515/The-worst-drivers-wheels-German-cars-like-BMWs-Audis-Mercs-study-reveals.html (accessed on 9 January 2023).

42. Davies, G.M.; Patel, D. The influence of car and driver stereotypes on attributions of vehicle speed, position on the road and culpability in a road accident scenario. *Leg. Criminol. Psychol.* **2005**, *10*, 45–62. [CrossRef]

43. Davies, G. Estimating the speed of vehicles: The influence of stereotypes. *Psych. Crime Law.* **2009**, *15*, 293–312. [CrossRef]

44. Fischer, J. *A Typology of Car Drivers.* Spiegel Institut Mannheim. Available online: http://www.spiegel-institut.de/en/presse/typology-car-drivers (accessed on 9 January 2023).

45. Perepjolkina, V. Aggressive driving: Associations between car's brand and driver's gender among young drivers in Latvia. Signum Temporis. *J. Pedag. Psych.* **2009**, *2*, 85–92. [CrossRef]

46. Piff, P.K.; Stancato, D.M.; Côté, S.; Mendoza-Denton, R.; Keltner, D. Higher social class predicts increased unethical behavior. *Proc. Natl. Acad. Sci. USA* **2012**, *109*, 4086–4091. [CrossRef]

47. Ghayad, A. Why BMW Drivers Are The Worst On The Road? Available online: https://engineerine.com/why-bmw-drivers-are-worst-on-road/ (accessed on 27 February 2023).

48. Stiles, D.A.; Sebben, D.J.; Gibbons, J.L.; Wiley, D.C. Why adolescent boys dream of becoming professional athletes. *Psychol. Rep.* **1999**, *84* (Suppl. S3), 1075–1085. [CrossRef]

49. Ferraro, T.; Rush, S. Why athletes resist sport psychology. *Athl. Insight J.* **2000**, *2*, 9–14.

50. Piepiora, P. Assessment of Personality Traits Influencing the Performance of Men in Team Sports in Terms of the Big Five. *Front. Psychol.* **2021**, *12*, 679724. [CrossRef] [PubMed]

51. Potkanowicz, E.; Mendel, R. The Case for Driver Science in Motorsport: A Review and Recommendations. *Sports Med.* **2013**, *43*, 565–574. [CrossRef]

52. Potkanowicz, E.S.; Ferguson, D.P.; Greenshields, J.T. Responses of Driver-Athletes to Repeated Driving Stints. *Med. Sci. Sports Exerc.* **2021**, *53*, 551–558. [CrossRef]

53. Filho, E.; Di Fronso, S.; Mazzoni, C.; Robazza, C.; Bortoli, L.; Bertollo, M. My heart is racing! Psychophysiological dynamics of skilled racecar drivers. *J. Sport. Scie.* **2015**, *33*, 945–959. [CrossRef]

54. Beaune, B.; Sylvain, D. Cardiac chronotropic adaptation to open-wheel racecar driving in young pilots. *Int. J. Perform. Anal. Sport.* **2011**, *11*, 325–334. [CrossRef]

55. Potkanowicz, E.S. A Real-Time Case Study in Driver Science: Physiological Strain and Related Variables. *Int. J. Sports Physiol. Perform.* **2015**, *10*, 1058–1060. [CrossRef] [PubMed]

56. Long, B.L.; Gillespie, A.I.; Tanaka, M.L. Mathematical Model to Predict Drivers' Reaction Speeds. *J. Appl. Biomech.* **2012**, *28*, 48–56. [CrossRef] [PubMed]

57. Klavora, P.; Heslegrave, R.J. Senior Drivers: An Overview of Problems and Intervention Strategies. *J. Aging Phys. Act.* **2002**, *10*, 322–335. [CrossRef]

58. Burtaverde, V.; Ene, C.; Chiriac, E.; Avram, E. Decoding the link between personality traits and resilience. Self-determination is the key. *Curr. Issues Pers. Psychol.* **2021**, *9*, 195–204. [CrossRef]

59. Topolewska, E.; Skimina, E.; Strus, W.; Cieciuch, J.; Rowiński, T. The short IPIP-BFM-20 questionnaire for measuring the Big Five. *Rocz. Psych.* **2017**, *17*, 385–402.

60. Donnellan, M.B.; Oswald, F.L.; Baird, B.M.; Lucas, R.E. The Mini-IPIP Scales: Tiny-yet-effective measures of the Big Five Factors of Personality. *Psych. Asses.* **2006**, *18*, 192–203. [CrossRef] [PubMed]

61. Piotrowski, A.; Makarowski, R.; Predoiu, R.; Predoiu, A.; Boe, O. Resilience and subjectively experienced stress among paramedics prior to and during the COVID-19 pandemic. *Fron. Psych.* **2021**, *12*, 664540. [CrossRef]

62. Makarowski, R.; Piotrowski, A.; Predoiu, R.; Görner, K.; Predoiu, A.; Mitrache, G.; Malinauskas, R.; Vicente-Salar, N.; Vazne, Z.; Cherepov, E.; et al. The English speaking, Hungarian, Latvian, Lithuanian, Romanian, Russian, Slovak, and Spanish adaptations of Makarowski's Stimulating and Instrumental Risk Questionnaire for martial arts athletes. *Archiv. Budo* **2021**, *17*, 1–33.

63. Makarowski, R.; Predoiu, R.; Piotrowski, A.; Görner, K.; Predoiu, A.; Oliveira, R.; Pelin, R.A.; Moanță, A.D.; Boe, O.; Rawat, S.; et al. Coping Strategies and Perceiving Stress among Athletes during Different Waves of the COVID-19 Pandemic—Data from Poland, Romania, and Slovakia. *Healthcare* **2022**, *10*, 1770. [CrossRef] [PubMed]

64. Faul, F.; Erdfelder, E.; Lang, A.-G.; Buchner, A. G*Power 3: A flexible statistical power analysis program for the social, behavioral, and biomedical sciences. *Behav. Res. Methods* **2007**, *39*, 175–191. [CrossRef]

65. Cohen, J. *Statistical Power Analysis for the Behavioral Sciences*; Lawrence Erlbaum Associates, Inc.: New York, NY, USA, 1988.

66. Tabachnick, B.G.; Fidell, L.S. *Using Multivariate Statistics*, 6th ed.; Pearson: London, UK, 2013.

67. Moran, M.; Baron-Epel, O.; Assi, N. Causes of road accidents as perceived by Arabs in Israel: A qualitative study. *Transp. Res. F: Traffic Psychol. Behav.* **2010**, *13*, 377–387. [CrossRef]
68. Thompson, E.R. Development and validation of an international English big-five mini-markers. *Pers. Individ. Differ.* **2008**, *45*, 542–548. [CrossRef]
69. Ishaq, G.; Rafique, R.; Asif, M. Personality Traits and News Addiction: Mediating Role of Self-Control. *J. Dow. Univ. Health Sci.* **2017**, *11*, 31–36. [CrossRef]
70. Merritt, C.J.; Tharp, I.J. Personality, self-efficacy and risk-taking in parkour (free-running). *Psychol. Sport Exerc.* **2013**, *14*, 608–611. [CrossRef]
71. Bowen, L.; Smith, A.P. Associations between job characteristics, mental health and driving: A secondary analysis. *J. Educ. Soc. Behav. Scien.* **2019**, *29*, 1–25. [CrossRef]
72. Erez, K.; Luria, G. The mediating role of smartphone addiction on the relationship between personality and young drivers' smartphone use while driving. *Transp. Res. F Traffic Psychol. Behav.* **2018**, *59*, 203–211. [CrossRef]
73. Swaminathan, V.; Kubat Dokumaci, U. Do all, or only some personality types engage in spreading negative WOM? An experimental study of negative WOM, big 5 personality traits and brand personality. *J. Glob. Schol. Mark. Scie.* **2021**, *31*, 260–272. [CrossRef]
74. Piepiora, P.; Witkowski, K. Personality profile of combat sports champions against neo-gladiators. *Arch. Budo* **2020**, *16*, 281–293.
75. Krahé, B.; Fenske, I. Predicting aggressive driving behavior: The role of macho personality, age, and power of car. *Aggress. Behav.* **2002**, *28*, 21–29. [CrossRef]
76. Zaitchik, M.C.; Mosher, D.L. Criminal justice implications of the macho personality constellation. *Crim. Justice Behav.* **1993**, *20*, 227–239. [CrossRef]
77. Sanjeev, M.; Sehrawat, A.; Santhosh, K. iPhone as a proxy indicator of adaptive narcissism: An empirical investigation. *Psychol. Mark.* **2019**, *36*, 895–904. [CrossRef]
78. Lönnqvist, J.E.; Ilmarinen, V.; Leikas, S. Not only assholes drive Mercedes Besides disagreeable men, also conscientious people drive high-status cars. *Int. J. Psychol.* **2020**, *55*, 572–576. [CrossRef]
79. Srivastava, A.; Dasgupta, S.A.; Ray, A.; Bala, P.K.; Chakraborty, S. Relationships between the "Big Five" personality types and consumer attitudes in Indian students toward augmented reality advertising. *Aslib J. Inf. Manag.* **2021**, *73*, 967–991. [CrossRef]
80. Managuran, K.D.; Varathan, K.D.; Al-Garadi, M.A.D. Can Music Likes in Facebook Determine Your Personality? *Malays. J. Comput. Sci.* **2021**, *34*, 61–81. [CrossRef]
81. Cashmore, E. *Sport and Exercise Psychology: The Key Concepts*, 2nd ed.; Routledge: Oxford, UK, 2008.
82. Predoiu, R.; Makarowski, R.; Görner, K.; Predoiu, A.; Boe, O.; Ciolacu, M.V.; Grigoroiu, C.; Piotrowski, A. Aggression in martial arts coaches and sports performance with the COVID-19 pandemic in the background—A dual processing analysis. *Arch. Budo* **2022**, *18*, 23–36.
83. Urzeală, C.; Teodorescu, S. Violence in sports. *Discobolul Phys. Educ. Sports Kinetothe. J.* **2018**, *53*, 17–24.
84. Szczypinska, M.; Samełko, A.; Guszkowska, M. Strategies for Coping with Stress in Athletes During the COVID-19 Pandemic and Their Predictors. *Front. Psychol.* **2021**, *12*, 624949. [CrossRef] [PubMed]
85. Gould, D.; Dieffenbach, K.; Moffet, A. Psychological characteristics and their development in Olympic champions. *J. Appl. Sport. Psychol.* **2020**, *14*, 172–204. [CrossRef]
86. Mansell, W.; Urmson, R.; Mansell, L. The 4Ds of Dealing with Distress—Distract, Dilute, Develop, and Discover: An Ul-tra-Brief Intervention for Occupational and Academic Stress. *Front. Psychol.* **2020**, *11*, 611156. [CrossRef]
87. LeUnes, A. *Sport Psychology*, 4th ed.; Psychology Press: New York, NY, USA, 2008.
88. Hernandez, J.; Anderson, K.B. Internal martial arts training and the reduction of hostility and aggression in martial arts students. *Psi. Chi. J. Psychol. Res.* **2015**, *20*, 169–176. [CrossRef]
89. di Fronso, S.; Montesano, C.; Costa, S.; Santi, G.; Robbaza, C.; Bertollo, M. Rebooting in sport training and competitions: Athletes' perceived stress levels and the role of interoceptive awareness. *J. Sports Sci.* **2022**, *40*, 542–549. [CrossRef] [PubMed]
90. Lyubomirsky, S.; King, L.A.; Diener, E. The benefits of frequent positive affect: Does happiness lead to success? *Psychol. Bull.* **2005**, *131*, 803–855. [CrossRef]
91. Vasile, C. Stress and aging perspectives. Psychological and neuropsychological interventions. *J. Educ. Sci. Psychol.* **2022**, *12*, 154–159.
92. Dymecka, J.; Gerymski, R.; Bidzan, M. Selected psychometric aspects of the Polish version of the Liverpool Self-efficacy Scale. *Curr. Issues Pers. Psychol.* **2020**, *8*, 339–351. [CrossRef]
93. Makarowski, R.; Piotrowski, A.; Görner, K.; Predoiu, R.; Predoiu, A.; Mitrache, G.; Malinauskas, R.; Vicente-Salar, N.; Vazne, Z.; Bochaver, K.; et al. The Hungarian, Latvian, Lithuanian, Polish, Romanian, Russian, Slovak, and Spanish, adaptation of the Makarowski's Aggression Questionnaire for martial arts athletes. *Archiv. Budo* **2021**, *17*, 75–108.
94. Buss, A.H.; Perry, M. The aggression questionnaire. *J. Pers. Soc. Psychol.* **1992**, *63*, 452–459. [CrossRef] [PubMed]

Article

Relationships between Sex and Adaptation to Physical Exercise in Young Athletes: A Pilot Study

Gabriella Pinto [1,2,†], Rosamaria Militello [3,†], Angela Amoresano [1,2], Pietro Amedeo Modesti [4], Alessandra Modesti [3] and Simone Luti [3,5,*]

1   Department of Chemical Sciences, University of Naples Federico II, 80126 Naples, Italy; gabriella.pinto@unina.it (G.P.); angela.amoresano@unina.it (A.A.)
2   INBB, Istituto Nazionale Biostrutture e Biosistemi, Consorzio Interuniversitario, 00136 Rome, Italy
3   Department of Biomedical, Experimental and Clinical Sciences "Mario Serio", University of Florence, 50134 Florence, Italy; rosamaria.militello@unifi.it (R.M.); alessandra.modesti@unifi.it (A.M.)
4   Department of Experimental and Clinical Medicine, University of Florence, 50134 Florence, Italy; pa.modesti@unifi.it
5   Institute for Sustainable Plant Protection, National Research Council of Italy, 50019 Sesto Fiorentino, Italy
*   Correspondence: simone.luti@unifi.it
†   These authors contributed equally to this work.

Abstract: The purpose of this study was to compare the redox, hormonal, metabolic, and lipid profiles of female and male basketball players during the seasonal training period, compared to their relative sedentary controls. 20 basketball players (10 female and 10 male) and 20 sedentary controls (10 female and 10 male) were enrolled in the study. Oxidative stress, adiponectin level, and metabolic profile were determined. Male and female athletes showed an increased antioxidant capacity (27% for males; 21% for females) and lactate level (389% for males; 460% for females) and reduced salivary cortisol (25% for males; 51% for females) compared to the sedentary controls. Moreover, a peculiar metabolite (in particular, amino acids and urea), hormonal, and lipidic profile were highlighted in the two groups of athletes. Female and male adaptations to training have several common traits, such as antioxidant potential enhancement, lactate increase, and activation of detoxifying processes, such as the urea cycle and arachidonic pathways as a response to inflammation. Moreover, we found different lipid and amino acid utilization related to sex. Deeper investigation could help coaches in developing training programs based on the athletes' sex in order to reduce the drop-out rate of sporting activity by girls and fight the gender stereotypes in sport that also have repercussions in social fields.

Keywords: sport metabolomics; oxidative stress; hormone signalling; adiponectin; basketball

Citation: Pinto, G.; Militello, R.; Amoresano, A.; Modesti, P.A.; Modesti, A.; Luti, S. Relationships between Sex and Adaptation to Physical Exercise in Young Athletes: A Pilot Study. *Healthcare* 2022, 10, 358. https://doi.org/10.3390/healthcare10020358

Academic Editors: João Paulo Brito and Rafael Oliveira

Received: 13 December 2021
Accepted: 7 February 2022
Published: 11 February 2022

Publisher's Note: MDPI stays neutral with regard to jurisdictional claims in published maps and institutional affiliations.

1. Introduction

During exercise and regular training, several metabolic changes occur in an organism, leading to the activation of adaptive mechanisms aimed at establishing a new dynamic equilibrium, especially at the metabolic level, which guarantees health and better performance in elite athletes [1]. Excessive training and effort could lead to chronic fatigue and muscle damage, with lower performance. Therefore, these adaptive mechanisms are the result of a fine balance between: (i) oxidative stress/inflammation induced by exercise that increases performance and health, and (ii) oxidative stress due to excessive effort that causes fatigue and muscle damage [2]. In this context, the role of oxidative stress and hormone signalling linked to inflammation appears to be crucial for adaptation [2]. Elite athletes have a very strong antioxidant defence system that guarantees free radical detoxification during acute exercise and recovery and, at the same time, the regulation of several hormones ensures adaptation [3,4].

Adiponectin is a signalling hormone with an anti-inflammatory effect, and its plasma level is modified by training [5]. It is a myokine induced by pro-inflammatory mediators (e.g., interferon γ and TNF) [6] produced during training, and it exerts a protective effect against oxidative stress [7]. It seems to play a role in adaptation to physical exercise, acting in an autocrine/paracrine manner. Several studies have indicated that adiponectin plasma concentration is determined by fat mass, with the additional independent effects of age and sex: adiponectin concentration is higher in females and it increases with age [8–10]. Some authors have suggested that adiponectin plasma levels are inversely associated with muscle strength [11], and most of this information has been gained through studies on male athletes, while less is known about female athletes. It could be of particular interest to evaluate training responses in relation to sex, since, between male and female athletes, different hormonal, physiological, and muscular statuses exist, which may influence these adaptive mechanisms. Comparison of these mechanisms could broaden our knowledge. Most of the sex-related differences in sport performance are ascribed to androgens; in particular, testosterone and its related metabolite, dihydrotestosterone [12–16]. This sex-based difference is the premise to explain why males acquire larger muscle mass and greater strength, larger and stronger bones, and higher circulating haemoglobin [16]. The observed differences in hormonal, physiological, and muscular status between women and men in physical exercise and adaptation may also be displayed at the metabolic level. There may be sex-independent or sex-dependent traits in adaptations to sport.

Basketball is an intermittent team sport played on a court [17] and characterized by frequent movement changes due to the transition between offense and defence [18]. These transitions differ in terms of movement pattern (e.g., running, jumping, or both), intensity, distance, frequency, and duration, while jumping (unlike in other team sports) approximately every minute [19,20]. Because of these frequent changes during the match, there are periods of high intensity activity interspersed with periods of low or moderate intensity, characterized by aerobic and anaerobic intermittent demands [19].

The aim of the present study is to evaluate plasma oxidative status, adiponectin, cortisol, and steroid hormones related to plasma metabolites and fatty acid levels between female and male basketball players, compared to relatively sedentary controls, with an effort to elucidate the different and common traits between male and female adaptations to physical exercise.

2. Materials and Methods

2.1. Design

This research was a preliminary cross-sectional study involving a total of 40 subjects: 10 female athletes, 10 female sedentary controls, 10 male athletes, and 10 male sedentary controls. We wanted to highlight the metabolomic, hormonal, and lipidomic profiles of young athletes attributable to training; therefore, we used female and male sedentary people as controls for the physiological differences found within the population. After initial screening, all participants received a complete explanation of the purpose, risks, and procedures of the study. A written informed consent was provided prior to enrolment in the study, which was conducted according to the policy statement set forth in the Declaration of Helsinki, and all the experiments were conducted according to established ethical guidelines. The study was approved by the local ethics committee of the University of Florence, Italy (AM Gsport 15840/CAM_BIO).

2.2. Participants

A cohort of 10 female professional basketball players, 10 male professional basketball players, and the respective controls, both 10 females and 10 males, were involved in this study.

The male and female athletes were recruited from the local sports club in Florence (Italy) "US AFFRICO-Firenze" and had been practicing this sport for more than 5 years.

The control groups were recruited among students of the degree course in Motor Sciences, Sport and Health at the University of Florence, and they did not practice any sports. All subjects volunteered to participate following an explanation of all the experimental procedures. They were all adults and of Caucasian ethnicity.

The participants completed a medical history and physical activity questionnaire in order to determine their eligibility. None of them used antioxidant or nutritional supplements. Subjects were selected because of non-smoking status, age, and stable body weight. All females were enrolled randomly with respect to menstrual cycle.

2.3. Training Protocol

All players attended the training sessions during the whole season from September until May, and the samples were collected during the twelfth week. The athletes trained at least 5 times a week, according to specific training programs, with sessions lasting 2 h per day. The training protocol involved both technical and aerobic exercise.

Team coaches tracked day-to-day training data for each player and for each training session over the entire season. To assess their adherence to the training plans, the subjects completed the physical activity questionnaires. In particular, we used the International Physical Activity Questionnaires (IPAQ) [21], and the subjects included in category 3, practicing 7 or more days of any combination of walking, moderate-intensity, or vigorous-intensity activities, achieving a minimum of at least 3000 MET minutes/week, were selected. In the present study, the athletic trainers used the rating of perceived exertion (RPE) system of Carl Foster to monitor training load (TL) to obtain the individual responses of athletes in different training sessions [22]. The RPEs of each player were registered across a whole season, and RPE was measured following the completion of each set of exercise and 30 min post-exercise (session RPE).

For both sexes, team coaches followed this classical training program: 30 min of low-moderate running, followed by interval training runs to improve speed and simulate different game situations. Then, as previously reported [23], 3 sets of 12 repetitions of exercises for the shoulder muscles and 3 sets of 15 repetitions of exercises for the abdominals (oblique and the rectus abdominis muscles) were performed.

2.4. Materials

Unless specified, all reagents were obtained from Bio-Rad Laboratories (Hercules, CA, USA). Solvents used for the sample preparation and LC–MS/MS analysis have >99.9% purity, as reported by the manufacturing companies. Acetonitrile (ACN) CHROMASOLV was obtained from Honeywell (Charlotte, NC, USA). Methanol (MeOH) and formic acid (HCOOH) were purchased from Sigma (St. Louis, MO, USA).

2.5. Procedures

2.5.1. Anthropometric Assessment

All subjects recruited were in the 18–30 age group. Anthropometric measurements were performed at the time of biological samples. Weight was measured to the nearest 0.1 kg and height (H) to the nearest 0.5 cm. All measurements were performed in resting conditions and by the same operator. Body mass index (BMI) was calculated from the ratio of body weight (kg) to body height (m^2).

2.5.2. Biological Samples

At the twelfth week of training of the athlete groups, a fasting capillary blood sample was taken, using a heparinized microvette (Sarstedt ref. 85.1018), from the forefinger of each volunteer after obtaining written and informed consent. The biological samples (capillary blood and saliva) were collected in morning, 48 h after the last competition and 24 h after the training session. Capillary collection was preferred to venous collection because of its reduced invasiveness, simple execution, and lower cost; moreover, its small volumes of samples were sufficient for the experiments carried out [23].

2.6. Plasma Oxidative Stress Measurements

The antioxidant capacity and levels of reactive oxygen metabolites on plasma were determined by using the BAP test (biological antioxidant potential) and the d-ROMs test (derivates of Reactive Oxygen Metabolite). The biomarkers d-ROM and BAP were selected based on their long-term stability. The measurements were carried out according to the manufacturer's instructions.

Analyses were performed using a free radical analyzer system (FREE Carpe Diem, DIACRON INTERNATIONAL s.r.l, Grosseto, Italy) that included a spectrophotometric device reader and a thermostatically regulated mini-centrifuge. The d-ROM results are expressed in arbitrary units (UCarr), one unit of which corresponds to 0.8 mg/L of hydrogen peroxide. The BAP results are expressed in µmol/L of the reduced ferric ions [24].

2.7. Adiponectin Western Blot Analysis

Plasma samples were clarified by centrifugation and total protein contents were obtained using a Bradford assay. An equal amount of each sample (12.5 µg of total proteins) was added to 4 × Laemmli buffer (0.5 M TrisHCl pH 6.8, 10% SDS, 20% glycerol, β-mercaptoethanol, and 0.1% bromophenol blue) and boiled for 5 min. Samples were separated on 12% SDS/PAGE and transferred onto a PVDF membrane using the Trans-Blot Turbo Transfer System (Bio-Rad Laboratories, Hercules, CA, USA). PVDF was probed with primary antibody (Acrp30 Santa Cruz) diluted 1:1000 in 2% milk, and then incubated overnight at 4 °C. After incubation with horseradish peroxidase (HRP)-conjugated anti-mouse IgG (1:10,000) (Santa Cruz Biotechnology, Dallas, TX, USA)), immune complexes were detected with the enhanced chemiluminescence (ECL) detection system (GE Healthcare, Chicago, IL, USA) and by Amersham Imager 600 (GE Healthcare). For quantification, the blot was subjected to densitometric analysis using the ImageJ 1.53 program. The intensity of the immunostained bands was normalized with the total protein intensities measured by Coomassie brilliant blue R-250 from the same PVDF membrane blot as previously reported [24].

2.8. Salivary Cortisol Measurement

The saliva samples of each participant were collected using a salivette cortisol (Sarstedt, Nümbrecht, Germany). All saliva samples were taken at the same hour of the day to avoid any variations due to circadian rhythm. Participants were instructed to not eat, drink, or brush their teeth for 30 min before saliva collection. The cotton sliver of the salivette was taken out and put in the sublingual for 1/2 min, and then put back into the salivette. The samples were centrifuged at 2000 rpm for 2 min. Saliva flow was collected from the salivette's bottom and stored at −20 °C for further laboratory analysis.

Salivary cortisol was detected using a specific commercial enzyme-linked immunosorbent assay kit (item no. 500360 Cayman chem, Ann Arbor, MI, USA). Saliva samples (50 µL) were used, and the assay was performed according to manufacturer-recommended procedures. The sensitivity provided by the manufacturer is approximately 35 pg/mL, with a detection range from 6.6–4000 pg/mL. The samples were analyzed in duplicate [25].

2.9. Gas Chromatography–MS (GC–MS) Analysis of Plasma Metabolites

Metabolite analysis of the plasma samples of the basketball players and control males and females were performed as reported in Militello et al. [23]. Briefly, 100 µL of pooled plasma samples were subjected to methanol/chloroform precipitation and the upper phase was evaporated at room temperature in a rotovapor. The obtained metabolites were analyzed in triplicate by the GC–MS technique after their derivatization with N-trimethylsilyl-N-methyl trifluoroacetamide (MSTFA). The MassHunter data processing tool (Agilent, Santa Clara, CA, USA) was used to obtain a global metabolic profiling using the Fiehn Metabolomics RTL library (Agilent G1676AA).

2.10. Extraction of Steroid Hormones and Standard Preparation

An aliquot (50 µL) of plasma from the male and female basketball players and from the control subjects weas added to four volumes of cold MeOH to precipitate the proteins and to extract steroid hormones. After a sonication step of 10 min, samples were centrifuged 10 min at 12,000 rpm at 4 °C, and supernatants were recovered and dried under vacuum.

A similar procedure was used to extract steroid hormones spiked in depleted plasma, in agreement with the guidelines of the certified Eureka kit (https://www.eurekakit.com, accessed on 1 March 2021). A list of nineteen analytes at six levels of concentration enclosed within physiological ranges was included in the datasheet of certified Eureka kit. Standard mixtures were used to perform the quantitative analysis by external standard method.

All dried extracts were then suspended in 50 µL of MeOH containing 0.1% HCOOH for the subsequent LC–MS/MS analysis.

The LC–MS/MS analysis was performed on an AB Sciex QTrap 4000 mass spectrometer, coupled to an ExpressHT™ Ultra HPLC system (Eksigent). Each extract (5 µL), cooled at 4 °C on a refrigerated auto sampler, was injected and separated on a Halo C18 1.0 × 50 mm, 2.7 µm column (at 40 °C) by using a flow rate of 40 µL/min. An aqueous solution containing 5 mM ammonium formate (Sigma Aldrich, Saint Luis, MO, USA) and 0.1% HCOOH was used as a mobile phase, while methanol acidified with 0.1% HCOOH was used as a phase B. A 5 min gradient (0–1 min 30% B, 1–5 min 30–95% B) was applied, with a further 2 min at 95% B to wash the column.

LC–MS/MS analysis was performed in MRM ion mode in positive ion mode (ESI$^+$). The instrumental settings included precursor ions (m/z) and product ions (m/z), and optimal collision energies as previously reported [24].

2.11. Transesterification of Fatty Acid in Plasma Samples

Methanol extracts deprived of the protein component (see above in paragraph "extraction of steroid hormones") were added with a methanolic solution containing 8% HCl, up to a final HCl concentration of 2%. The reaction of transesterification was performed overnight at 95 °C within a Reacti-Therm™ (Thermo Fisher Scientific™, Waltham, MA, USA) system. At the end of the reaction, the mixture was added to an aqueous solution (1 mL) and fatty acid methyl esters (FAME) were extracted by hexane (1 mL).

The same reaction was performed on a standard mixture containing palmitoleic acid, palmitic acid, linoleic acid, and oleic acid (Sigma Aldrich) for the quantification by external standard method.

After centrifugation, the hexane layer of all samples containing FAMEs was placed into a gas chromatography vial, capped, and directly subjected to a GC analysis performed as reported in Table 1.

Table 1. GC–MS parameters used for fatty acids analysis.

	Rate (°C/min)	Value (°C)	Hold Time (min)	Run Time (min)
(initial)		90	1	1
Ramp 1	10	140	2	8
Ramp 2	5	180	0	16
Ramp 3	10	280	5	31

2.12. Gas Chromatography–Mass Spectrometry (GC–MS) Analysis of Plasma Fatty Acid Methyl Esters (FAMEs)

The gas chromatography (GC) analyses were performed using Agilent GC 6890 coupled with a 5973 MS detector. Each hexane extract (1 µL) was injected into the GC–MS and the analytes were separated on an HP-5 capillary column (30 m × 0.25 mm, 0.25 mM, 5% polisilarilene, and 95% polydimethylsiloxane). Helium was used as the carrier gas at a rate of 1.0 mL min^{-1}. The GC injector was maintained at 230 °C, while the oven temperature was held at 60 °C for 3 min, and then increased to 150 °C at 10 °C/min, increasing to 180 °C at 5 °C/min, and finally to 280 °C at 10 °C/min and held for 5 min, for a total separation time of 30 min. The analyzer temperature was kept at 250 °C. The collision energy was set to a value of 70 eV and the fragment ions generated were analyzed at a mass range of 20–500 m/z.

The identification of each compound was based on the comparison of retention time with the relative standard and fragmentation spectra matching those collected into the NIST 05 Mass Spectral Library.

2.13. Statistical Analysis

Data are presented as means $+/-$ standard deviation (SD) from at least three experiments. Statistical analysis was performed by two-way ANOVA (Tukey's multiple comparisons test) using GraphPad Prism 6.01. The Tukey test compared every mean with every other mean, computing a confidence interval for multiple comparisons of 95% confidence. Significance was defined as $p < 0.05$. We checked the normality and homogeneity distribution of our data by the D'Agostino–Pearson and Brown–Forsythe tests, respectively. For Western blot quantification, the blot was subjected to densitometric analysis using the ImageJ program. As measure of the effect size, we reported the eta-squared (η^2) calculated in ANOVA (Tukey's multiple comparisons test) performed on all groups together using GraphPad Prism 8, considering the following intervals: 0.01–0.20 = small effect; 0.21–0.60 = moderate effect; and 0.61–0.99 = high effect.

3. Results

3.1. Participants' Characteristics

Descriptive characteristics of the participants as mean $\pm$ SD are presented in Table 2. The mean age of the participants was 24.4 ± 4.2 years.

Table 2. Participants' characteristics.

Characteristics	Mean (SD)				Tukey's Test [a]			
	Basket Male	Control Male	Basket Female	Control Female	BM vs. CM	BF vs. CF	BM vs. BF	CM vs. CF
Age (year)	21 ± 2.2	26.1 ± 4.1	25.1 ± 5.5	26.9 ± 2.2	0.01 *	0.77	0.08	0.97
Weight (kg)	81.5 ± 10.2	73 ± 8.7	68.7 ± 11.9	58.7 ± 5.8	0.17	0.17	0.03 *	0.02 *
Height (cm)	186 ± 0.06	178.7 ± 0.06	175.6 ± 0.08	163.4 ± 0.06	0.06	0.003 **	0.006 **	<0.0001 ****
BMI (kg/m^2)	23.6 ± 2.7	22.9 ± 2.9	22.1 ± 2.05	22 ± 2.3	0.64	1.00	0.18	0.72

[a] Tukey's test was performed by GraphPad Prism 8.0 software between the male basketball group (BM), male control group (CM), female basketball group (BF), and female control group (CF) (* $p < 0.05$; ** $p < 0.01$; and **** $p < 0.0001$).

Significant differences emerged between the heights of female basketball players in comparison with the respective controls, and between males and females in both groups (athletes and controls). As expected, males were taller and heavier than females. Despite this, BMI was calculated, and no significant differences were found between the player groups and the control groups ($p = 0.64$ for males, $p = 1$ for females; $\eta^2 = 0.084$) and between males and females ($p = 0.18$ for athletes, $p = 0.72$ for controls; $\eta^2 = 0.084$). Although a significant difference is evident in the age of male athletes compared to the controls, we did not consider it relevant because they were still young males.

Plasma oxidative stress, adiponectin level, metabolite analysis, and salivary cortisol measurements in females were derived from our previous study [23].

3.2. Plasma Oxidative Stress and Salivary Cortisol Measurements

We measured antioxidant capacity (BAP) and the levels of oxidative species (d-ROM) from all the participants, and the results are reported in Figure 1. In a detailed analysis of antioxidant capacity (Figure 1A), we found a significant increase (25.2%, $p < 0.001$; $\eta^2 = 0.655$) in the male basketball players compared to the male control group, and the BAP mean value was 2153.14 ± 211 μmol/L for the athletes in comparison with the controls (1720.33 ± 216 μmol/L). Moreover, a significant increase (21.7%, $p < 0.001$; $\eta^2 = 0.655$) was observed in the female basketball players (1764.7 ± 163 μmol/L) compared to the female control group (1450 ± 139 μmol/L).

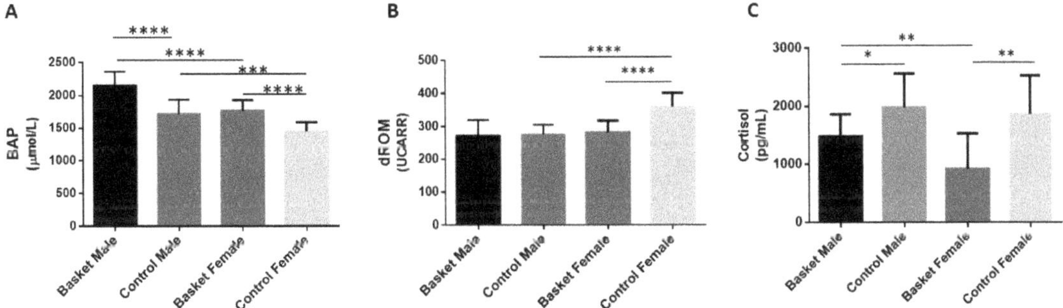

Figure 1. Plasma oxidative stress and salivary cortisol determination. (**A**) The antioxidant capacity was evaluated using the BAP test (biological antioxidant potential). (**B**) The levels of reactive oxygen metabolites using the d-ROM test (derivates of Reactive Oxygen Metabolite) by a free radical analyser system (FREE Carpe Diem, DIACRON INTERNATIONAL s.r.l). (**C**) The cortisol levels measured using an enzyme-linked immunosorbent assay kit. All the measurements ($n = 10$) were performed in triplicate and are reported in the histograms as mean ± SD. Female data are from Militello et al., 2021. The statistical analysis was carried out by two-way ANOVA using GraphPad Prism 8 (* $p < 0.05$; ** $p < 0.01$; *** $p < 0.001$; and **** $p < 0.0001$).

Significant differences were also found between males and females in both the basketball and control groups; in particular, male athletes showed an increase (22%; $p < 0.001$; $\eta^2 = 0.655$) compared to female athletes and an analog difference was found between control groups where, in fact, males showed an increase (18.6%; $p < 0.05$; $\eta^2 = 0.655$) compared to females.

Regarding d-ROM values (Figure 1B), the results showed that there is no significant difference between the male athletes (271.9 ± 47 UCarr) and the respective controls (275.5 ± 29 UCarr). On the contrary, the female athletes (281.1 ± 36 UCarr) showed a decrease of d-ROM values of about 21.7% ($p < 0.0001$; $\eta^2 = 0.452$) compared to the female controls (358.8 ± 42 UCarr).

A significant difference was also shown in the control groups between males and females, with a decrease in male subjects of 23.2% ($p < 0.0001$; $\eta^2 = 0.452$) compared to females. The same difference was not found between the male and female basketball groups.

We determined the concentration of salivary cortisol (Figure 1C) and we found a significant decrease (51%; $p = 0.009$; $\eta^2 = 0.385$) in the female basketball group in comparison to the female control group: the cortisol mean value was 1867.68 ± 659.83 pg/mL for female controls and 915.09 ± 613.69 pg/mL for female basketball players. A significant difference was found also in the male groups; in fact, the athletes showed a reduction of 25% (1478.3 ± 385 pg/mL, $p = 0.017$; $\eta^2 = 0.385$) compared to the control values (1979.65 ± 583.9 pg/mL).

There was no significant difference between control groups, but between the athlete groups, the male basketball players had shown an increase of 61.5% ($p = 0.004$; $\eta^2 = 0.385$) in comparison with the female athletes.

3.3. Plasma Adiponectin Determination

The plasma adiponectin level was determined by Western blot analysis as previously reported [23]. In the female groups, no significant differences were found between athletes and controls. Regarding the male groups, adiponectin levels showed a significant increase of 35.4% ($p < 0.0001$, $\eta^2 = 0.230$) in basketball players in comparison with the respective controls (Figure 2). In addition, between the controls, we observed a significant decrease of 31.4% ($p < 0.0001$, $\eta^2 = 0.230$) in males compared to females.

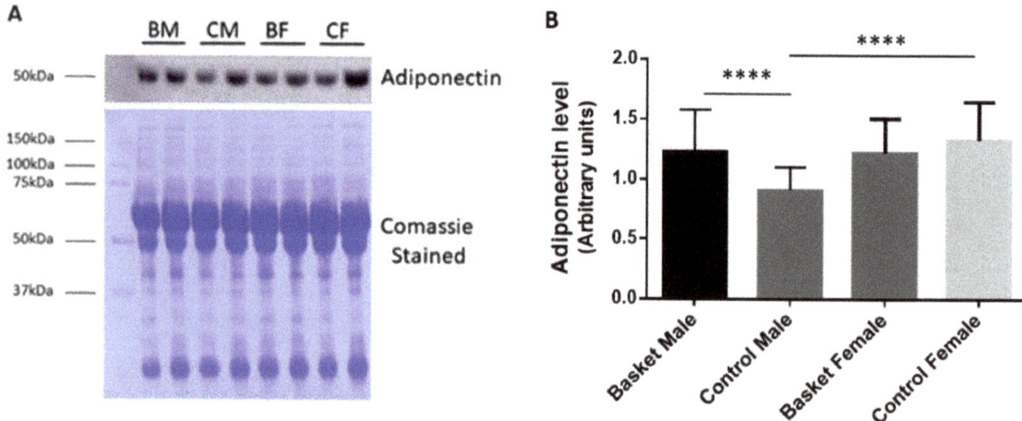

Figure 2. Plasma adiponectin levels. (**A**) A representative immunoblot of adiponectin with the corresponding Coomassie-stained PVDF membrane. (**B**) Relative quantification of adiponectin carried out by the Image J program. Female data are from Militello et al., 2021. Statistical analysis was performed by two-way ANOVA (Tukey's multiple comparisons test) using GraphPad Prism 8. Bars represent the mean ± SD ($n = 10$; **** $p < 0.0001$).

3.4. GC–MS Metabolomic Analysis of Plasma Samples

Intending to highlight the metabolic similarity and differences between female and male basketball athletes, we determined their plasma metabolic profiles through GC–MS. A total of 50 different compounds were identified (Supplementary Table S1).

In female athletes, compared to female controls, we found six statistically different concentrated metabolites (Figure 3A). Among them, lactic acid (460%, $p = 0.0013$; $\eta^2 = 0.937$), urea (220%, $p = 0.0361$; $\eta^2 = 0.928$), and ornithine (130%, $p = 0.0467$; $\eta^2 = 0.815$) showed an increase in athletes compared to controls, while hydroxybutyric acid (75%, $p= 0.0258$; $\eta^2 = 0.641$), L-glutamic acid (83%, $p = 0.0353$; $\eta^2 = 0.637$), and L-asparagine (82%, $p = 0.0498$; $\eta^2 = 0.663$) were statistically decreased.

In males, as reported in Figure 3B, we found nine metabolites differentially concentrated in the basketball players compared to the controls. Among them, lactic acid (389%, $p = 0.0018$; $\eta^2 = 0.937$), acetohydroxamic acid (158%, $p = 0.0081$; $\eta^2 = 0.823$), L-ornithine (132%, $p = 0.0294$; $\eta^2 = 0.815$), and L- valine (218%, $p = 0.0002$; $\eta^2 = 0.936$) were increased. On the contrary, L-glutamic acid (79%, $p = 0.0467$; $\eta^2 = 0.865$), L-glycine (15%, $p = 0.0001$; $\eta^2 = 0.970$), methyl-beta-D-galactopyranoside (63%, $p = 0.0206$; $\eta^2 = 0.660$), urea (8%, $p = 0.0005$; $\eta^2 = 0.928$), and uric acid (61%, $p = 0.0091$; $\eta^2 = 0.726$) were decreased. Interestingly, L-isoleucine was found only in the male athletes.

Comparing the athletes, we found eight metabolites differentially concentrated in plasma. The amino acids L-ornithine (129%, $p = 0.0472$; $\eta^2 = 0.815$), L-proline (149%, $p = 0.0011$; $\eta^2 = 0.911$), L-threonine (130%, $p = 0.0435$; $\eta^2 = 0.643$), and L-valine (247%, $p = 0.0001$; $\eta^2 = 0.936$) were higher in males. Moreover, sorbitol (142%, $p = 0.0128$; $\eta^2 = 0.730$) and acetohydroxamic acid (191%, $p = 0.0017$; $\eta^2 = 0.823$) were more abundant in males. On the contrary, L-glycine (14%, $p = 0.0001$; $\eta^2 = 0.970$) and urea (7%, $p = 0.0004$; $\eta^2 = 0.928$) were lower in males compared to females.

With respect to the non-athlete subjects, we found that only L-proline (154%, $p = 0.0021$; $\eta^2 = 0.911$) and urea (191%, $p = 0.0201$; $\eta^2 = 0.928$) were differentially concentrated, and both were more abundant in males.

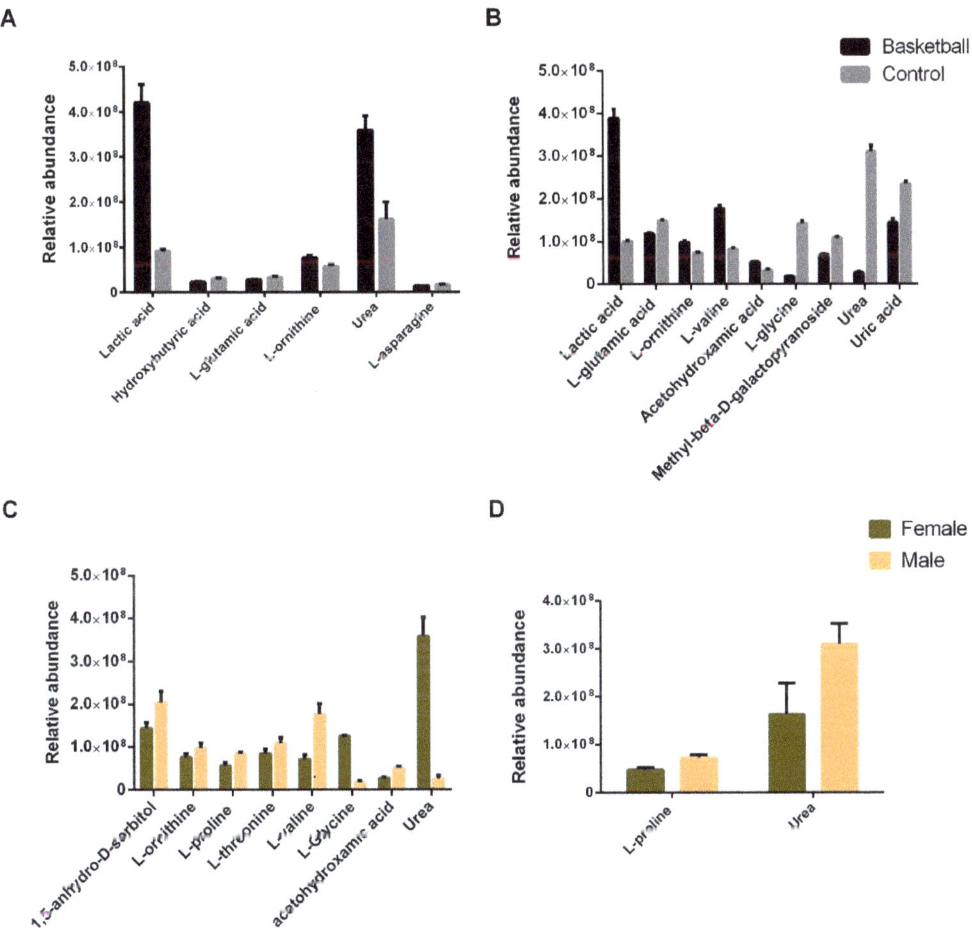

Figure 3. Plasma metabolomic profile of female and male basketball players using gas chromatography–mass spectrometry (GC–MS). (**A**) Histogram representation of plasma metabolites whose relative abundance is statistically different between female basketball athletes and controls ($p < 0.05$) (Militello et al., 2021). (**B**) Histogram representation of plasma metabolites whose relative abundance is statistically different between male basketball athletes and controls ($p < 0.05$). (**C**) Histogram representation of plasma metabolites whose relative abundance is statistically different between female and male basketball athletes ($p < 0.05$). (**D**) Histogram representation of plasma metabolites whose relative abundance is statistically different between female and male controls ($p < 0.05$). Statistical analysis was performed by two-way ANOVA (Tukey's multiple comparisons test) using GraphPad Prism 8.

3.5. Steroid Hormones Evaluation in Plasma Samples

Nine steroid hormones were quantified in a LC–MS/MS single run by using an MRM mode on plasma samples. The hormonal levels of both male and female of athletes were compared to those of the relative control groups. The quantification of each steroid hormone was obtained by the interpolation of extracted ion chromatogram peak area onto the relative calibration curves built up for each analyte.

Our samples indicated that dehydroepiandrosterone (DHEA) and its sulphate (DHEA-S; 182% for males, $p = 0.0002$; 169% for females, $p = 0.0012$; $\eta^2 = 0.9277$) were enhanced in both male and female subjects who practice basketball, compared to the control groups (Figure 4A,B). Such increases enabled us to quantify the DHEA in the plasma of athletes after basketball activity, which was otherwise under the detection limit in the control subjects (Figure 4B). An analogue trend is observed for estrone (252% for males, $p = 0.0001$; 242% for females, $p = 0.0002$; $\eta^2 = 0.9549$), testosterone (364% for males, $p \leq 0.05$), and 17-OH-progesterone (303% for males, $p = 0.0001$; 451% for females, $p = 0.0001$; $\eta^2 = 0.9954$), displaying a significant increase following physical activity (Figure 4C). Interestingly, the level of testosterone in the control females was below the limit of instrumental detection, whereas it was detected in females as a consequence of physical activity. Finally, cortisol, 21-deoxycortisol, cortisone, and estradiol showed a different trend in relation to sex: cortisol (78%, $p = 0.0075$; $\eta^2 = 0.8084$), cortisone (70%, $p = 0.0197$; $\eta^2 = 0.7490$), and estradiol (62%, $p = 0.0016$; $\eta^2 = 0.8298$) decreased only in female basketball players, while there was no change in males, and 21-deoxycortisol (152%, $p = 0.0094$; $\eta^2 = 0.7197$) increased in male athletes, but not in females ($p \leq 0.05$; Figure 4A,B).

Comparing the athletes, cortisol (124%, $p = 0.0161$; $\eta^2 = 0.8084$), cortisone (139%, $p = 0.0252$; $\eta^2 = 0.7490$), 17-OH-progesterone (113%, $p = 0.0038$; $\eta^2 = 0.9954$), and estradiol (133%, $p = 0.0396$; $\eta^2 = 0.8298$) were higher in males.

Finally, the male controls had higher 17-OH-progesterone (166%, $p = 0.0019$; $\eta^2 = 0.9954$) compared to the female controls.

3.6. Fatty Acids Evaluation in Plasma Samples

The quantification of plasma fatty acids was performed by the reaction of transesterification, converting triglycerides and free fatty acid in the corresponding FAMEs, followed by GC–MS analysis.

Our results showed that there was a general trend of the identified fatty acids to decrease with physical activity, independent of sex. In particular, palmitoleic acid (36% for males, $p = 0.0001$; 65%, $p = 0.0369$ for females; $\eta^2 = 0.9680$), palmitic acid (34% for males $p = 0.001$; 84% for females, $p = 0.003$; $\eta^2 = 0.9533$), linoleic acid (35%, $p = 0.0084$ for males; 80%, $p = 0.0332$ for females; $\eta^2 = 0.9742$), oleic acid (48% for males, $p = 0.0001$; 78%, $p = 0.0157$ for females; $\eta^2 = 0.9343$), arachidonic acid (26%, $p = 0.0047$ for males; 59%, $p = 0.0001$ for females; $\eta^2 = 0.9788$), dihomo γ-linolenic acid (41%, $p = 0.0001$ for males; 14%, $p = 0.0001$ for females; $\eta^2 = 0.9764$), 15-tetracosenoic (nervonic) acid (39% for females), and tetracosanoic (lignoceric) acid (30% for females) decreased in both female and male basketball players compared to the control groups ($p \leq 0.05$; Figure 5). Even two C24 fatty acids, e.g., 15-tetracosenoic (nervonic) acid and tetracosanoic (lignoceric) acid, decreased because of the physical activity to concentrations lower than the limit of instrumental detection in the male basketball players group. Interestingly, docosenoic acid was slightly more abundant in females than males, but it did not change in relation to sport (Figure 5), and stearic acid exhibited a different tendency in relation to physical activity and sex; in fact, it decreased in the male basketball players (26%, $p = 0.0001$; $\eta^2 = 0.9714$) but increased in the female athletes (131%, $p = 0.132$; $\eta^2 = 0.9714$) (Figure 5).

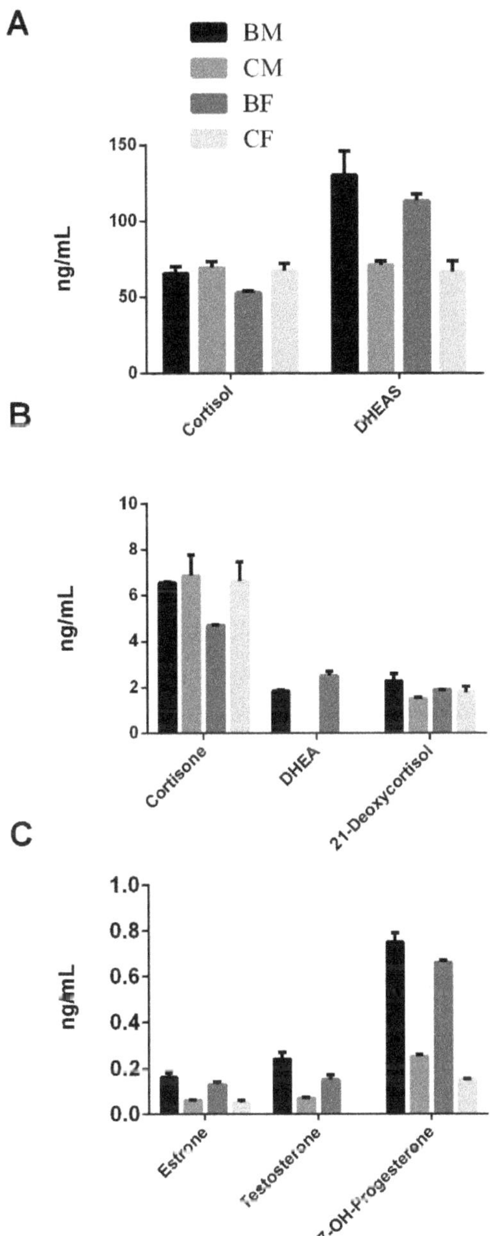

Figure 4. Measurements of steroid hormones quantified in the plasma of both male and female basketball players vs their respective controls. The histogram representations were separated in three panels (**A–C**) in agreement with the order of magnitude expressed in ng/mL. The statistical analysis was carried out by two-tailed t-test using GraphPad Prism 8, and only statistically significant differences were reported ($p < 0.05$).

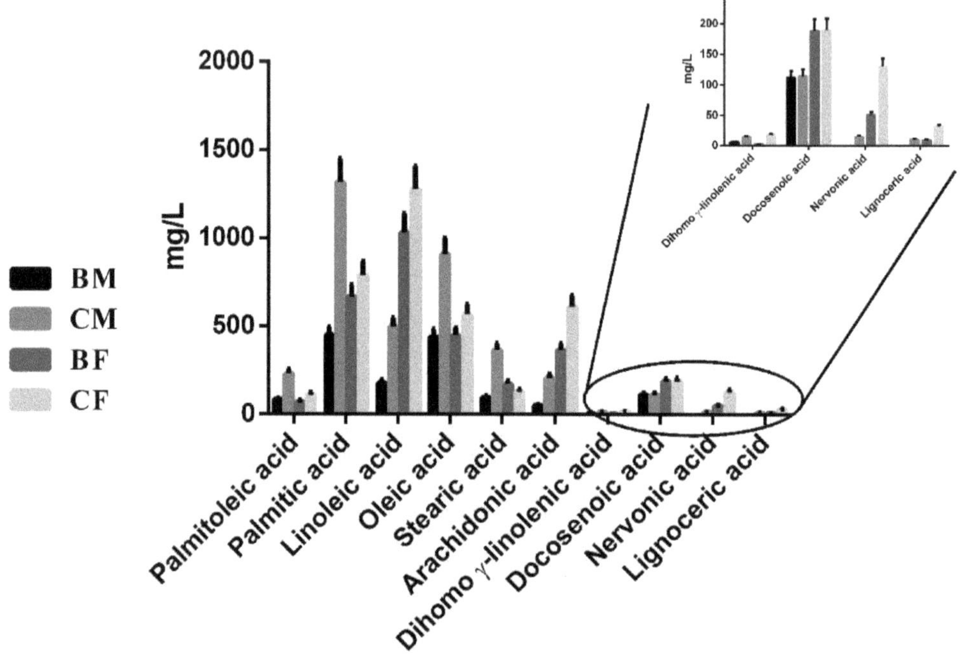

Figure 5. Comparison of the concentrations (expressed in mg/L) of FAME obtained by acid transesterification of the plasma from both male and female basketball players and from the control subjects. The statistical analysis was carried out by two-tailed *t*-test using GraphPad Prism 8 and only statistically significant differences were reported ($p < 0.05$).

Comparing the athletes, linoleic acid (17%, $p = 0.001$; $\eta^2 = 0.9742$), stearic acid (57%, $p = 0.0123$) $\eta^2 = 0.9714$), arachidonic acid (15%, $p = 0.0001$; $\eta^2 = 0.9788$), and docosenoic acid (59%, $p = 0.0015$; $\eta^2 = 0.8990$) were lower in males.

There were also differences in the sedentary controls where, in fact, in male subjects, palmitoleic acid (206%, $p = 0.0001$; $\eta^2 = 0.9680$), palmitic acid (166%, $p = 0.003$; $\eta^2 = 0.9533$), oleic acid (160%, $p = 0.0007$; $\eta^2 = 0.9343$), and stearic acid (278%, $p = 0.0001$; $\eta^2 = 0.9714$) were higher than in female subjects. Instead, linoleic acid (30%, $p = 0.0001$; $\eta^2 = 0.9742$), arachidonic acid (33%, $p = 0.001$; $\eta^2 = 0.9788$), and docosenoic acid (60%; $p = 0.0016$; $\eta^2 = 0.8990$) showed the opposite trend, decreasing in the male subjects.

3.7. Metabolic Pathways Analysis

To have an overview of the metabolic process involved in adaptation to physical activity, we used the plasma metabolites identified in this study, showing a statistically significant increase/decrease level in female and male basketball players in comparison with the relative sedentary controls. We used MetScape (http://metscape.med.umich.edu, accessed on 10 March 2021—from Cytoscape), which provides a bioinformatics tool for the interpretation of metabolomic data [26]. The results obtained are reported in Table 3. We identified 14 metabolic pathways in common between female and male athletes, including, among others, amino acids, fatty acids, and glycidic metabolism; 4 pathways typical of the female metabolism involving mainly fatty acids; and 6 pathways characteristic of males, including amino acids, fatty acids, and porphyrin metabolism (for details see Table 3).

Table 3. The selected metabolic pathways obtained using plasma metabolites identified in this study showing a statistically significant increase/decrease in the female and male basketball players in comparison with the relative sedentary controls. The pathways analysis was carried out using the MetScape 3 App for Cytoscape (http://metscape.med.umich.edu, accessed on 10 March 2021).

Metabolic Pathways Involved in Females and Males
Androgen and estrogen biosynthesis and metabolism
Arachidonic acid metabolism
Bile acid biosynthesis
C21-steroid hormone biosynthesis and metabolism
De novo fatty acid biosynthesis
Di-unsaturated fatty acid beta-oxidation
Glycolysis and gluconeogenesis
Histidine metabolism
Leukotriene metabolism
Linoleate metabolism
Omega-6 fatty acid metabolism
Purine metabolism
Urea cycle and metabolism of arginine, proline, glutamate, aspartate, and asparagine
Vitamin B9 (folate) metabolism
Metabolic pathways involved only in females
Butanoate metabolism
De novo fatty acid biosynthesis
Omega-6 fatty acid metabolism
Squalene and cholesterol biosynthesis
Metabolic pathways involved only in males
Glycine, serine, alanine, and threonine metabolism
Lysine metabolism
Porphyrin metabolism
Prostaglandin formation from arachidonate
Saturated fatty acids beta-oxidation
Valine, leucine, and isoleucine degradation

4. Discussion

The purpose of this study was to compare the redox, hormonal, metabolic, and lipid profiles of female and male basketball players during the seasonal training period, compared to the relative sedentary control groups. Further, this investigation aimed to be descriptive and exploratory to deepen the understanding of biochemical alterations that occur in adaptation to physical exercise in female and male athletes, underlining their common and different traits. As the control, we planned to use untrained subjects because they represent the reference value of the analyzed parameters, trying to highlight the changes due only to regular physical activity.

In our experimental conditions, in female athletes with a body mass index comparable to controls, the antioxidant capacity showed a significant increase and a reduction in oxidative species. In males, we found a similar increase in the antioxidant capacity, although the oxidative species are stackable with the relative controls. Therefore, it is known that females had a greater oxidative status than males [27]. In our experimental conditions, we observed in both sexes a general increase in antioxidant capacity. As previously reported in our study, during physical activity there is an increase in ROS production due to the stress of physical exercise [2]. Consequently, the high levels of antioxidant capacity of female and male athletes attenuate this oxidant production. Several studies reported that in basketball players, oxidative stress at the end of the season is not severe, despite the high number of matches played, and this is probably thanks to the increases in the antioxidant defense mechanisms [28,29]. The same trends were observed for other sports, such as soccer, cycling, and swimming [2]. In this context, we suggest that in both sexes, adaptation to physical exercise leads to an increase in antioxidant capacity that is able to counteract the oxidative stress produced during exercise, guaranteeing health and better performance. However, high effort and excessive levels of oxidative stress can lead to a rise in inflammatory markers and overtraining syndrome [5].

Resistance exercise and/or training elicits a milieu of acute physiological responses and chronic adaptations that are critical for increasing muscular strength, power, hypertrophy, and local muscular endurance [30]. In our study, we analyzed several hormones, such as salivary cortisol, plasma adiponectin, and several steroid hormones. Salivary cortisol quantification revealed that both in females and in males, this hormone decreases with athletic activity. This agrees with other studies that reported a reduction of cortisol for both females and males after a match [31]. We also analyzed the same hormone in plasma, finding a comparable trend for females and males, although in males, this difference is not statistically significant. This divergence could be explained by the different techniques used and/or by the biological samples analyzed; in fact, other authors have reported that cortisol changes occurred simultaneously in plasma and saliva, but the timing of post-exercise hormone moving varied between trials and individuals [32]. Moreover, specific proteins binding cortisol affecting the circulating pool of bioactive free cortisol in plasma have been described [33].

In this study, we evaluated the levels of adiponectin by Western blot, highlighting its increase in male players. Similarly, Jurimae and colleagues found that in highly trained athletes, during the recovery period, adiponectin levels increased [34,35]. Contrarily, in females, we found that the plasma levels of this adipokine were not changed. Nevertheless, the adiponectin levels that we found in control females was higher than in control males, confirming the high levels of this adipokine in females [36,37]. Basketball is a sport more associated with fast twitch muscular fibres (type II) mainly involved in quick bursts with great power [16,38], than it is with slow twitch fibres (type I), which in turn are mainly involved in endurance sports. Interestingly, males expressed 42% more of this type of fibers than did females, which, in addition to showing a lower expression of AdipoR1 [37], suggests a sex-related difference in adiponectin susceptibility.

Finally, we evaluated the levels of several steroid hormones. We found an increase of testosterone, DHEA, estrone, and 17-OH-progesterone both in females and males. Testosterone increases muscle mass and strength over weeks to months, with a strong dose-response evident from below to above physiological testosterone doses and concentrations [16], and several reports have indicated its increase after exercise [39]. Likewise, other authors have recorded a significant increase of DHEA levels monitored in the sera of older athletes immediately after exercise [40]. In a similar manner, estrogens have been shown to reduce bone resorption and muscle damage, which may have important ramifications for musculoskeletal adaptations to resistance training [30]. On the other hand, estradiol decreases in athletes, regardless of sex, as reported by others, emphasizing the benefit of physical activity, especially for females to reduce the main risk factor for estrogen-responsive breast cancer [41].

To have a further overview of the process involved in female and male adaptation to physical exercise, we analyzed the plasma metabolic and lipid profiles of participants. Comparing players versus sedentary subjects, our results showed that there was a general trend of fatty acids decreasing with physical activity, independent of sex. This finding is in agreement with the activation of fatty acid metabolism during physical activity, where energy reservoirs of triacylglycerols stored within the adipose tissues were used as fuel for working muscles [42].

The MetScape program (version 3.1.3) allowed us to identify the main metabolic pathways implicated in physical exercise in females and males. These include amino acids, fatty acids, and glycidic metabolism, but also arachidonic acid metabolism as a response to the inflammation process [43]. Moreover, we identified the urea cycle, which is the main system for ammonia elimination and detoxification, for reducing fatigue and improving exercise endurance capacity [44]. Considering the pathways differentially involved among the sexes, in females we observed mainly lipid metabolism (such as butanoate, squalene, and others), while in males, we observed primarily amino acids metabolism, in particular the branched chain. Increases in serum BCAAs (branched chain amino acids) during or after exercise may indicate their mobilization from either liver or muscle [45]. In skeletal muscle, they promote glucose uptake and protein synthesis, thus playing an important role during exercise and especially in post-exercise recovery. Moreover, exercise increases the ability of mitochondria from skeletal muscle to oxidize BCAAs as an alternative source of energy without inducing insulin resistance [46].

Already, other researchers have highlighted significant sex-related differences in energy and macronutrient metabolism within many tissues and organs, resulting in a distinct metabolic profile between females and males and, in particular, the use of nutritional fats and the distribution of fat depots, glucose homeostasis, amino acid transport, and protein metabolism [47]. Moreover, sex differences in substrate metabolism during whole body moderate-to-high intensity endurance exercise have also been reported, with females shown to oxidize more fat and less carbohydrate than males when exercising at the same relative intensity [48–50]. Our obtained data suggest that adaptation to training increases this difference between male and female metabolisms.

Our analysis was performed 24 h after training sessions, and, interestingly, we observed a relevant increase in lactate in athletes. Lactate, once recognized as merely a product of glycogenolysis, is now seen as a carbohydrate fuel source [51,52] and a signaling molecule contributing to mitochondrial adaptations in skeletal muscle [53,54]. We think that it is a key regulator of adaptation, and, probably, the ability to produce and tolerate high levels of blood lactate during effort is associated with a successful performance.

Further investigations are compelling to elucidate the role of lactate in adaptation to physical exercise.

5. Limitation

The main limitation of this study is the lower number of participants belonging to the same teams. Furthermore, our data relates exclusively to basketball, as it is characterized by specific metabolic demands, and it cannot be transferred to all other sports. Therefore, this should be considered a pilot study. It will be interesting to evaluate if the same differences are also evident for sports with metabolic characteristics different from those of basketball as analyzed in this study.

Moreover, we did not consider the menstrual cycle phase of female participants. There are conflicting results on menstrual cycle phase effects on physical activity [55], with studies reporting improved performance outcomes during the early follicular [56], ovulatory [57], and mid-luteal [58] phases; whereas others have shown no changes in exercise performance between menstrual cycle phases [59–61].

6. Conclusions

Despite its limitations, our study indicates that female and male adaptations to training have several common traits, such as antioxidant potential enhancement, lactate elevation, and activation of detoxifying process such as the urea cycle and arachidonic pathways as a response to inflammation. We also highlighted some different sex-related behaviors, including dissimilar lipid and amino acids utilization, which need further investigation.

Supplementary Materials: The following supporting information can be downloaded at: https://www.mdpi.com/article/10.3390/healthcare10020358/s1; Table S1: List of Metabolites identified in plasma samples of the subject utilized in this study by Gas Chromatography mass spectrometry (GC-MS) analysis.

Author Contributions: Conceptualization, A.M. and S.L.; data curation, G.P., R.M. and A.A.; formal analysis, G.P., R.M., A.A. and S.L.; funding acquisition, A.M.; investigation, G.P., R.M., A.M. and S.L.; methodology, G.P. and A.A.; project administration, P.A.M. and A.M.; resources, P.A.M.; supervision, A.M.; validation, G.P., R.M., A.A., P.A.M. and S.L.; writing—original draft, G.P., R.M., A.A., P.A.M., A.M. and S.L. All authors have read and agreed to the published version of the manuscript.

Funding: This research was funded by Ricerca di Ateneo ex 60% 2020 from the University of Florence.

Institutional Review Board Statement: The study was approved by the local ethics committee of the University of Florence, Italy (AM_Gsport 15840/CAM_BIO).

Informed Consent Statement: A written informed consent was provided by each subject prior to enrolment in the study. The study was conducted according to the policy statement set forth in the Declaration of Helsinki, and all the experiments were conducted according to established ethical guidelines.

Data Availability Statement: The data presented in this study are available on request from the authors.

Conflicts of Interest: The authors declare no conflict of interest.

References

1. Hawley, J.A.; Lundby, C.; Cotter, J.D.; Burke, L.M. Maximizing Cellular Adaptation to Endurance Exercise in Skeletal Muscle. *Cell Metab.* **2018**, *27*, 962–976. [CrossRef] [PubMed]
2. Luti, S.; Modesti, A.; Modesti, P.A. Inflammation, Peripheral Signals and Redox Homeostasis in Athletes Who Practice Different Sports. *Antioxidants* **2020**, *9*, 1065. [CrossRef] [PubMed]
3. Hadžović-Džuvo, A.; Valjevac, A.; Lepara, O.; Pjanić, S.; Hadžimuratović, A.; Mekić, A. Oxidative stress status in elite athletes engaged in different sport disciplines. *Bosn. J. Basic Med. Sci.* **2014**, *14*, 56. [CrossRef] [PubMed]
4. Mangine, G.T.; Hoffman, J.R.; Gonzalez, A.M.; Townsend, J.R.; Wells, A.J.; Jajtner, A.R.; Beyer, K.S.; Boone, C.H.; Wang, R.; Miramonti, A.A.; et al. Exercise-Induced Hormone Elevations Are Related to Muscle Growth. *J. Strength Cond. Res.* **2017**, *31*, 45–53. [CrossRef]
5. Magherini, F.; Fiaschi, T.; Marzocchini, R.; Mannelli, M.; Gamberi, T.; Modesti, P.A.; Modesti, A. Oxidative stress in exercise training: The involvement of inflammation and peripheral signals. *Free Radic. Res.* **2019**, *53*, 1155–1165. [CrossRef]
6. Martinez-Huenchullan, S.F.; Tam, C.S.; Ban, L.A.; Ehrenfeld-Slater, P.; Mclennan, S.V.; Twigg, S.M. Skeletal muscle adiponectin induction in obesity and exercise. *Metabolism* **2020**, *102*, 154008. [CrossRef]
7. Ren, Y.; Li, Y.; Yan, J.; Ma, M.; Zhou, D.; Xue, Z.; Zhang, Z.; Liu, H.; Yang, H.; Jia, L.; et al. Adiponectin modulates oxidative stress-induced mitophagy and protects C2C12 myoblasts against apoptosis. *Sci. Rep.* **2017**, *7*, 3209. [CrossRef]
8. Cnop, M.; Havel, P.J.; Utzschneider, K.M.; Carr, D.B.; Sinha, M.K.; Boyko, E.J.; Retzlaff, B.M.; Knopp, R.H.; Brunzell, J.D.; Kahn, S.E. Relationship of adiponectin to body fat distribution, insulin sensitivity and plasma lipoproteins: Evidence for independent roles of age and sex. *Diabetologia* **2003**, *46*, 459–469. [CrossRef]
9. Böttner, A.; Kratzsch, J.; Müller, G.; Kapellen, T.M.; Blüher, S.; Keller, E.; Blüher, M.; Kiess, W. Gender differences of adiponectin levels develop during the progression of puberty and are related to serum androgen levels. *J. Clin. Endocrinol. Metab.* **2004**, *89*, 4053–4061. [CrossRef]
10. Gamberi, T.; Magherini, F.; Fiaschi, T. Adiponectin in Myopathies. *Int. J. Mol. Sci.* **2019**, *20*, 1544. [CrossRef]
11. Agostinis-Sobrinho, C.; Santos, R.; Moreira, C.; Abreu, S.; Lopes, L.; Oliveira-Santos, J.; Rosário, R.; Póvoas, S.; Mota, J. Association between serum adiponectin levels and muscular fitness in Portuguese adolescents: LabMed Physical Activity Study. *Nutr. Metab. Cardiovasc. Dis.* **2016**, *26*, 517–524. [CrossRef]
12. Senefeld, J.W.; Clayburn, A.J.; Baker, S.E.; Carter, R.E.; Johnson, P.W.; Joyner, M.J. Sex differences in youth elite swimming. *PLoS ONE* **2019**, *14*, e0225724. [CrossRef]
13. Handelsman, D.J.; Hirschberg, A.L.; Bermon, S. Circulating Testosterone as the Hormonal Basis of Sex Differences in Athletic Performance. *Endocr. Rev.* **2018**, *39*, 803–829. [CrossRef]
14. Vesper, H.W.; Wang, Y.; Vidal, M.; Botelho, J.C.; Caudill, S.P. Serum Total Testosterone Concentrations in the US Household Population from the NHANES 2011–2012 Study Population. *Clin. Chem.* **2015**, *61*, 1495–1504. [CrossRef]
15. Handelsman, D.J.; Sikaris, K.; Ly, L.P. Estimating age-specific trends in circulating testosterone and sex hormone-binding globulin in males and females across the lifespan. *Ann. Clin. Biochem.* **2016**, *53*, 377–384. [CrossRef]
16. Handelsman, D.J. Sex differences in athletic performance emerge coinciding with the onset of male puberty. *Clin. Endocrinol.* **2017**, *87*, 68–72. [CrossRef]
17. Hoffman, J.R.; Bar-Eli, M.; Tenenbaum, G. An examination of mood changes and performance in a professional basketball team. *J. Sports Med. Phys. Fit.* **1999**, *39*, 74–79.
18. McInnes, S.E.; Carlson, J.S.; Jones, C.J.; McKenna, M.J. The physiological load imposed on basketball players during competition. *J. Sports Sci.* **1995**, *13*, 387–397. [CrossRef]
19. Ben Abdelkrim, N.; El Fazaa, S.; El Ati, J. Time-motion analysis and physiological data of elite under-19-year-old basketball players during competition. *Br. J. Sports Med.* **2007**, *41*, 69–75; discussion 75. [CrossRef]
20. Scanlan, A.; Dascombe, B.; Reaburn, P. A comparison of the activity demands of elite and sub-elite Australian men's basketball competition. *J. Sports Sci.* **2011**, *29*, 1153–1160. [CrossRef]
21. Hagströmer, M.; Oja, P.; Sjöström, M. The International Physical Activity Questionnaire (IPAQ): A study of concurrent and construct validity. *Public Health Nutr.* **2006**, *9*, 755–762. [CrossRef]
22. Foster, C.; Florhaug, J.A.; Franklin, J.; Gottschall, L.; Hrovatin, L.A.; Parker, S.; Doleshal, P.; Dodge, C. A new approach to monitoring exercise training. *J. Strength Cond. Res.* **2001**, *15*, 109–115.

23. Militello, R.; Luti, S.; Parri, M.; Marzocchini, R.; Soldaini, R.; Modesti, A.; Modesti, P.A. Redox Homeostasis and Metabolic Profile in Young Female Basketball Players during in-Season Training. *Healthcare* **2021**, *9*, 368. [CrossRef]

24. Luti, S.; Fiaschi, T.; Magherini, F.; Modesti, P.A.; Piomboni, P.; Governini, L.; Luddi, A.; Amoresano, A.; Illiano, A.; Pinto, G.; et al. Relationship between the metabolic and lipid profile in follicular fluid of women undergoing in vitro fertilization. *Mol. Reprod. Dev.* **2020**, *87*, 986–997. [CrossRef]

25. McGuigan, M.R.; Egan, A.D.; Foster, C. Salivary Cortisol Responses and Perceived Exertion during High Intensity and Low Intensity Bouts of Resistance Exercise. *J. Sports Sci. Med.* **2004**, *3*, 8–15.

26. Gao, J.; Tarcea, V.G.; Karnovsky, A.; Mirel, B.R.; Weymouth, T.E.; Beecher, C.W.; Cavalcoli, J.D.; Athey, B.D.; Omenn, G.S.; Burant, C.F.; et al. Metscape: A Cytoscape plug-in for visualizing and interpreting metabolomic data in the context of human metabolic networks. *Bioinformatics* **2010**, *26*, 971–973. [CrossRef]

27. Wiecek, M.; Maciejczyk, M.; Szymura, J.; Szygula, Z. Changes in Oxidative Stress and Acid-Base Balance in Men and Women Following Maximal-Intensity Physical Exercise. *Physiol. Res.* **2015**, *64*, 93–102. [CrossRef]

28. Spanidis, Y.; Goutzourelas, N.; Stagos, D.; Mpesios, A.; Priftis, A.; Bar-Or, D.; Spandidos, D.A.; Tsatsakis, A.M.; Leon, G.; Kouretas, D. Variations in oxidative stress markers in elite basketball players at the beginning and end of a season. *Exp. Ther. Med.* **2016**, *11*, 147–153. [CrossRef]

29. Yilmaz, N.; Erel, Ö.; Hazer, M.; Bagci, C.; Namiduru, E.; Giil, E. Biochemical assessments of retinol, α-tocopherol, pyridoxal—5-phosphate oxidative stress index and total antioxidant status in adolescent professional basketball players and sedentary controls. *Int. J. Adolesc. Med. Health* **2007**, *19*, 177–186. [CrossRef]

30. Kraemer, W.J.; Ratamess, N.A. Hormonal responses and adaptations to resistance exercise and training. *Sports Med.* **2005**, *35*, 339–361. [CrossRef]

31. Casanova, N.; Palmeira-DE-Oliveira, A.; Pereira, A.; Crisóstomo, L.; Travassos, B.; Costa, A.M. Cortisol, testosterone and mood state variation during an official female football competition. *J. Sports Med. Phys. Fit.* **2016**, *56*, 775–781.

32. Tanner, A.V.; Nielsen, B.V.; Allgrove, J. Salivary and plasma cortisol and testosterone responses to interval and tempo runs and a bodyweight-only circuit session in endurance-trained men. *J. Sports Sci.* **2014**, *32*, 680–689. [CrossRef] [PubMed]

33. Schwinn, A.-C.; Sauer, F.J.; Gerber, V.; Bruckmaier, R.M.; Gross, J.J. Free and bound cortisol in plasma and saliva during ACTH challenge in dairy cows and horses. *J. Anim. Sci.* **2018**, *96*, 76–84. [CrossRef] [PubMed]

34. Jurimae, J.; Purge, P.; Jurimae, T. Adiponectin is altered after maximal exercise in highly trained male rowers. *Eur. J. Appl. Physiol.* **2005**, *93*, 502–505. [CrossRef] [PubMed]

35. Plinta, R.; Olszanecka-Glinianowicz, M.; Drosdzol-Cop, A.; Chudek, J.; Skrzypulec-Plinta, V. The effect of three-month pre-season preparatory period and short-term exercise on plasma leptin, adiponectin, visfatin, and ghrelin levels in young female handball and basketball players. *J. Endocrinol. Investig.* **2012**, *35*, 595–601. [CrossRef]

36. Waki, H.; Yamauchi, T.; Kamon, J.; Ito, Y.; Uchida, S.; Kita, S.; Hara, K.; Hada, Y.; Vasseur, F.; Froguel, P.; et al. Impaired multimerization of human adiponectin mutants associated with diabetes. Molecular structure and multimer formation of adiponectin. *J. Biol. Chem.* **2003**, *278*, 40352–40363. [CrossRef]

37. Høeg, L.D.; Sjøberg, K.A.; Lundsgaard, A.-M.; Jordy, A.B.; Hiscock, N.; Wojtaszewski, J.F.P.; Richter, E.A.; Kiens, B. Adiponectin concentration is associated with muscle insulin sensitivity, AMPK phosphorylation, and ceramide content in skeletal muscles of men but not women. *J. Appl. Physiol.* **2013**, *114*, 592–601. [CrossRef]

38. Senefeld, J.; Joyner, M.J.; Stevens, A.; Hunter, S.K. Sex differences in elite swimming with advanced age are less than marathon running. *Scand. J. Med. Sci. Sports* **2016**, *26*, 17–28. [CrossRef]

39. Vingren, J.L.; Kraemer, W.J.; Ratamess, N.A.; Anderson, J.M.; Volek, J.S.; Maresh, C.M. Testosterone physiology in resistance exercise and training: The up-stream regulatory elements. *Sports Med.* **2010**, *40*, 1037–1053. [CrossRef]

40. Heaney, J.L.J.; Carroll, D.; Phillips, A.C. DHEA, DHEA-S and cortisol responses to acute exercise in older adults in relation to exercise training status and sex. *Age* **2013**, *35*, 395–405. [CrossRef]

41. Ennour-Idrissi, K.; Maunsell, E.; Diorio, C. Effect of physical activity on sex hormones in women: A systematic review and meta-analysis of randomized controlled trials. *Breast Cancer Res. BCR* **2015**, *17*, 139. [CrossRef]

42. Mika, A.; Macaluso, F.; Barone, R.; Di Felice, V.; Sledzinski, T. Effect of Exercise on Fatty Acid Metabolism and Adipokine Secretion in Adipose Tissue. *Front. Physiol.* **2019**, *10*, 26. [CrossRef]

43. Wang, T.; Fu, X.; Chen, Q.; Patra, J.K.; Wang, D.; Wang, Z.; Gai, Z. Arachidonic Acid Metabolism and Kidney Inflammation. *Int. J. Mol. Sci.* **2019**, *20*, 3683. [CrossRef]

44. Chen, S.; Minegishi, Y.; Hasumura, T.; Shimotoyodome, A.; Ota, N. Involvement of ammonia metabolism in the improvement of endurance performance by tea catechins in mice. *Sci. Rep.* **2020**, *10*, 6065. [CrossRef]

45. Shimomura, Y.; Murakami, T.; Nakai, N.; Nagasaki, M.; Harris, R.A. Exercise promotes BCAA catabolism: Effects of BCAA supplementation on skeletal muscle during exercise. *J. Nutr.* **2004**, *134*, 1583S–1587S. [CrossRef]

46. Shou, J.; Chen, P.-J.; Xiao, W.-H. The Effects of BCAAs on Insulin Resistance in Athletes. *J. Nutr. Sci. Vitaminol.* **2019**, *65*, 383–389. [CrossRef]

47. Comitato, R.; Saba, A.; Turrini, A.; Arganini, C.; Virgili, F. Sex Hormones and Macronutrient Metabolism. *Crit. Rev. Food Sci. Nutr.* **2015**, *55*, 227–241. [CrossRef]

48. Carter, S.L.; Rennie, C.; Tarnopolsky, M.A. Substrate utilization during endurance exercise in men and women after endurance training. *Am. J. Physiol. Endocrinol. Metab.* **2001**, *280*, E898–E907. [CrossRef]

49. Roepstorff, C.; Steffensen, C.H.; Madsen, M.; Stallknecht, B.; Kanstrup, I.-L.; Richter, E.A.; Kiens, B. Gender differences in substrate utilization during submaximal exercise in endurance-trained subjects. *Am. J. Physiol. Endocrinol. Metab.* **2002**, *282*, E435–E447. [CrossRef]
50. Roepstorff, C.; Thiele, M.; Hillig, T.; Pilegaard, H.; Richter, E.A.; Wojtaszewski, J.F.P.; Kiens, B. Higher skeletal muscle alpha2AMPK activation and lower energy charge and fat oxidation in men than in women during submaximal exercise. *J. Physiol.* **2006**, *574*, 125–138. [CrossRef]
51. Brooks, G.A. The Science and Translation of Lactate Shuttle Theory. *Cell Metab.* **2018**, *27*, 757–785. [CrossRef]
52. Ferguson, B.S.; Rogatzki, M.J.; Goodwin, M.L.; Kane, D.A.; Rightmire, Z.; Gladden, L.B. Lactate metabolism: Historical context, prior misinterpretations, and current understanding. *Eur. J. Appl. Physiol.* **2018**, *118*, 691–728. [CrossRef]
53. Kitaoka, Y.; Takeda, K.; Tamura, Y.; Hatta, H. Lactate administration increases mRNA expression of PGC-1α and UCP3 in mouse skeletal muscle. *Appl. Physiol. Nutr. Metab.* **2016**, *41*, 695–698. [CrossRef]
54. Hashimoto, T.; Hussien, R.; Oommen, S.; Gohil, K.; Brooks, G.A. Lactate sensitive transcription factor network in L6 cells: Activation of MCT1 and mitochondrial biogenesis. *FASEB J.* **2007**, *21*, 2602–2612. [CrossRef]
55. McNulty, K.L.; Elliott-Sale, K.J.; Dolan, E.; Swinton, P.A.; Ansdell, P.; Goodall, S.; Thomas, K.; Hicks, K.M. The Effects of Menstrual Cycle Phase on Exercise Performance in Eumenorrheic Women: A Systematic Review and Meta-Analysis. *Sports Med.* **2020**, *50*, 1813–1827. [CrossRef]
56. Pallavi, L.C.; Souza, U.J.D.; Shivaprakash, G. Assessment of Musculoskeletal Strength and Levels of Fatigue during Different Phases of Menstrual Cycle in Young Adults. *J. Clin. Diagn. Res. JCDR* **2017**, *11*, CC11–CC13. [CrossRef]
57. Bambaeichi, E.; Reilly, T.; Cable, N.T.; Giacomoni, M. The isolated and combined effects of menstrual cycle phase and time-of-day on muscle strength of eumenorrheic females. *Chronobiol. Int.* **2004**, *21*, 645–660. [CrossRef]
58. Oosthuyse, T.; Bosch, A.N.; Jackson, S. Cycling time trial performance during different phases of the menstrual cycle. *Eur. J. Appl. Physiol.* **2005**, *94*, 268–276. [CrossRef]
59. McLay, R.T.; Thomson, C.D.; Williams, S.M.; Rehrer, N.J. Carbohydrate loading and female endurance athletes: Effect of menstrual-cycle phase. *Int. J. Sport Nutr. Exerc. Metab.* **2007**, *17*, 189–205. [CrossRef]
60. Janse DE Jonge, X.A.K.; Thompson, M.W.; Chuter, V.H.; Silk, L.N.; Thom, J.M. Exercise performance over the menstrual cycle in temperate and hot, humid conditions. *Med. Sci. Sports Exerc.* **2012**, *44*, 2190–2198. [CrossRef]
61. Vaiksaar, S.; Jürimäe, J.; Mäestu, J.; Purge, P.; Kalytka, S.; Shakhlina, L.; Jürimäe, T. No effect of menstrual cycle phase and oral contraceptive use on endurance performance in rowers. *J. Strength Cond. Res.* **2011**, *25*, 1571–1578. [CrossRef] [PubMed]

Article

Chronic Training Induces Metabolic and Proteomic Response in Male and Female Basketball Players: Salivary Modifications during In-Season Training Programs

Simone Luti [1,*,†], Rosamaria Militello [1,†], Gabriella Pinto [2,3], Anna Illiano [2,3], Angela Amoresano [2,3], Giovanni Chiappetta [4], Riccardo Marzocchini [1], Pietro Amedeo Modesti [5,6], Simone Pratesi [5], Luigia Pazzagli [1], Alessandra Modesti [1] and Tania Gamberi [1]

[1] Department of Biomedical, Experimental and Clinical Sciences "Mario Serio", University of Florence, Viale G. Morgagni, 50, 50134 Florence, Italy
[2] Istituto Nazionale Biostrutture e Biosistemi, Viale delle Medaglie d'Oro, 305, 00136 Rome, Italy
[3] Department of Chemical Sciences, Polytechnic and Basic Sciences School, University of Naples Federico II, Via Cintia, 21, 80126 Napoles, Italy
[4] Biological Mass Spectrometry and Proteomics Group, SMBP, PDC CNRS UMR, 8249, ESPCI Paris, Université PSL, 10 rue Vauquelin, 75005 Paris, France
[5] Department of Experimental and Clinical Medicine, University of Florence, Largo Brambilla, 3, 50134 Florence, Italy
[6] Sport Medicine Unit, Careggi University Hospital, University of Florence, Largo Brambilla 3, 50134 Florence, Italy
* Correspondence: simone.luti@unifi.it
† These authors contributed equally to this work.

Citation: Luti, S.; Militello, R.; Pinto, G.; Illiano, A.; Amoresano, A.; Chiappetta, G.; Marzocchini, R.; Modesti, P.A.; Pratesi, S.; Pazzagli, L.; et al. Chronic Training Induces Metabolic and Proteomic Response in Male and Female Basketball Players: Salivary Modifications during In-Season Training Programs. *Healthcare* **2023**, *11*, 241. https://doi.org/10.3390/healthcare11020241

Academic Editors: João Paulo Brito and Rafael Oliveira

Received: 5 December 2022
Revised: 9 January 2023
Accepted: 9 January 2023
Published: 12 January 2023

Abstract: The aim of this study was to characterize the salivary proteome and metabolome of highly trained female and male young basketball players, highlighting common and different traits. A total of 20 male and female basketball players (10 female and 10 male) and 20 sedentary control subjects (10 female and 10 male) were included in the study. The athletes exercised at least five times per week for 2 h per day. Saliva samples were collected mid-season, between 9:00 and 11:00 a.m. and away from sport competition. The proteome and metabolome were analyzed by using 2DE and GC–MS techniques, respectively. A computerized 2DE gel image analysis revealed 43 spots that varied in intensity among groups. Between these spots, 10 (23.2%) were differentially expressed among male athletes and controls, 22 (51.2%) between female basketball players and controls, 11 spots (25.6%) between male and female athletes, and 13 spots (30.2%) between male and female controls. Among the proteins identified were Immunoglobulin, Alpha-Amylase, and Dermcidin, which are inflammation-related proteins. In addition, several amino acids, such as glutamic acid, lysine, ornithine, glycine, tyrosine, threonine, and valine, were increased in trained athletes. In this study, we highlight that saliva is a useful biofluid to assess athlete performance and confirm that the adaptation of men and women to exercise has some common features, but also some different sex-specific behaviors, including differential amino acid utilization and expression of inflammation-related proteins, which need to be further investigated. Moreover, in the future, it will be interesting to examine the influence of sport-type on these differences

Keywords: physical exercise; basketball; saliva; metabolomics; proteomics; training; sport; sex

1. Introduction

Exercise is an important regulator of cellular metabolic pathways, and several studies have focused on blood sampling to measure metabolites to investigate adaptations and responses to acute and chronic exercise [1]. Acute and chronic training are two very different exercise paradigms. Chronic exercise or exercise training is a repeated number of exercise sessions over a short or long-term period, while acute is defined as a single session of exercise [2–4]. However, it is known that acute exercise, if continued, can lead to muscle tissue damage, releasing proteins such as creatine kinase, myoglobin, or troponin into the plasma, which can reach the saliva through active or passive flow [5]. Recently, Franco-Martínez et al. [6] found sex-specific differences between men and women in the salivary proteome at rest and after acute exercise and concluded that sex is the factor that most strongly modulates salivary protein content. For these reasons, saliva is a useful alternative to blood to study the molecules involved in exercise adaptation [7].

The measurement of total salivary content may provide a non-invasive, low-stress method for assessing exercise-associated metabolic, hormonal, and immunologic status and for evaluating exercise load while avoiding upper-respiratory-tract infections. Recently, many studies have used saliva to assess concentrations of these compounds in response to exercise and training [8], and its composition reflects systemic health status [9]. Saliva is a hypotonic fluid composed mainly of water; electrolytes; and biomolecules such as proteins [10], cellular and bronchoalveolar debris, nasal secretions [11], and secretions from the salivary glands, which are composed of serous and mucous cells (acinar cells) and different types of duct cells, and these components contribute differently to the composition of saliva [12]. However, the major limitation in salivary samples is the inter-individual variability, which is determined by various factors, such as sex, age, and circadian rhythms, that may affect salivary composition, thus making a comparison between patients difficult [13,14].

Despite these limitations, it is interesting to monitor the concentration changes of salivary metabolites that occur during strenuous exercise. Pitti et al. [15] reported an altered concentration of 56 metabolites after a football match in the saliva of female athletes. They concluded that these modifications could be due to changes in water content rather than metabolite synthesis, especially after intense effort. In our previous studies, to delve deeper into sex-specific adaptations to chronic in-season training, we examined oxidative status, adiponectin, cortisol, steroid hormones, and fatty acid levels in the plasma of young male and female athletes compared to matched sedentary controls [16,17]. We concluded that the skeletal muscles of female athletes who train regularly during the competitive season have higher plasma concentrations of urea than those of their male counterparts, indicating increased protein catabolism that may lead to fatigue and overtraining. Females and males have different muscle properties, such as a different size, number, and diameter of muscle fibers, and this mainly explains the differences in maximum strength. This partly explains the sex differences in fatigability [18].

The aim of the present study is to investigate the sex-specific changes in the salivary proteome and metabolome of female and male basketball players during chronic training and away from competitions. Since many studies have already been conducted on the salivary proteome and metabolome in response to acute exercise, our first objective was to determine the effects during continuous chronic training [19]. We also wanted to investigate the influence of sex in sport adaptations, since male and female players underwent the same type of training. Interestingly, in a previous study we analyzed the plasma proteome and metabolome of the same highly trained athletes [16,17]. Overall, we hypothesized that the salivary proteome and metabolome may reflect the sex-specific sport adaptation identified in plasma samples. Indeed, in plasma, we found that regular exercise reduced the concentrations of proteins involved in chronic inflammation in females [17]; similarly, in this study, we found lower concentrations of inflammation-related proteins, such as Immunoglobulin A, Alpha-Amylase, and Dermcidin, in female athletes, compared to male players.

2. Materials and Methods

Unless specified, all reagents were obtained from Bio-Rad Laboratories (Hercules, CA, USA). Solvents used for the sample preparation and LC–MS/MS analysis have >99.9% purity, as reported by the manufacturing companies. Acetonitrile (ACN) CHROMASOLV was obtained from Honeywell (Charlotte, NC, USA). Methanol (MeOH) and formic acid (HCOOH) were purchased from Sigma (St. Louis, MO, USA).

2.1. Participants

Students from the University of Florence enrolled in a degree course in Motor Sciences, Sport, and Health were recruited as sedentary subjects, while basketball players from local sports clubs, "US AFFRICO Firenze", were selected. All participants were adults of Caucasian origin aged between 18 and 30 years. Women were randomly selected in relation to menstrual cycle. All subjects completed a physical activity questionnaire to determine their eligibility, using, in particular, the International Physical Activity Questionnaires (IPAQs) [20].

All participants were selected based on their health status, age, and stable body weight. Individuals who smoke and use antioxidants or dietary supplements were excluded from the study. All players attended the training sessions throughout the season, from September to May, and samples were collected during the 12th week. Athletes trained at least 5 times per week, with sessions of 2 h per day, as previously reported [17]. Briefly, the training protocol included both technical and aerobic exercises: 30 min of low–moderate running, followed by interval training runs to improve speed and simulate different game situations, 3 sets of 15 repetitions of exercises for the abdominals (oblique and the rectus abdominis muscles) and 3 sets of 12 repetitions of exercises for the shoulder muscles. Both male and female athletes followed the same training protocol [16].

Written informed consent was obtained from the participants; the research was carried out according to the policy statement set forth in the Declaration of Helsinki of 2013 and authorized by the local Ethics Committee of the University of Florence, Italy (AM_Gsport 15840/CAM_BIO).

2.2. Saliva Samples Collection and Preparation

Subjects completed a questionnaire about their lifestyle habits and other general information. Saliva samples (3 mL) were collected in the morning, between 9:00 and 11:00 a.m., to avoid fluctuations due to circadian rhythms, before daily training sessions and outside of competition. Participants were instructed not to eat, drink, or brush their teeth for 30 min prior to collection. Saliva samples from each participant were collected by using Salivette Cortisol (Sarstedt AG and Co., Nümbrecht, Germany) and were stored in a freezer at $-80\ ^\circ$C until their use. The cotton sliver of the Salivette was taken out and placed in the sublingual for 1/2 min and then put back into the Salivette. Samples were then centrifuged at 2000 rpm for 2 min. Saliva samples were grouped in four different pools, i.e., one for each group involved in the study; therefore, each pool consisted of the saliva samples of 10 subjects. For proteomic analysis, saliva pools were precipitated overnight at $-20\ ^\circ$C with ice cold acetone (1:5) and then centrifuged for 30 min at 9000 rpm. The acetone was then removed, and the residuals were left to evaporate. A total of 150 µL of rehydration solution composed of 8 M urea, 4% (w/v) 3-((3-cholamidopropyl) dimethylammonio)-1-propanesulfonate (CHAPS), and 50 mM Dithiothreitol (DTT) was used to suspend the pellets. A Bradford assay was carried out to assess the total protein contents.

2.3. Two-Dimensional Polyacrylamide Gel Electrophoresis (2DE)

A total of 30 µg of protein samples was loaded on 11 cm IPG strips, pH 3–10 NL (Bio Rad Laboratories, Hercules, CA, USA), that were actively rehydrated for 16 h (at 50 V) in 200 µL of rehydration solution supplemented with 0.5% (v/v) carrier ampholyte (Bio-Rad Laboratories, CA, USA) and a trace of bromophenol blue. Isoelectric focusing (IEF) was carried as follow: 250 V for 20 min (rapid), from 250 to 8000 V for 1 h, and then 8000 V until a total of 23,000 V/h was reached, with a limiting current of 50 µA/strip. Strips were then equilibrated for 15 min in 6M urea, 30% (v/v) glycerol, 2% (w/v) SDS, and 2% (w/v) DTT in 0.05 M Tris-HCl buffer (pH 6.8), and for 15 min in the same solution with 2.5% (w/v) iodoacetamide instead of DTT. The strips were moved on 9–16% polyacrylamide linear gradient gels, and SDS–PAGE was performed at 200 V until the dye front reached the bottom of the gel. Ammoniacal silver nitrate stain was used for analytical gels, as previously described [21]. For preparative gel, 200 μg of proteins (50 μg from each pool) was used, and the gel was stained with colloidal Coomassie blue G-250 [22].

2.4. Protein Identification by Mass Spectrometry

2.4.1. In Situ Digestion of 2DE Spots

Each 2DE spot was manually picked up from Coomassie-stained preparative gels and subjected to in situ digestion, as previously described [17]. Peptide mixtures were eluted from each gel and recovered after several washings by using 0.1% formic acid (HCOOH) and acetonitrile (ACN). Finally, the peptide mixture was vacuum-dried and resuspended in 2% ACN containing 0.1% HCOOH for the subsequent liquid chromatography–tandem mass spectrometry (LC–MS/MS) analysis.

2.4.2. LC–MS/MS Analysis

A 6520 Accurate-Mass Q-TOF LC/MS system (Agilent Technologies, Santa Clara, CA, USA) equipped with a 1200 HPLC system and a chip cube (Agilent Technologies, Santa Clara, CA, USA) was used to analyze the peptide mixtures. The peptide mixture (1 μL) was automatically injected by an autosampler and desalted at a flow rate of 4 μL/min in a 40 nL enrichment column with 0.1% HCOOH as an eluent. A C18 reverse-phase capillary column (75 mm × 43 mm) included into an Agilent Technologies chip (Santa Clara, CA, USA) was used to fractionate the sample at a flow rate of 400 nL/min, with a linear gradient of eluent B (0.1% HCOOH in 95% ACN) in A (0.1% HCOOH in 2% ACN) from 5 to 80% in 50 min.

An analysis of the peptide was carried out by using the data-dependent acquisition of one MS scan (mass range m/z 300–2400), followed by an MS/MS scan of the five most abundant ions in each MS scan. MS/MS spectra were measured automatically when the MS signal was greater than the threshold of 50,000 counts. Charge ions preferably isolated were double, triple, and quadruple, and they were fragmented over singly charged ions. The acquired MS and MS/MS data were converted in a .mgf file, using Mass Hunter software (Agilent Technologies, Santa Clara, CA, USA), in order to be used for protein identification, with a licensed version of Mascot Software (London, UK). Searches were performed by using UniProtKB and the NCBI protein database with *Homo sapiens* taxonomy. Mascot search parameters were as follows: trypsin as an enzyme, allowed number of missed cleavage 3; carbamidomethyl, C as fixed modifications; oxidation of methionine (Met), pyro-Glu (N-terminal Gln and Glu) as variable modifications and peptide charge from +2 to +4. A mass tolerance value of 10 ppm was set for the precursor, and 0.2 was set for the fragment ions. Proteins identified with at least 1 peptide displaying a p-value < 0.05 were selected as significant.

2.5. Western Blot Analysis

An equal amount of four different pools of saliva samples (10 μg of total proteins) was combined with 4× Laemmli buffer (0.5 M TrisHCl pH 6.8, 10% sodium dodecyl sulfate (SDS), 20% glycerol, β-mercaptoethanol, and 0.1% bromophenol blue) and boiled for 5 min. After that, 12% SDS/ polyacrylamide gel was used to separate samples that ware then transferred onto polyvinylidene fluoride membrane (PVDF), using the Trans-Blot Turbo Transfer System (BIO-RAD Laboratories, CA, USA). The PVDF was probed with primary antibody Alpha-Amylase (sc-46657), Immunoglobulin A (IgA) (sc-166334), and Dermcidin (sc-33656), all provided by Santa Cruz Biotechnology (Santa Cruz Biotechnology, Santa Cruz, CA, USA), and Cystatin A (GT2264), which was provided by GeneTex (GeneTex, Irvine, CA, USA); diluted 1:1000 in 2% milk; and incubated overnight at 4 °C. An enhanced chemiluminescence (ECL) detection system (GE Healthcare, Chicago, IL, USA) and Amersham Imager 600 (GE Healthcare, IL, USA) were used to detect the bands after incubation with horseradish peroxidase (HRP)-conjugated anti-mouse IgG (1:10,000) (Santa Cruz Laboratories, Santa Cruz, CA, USA). The PVDF membrane was stained with Coomassie brilliant blue R-250 blot, and the total protein intensities were used to normalize the intensity of the immuno-stained bands. The blot was analyzed by using the ImageJ program, as described previously [23].

2.6. Analysis of Salivary Samples by Gas Chromatography–Mass Spectrometry (GC–MS)

Metabolite analyses were performed on four pools of salivary samples, one for each group (BM, BF, CM, and CF). The same volume of saliva from each participant was taken to make each pool in order to reduce intraindividual variability of metabolites levels. The GC–MS analysis was carried out on 150 µL of saliva, as reported by Luti et al. 2020 [18], but with slight modifications. Briefly, 150 µL of the samples was mixed with 150 µL ice-cold MeOH (#34854-1; Sigma Aldrich, Darmstadt, Germany) containing 20 µM norvaline (#N7502; Sigma Aldrich), which served as an internal standard, and 150 µL of chloroform (ratio 1:1:1; #900688-1; Sigma Aldrich). Then samples were vortexed at 4 °C for 20 min, centrifuged at 8000× g for 10 min at 4 °C, and the upper phase was collected in a new tube and evaporated at room temperature in a rotavapor.

For derivatization, dried polar metabolites were dissolved in 30 µL of 2% methoxyamine hydrochloride (#226904; Sigma Aldrich) in pyridine (25104; Thermo Fisher Scientific, Waltham, MA, USA) and held at 30 °C for 90 min. After dissolution and reaction, 45 mL MSTFA + 1% TMCS (69478-10x; Sigma Aldrich) was added, and samples were incubated at 37 °C for 60 min. Gas chromatographic runs were performed with helium as carrier gas at 0.6 mL/min at an inlet temperature of 250 °C. The GC oven temperature ramp was from 60 to 325 °C, at 10 °C/min. The data acquisition rate was 10 Hz. For the Quadrupole, an EI source (70 eV) was used, and full-scan spectra (mass range from 50 to 600) were recorded in the positive ion mode. The injection volume was 1 µL, and a split ratio of 1:10 was used.

Global metabolic profiling was obtained by using a MassHunter data-processing tool (Agilent). Metabolite identification was performed at level 1, as proposed by the metabolomics standards initiative [24], using a retention index and mass spectrum as the two independent and orthogonal data required for identification. A Fiehn Metabolomics RTL library (Agilent G1676AA) was used as the reference library of compounds [25]. For the identification of significant metabolic pathways involved in male and female in-season chronic training, we used MetaboAnalyst 4.0 (http://www.metaboanalyst.ca, date: 10 April 2022). A p-value < 0.05 and a false discovery rate (FDR) < 0.1 were used to select the involved pathways [26].

2.7. Image and Statistical Analysis

Data are presented as means +/− standard deviation (SD) from at least three experiments. Amersham Imager 600 (GE Healthcare) was used to scan the 2DE images which were saved with a resolution of 300 dpi and in 16-bit TIFF format. Progenesis SameSpots software v4.0 (Nonlinear Dynamics, UK) was used to perform image analysis after spots detection and background subtraction. We selected as the reference image the gel scan showing the best protein pattern, and its spots were used to match, quantify, and normalize the spots volume across all the other gels. The default parameters of the Progenesis SameSpots Stat module were used to perform the statistical analysis and principal component analysis. The differentially expressed spots were identified by using one-way ANOVA (p value < 0.05), and to find out the significant differences between groups, Tukey's multiple comparisons test, using GraphPad Prism v8.0 software, was performed. The Tukey test compared every mean with every other mean, computing a confidence interval for multiple comparisons of 95% confidence. Significance was defined as p-value < 0.05.

We checked the normality and homogeneity distribution of our data by the D'Agostino–Pearson and Brown–Forsythe tests, respectively. As a measure of the effect size, we reported the eta-squared (η^2) calculated in ANOVA (Tukey's multiple comparisons test) performed on all groups together, using GraphPad Prism 8, considering the following intervals: 0.01–0.20 = small effect, 0.21–0.60 = moderate effect, and 0.61–0.99 = high effect [16]. The power analysis performed post hoc with GPower 3.1 confirmed that the sample size is satisfactory for a power of 0.90 and $p < 0.05$ for each variable analyzed, with the only exception being L-glutamic acid, for which the power is 0.5.

3. Results

The characteristics of 20 professional basketball players (10 female and 10 male) and 20 sedentary participants (10 male and 10 female) as controls are reported in Table 1.

Table 1. Participants' characteristics.

Characteristics	Mean (SD)				Tukey's Test [a]			
	BM	CM	BF	CF	BM vs. CM	BF vs. CF	BM vs. BF	CM vs. CF
Age (year)	21 ± 2.2	26.1 ± 4.1	25.1 ± 5.5	26.9 ± 2.2	0.01 *	0.77	0.08	0.97
Weight (kg)	81.5 ± 10.2	73 ± 8.7	68.7 ± 11.9	58.7 ± 5.8	0.17	0.17	0.03 *	0.02 *
Height (cm)	186 ± 0.06	178.7 ± 0.06	175.6 ± 0.08	163.4 ± 0.06	0.06	0.003 **	0.006 **	<0.0001 ****
BMI (kg/m^2)	23.6 ± 2.7	22.9 ± 2.9	22.1 ± 2.05	22 ± 2.3	0.64	1	0.18	0.72

[a] Tukey's test was performed by GraphPad Prism 8.0 software between male basket group (BM), male control group (CM), female basket group (BF), and female control group (CF): (* p-value < 0.05), (** p-value < 0.01), and (**** p-value < 0.0001).

In summary, the mean age of all participants was 24.4 ± 4.2 years, and no significant differences were found in the Body Mass Index (BMI) between the player groups and the control groups (p-value = 0.64 for males, and p-value = 1 for females; $\eta^2 = 0.084$) and between males and females (p-value = 0.18 for athletes, and p-value = 0.72 for controls; $\eta^2 = 0.084$), despite the fact that males were taller and heavier than females in both groups (athletes and controls) and female athletes were taller than the respective control. Moreover, all the professional basketball players, both male and female, trained at least five times a week, according to the specific training programs reported previously [16].

3.1. Overview of Saliva Protein Profiles

The 2DE silver-stained gel images were analyzed by Progenesis SameSpots software 4.0 (Nonlinear Dynamics, UK), using default parameters. A representative gel for each group was reported in Figure 1a. Each saliva sample was run in triplicate to obtain statistically significant results. After automatic spot identification, an average of about 1157 protein spots was detected in each gel. The computational 2DE gel image analysis pointed out 72 differentially expressed protein spots (ANOVA p-value < 0.05), while the Tukey's test showed that 43 of them were differentially expressed in groups that can be compared to each other, as reported in Supplementary Table S1: the terms of comparison chosen were the practice of sport and sex. Among these spots, 10 (23.2%) were differentially expressed between male athletes and controls, 22 (51.2%) between female basketball players and controls, 11 spots (25.6%) were differentially expressed between male and female athletes, and 13 spots (30.2%) between male and female controls.

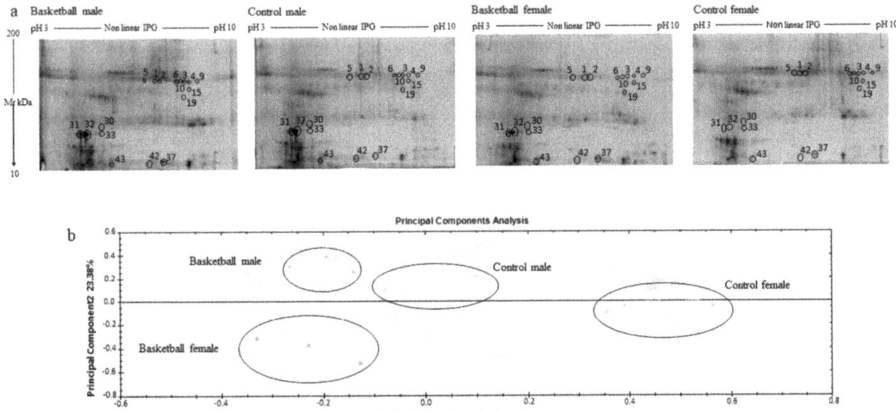

Figure 1. Proteomic profile of Basketball players and controls. (**a**) Representative 2DE images of silver-stained gels of saliva proteins run on NL pH 3–10 IP strip and in 9–16% polyacrylamide linear gradient. Circles and numbers indicate statistically differentially abundant proteins between the four groups analyzed, as reported in Table 2. (**b**) Multivariate analysis of the 2DE gel images results using Principal Components Analysis (PCA) performed by Progenesis SameSpots 4.0 software (Nonlinear Dynamic, UK).

3.2. Principal Component Analysis

A multivariate analysis PCA (principal component analysis) was carried out to obtain an overview of the proteomic data for overall trends in all groups. In the PCA biplot shown in Figure 1b, each dot describes the collective expression profiles of one sample; gels were grouped according to the difference of protein spot abundance, and the plot demonstrates consistent reproducibility among repeated samples within each group. The PCA biplot reveals four distinct main protein profile groups corresponding to the (i) basketball male group (pink circle), (ii) basketball female group (violet circle), (iii) control male group (blue circle), and (iv) control female group (orange circle). The first principal component, which distinguished 37.68% of the variance, clearly separates the proteome data of female controls group from the other groups, while the second component, with an additional 23.38% of variance, clearly distinguished the basketball male group from the basketball female group. The PCA plot suggests that sport training drastically affects the protein pattern; in fact, the greater differences are evident between the athletes than between the controls.

3.3. Proteins Differentially Modulated by Exercise and Sex

The proteins modulated by chronic in-season training or sex identified by the mass spectrometry approach are reported in Table 2. By comparing the proteins differentially expressed between athletes and sedentary controls, we found the protein Collagen-1(I) chain (P02452) to be downregulated in both male and female athletes (fold change −2.9 in male and −2.6 in female). We found three proteins, Immunoglobulin heavy constant alpha 1 and alpha 2 (P01876 and P01877, respectively) and the protein Alpha-Amylase 1A (P0DTE8) detected in multiple spots within the gel and with different fold changes, and it is likely that different forms of the same protein, processed or post-translationally modified, migrate differently. Post-translational modifications of Immunoglobulins are very frequent and often are associated with physiological and pathological conditions [27], while proteomic studies reported that Alpha-Amylase could be detected in more than twenty spots on 2DE gels. Indeed, posttranslational modifications can give rise to a pattern of isozymes [28,29].

Table 2. Differentially expressed protein spots from 2DGE identified by LC–MS/MS analysis.

Spot No. [a]	AC [b]	Gene Name	Protein Name	Score [c]	Protein Mass	Protein Cover [d]	Tukey's Test [e]/Fold Change [f]			
							BM vs. CM	BF vs. CF	BM vs. BF	CM vs. CF
1	P0DTE8	AMY1A	Alpha-Amylase 1A	791	56,484	52.2	ns	*/−2.6	ns	ns
	P01876	IGHA1	Immunoglobulin heavy constant alpha 1	243	38,486	22.9	ns	*/−2.6	ns	ns
	P01877	IGHA2	Immunoglobulin heavy constant alpha 2	170	37,366	19	ns	*/−2.6	ns	ns
2	P0DTE8	AMY1A	Alpha-Amylase 1A	628	56,484	31.9	ns	*/−2.6	ns	ns
	P01876	IGHA1	Immunoglobulin heavy constant alpha	141	38,486	14.7	ns	*/−2.6	ns	ns
	P01877	IGHA2	Immunoglobulin heavy constant alpha 2	121	37,366	13.9	ns	*/−2.6	ns	ns
3	P0DTE8	AMY1A	Alpha-Amylase 1A	1120	56,484	52.8	*/−2.9	**/−2.6	ns	*/−1.7
	P02452	COL1A1	Collagen alpha-1(I) chain	117	37,077	11.3	*/−2.9	**/−2.6	ns	*/−1.7
	P01876	IGHA1	Immunoglobulin heavy constant alpha 1	223	38,486	20.2	*/−2.9	**/−2.6	ns	*/−1.7
	P01877	IGHA2	Immunoglobulin heavy constant alpha 2	176	37,366	16.2	*/−2.9	**/−2.6	ns	*/−1.7
4	P0DTE8	AMY1A	Alpha-Amylase 1A	1073	56,484	57.1	ns	**/−3.3	ns	ns
	P01876	IGHA1	Immunoglobulin heavy constant alpha 1	194	38,486	16.6	ns	**/−3.3	ns	ns
	P01877	IGHA2	Immunoglobulin heavy constant alpha 2	143	37,366	12.6	ns	**/−3.3	ns	ns
	A0N4V7	Tcr-alpha	putative T-cell receptor beta	39	2269	38.1	ns	**/−3.3	ns	ns
5	P0DTE8	AMY1A	Alpha-Amylase 1A	478	56,484	30	**/1.4	**/−1.5	****/1.9	ns
6	P0DTE8	AMY1A	Alpha-Amylase 1A	1011	56,484	48.8	ns	*/−2.4	ns	ns
	P01876	IGHA1	Immunoglobulin heavy constant alpha 1	254	38,486	25.7	ns	*/−2.4	ns	ns
	P01877	IGHA2	Immunoglobulin heavy constant alpha 2	212	37,366	21.9	ns	*/−2.4	ns	ns
9	P0DTE8	AMY1A	Alpha-Amylase 1A	904	58,415	43.2	ns	*/−3.2	ns	ns
	P68871	HBB	Hemoglobin subunit beta	147	16,102	35.4	ns	*/−3.2	ns	ns
	P01876	IGHA1	Immunoglobulin heavy constant alpha 1	240	38,486	21.5	ns	*/−3.2	ns	ns
	P01877	IGHA2	Immunoglobulin heavy constant alpha 2	179	37,366	16.2	ns	*/−3.2	ns	ns
10	P0DTE8	AMY1A	Alpha-Amylase 1A	911	58,415	41.1	*/−2.3	ns	ns	ns
	P01876	IGHA1	Immunoglobulin heavy constant alpha 1	211	38,486	17.3	*/−2.3	ns	ns	ns
	P01877	IGHA2	Immunoglobulin heavy constant alpha 2	175	37,366	16.2	*/−2.3	ns	ns	ns
15	P0DTE8	AMY1A	Alpha-Amylase 1A	375	58,415	18	ns	*/−2.7	ns	ns
	P25311	AZGP1	Zinc-alpha-2-glycoprotein	390	34,465	37.6	ns	*/−2.7	ns	ns
	Q01469	FABP5	Fatty acid–binding protein 5	147	15,497	28.1	ns	*/−2.7	ns	ns

Table 2. Cont.

Spot No. [a]	AC [b]	Gene Name	Protein Name	Score [c]	Protein Mass	Protein Cover [d]	Tukey's Test [e]/Fold Change [f]			
							BM vs. CM	BF vs. CF	BM vs. BF	CM vs. CF
19	P0DTE8	AMY1A	Alpha-Amylase 1A	831	58,415	44.2	ns	**/−2	ns	**/−2.2
30	P02768	ALB	Albumin	269	71,317	15.1	ns	ns	ns	*/2.1
	P07355	ANXA2	Annexin A2	249	38,808	20.6	ns	ns	ns	*/2.1
	P81605	DCD	Dermcidin	208	11,391	38.2	ns	ns	ns	*/2.1
31	P01040	CSTA	Cystatin A	246	11,000	98	ns	ns	ns	*/3.8
	P81605	DCD	Dermcidin	177	11,391	35.5	ns	ns	ns	*/3.8
32	P01040	CSTA	Cystatin A	348	11,000	94.9	ns	ns	ns	*/4.3
	P81605	DCD	Dermcidin	73	11,391	30.9	ns	ns	ns	*/4.3
33	P02768	ALB	Albumin	675	71,317	28.1	ns	ns	**/2.7	ns
	P81605	DCD	Dermcidin	150	11,391	30.9	ns	ns	**/2.7	ns
	P01876	IGHA1	Immunoglobulin heavy constant alpha 1	214	38,486	15.3	ns	ns	**/2.7	ns
42	P02768	ALB	Albumin	190	68,408	20.5	ns	ns	ns	*/2.1
	P31151	S100A7	Psoriasin	115	11,564	62.4	ns	ns	ns	*/2.1
43	P81605	DCD	Dermcidin	143	11,391	37.3	ns	ns	ns	*/3

[a] Spot numbers reported in the representative 2DE images shown in Figure 1. [b] Accession number in Swiss-Prot/UniProtKB (http://www.uniprot.org/). [c] MASCOT MS score (Matrix Science, London, UK; http://www.matrixscience.com). MS matching score greater thank 56 was required for a significant MS hit (p-value < 0.05). [d] Sequence coverage = (number of the identified residues/total number of amino acid residues in the protein sequencer) × 100%. [e] Tukey's post hoc test was performed on ANOVA p-values by GraphPad Prism 6.0 software (* p-value < 0.05), (** p-value < 0.01), (**** p-value < 0.0001), and (ns = not significant). [f] Fold change was calculated by GraphPad Prism 6.0 software. It is the ratio of the mean normalized spot volumes of male basket group (BM), male control group (CM), female basket group (BF), and female control group (CF). It was reported only for statistically significant values.

As reported above, we observed that exercise's effects on saliva proteins were not similar for males and females. Moreover, we also evaluated the possible sex-related differences in sedentary controls. In the male controls, we found increased levels of the proteins Dermcidin (P81605) (Spots 30, 31, 32, and 43) and Cystatin-A (P01040), which were both found in more of one spot (Spots 31 and 32), and Albumin (P02768) (42). Dermcidin was also found in Spot 33, where its modulation is related to sport and sex, and it is upregulated in male athletes (fold change 2.7).

3.4. Saliva IgA, Alpha-Amylase, Dermcidin, and Cystatin a Determination

To confirm the differential expression levels of the identified proteins, we performed a Western blot analysis of IgA and Alpha-Amylase, whose expression increased in male athletes compared with female players (Figure 2a,b). The Western blot analysis confirms the results obtained by the proteomic analyses; in fact, IgA was lower in female athletes in comparison to male players, as reported in Spot 33, and in comparison to control females, as reported in Spots 1, 2, 3, 4, and 6. For Alpha-Amylase, the Western blot analysis confirms the trend observed in Spot 5, in which only this protein was identified. However, it does not validate the decreased level in basketball female players in comparison to their controls observed also in other spots, such as Spots 1, 2, 3, 4, 6, 9, and 15. It is important to underline that, in all the spots, more than one protein was identified, and the fold changes reported in Table 2 were calculated through an analysis performed on silver stain gel. The presence of different proteins in the same spot could explain why these results were not confirmed by the Western blot. Furthermore, IgA and Alpha-Amylase are abundant proteins in saliva, and, like albumin in blood, they appear over a wide range of pH values and molecular weights, they so were found in different spot interfering with proteomic studies.

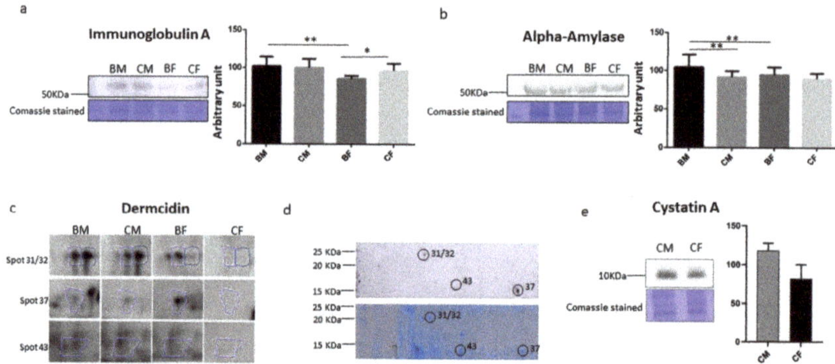

Figure 2. Validation of proteomic results. Histograms and representative immunoblot images of (**a**) Immunoglobulin A and (**b**) Alpha-Amylase in BM (basketball male group), CM (control male group), BF (basketball female group), and CF (control female group). Normalization of immunoblot was performed on Coomassie-stained PVDF membrane. The statistical analysis was carried out by two-tailed *t*-test, using Graphpad Prism 6 (* $p < 0.05$; ** $p < 0.01$). (**c**) Representative image of 2DE silver-stain spots, (**d**) 2DE Western blot for Dermcidin, and (**e**) Cystatin A.

Dermcidin was found to be increased in the control males compared to the females in more than one spot (Spots 30, 31, 32, and 43; Figure 2c), so we decided to perform a Western blot analysis on 2DE gel, as reported in Figure 2d. We confirmed the presence of Dermcidin in Spots 31, 32, and 43, and we also identified Dermcidin in Spot 37 (Figure 2d) that was further validated by mass spectrometry, confirming the presence of this protein as reported in Table 3. In addition, the identification of Spot 43 shows Dermcidin as a unique protein, but its expression is modified only in the controls (increased in male vs. female). Spots 37 and 43 show a similar size but different isoelectric points, as is evident from the gel in Figure 1a. Dermcidin is secreted into sweat, where it is proteolytically processed and also post-translationally modified in order to give rise to antimicrobial peptides, thus explaining its presences in several spots [30]. Furthermore, it is also released by skeletal muscles, and the full-length protein stimulates apoptosis under hypoxic conditions [31].

Table 3. Dermcidin validation by mass spectrometry.

Spot No. [a]	Accession No. [b]	Description	Coverage (%) [c]	Unique Peptides [d]	Score Mascot [e]	Score Sequest HT [f]
31–32	P81605	Dermcidin	21	2	64	2.22
37	P81605	Dermcidin	21	2	70	5.09
43	P81605	Dermcidin	25	4	147	14.15

[a] Spot numbers reported in the 2DE Western blot shown in Figure 3d. [b] Accession number in Swiss-Prot/UniprotKB. [c] Percentage of the protein sequence covered by the peptides. [d] Total number of peptide sequences unique to the protein group. [e] Cumulative protein score based on summing the ion scores of the unique peptides identified. [f] Sum of the scores of the individual peptides from the Sequest HT search.

Moreover, in Spots 31 and 32, another protein, Cystatin A, was identified with a higher protein coverage and score. In Figure 2e, the immunoblot quantifications were reported, confirming the results of Cystatin A modulation.

3.5. Metabolites Differentially Modulated by Exercise and by Sex

The metabolic modulation of salivary samples due to the chronic training was evaluated through GC–MS, and each compound obtained was identified by the Fiehn library, which allows users to identify several salivary compounds differentially expressed between the groups (Figure 3 and Table 4). The overall identified metabolites were reported in Supplementary Table S2. Metabolite concentrations were normalized by using both saliva volume (for each group, 150 μL) and total metabolites concentration according to Pitti et al. [15] with similar results.

Table 4. Differentially expressed metabolites identified in saliva samples by gas chromatography–mass spectrometry (GC–MS) analysis.

Compound Name	CAS Number *	Formula	KEGG ID °	Fold Change/Tukey's Test			
				BM vs. CM	BF vs. CF	BM vs. BF	CM vs. CF
D-allose	579-36-2	$C_6H_{12}O_6$	C01487	−3.1/***	−1.8/*	ns	ns
D-mannitol	87-78-5	$C_6H_{12}O_6$	C00392	ns	2.4/***	−2.1/***	ns
L-glutamic acid	56-86-0	$C_5H_9NO_4$	C00025	3.2/****	2.2/***	ns	ns
L-lysine	56-87-1	$C_6H_{14}N_2O_2$	C00047	5.0/****	3.0/*	ns	ns
L-ornithine	70-26-8	$C_5H_{12}N_2O_2$	C00077	5.2/**	4.0/*	ns	ns
L-threonine	72-19-5	$C_4H_9NO_3$	C00188	8.9/**	ns	ns	ns
L-valine	72-18-4	C_5HNO_2	C00183	5.0/*	ns	ns	ns
O phosphocolamine	1071-23-4	$C_2H_8NO_4P$	C00346	−2.2/****	−2.3/**	ns	ns
Citric acid	5949-29-1	$C_6H_8O_7$	C00158	ns	−1.8/**	−2.5/*	−2.7/***
Glycine	56-40-6	$C_2H_5NO_2$	C00037	4.9/*	2.5/*	ns	ns
Methyl-beta-D-galactopyranoside	1824-94-8	$C_7H_{14}O_6$	C03619	ns	−5.8/****	ns	−8.2/****
N-acetylneuraminic acid	131-48-6	$C_{11}H_{19}NO_9$	C00270	2.8/*	2.0/*	ns	ns
Taurine	107-35-7	$C_2H_7NO_3S$	C00245	ns	ns	2.6/*	ns
Tyrosine	60-18-4	$C_9H_{11}NO_3$	C00082	2.3/**	1.6/*	ns	ns

* Chemical Abstract Service number ° KEGG identifier (http://www.genome.jp/kegg/). Tukey's post hoc test was performed on ANOVA *p*-values by GraphPad Prism 6.0 software: (* $p < 0.05$), (** $p < 0.01$), (*** $p < 0.001$), (**** $p < 0.0001$), and (ns = not significant). Fold change was calculated by GraphPad Prism 6.0 software and was reported only for statistically significant values.

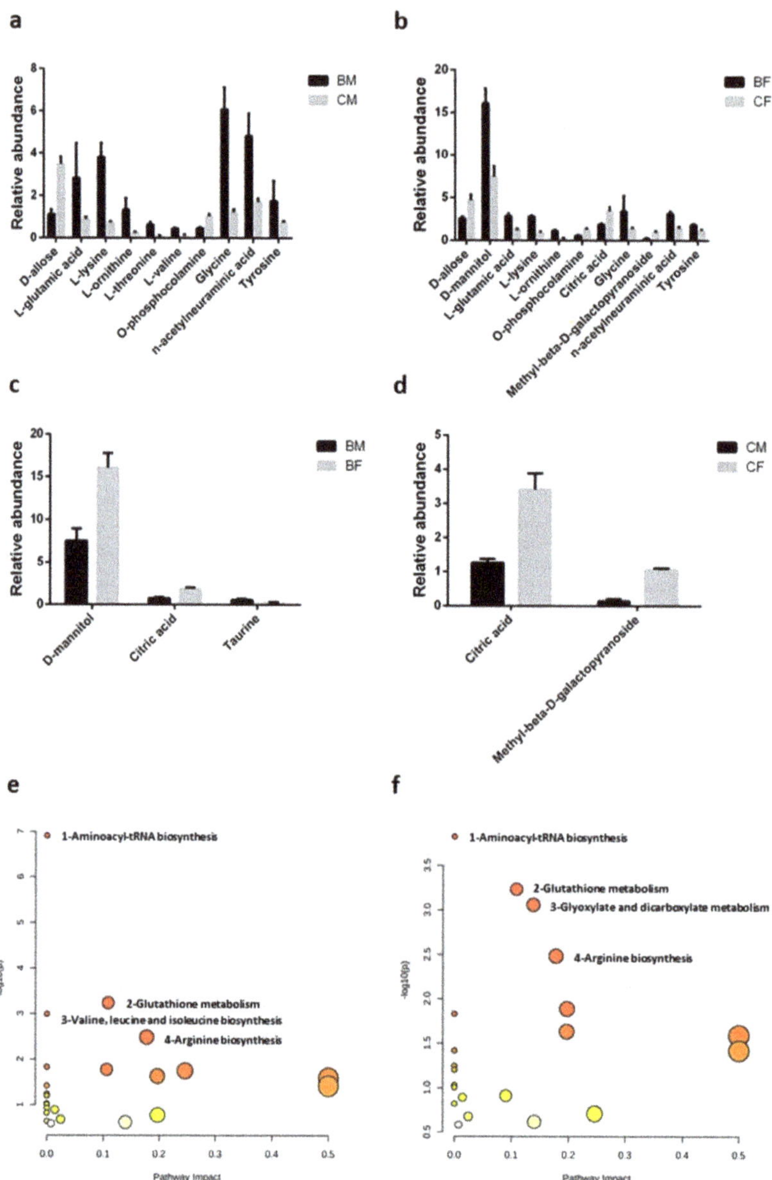

Figure 3. Plasma metabolomic profile of female and male basketball players. Histogram representation of saliva metabolites whose relative abundance is statistically different ($p < 0.05$) between (**a**) male basketball athletes and controls; (**b**) female basketball athletes and controls; (**c**) female and male basketball athletes; and (**d**) female and male controls (CM, control male group; BF, basketball female group; and CF, control female group). Statistical analysis was performed by two-way ANOVA (Tukey's multiple comparisons test), using GraphPad Prism 6. Representation of the metabolic pathways involved in (**e**) male and (**f**) female in-season chronic training, using the differentially abundant metabolites reported in Table 4. For analysis, we used MetaboAnalyst 4.0, setting a p-value < 0.05 and a false discovery rate (FDR) < 0.1 to select the involved pathways.

Comparing both male and female athletes with their respective sedentary controls, we identified several amino acids modulated by training and, more precisely, showing a statistically significant increase: L-glutamic acid ($\eta^2 = 0.3589$), L-lysine ($\eta^2 = 0.8801$), L-ornithine ($\eta^2 = 0.8508$), glycine ($\eta^2 = 0.8007$), and tyrosine ($\eta^2 = 0.7897$). In addition to these, in male basketball players, two more amino acids, L-threonine ($\eta^2 = 0.8120$) and L-valine ($\eta^2 = 0.5985$), showed a significant increase compared to the sedentary controls. Again, when comparing both male and female athletes with the controls, we found a significant increase in the N-acetylneuraminic acid salivary level ($\eta^2 = 0.7202$). On the contrary, other metabolites, such as D-allose ($\eta^2 = 0.8869$) and O-phosphocolamine ($\eta^2 = 0.9167$), decreased (Figure 3a,b and Table 4).

As with proteins, we observed that some metabolites were sex-specific during chronic exercise, suggesting a possible difference between the male and female athletes and the controls. When comparing male and female athletes, in the males, we found a significant decrease in D-mannitol ($\eta^2 = 0.8319$) and citric acid ($\eta^2 = 0.9044$) and a great increase in taurine ($\eta^2 = 0.7411$) (Figure 3c; Table 4). With regard to citric acid, it is interesting to note that in sedentary controls, the increases are significant in females compared to males. Moreover, male controls showed a reduced value of methyl-beta-D-galactopyranoside ($\eta^2 = 0.9632$) compared to the female controls (Figure 3d and Table 4).

We used metabolites showing significant modifications (Table 4) to identify the metabolic pathway related to chronic training and sex. These analyses were conducted by using MetaboAnalyst 4.0. From a total of 24 pathways indicated comparing athletes versus sedentary controls, 4 for male and 4 for female athletes were selected by following the criteria of $p < 0.05$ and FDR < 0.1 (Figure 3e,f). From all observed significant pathways, three of them were involved in both males and females. These pathways are as follows:

(i) Aminoacyl-tRNA biosynthesis—in detail, for males, glycine, L-valine, L-lysine, L-threonine, L-tyrosine, and L-glutamate (number 1 in Figure 3e), and for females, glycine, L-lysine, L-tyrosine, and L-glutamate (number 1 in Figure 3f);

(ii) Glutathione metabolism—for both males and females, glycine, L-glutamate, and L-ornithine (number 2 in Figure 3e,f)

(iii) Arginine biosynthesis; for both males and females, L-glutamate and L-ornithine (number 4 in Figure 3e,f).

On the other hand, valine, leucine, and isoleucine biosynthesis were significant only in males (L-valine and L-threonine in pathway number 3 panel in Figure 3e), while glyoxylate and dicarboxylate metabolism were significant only in females (citric acid, glycine, and L-glutamate in pathway number 3 in Figure 3f).

4. Discussion

In the present study, we investigated the sex-specific changes in the salivary proteome and metabolome in highly trained athletes during continuous chronic training. Several studies have been conducted on the salivary proteome and metabolome in response to acute training; very little is known about the repeated number of sessions [6,19,32]. Overall, we found reduced levels of proteins involved in chronic inflammation, such as Immunoglobulin A, Alpha-Amylase, and Dermcidin, in females, thus confirming the anti-inflammatory effect of exercise training. In addition, we found a training effect on metabolism related to glutathione biosynthesis in both male and female players, whereas an increase in branched chain amino acids was observed only in male players.

In a recent paper, McKetney et al. [33] reported a multiomics analysis in the saliva of soldiers before, during, and after a military deployment in which a military attack was simulated to examine the acute and chronic exercise response. However, in their experimental model, a stress condition due to the period of the military mission is foreseen.

This study focused on determining the effects of chronic training on the salivary proteome and metabolome during the season. Both male and female basketball athletes were evaluated to highlight possible gender differences. Among proteins, significant modulations were observed in molecules with antimicrobial activity involved in mucosal immunity, such as salivary IgA and Alpha-Amylase [7].

In our experimental model, decreased expression of salivary IgA was observed in female athletes, suggesting that highly trained athletes may experience an immunosuppressive state during in-season training periods (Figure 2a). The same evidence was previously obtained in plasma samples from the same subjects [17]: indeed, we showed a reduction of several plasma proteins associated with chronic inflammation, enhancing the anti-inflammatory effect of regular training in females. Several studies reported lower salivary IgA levels as an acute effect of exercise [34,35], while other studies found no change [36] or even an increase [7,37], suggesting a possible role of exercise in modulating this protein. In addition, the salivary IgA level may be helpful to detect excessive exercise load, which may determine the risk of respiratory infections in elite athletes. However, in our results, we found that all identified isoforms were less expressed in female athletes compared to their controls, suggesting that the effect is more pronounced in females than in males [38,39].

The same trend was observed in the modulation of salivary Alpha-Amylase between male and female athletes. The highlighted data may indicate that the intensity, workload, and duration of training during season must take into account the sex difference. Specifically, we observed a lower level of salivary Alpha-Amylase in female athletes (Figure 2b) [40]. These proteins are an indicator of sympathetic nervous system activity under exercise conditions, and it is observed to decrease in saliva from elite athletes after a heavy training season, indicating a non-functional overreaching [41,42]. This decrease is more pronounced in the athlete groups than in the control groups (Figure 2b). The main interest of investigators in the studying muscle adaptation has been changes in the levels of this enzyme after acute exercise, but there are few data on its modulation after chronic training. As with salivary IgA levels, it is suggested that the stress response to physical exercise can be assessed by monitoring salivary Alpha-Amylase, since the salivary levels of both enzymes are altered in response to sympathetic and parasympathetic nervous system activation after exercise [7] and are known to be higher in most competent and experienced male athletes than in females [42,43].

In our condition, training protocols for male and female athletes were identical. This altered female-specific salivary IgA and Alpha-Amylase response could be due to an inadequate training workload. Consequently, training that is not specifically planned and modulated for female athletes could affect female athletic performance and, at best, result in lower performance. In summary, we demonstrated a stress response of the body in female athletes to chronically inadequate training.

In our previous study, we found reduced salivary cortisol levels and absence of oxidative stress in both female and male basketball players under the same conditions [16]. Kivlighan et al. [42] observed an inverse relationship between salivary cortisol and Alpha-Amylase during competition in female athletes and not in male athletes. Indeed, salivary cortisol levels increased in response to acute exercise in females. Moreover, Cadegiani et al. [44] reported that high resting cortisol levels are associated with overtraining. Both cortisol and Alpha-Amylase can be signs of stress in response to exercise [40,45], but they cannot be considered equivalent because Alpha-Amylase is produced locally in the salivary glands and is therefore a more direct and sensitive marker of exercise-induced stress than cortisol, which is transported from blood to saliva [46,47].

Regarding Dermcidin, our results have shown that this protein is present in saliva and that it is modulated during training. Dermcidin is expressed as a precursor that can be proteolytically processed to obtain various antimicrobial peptides. The peptides produced provide a line of defense and are abundant in body fluids. Moreover, this protein is expressed in various cell types, but its function remains to be elucidated. Recently Esposito et al. [31] identified Dermcidin as a novel myokine that modulates cardiac myocytes and their function. The authors proposed that this protein is a signaling molecule released from skeletal muscles. We can suggest that this myokine may be a factor in inter-organ communication in male athletes with higher skeletal muscle mass compared to females, affecting nearby or distant organs and stimulating various mechanisms, including inflammatory processes that may lead to muscle fatigue.

We found a higher saliva concentration of Cystatin A in male subjects compared with female controls; Cystatin A is a protein belonging to the Cystatin family, a group of cysteine proteases produced mainly in the submandibular glands. Under our experimental conditions, this protein was not increased during chronic exercise, whereas several authors reported that acute exercise significantly increased its salivary levels. We conclude that the lack of modulation in Cystatin A observed during chronic exercise is related to the time elapsed after the end of acute exercise [48]. However, this protein family seems to be linked to an increase in oxidative stress, as reported by some authors [49], and this could be related to our results in which all the athletes analyzed, both male and female, had a low level of this type of stress.

The metabolites identified as significantly associated with chronic training in both male and female athletes are related to amino acid biosynthesis and translation, as evidenced by modulation of the aminoacyl-tRNA biosynthesis pathway. Moreover, both male and female athletes exhibited metabolites related to glutathione biosynthesis, indicating an increase in antioxidant defenses [50]. Regarding threonine and ornithine matching in the glutathione biosynthesis pathway, the first is converted to pyruvate to produce glucose, and its salivary level could be due to a mechanism to maintain glycemia in athletes. Ornithine is involved in the urea cycle, and its increase could indicate a rise in the urea cycle to eliminate ammonia produced by amino acid degradation during chronic training [51]. In males, but not in females, branched chain amino acids have been associated with an increase in body mass, as reported by Rodriguez-Carmona et al. [52], and in our experimental conditions, we found a significant increase in L-valine and L-threonine in the saliva of male athletes only. As expected, in male athletes, a training-related metabolic pathway is associated with valine, leucine, and isoleucine biosynthesis, considering that branched chain amino acids are associated with an increase in skeletal muscle mass, which is related to sex-specific differences in whole body muscle mass [53–55].

In our previous manuscript [16], we found increased amino acid metabolism in the plasma of male athletes, so based on these further results, we can hypothesize that the changes in these amino acid levels in saliva may reflect the changes in blood. Additionally, males exhibited significant differences in amino acids levels compared with females, again suggesting differences in amino acids' utilization.

The main limitation of the study is the number of participants, so it should be considered a pilot study. Furthermore, the findings reported relate to basketball, which is characterized by specific metabolic demands and cannot be generalized to all other sports. Future studies should include a larger number of participants practicing different sports. Finally, we did not consider the menstrual phase of the participants. There are controversial data on the influence of the menstrual cycle on physical activity [56], and it is very difficult to evaluate this.

5. Conclusions

In this study, we highlighted that saliva is a useful biofluid to evaluate the performance of athletes, because we found similar results to our previous work involving a plasma analysis of the same subjects [16]. Saliva reflects the cues found in plasma, so it can be used as a valid tool to monitor training and investigate differences between the sexes.

Regarding adaptation to training, we confirm that males and females share some common characteristics, but also some different sex-specific behaviors, including dissimilar amino acid utilization. This aspect has often been overlooked because women are still underrepresented in sports medicine research compared to men, and further studies should be conducted to fully understand female physiology and metabolism during exercise. Women and men share many gender-specific anthropometric and physiological characteristics that may affect adaptive mechanisms during exercise, and future research should focus on uncovering these differences to allow for more targeted interventions to maintain health, fitness, and performance and treat diseases from a gender perspective. In the future, it may be interesting to examine the effects of different sports that require different levels of effort and types of exercise on the gender differences found in this study.

Supplementary Materials: The following supporting information can be downloaded at https://www.mdpi.com/article/10.3390/healthcare11020241/s1. Table S1: Quantitative data and statistical analyses of protein spots whose intensity levels significantly differed among saliva of the four groups. Table S2: List of Metabolites identified in saliva samples of the subject utilized in this study by gas chromatography–mass spectrometry (GC–MS) analysis.

Author Contributions: Conceptualization, A.A., P.A.M., A.M. and T.G.; data curation, R.M. (Rosamaria Militello), G.P. and R.M. (Riccardo Marzocchini); formal analysis, S.L., R.M. (Rosamaria Militello), G.P., A.I., G.C., R.M. (Riccardo Marzocchini) and L.P.; funding acquisition, A.M.; investigation, S.P.; methodology, S.L., R.M. (Rosamaria Militello), A.I., G.C., S.P. and L.P.; project administration, A.M. and T.G.; resources, R.M. (Riccardo Marzocchini) and P.A.M.; software, G.P., A.I., A.A. and G.C.; supervision, P.A.M., A.M. and T.G.; validation, A.A.; visualization, S.P., L.P. and T.G.; writing—original draft, S.L. and A.M.; writing—review and editing, S.L., P.A.M., A.M. and T.G. All authors have read and agreed to the published version of the manuscript.

Funding: This research received no external funding.

Institutional Review Board Statement: The study was conducted in accordance with the Declaration of Helsinki, and approved by the Ethics Committee of University of Florence (AM_Gsport 15840/CAM_BIO).

Informed Consent Statement: Informed consent was obtained from all subjects involved in the study.

Data Availability Statement: The data presented in this study are available upon request from the authors.

Conflicts of Interest: The authors declare no conflict of interest.

References

1. Finsterer, J. Biomarkers of peripheral muscle fatigue during exercise. *BMC Musculoskelet. Disord.* **2012**, *13*, 218. [CrossRef] [PubMed]
2. Williams, M.A.; Haskell, W.L.; Ades, P.A.; Amsterdam, E.A.; Bittner, V.; Franklin, B.A.; Gulanick, M.; Laing, S.T.; Stewart, K.J. Resistance Exercise in Individuals with and without Cardiovascular Disease: 2007 Update. *Circulation* **2007**, *116*, 572–584. [CrossRef] [PubMed]
3. Luti, S.; Modesti, A.; Modesti, P.A. Inflammation, peripheral signals and redox homeostasis in athletes who practice different sports. *Antioxidants* **2020**, *9*, 1065. [CrossRef] [PubMed]
4. Hwang, Y.; Park, J.; Lim, K. Effects of Pilates Exercise on Salivary Secretory Immunoglobulin A Levels in Older Women. *J. Aging Phys. Act.* **2016**, *24*, 399–406. [CrossRef]
5. Lindsay, A.; Costello, J.T. Realising the Potential of Urine and Saliva as Diagnostic Tools in Sport and Exercise Medicine. *Sport. Med.* **2017**, *47*, 11–31. [CrossRef]
6. Franco-Martínez, L.; González-Hernández, J.M.; Horvatić, A.; Guillemin, N.; Cerón, J.J.; Martínez-Subiela, S.; Sentandreu, M.Á.; Brkljačić, M.; Mrljak, V.; Tvarijonaviciute, A.; et al. Differences on salivary proteome at rest and in response to an acute exercise in men and women: A pilot study. *J. Proteom.* **2020**, *214*, 103629. [CrossRef]
7. Papacosta, E.; Nassis, G.P. Saliva as a tool for monitoring steroid, peptide and immune markers in sport and exercise science. *J. Sci. Med. Sport* **2011**, *14*, 424–434. [CrossRef]
8. Ntovas, P.; Loumprinis, N.; Maniatakos, P.; Margaritidi, L.; Rahiotis, C. The Effects of Physical Exercise on Saliva Composition: A Comprehensive Review. *Dent. J.* **2022**, *10*, 7. [CrossRef]
9. Carpenter, G.H. The Secretion, Components, and Properties of Saliva. *Annu. Rev. Food Sci. Technol.* **2013**, *4*, 267–276. [CrossRef]
10. Miller, C.S.; Foley, J.D.; Bailey, A.L.; Campell, C.L.; Humphries, R.L.; Christodoulides, N.; Floriano, P.N.; Simmons, G.; Bhagwandin, B.; Jacobson, J.W.; et al. Current developments in salivary diagnostics. *Biomark. Med.* **2010**, *4*, 171–189. [CrossRef]
11. Kaufman, E.; Lamster, I.B. The diagnostic applications of saliva—A review. *Crit. Rev. Oral Biol. Med.* **2002**, *13*, 197–212. [CrossRef] [PubMed]
12. Jasim, H.; Olausson, P.; Hedenberg-Magnusson, B.; Ernberg, M.; Ghafouri, B. The proteomic profile of whole and glandular saliva in healthy pain-free subjects. *Sci. Rep.* **2016**, *6*, 39073. [CrossRef] [PubMed]
13. Fleissig, Y.; Reichenberg, E.; Redlich, M.; Zaks, B.; Deutsch, O.; Aframian, D.J.; Palmon, A. Comparative proteomic analysis of human oral fluids according to gender and age. *Oral Dis.* **2010**, *16*, 831–838. [CrossRef] [PubMed]

14. Soares Nunes, L.A.; Mussavira, S.; Sukumaran Bindhu, O. Clinical and diagnostic utility of saliva as a non-invasive diagnostic fluid: A systematic review. *Biochem. Med.* **2015**, *25*, 177–192. [CrossRef]
15. Pitti, E.; Petrella, G.; Di Marino, S.; Summa, V.; Perrone, M.; D'Ottavio, S.; Bernardini, A.; Cicero, D.O. Salivary Metabolome and Soccer Match: Challenges for Understanding Exercise induced Changes. *Metabolites* **2019**, *9*, 141. [CrossRef]
16. Pinto, G.; Militello, R.; Amoresano, A.; Modesti, P.A.; Modesti, A.; Luti, S. Relationships between Sex and Adaptation to Physical Exercise in Young Athletes: A Pilot Study. *Healthcare* **2022**, *10*, 358. [CrossRef]
17. Militello, R.; Pinto, G.; Illiano, A.; Luti, S.; Magherini, F.; Amoresano, A.; Modesti, P.A.; Modesti, A. Modulation of Plasma Proteomic Profile by Regular Training in Male and Female Basketball Players: A Preliminary Study. *Front. Physiol.* **2022**, *13*, 813447. [CrossRef]
18. Costill, D.L.; Daniels, J.; Evans, W.; Fink, W.; Krahenbuhl, G.; Saltin, B. Skeletal muscle enzymes and fiber composition in male and female track athletes. *J. Appl. Physiol.* **1976**, *40*, 149–154. [CrossRef]
19. Bongiovanni, T.; Pintus, R.; Dessì, A.; Noto, A.; Sardo, S.; Finco, G.; Corsello, G.; Fanos, V. Sportomics: Metabolomics applied to sports. The new revolution? *Eur. Rev. Med. Pharmacol. Sci.* **2019**, *23*, 11011–11019. [CrossRef]
20. Hagströmer, M.; Oja, P.; Sjöström, M. The International Physical Activity Questionnaire (IPAQ): A study of concurrent and construct validity. *Public Health Nutr.* **2006**, *9*, 755–762. [CrossRef]
21. Guidi, F.; Magherini, F.; Gamberi, T.; Bini, L.; Puglia, M.; Marzocchini, R.; Ranaldi, F.; Modesti, P.A.; Gulisano, M.; Modesti, A. Plasma protein carbonylation and physical exercise. *Mol. Biosyst.* **2011**, *7*, 640–650. [CrossRef] [PubMed]
22. Pietrovito, L.; Cano-Cortés, V.; Gamberi, T.; Magherini, F.; Bianchi, L.; Bini, L.; Sánchez-Martín, R.M.; Fasano, M.; Modesti, A. Cellular response to empty and palladium-conjugated amino-polystyrene nanospheres uptake: A proteomic study. *Proteomics* **2015**, *15*, 34–43. [CrossRef] [PubMed]
23. Luti, S.; Fiaschi, T.; Magherini, F.; Modesti, P.A.; Piomboni, P.; Governini, L.; Luddi, A.; Amoresano, A.; Illiano, A.; Pinto, G.; et al. Relationship between the metabolic and lipid profile in follicular fluid of women undergoing in vitro fertilization. *Mol. Reprod. Dev.* **2020**, *87*, 986–997. [CrossRef] [PubMed]
24. Sumner, L.W.; Amberg, A.; Barrett, D.; Beale, M.H.; Beger, R.; Daykin, C.A.; Fan, T.W.-M.; Fiehn, O.; Goodacre, R.; Griffin, J.L.; et al. Proposed minimum reporting standards for chemical analysis. *Metabolomics* **2007**, *3*, 211–221. [CrossRef]
25. Kind, T.; Wohlgemuth, G.; Lee, D.Y.; Lu, Y.; Palazoglu, M.; Shahbaz, S.; Fiehn, O. FiehnLib: Mass Spectral and Retention Index Libraries for Metabolomics Based on Quadrupole and Time-of-Flight Gas Chromatography/Mass Spectrometry. *Anal. Chem.* **2009**, *81*, 10038–10048. [CrossRef]
26. Benjamini, Y.; Hochberg, Y. Controlling the False Discovery Rate: A Practical and Powerful Approach to Multiple Testing. *J. R. Stat. Soc. Ser. B* **1995**, *57*, 289–300. [CrossRef]
27. Harb, J.; Wilson, B.S.; Hermouet, S. Editorial: Structure, Isotypes, Targets, and Post-translational Modifications of Immunoglobulins and Their Role in Infection, Inflammation and Autoimmunity. *Front. Immunol.* **2020**, *11*, 1761. [CrossRef]
28. Hirtz, C.; Chevalier, F.; Centeno, D.; Rofidal, V.; Egea, J.-C.; Rossignol, M.; Sommerer, N.; Deville de Périère, D. MS characterization of multiple forms of alpha-amylase in human saliva. *Proteomics* **2005**, *5*, 4597–4607. [CrossRef]
29. Bank, R.A.; Hettema, E.H.; Arwert, F.; Amerongen, A.V.N.; Pronk, J.C. Electrophoretic characterization of posttranslational modifications of human parotid salivary α-amylase. *Electrophoresis* **1991**, *12*, 74–79. [CrossRef]
30. Schittek, B. The multiple facets of dermcidin in cell survival and host defense. *J. Innate Immun.* **2012**, *4*, 349–360. [CrossRef]
31. Esposito, G.; Schiattarella, G.G.; Perrino, C.; Cattaneo, F.; Pironti, G.; Franzone, A.; Gargiulo, G.; Magliulo, F.; Serino, F.; Carotenuto, G.; et al. Dermcidin: A skeletal muscle myokine modulating cardiomyocyte survival and infarct size after coronary artery ligation. *Cardiovasc. Res.* **2015**, *107*, 431–441. [CrossRef] [PubMed]
32. Puigarnau, S.; Fernández, A.; Obis, E.; Jové, M.; Castaner, M.; Pamplona, R.; Portero-Otin, M.; Camerino, O. Metabolomics reveals that fittest trail runners show a better adaptation of bioenergetic pathways. *J. Sci. Med. Sport* **2022**, *25*, 425–431. [CrossRef] [PubMed]
33. McKetney, J.; Jenkins, C.C.; Minogue, C.; Mach, P.M.; Hussey, E.K.; Glaros, T.G.; Coon, J.; Dhummakupt, E.S. Proteomic and metabolomic profiling of acute and chronic stress events associated with military exercises. *Mol. Omi.* **2022**, *18*, 279–295. [CrossRef]
34. Tomasi, T.B.; Trudeau, F.B.; Czerwinski, D.; Erredge, S. Immune parameters in athletes before and after strenuous exercise. *J. Clin. Immunol.* **1982**, *2*, 173–178. [CrossRef] [PubMed]
35. Walsh, N.P.; Bishop, N.C.; Blackwell, J.; Wierzbicki, S.G.; Montague, J.C. Salivary IgA response to prolonged exercise in a cold environment in trained cyclists. *Med. Sci. Sports Exerc.* **2002**, *34*, 1632–1637. [CrossRef]
36. McDowell, S.L.; Chaloa, K.; Housh, T.J.; Tharp, G.D.; Johnson, G.O. The effect of exercise intensity and duration on salivary immunoglobulin A. *Eur. J. Appl. Physiol. Occup. Physiol.* **1991**, *63*, 108–111. [CrossRef]
37. Allgrove, J.E.; Gomes, E.; Hough, J.; Gleeson, M. Effects of exercise intensity on salivary antimicrobial proteins and markers of stress in active men. *J. Sports Sci.* **2008**, *26*, 653–661. [CrossRef]
38. Gleeson, M. Mucosal immunity and respiratory illness in elite athletes. *Int. J. Sports Med.* **2000**, *21* (Suppl. S1), S33–S43. [CrossRef]
39. Hackney, A.C.; Koltun, K.J. The immune system and overtraining in athletes: Clinical implications. *Acta Clin. Croat.* **2012**, *51*, 633–641.
40. Nicolette, C.B.; Bishop, N.C.; Gleeson, M. Acute and chronic effects of exercise on markers of mucosal immunity. *Front. Biosci.* **2009**, *14*, 4444–4456. [CrossRef]

41. Halson, S.L.; Jeukendrup, A.E. Does overtraining exist? An analysis of overreaching and overtraining research. *Sports Med.* **2004**, *34*, 967–981. [CrossRef] [PubMed]

42. Kivlighan, K.T.; Granger, D.A. Salivary alpha-amylase response to competition: Relation to gender, previous experience, and attitudes. *Psychoneuroendocrinology* **2006**, *31*, 703–714. [CrossRef] [PubMed]

43. Alix-Sy, D.; Le Scanff, C.; Filaire, E. Psychophysiological responses in the pre-competition period in elite soccer players. *J. Sports Sci. Med.* **2008**, *7*, 446–454. [PubMed]

44. Cadegiani, F.A.; Kater, C.E. Hormonal aspects of overtraining syndrome: A systematic review. *BMC Sports Sci. Med. Rehabil.* **2017**, *9*, 14. [CrossRef]

45. de Oliveira, V.N.; Bessa, A.; Lamounier, R.P.M.S.; de Santana, M.G.; de Mello, M.T.; Espindola, F.S. Changes in the Salivary Biomarkers Induced by an Effort Test. *Int. J. Sports Med.* **2010**, *31*, 377–381. [CrossRef]

46. Lee, D.Y.; Kim, E.; Choi, M.H. Technical and clinical aspects of cortisol as a biochemical marker of chronic stress. *BMB Rep.* **2015**, *48*, 209–216. [CrossRef]

47. Nater, U.M.; Rohleder, N. Salivary alpha-amylase as a non-invasive biomarker for the sympathetic nervous system: Current state of research. *Psychoneuroendocrinology* **2009**, *34*, 486–496. [CrossRef]

48. Sant'Anna, M.d.L.; Oliveira, L.T.; Gomes, D.V.; Marques, S.T.F.; Provance, D.W.; Sorenson, M.M.; Salerno, V.P. Physical exercise stimulates salivary secretion of cystatins. *PLoS ONE* **2019**, *14*, e0224147. [CrossRef]

49. Jain, S.; Ahmad, Y.; Bhargava, K. Salivary proteome patterns of individuals exposed to High Altitude. *Arch. Oral Biol.* **2018**, *96*, 104–112. [CrossRef]

50. Khodagholi, F.; Zareh Shahamati, S.; Maleki Chamgordani, M.; Mousavi, M.A.; Moslemi, M.; Salehpour, M.; Rafiei, S.; Foolad, F. Interval aerobic training improves bioenergetics state and mitochondrial dynamics of different brain regions in restraint stressed rats. *Mol. Biol. Rep.* **2021**, *48*, 2071–2082. [CrossRef]

51. Sugino, T.; Shirai, T.; Kajimoto, Y.; Kajimoto, O. L-ornithine supplementation attenuates physical fatigue in healthy volunteers by modulating lipid and amino acid metabolism. *Nutr. Res.* **2008**, *28*, 738–743. [CrossRef] [PubMed]

52. Rodríguez-Carmona, Y.; Meijer, J.L.; Zhou, Y.; Jansen, E.C.; Perng, W.; Banker, M.; Song, P.X.K.; Téllez-Rojo, M.M.; Cantoral, A.; Peterson, K.E. Metabolomics reveals sex-specific pathways associated with changes in adiposity and muscle mass in a cohort of Mexican adolescents. *Pediatr. Obes.* **2022**, *17*, e12887. [CrossRef]

53. Low, S.; Wang, J.; Moh, A.; Ang, S.F.; Ang, K.; Shao, Y.-M.; Ching, J.; Wee, H.N.; Lee, L.S.; Kovalik, J.-P.; et al. Amino acid profile of skeletal muscle loss in type 2 diabetes: Results from a 7-year longitudinal study in asians. *Diabetes Res. Clin. Pract.* **2022**, *186*, 109803. [CrossRef] [PubMed]

54. Janssen, I.; Heymsfield, S.B.; Wang, Z.M.; Ross, R. Skeletal muscle mass and distribution in 468 men and women aged 18–88 yr. *J. Appl. Physiol.* **2000**, *89*, 81–88. [CrossRef] [PubMed]

55. Yamamoto, K.; Ishizu, Y.; Honda, T.; Ito, T.; Imai, N.; Nakamura, M.; Kawashima, H.; Kitaura, Y.; Ishigami, M.; Fujishiro, M. Patients with low muscle mass have characteristic microbiome with low potential for amino acid synthesis in chronic liver disease. *Sci. Rep.* **2022**, *12*, 3674. [CrossRef] [PubMed]

56. McNulty, K.L.; Elliott-Sale, K.J.; Dolan, E.; Swinton, P.A.; Ansdell, P.; Goodall, S.; Thomas, K.; Hicks, K.M. The Effects of Menstrual Cycle Phase on Exercise Performance in Eumenorrheic Women: A Systematic Review and Meta-Analysis. *Sport. Med.* **2020**, *50*, 1813–1827. [CrossRef]

healthcare

Article

The Association between Rapid Weight Loss and Body Composition in Elite Combat Sports Athletes

Marius Baranauskas [1,*] , Ingrida Kupčiūnaitė [1] and Rimantas Stukas [2]

[1] Faculty of Biomedical Sciences, Panevėžys University of Applied Sciences, 35200 Panevėžys, Lithuania; ingrida.kupciunaite@panko.lt
[2] Department of Public Health, Institute of Health Sciences, Faculty of Medicine, Vilnius University, 01513 Vilnius, Lithuania; rimantas.stukas@mf.vu.lt
* Correspondence: baranauskas.marius@panko.lt

Abstract: Rapid Weight Loss (RWL) is a rapid reduction in weight over a short period of time seeking to attain the norm required for a competition in a particular weight category. RWL has a negative health impact on athletes including the significant muscle damage induced by RWL. This study aimed to identify the association between RWL and body composition among competitive combat athletes (n = 43) in Lithuania. Our focus was laid on the disclosure of their RWL practice by using a previously standardized RWL Questionnaire. The body composition of the athletes was measured by means of the standing-posture 8-12-electrode multi-frequency bioelectrical impedance analysis (BIA) and the electrical signals of 5, 50, 250, 550 and 1000 kHz. This non-experimental cross-sectional study resulted in preliminary findings on the prevalence and profile of RWL among combat athletes in Lithuania. 88% of the athletes surveyed in our study had lost weight in order to compete, with the average weight loss of 4.6 ± 2% of the habitual body mass. The athletes started to resort to weight cycling as early as 9 years old, with a mean age of 12.8 ± 2.1 years. The combination of practiced weight loss techniques such as skipping meals (adjusted Odd Ratio (AOR) 6.3; 95% CI: 1.3–31.8), restricting fluids (AOR 5.5; 95% CI: 1.0–31.8), increased exercise (AOR 3.6; 95% CI: 1.0–12.5), training with rubber/plastic suits (AOR 3.2; 95% CI: 0.9–11.3) predicted the risk of RWL aggressiveness. RWL magnitude potentially played an important role in maintaining the loss of muscle mass in athletes during the preparatory training phase (β –0.01 kg, $p < 0.001$). Therefore, an adequate regulatory programme should be integrated into the training plans of high-performance combat sports athletes to keep not only the athletes but also their coaches responsible for a proper weight control.

Keywords: combat sports; elite athletes; rapid weight loss; extreme weight loss; weight management

Citation: Baranauskas, M.; Kupčiūnaitė, I.; Stukas, R. The Association between Rapid Weight Loss and Body Composition in Elite Combat Sports Athletes. *Healthcare* **2022**, *10*, 665. https://doi.org/10.3390/healthcare10040665

Academic Editors: João Paulo Brito, Rafael Oliveira and José Carmelo Adsuar Sala

Received: 28 February 2022
Accepted: 31 March 2022
Published: 1 April 2022

Publisher's Note: MDPI stays neutral with regard to jurisdictional claims in published maps and institutional affiliations.

1. Introduction

The athletes involved in any of the Olympic combat sports branches such as judo, wrestling, boxing and taekwondo are categorised by body weight (BW) into weight classes seeking to mitigate the disparities in size and strength [1,2]. However, the majority of combat sport athletes take part in competitions of a weight category which is below their usual body weight [3] in order to attain the advantage over their weaker or smaller opponents [4–8]. Rapid Weight Loss (RWL) is a rapid reduction of weight over a brief time period in an attempt of attaining the norm meeting the requirements for a competition in a particular weight category. It is described as a temporary weight loss of up to 5% of a person's weight over a short period of time (commonly within one week) [9]. Data obtained by many studies suggest that the prevalence of RWL practice among combat athletes ranged from 42% to 90% [5,10–14]. Before each competition, usually 2 to 3 days prior to the weigh-in procedure, combat sport athletes usually lowered their body weight by approximately 2% to 10% [15]. These practices appeared to be most common among athletes competing at higher contest levels where weight was managed by more aggressive strategies. Some most frequently deployed RWL techniques involve more intensive exercise; dehydration; use of sauna and rubber or plastic suits; reduction of energy intake; use of rubber or plastic suits for training; low intake of carbohydrates; limited consumption of fats; fasting; vomiting; diet pills, laxatives, and diuretics, all of which may have a negative impact on the performance of athletes or at least increase the risk of injury [16–19]. It was also well documented that RWL practice negatively impacts the health of athletes, leading to physical and psychological damage and may result in reduced bone density; worse performance of muscles; mood swings; dehydrated body condition; accelerated heart rate; poorer short-term memory, cognitive and mental function; or mounting anxiety, bad temper, weakness, depression, and a feeling of isolation [4,16,20]. Extreme cases of RWL have led to deaths as a result of dehydration and hyperthermia and myocardial infarction [21,22]. The threat of RWL has been recognised by sports bodies and position statements have been issued to reflect their stand [3,23]. In line with the criteria for prohibited methods stipulated in the World Anti-Doping Agency Code, Artioli et al. [15] even recommended to prohibit RWL practices in combat sports.

In summary, there are sufficient findings in scientific literature about the negative effects of the RWL magnitude not only on physical but also on mental health conditions among combat sports athletes. On the other hand, there is little evidence to generalise the impact of RWL magnitude on athletes' body composition. To our knowledge, RWL decreases performance, bone mineral density, and muscle mass in both, males and females [24]. In addition, it was revealed that an acute restriction of food and fluid intake appeared to negatively affect fat-free mass and the indices of kidney function in combat sports athletes including a significant muscle damage induced by RWL [25,26]. However, up until now, the association between the RWL magnitude and the habitual weight status or body composition of elite combat athletes in the training process have not yet been evaluated.

Taking into account the international context of research, the available evidence on RWL among Lithuanian athletes in weight-sensitive sports is limited. Therefore, the requirement to identify the prevalence, magnitude and methods of RWL among athletes seems to be of primary importance. In the next step, this study aimed to identify the association between RWL and body composition among competitive combat athletes in Lithuania. We identified the following challenges: (1) the primary objective of this study was to investigate the prevalence, magnitude and self-reported methods of RWL; (2) the secondary objective of the study was to identify and evaluate the body composition in a sample of high-performance athletes; (3) the third objective of the study was to reveal the relationship between RWL and the individual weight components among combat sports athletes.

2. Materials and Methods

2.1. Participants and Procedures

In March and May 2018, an observational analytical non-experimental cross-sectional study was carried out. The target population of the survey was the elite combat sport athletes belonging to the Lithuanian Sports Centre (LSC) based in Vilnius, Lithuania, whose lists were approved by the Lithuanian National Olympic Committee (LNOC) of Lithuania. Our investigation included only the athletes with the history of attainment of an Olympic qualification quota place or those with a history of participation in the European Athletics Championships (ECh) and/or the World Athletics Championships (WCh) to obtain Olympic qualification. On the basis of the above mentioned inclusion criteria, during the preparatory training phase, 43 high-performance athletes of combat sports were selected and examined, who represented 84% of the candidates for the Lithuanian Olympic team. The pool of respondents included judo (n = 11), Greco-Roman wrestling (n = 19) and boxing (n = 13) athletes. The qualification standards the athletes had met before the study, served as major inclusion criteria.

2.2. Measures

2.2.1. Anthropometric Measures

The athletes' height was measured at the Lithuanian Sports Medicine Centre (LSMC) by means of a stadiometer ($\pm$0.01 m). The BW, the individual weight components (lean body mass (LBM) (in kg and %), muscle mass (MM) (in kg and %) and body fat (BF) (in kg and %) were measured at the LSC by means of the standing-posture 8-12-electrode multi-frequency bioelectrical impedance analysis (BIA) and using electrical signals of 5, 50, 250, 550 and 1000 kHz (X-SCAN body composition analyzer with the certification EN ISO (the European Union has adopted an international standard) 13,488; Kyungsan City, South Korea) [27,28]. The muscle and fat mass index (MFMI) of each athlete was determined by dividing the muscle mass (in kg) by body fat (in kg). Lean body mass, body fat, muscle mass and MFMI were assessed according to the norms identified for high-performance athletes (male and female) and validated by the scientists in Lithuania [27].

2.2.2. Rapid Weight Loss Questionnaire

A questionnaire in line with the one deployed by Artioli et al. but adapted to the specifics of our study was used to explore the RWL practice [29]. The data were collected using a face-to-face questionnaire, translated to Lithuanian using back translation method. The Rapid Weight Loss Questionnaire (RWLQ) consisted of 18 items, and a special scoring system was used. The subjects with RWL practice records were assigned 3 points, while those without were assigned 0 points. When athletes reported the highest weight ever lost during their sporting career before the competition, 0.5 point was awarded for each kilogram of the weight lost. In addition, one point was awarded for 1 kg lost for those with the usual weight cut before the competition. The respondents were awarded scores of 5, 4, 3, 2, 1, and 0 after they had lost weight during the shortest period of time (from one to three days), and over increasingly longer periods of time, i.e., from 4 to 5 days, from 6 to 7 days, from 8 to 10 days, from 11 to 14 days, and from 14 days and longer, accordingly. One point was assigned per kilogram for the weight gained over one week after the competition. As for the methods deployed by athletes in losing weight, according to their responses such as "never", "not anymore", "almost never", "sometimes", and "frequently", we awarded the scores 0, 0.5, 1, 2, and 3 accordingly. After calculating the total scores, it was obvious that higher scores corresponded to more aggressive weight loss utilising harmful methods and an increased risk of losing weight too rapidly. The self-reported history of the weight loss section highlighted the self-reported histories of athletes and was formed of three questions. Firstly, the athletes were enquired about the weight category they were competing in, secondly, if the athlete had moved from one weight category to another, thirdly, what weight class they belonged to over the previous off-season.

2.3. Statistical Analysis

The statistical analysis was performed using SPSS V.25 for Windows (Corporate headquarters 1 New Orchard Road, Armonk, NY, USA). Standard descriptive summary statistics were applied to characterize the responses. All the normally distributed continuous variables were presented as means $\pm$ standard deviations (SD), whereas the qualitative variables were presented as relative frequencies (in %). When normality was confirmed, independent and paired t-tests were used to assess the differences between some groups. Pearson correlations (parametric tests) were used to assess the relationship between the RWL score and the variables of interest (weight cut for competition (in kg) and weight regain in a week after competition (in kg)).

The multinomial logistic regression analysis was used to relate the RWL scores and weight loss methods, later the results were presented by Odds Ratios (ORs) with 95% Confidence Interval (CI). As dependent variables must be categorical, continuous variables must be transformed, mostly via the classification of quartiles or percentiles. Thus, the dependent variable (RLWS (RWL score)) in this study was changed into the ordinal scale measurements by simply ranking the observations and using the values as a cut-off point of two subscales (20.7 $\geq$ RWLS > 20.7). The logistic model was adjusted for sports type, gender, and age. In this study, the parameters were estimated by means of the maximum likelihood method, also the appropriateness, adequacy and usefulness of the model were evaluated by the Wald (W) statistic, the estimated coefficient (β), with standard error (SE) (<5) and Nagelkerke R^2 statistic. The multiple linear regression analysis was used to determine the association between the aggressive RWL methods (in score) and the individual weight components (MM (in kg) and BF (in kg)). The model was adjusted for sex, age and type of sport. The significance level $p < 0.05$ was determined for all statistical tests.

3. Results

3.1. Characteristics of the Subjects

The sample under analysis included 86% ($n = 37$) men and 14% ($n = 6$) women. The combat athletes' age was within the range of 16 to 29. All the subjects participated in the National (Lithuanian) Championships and held prizes, 49% of combat athletes participated in the ECh (9% won medals), 26% of combat athletes participated in the WCh (12% won medals) and 5% of sportsmen participated in the Olympic Games (OG) (did not win medals).

The athletes' training workload dimensions were totally consistent with the training plans that had been approved by the LSC and the LNOC. The Tokyo 2020 programme provided specifications of the training plans. Over the study period the athletes underwent testing which revealed their training period range of 7.4 $\pm$ 3.9 years, workouts performed 5.5 $\pm$ 0.9 days per week, an average duration of workout 147.5 $\pm$ 40.4 min per day.

Table 1 shows the distribution of combat athletes (in percentage) by the current weight class of participation in competitions in line with the weight classes being contested in judo, Greco-Roman wrestling and boxing qualification events for Tokyo 2020.

Table 1. Weight classes of combat sports athletes.

Sport	Male		Female		Weigh-In Procedures
	Weight Classes	% of Athletes	Weight Classes	% of Athletes	
Judo	<60 kg	80.0	<48 kg	66.7	Have a trial 1 h weight-in period the day before the competition in the evening, and the next day during the checkweigher.
	66 kg	20.0	52 kg	33.3	
	73 kg	-	57 kg	-	
	81 kg	-	63 kg	-	
	90 kg	-	70 kg	-	
	100 kg	-	78 kg	-	
	+100 kg	-	+78 kg	-	
Greco-Roman Wrestling	<60 kg	5.3	-	-	The weigh-in for each category always takes place on the day before the beginning of the competition concerned and lasts 30 min.
	67 kg	36.8	-	-	
	77 kg	36.8	-	-	
	87 kg	10.5	-	-	
	97 kg	5.3	-	-	
	130 kg	5.3	-	-	
Boxing	Flyweight (<52 kg)	15.4	48–51 kg	-	There is a weight-in of all competitors every morning throughout the competition (several days) and must keep their weight class limit.
	Featherweight (57 kg)	30.8	54–57 kg	-	
	Lightweight (63 kg)	23.1	57–60 kg	-	
	Welterweight (69 kg)	7.7	64–69 kg	-	
	Middleweight (75 kg)	7.7	69–75 kg	-	
	Light heavyweight (81 kg)	15.4	-	-	
	Heavyweight (81–91 kg)	-	-	-	
	Super heavyweight (91+ kg)	-	-	-	

3.2. The Body Composition of Athletes

The body composition of combat athletes was assessed as indicated in Table 2. The BW, LBM and MM in subjects varied within the normal range. Nonetheless, the mean of LBM ($83.9 \pm 5.5\%$) in male athletes met the maximum (max.) limit (85%). Δ LBM$_{in\,\%}$ (actual LBM$_{in\,\%}$—max. recommended LBM$_{in\,\%}$) was—$1.1 \pm 0.8\%$ (95% PI: -0.9–0.7). The LBM ($78 \pm 2.7\%$) of female athletes was not different from the max. limit (80%). Δ LBM$_{in\,\%}$ (actual LBM$_{(in\,\%)}$—max. recommended LBM$_{in\,\%}$) was—$2 \pm 1.1\%$ (95% PI: -4.8–0.9).

Table 2. Body composition of combat sports athletes.

Variables	Judo		Greco-Roman Wrestling	Boxing	Normative Male/Female
	Male	**Female**	**Male**	**Male**	
Height (m)	1.63 ± 0.17	1.64 ± 0.07	1.77 ± 0.08	1.73 ± 0.08	
BW (kg)	59.4 ± 25.7	55.8 ± 2.3	75.1 ± 15.7	66.5 ± 14.2	
LBM (in kg)	48.7 ± 14.5	43.6 ± 2.8	62.3 ± 11.2	55.4 ± 7.7	
LBM (% of BW)	85.1 ± 9.0	78.0 ± 2.7	83.1 ± 3.8	84.3 ± 6.2	75–85/70–80
MM (in kg)	45.3 ± 13.1	40.4 ± 2.7	57.7 ± 10.1	51.6 ± 6.7	
MM (% of BW)	79.3 ± 8.9	72.2 ± 2.7	76.5 ± 4.6	78.7 ± 6.2	74–80/64–80
BF (in kg)	10.7 ± 11.5	12.2 ± 1.4	13.2 ± 5.3	11.2 ± 6.7	
BF (% of BW)	14.9 ± 9.0	22.1 ± 2.6	16.7 ± 3.9	15.7 ± 6.2	15–19 [1]/20–24 [2]
MFMI	6.9 ± 3.5	3.4 ± 0.6	4.8 ± 1.4	6.4 ± 3.9	4.7–6 [1]/3–3.99 [3]

[1]—a large (acceptable) BF (% of BW); [2]—an average (optimal) BF (% of BW); [3]—an average MFMI; The data are presented as means $\pm$ standard deviation (SD). BW–body weight; LBM–lean body mass; MM–muscle mass, BF–body fat; MFMI–muscle and fat mass index.

The mean of BF in male athletes involved in sports such as judo, Greco-Roman wrestling and boxing ($16 \pm 5.4\%$ of BW) was acceptable and met the average (avg.) recommended limit (17% of BW). Δ BF$_{in\,\%}$ (actual BF$_{in\,\%}$—avg. recommended BF$_{in\,\%}$) was $-1 \pm 0.9\%$ (95% PI: -2.8–0.8). Meanwhile, BF of female athletes involved in judo was $22.1 \pm 2.6\%$ of body weight, which was acceptable (20–24% of BW) and the BF was not different from the avg. recommended rate (22%): Δ BF$_{in\,\%}$ (actual BF$_{in\,\%}$—avg. recommended BF$_{in\,\%}$) was $0.1 \pm 1.1\%$ (95% PI: -2.6–2.8). Apart from that, the female athletes' MFMI (3.4 ± 0.6) was consistent with the avg. one (3–3.9) (Δ MFMI (actual MFMI—minimum recommended MFMI) was 0.4 ± 0.2; 95% PI: -0.2–2.9). MFMI in male athletes was high and there was no statistically significant difference from the max. range of 4.7 to 6. The paired sample *t*-test between the MFMI in male athletes (5.7 ± 2.8) and the max. MFMI limit were found without statistical significance (Δ MFMI (actual MFMI—max. recommended MFMI) was -0.3 ± 0.4; 95% PI: -1.2–0.6).

3.3. Prevalence, Magnitude and Methods of RWL

The prevalence of the reported RWL among elite combat sports athletes was 88%, which means that 38 out of 43 interviewed subjects reported practicing RWL (Figure 1).

Table 3 presents the profile of sport practice and weight loss history reported by the combat sport athletes. On average, the athletes began participating in combat sports before turning 10 years old and had been participating competitively after they reached the age of 11 years. In addition, the athletes began to resort to weight-cutting as early as 9 years old, with the mean age of 12.8 ± 2.1 years. The athletes were engaged in weight loss practices that took a short duration of time (4.9 ± 3.3 days) and they had 7.1 ± 5.6 episodes of weight reductions in the previous year. Their current mean weight (68.1 ± 16.7 kg) was almost similar to the previous off-season weight (67.8 ± 16.7 kg). However, the average ever lost most weight (3.1 ± 1.5 kg) was less than 4 kg, accounting for almost 5% of their body weight, and 104.1 ± 28.3 kg of this weight loss most frequently recovered a week later.

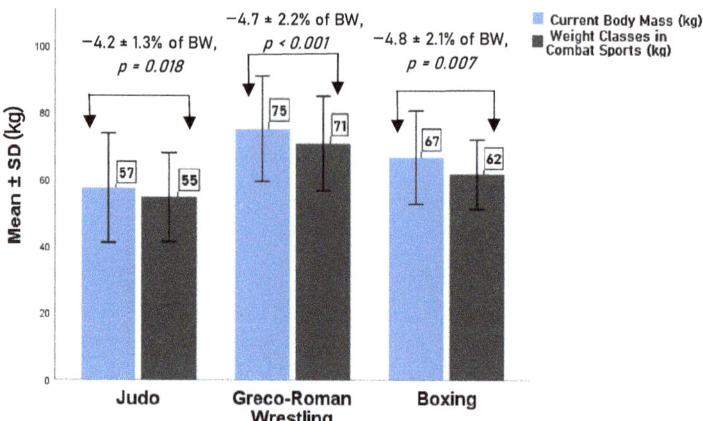

Figure 1. Prevalence of RWL (in kg and % of BW) among elite combat sports athletes (*t*-test was used).

Table 3. Profile of sport practice and weight loss history reported by the combat sports athletes.

Profile of Sport Practice and Weight Loss History	Judo	Greco-Roman Wrestling	Boxing	Total
Prevalence of the reported RWL (%)	73	95	92	88
Total RWLS	20.3 ± 6.2	20.9 ± 3.6	23.5 ± 4.8	21.6 ± 4.6
Age at the start of sport practice (years)	8.7 ± 3.1	9.8 ± 2.2	10.9 ± 2.9	9.9 ± 2.7
Age at the start of competition (years)	9.2 ± 2.4	10.9 ± 2.1	10.2 ± 4.9	10.4 ± 4.7
Fights over previous 12 months	10.4 ± 4.2	15.8 ± 9.5	13.3 ± 5.2	13.7 ± 7.4
Age at the start of weight cut (years)	12.5 ± 1.1	12.9 ± 2.5	13.2 ± 1.9	12.8 ± 2.1
Off-season weight (kg)	57.1 ± 16	75.3 ± 15.8	66.3 ± 13.7	67.9 ± 16.8
Frequency of weight reductions in previous year	5.6 ± 3	8.2 ± 7.3	6.5 ± 3.7	7.1 ± 5.6
Duration of weight reduction (days)	5.3 ± 4	4.7 ± 3.3	4.9 ± 2.8	4.9 ± 3.3
BW regain in a week after fight (kg)	2.6 ± 0.5	3.4 ± 1.6	2.8 ± 1.5	3.1 ± 1.4
BW regain in a week after fight (% of weight cut)	117.7 ± 23.3	106.7 ± 30.6	91.4 ± 24.2	104.1 ± 28.3

BW—body weight; RLWS—rapid weight loss score. The data are presented as means ± standard deviation (SD).

In this sample of combat athletes, the RWLS was 21.6 ± 4.6. Multivariate logistic regression was constructed to obtain how the predictors such as different self-reported RWL methods (skipping one or two meals, restricting fluids, increased exercise, training with rubber/plastic suits) may predict the magnitude of RWL (RWLS ≥ 20.7). ORs of covariates in multivariate logistic regression model were adjusted for the sports type, sex, and age. The results of multivariate analysis were displayed in Table 4. The regression of model identified that the athletes at high risk of RWL were significantly more likely to skip one or two meals (AOR 6.3; 95% CI: 1.3–31.8), to increase exercise (AOR 3.6; 95% CI: 1.0–12.5) and to train with rubber/plastic suits (AOR 3.2; 95% CI: 0.9–11.3). Even though the use of energy restriction strategies such as skipping one or two meals is a common practice, the methods to diminish body water stores (i.e., restrict fluid intake) were also frequently practised for RWL by this cohort (AOR 5.5; 95% CI: 1.0–31.8). In all, 92.1% of the athletes used fluid restriction "sometimes" or "always", and 68.4% of them practised this method between 1 and 24 h before weigh-ins. More specifically, 68.4% of the weight-reducing athletes drank less than 500 mL of drinks or water on the last day before the competition, while 23.7% of the athletes sustained from drinking fluids at all. According to our study, the athletes experienced dehydration due to the reduced fluid intake which corresponded to 1.4 ± 0.7% of BW. A more detailed analysis of the frequency of the weight loss methods deployed by the combat sports representatives is given in Table S1.

Table 4. The association between the RWL methods deployed by the combat sports representatives and the RWL score.

RWL Score [a] (Score > 20.7)	β	SE	W	*p*	AOR (95% CI)
Skipping one or two meals	1.8	0.8	5.0	*0.025*	6.3 (1.3, 31.8)
Restricting fluids	1.7	0.9	3.6	*0.05*	5.5 (1.0, 31.8)
Increased exercise	1.3	0.6	4.1	*0.043*	3.6 (1.0, 12.5)
Training with rubber/plastic suits	1.2	0.6	3.3	*0.049*	3.2 (0.9, 11.3)
Constant	−10.4	3.4	9.2	*0.002*	0

[a]—reference category is RWL score (score ≤ 20.7); β—the estimated coefficient; SE—the standard error (SE) of β; W—the Wald statistic; OR—Odds Ratio; CI—confidence interval; Nagelkerke R^2 = 0.66; ORs in the logistic model was adjusted for the sports type, sex and age (AORs).

According to our study, a correlation has been found between the RWLS and the weight cut/lost for the competition (r = 0.39, p = 0.016), and the weight regained in a week after the competition (r = 0.35, p = 0.033) in a sample of combat athletes.

3.4. The Association between RWL and Body Composition

Multivariate liner regressions were constructed to obtain how the predictor RWL magnitude may predict the body composition (MM (in kg), BF (in kg), MFMI) of combat sports athletes. Multivariate linear regression model was adjusted for athlete sport, sex and age. The results of multivariate analysis were displayed in Table 5. After using a multivariate linear regression method, we found that with a 95% confidence level the MM decreased from 0 to −0.03 kg depending on increased RWLS for 1 point (β −0.01, p < 0.001). Meanwhile, no association has been established between the current MM status (in kg), MFMI and the application of more aggressive RWL methods (in score) (Table 5).

Table 5. The association between MM, BF, MFMI and RWL score.

RWL (Score)	β	95% CI	*p*
MM (kg)	−0.01	(−0.03; 0)	*<0.001*
BF (kg)	0.06	(0.04; 0,1)	*0.087*
MFMI	0.08	(0.03; 0.1)	*0.121*

The association between MM, BF, MFMI and RWL score was estimated controlling for athlete sport, sex, and age (adjusted for sports type, sex and age). F (6, 36) = 7.2, p < 0.0001, R^2 = 0.56. MM—muscle mass; BF—body fat; MFMI–muscle and fat mass index.

4. Discussion

4.1. The Prevalence of RWL

This study has been an initial effort to explore the scope of prevalence of the RWL practices in the Lithuanian elite combat athletes which revealed the real picture of a rapid weight loss practice observed in 88% of the subjects (73% judokas, 95% Greco-Roman wrestlers, and 92% boxing Olympic athletes) over the previous year. This figure stood at a similar level in comparison to that observed for athletes practising similar combat sports such as wrestling (42–90%) [11,14,18], judo (63–90%) [12,16,30] and boxing (50–100%) [1], and it showed the prevalence of the ongoing practice of RWL among combat sports representatives, without regard to the sport branch.

Additionally, according to our study, the athletes began to adjust their bodies to weight-cutting being as young as 9 years of age, with the mean age of 12.8 ± 2.1. Similar results were highlighted in scientific literature with evidence on about 60% of judo athletes starting weight cycling before fights at the age of 12–15 and wrestlers at the age of 15.5 ± 2.4 [1,5,12]. It was also shown that the normal growth and development of adolescents were interrupted due to a disturbance in hormone levels induced by continuous by RWL practices [11,15,31–33].

4.2. The Magnitude of RWL

Combat sports athletes most commonly lose $\geq$ 5% of BW over the seven days prior to weigh-in [1]. According to our data, the average BW loss found before the competition in a sample of the Lithuanian combat athletes varied within the range of 3.1 $\pm$ 1.5 kg (4.6 $\pm$ 2.0% of BW). It should be taken into consideration that under some conditions, it is allowed to reach from 5% to 8% of BW loss with an acceptable slight effect on the health condition and performance of athletes [34,35]. As the National Collegiate Athletic Association (NCAA) weight loss guidelines stipulate, the Minimum Wrestling Weight (MWW) is grounded on 5% of BF as the lowest BF percentage allowed and the levels of safe weight loss attainable before the first weigh-in procedure at the competition— namely, losing no more than 1.5% of BW per week [36]. In line with the data from our study, male athletes involved in sports such as judo, Greco-Roman wrestling and boxing had a high BF percentage (16 $\pm$ 5.4%). Therefore, before competing, the BW of combat athletes can be adjusted as much as possible. On the other hand, the athletes' LBM was relatively high and led to a high ratio between muscle mass and fat mass (5.7 $\pm$ 2.8). Therefore, BW loss should not be a priority for elite athletes in the run-up to the competition.

4.3. The Methods of RWL

According to the scholarly literature, the most frequently practised methods to start RWL were fluid restriction [5,6,20,25,37–44], energy restriction [25,37,40,45], wearing rubber/plastic suits [5,37,39,41], increased exercise [37,40,42], heated room training [37,39], use of sauna [37,41], spitting [5], using laxatives [41] and gradual dieting [37]. However, according to our study, aggressive RWL methods among the Lithuanian combat athletes were less common. Additionally, emphasis can be laid on the fact that in a sample of Lithuanian athletes, those at high risk of RWL were significantly more likely to skip one or two meals, to increase exercising and to train with rubber/plastic suits.

On the other hand, the results obtained from our study were in line with the findings of other studies carried out by previous researchers and showed that among the most prevailing methods employed by the cohort of combat athletes in Lithuania for RWL was the method of reducing the body water stores (i.e., fluid restriction). The human body of the general population is composed of ~60% of water and most probably larger amount in athletes with higher LBM levels [46]. Taking into account the size of the body and the short time frame for the occurrence of the possible fluctuations, it is not surprising that dehydration appeared to be the primary RWL strategy practiced by a number of combat sports representatives. According to our study, fluid restriction was used by 92.1% of the combat athletes in Lithuania, while 68.4% of weight-reducing athletes consumed less than 500 mL of drinks or water on the last day before the competition, while 23.7% of athletes did not consume any fluids at all. Therefore, the athletes experienced dehydration as a result of reduced fluid intake which corresponded to 1.4 $\pm$ 0.7% of BW. On the one hand, mild dehydration (<2% BW) tends not to affect the performance, while on the other hand, a larger scope may pose problems, especially during the limited time to restore the rehydrated post weigh-in [34,35]. In other words, the athletes can take a choice from two methods available to decrease their bodily water, either to take in a lower amount of fluids and/or excrete a larger amount of fluids. The restriction in fluid intake that lasts for 24 h (<300 mL) may result in 1.5–2.0% BW loss [47]; however, the athletes most frequently achieve a loss larger than this as a result of the accelerated sweating rate via either active (as a result of intensive exercise) and/or passive techniques deployed (i.e., sauna rooms, heated environments, the use of sweat suits, etc.) [1,7]. This data was in line with the data obtained from our study showing that elite combat athletes from Lithuania were combining not only an extremely low fluid intake, but were also exercising more intensively and training with rubber/plastic suits. Therefore, future studies should focus and verify more accurate levels of dehydration observed in athletes. In addition, it should be highlighted that severe (or even moderate) dehydration employed as a tool to lose weight in weight controlled sports accelerates the risk of acute cardiovascular ailments [15,48]. The athletes who use thermal exposure seeking to achieve such dehydration run an increased risk suffering from heat stroke or heat illness [15]. Additionally, acute dehydration may significantly impact the electrolyte concentration composition which could impact the cell fluid balance, metabolism and result in the impairment of the neuromuscular function [49,50]. Scholarly works suggest that in-depth dehydration is also likely to induce alterations in the morphology of the brain and pose the risk of potential brain injuries as a consequence of head traumas triggered by strikes [51,52].

What is more, typically in a wide range of sports, prior to competitions, the nutritional status includes "fine tuning" procedures consistent to satisfying the needs of comfort, repetitive and habitual practices rather than clinging to excessive weight adjusting patterns. Even though there is insufficient evidence of research into the specific dietary patterns the athletes select to achieve RWL [53], the most frequently observed acute caloric restriction tends to impact the performance through the depleted glycogen depots [35]. Our study showed that energy restriction strategy (skip one or two meals) was frequently used by cohort of combat sports athletes. Therefore, this strategy of energy restriction had the association with the magnitude of RWL. The authors also point out that earlier research carried out by them in Lithuania found that the wrestlers during the RWL period followed low-energy and low-carbohydrate diet [54]. More specifically, low total energy intake contributed to the insufficiency of the intake of vitamins A, B_1, B_2, PP, D, E, B_6, folic acid, and the intakes of minerals such as potassium, calcium, phosphorus, magnesium, iron, zinc during the rapid bodyweight reduction. Greenhaff et al. [55] claimed that 'a diet low in carbohydrates even when the food consumption is sufficient, could impact the buffering properties of the blood'. The interacting effect of RWL and the muscle hydrogen ion efflux could be regulated down by a low carbohydrate intake and can result in increased fatigue during the intense muscular contractions [45].

4.4. The Body Composition and RWL

The combat athletes were engaged in RWL practices taking a short period of time and had 7 episodes of weight reductions in the previous year. The average most weight ever lost accounted for an almost 5% of their body weight, and 104% of this weight loss was usually regained in the week after a fight. Additionally, we found the correlation between RWLS and the weight cut for the competition, and the weight regained in a week after the competition in a sample of combat athletes. It can be explained that applying more aggressive RWL methods before the competition relates to how much weight combat athletes restore after the competition. This type of weight cycling may lead to MM damage and catabolism [24–26]. In this context the results of our study were interesting as the relationship was found between RWL and body composition. Utilising more aggressive and harmful RWL methods related to decreased MM during the preparatory training phase in a sample of Lithuanian combat sports athletes. Further research will be necessary to identify the effect of RWL magnitude on the athlete body composition in the relation to manipulations with the athlete's body mass.

4.5. Limitations of the Study

Our study had a few limitations. The study was carried out by a single centre and the size of the sample was small. A larger sample and an international multi-centre size could represent the findings of the quantitative data more accurately. In addition, the survey tools were unable to determine and provide a cut-off to indicate where the weight management behaviour was dangerous and could potentially cause harm. In the view of the above, scholars have now developed a new and safe (in terms of hyponatremia) method of 'water loading' that promotes dehydration [56–60]. In connection with this new RWL method, the limitation of our study was the absence of additional questions on the mentioned method in the questionnaire (RWLQ) we used [29]. Moreover, some findings were based on the data that was self-reported retrospectively; thus, some of the biased responses on behalf of the athletes were unavoidable.

5. Conclusions

This study proves high levels of RWL practices in the Lithuanian athletes of elite combat sports (88%) five days prior to the competition, resulting in the reduction of the average of 3.1 kg (4.6%) of their body weight. The athletes began to resort to weight-cutting as early as 9 years old, with a mean age of 12.8 $\pm$ 2.1 years. The combination of practiced weight loss techniques such as skipping one or two meals, restricting fluids, increased exercise, training with rubber/plastic suits may predict the magnitude of RWL. Utilising more aggressive and harmful RWL methods before the competition is associated to on how much weight combat athletes are cutting for the competition and regaining after the event. RWL being a widely spread practice among athletes of combat sports, can trigger negative consequences of the athletes' body composition. RWL magnitude potentially plays an important role in maintaining the loss of muscle mass during the preparatory training phase. Therefore, an adequate regulatory programme should be integrated into the training plans for high-performance combat sports athletes to keep not only the athletes but also their coaches responsible for a proper weight control.

Supplementary Materials: The following supporting information can be downloaded at: https://www.mdpi.com/article/10.3390/healthcare10040665/s1, Table S1: Frequency analysis of the weight loss methods reported by the combat sport athletes; Table S2: Acronyms and terminology.

Author Contributions: Conceptualization, M.B. and R.S.; methodology, M.B.; software, M.B.; validation, M.B., I.K. and R.S.; formal analysis, M.B.; investigation, M.B.; resources, M.B. and I.K.; data curation, I.K.; writing—original draft preparation, M.B.; writing—review and editing, I.K.; visualization, M.B.; supervision, R.S.; project administration, I.K. All authors have read and agreed to the published version of the manuscript.

Funding: This research received no external funding.

Institutional Review Board Statement: The study was conducted in accordance with the Declaration of Helsinki, and approved by the Vilnius Regional Committee of Biomedical Research Ethics (protocol code 158200-17-898-419, protocol approved on 11 April 20170). The clinical investigations were carried out under the guidance of the principles stipulated by the Declaration of Helsinki.

Informed Consent Statement: Informed consent was obtained from all subjects involved in the study.

Data Availability Statement: Data are available on request.

Acknowledgments: We are grateful to the Lithuanian Sports Centre and Lithuanian Sports Medicine Centre for their help in conducting the measurements of the body composition in combat sports athletes.

Conflicts of Interest: The authors declare no conflict of interest.

References

1. Franchini, E.; Brito, C.J.; Artioli, G.G. Weight loss in combat sports: Physiological, psychological and performance effects. *J. Int. Soc. Sports Nutr.* **2012**, *9*, 52. [CrossRef] [PubMed]
2. Jetton, A.M.; Lawrence, M.M.; Meucci, M.; Haines, T.L.; Collier, S.R.; Morris, D.M.; Utter, A.C. Dehydration and acute weight gain in mixed martial arts fighters before competition. *J. Strength Cond. Res.* **2013**, *27*, 1322–1326. [CrossRef] [PubMed]
3. Oppliger, R.; Case, H.; Horsewill, C.; Landry, L.; Shelter, A. American College of Sports Medicine position stand. Weight loss in wrestlers. *Med. Sci. Sports Exerc.* **1996**, *28*, 9–12. [CrossRef]
4. Malliaropoulos, N.; Rachid, S.; Korakakis, V.; Fraser, S.A.; Bikos, G.; Maffulli, N.; Angioi, M. Prevalence, techniques and knowledge of rapid weight loss amongst adult British judo athletes: A questionnaire based study. *Muscles Ligaments Tendons J.* **2017**, *7*, 459–466. [CrossRef]
5. Artioli, G.G.; Gualano, B.; Franchini, E.; Scagliusi, F.B.; Takesian, M.; Fuchs, M.; Antonio Herbert Lancha, J.R. Prevalence, magnitude, and methods of rapid weight loss among judo competitors. *Med. Sci. Sports Exerc.* **2010**, *42*, 436–442. [CrossRef]
6. Barley, O.R.; Chapman, D.W.; Abbiss, C.R. Weight loss strategies in combat sports and concerning habits in mixed martial arts. *Int. J. Sports Physiol. Perform.* **2018**, *13*, 933–939. [CrossRef]
7. Reale, R.; Slater, G.; Burke, L.M. Weight management practices of Australian Olympic combat sport athletes. *Int. J. Sports Physiol. Perform.* **2018**, *13*, 459–466. [CrossRef]

8. Kiningham, R.B.; Gorenflo, D.W. Weight loss methods of high school wrestlers. *Med. Sci. Sports Exerc.* **2001**, *33*, 810–813. [CrossRef]

9. Khodaee, M.; Olewinski, L.; Shadgan, B.; Kiningham, R.R. Rapid weight loss in sports with weight classes. *Cur. Sports Med. Rep.* **2015**, *14*, 435–441. [CrossRef]

10. Kazemi, M.; Rahman, A.; De Ciantis, M. Weight cycling in adolescent taekwondo athletes. *J. Can. Chiropr. Assoc.* **2011**, *55*, 318–324.

11. Steen, S.N.; Brownell, K.D. Patterns of weight loss and regain in wrestlers: Has the tradition changed? *Med. Sci. Sports Exerc.* **1990**, *22*, 762–768. [CrossRef]

12. Berkovich, B.E.; Stark, A.H.; Eliakim, A.; Nemet, D.; Sinai, T. Rapid weight loss in competitive judo and taekwondo athletes: Attitudes and practices of coaches and trainers. *Int. J. Sport Nutr. Exerc. Metab.* **2019**, *29*, 532–538. [CrossRef] [PubMed]

13. Figlioli, F.; Bianco, A.; Thomas, E.; Stajer, V.; Korovljev, D.; Trivic, T.; Maksimovic, N.; Drid, P. Rapid weight loss habits before a competition in sambo athletes. *Nutrients* **2021**, *13*, 1063. [CrossRef] [PubMed]

14. Viveiros, L.; Moreira, A.; Zourdos, M.C.; Aoki, M.S.; Capitani, C.D. Pattern of weight loss of young female and male wrestlers. *J. Strength Cond. Res.* **2015**, *29*, 3149–3155. [CrossRef] [PubMed]

15. Artioli, G.G.; Saunders, B.; Iglesias, R.T.; Franchini, E. It is time to ban rapid weight loss from combat sports. *Sports Med.* **2016**, *46*, 1579–1584. [CrossRef] [PubMed]

16. Brito, C.J.; Roas, A.F.C.M.; Brito, I.S.S.; Marins, J.C.B.; Cordova, C.; Franchini, E. Methods of body-mass reduction by combat sport athletes. *Int. J. Sport Nutr. Exerc. Metab.* **2012**, *22*, 89–97. [CrossRef] [PubMed]

17. Langan-Evans, C.; Close, G.; Morton, J. Making weight in combat sports. *Strength Cond. J.* **2011**, *33*, 25–39. [CrossRef]

18. Oppliger, R.A.; Steen, S.N.; Scott, J.R. Weight loss practice in college wrestling. *Int. J. Sport Nutr. Exerc. Metab.* **2003**, *13*, 29–46. [CrossRef] [PubMed]

19. Kim, J.C.; Park, K.J. Injuries and rapid weight loss in elite Korean wrestlers: An epidemiological study. *Physician Sportsmed* **2020**, *49*, 308–315. [CrossRef]

20. Degoutte, F.; Jouanel, P.; Begue, R.J.; Colombier, M.; Lac, G.; Pequignot, J.M.; Filaire, E. Food restriction, performance, biochemical, psychological, and endocrine changes in judo athletes. *Int. J. Sports Med.* **2006**, *27*, 9–18. [CrossRef] [PubMed]

21. Artioli, G.G.; Franchini, E.; Nicastro, H.; Sterkowicz, S.; Solis, M.Y.; Lancha, A.H., Jr. The need of a weight management control program in judo: A proposal based on the successful case of wrestling. *J. Int. Soc. Sports Nutr.* **2010**, *7*, 15. [CrossRef] [PubMed]

22. Sundgot-Borgen, J.; Garthe, I. Elite athletes in aesthetic and Olympic weight-class sports and the challenge of body weight and body compositions. *J. Sports Sci.* **2011**, *29*, S101–S114. [CrossRef] [PubMed]

23. Turocy, P.S.; DePalma, B.F.; Horswill, C.A.; Laquale, K.M.; Martin, T.J.; Perry, A.C.; Somova, M.J.; Utter, A.C. National Athletic Trainers' Association position statement: Safe weight loss and maintenance practices in sport and exercise. *J. Athl. Train.* **2011**, *46*, 322–336. [CrossRef]

24. Bongiovanni, T.; Mascherini, G.; Genovesi, F.; Pasta, G.; Iaia, F.M.; Trecroci, A.; Ventimiglia, M.; Alberti, G.; Campa, F. Bioimpedance vector references need to be period-specific for assessing body composition and cellular health in elite soccer players: A brief report. *J. Funct. Morphol. Kinesiol.* **2020**, *5*, 73. [CrossRef]

25. Drid, P.; Krstulovic, S.; Erceg, M.; Trivic, T.; Stojanovic, M.; Ostojic, S. The effect of rapid weight loss on body composition and circulating markers of creatine metabolism in judokas. *Kinesiology* **2019**, *51*, 10–13. [CrossRef]

26. Roklicer, R.; Lakicevic, N.; Stajer, V.; Trivic, T.; Bianco, A.; Mani, D.; Milosevic, Z.; Maksimovic, N.; Paoli, A.; Drid, P. The effects of rapid weight loss on skeletal muscle in judo athletes. *J. Transl. Med.* **2020**, *18*, 142. [CrossRef] [PubMed]

27. Skernevičius, J.; Milašius, K.; Raslanas, A.; Dadelienė, R. Sporto treniruotė (*Fitness Training*). In *Sportininkų Gebėjimai ir jų Ugdymas (Athlete Skills and Training Them)*, 1st ed.; Čepulėnas, A., Saplinskas, J., Paulauskas, R., Eds.; Lithuanian University of Educational Sciences Press: Vilnius, Lithuania, 2011; pp. 165–217.

28. Lukaski, H.C.; Bolonchuk, W.W. (Eds.) *Theory and Validation of the Tetrapolar Bioelectrical Impedance Method to Assess Human Body Composition*; Institute of Physical Science and Medicine: London, UK, 1987.

29. Artioli, G.G.; Scagliusi, F.; Kashiwagura, D.; Franchini, E.; Gualano, B.; Junior, A.L. Development, validity and reliability of a questionnaire designed to evaluate rapid weight loss patterns in judo players. *Scand. J. Med. Sci. Sports* **2010**, *20*, e177–e187. [CrossRef] [PubMed]

30. Artioli, G.G.; Gualano, B.; Franchini, E.; Batista, R.N.; Polacow, V.O.; Lancha, A.H. Physiological, performance, and nutritional profile of the Brazilian Olympic wushu (kung-fu) team. *J. Strength Cond. Res.* **2009**, *23*, 20–25. [CrossRef] [PubMed]

31. Loucks, A.B.; Kiens, B.; Wright, H.H. Energy availability in athletes. *J. Sports Sci.* **2011**, *29*, S7–S15. [CrossRef]

32. Saarni, S.E.; Rissanen, A.; Sarna, S.; Koskenvuo, M.; Kaprio, J. Weight cycling of athletes and subsequent weight gain in middle age. *Int. J. Obes.* **2006**, *30*, 1639–1644. [CrossRef]

33. Roemmich, J.N.; Sinning, W.E. Weight loss and wrestling training: Effects on growth-related hormones. *J. Appl. Physiol.* **1997**, *82*, 1760–1764. [CrossRef]

34. Reale, R.; Slater, G.; Burke, L.M. Acute weight loss strategies for combat sports and applications to Olympic success. *Int. J. Sports. Physiol. Perf.* **2016**, *12*, 142–151. [CrossRef]

35. Reale, R.; Slater, G.; Burke, L.M. Individualised dietary strategies for Olympic combat sports: Acute weight loss, recovery and competition nutrition. *Eur. J. Sports. Sci.* **2017**, *17*, 727–740. [CrossRef] [PubMed]

36. Gibbs, A.E.; Pickerman, J.; Sekiya, J.K. Weight management in amateur wrestling. *Sports Health* **2009**, *1*, 227–230. [CrossRef] [PubMed]

37. Berkovich, B.E.; Eliakim, A.; Nemet, D.; Stark, A.H.; Sinai, T. Rapid weight loss among adolescents participating in competitive judo. *Int. J. Sport Nutr. Exerc. Metab.* **2016**, *26*, 276–284. [CrossRef] [PubMed]

38. Finaud, J.; Degoutte, F.; Scislowski, V.; Rouveix, M.; Durand, D.; Filaire, E. Competition and food restriction effects on oxidative stress in judo. *Int. J. Sports Med.* **2006**, *27*, 834–841. [CrossRef]

39. Artioli, G.G.; Iglesias, R.T.; Franchini, E.; Gualano, B.; Kashiwagura, D.B.; Solis, M.Y.; Benatti, F.; Fuchs, M.; Junior, A.H.L. Rapid weight loss followed by recovery time does not affect judo-related performance. *J. Sports Sci.* **2010**, *28*, 21–32. [CrossRef] [PubMed]

40. Coufalova, K.; Cochrane, D.J.; Maly, T.; Heller, J. Changes in body composition, anthropometric indicators and maximal strength due to weight reduction in judo. *Arch. Budo* **2014**, *10*, 161–168.

41. Fortes, L.S.; Lira, H.A.A.S.; Ferreira, M.E.C. Effect of rapid weight loss on decision-making performance in judo athletes. *J. Phys. Educ.* **2017**, *28*, 2817.

42. Kons, R.; Athayde, M.; Follmer, B.; Detanico, D. Methods and magnitudes of rapid weight loss in judo athletes over pre-competition periods. *Hum. Mov.* **2017**, *18*, 49–55. [CrossRef]

43. Morales, J.; Ubasart, C.; Solana-Tramunt, M.; Villarrasa-Sapina, I.; Gonzalez, L.M.; Fukuda, D.; Franchini, E. Effects of rapid weight loss on balance and reaction time in elite judo athletes. *Int. J. Sport Physiol.* **2018**, *13*, 1371–1377. [CrossRef]

44. Hillier, M.; Sutton, L.; James, L.; Mojtahedi, D.; Keay, N.; Hind, K. High prevalence and magnitude of rapid weight loss in mixed martial arts athletes. *Int. J. Sport Nutr. Exerc. Metab.* **2019**, *29*, 512–517. [CrossRef] [PubMed]

45. Filaire, E.; Maso, F.; Degoutte, F.; Jouanel, P.; Lac, G. Food restriction, performance, psychological state and lipid values in judo athletes. *Int. J. Sports Med.* **2001**, *22*, 454–459. [CrossRef]

46. Sawka, M.N.; Cheuvront, S.N.; Carter, R. Human water needs. *Nutr. Rev.* **2005**, *63*, S30–S39. [CrossRef]

47. James, L.J.; Shirreffs, S.M. Fluid and electrolyte balance during 24-h fluid and/or energy restriction. *Int. J. Sport Nutr. Exerc. Metab.* **2013**, *23*, 545–553. [CrossRef] [PubMed]

48. Lowe, G.; Lee, A.; Rumley, A.; Price, J.; Fowkes, F. Blood viscosity and risk of cardiovascular events: The Edinburgh artery study. *Br. J. Haematol.* **1997**, *96*, 168–173. [CrossRef]

49. Cheuvront, S.N.; Kenefick, R.W. Dehydration: Physiology, assessment, and performance effects. *Compr. Physiol.* **2014**, *4*, 257–285. [PubMed]

50. Sjøgaard, G. Water and electrolyte fluxes during exercise and their relation to muscle fatigue. *Acta Physiol. Scand. Suppl.* **1985**, *556*, 129–136.

51. Crighton, B.; Close, G.L.; Morton, J.P. Alarming weight cutting behaviours in mixed martial arts: A cause for concern and a call for action. *Br. J. Sports Med.* **2015**, *50*, 446–447. [CrossRef]

52. Kempton, M.J.; Ettinger, U.; Schmechtig, A.; Winter, E.M.; Smith, L.; McMorris, T.; Wilkinson, I.D.; Williams, S.C.; Smith, M.S. Effects of acute dehydration on brain morphology in healthy humans. *Hum. Brain Mapp.* **2009**, *30*, 291–298. [CrossRef]

53. Barley, O.R.; Chapman, D.W.; Abbiss, C.R. The current state of weight-cutting in combat sports. *Sports* **2019**, *7*, 123. [CrossRef] [PubMed]

54. Baranauskas, M.; Tubelis, L.; Stukas, R.; Švedas, E.; Samsonienė, L.; Karanauskienė, D. The influence of short-term hipocaloric nutrition on body weight reduction in Lithuanian Olympic team wrestlers. *Balt. J. Sport Health Sci.* **2011**, *4*, 5–12.

55. Greenhaff, P.L.; Gleeson, M.; Maughan, R.J. The effects of dietary manipulation on blood acid-base status and the performance of high intensity exercise. *Eur. J. Appl. Physiol. Occup. Physiol.* **1987**, *56*, 331–337. [CrossRef] [PubMed]

56. Reale, R.; Slater, G.; Cox, G.R.; Dunican, I.C.; Burke, L.M. The effect of water loading on acute weight loss following fluid restriction in combat sports athletes. *Int. J. Sport. Nutr. Exerc. Metab.* **2018**, *28*, 565–573. [CrossRef] [PubMed]

57. Verbalis, J.G. Disorders of body water homeostasis. *Best Pract. Res. Clin. Endocrinol. Metab.* **2003**, *17*, 471–503. [CrossRef]

58. Robertson, G.L.; Shelton, R.L.; Athar, S. The osmoregulation of vasopressin. *Kidney Int.* **1976**, *10*, 25–37. [CrossRef]

59. Lankford, S.P.; Chou, Y.; Terada, S.M.; Wall, J.B.; Knepper, W.; Knepper, M.A. Regulation of collecting duct water permeability independent of cAMP-mediated AVP response. *Am. J. Physiol. Renal Physiol.* **1991**, *261*, F554–F566. [CrossRef]

60. Kishore, B.K.; Knepper, W.; Knepper, M.A. Quantitation of aquaporin-2 abundance in micro dissected collecting ducts: Axial distribution and control by AVP. *Am. J. Physiol. Renal Physiol.* **1996**, *271*, F62–F70. [CrossRef]

Article

The Performance, Physiology and Morphology of Female and Male Olympic-Distance Triathletes

Paulo J. Puccinelli [1], Claudio A. B. de Lira [2], Rodrigo L. Vancini [3], Pantelis T. Nikolaidis [4], Beat Knechtle [5,6,*], Thomas Rosemann [6] and Marilia S. Andrade [1]

[1] Programa de Pós-Graduação em Medicina Translacional, Department of Physiology, Federal University of São Paulo, São Paulo 04021-001, Brazil; paulopuccinelli@hotmail.com (P.J.P.); marilia1707@gmail.com (M.S.A.)
[2] Human and Exercise Physiology Division, Faculty of Physical Education and Dance, Federal University of Goiás, Goiânia 74690-900, Brazil; andre.claudio@gmail.com
[3] Center of Physical Education and Sports, Federal University of Espírito Santo, Vitória 29075-910, Brazil; rodrigoluizvancini@gmail.com
[4] School of Health and Caring Sciences, University of West Attica, 12243 Athens, Greece; pademil@hotmail.com
[5] Medbase St. Gallen Am Vadianplatz, St. Gallen and Institute of Primary Care, 9100 St. Gallen, Switzerland
[6] Institute of Primary Care, University of Zurich, 8091 Zurich, Switzerland; thomas.rosemann@usz.ch
* Correspondence: beat.knechtle@hispeed.ch; Tel.: +41-71-226-93-00

Abstract: Sex differences in triathlon performance have been decreasing in recent decades and little information is available to explain it. Thirty-nine male and eighteen female amateur triathletes were evaluated for fat mass, lean mass, maximal oxygen uptake (VO_2 max), ventilatory threshold (VT), respiratory compensation point (RCP), and performance in a national Olympic triathlon race. Female athletes presented higher fat mass ($p = 0.02$, $d = 0.84$, power = 0.78) and lower lean mass ($p < 0.01$, $d = 3.11$, power = 0.99). VO_2 max ($p < 0.01$, $d = 1.46$, power = 0.99), maximal aerobic velocity (MAV) ($p < 0.01$, $d = 2.05$, power = 0.99), velocities in VT ($p < 0.01$, $d = 1.26$, power = 0.97), and RCP ($p < 0.01$, $d = 1.53$, power = 0.99) were significantly worse in the female group. VT (%VO_2 max) ($p = 0.012$, $d = 0.73$, power = 0.58) and RCP (%VO_2 max) ($p = 0.005$, $d = 0.85$, power = 0.89) were higher in the female group. Female athletes presented lower VO_2 max value, lower lean mass, and higher fat mass. However, females presented higher values of aerobic endurance (%VO_2 max), which can attenuate sex differences in triathlon performance. Coaches and athletes should consider that female athletes can maintain a higher percentage of MAV values than males during the running split to prescribe individual training.

Keywords: triathlon; sports medicine; sports physiology; female athlete; VO_2 max

Citation: Puccinelli, P.J.; de Lira, C.A.B.; Vancini, R.L.; Nikolaidis, P.T.; Knechtle, B.; Rosemann, T.; Andrade, M.S. The Performance, Physiology and Morphology of Female and Male Olympic-Distance Triathletes. *Healthcare* 2022, 10, 797. https://doi.org/10.3390/healthcare10050797

Academic Editors: Parisi Attilio, João Paulo Brito and Rafael Oliveira

Received: 9 March 2022
Accepted: 21 April 2022
Published: 25 April 2022

Publisher's Note: MDPI stays neutral with regard to jurisdictional claims in published maps and institutional affiliations.

1. Introduction

The participation of women in amateur and elite endurance sports events, including triathlon, has increased and their performance has improved during the last three decades [1–6]. Factors that are possibly associated with the increasing participation of women are the acceptance of female athletes in society, the importance of regular physical activity for the prevention and treatment of noncommunicable diseases, and the feeling of well-being that comes from a more active lifestyle [7].

Sex differences in triathlon performance seem to be decreasing, and currently vary between 12 and 18% [8,9], which seems to be influenced by distance, the level of competition, and the participation of the athletes [10,11].

Longer triathlon events, such as the Ironman (3.8 km of swimming, 180 km of cycling, and 42.2 km of running), or ultra-triathlons such as the Double Iron ultra-triathlon (7.6 km of swimming, 360 km of cycling, and 84.4 km of running) seem to be associated with decreased sex differences in performance, compared to shorter triathlon distances [12].

In addition, a lower tendency to sex difference was observed for elite athletes when compared to amateur athletes [12].

Some morphological sex differences related to body composition, such as lower fat mass percentage and higher muscle mass in the male sex [13–15], seem to be associated with better male performance [16].

Regarding the physiological factors that influence endurance performance, maximal oxygen uptake (VO_2 max), ventilatory threshold (VT), and running economy (RE) are variables commonly investigated to predict aerobic performance [17].

In terms of VO_2 max values, average values for females are approximately 75% of the values for males [18]. However, among athletes, these differences may be lower [19]. Lower blood hemoglobin concentrations typically found in women, as well as lower red cell mass and hematocrit level, which result in lower arterial oxygen content (CaO_2) and lower O_2 delivery to muscles during exercise [20,21] are the main factors responsible for these VO_2 max gender differences.

Differently from VO_2 max, data about sex differences according to the %VO_2 max at VT seems to be contradictory. Female athletes have 7 to 23% more type I muscle fibers than men [22–24]. This difference in muscle fiber composition means that women have a greater oxidation of fat [25] and faster oxygen consumption kinetics [26], which should directly impact the VT, since this is dependent on the oxidative capacity of the muscles during exercise [27]. In addition, female athletes also have a higher rate of mitochondrial respiration [28]. These differences impact muscle metabolism, making women more apt to resynthesize ATP through the oxidative metabolism. Considering these sex differences, higher VT could be expected for female athletes; however, literature data show conflicting results. [17–20]. Moreover, as for VO_2 max, there are very few data for VT in Olympic-distance triathletes [19].

Women's participation in amateur triathlon events has increased in recent years. As a result, women's performance has also improved, and sex differences have decreased [19]. So far, it is not possible to define whether the difference is associated with training volume or physiological limitations. Therefore, understanding physiological differences between the sexes can help clarify this issue.

Considering the importance of understanding the differences between sexes in endurance sports performance and the lack of data regarding both Olympic-distance triathletes and amateur athletes, we compared the physiological and morphological characteristics of male and female amateur triathletes of the same mean age who competed in an Olympic-distance event. Better knowledge about gender differences and female characteristics can explain the narrowing performance gap between the sexes of amateur triathlon athletes, and may help women reach their best performance.

We hypothesized that male triathletes would present higher VO_2 max, higher lean mass, and lower fat mass than female triathletes, but that there would be no sex differences according to VT. Because of their higher VO_2 max levels and better body composition, we hypothesized that men would present lower overall race time and split times than female triathletes in the Olympic triathlon race.

2. Materials and Methods

2.1. Ethical Approval

All experimental procedures were approved by the Human Research Ethics Committee of the Federal University of Sao Paulo (approval number 1659697) and conformed to the principles outlined in the Declaration of Helsinki. The study was conducted in accordance with recognized ethical standards and national/international laws. After receiving instructions regarding the experimental procedures, their possible risks and benefits, the objectives and justification of the research, and the principles of respect for persons involved, which encompassed a guarantee of privacy, confidentiality, and anonymity rights, the athletes signed the consent form.

2.2. Participants

Ninety-three athletes who had applied for Olympic-distance triathlon races accepted an invitation to participate in the study. However, thirty-six did not meet the inclusion criteria. Therefore, fifty-seven athletes participated in the study.

The inclusion criteria to participate in the study included having participated in at least one Olympic-distance triathlon race, with at least one year of triathlon practice. The exclusion criteria included having no medical approval for maximum effort, being pregnant, having acute pain in the lower limbs, edema, or not finishing the race.

The main reasons for the thirty-six exclusions were giving up on participating in the Olympic-distance triathlon race (n = 21), not finishing the race (n = 6), having injuries during the training period (n = 4), absence on the day scheduled for laboratory evaluations (n = 4), and one woman got pregnant.

Characterization of the sample according to the age and training habits are presented in Table 1.

Table 1. Characteristics of participants.

	Male Triathletes (n = 39)	Female Triathletes (n = 18)	p Value
Age (years)	38.8 ± 6.9	41.3 ± 6.68.4	0.210
Triathlon experience (years)	2.7 ± 1.7	3.3 ± 1.6	0.232
Training per week (hours)	13.2 ± 4.1	14.4 ± 3.5	0.287

Data are presented as mean ± standard deviations.

As the number of female athletes who participated in the study was smaller than the number of male athletes, the power of the statistical analysis is shown with the p-value. This was employed to identify the possible lack of statistical difference between the groups due to the small sample size.

2.3. Procedures

Each participant reported to the laboratory for one day, in which they answered a questionnaire about training habits. Afterwards, anthropometric data measurement and a cardiorespiratory maximal test on a treadmill were performed. The organizing committee of the races provided the overall triathlon race time and split times. Thirty-nine male and six female amateur triathletes participated in the same race.

2.4. Assessments

2.4.1. Questionnaires

The athletes answered a questionnaire about training habits with the four following questions: (1) How many years have you been practicing triathlon? (2) How many hours a week do you train swimming? (3) How many hours a week do you train cycling? (4) How many hours a week do you train running?

2.4.2. Body Composition and Anthropometry

The height and body mass of the participants were assessed using a calibrated stadiometer and were measured to the nearest 0.1 kg and 0.1 cm, respectively. Dual energy X-ray absorptiometry (DXA, software version 12.3, Lunar DPX, GE Healthcare, Madison, WI, USA) was used to assess body composition (lean and fat mass). Athletes were instructed to follow their normal ad libitum hydration habits. They were evaluated after bladder voiding; no fasting or other limitations on their usual activities were implemented [29]. This method has been previously demonstrated as a reliable technique for body composition assessments [30,31].

2.4.3. Cardiorespiratory Maximal Test on a Treadmill

Cardiopulmonary exercise testing (CPET) was conducted on a motorized treadmill (Inbrasport, ATL, Porto Alegre, Brazil) using a computer-based metabolic analyzer (Quark, Cosmed, Italy). The calibration procedure was performed prior to each test, according to the manufacturer's guidelines. CPET was used to measure VO_2 max, VT, respiratory compensation point (RCP), and maximal aerobic velocity (MAV). The VO_2 max was determined as the stabilization of VO_2 (increase lower than 2.1 $mL \cdot kg^{-1} \cdot min^{-1}$) even after increasing the treadmill velocity during the last stage of the CPET [32]. All the volunteers reached VO_2 max. VT was determined based on the following criteria: an increase in the ventilatory equivalent for oxygen without an increase in the ventilatory equivalent for carbon dioxide, and an increase in the partial pressure of exhaled oxygen. RCP was determined based on the increase in the ventilatory equivalent for carbon dioxide and the decrease in the partial pressure of exhaled carbon dioxide [33]. VT and RCP were determined separately by two experienced investigators; a third investigator was asked in cases of discordance. MAV was determined as the minimal velocity eliciting the VO_2 max [34]. The percentage of MAV that the athlete maintained during the running split was also calculated.

Athletes warmed up for 4 min at 10 $km \cdot h^{-1}$ (males) and 9 $km \cdot h^{-1}$ (females). After the warm-up period, the running velocity was increased by 1 $km \cdot h^{-1}$ every minute until voluntary exhaustion [35]. The entire CPET lasted between 8 and 12 min and treadmill grade was set at 1% to simulate the energetic cost of outdoor running [36]. The heart rate was measured by a monitor (Ambit 2S, Suunto, Finland) throughout the entire test, and perceived exertion was rated according to the Borg scale (a 10-point scale) [37].

2.5. Statistical Analysis

Data are presented as the mean and the standard deviations. All variables presented normal distribution and homogeneous variability according to the Shapiro–Wilk and Levene tests, respectively. In order to compare the triathlon race times and morphological and physiological characteristics of the sexes, Welch's unequal variances t-test was used. This test was chosen because it is more reliable when the two samples have unequal sample sizes [38]. The measures of the effect size for differences between sexes were determined by calculating the mean difference between the two sexes, and then dividing the result by the pooled standard deviation. Calculating effect sizes, the magnitude of any change was judged according to the following criteria: $d = 0.2$ was considered a "small" effect size; 0.5 represented a "medium" effect size; and 0.8 a represented "large" effect size [39]. Considering that the study had a convenience sample, the power of all between-sex comparisons were calculated. Power analysis was performed using G*Power software [40]. The power of the tests varied from 0 to 1. Usually, researchers use 0.80 as the power level of the test [41]. Therefore, the same values were considered in this study to interpret the results. The level of significance was set at $p < 0.05$.

3. Results

Female athletes presented significantly lower body mass ($p < 0.01$, d = 2.00, power = 0.99) and height ($p < 0.01$, d = 1.80, power = 0.99) than male athletes. There was no difference in mean age between the groups ($p = 0.21$, d = 0.35, power = 0.65). Overall race time and split times were compared for sexes who participated in the same triathlon event. Regarding performance, female athletes presented higher race times for swimming (+11%), cycling (+7.5%), running (+7%), and overall race time (+8%). According to morphologic characteristics, male athletes presented higher lean body mass (kg) ($p < 0.01$, d = 3.11, power = 0.99). According to fat mass distribution, the percentage of trunk fat mass was not different between sexes ($p = 0.522$, $d = 0.17$, power = 0.73), nor was the percentage of android fat mass ($p = 0.921$, $d = 0.02$, power = 0.74), but the percentage of gynoid fat mass was higher in female athletes ($p < 0.01$, $d = 1.37$, power = 0.98). VO_2 max ($p < 0.01$, $d = 1.46$, power = 0.99), MAV ($p < 0.01$, $d = 2.05$, power = 0.99), and velocities associated with VT ($p = 0.02$, $d = 1.26$, power = 0.97) and RCP ($p < 0.01$, $d = 1.53$, power = 0.99) were significantly higher in the male group. %VO$_2$ max at VT (p = 0.012, d = 0.73, power = 0.58) and %VO$_2$ max at RCP (p = 0.005, d = 0.85, power = 0.89) were higher in the female group. During the running split, female athletes were running at a higher percentage of MAV (75 ± 8%) than males (62 ± 6%) (p < 0.01, d = 1.83, power = 0.99) (Table 2).

Table 2. Descriptive characteristics of the triathletes and comparison between the sexes.

	Male (n = 39)	Female (n = 18)	p Value	d Value	Power (1-Beta)
Anthropometric profile					
Age (years)	38.9 ± 6.9	41.3 ± 6.6	0.21	0.35	0.65
Body mass (kg)	74.3 ± 8.8 *	59.5 ± 5.6	<0.01	2.00	0.99
Height (cm)	174.8 ± 6.5 *	164.5 ± 4.8	<0.01	1.80	0.99
Fat mass (%)	16.8 ± 5.6 *	23.2 ± 9.2	0.02	0.84	0.78
Lean Mass (kg)	59.0 ± 5.7 *	43.0 ± 4.5	<0.01	3.11	0.99
Trunk fat mass (%)	19.8 ± 6.8	21.3 ± 10.2	0.52	0.17	0.73
Android fat mass (%)	22.7 ± 8.6	22.4 ± 12.0	0.92	0.02	0.74
Gynoid fat mass (%)	21.9 ± 6.2 *	33.2 ± 9.8	<0.01	1.37	0.98
Maximal graded exercise test					
VO$_2$ max (ml·kg^{-1}·min^{-1})	59.9 ± 6.3 *	49.5 ± 7.8	<0.01	1.46	0.99
VT (%VO$_2$ max)	74.4 ± 5.6 *	78.7 ± 6.1	0.01	0.73	0.58
Velocity at VT (km·h^{-1})	12.4 ± 1.4 *	10.5 ± 1.6	<0.01	1.26	0.97
RCP (%VO$_2$ max)	87.5 ± 4.6 *	91.2 ± 4.1	0.01	0.85	0.89
Velocity at RCP (km·h^{-1})	14.8 ± 1.5 *	12.5 ± 1.5	<0.01	1.53	0.99
MAV (km·h^{-1})	17.8 ± 1.4 *	14.6 ± 1.7	<0.01	2.05	0.99
Running split					
%MAV	62 ± 6 *	75 ± 8	<0.01	1.83	0.99
Velocity (km·h^{-1})	11.0 ± 1.0 *	11.0 ± 1.8	0.99	0.00	0.99

Data are presented as mean ± standard deviations. d value: Effect size (Cohen's D). VO$_2$ max: maximal oxygen uptake. VT: ventilatory threshold. RCP: respiratory compensation point (RCP). MAV: maximal aerobic velocity. * significant difference between sexes (p < 0.05).

4. Discussion

The primary aim of this study was to compare the sex differences of amateur Olympic-distance triathletes in relation to performance and physiological and morphological characteristics. The main findings were that: (i) the sex differences in performance were 8.0% for overall race time, 11% for swimming, 7.5% for cycling, and 7% for running; (ii) female athletes presented a lower VO$_2$ max and a higher %VO$_2$ max at VT and RCP than male athletes; (iii) female athletes presented lower lean mass than males; and (iv) female athletes presented higher total fat mass and gynoid fat mass than males, but the same android and trunk fat masses.

The sex differences in 1.5 km of swimming, 40 km of cycling, 10 km of running, and overall race time were 11.0, 7.5, 7.0, and 8.0%, respectively. Higher sex differences were previously shown for the top 10 athletes of each age group of the World Championship from 2009 to 2011, with a 13.3% performance difference in swimming, 10.7% difference in cycling, 7.5% difference in running, and 12% difference in overall race time [42]. Higher sex difference between the top five athletes from the "Zurich triathlon", which occurs in Zurich, Switzerland, in each category have also been shown (18.5% in swimming, 15.5% in cycling, 18.5% in running, and 17.1% in overall race time) [6]. Therefore, it is evident that the sex differences in a given performance depend on the race level (world, national or regional championship). In the present study, minor differences were found between the sexes; however, only amateur athletes were studied, which differs from the studies mentioned above that evaluated elite athletes.

As expected, female athletes presented lower VO$_2$ max and MAV values (49.5 ± 7.8 mL·kg^{-1}·min^{-1} and 14.6 ± 1.7 km·h^{-1}, respectively) than male athletes (59.9 ± 6.3 mL·kg^{-1}·min^{-1} and 17.8 ± 1.0 km·h^{-1}, respectively), showing a sex difference of approximately 19%. Similar sex difference have previously been shown for elite younger triathletes, reporting 20% lower values for females than for males (56.1 and 67.9 mL·kg^{-1}·min^{-1}) [43]. However, VO_2 max values for ultra-endurance triathletes seem to be more similar between the sexes (68.8 and 65.9 mL·kg^{-1}·min^{-1} for males and females, respectively), evidencing a sex difference of only 4.4% [44].

Besides maximal capacity for oxygen uptake, endurance performance also depends on VT. It has been suggested that 70% of success in endurance running depends on these physiological parameters [17]. An important new finding from the present study is that the female athletes presented higher values for VT (78.7 $\pm$ 6.1% for females and 74.4 $\pm$ 5.6% of VO_2 max for males) and RCP (91.2 $\pm$ 4.1% for females and 87.5 $\pm$ 4.6% VO_2 max for males) than male athletes. In addition, female athletes maintained a velocity corresponding to 75% of the MAV during the running split, which is higher than the value for males, who maintained 62% of their MAV [34]. The VT is limited by the peripheral conditions (i.e., mitochondrial volume, capillary density, oxidative enzyme capacity) [45,46]. Considering this context, females present different metabolic (greater proportional area of type I fibers [22–24], greater whole-muscle oxidative capacity [26], and greater mitochondrial oxidative function [28]), contractile (Ca^{2+} transients were smaller in magnitude and longer in duration in females [47]), and hemodynamic (greater vasodilatory responses of the arteries to muscles and higher density of capillaries per unit of skeletal muscle [22]) properties of skeletal muscles than males, favoring ATP resynthesis from oxidative phosphorylation during exercise [48,49], which could contribute to a higher VT.

Triathlon performance is also associated with body composition [16,50]. In this study, female athletes presented lower lean mass than males and higher total fat mass and gynoid fat mass percentage. The android and trunk body mass did not differ between the sexes. Moreover, fat mass values for both sexes were higher than those reported for elite athletes (<13% for female and <5% for males) [43]. Therefore, female body composition seems to be disadvantageous for athletic performance [13,14,51].

Regarding the limitations of this study, the test measurements cited were performed only on a treadmill. Thus, as the physiologic characteristics were only measured during a running activity using a treadmill, it would be very interesting to identify sex differences with the same measurements in tests performed during cycling or swimming activities. The inclusion of amateur athletes rather than elite athletes was another study limitation. Furthermore, this was a cross-sectional study, which prevented us from the studying the performance difference between sexes over time. Considering the increased popularity of Olympic-distance triathlon, especially among women, who were underrepresented in this sport until recently [52], the findings of the present study have practical applications for training monitoring. Strength and conditioning coaches working with triathletes might develop separate exercise programs for each sex.

Thus, an awareness of physiological sex differences related to performance would help coaches to prescribe sex-tailored training. In this context, the main finding from the present study was that the female athletes presented higher values of aerobic endurance (%VO_2 max) than male athletes. These findings suggest that female athletes can maintain a higher percentage of MAV values than males during the running split; therefore, coaches could consider these findings to prescribe individual training.

5. Conclusions

In summary, female athletes present lower VO_2 max and lean mass, and higher fat mass. However, they present higher values of aerobic endurance (%VO_2 max), which can attenuate sex differences in triathlon performance. However, the sex differences in VT require further investigation, as there are few data about this variable in the literature.

Author Contributions: Conceptualization, P.J.P. and M.S.A.; methodology, P.J.P. and C.A.B.d.L.; software, R.L.V.; validation, C.A.B.d.L. and M.S.A.; formal analysis, P.J.P. and R.L.V.; investigation, M.S.A.; data curation, P.T.N.; writing—original draft preparation, C.A.B.d.L.; writing—review and editing, P.T.N., B.K. and T.R.; visualization, B.K. and T.R.; supervision, M.S.A. All authors have read and agreed to the published version of the manuscript.

Funding: This research received no external funding.

Institutional Review Board Statement: The study was conducted according to the guidelines of the Declaration of Helsinki, and approved by the Human Research Ethics Committee of the University (approval number 1659697).

Informed Consent Statement: Informed consent was obtained from all subjects involved in the study.

Data Availability Statement: Data supporting reported results can be asked to corresponding author.

Acknowledgments: We would like to thank all of the participants who volunteered their time to participate in the study, the Olympic Training and Research Center (Centro Olímpico de Treinamento e Pesquisa, COTP, São Paulo, Brazil).

Conflicts of Interest: The authors declare no conflict of interest.

References

1. Jokl, P.; Sethi, P.M.; Cooper, A.J. Master's performance in the New York City Marathon 1983–1999. *Br. J. Sports Med.* **2004**, *38*, 408–412. [CrossRef]
2. Leyk, D.; Erley, O.; Ridder, D.; Leurs, M.; Rüther, T.; Wunderlich, M.; Sievert, A.; Baum, K.; Essfeld, D. Age-related changes in marathon and half-marathon performances. *Int. J. Sports Med.* **2007**, *28*, 513–517. [CrossRef]
3. Hoffman, M.D.; Wegelin, J.A. The western states 100-mile endurance run: Participation and performance trends. *Med. Sci. Sports Exerc.* **2009**, *41*, 2191–2198. [CrossRef]
4. Hoffman, M.D. Performance trends in 161-km ultramarathons. *Int. J. Sports Med.* **2010**, *31*, 31–37. [CrossRef] [PubMed]
5. Lepers, R.; Cattagni, T. Do older athletes reach limits in their performance during marathon running? *Age* **2012**, *34*, 773–781. [CrossRef] [PubMed]
6. Etter, F.; Knechtle, B.; Bukowski, A.; Rüst, C.A.; Rosemann, T.; Lepers, R. Age and gender interactions in short distance triathlon performance. *J. Sports Sci.* **2013**, *31*, 996–1006. [CrossRef]
7. Deaner, R.O. Distance running as an ideal domain for showing a sex difference in competitiveness. *Arch. Sex. Behav.* **2013**, *42*, 413–428. [CrossRef]
8. Lepers, R.; Knechtle, B.; Stapley, P.J. Trends in triathlon performance: Effects of sex and age. *Sport. Med.* **2013**, *43*, 851–863. [CrossRef]
9. Stiefel, M.; Rüst, C.A.; Rosemann, T.; Knechtle, B. A comparison of participation and performance in age-group finishers competing in and qualifying for Ironman Hawaii. *Int. J. Gen. Med.* **2013**, *6*, 67–77. [CrossRef]
10. Lepers, R.; Rüst, C.A.; Stapley, P.J.; Knechtle, B. Relative improvements in endurance performance with age: Evidence from 25 years of Hawaii Ironman racing. *Age* **2013**, *35*, 953–962. [CrossRef]
11. Knechtle, B.; Zingg, M.; Rosemann, T.; Rust, C. Sex difference in top performers from Ironman to double deca iron ultra-triathlon. *Open Access J. Sport. Med.* **2014**, *5*, 159–172. [CrossRef]
12. Joyner, M.J. Physiological limits to endurance exercise performance: Influence of sex. *J. Physiol.* **2017**, *595*, 2949–2954. [CrossRef] [PubMed]
13. Fleck, S.J. Body composition of elite American athletes. *Am. J. Sports Med.* **1983**, *11*, 398–403. [CrossRef]
14. Knechtle, B.; Wirth, A.; Baumann, B.; Knechtle, P.; Rosemann, T. Personal best time, percent body fat, and training are differently associated with race time for male and female ironman triathletes. *Res. Q. Exerc. Sport* **2010**, *81*, 62–68. [CrossRef]
15. Heydenreich, J.; Kayser, B.; Schutz, Y.; Melzer, K. Total Energy Expenditure, Energy Intake, and Body Composition in Endurance Athletes Across the Training Season: A Systematic Review. *Sport. Med.-Open* **2017**, *3*, 8. [CrossRef] [PubMed]
16. Landers, G.J.; Blanksby, B.A.; Ackland, T.R.; Smith, D. Morphology and performance of world championship triathletes. *Ann. Hum. Biol.* **2000**, *27*, 387–400. [CrossRef]
17. Millet, G.P.; Vleck, V.E.; Bentley, D.J. Physiological requirements in triathlon. *J. Hum. Sport Exerc.* **2011**, *6*, 184–204. [CrossRef]
18. Sharma, H.B.; Kailashiya, J. Gender difference in aerobic capacity and the contribution by body composition and haemoglobin concentration: A study in young Indian National hockey players. *J. Clin. Diagnostic Res.* **2016**, *10*, CC09–CC13. [CrossRef] [PubMed]
19. Lepers, R. Sex difference in triathlon performance. *Front. Physiol.* **2019**, *24*, 973. [CrossRef] [PubMed]
20. Cureton, K.; Bishop, P.; Hutchinson, P.; Newland, H.; Vickery, S.; Zwiren, L. Sex difference in maximal oxygen uptake—Effect of equating haemoglobin concentration. *Eur. J. Appl. Physiol. Occup. Physiol.* **1986**, *54*, 656–660. [CrossRef]

21. Ansdell, P.; Thomas, K.; Hicks, K.M.; Hunter, S.K.; Howatson, G.; Goodall, S. Physiological sex differences affect the integrative response to exercise: Acute and chronic implications. *Exp. Physiol.* **2020**, *105*, 2007–2021. [CrossRef] [PubMed]

22. Roepstorff, C.; Thiele, M.; Hillig, T.; Pilegaard, H.; Richter, E.A.; Wojtaszewski, J.F.P.; Kiens, B. Higher skeletal muscle α2AMPK activation and lower energy charge and fat oxidation in men than in women during submaximal exercise. *J. Physiol.* **2006**, *574*, 125–138. [CrossRef] [PubMed]

23. Simoneau, J.A.; Bouchard, C. Human variation in skeletal muscle fiber-type proportion and enzyme activities. *Am. J. Physiol.-Endocrinol. Metab.* **1989**, *257*, E567–E572. [CrossRef]

24. Staron, R.S.; Hagerman, F.C.; Hikida, R.S.; Murray, T.F.; Hostler, D.P.; Crill, M.T.; Ragg, K.E.; Toma, K. Fiber type composition of the vastus lateralis muscle of young men and women. *J. Histochem. Cytochem.* **2000**, *48*, 623–629. [CrossRef]

25. Tarnopolsky, M.A. Sex differences in exercise metabolism and the role of 17-beta estradiol. *Med. Sci. Sports Exerc.* **2008**, *40*, 648–654. [CrossRef]

26. Beltrame, T.; Rodrigo, V.; Hughson, R.L. Sex differences in the oxygen delivery, extraction, and uptake during moderate-walking exercise transition. *Appl. Physiol. Nutr. Metab.* **2017**, *42*, 994–1000. [CrossRef]

27. Weltman, A. *The Blood Lactate Response to Exercise*; Human Kinetics: Champaign, IL, USA, 1995; ISBN 0873227697 9780873227698.

28. Cardinale, D.A.; Larsen, F.J.; Schiffer, T.A.; Morales-Alamo, D.; Ekblom, B.; Calbet, J.A.L.; Holmberg, H.C.; Boushel, R. Superior intrinsic mitochondrial respiration in women than in men. *Front. Physiol.* **2018**, *9*, 1133. [CrossRef] [PubMed]

29. Sanfilippo, J.; Krueger, D.; Heiderscheit, B.; Binkley, N. Dual-Energy X-ray Absorptiometry Body Composition in NCAA Division I Athletes: Exploration of Mass Distribution. *Sports Health* **2019**, *11*, 453–460. [CrossRef] [PubMed]

30. Hill, A.M.; LaForgia, J.; Coates, A.M.; Buckley, J.D.; Howe, P.R.C. Estimating abdominal adipose tissue with DXA and anthropometry. *Obesity* **2007**, *15*, 504. [CrossRef] [PubMed]

31. Choi, Y.J.; Seo, Y.K.; Lee, E.J.; Chung, Y.S. Quantification of visceral fat using dual-energy x-ray absorptiometry and its reliability according to the amount of visceral fat in Korean Adults. *J. Clin. Densitom.* **2015**, *18*, 192–197. [CrossRef] [PubMed]

32. Longstreet Taylor, H.; Buskirk, E.; Henschel, A. Maximal oxygen intake as an objective measure of cardio-respiratory performance. *J. Appl. Physiol.* **1955**, *8*, 73–80. [CrossRef]

33. Whipp, B.J.; Ward, S.A.; Wasserman, K. Respiratory markers of the anaerobic threshold. *Adv. Cardiol.* **1986**, *35*, 47–64. [CrossRef] [PubMed]

34. Puccinelli, P.J.; Lima, G.H.O.; Pesquero, J.B.; de Lira, C.A.B.; Vancini, R.L.; Nikolaids, P.T.; Knechtle, B.; Andrade, M.S. Previous experience, aerobic capacity and body composition are the best predictors for Olympic distance triathlon performance: Predictors in amateur triathlon. *Physiol. Behav.* **2020**, *225*, 113110. [CrossRef] [PubMed]

35. de Lira, C.A.B.; Peixinho-Pena, L.F.; Vancini, R.L.; de Freitas Guina Fachina, R.J.; de Almeida, A.A.; Andrade, M.D.S.; da Silva, A.C. Heart rate response during a simulated Olympic boxing match is predominantly above ventilatory threshold 2: A cross sectional study. *Open Access J. Sport. Med.* **2013**, *4*, 175–182. [CrossRef]

36. Jones, A.M.; Doust, J.H. A 1% treadmill grade most accurately reflects the energetic cost of outdoor running. *J. Sports Sci.* **1996**, *14*, 321–327. [CrossRef] [PubMed]

37. Noble, B.J.; Borg, G.A.V.; Jacobs, I.; Ceci, R.; Kaiser, P. A category-ratio perceived exertion scale: Relationship to blood and muscle lactates and heart rate. *Med. Sci. Sports Exerc.* **1983**, *15*, 523–528. [CrossRef] [PubMed]

38. Ruxton, G.D. The unequal variance *t*-test is an underusedalternative to Student's *t*-test and the Mann–Whitney *U* test. *Behav. Ecol.* **2006**, *17*, 688–690. [CrossRef]

39. Cohen, J. *Statistical Power for the Behavioral Sciences*, 2nd ed.; Lawrence Erlbaum Associates: Hillsdale, NJ, USA, 1988; ISBN 0-8058-0283-5.

40. Kang, H. Sample size determination and power analysis using the G*Power software. *J. Educ. Eval. Health Prof.* **2021**, *18*, 17. [CrossRef] [PubMed]

41. Banerjee, A.; Chitnis, U.; Jadhav, S.; Bhawalkar, J.; Chaudhury, S. Hypothesis testing, type I and type II errors. *Ind. Psychiatry J.* **2009**, *18*, 127–131. [CrossRef]

42. Stevenson, J.L.; Song, H.; Cooper, J.A. Age and sex differences pertaining to modes of locomotion in triathlon. *Med. Sci. Sports Exerc.* **2013**, *45*, 976–984. [CrossRef]

43. Bunc, V.; Heller, J.; Horcic, J.; Novotny, J. Physiological profile of best Czech male and female young triathletes. *J. Sports Med. Phys. Fitness* **1996**, *36*, 265–270. [PubMed]

44. O'toole, M.L.; Douglas, W.; Crosby, L.O.; Douglas, P.S. The ultraendurance triathlete: A physiological profile. *Med. Sci. Sports Exerc.* **1987**, *19*, 45–50. [CrossRef] [PubMed]

45. Holloszy, J.O.; Coyle, E.F. Adaptations of skeletal muscle to endurance exercise and their metabolic consequences. *J. Appl. Physiol. Respir. Environ. Exerc. Physiol.* **1984**, *56*, 831–838. [CrossRef]

46. Coyle, E.F.; Coggan, A.R.; Hopper, M.K.; Walters, T.J. Determinants of endurance in well-trained cyclists. *J. Appl. Physiol.* **1988**, *64*, 2622–2630. [CrossRef]

47. Wasserstrom, J.A.; Kapur, S.; Jones, S.; Faruque, T.; Sharma, R.; Kelly, J.E.; Pappas, A.; Ho, W.; Kadish, A.H.; Aistrup, G.L. Characteristics of intracellular Ca^{2+} cycling in intact rat heart: A comparison of sex differences. *Am. J. Physiol.-Heart Circ. Physiol.* **2008**, *295*, H1895–H1904. [CrossRef]

48. Mitchell, E.A.; Martin, N.R.W.; Bailey, S.J.; Ferguson, R.A. Critical power is positively related to skeletal muscle capillarity and type I muscle fibers in endurance-trained individuals. *J. Appl. Physiol.* **2018**, *125*, 737–745. [CrossRef] [PubMed]

49. Vanhatalo, A.; Black, M.I.; DiMenna, F.J.; Blackwell, J.R.; Schmidt, J.F.; Thompson, C.; Wylie, L.J.; Mohr, M.; Bangsbo, J.; Krustrup, P.; et al. The mechanistic bases of the power-time relationship: Muscle metabolic responses and relationships to muscle fibre type. *J. Physiol.* **2016**, *594*, 4407–4423. [CrossRef]
50. Ackland, T.R.; Blanksby, B.A.; Landers, G.; Smith, D. Anthropmetrlc profiles of elite trlathletes. *J. Sci. Med. Sport* **1998**, *1*, 52–56. [CrossRef]
51. Knechtle, B.; Knechtle, P.; Rosemann, T. Similarity of anthropometric measures for male ultra-triathletes and ultra-runners. *Percept. Mot. Skills* **2010**, *111*, 805–818. [CrossRef]
52. Wonerow, M.; Rüst, C.A.; Nikolaidis, P.T.; Rosemann, T.; Knechtle, B. Performance trends in age group triathletes in the olympic distance triathlon at the world championships 2009–2014. *Chin. J. Physiol.* **2017**, *60*, 137–150. [CrossRef] [PubMed]

Article

Physiological Features of Olympic-Distance Amateur Triathletes, as Well as Their Associations with Performance in Women and Men: A Cross–Sectional Study

José Geraldo Barbosa [1], Claudio Andre Barbosa de Lira [2], Rodrigo Luiz Vancini [3], Vinicius Ribeiro dos Anjos [1], Lavínia Vivan [1], Aldo Seffrin [1], Pedro Forte [4,5,6], Katja Weiss [7], Beat Knechtle [7,8,*] and Marilia Santos Andrade [1]

1. Department of Physiology, Federal University of São Paulo, São Paulo 04021-001, Brazil
2. Human and Exercise Physiology Division, Faculty of Physical Education and Dance, Federal University of Goiás, Goiânia 74690-900, Brazil
3. Center for Physical Education and Sports, Federal University of Espírito Santo, Vitória 29075-210, Brazil
4. Department of Sports, Higher Institute of Educational Sciences of the Douro, 4560-547 Porto, Portugal
5. Department of Sports Sciences, Instituto Politécnico de Bragança, 5300-253 Bragança, Portugal
6. Research Center in Sports, Health and Human Development, 7000-671 Covilhã, Portugal
7. Institute of Primary Care, University of Zurich, 8091 Zurich, Switzerland
8. Medbase St. Gallen Am Vadianplatz, 9000 St. Gallen, Switzerland
* Correspondence: beat.knechtle@hispeed.ch

Abstract: The purpose of this study was to verify the physiological and anthropometric determinants of triathlon performance in female and male athletes. This study included 40 triathletes (20 male and 20 female). Dual-energy X-ray absorptiometry (DEXA) was used to assess body composition, and an incremental cardiopulmonary test was used to assess physiological variables. A questionnaire about physical training habits was also completed by the athletes. Athletes competed in the Olympic-distance triathlon race. For the female group, the total race time can be predicted by $\dot{V}O_2max$ ($\beta = -131$, t $= -6.61$, $p < 0.001$), lean mass ($\beta = -61.4$, t $= -2.66$, $p = 0.018$), and triathlon experience ($\beta = -886.1$, t $= -3.01$, $p = 0.009$) ($r^2 = 0.825$, $p < 0.05$). For the male group, the total race time can be predicted by maximal aerobic speed ($\beta = -294.1$, t $= -2.89$, $p = 0.010$) and percentage of body fat ($\beta = 53.6$, t $= 2.20$, $p = 0.042$) ($r^2 = 0.578$, $p < 0.05$). The variables that can predict the performance of men are not the same as those that can predict the triathlon performance of women. These data can help athletes and coaches develop performance-enhancing strategies.

Keywords: triathlon; ports physiology; performance; women; maximal aerobic speed

Citation: Barbosa, J.G.; de Lira, C.A.B.; Vancini, R.L.; dos Anjos, V.R.; Vivan, L.; Seffrin, A.; Forte, P.; Weiss, K.; Knechtle, B.; Andrade, M.S. Physiological Features of Olympic-Distance Amateur Triathletes, as Well as Their Associations with Performance in Women and Men: A Cross–Sectional Study. *Healthcare* **2023**, *11*, 622. https://doi.org/10.3390/healthcare11040622

Academic Editors: João Paulo Brito and Rafael Oliveira

Received: 14 December 2022
Revised: 16 February 2023
Accepted: 18 February 2023
Published: 20 February 2023

1. Introduction

Although female participation in triathlon is still lower than male participation (25–40%), there has been a significant increase in female participation in this sport since 1990 [1–3]. Female participation has increased not only in triathlon, but also in several other sports. In running, female participation reached the same percentage as male participation in 2018, and at the most recent Olympic Games (Tokyo 2020), female participation set a new record, reaching 49% [4].

Despite a recent surge of female participation in sports, most scientific studies on sports sciences continue to focus on men [4]. Therefore, the results of the studies conducted on male athletes on sports training are applied to both male and female athletes, despite the lack of a reasonable scientific justification [5,6].

The literature's consensus that maximal oxygen uptake ($\dot{V}O_2max$), the percentage of $\dot{V}O_2max$ that can be sustained for an extended period of time, the running economy, and body composition are important variables associated with performance in long-distance events [7]. However, the relative importance of each one can differ for male or female performance, once there are several physiological (red cell mass, hemoglobin, muscular fiber type percentage, muscle capillarization, vasodilatory capacity, and energetic substrate use) and body composition (fat mass percentage, lean mass) differences between sexes [8]. The variables associated with triathlon performance have previously been studied. Therefore, prediction equations for triathlon performance have also been developed; however, previous studies have only included male athletes, or when female athletes were included, the sex was not evaluated as a biological variable [9,10]. The knowledge of the main predictive variables of performance for each sex separately, can help coaches in directing the training sessions to obtain adaptive responses of the most important predictive variables of performance and optimize the improvement of sports performance.

Therefore, the purpose of this study was to confirm the levels of association between physiological and body composition variables and performance in an Olympic-distance triathlon test for each sex and, later, to describe a performance prediction equation for the modality based on the measured variables for each sex. Second, the study aimed to compare physiological and body composition variables between sexes. We hypothesized that the level of association between the measured variables and triathlon race results would differ between sexes, and thus the Olympic-distance triathlon race prediction equations would differ for each sex.

2. Materials and Methods

2.1. Participants

Triathletes were invited to take part in the study via social media (WhatsApp, Instagram, and email), as well as folders distributed at triathlon competitions. The inclusion criteria for participating in the study were being enrolled in the 30th Santos International Olympic Triathlon in February 2022, between the ages of $\geq$18 and no more than 61 years, both sexes, training triathlon regularly for at least 6 months, and having a medical allowance. The exclusion criteria included being pregnant, having competed in the alternate modality, failure to finish the race, or failure to submit to laboratory tests for any reason. Initially, 42 athletes were selected to participate in the study (22 men and 20 women). Two male athletes were excluded from the study, one was due to the fact that he did not finish the race and one was due to the fact that he failed to submit to laboratory tests. As a result, the study includes 20 male athletes and 20 female athletes. Data were collected during the pre-season. Table 1 shows the descriptive characteristics of the sample.

Table 1. Anthropometric data for men and women.

	Women (n = 20)	Men (n = 20)	*p*-Value	Cohen's *d*	Power
Age (years)	42.7 ± 7.3 (3.9–45.9)	43.7 ± 9.3 (38.2–46.4)	0.880	0.05	0.052
Body mass (kg)	58.8 ± 6.7 (55.9–61.8)	74.8 ± 6.9 (71.8–77.9)	<0.001	2.35	0.999
Height (cm)	165.0 ± 5.7 (163.0–168.0)	175.0 ± 8.2 (171.0–178.0)	<0.001	1.35	0.986
Fat mass (kg)	13.3 ± 7.2 (9.9 16.9)	12.8 ± 5.0 (10.6–15.0)	0.826	0.07	0.055
Lean mass (kg)	42.2 ± 6.5 (45.1–39.4)	58.3 ± 5.8 (55.8–60.8)	<0.001	2.60	1.000
% Body fat	23.3 ± 11.3 (18.3–28.2)	17.8 ± 6.3 (15.0–20.6)	0.066	0.60	0.453
% Gynoid fat	53.2 ± 6.0 (50.6–55.9)	39.1 ± 5.1 (36.8–41.3)	<0.001	2.53	1.000
% Android fat	42.9 ± 6.6 (40.0–45.8)	57.1 ± 5.4 (54.7–59.5)	<0.001	2.36	0.999

Mean ± standard deviation. Confidence interval: 95%.

2.2. Procedures

All experimental procedures followed the Declaration of Helsinki and the Recommendations for the Conduct, Reporting, Editing, and Publication of Scholarly Work in Medical Journals. The study was approved by the Human Ethics Committee of the University Federal of São Paulo-UNIFESP (approval number 5.059.538, 25 October 2021). The participants were given information about the purpose of the research, all of the proposed physiological laboratory tests, and the risks and benefits. The researchers justified the principles of respect for the volunteers, as well as the guarantee of privacy, confidentiality, and anonymity rights. After all, every participant signed the informed consent form. Initially, the volunteers completed an online questionnaire from the Google Forms Platform about their training habits and medical conditions. Then, they attend the Exercise Physiology Laboratory at UNIFESP once, during the morning period. During the visit, volunteers were measured for height, body mass, and body composition. Thereafter, participants were submitted to a running economy evaluation. After 30 min of rest, participants were subjected to a cardiorespiratory maximal treadmill test. They were instructed to abstain from strenuous training in the last 24 h before the test and not consume hyper-stimulating foods on the day (e.g., caffeine). Wearing light clothes and comfortable running shoes was also recommended.

The organizers provided the results of the total race time and split times of the competition, which were taken from the official website of the event (https://www.internacionaldesantos.com.br, accessed on 5 March 2022). All the data were collected in January 2022 (pre-competition phase), 1 month before the 30th Santos International Olympic Triathlon.

2.3. Questionnaire

The questionnaire includes two open questions about their medical condition: Do you have any chronic diseases? Do you take any medications? The questionnaire also includes four open questions about their training habits: How many hours per week do you cycle train? How many hours per week do you train for running? How many hours per week do you train for swimming? How many days, months, or years have you been training for a triathlon?

2.4. Morphological Variables

Body mass and height were measured to the nearest 0.1 kg and 0.1 cm using a calibrated stadiometer Filizola® PL (Filizola, São Paulo, SP, Brazil), respectively. Body composition was determined using dual-energy X-ray absorptiometry (DEXA, software version 12.3, Lunar DPX, GE Healthcare, Madison, WI, USA). The volunteers were instructed to drink water ad libitum and were not given any instructions about fasting or taking any specific feedings prior to the procedure. They were all evaluated after bladder voiding [11]. These procedures had been shown to be a reliable method for assessing body composition [12,13].

2.5. Running Economy Test

The volunteers were subjected to a treadmill running test for 4 min on a motorized treadmill (Inbrasport, ATL, Porto Alegre, Brazil) using a computer-based breath, via a breath gas exchange analyzer (Quark, Comedy, Italy) at a constant speed of 8 km/h, which was below the ventilatory threshold (VT) for all volunteers. Prior to each test, the calibration procedure was carried out according to the manufacturer's instructions. The last minute was taken into account when calculating the average oxygen uptake, CO_2 production, and respiratory exchange rate (RER). The RER should be lower than 1.0., as all the participants were exercising lower than the ventilatory threshold intensity. According to Silva et al. [14], these variables were used to calculate the oxygen cost and the energy cost of running.

2.6. Cardiorespiratory Maximal Treadmill Test

After a 30-min recovery period from the running economy test, which was enough to return all volunteers' heart rates to rest levels, they were subjected to the cardiorespiratory maximal treadmill test. The same computer-based metabolic analyzer (Quark, Comedy, Italy) was used to measure $\dot{V}O_2$max, VT, and respiratory compensation point (RCP). The maximal aerobic speed (MAS) was also measured. $\dot{V}O_2$max was defined as a stable increase in oxygen uptake (less than 2.1 mL/kg/min) even after increasing exercise intensity [15]. VT was calculated using the following criteria: An increase in the ventilatory equivalent for oxygen without an increase in the equivalent for carbon dioxide and an increase in end-tidal pressure of oxygen. The RCP was determined by increasing the CO_2 equivalent ventilatory and decreasing the end-tidal pressure of CO_2 [16]. Two independent investigators determined VT and RCP, and a third researcher was consulted in the event of disagreement. The MAS was defined as the lowest exercise intensity that produced $\dot{V}O_2$max [10].

2.7. Total Race and Split Time Results

Total race and split time results were provided by the organizers of the 30th Santos International Olympic Triathlon, which were accessed via the official website of the event (https://www.internacionaldesantos.com.br/, accessed on 5 March 2022).

2.8. Statistical Analysis

The data were presented in the form of mean and standard deviations. According to the Kolmogorov–Smirnov and Levene's tests, all variables had a normal distribution and homogeneous variability. A Student's *t*-test for independent samples was used to compare the variables of male and female athletes. To determine the magnitude of the differences, between group effect sizes were computed for each outcome. Using Cohen's effect sizes, the magnitude of any change was judged according to the following criteria: $d < 0.2$ was considered as ignored; $0.2 \leq d < 0.5$ was considered as a "small" effect size; $0.5 \leq d < 0.8$ represented a "moderate" effect size; $0.8 \leq d < 1.3$ a "large" effect size; and $d \geq 1.3$ a "very large" effect [17]. The Pearson linear correlation coefficient and dispersion diagrams were used to validate the level of association between each split time and total race time with other measured variables. To compare the triathlon race time to the previous triathlon experience, a one-way ANOVA was used. Thereafter, we considered the significant correlations for the stepwise adjustment of the multiple linear regression model. The formula of the regression model is $x = \alpha + \beta \cdot y + E$, where x is the dependent variable, y is the independent variable, α is the intercept, β is the slope, and E is the residual. For each regression equation, the coefficient of determination (r^2), a number that measures how well a statistical model predicts an outcome, was presented. For all regression models presented, Durbin–Watson Test (to detect autocorrelation), variance inflation factor (VIF) and tolerance (to detect multicollinearity), the normality of the distribution of residuals, and Q-Q plot (to detect homoscedasticity) were presented. The G*Power version 3.1.9.2 (Franz, Universität Kiel, Germany) was used to determine the sample size and analyze the test power level. A sample size calculation for regression analysis for overall race time with two predictors, using previous published data from Puccinelli et al. [10] ($r^2 = 0.607$), showed that 20 athletes were needed to detect a relevant difference with 80% power and a significance level of 5%. The powers of the analyses were also calculated. The analyses were carried out using the IBM SPSS Statistics (version 22, USA) software, with the level of significance set at $p < 0.05$.

3. Results

In terms of physiological variables, men had significantly higher values for $\dot{V}O_2$max (L/min) ($p < 0.001$, $d = 2.74$), $\dot{V}O_2$max (mL/kg/min) ($p = 0.020$, $d = 0.769$), MAS ($p < 0.001$, $d = 1.30$), VT speed ($p = 0.011$, $d = 0.843$), and RCP speed ($p < 0.001$, $d = 1.78$). The results showed no significant difference in the percentage of $\dot{V}O_2$max at VT ($p = 0.129$, $d = 0.490$)

or RCP (p = 0.558, d = 0.173) between men and women. Running economy, as measured by oxygen cost or energy cost, did not differ significantly between sexes (p = 0.540, d = 0.196 and p = 0.600, d = 0.167, respectively) (Table 2).

Table 2. Measured variables in the cardiorespiratory maximal test and performance for each sex.

	Women (n = 20)	Men (n = 20)	p-Value	Cohen's d	Power
Cardiorespiratory maximal test					
$\dot{V}O_2$max (L/min)	2.90 ± 0.39 (2.72–3.07)	4.08 ± 0.47 (3.88–4.28)	<0.001	2.74	1.00
$\dot{V}O_2$max (mL/kg/min)	49.7 ± 7.6 (46.3–53.0)	54.6 ± 5.0 (52.4–56.8)	0.020	0.769	0.659
MAS (km/h)	14.9 ± 1.8 (14.1–15.7)	17.1 ± 1.6 (16.4–17.8)	<0.001	1.30	0.979
$\dot{V}O_2$ at VT (mL/kg/min)	38.0 ± 6.3 (35.3–40.0)	40.4 ± 4.0 (38.6–42.2)	0.164	0.499	0.336
% $\dot{V}O_2$max at VT	76.2 ± 5.3 (73.9–78.6)	73.8 ± 4.7 (71.7–75.9)	0.129	0.490	0.326
Speed at VT (km/h)	10.7 ± 1.6 (10.0–11.4)	11.8 ± 1.1 (11.4–12.3)	0.011	0.843	0.738
$\dot{V}O_2$ at RCP (mL/kg/min)	45.1 ± 6.9 (42.1–48.1)	49.1 ± 4.7 (47.4–51.1)	0.025	0.736	0.621
% $\dot{V}O_2$max at RCP	90.5 ± 3.9 (88.7–92.2)	89.8 ± 3.7 (88.2–91.4)	0.588	0.173	0.083
Speed at RCP (km/h)	12.8 ± 1.6 (12.0–13.5)	14.4 ± 1,2 (13.9–15.0)	<0.001	1.78	0.999
Oxygen cost (mL/kg/km)	226.0 ± 20.3 (217.0–235.0)	222.0 ± 18.1 (214.0–230.0)	0.540	0.196	0.092
Energy cost (Kcal/kg/km)	1.11 ± 0.09 (1.07–1.15)	1.09 ± 0.09 (1.05–1.13)	0.600	0.167	0.080
Race performance					
Swimming (seconds)	2072 ± 518 (1844–2219)	1796 ± 265 (1680–1912)	0.041	0.670	0.670
Cycling (seconds)	4274 ± 405 (4097–4452)	3969 ± 329 (3825–4114)	0.013	0.826	0.720
Running (seconds)	3143 ± 546 (2904–3382)	3060 ± 371 (2853–3266)	0.608	0.163	0.079
Total race time (seconds)	9489 ± 1357 (8894–10083)	8825 ± 920 (8421–9228)	0.078	0.573	0.423

Mean ± standard deviation. Confidence interval: 95%. MAS: Maximal aerobic speed; $\dot{V}O_2$ at VT: $\dot{V}O_2$ at ventilatory threshold; % $\dot{V}O_2$max at VT: % $\dot{V}O_2$max at ventilatory threshold; $\dot{V}O_2$ at RCP: $\dot{V}O_2$ at respiratory compensation point; % $\dot{V}O_2$max at RCP: % $\dot{V}O_2$max at respiratory compensation point.

Men were significantly faster in the swimming split (p = 0.041, d = 0.670) and cycling split (p = 0.013, d = 0.826) of the Olympic-distance triathlon race, but there was no significant difference in the running split (p = 0.608, d = 0.163) or total race time (p = 0.078, d = 0.573) (Table 2). There was no significant difference in the years of triathlon experience (p = 0.807) between women [3(1–3)] and men [3(1–3)]. The level of association between each split and total race performance with body composition or physiologic variables was presented in Table 3.

Table 3. Pearson's correlation coefficient between performance in swimming, cycling, running, total race time, and measured variables for both sexes.

	Swimming (n = 20)	Cycling (n = 20)	Running (n = 20)	Total Time (n = 20)
Fat mass (kg)	W: r = 0.278 M: r = 0.475 *	W: r = 0.604 * M: r = 0.245	W: r = 0.612 * M: r = 0.657 *	W: r = 0.533 M: r = 0.560 *
Lean mass (Kg)	W: r= −0.345 M: r= −0.257	W: r= −0.530 * M: r= −0.210	W: r= −0.413 M: r= −0.382	W: r= −0.456 * M: r= −0.344
% Body fat	W: r = 0.353 M: r = 0.495 *	W: r = 0.653 * M: r = 0.300	W: r = 0.648 * M: r = 0.702 *	W: r = 0.590 * M: r = 0.609 *
% Gynoid fat	W: r = 0.188 M: r= −0.541 *	W: r= −0.037 M: r= −0.597 *	W: r= −0.035 M: r= −0.258	W: r = 0.046 M: r= −0.501 *
% Android fat	W: r= −0.138 M: r = 0.580 *	W: r = 0.108 M: r = 0.598 *	W: r = 0.110 M: r = 0.353	W: r = 0.024 M: r = 0.561*
$\dot{V}O_2$max (L/min)	W: r= −0.713 * M: r= −0.169	W: r= −0.776 * M: r= −0.154	W: r= −0.711 * M: r= −0.361	W: r= −0.790 * M: r= −0.288
$\dot{V}O_2$max (mL/kg/min)	W: r= −0.634 * M: r= −0.375	W: r= −0.781 * M: r= −0.237	W: r= −0.828 * M: r= −0.678 *	W: r= −0.808 * M: r= −0.539 *
MAS (km/h)	W: r= −0.633 * M: r= −0.442	W: r= −0.690 * M: r= −0.406	W: r= −0.845 * M: r= −0.790 *	W: r= −0.788 * M: r= −0.676 *
$\dot{V}O_2$ at VT (mL/kg/min)	W: r= −0.659 * M: r= −0.475 *	W: r= −0.794 * M: r= −0.343	W: r= −0.802 * M: r= −0.614 *	W: r= −0.811 * M: r= −0.573 *
% $\dot{V}O_2$max at VT (%)	W: r= −0.166 M: r= −0.278	W: r= −0.141 M: r= −0.235	W: r= −0.089 M: r= −0.036	W: r= −0.141 M: r= −0.182
Speed at VT (km/h)	W: r= −0.490 * M: r= −0.591 *	W: r= −0.701 * M: r= −0.385	W: r= −0.787 * M: r= −0.803 *	W: r= −0.713 * M: r= −0.719 *
$\dot{V}O_2$max at RCP (mL/kg/min)	W: r= −0.581 * M: r= −0.548 *	W: r= −0.757 * M: r= −0.298	W: r= −0.793 * M: r= −0.695 *	W: r= −0.581 * M: r= −0.620 *
% $\dot{V}O_2$max at RCP	W: r = 0.208 M: r= −0.386	W: r = 0.107 M: r= −0.240	W: r = 0.149 M: r= −0.050	W: r = 0.171 M: r = 0.224
Speed at RCP (km/h)	W: r= −0.576 * M: r= −0.556 *	W: r= −0.731 * M: r= −0.416	W: r= −0.842 * M: r= −0.851 *	W: r= −0.777 * M: r= −0.744 *
Oxygen cost (mL/kg/min)	W: r= −0.065 M: r = 0.034	W: r= −0.510 * M: r= −0.60	W: r = 0.292 M: r = 0.141	W: r= −0.294 M: r = 0.061
Energy cost (kcal/kg/km)	W: r = 0.014 M: r = 0.070	W: r= −0.453 * M: r= −0.040	W: r= −0.207 M: r = 0.200	W: r= −0.213 M: r = 0.109

W: Women; M: Men; r: Pearson's correlation coefficient; p-value between parenthesis; * $p < 0.05$; mean $\pm$ standard deviation. Confidence interval: 95%. MAS: Maximal aerobic speed; $\dot{V}O_2$ at VT: $\dot{V}O_2$ at ventilatory threshold; % $\dot{V}O_2$max at VT: % $\dot{V}O_2$max at ventilatory threshold; $\dot{V}O_2$ at RCP: $\dot{V}O_2$ at respiratory compensation point; % $\dot{V}O_2$max at RCP: % $\dot{V}O_2$max at respiratory compensation point.

When female ($p = 0.113$) and male ($p = 0.217$) athletes with less than 1 year, 1–3 years, and more than 3 years of triathlon training experience were compared, no significant difference in the triathlon race time was found. There was also no significant difference in swimming, cycling, and running split times between those with less than 1 year, 1–3 years, and more than 3 years of triathlon training experience in the female group ($p = 0.053$, $p = 0.397$, $p = 0.137$, respectively) and male group ($p = 0.077$, $p = 0.248$, $p = 0.380$, respectively).

Multiple linear regression adjusted models were fitted to determine which measured body composition and physiological variables can better predict the results by the stepwise method in each split time and the overall race time for women and men. The statistical models resulting from the analyses for the female sample are presented in Table 4, while those for the male sample are presented in Table 5.

Table 4. Multiple linear regression models for estimating performance in swimming, cycling, running, and total race time for female athletes.

Modality	r^2	Z	Df	*p*-Value	SEE	Tolerance	VIF	Durbin–Watson	Power
			Time 1500 m (s) = 4817 − 877 (absol. $\dot{V}O_2$max) − 378 (T.E.)						
Swimming	0.634	13.8	2.16	<0.001	304	>0.997	1	2.17	0.99
			Time 40 Km (s) = 6347 − 41.8 (relat. $\dot{V}O_2$max)						
Cycling	0.610	28.2	1.18	<0.001	246	0.961	1	1.59	0.99
			Time 10 Km (s) = 6766 − 284 (speed at RCP)						
Running	0.709	43.8	1.18	<0.001	287	1	1	2.35	0.99
			Total time (s) = 19077.4 − 131 (relat. $\dot{V}O_2$max) − 61.4 (lean mass) − 886.1 (T.E)						
Total Time	0.825	23.6	3.15	<0.001	561	1	<1.08	2.15	1.00

Absol. $\dot{V}O_2$max: Absolute $\dot{V}O_2$max (L/m); T.E.: Triathlon experience > 3 years; Relat. $\dot{V}O_2$max: $\dot{V}O_2$max relative to body mass (mL/kg/min); speed at RCP: Speed at respiratory compensation point.

Table 5. Multiple linear regression models for estimating performance in swimming, cycling, running, and total race time for men.

Modality	R^2	Z	Df	*p*-Value	SEE	Tolerance	VIF	Durbin–Watson	Power
			Time 1500 m (s) = 1836.8 − 89.6 (speed at RCP) + 22 (% Android fat)						
Swimming	0.494	8.31	2.17	0.003	183	0.907	1.1	1.82	0.95
			Time 40 Km (s) = 1907.2 + 36.1 (% Android fat)						
Cycling	0.357	10	2.18	0.005	257	1	1	2.17	0.88
			Time 10 km (s) = 7750 − 325 (speed at RCP)						
Running	0.724	47.3	1.18	<0.001	241	1	1	1.71	0.99
			Total time (s) = 12850 − 294.1 (MAS) + 56.3 (% Body Fat)						
Total Time	0.578	11.7	2.17	<0.001	582	0.808	<1.24	1.60	0.99

Speed at VT-2: Speed at ventilatory threshold-2 (km/h); MAS: Maximal aerobic speed (km/h).

For the women, in swimming, the variables that better adjusted to the model were absolute $\dot{V}O_2$max ($\beta = -877$, t $= -4.50$, $p < 0.001$) and experience in triathlon competition ($\beta = -378$, t $= -245$, $p = 0.026$), and both explain 63.4% of the split time performance in swimming. In cycling, the best model used only one variable: $\dot{V}O_2$max relative to body mass ($\beta = -41.8$, t $= -5.31$, $p < 0.001$), and it can explain 61% of the cycling split time. Similarly, the best model for running used only one variable: The speed at RCP ($\beta = -325$, t $= -6.62$, $p < 0.001$), and it can explain 70.9% of the running split time. Finally, for total race time, the best model includes $\dot{V}O_2$max ($\beta = -131$, t $= -6.61$, $p < 0.001$), lean mass ($\beta = -61.4$, t $= -2.66$, $p = 0.018$), and triathlon experience ($\beta = -886.1$, t $= -3.01$, $p = 0.009$), and they can explain 82.5% of the overall race time (Table 4).

For the men, in swimming, the best variables were speed at RCP ($\beta = -89.6$, t $= -2.31$, $p = 0.034$) and percentage of android fat ($\beta = 22.0$, t $= 2.50$, $p = 0.023$), and both can explain 49.4% of the swimming performance. In cycling, the best equation used only one variable: Percentage of android fat ($\beta = 36.1$, t $= 3.16$, $p = 0.005$), and it can explain 35.7% of the cycling split time. Similarly, in running, the best model used only one variable: Speed at RCP ($\beta = -325$, t $= -6.88$, $p < 0.001$), and it can predict 72.4% of the running performance. For the total race time, the variables that best adjusted to the model were MAS ($\beta = -294.1$, t $= -2.89$, $p = 0.010$) and percentage of body fat ($\beta = 53.6$, t $= 2.20$, $p = 0.042$), and both can predict 57.8% of the performance (Table 5).

4. Discussion

The main findings of this study were as follows: (I) The variables that better adjusted to the regression models for triathlon performance were different for male and female athletes; (II) for women, $\dot{V}O_2$max was part of the prediction equations for performance in swimming, cycling, and overall race time; (III) for women, the triathlon experience time was part of the prediction equations for performance in swimming and overall race time; (IV) for women, the lean mass was part of the prediction equations for the overall race time; (V) for men, body composition (android fat mass percentage or body fat percentage) was part of the prediction equations for performance in swimming, cycling, and overall race time; and (VI) the model for predicting performance in running split, which is the speed at RCP, was the only one that was found to be very similar between sexes. The current study's findings confirmed this initial hypothesis, as the regression models for performance in swimming, cycling, and overall race time differed between sexes.

4.1. Physiological, Body Composition, and Performance Sex Differences

In terms of body composition, the male sample had significantly more lean mass than the female sample. However, there was no difference in fat mass (kg) between sexes. The % fat mass was also not significantly different ($p = 0.066$) between sexes; however, the effect size of the difference was 0.6 (medium effect), and the power of this analysis was very low (power = 0.453); therefore, caution should be exercised when claiming that there is no difference in % fat mass between sexes. Despite the similarity of the total fat mass (kg), the distribution of body fat differed significantly between sexes. Female athletes had a higher gynoid percentage, while male athletes had a higher android percentage. These data are consistent with the previous study [18] for non-athletes and athletes [19]. Moreover, these findings are especially important given that the regional fat distribution in gynoid or android regions is associated with performance [19] and lipid profile [18].

In terms of physiological variables, male athletes had higher $\dot{V}O_2max$ and MAS than female athletes. However, the percentage of $\dot{V}O_2max$ at VT and RCP did not differ between sexes. Although the literature agrees that men have higher maximum oxygen consumption values [8], data on VTs are contradictory. Puccinelli et al. [10] discovered that female athletes had a higher percentage of $\dot{V}O_2max$ at VT and RCP than male athletes. These contradictory findings could be attributed to the varying levels of training of the athletes. Moreover, Puccinelli et al. [10] reported that the $\dot{V}O_2max$ values for male athletes were 59.9 ± 6.3 mL/kg/min and our male sample had a lower value of 54.6 ± 5.0 mL/kg/min, whereas female $\dot{V}O_2max$ values were comparable between the two studies (49.5 ± 7.8 and 49.7 ± 7.6 mL/kg/min, respectively).

The running economy did not differ between sexes. This finding is consistent with previous literature data [4]. Men outperformed women in cycling and swimming, with the difference being more evident in cycling. In the running split and the overall race time, there was no significant difference between sexes; however, due to the fact that the power of this statistical analysis was low (power = 0.079; power = 0.423, respectively), caution is advised in interpreting this lack of significant difference between sexes.

4.2. Predictors of Triathlon Overall and Split Race Times for the Female Sample

For the female sample, absolute (L/min) and relative (mL/kg/min) $\dot{V}O_2max$ values are strongly related to all the split and overall race times. This is an expected result since the higher the individual's oxygen consumption capacity, the greater the exercise intensity they can sustain [9,10]. The $\dot{V}O_2max$ by the regression models for swimming split time, cycling split time, and overall split time, demonstrate the importance of this variable in female performance.

Triathlon experience appears to be an important variable in predicting performance for female athletes, as evidenced by the prediction equations for swimming split time and overall race time. Moreover, previous research has shown that triathlon experience is an important predictor of performance in splits, particularly in swimming and total race time [9,20–23]. Swimming in the sea, bay, lake, or river (current, temperature, navigation, buoyancy, etc.) can have very different environmental conditions than swimming in a pool, where most athletes train, and these differences can make triathlon training experience especially important for swimming performance. Previous studies have already demonstrated the positive impact of previous experience on performance in the triathlon's total race time [20,21]. In the same direction, sports practice during childhood, even non-specific physical activity, has previously been shown to be highly correlated to better performances in swimming and total race times in the Olympic triathlon [24].

Apart from swimming, regarding body composition, body fat mass (kg or %) showed a significant correlation with cycling and running split time and total race time for women. These findings are consistent with previous literature data [5,9,25]. The lack of association between body fat and swimming split time could be attributed to fatter individuals having better body buoyancy due to lower body density, which contributes to passive floating and gliding [25,26]. Even though there is a significant association between body fat and running or cycling performance, it is not included in the regression models. The absence of body fat in the equation can be explained by the fact that this variable is associated with $\dot{V}O_2$max, and since $\dot{V}O_2$max composes the equation, body fat could also not compose the equation to avoid the multicollinearity effect.

The cycling performance was strongly associated with lean mass. The important relationship that exists between muscle mass and cycling performance can be explained by a greater ability to apply force on the pedals, which will generate greater power and displacement speed in cycling [27]. In contrast, the present study found no significant association between muscle mass and swimming split performance. Despite the fact that muscle mass contributes to propulsive force during swimming, it also contributes to an increase in body density and the tendency for body sinking, which increases swimming drag [28,29]. Therefore, when considering swimming speed as the function of drag and propulsion forces, there is no agreement in the previous literature on whether muscle mass has a positive or negative effect on swimming performance [24,26].

In terms of overall race time, muscle mass appears to be very important for female athletes, as this variable composes the regression model used to predict the overall race time. The speed at RCP, MAS, and $\dot{V}O_2$max (mL/kg/min) represented the measured variables that showed the strongest association with running performance, which is consistent with previous literature data [22,30,31]. Regarding the regression model for running split time, the speed at RCP has been used to build the model.

4.3. Predictors of Overall Triathlon and Each Split Race Time for the Male Sample

Regarding the physiological variables, the $\dot{V}O_2$max (mL/kg/min) presented a significant correlation with running split time and overall split time, similar to the MAS. At the same time, the speed and $\dot{V}O_2$ measured at the VT and RCP had a significant correlation with swimming and running split times, as well as overall race time. The significance of these variables can be seen in the fact that the speed at RCP composes the regression models used to predict the swimming and running time, whereas MAS composes the model used to predict the overall race time. The inclusion of the MAS rather than the $\dot{V}O_2$max in the prediction model of the overall race time can be justified since the MAS is known to be dependent on the $\dot{V}O_2$max and running economy [10]. The lack of association of $\dot{V}O_2$max and VT measurements with cycling performance may be due to some factors, such as the lack of specificity of the assessment method. The test used to determine $\dot{V}O_2$max and VTs was developed on a treadmill, and it is possible that an assessment performed on a bicycle will be more effective in identifying the variables associated with cycling split performance. Another complicating factor to consider is the tactical dimension in cycling split, such as drafting, pacing, and contextual factors on race dynamics [32]. Moreover, the results showed no significant association between running economy and performance. It is already expected, since running economy is regarded as a particularly important variable for distinguishing performance only among athletes with similar values for $\dot{V}O_2$max [33], which is not the case in the present study.

In terms of body composition variables, lean mass (kg) was not related to any split or overall race time. In contrast, fat mass (%) was associated with swimming, running, and overall race time. In addition, the fat mass percentage was used to create regression models to predict overall race time. Even though the total fat mass (%) was not associated with cycling split time, the distribution of fat mass affects performance in cycling and swimming split times. The higher the percentage of fat mass in the android, the worse the performance. With increased abdominal volume, it can be more difficult to assume an aerodynamic posture on a time trial bike, which can impair cycling performance. In a time trial bike position, the increase in abdominal volume can also make breathing mechanics difficult, where small shifts in cyclist positions may have a meaningful impact on performance [34]. A higher android fat percentage can also impair optimal body posture while swimming. McLean et al. [26] demonstrated that a higher android fat mass contributes to decreased buoyancy, which increases drag forces and reduces swimming speed [25]. Therefore, fat mass distribution is a very important variable associated with performance, and it was used to construct the regression model for swimming and cycling split times.

The regression model used in this study to predict swimming split time was composed of RCP speed and android fat mass (%). The android fat mass was the unique variable that composed the regression model for cycling split time. It is possible that evaluating the physiological variables in a more specific ergometer (bike) will yield regression models with better predictive values. In this direction, for the running split time, only the RCP speed is used to build the regression model. Finally, MAS and total body fat (%) composed the regression model for overall race time. As a result, it is demonstrated that only a treadmill assessment of MAS and a body composition assessment can predict more than 50% of the overall race time in an Olympic-distance triathlon.

4.4. Limitations and Strengh of the Study

The lack of physiological measurements in swimming or cycling activities is one of the study's limitations. It is possible that with these additional evaluations (cardiorespiratory maximal tests performed at cycle ergometer or at swimming pool), better equations for predicting triathlon performance can be presented. The authors propose that future studies should be designed with this goal in mind. The presentation of triathlon performance prediction equations for female athletes is an important strength of the study, as women are understudied, and the factors associated with performance differ between sexes. In addition, the sample size can be considered as very adequate since it reached a high level of power. As the power of a hypothesis test is 1 minus the probability of a type II error, the probability of making a type II error was exceedingly small. Finally, another strength of the study was using reliable and valid instruments, such as DEXA and breath, via the breath gas exchange analyzer.

5. Conclusions

For women's endurance performance, there are strong correlations between physiological variables usually measured in a laboratory. Moreover, it was possible to develop significantly predictive performance equations for triathlon total race time and splits. The physiological measures evaluated in the incremental treadmill test and body composition variables can predict more than 50% of the performance in the total time of the Olympic triathlon event. In addition, regression models for predicting female performance can predict a higher percentage of performance than models for predicting male performance. Finally, and perhaps most importantly, the variables capable of predicting male performance are not the same as those capable of predicting female performance, justifying the need to evaluate and study each gender separately.

Author Contributions: Conceptualization, M.S.A., J.G.B. and C.A.B.d.L.; methodology, J.G.B., L.V. and V.R.d.A.; software, A.S.; validation, M.S.A., R.L.V. and C.A.B.d.L.; formal analysis, M.S.A. and J.G.B.; investigation, L.V. and J.G.B.; data curation, C.A.B.d.L. and R.L.V.; writing—original draft preparation, J.G.B.; writing—review and editing, M.S.A. and P.F.; visualization, K.W. and B.K.; supervision, M.S.A. All authors have read and agreed to the published version of the manuscript.

Funding: This research was funded by FAPESP, grant number 2021/08114-1.

Institutional Review Board Statement: The study was conducted in accordance with the Declaration of Helsinki, and approved by the Institutional Review Board of UNIFESP (protocol code 5.059.538).

Informed Consent Statement: Informed consent was obtained from all subjects involved in the study.

Data Availability Statement: Data supporting the reported results can be requested from the corresponding author.

Acknowledgments: We would like to thank all of the participants who volunteered their time to participate in the study, the Olympic Training and Research Center (Centro Olímpico de Treinamento e Pesquisa, COTP, São Paulo, Brazil), and the Medicina Translacional post-graduation program, UNIFESP, Brazil.

Conflicts of Interest: The authors declare no conflict of interest.

References

1. Lepers, R.; Rüst, C.A.; Stapley, P.J.; Knechtle, B. Relative Improvements in Endurance Performance with Age: Evidence from 25 Years of Hawaii Ironman Racing. *Age* **2013**, *35*, 953–962. [CrossRef] [PubMed]
2. Wonerow, M.; Rüst, C.A.; Nikolaidis, P.T.; Rosemann, T.; Knechtle, B. Performance Trends in Age Group Triathletes in the Olympic Distance Triathlon at the World Championships 2009–2014. *Chin. J. Physiol.* **2017**, *60*, 137–150. [CrossRef] [PubMed]
3. Joyner, M.J. Physiological Limits to Endurance Exercise Performance: Influence of Sex. *J. Physiol.* **2017**, *595*, 2949–2959. [CrossRef]
4. Besson, T.; Macchi, R.; Rossi, J.; Morio, C.Y.M.; Kunimasa, Y.; Nicol, C.; Vercruyssen, F.; Millet, G.Y. Sex Differences in Endurance Running. *Sport. Med.* **2022**, *52*, 1235–1257. [CrossRef] [PubMed]
5. Lepers, R. Sex Difference in Triathlon Performance. *Front. Physiol.* **2019**, *10*, 973. [CrossRef]
6. Ansdell, P.; Thomas, K.; Hicks, K.M.; Hunter, S.K.; Howatson, G.; Goodall, S. Physiological Sex Differences Affect the Integrative Response to Exercise: Acute and Chronic Implications. *Exp. Physiol.* **2020**, *105*, 2007–2021. [CrossRef]
7. Lanferdini, F.J.; Diefenthaeler, F.; Ardigò, L.P.; Peyré-Tartaruga, L.A.; Padulo, J. Editorial: Structural and Mechanistic Determinants of Endurance Performance. *Front. Physiol.* **2022**, *13*, 1035583. [CrossRef]
8. de Araújo Moury Fernandes, G.C.; Barbosa Junior, J.G.G.; Seffrin, A.; Vivan, L.; de Lira, C.A.B.; Vancini, R.L.; Weiss, K.; Knechtle, B.; Andrade, M.S. Amateur Female Athletes Perform the Running Split of a Triathlon Race at Higher Relative Intensity than the Male Athletes: A Cross-Sectional Study. *Healthcare* **2023**, *11*, 418. [CrossRef]
9. Knechtle, B.; Zingg, M.; Stiefel, M.; Rosemann, T.; Rüst, C. What Predicts Performance in Ultra-Triathlon Races?—A Comparison between Ironman Distance Triathlon and Ultra-Triathlon. *Open Access J. Sport. Med.* **2015**, *6*, 149. [CrossRef]
10. Puccinelli, P.J.; Lima, G.H.O.; Pesquero, J.B.; de Lira, C.A.B.; Vancini, R.L.; Nikolaids, P.T.; Knechtle, B.; Andrade, M.S. Predictors of Performance in Amateur Olympic Distance Triathlon: Predictors in Amateur Triathlon. *Physiol. Behav.* **2020**, *225*, 113110. [CrossRef]
11. Sanfilippo, J.; Krueger, D.; Heiderscheit, B.; Binkley, N. Dual-Energy X-ray Absorptiometry Body Composition in NCAA Division I Athletes: Exploration of Mass Distribution. *Sport. Health* **2019**, *11*, 453–460. [CrossRef]
12. Choi, Y.J.; Seo, Y.K.; Lee, E.J.; Chung, Y.S. Quantification of Visceral Fat Using Dual-Energy X-ray Absorptiometry and Its Reliability According to the Amount of Visceral Fat in Korean Adults. *J. Clin. Densitom.* **2015**, *18*, 192–197. [CrossRef]
13. Borga, M.; West, J.; Bell, J.D.; Harvey, N.C.; Romu, T.; Heymsfield, S.B.; Leinhard, O.D. Advanced Body Composition Assessment: From Body Mass Index to Body Composition Profiling. *J. Investig. Med.* **2018**, *66*, 887–895. [CrossRef]
14. Silva, W.A.; de Lira, C.A.B.; Vancini, R.L.; Andrade, M.S. Hip Muscular Strength Balance Is Associated with Running Economy in Recreationally-Trained Endurance Runners. *PeerJ* **2018**, *6*, e5219. [CrossRef]
15. Molinari, C.A.; Edwards, J.; Billat, V. Maximal Time Spent at VO2max from Sprint to the Marathon. *Int. J. Environ. Res. Public Health* **2020**, *17*, 9250. [CrossRef]
16. Kominami, K.; Imahashi, K.; Katsuragawa, T.; Murakami, M. The Ratio of Oxygen Uptake From Ventilatory Anaerobic Threshold to Respiratory Compensation Point Is Maintained During Incremental Exercise in Older Adults. *Front. Physiol.* **2022**, *13*, 769387. [CrossRef]
17. Hopkins, W.G.; Marshall, S.W.; Batterham, A.M.; Hanin, J. Progressive statistics for studies in sports medicine and exercise science. *Med. Sci. Sport. Exerc.* **2009**, *41*, 3–13. [CrossRef]
18. Min, K.B.; Min, J.Y. Android and Gynoid Fat Percentages and Serum Lipid Levels in United States Adults. *Clin. Endocrinol.* **2015**, *82*, 377–387. [CrossRef]
19. Puccinelli, P.; de Lira, C.A.; Vancini, R.L.; Nikolaidis, P.T.; Knechtle, B.; Andrade, M.S. Distribution of Body Fat Is Associated with Physical Performance of Male Amateur Triathlon Athletes. *J. Sport. Med. Phys. Fit.* **2022**, *62*, 215–222. [CrossRef]
20. Rüst, C.A.; Knechtle, B.; Knechtle, T.; Rosemann, T.; Lepers, R. Personal Best Times in an Olympic Distance Triathlon and in a Marathon Predict Ironman Race Time in Recreational Male Triathletes. *Open Access J. Sport. Med.* **2011**, *22*, 121–129. [CrossRef] [PubMed]
21. Knechtle, B.; Zingg, M.A.; Rosemann, T.; Rüst, C.A. The Aspect of Experience in Ultra-Triathlon Races. *Springerplus* **2015**, *19*, 278. [CrossRef] [PubMed]

22. Joyner, M.J.; Coyle, E.F. Endurance Exercise Performance: The Physiology of Champions. *J. Physiol.* **2008**, *586*, 35–44. [CrossRef] [PubMed]
23. Sinisgalli, R.; de Lira, C.A.B.; Vancini, R.L.; Puccinelli, P.J.G.; Hill, L.; Knechtle, B.; Nikolaidis, P.T.; Andrade, M.S. Impact of Training Volume and Experience on Amateur Ironman Triathlon Performance. *Physiol. Behav.* **2021**, *232*, 113344. [CrossRef]
24. Hodges, N.J.; Starkes, J.L.; Nananidou, A.; Kerr, T.; Weir, P.L. Predicting Performance Times from Deliberate Practice Hours for Triathletes and Swimmers: What, When, and Where Is Practice Important? *J. Exp. Psychol. Appl.* **2004**, *10*, 219–237. [CrossRef]
25. Cortesi, M.; Gatta, G.; Michielon, G.; di Michele, R.; Bartolomei, S.; Scurati, R. Passive Drag in Young Swimmers: Effects of Body Composition, Morphology and Gliding Position. *Int. J. Environ. Res. Public Health* **2020**, *17*, 2002. [CrossRef]
26. McLean, S.P.; Hinrichs, R.N. Buoyancy, Gender, and Swimming Performance. *J. Appl. Biomech.* **2000**, *16*, 248–263. [CrossRef]
27. Rønnestad, B.R.; Mujika, I. Optimizing Strength Training for Running and Cycling Endurance Performance: A Review. *Scand. J. Med. Sci. Sport.* **2014**, *24*, 603–612. [CrossRef]
28. Papic, C.; McCabe, C.; Gonjo, T.; Sanders, R. Effect of Torso Morphology on Maximum Hydrodynamic Resistance in Front Crawl Swimming. *Sport. Biomech.* **2020**, *7*, 1–15. [CrossRef]
29. Wang, A.X.G.; Kabala, Z.J. Body Morphology and Drag in Swimming: CFD Analysis of the Effects of Differences in Male and Female Body Types. *Fluids* **2022**, *7*, 332. [CrossRef]
30. Støa, E.M.; Helgerud, J.; Rønnestad, B.R.; Hansen, J.; Ellefsen, S.; Støren, Ø. Factors Influencing Running Velocity at Lactate Threshold in Male and Female Runners at Different Levels of Performance. *Front. Physiol.* **2020**, *11*, 585267. [CrossRef]
31. McLaughlin, J.E.; Howley, E.T.; Bassett, D.R.; Thompson, D.L.; Fitzhugh, E.C. Test of the Classic Model for Predicting Endurance Running Performance. *Med. Sci. Sport. Exerc.* **2010**, *42*, 991–997. [CrossRef] [PubMed]
32. Phillips, K.E.; Hopkins, W.G. Determinants of Cycling Performance: A Review of the Dimensions and Features Regulating Performance in Elite Cycling Competitions. *Sport. Med. Open* **2020**, *6*, 23. [CrossRef]
33. Bassett, D.R.; Howley, E.T. Limiting Factors for Maximum Oxygen Uptake and Determinants of Endurance Performance. *Med. Sci. Sport. Exerc.* **2000**, *32*, 70–84. [CrossRef]
34. Faulkner, S.H.; Jobling, P. The Effect of Upper-Body Positioning on the Aerodynamic–Physiological Economy of Time-Trial Cycling. *Int. J. Sport. Physiol. Perform.* **2020**, *16*, 51–58. [CrossRef]

Article

Effect of Low-Intensity Aerobic Training Combined with Blood Flow Restriction on Body Composition, Physical Fitness, and Vascular Responses in Recreational Runners

Hyoung Jean Beak [1], Wonil Park [2], Ji Hye Yang [3] and Jooyoung Kim [1,4,*]

1 Sports Convergence Institute, Konkuk University, Chungju 27478, Korea
2 Department of Physical Education, College of Education, Chung-Ang University, Seoul 06974, Korea
3 Department of Medicine, CHA University, Pocheon 11160, Korea
4 College of Liberal Arts, Konkuk University, Chungju 27478, Korea
* Correspondence: hirase1125@hanmail.net; Tel.: +82-43-840-3520; Fax: +82-43-840-3988

Abstract: This study investigated the effect of low-intensity aerobic training combined with blood flow restriction (LABFR) on body composition, physical fitness, and vascular functions in recreational runners. The participants were 30 healthy male recreational runners, randomized between the LABFR ($n = 15$) and control ($n = 15$) groups. The LABFR group performed five sets of a repeated pattern of 2 min running at 40% VO_{2max} and 1 min passive rest, while wearing the occlusion cuff belts on the proximal end of the thigh. The frequency was three times a week for the period of eight weeks. The control group performed the identical running protocol without wearing the occlusion cuff belts. At the end of the training, the participants' body composition (fat mass, body fat, muscle mass, and right and left thigh circumference), physical fitness (power and VO_{2max}), and vascular responses (flow-mediated dilation (FMD), brachial ankle pulse wave velocity (baPWV), ankle brachial index (ABI), systolic blood pressure (SBP) and diastolic blood pressure (DBP)) were measured. The results showed a significant time $\times$ group interaction effect on muscle mass (F = 53.242, $p = 0.001$, $\eta_p^2 = 0.664$) and right thigh circumference (F = 4.544, $p = 0.042$, $\eta_p^2 = 0.144$), but no significant variation in any other factors, including fat mass, body fat, left thigh circumference, FMD, baPWV, ABI, SBP, and DBP ($p > 0.05$). Overall, our results suggested that eight-week LABFR exerted a positive effect on the body composition, especially muscle mass and thigh circumference, of recreational runners.

Keywords: blood flow restriction; body composition; low-intensity aerobic training; recreational runner; physical fitness; vascular responses

Citation: Beak, H.J.; Park, W.; Yang, J.H.; Kim, J. Effect of Low-Intensity Aerobic Training Combined with Blood Flow Restriction on Body Composition, Physical Fitness, and Vascular Responses in Recreational Runners. *Healthcare* **2022**, *10*, 1789. https://doi.org/10.3390/healthcare10091789

Academic Editors: João Paulo Brito and Rafael Oliveira

Received: 20 July 2022
Accepted: 15 September 2022
Published: 16 September 2022

Publisher's Note: MDPI stays neutral with regard to jurisdictional claims in published maps and institutional affiliations.

1. Introduction

Recent interest in the application of blood flow restriction (BFR) has focused on the effect of training in general training adaptations during periods of reduced blood flow [1]. In Japan, Dr. Yoshiaki Sato promoted "kaatsu training", involving "training with added pressure" through which BFR became widely known to the general public [2]. In typical BFR training, a cuff/tourniquet system is used to apply a partial restriction to the arterial inflow in the working musculature during exercise, while the venous outflow is completely restricted [3] so that the consequent blood pooling allows for an increased training effect [4]. Recent evidence has indicated the superiority of the reinforced training stimuli using the combined BFR training compared to the same exercise without BFR training [5].

The BFR method is usually used during low-load resistance exercise, and has been shown to be effective in enhancing long-term hypertrophic and strength responses in both clinical and athletic populations [6]. According to several studies on BFR training, an increase in muscle strength and hypertrophy could be expected from the conventional resistance training only via high-intensity training using an approximately 70–85% load of one-repetition maximum (1 RM), whereas the BFR training exerted a similar level of effect using a 20–50% load of 1 RM [5–7]. In fact, BFR training has been shown to offer significant benefits in the change in skeletal muscles in a number of studies, for example, by increasing the local muscle mass, strength, and endurance, despite the training being performed with lower resistance [8]. Moreover, several recent studies have reported a positive effect of BFR training in controlling the arterial compliance compared to the absence of BFR training, as well as stronger advantages in the change in vascular functions through four or more weeks of BFR training compared to conventional resistance training, thus, implicating a potential contribution in cardiovascular health [9].

Meanwhile, changes induced by low-intensity aerobic training combined with BFR (LABFR) have been investigated in recent studies, with the results showing the advantages of BFR training as a single mode of training based on the simultaneous increase in aerobic fitness and muscular strength, in addition to the increased maximum oxygen uptake (VO_{2max}), delayed onset of blood lactate accumulation, and enhanced economy of motion [10,11]. For these reasons, LABFR has been applied to individuals with a low level of training or a handicap in certain training, such as those recovering from an injury or those seeking assistive training to add new stimuli to aerobic training, and the reported effects were positive in inducing a variety of physiological changes [12–14]. In fact, Abe et al. [15] reported that, although the intensity of slow-walk training with BFR was set to a low level, secretion of growth hormone after acute exercise was increased and, after three weeks, the thigh muscle cross-sectional area and volume were increased by 4–7%. In a recent study by Pinheiro et al. [12], trained cyclists with knee osteoarthritis performed low-intensity aerobic training combined with BFR for nine weeks, in which a positive change was found, not only for aerobic fitness such, as with 20 km cycling time-trial performance and peak oxygen consumption (VO_{2peak}), but also for right and left leg maximal strength and the cross-sectional area of the vastus lateralis.

These studies on LABFR involved individuals who were injured or poorly trained. However, in some cases, low-intensity aerobic training is one of the training strategies used by recreational runners. High-intensity interval training can increase catabolic hormones, such as cortisol, which inhibit muscle protein synthesis, reducing muscle hypertrophy and muscle strength development [16,17]. In addition, BFR changes acute physiological stressors, such as local muscle oxygen availability and vascular shear stress, which can lead to adaptations that cannot be easily obtained with traditional training [1]. These advantages can provide positive effects for recreational runners. However, research on the positive effects of LABFR for recreational runners is limited. Thus, this study aimed to investigate the effect of an eight-week protocol of LABFR on body composition, physical fitness, and vascular responses in recreational runners. We hypothesized that regular LABFR training will increase the body composition, physical fitness, and vascular responses in recreational runners.

2. Methods

2.1. Subjects

Thirty health male recreational runners participated in this study. A recreational runner was defined as an individual who had regularly participated in running three times a week for a minimum of six months. The sample size was estimated by setting the effect size to 0.25 with an alpha level of $p < 0.05$ and statistical power > 0.80. As a result, the sample size required for this study was $n = 24$. Considering the drop-out rate, a total of 30 individuals were recruited. Individuals who received an operation related to a musculoskeletal disorder in the past six months, those with an injury history, and those with a vascular health problem (e.g., hypertension or deep vein thrombosis, or distended varicose veins) or a history of medical treatment or drug administration, were excluded. Prior to the experiment, the participants were given detailed explanations on the purpose, procedures and methods of the experiment, and all participants voluntarily signed an informed consent form. The 30 participants were randomized between the LABFR group ($n = 15$) and the control group ($n = 15$), but 1 from the LABFR group dropped out of the study for a personal reason, leaving 14 individuals in the LABFR group and 15 individuals in the control group at the end of the experiment (Table 1). An independent t-test found no significant differences between groups for the variables of age, height, weight, body mass index (BMI), and body fat ($p > 0.05$). The study procedures and methods, and the consent form, were approved by the university's Institutional Review Board. The study was conducted between April 2021 and January 2022.

Table 1. Physical characteristics of the subjects.

	LABFR ($n = 14$)	**CON ($n = 15$)**
Age (years)	30.21 ± 4.93	29.67 ± 3.06
Height (cm)	175.07 ± 8.46	172.27 ± 5.66
Weight (kg)	75.23 ± 9.08	73.59 ± 10.68
BMI (kg/m^2)	24.44 ± 2.32	24.63 ± 2.77
Body fat (%)	12.54 ± 4.27	15.33 ± 5.05

Data are presented as mean ± standard deviation (SD); BMI, body mass index; LABFR, low-intensity aerobic training combined with blood flow restriction group; CON, control group.

2.2. Low-Intensity Aerobic Training Combined with Blood Flow Restriction

Participants visited a laboratory with BFR equipment to perform training. Here, VO_{2max} was measured for LABFR. Before the VO_{2max} measurement, the participants attended an education session on relevant procedures and precautions. Afterwards, each participant performed a warm-up on the treadmill, followed by exercise at 1.7 mph with 10% slope. At every 3 min interval, the speed was increased by 0.8–0.9 mph, and the slope was increased by 2% as part of the Bruce protocol to measure the VO_{2max}. In the measurements, gas analyzers (Quark b, Cosmed, Rome, Italy) were concurrently applied, after adjustments for the daily volume, humidity, and temperature. The criteria of VO_{2max} were (i) a change in $VO_2 \leq 2$ mL·kg^{-1}·min^{-1}; (ii) respiratory exchange ratio ≥ 1.1; (iii) age-predicted maximal heart rate $\geq 85\%$. Before performing LABFR, 10 min of light jogging was performed as a warm-up. Afterwards, each participant wore an occlusion cuff belt (KAATSU NANO, KAATSU JAPAN, Tokyo, Japan) at the proximal end of both thighs (Figure 1). For LABFR, each participant repeated a pattern of 2 min running on a treadmill (HERA-9000, Health-One, Goyang, Republic of Korea) with the exercise intensity set to 40% VO_{2max}, followed by a 1 min passive rest, performing a total of five sets. The total time taken for exercise was 15 min. The pressure range during LABFR was set to 160–240 mmHg, because the use of LABFR with higher occlusion pressures (≥ 130 mmHg) has been shown to improve both the aerobic fitness and aerobic performance in young adults [18]. Participants performed LABFR three times per week for a period of eight weeks. They were instructed to take sufficient recovery time after training, and were also asked to avoid consuming caffeine-containing foods and beverages for 24 h before training. The LABFR protocol in this study was designed with reference to previous studies [2,18]. The LABFR protocol used in this study was registered at https://www.protocols.io/ (accessed on 20 July 2022). The control group performed the same protocol as the LABFR group without wearing the occlusion cuff belts for blood flow restriction.

2.3. Body Composition

The body composition (fat mass, body fat, and muscle mass) of the participants was measured using a bioelectrical impedance analysis device (InBody-270, Biospace, Seoul, Republic of Korea) [19]. For accurate measurements, the participants were prohibited from performing excessive physical activity or exercise on the day prior to the day of measurement. They were also advised not to drink caffeinated or alcoholic beverages with a potential effect on moisture imbalance, although they were allowed to drink an adequate quantity of water. After a minimum of 8 h of fasting, the measurements were taken during the morning hours. The thigh circumference was also measured to examine the changes in muscle hypertrophy [20]. For this, while the participant's leg was in a natural extension state, a spot 18 cm from the knee joint was marked with ink and the circumference around the spot was measured using a fiberglass tape (Gulick II, Country Technology Inc., Gays Mills, WI, USA).

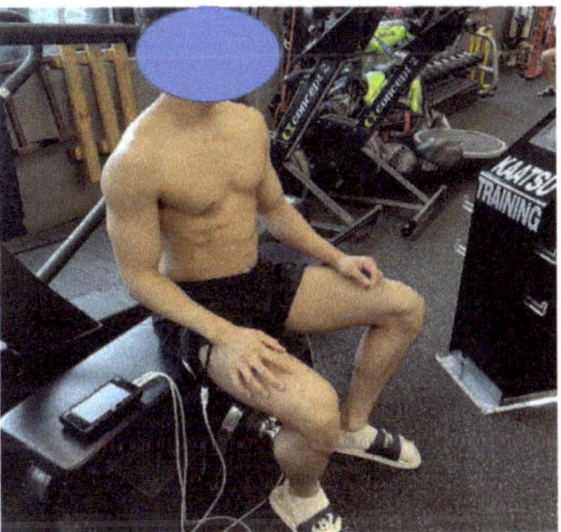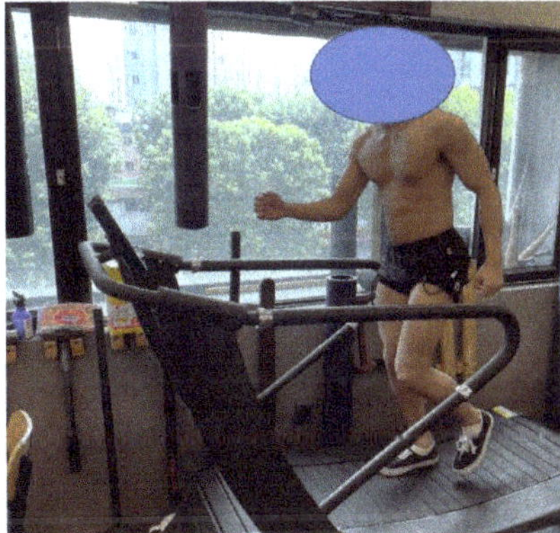

Figure 1. Occlusion cuff belts used in this study for blood flow restriction.

2.4. Physical Fitness

The physical fitness was measured and subdivided into power and cardiorespiratory fitness. Power was measured using the vertical jump. For this, a jump-MD (TKK-5406, TAKEI, Niigata, Japan) was used. While standing on a mat with the feet aligned with each respective shoulder, the participant made as high a vertical jump as possible through the instant flexion and extension of the knees at the investigator's signal. The measurements were taken twice, and the higher of the two values was recorded. The participant was cautioned against jumping through rebounding. For cardiorespiratory fitness, the VO_{2max} was measured. The measurement procedure for VO_2max has already been described in detail in Section 2.2.

2.5. Vascular Responses

Flow-mediated dilation (FMD) is a method of assessing the endothelial functions based on the level of expansion of the brachial artery in response to the nitric oxide produced by vascular endothelial cells. To measure the FMD, an ECG-guided high-resolution B-mode ultrasound system (UNEX-EF-38G, UNEX Corp., Nagoya, Japan) was used to measure the baseline diameter of the brachial artery at rest, and the maximum diameter after applying shear stress to express the change in brachial arterial diameter as a ratio was determined [21]. The respective equation for FMD is as follows: FMD = [(maximal diameter–baseline diameter)/baseline diameter × 100]. The brachial ankle pulse wave velocity (baPWV) of each participant in the supine position was measured using a non-invasive vascular screening device (VP-1000 Plus, Omron, Kyoto, Japan) after at least 5 min of rest. The sampling time for one pulse wave recording was 10 s. For each participant, two consecutive measurements were taken and the mean of two values was used in the analysis. The factors that determine a pulse wave are time difference (ΔT), distance between two measured spots (L), and participant's height (cm). The pulse wave (L/ΔT, cm/s) for an arterial segment was automatically calculated by the device [22]. Ankle brachial index (ABI) was measured using the same device as in the baPWV measurements. The cuff was applied to the limbs of each participant in the supine position, after at least 10 min of rest. The systolic blood pressure (SBP) was measured at the left and right brachial artery and the left and right posterior tibial artery, and by dividing the highest posterior tibial arterial pressure by the highest brachial arterial pressure, the ratio for ABI was obtained [23]. The respective equation for ABI was as follows: ABI = (highest left and right ankle systolic blood pressure)/(highest left and right arm systolic blood pressure).

2.6. Statistical Analysis

All data in this study were presented as the mean and standard deviation. The normality and homogeneity were tested using the Kolmogorov–Smirnov and Levene's test, respectively. To analyze the time $\times$ group interaction effect, a repeated measure ANOVA was applied, and Bonferroni procedures were used for post hoc comparison. The partial eta squared (η_p^2) was used to determine the effect size in the mixed ANOVA. All analyses were performed using the SPSS software (SPSS Statistics 25.0, IBM, Armonk, NY, USA). The level of significance was set to 0.05.

3. Results

Table 2 presents the changes in body composition after performing the eight-week LABFR protocol. All body composition variables were normally distributed, as assessed using the Kolmogorov–Smirnov test ($p > 0.05$), except for pre- and post-body fat and pre- and post-right thigh circumference ($p < 0.05$). Levene's test showed no significant difference in variances for all body composition variables ($p > 0.05$). The results indicated no significant time $\times$ group interaction effect on fat mass (F = 0.265, $p = 0.611$, $\eta_p^2 = 0.010$) and body fat (F = 0.490, $p = 0.490$; $\eta_p^2 = 0.018$). In contrast, significant time $\times$ group interactions effects were found on muscle mass (F = 53.242, $p = 0.001$, $\eta_p^2 = 0.664$) and right thigh circumference (F = 4.544, $p = 0.042$, $\eta_p^2 = 0.144$). Compared to the control group, the LABFR group exhibited an increase in muscle mass and right thigh circumference, whereas no interaction effect was found in left thigh circumference (F = 3.171, $p = 0.086$, $\eta_p^2 = 0.105$).

Table 2. Change in body composition after low-intensity aerobic training combined with BFR.

	Group	Pre	Post	F	p	η_p^2
Fat mass (kg)	BFR ($n = 14$)	12.54 ± 4.27	11.56 ± 3.09	0.265	0.611	0.010
	CON ($n = 15$)	15.33 ± 5.05	14.52 ± 4.77			
Body fat (%)	BFR ($n = 14$)	16.90 ± 6.42	15.82 ± 6.63	0.490	0.490	0.018
	CON ($n = 15$)	20.92 ± 6.88	20.18 ± 6.76			
Muscle mass (kg)	BFR ($n = 14$)	35.59 ± 6.08	36.06 ± 5.91	53.242	0.001 ***	0.664
	CON ($n = 15$)	32.98 ± 6.13	32.46 ± 6.08			
Thigh circumference (right, cm)	BFR ($n = 14$)	56.42 ± 4.23	57.10 ± 4.32	4.544	0.042 *	0.144
	CON ($n = 15$)	55.26 ± 5.43	55.36 ± 5.54			
Thigh circumference (left, cm)	BFR ($n = 14$)	56.07 ± 3.94	57.10 ± 4.27	3.171	0.086	0.105
	CON ($n = 15$)	55.00 ± 5.53	55.46 ± 5.44			

Data are presented as mean ± standard deviation (SD); * $p < 0.05$, *** $p < 0.001$; BFR, blood flow restriction; LABFR, low-intensity aerobic training combined with blood flow restriction group; CON, control group.

Table 3 presents the changes in physical fitness after the eight-week LABFR protocol. All physical fitness variables were normally distributed, as assessed using the Kolmogorov–Smirnov test ($p > 0.05$). Levene's test showed that pre- and post-power were significantly different variances ($p < 0.05$). The results indicated no significant time $\times$ group interaction effect on either power (F = 0.624, $p = 0.437$, $\eta_p^2 = 0.023$) or VO_{2max} (F = 0.258, $p = 0.616$, $\eta_p^2 = 0.009$). Lastly, Table 4 presents the changes in vascular responses after the eight-week LABFR protocol. All vascular responses variables were normally distributed, as assessed using Kolmogorov–Smirnov test ($p > 0.05$). Levene's test showed no significant difference in variances for all vascular response variables ($p > 0.05$). The results showed no significant time $\times$ group interaction effect on FMD (F = 0.042, $p = 0.840$, $\eta_p^2 = 0.002$) or any other factors, as follows: baPWV (F = 0.073, $p = 0.789$, $\eta_p^2 = 0.003$) and ABI (F = 0.188, $p = 0.668$, $\eta_p^2 = 0.007$), SBP (F = 0.039, $p = 0.845$, $\eta_p^2 = 0.001$), and DBP (F = 2.354, $p = 0.137$, $\eta_p^2 = 0.080$).

Table 3. Change in physical fitness after low-intensity aerobic training combined with BFR.

	Group	Pre	Post	F	*p*	η_p^2
Power (cm)	BFR (*n* = 14)	57.57 ± 8.21	62.71 ± 8.38	0.624	0.437	0.023
	CON (*n* = 15)	48.93 ± 14.16	53.06 ± 12.10			
VO$_{2max}$ (mL/kg/min)	BFR (*n* = 14)	48.59 ± 8.56	51.74 ± 9.45	0.258	0.616	0.009
	CON (*n* = 15)	44.01 ± 6.80	46.48 ± 6.58			

Data are presented as mean ± standard deviation (SD); BFR, blood flow restriction; LABFR, low-intensity aerobic training combined with blood flow restriction group; CON, control group; VO$_{2max}$, maximum oxygen uptake.

Table 4. Change in vascular responses after low-intensity aerobic training combined with BFR.

	Group	Pre	Post	F	*p*	η_p^2
FMD (%)	BFR (*n* = 14)	7.27 ± 2.96	7.39 ± 2.02	0.042	0.840	0.002
	CON (*n* = 15)	6.73 ± 1.88	7.01 ± 1.95			
baPWV (cm/s)	BFR (*n* = 14)	1216.07 ± 139.47	1141.89 ± 154.45	0.073	0.789	0.003
	CON (*n* = 15)	1179.93 ± 133.21	1110.76 + 139.25			
ABI	BFR (*n* = 14)	1.11 ± 0.071	1.01 ± 0.115	0.188	0.668	0.007
	CON (*n* = 15)	1.09 ± 0.074	1.00 ± 0.082			
SBP (mmHg)	BFR (*n* = 14)	126.71 ± 7.31	118.35 ± 7.93	0.039	0.845	0.001
	CON (*n* = 15)	125.46 ± 9.50	117.53 ± 6.34			
DBP (mmHg)	BFR (*n* = 14)	78.00 ± 9.79	67.21 ± 5.82	2.354	0.137	0.080
	CON (*n* = 15)	76.13 ± 9.14	69.93 ± 8.22			

Data are presented as mean ± standard deviation (SD); BFR, blood flow restriction; LABFR, low-intensity aerobic training combined with blood flow restriction group; CON, control group; ABI, ankle brachial index; baPWV, brachial ankle pulse wave velocity; FMD, flow-mediated dilation; SBP, systolic blood pressure; DBP, diastolic blood pressure.

4. Discussion

The purpose of this study was to investigate the effect of an eight-week LABFR protocol on the body composition, physical fitness, and vascular responses in recreational runners. The results indicated a higher level of increase in muscle mass and thigh circumference in the LABFR group compared to the control group, which agrees with the results of a number of previous studies. Abe et al. [13] and Kim et al. [14] had reported a positive effect of LABFR on the cross-sectional area and volume of thigh and quadriceps muscles. In Abe et al. [13], when young adults who did not regularly participate in aerobic training performed LABFR for eight weeks, the cross-sectional area and volume of the thigh and quadriceps muscles showed a significant increase compared to the control group. In Kim et al. [14], healthy undergraduates performed LABFR for six weeks, as a result of which the leg muscle mass significantly increased, and the knee flexion muscle strength increased to a similar level as in vigorous-intensity cycling.

A number of studies suggested that the BFR training may promote muscle growth by activating the intracellular signaling pathways for muscle protein synthesis, such as the mTOR pathway, through the creation of a hypoxic environment inside the muscle [24], and by altering the genetic regulation of muscle satellite cells [25]. Another study claimed that the muscle hypertrophy induced through BFR could include increased mechanical tension and metabolic stress and muscle growth as a result of autocrine and/or paracrine actions [26]. However, it should be noted that the BFR-mediated mechanisms related to increased muscle mass or hypertrophy had mainly been based on a resistance training model. Abe et al. [13] and Kim et al. [14], likewise, could not clearly identify the potential mechanism of LABFR, but suggested only a conjectured mechanism based on resistance training models used in previous studies despite the reported positive effect of LABFR. Thus, further studies should investigate the potential mechanisms involved in the post-LABFR changes. An interesting finding in this study was the lack of change in power despite the increased muscle mass and thigh circumference. While certain previous studies have reported the possibility of a positive relation between increasing the muscle mass and improving the muscle force, others reported a very weak or lack of a relationship [27,28]. It is also noteworthy that only the right thigh circumference showed a significant increase after LABFR in this study. The reason for this change is not clear and should be investigated in future research.

Likewise, in this study, we hypothesized that there would be a significant change in physical fitness and vascular responses based on several previous studies suggesting a potential additional training effect from ischemia–reperfusion caused by cuff pressure on oxidative metabolism and angiogenesis [29,30]. However, the eight-week LABFR protocol had not exerted any effect on either physical fitness measure, including VO_{2max} or vascular responses, such as FMD, baPWV, ABI, SBP, and DBP, in contrast to recent reports. The study of Karabulut et al. [31], in which healthy and active young male adults performed six-weeks of LABFR training on a treadmill, reported that the VO_{2max} was higher in the LABFR group than in the control group. In Pinheiro et al. [12], where trained cyclists performed a 9-week course of LABFR training, it was also reported that LABFR contributed to improving the VO_{2peak}. Likewise, in de Oliveira et al. [32], four-week LABFR training could improve the onset of blood lactate accumulation in active males.

However, the participants in Karabulut et al. [31] and de Oliveira et al. [32] were males from the general population, and did not regularly participate in endurance training programs, while the participant in Pinheiro et al. [12] was a trained cyclist with knee osteoarthritis, which deviated from the present investigation. Moreover, Pinheiro et al. [12] was a case report on one trained cyclist. The difference in subject characteristics is likely to have had an influence on vascular responses. The participants in this study were recreational runners without any physical problems, who had been performing regular running for a long period of time before participating in this study, and it is likely to have been difficult to induce a significant change in vascular responses.

This study has a few limitations. First, the subjects of this study included only recreational runners who regularly participated in running. Therefore, the effect of LABFR on individuals regularly participating in other types of exercise remains unknown. In a follow-up study, the effect of LABFR should be investigated in different exercise models. Second, a bioelectrical impedance analysis device and a fiberglass tape were used to measure the changes in muscle mass and thigh circumference. In addition, power was measured through vertical jump. A greater variety of instruments could not be used due to the limitations of laboratory conditions. In the future, the changes in skeletal muscles should be monitored using a magnetic resonance imager or an isokinetic dynamometer to extend the scope of the instruments.

5. Conclusions

To conclude, our results indicated that the eight-week LABFR protocol had a potential positive effect on the body composition, especially muscle mass and thigh circumference, in recreational runners. These findings suggest that LABFR can be a useful training strategy for recreational runners seeking muscle-related changes. However, the lack of significant changes in indicators, such as VO_{2max} or vascular responses that are more closely associated with the runner, raises a question regarding the practical effectiveness of LABFR, while implying that care should be taken in result interpretations. The fact that only a few studies similar to the present investigation have been conducted, as well as the limitations of the present investigation, should also be considered. In the future, a greater number of studies investigating the effect of LABFR in recreational runners using a reliable design and more varied indicators should be conducted.

Author Contributions: Conceptualization, H.J.B. and J.K.; methodology, H.J.B. and J.H.Y.; formal analysis, H.J.B. and W.P.; investigation, H.J.B. and W.P.; data curation, J.H.Y.; writing—original draft preparation, J.K.; writing—review and editing, J.K; supervision, J.K. All authors have read and agreed to the published version of the manuscript.

Funding: This paper was supported by Konkuk University in 2021.

Institutional Review Board Statement: The study was conducted in accordance with the Declaration of Helsinki, and approved by the Institutional Review Board of CHA University (1044308-202009-HR-042-02).

Informed Consent Statement: Informed consent was obtained from all participants involved in the study and written informed consent has been obtained from the participants to publish this paper.

Data Availability Statement: The data are not publicly available due to privacy or ethical reasons.

Conflicts of Interest: The authors declare no conflict of interest.

References

1. Pignanelli, C.; Christiansen, D.; Burr, J.F. Blood flow restriction training and the high-performance athlete: Science to application. *J. Appl. Physiol.* **2021**, *130*, 1163–1170. [CrossRef] [PubMed]
2. Patterson, S.D.; Hughes, L.; Warmington, S.; Burr, J.; Scott, B.R.; Owens, J.; Abe, T.; Nielsen, J.L.; Libardi, C.A.; Laurentino, G.; et al. Blood Flow Restriction Exercise: Considerations of Methodology, Application, and Safety. *Front. Physiol.* **2019**, *10*, 533. [CrossRef] [PubMed]
3. Wortman, R.J.; Brown, S.M.; Savage-Elliott, I.; Finley, Z.J.; Mulcahey, M.K. Blood Flow Restriction Training for Athletes: A Systematic Review. *Am. J. Sports Med.* **2021**, *49*, 1938–1944. [CrossRef]
4. Luebbers, P.E.; Fry, A.C.; Kriley, L.M.; Butler, M.S. The effects of a 7-week practical blood flow restriction program on well-trained collegiate athletes. *J. Strength Cond. Res.* **2014**, *28*, 2270–2280. [CrossRef] [PubMed]
5. Miller, B.C.; Tirko, A.W.; Shipe, J.M.; Sumeriski, O.R.; Moran, K. The Systemic Effects of Blood Flow Restriction Training: A Systematic Review. *Int. J. Sports Phys. Ther.* **2021**, *16*, 978–990. [CrossRef]
6. Gawel, D.; Jarosz, J.; Matykiewicz, P.; Kaszuba, M.; Trybulski, R. Acute impact of blood flow restriction during resistance exercise—Review. *Trends Sport Sci.* **2021**, *28*, 83–92.
7. Alves, A.R.; Marinho, D.A.; Pecego, M.; Ferraz, R.; Marques, M.C.; Neiva, H.P. Strength training combined with high-intensity interval aerobic training in young adults' body composition. *Trends Sport Sci.* **2021**, *28*, 187–193.
8. Lixandrão, M.E.; Roschel, H.; Ugrinowitsch, C.; Miquelini, M.; Alvarez, I.F.; Libardi, C.A. Blood-Flow Restriction Resistance Exercise Promotes Lower Pain and Ratings of Perceived Exertion Compared With Either High- or Low-Intensity Resistance Exercise Performed to Muscular Failure. *J. Sport Rehabil.* **2019**, *28*, 706–710. [CrossRef]
9. Liu, Y.; Jiang, N.; Pang, F.; Chen, T. Resistance Training with Blood Flow Restriction on Vascular Function: A Meta-analysis. *Int. J. Sports Med.* **2021**, *42*, 577–587. [CrossRef]
10. Freitas, E.; Miller, R.M.; Heishman, A.D.; Aniceto, R.R.; Larson, R.; Pereira, H.M.; Bemben, D.; Bemben, M.G. The perceptual responses of individuals with multiple sclerosis to blood flow restriction versus traditional resistance exercise. *Physiol. Behav.* **2021**, *229*, 113219. [CrossRef]
11. Smith, N.; Scott, B.R.; Girard, O.; Peiffer, J.J. Aerobic Training With Blood Flow Restriction for Endurance Athletes: Potential Benefits and Considerations of Implementation. *J. Strength Cond. Res* **2021**. *Online ahead of print*. [CrossRef]
12. Pinheiro, F.A.; Pires, F.O.; Rønnestad, B.R.; Hardt, F.; Conceição, M.S.; Lixandrão, M.E.; Berton, R.; Tricoli, V. The Effect of Low-intensity Aerobic Training Combined with Blood Flow Restriction on Maximal Strength, Muscle Mass, and Cycling Performance in a Cyclist with Knee Displacement. *Int. J. Environ. Res. Public Health* **2022**, *19*, 2993. [CrossRef]
13. Abe, T.; Fujita, S.; Nakajima, T.; Sakamaki, M.; Ozaki, H.; Ogasawara, R.; Sugaya, M.; Kudo, M.; Kurano, M.; Yasuda, T.; et al. Effects of Low-Intensity Cycle Training with Restricted Leg Blood Flow on Thigh Muscle Volume and VO_{2max} in Young Men. *J. Sports Sci. Med.* **2010**, *9*, 452–458.
14. Kim, D.; Singh, H.; Loenneke, J.P.; Thiebaud, R.S.; Fahs, C.A.; Rossow, L.M.; Young, K.; Seo, D.I.; Bemben, D.A.; Bemben, M.G. Comparative Effects of Vigorous-Intensity and Low-Intensity Blood Flow Restricted Cycle Training and Detraining on Muscle Mass, Strength, and Aerobic Capacity. *J. Strength Cond. Res.* **2016**, *30*, 1453–1461. [CrossRef]
15. Abe, T.; Kearns, C.F.; Sato, Y. Muscle size and strength are increased following walk training with restricted venous blood flow from the leg muscle, Kaatsu-walk training. *J. Appl. Physiol.* **2008**, *104*, 1255. [CrossRef]
16. Astorino, T.A.; Allen, R.P.; Roberson, D.W.; Jurancich, M. Effect of high-intensity interval training on cardiovascular function, VO_2max, and muscular force. *J. Strength Cond. Res.* **2012**, *26*, 138–145. [CrossRef]
17. Tanner, A.V.; Nielsen, B.V.; Allgrove, J. Salivary and plasma cortisol and testosterone responses to interval and tempo runs and a bodyweight-only circuit session in endurance trained men. *J. Sports Sci.* **2014**, *32*, 680–689. [CrossRef]
18. Bennett, H.; Slattery, F. Effects of Blood Flow Restriction Training on Aerobic Capacity and Performance: A Systematic Review. *J. Strength Cond. Res.* **2019**, *33*, 572–583. [CrossRef]
19. Yoon, E.J.; Kim, J. Effect of Body Fat Percentage on Muscle Damage Induced by High-Intensity Eccentric Exercise. *Int. J. Environ. Res. Public Health* **2020**, *17*, 3476. [CrossRef]
20. Alves, E.D.; Salermo, G.P.; Panissa, V.; Franchini, E.; Takito, M.Y. Effects of long or short duration stimulus during high-intensity interval training on physical performance, energy intake, and body composition. *J. Exerc. Rehabil.* **2017**, *13*, 393–399. [CrossRef]

21. de Oliveira, G.V.; Mendes Cordeiro, E.; Volino-Souza, M.; Rezende, C.; Conte-Junior, C.A.; Silveira Alvares, T. Flow-Mediated Dilation in Healthy Young Individuals Is Impaired after a Single Resistance Exercise Session. *Int. J. Environ. Res. Public Health* **2020**, *17*, 5194. [CrossRef] [PubMed]

22. Joo, H.J.; Cho, S.A.; Cho, J.Y.; Lee, S.; Park, J.H.; Hwang, S.H.; Hong, S.J.; Yu, C.W.; Lim, D.S. Brachial-Ankle Pulse Wave Velocity is Associated with Composite Carotid and Coronary Atherosclerosis in a Middle-Aged Asymptomatic Population. *J. Atheroscler. Thromb.* **2016**, *23*, 1033–1046. [CrossRef]

23. Mahé, G.; Lanéelle, D.; Le Faucheur, A. Postexercise Ankle-Brachial Index Testing to Diagnose Peripheral Artery Disease. *JAMA* **2021**, *325*, 89. [CrossRef] [PubMed]

24. Hwang, P.S.; Willoughby, D.S. Mechanisms Behind Blood Flow-Restricted Training and its Effect Toward Muscle Growth. *J. Strength Cond. Res.* **2019**, *33*, 167–179. [CrossRef] [PubMed]

25. Torma, F.; Gombos, Z.; Fridvalszki, M.; Langmar, G.; Tarcza, Z.; Merkely, B.; Naito, H.; Ichinoseki-Sekine, N.; Takeda, M.; Murlasits, Z.; et al. Blood flow restriction in human skeletal muscle during rest periods after high-load resistance training down-regulates miR-206 and induces Pax7. *J. Sport Health Sci.* **2021**, *10*, 470–477. [CrossRef]

26. Pearson, S.J.; Hussain, S.R. A review on the mechanisms of blood-flow restriction resistance training-induced muscle hypertrophy. *Sports Med.* **2015**, *45*, 187–200. [CrossRef] [PubMed]

27. Erskine, R.M.; Fletcher, G.; Folland, J.P. The contribution of muscle hypertrophy to strength changes following resistance training. *Eur. J. Appl. Physiol.* **2014**, *114*, 1239–1249. [CrossRef]

28. Reggiani, C.; Schiaffino, S. Muscle hypertrophy and muscle strength: Dependent or independent variables? A provocative review. *Eur. J. Transl. Myol.* **2020**, *30*, 9311. [CrossRef]

29. Clanton, T.L.; Zuo, L.; Klawitter, P. Oxidants and skeletal muscle function: Physiologic and pathophysiologic implications. *Proc. Soc. Exp. Biol. Med.* **1999**, *222*, 253–262. [CrossRef]

30. Zhao, Y.; Li, J.; Lin, A.; Xiao, M.; Xiao, B.; Wan, C. Improving angiogenesis and muscle performance in the ischemic limb model by physiological ischemic training in rabbits. *Am. J. Phys. Med. Rehabil.* **2011**, *90*, 1020–1029. [CrossRef]

31. Karabulut, M.; Esparza, B.; Dowllah, I.M.; Karabulut, U. The impact of low-intensity blood flow restriction endurance training on aerobic capacity, hemodynamics, and arterial stiffness. *J. Sports Med. Phys. Fitness* **2021**, *61*, 877–884. [CrossRef] [PubMed]

32. de Oliveira, M.F.; Caputo, F.; Corvino, R.B.; Denadai, B.S. Short-term low-intensity blood flow restricted interval training improves both aerobic fitness and muscle strength. *Scand. J. Med. Sci. Sports* **2016**, *26*, 1017–1025. [CrossRef] [PubMed]

Article

Evaluation of the Association of *VDR* rs2228570 Polymorphism with Elite Track and Field Athletes' Competitive Performance

Celal Bulgay [1], Işık Bayraktar [2], Hasan Huseyin Kazan [3], Damla Selin Yıldırım [4], Erdal Zorba [5], Onur Akman [6], Mehmet Ali Ergun [7], Mesut Cerit [4], Korkut Ulucan [8], Özgür Eken [9,*], Halil İbrahim Ceylan [10,*], Georgian Badicu [11], Wilhelm Robert Grosz [11] and Raluca Mijaică [11]

[1] Sports Science Faculty, Bingol University, Bingol 12000, Turkey
[2] Faculty of Sports Sciences, Alanya Alaaddin Keykubat University, Alanya 07450, Turkey
[3] Faculty of Medicine, Near East University, 1010-1107 Nicosia, Cyprus
[4] Sports Science Faculty, Lokman Hekim University, Ankara 06510, Turkey
[5] Sports Science Faculty, Gazi University, Ankara 06560, Turkey
[6] Sports Science Faculty, Bayburt University, Bayburt 69000, Turkey
[7] Faculty of Medicine, Gazi University, Ankara 06560, Turkey
[8] Department of Medical Biology and Genetics, Marmara University, Istanbul 34722, Turkey
[9] Department of Physical Education and Sport Teaching, Inonu University, Malatya 44000, Turkey
[10] Physical Education and Sports Teaching Department, Kazim Karabekir Faculty of Education, Ataturk University, Erzurum 25240, Turkey
[11] Department of Physical Education and Special Motricity, Faculty of Physical Education and Mountain Sports, Transilvania University of Braşov, 500068 Braşov, Romania
* Correspondence: ozgureken86@gmail.com (Ö.E.); halil.ibrahimceylan60@gmail.com (H.İ.C.)

Abstract: The present study aimed to examine the vitamin D receptor (*VDR*), rs2228570 polymorphism, and its effect on elite athletes' performance. A total of 60 elite athletes (31 sprint/power and 29 endurance) and 20 control/ physically inactive, aged 18–35, voluntarily participated in the study. The International Association of Athletics Federations (IAAF) score scale was used to determine the performance levels of the athletes' personal best (PB). Whole exome sequencing (WES) was performed by the genomic DNA isolated from the peripheral blood of the participants. Sports type, sex, and competitive performance were chosen as the parameters to compare within and between the groups by linear regression models. The results showed no statistically significant difference between the CC, TC, and TT genotypes within and between the groups ($p > 0.05$). Additionally, our results underlined that there were no statistically significant differences for the association of rs2228570 polymorphism with PBs within the groups of the ($p > 0.05$) athletes. The genetic profile in the selected gene was similar in elite endurance, sprint athletes, and in controls, suggesting that rs2228570 polymorphism does not determine competitive performance in the analyzed athlete cohort

Keywords: athletics; sprint/power athletes; endurance athletes; *VDR*; rs2228570 polymorphism

Citation: Bulgay, C.; Bayraktar, I.; Kazan, H.H.; Yıldırım, D.S.; Zorba, E.; Akman, O.; Ergun, M.A.; Cerit, M.; Ulucan, K.; Eken, Ö.; et al. Evaluation of the Association of *VDR* rs2228570 Polymorphism with Elite Track and Field Athletes' Competitive Performance. *Healthcare* **2023**, *11*, 681. https://doi.org/10.3390/healthcare11050681

Academic Editor: Rahman Shiri

Received: 26 January 2023
Revised: 18 February 2023
Accepted: 21 February 2023
Published: 25 February 2023

1. Introduction

From the inception of humanity to the present date, the differences created by physical performance and related variables have been regarded as very important parameters in terms of the survival and continuity of generations. Particularly within the last century, in addition to factors that affect physical performance such as lifestyle and environmental-epigenetic interactions, changes in which polymorphisms resulting from genetics also play a role have become the focal point of the scientific world. Since new findings and locations of interaction have been revealed along with their causes as a result of the gradual decoding of human genome codes in the past 25 years, new methods and applications that will contribute to the scientific world within the framework of improving individual effort limits have started to gain momentum as well.

Elite athletes easily adapt to the interaction of genomic and epigenomic features; training applications; and the changes caused in organisms by nutrition, lifestyle, and environmental factors. In addition to genetic adaptation, the reshaping of muscles as a result of training and positive responses to training stimuli are the most apparent reflections of the ability to adapt. When used in athletes who exhibit elite levels of performance, candidate genes that reveal athletic ability serve as important clues that can reveal the reflections of the scope and severity of training applications, load, rest and recovery processes, nutrition, and injury risks on the field or podium [1,2].

In contemporary medicine, a new era titled "personalized medicine" is beginning. The necessity to address a variety of metabolic diseases and support individuals with an approach toward sports activities that are specified based on their genetic infrastructure has been increasingly emphasized in many countries and scientific communities [3]. Within this scope, with the discovery of approximately 250 candidate genes that affect athletic performance up to date, the regulation of methods and applications affecting competitive performance in consideration of the genetic features of individuals and particularly elite athletes, and the implementation of application-oriented exercises within this framework, have started to become increasingly common among trainers and athletes as well [4,5]. In this context, one of the candidate genes that has been considered by researchers in recent years in relation to the improvement of physical effort and athlete health is the vitamin D receptor (*VDR*) gene. The vitamin D receptor (VDR) is a member of the nuclear receptor group. The gene coding VDR, a member of the nuclear hormone receptor gene family, was defined for the first time in 1969. It is localized in 12q13.11 and has a size of 100 kb, and more than 100 polymorphisms were defined [6,7]. Exons 2 and 3 of the gene, which consist of 11 exons, code the DNA binding site, whereas exons 4 and 9 code the ligand binding site.

There are four functionally important polymorphic regions within the *VDR* gene. The most significant variations of these are polymorphisms such as rs1544410 (intron 8), rs2228570 (exon 2), rs731236 (intron 8), and rs7975232 (exon 9). These variations can cause various metabolisms [8,9]. The Fok1 polymorphism (rs2228570) is located at the $5'$ end of exon 2 of *VDR* and corresponds to the start codon. As a result of the transition of T to C, ATG–ACG conversion occurs in the first codon and a 424 amino acid variant polypeptide is produced [6,9–12]. In the absence of this variation, a 427 amino acid polypeptide, which is regarded as normal, is synthesized. It is stated that the VDR protein has effects on parathyroid cells, hematopoietic cells, keratinocytes, pancreatic islet cells, reproductive organs, and the immune system. VDR also exhibits a wide tissue distribution involving endothelial, vascular smooth muscles, and cardiomyocytes [13].

Vitamin D, which carries out its functions through VDR, is a fat-soluble molecule derived from the steroidal hormone family [14]. Vitamin D metabolites adjust calcium homeostasis with the parathyroid hormone D [15]. Vitamin D synthesized in the human body from cholesterol plays a role in mechanisms affecting skeletal (mineral density, stress fractures, etc.) and muscular (resistance and power) structure [16] and is stored in fat cells to be transferred into circulation when needed. Vitamin D plays an important role in the maintenance of mineral balance [17] and the regulation of calcium metabolism and protein synthesis within muscle cells by enabling the absorption of phosphorus and calcium in the kidney and intestines [18]. In the functioning of vitamin D metabolism, it is thought that the functional polymorphisms on the receptor have different levels of impact on individuals [8,18]. Vitamin D plays an important role in the healthy development of muscles and bones and neuromuscular functions [19–21]. Vitamin D triggers muscle cells to receive the inorganic phosphate obtained from energy-rich phosphate compounds (ATP and CrP) that play a key role in muscular contraction mechanisms, whereas specific VDRs localized in the cell membrane enable the distribution and regulation of intracellular calcium [22,23].

It is known that an increase in vitamin D levels in an athlete has positive effects on the musculoskeletal system. As a result of increased vitamin D levels, ATP regeneration, protein synthesis [18], and a positive acceleration in the development of physical performance aspects such as explosive muscle strength are observed, as well as a decrease in the risk of stress fractures. On the other hand, in the case of vitamin D deficiency, muscular atrophy linked with different sports branches, a decrease in muscular contraction rate, chronic muscle pain, and prolonged post-training muscle recovery time are observed. In consideration of the role of vitamin D in muscular contraction strength and stimulation processes, it can also have a direct effect on functional conditions [18,24,25]. Recent studies suggest that the 25 (OH) D level in athletes is above 40 ng/mL [26–28]. In another similar study, it was found that high serum 25 (OH) D3 levels were directly related to speed, power, vertical jump, and muscle strength [26,28,29]. In long-distance runners whose vitamin D concentrations were found to be below 32 ng/mL, it was observed that the tumor necrosis factor significantly increased following extensive training loads, and low levels of vitamin D as well as anti-inflammatory and pro-inflammatory cytokines that ensure recovery were found.

Vitamin D deficiency is caused by insufficient time and frequency of sunlight exposure [30]. However, in studies conducted in various climates and regions, cases of vitamin D deficiency were reported despite sufficient time and frequency of sunlight exposure. In certain studies, it was stated that even athletes who trained outdoors had low vitamin D levels [31], which could lead to over-training syndrome induced by overload, malnourishment, and insufficient rest in the later stages of training [32].

VDR polymorphisms can affect the vitamin D demands of the body at varying levels. Essentially, vitamin D can be acquired through dietary sources, but is predominantly synthesized endogenously from the ultraviolet-B radiation of the sun, and it reinforces muscle mass and strength levels by increasing the hormone levels in the body within the framework of enzymatic processes [33]. In a study investigating the effects of vitamin D intake on athletic performance under different seasonal and climatic conditions, the performances of athletes during the summer and winter seasons were compared, and it was stated that the physical efforts exhibited during the winter season were more efficient [34]. This situation explains the fact that vitamin D deficiency does not only occur through being deprived of solar rays, but genetic and epigenetic interactions can play a key role in vitamin D deficiency as well.

The present study aimed to examine the *VDR* rs2228570 polymorphism and its effects on the performance of elite athletes.

2. Methods

2.1. Participants

This study involved 60 elite athletes (sprint/power: 11 females (35.5%) and 20 males (64.5%); endurance: 10 females (34.5%) and 19 males (65.5%)) licensed in different clubs and affiliated to the Turkish Athletics Federation. The number of the controls (non-athletes) were 20 (female 6 (30.0%) and male 14 (70.0%)); they were healthy unrelated citizens of Turkey without any competitive sports experience. All the athletes and controls were of Caucasian ancestry. The individuals with any known and declared diseases were excluded from this study. The athletes were categorized as either sprint/power or endurance athletes, as determined by the distance, duration, and energy requirements of their events. All athletes were ranked in the top 10 in their sport disciplined nationally. Those in the elite group had participated in international competitions such as the Olympic Games, European Championships, Universiade, Mediterranean Games, and Balkan Championship. The sprint/power group included sprint and power athletes whose events demand predominantly anaerobic energy production. The athletes in this group were 100–400 m runners, jumpers, and throwers. The endurance athlete group included athletes competing in long-distance events demanding predominantly aerobic energy production. This group included 3000 m, 5000 m, 10,000 m, and marathon runners. The athletes in this group (n = 31) were 100–400 m 123 runners (n = 9), jumpers (n = 3), and throwers (n = 19). The endurance athlete group (n = 29) included athletes competing in long-distance events demanding predominantly aerobic energy production. This group included 3000 m (n = 12); 5000 m (n = 5); 10,000 m (n = 4); and marathon (n = 8) runners.

2.2. Study Design

This study was carried out in accordance with the Declaration of Helsinki and approval was obtained from the Gazi University Non-Interventional Clinical Research Ethics Committee with the decision dated April 05, 2021 and numbered 09. The informed voluntary consent and demographic information forms were applied for the athletes and control groups before the measurements.

2.3. Personal Best (PB)

The International Association of Athletics Federations (IAAF; World Athletics) score scale was used to determine the performance levels of the athletes depending on their personal best/competitive performance [35]. For instance, the IAAF score scale of a male athlete who runs 100 m in 10.05 sec is 1189, whereas that of a marathon runner who completes the race in 2 h 20 min 11 sec is 997. Thus, the performance scale of the marathon runner is less than that of the 100 m runner. The IAAF scales are useful for the determination of performances of athletes from diverse athletics events and genders.

2.4. Whole Exome Sequencing

Total genomic DNA was isolated from the peripheral venous blood of the participants (4cc) for further genetic screenings using a DNeasy Blood and Tissue Kit (Qiagen, Germany) according to the supplier's instructions. The quality of the isolated DNA was verified using 1% agarose gel electrophoresis and NanoDrop (NanoDrop 1000 Spectrophotometer; Thermo Scientific, USA) according to optical density ratios, and the concentration was determined by NanoDrop.

Whole exome sequencing (WES) was performed after library preparation by Twist Human Comprehensive Exome Panel (Twist Biosciences, USA) according to the instructions of the supplier. Briefly, DNA was fragmented enzymatically, size selection was carried out, and hybridization was applied using Twist Hybridization probes and Dynabeads™ MyOne™ Streptavidin T1 (Invitrogen, USA), and the library was enriched by polymerase chain reaction (PCR). The concentration and size of the libraries were determined, and sequences were performed using Illumina NextSeq500 according to the manufacturer's standard protocol.

Raw data were processed to by the genome analysis toolkit (GATK) [36]. The HaplotypeCaller program was used to obtain binary alignment map (BAM) files and subsequently produce an output variant call format (VCF) file via the GRCh38/hg38 reference genome. Variants were annotated by ANNOVAR [37] and each single nucleotide polymorphism (SNPs) was analyzed manually. All molecular analyzes were performed in Gazi University Medical Genetics Laboratory.

2.5. Statistical Analyses

The SPSS statistical package version 25.0 for Mac was used to perform all statistical analyses. In the evaluation of the data, descriptive statistical methods (number, percentage, and mean) were used. Before performing any analysis on the data, the study determined whether they met the requirements for parametric tests. To that end, the variables were tested for normality, whereas Kolmogorov–Smirnov and Shapiro–Wilk ($p = 0.200$; 0.785, respectively) tests were used for homogeneity of variance. As result of these tests, parametric tests were performed for the variables distributed. Genotype and allele frequencies were calculated for the polymorphism, and the Hardy–Weinberg equilibrium (HWE) was assessed using the chi-squared (χ^2) test. Allele and genotype frequencies and any associations between the polymorphism and the athletic parameters were assessed by SNPStats (https://www.snpstats.net/start.htm, acceded on 11 October 2022) [38] using linear regression models, which were used to assess the score of variation in PB with linear regression multiple inheritance models: co-dominant, dominant, recessive, over-dominant, and additive. To confirm the results obtained using the linear regression models, we also analyzed the data by means of the one-way analysis of covariance (ANCOVA), adjusting for sex and sports experience. Data were significant when $p < 0.05$.

3. Results

The present study aims to decipher any possible association of the rs2228570 polymorphism with the competitive performances of a group of elite athletes (mean age $\pm$ SD: 25.1 ± 4.8; height (cm): 174.97 ± 7.9; body weight (kg) 72.5 ± 22.4; sports experience (year) 9.4 ± 4.8; personal-best (PB) = 1005.63 ± 94.55) in the presence of a control group (mean age $\pm$ SD: 23.5 ± 7.1). Three groups (speed/throw/jump) and long-distance athletes and controls have been chosen to assess this aim.

The genotype and allele frequencies were determined. According to the results, there were not any statistically significant different between the CC, TC, and TT genotypes within and between the groups (Table 1; $p > 0.05$). For allele frequencies, although the number of T alleles (n = 112) was higher than the C allele (n = 48), it was not statistically significant ($p > 0.05$; Table 1).

Table 1. Genotype and allele frequencies of *VDR* rs2228570 polymorphism in elite athletes and controls.

	Genotype			*p*-Value	Allele		*p*-Value
	CC	TC	TT		C	T	
Sprint/Power	1(3.2%)	17(54.8%)	13(42.0%)		19(30.6%)	43(69.4%)	
Endurance	1(3.4%)	13(44.8%)	15(51.7%)	0.717	15(25.9%)	43(74.1%)	0.618
Controls	2(10.0%)	10(50.0%)	8(40.0%)		14(29.6%)	26(70.4%)	

Statistically significant differences ($p < 0.05$); T: thymine, C: cytosine.

Additionally, when we compare the genotype or allele distribution for both sex and sports experience, no significant difference was detected (data not shown). Therefore, to increase the power of the statistical results, we decided to pool the sample.

VDR rs2228570 polymorphism was also evaluated to determine whether it was associated with personal bests (PBs) using different genetic models, codominant, dominant, recessive, and over-dominant. Our results underlined that there was not any significance for the association of rs2228570 polymorphism with PBs within the athlete group ($p > 0.05$; Table 2).

Table 2. Association analysis of the *VDR* rs2228570 polymorphism with competitive performance.

Model	Genotype	n	Mean Score (PB)	Difference (95% CI)	*p*-Value
	CC	28	1011	0.00	
Co-dominant	TC	30	997	−18 (−65 to 28)	0.72
	TT	2	1009	8 (−122 to 138)	
	CC	28	1011	0.00	
Dominant	TC-TT	32	997	−17 (−62 to 28)	0.48
	CC-TC	58	1003	0.00	
Recessive	TT	2	1009	17 (−110 to 145)	0.79
	CC-TT	30	1010	0.00	
Over-dominant	TC	30	997	−19(−64 to 27)	0.42

Statistically significant differences ($p < 0.05$); adjusted by sports experience + sex.

4. Discussion

A predisposition towards sports activities in humans is a large variable factor. Although muscle mass and function were initially listed among the main reasons behind this variability, it has been stated within the framework of the scientific data reported in recent years that elite athletic performance ability originates from genetics. In this context, it is thought that high levels of effort can be achieved through the regulation of variables such as the type of nutrition, as well as the severity and duration of training programs planned in line with the properties of the related candidate genes in addition to the existing genetic infrastructure of the individual. In recent years, genetic tests and applications that reveal branch-specific athletic ability have been among the most frequently emphasized concepts, particularly in the field of the sports industry. In tests carried out within this scope, the presence of candidate genes that reveal capability is regarded as indicating the capability of individuals, and outstanding candidate genes are taken into consideration, particularly in elite athletes. Genetic differences allow certain individuals to reach their goals in a very short period of time, whereas others can achieve high performance following a very long period. Although there are various body types among humans, body type and physiological characteristics depend on the distribution in the genetic structure, which varies based on environmental and epigenetic factors. These differences determine the outcome for elite athletes [39].

Similarly, the type and efficiency of nutrition are also affected by genetic factors. For example, the microbiome is heavily affected by host genetics [40]. Polymorphisms in various genes, including *VDR* rs2228570, can differentiate individual responses to training loads by influencing muscle strength and resistance [41]. VDR can affect the vitamin D needs of an individual. Within this framework, differences between individuals that result from genetics, common environment, and epigenetic factors determine the functionality of primary nutritional sources and supplements within the metabolism. Similarly, vitamin D absorption levels originating from *VDR* gene polymorphisms differ among individuals. Therefore, functional polymorphisms on the receptor play a very significant role in terms of understanding the effects of vitamin D metabolites on athletic performance and a healthy lifestyle [42].

Vitamin D is among the bioactive molecules required for bone mineral density, as well as muscular contraction and regeneration. Additionally, it accelerates regeneration time by reducing the level of inflammation that occurs after exhaustive training loads. Within this framework, it was stated in previous studies that the daily level of vitamin D required for athletes should be at high values such as approximately 32–40 ng/mL (20 ng/mL for sedentary individuals). However, variables originating from individual differences (climate; geography; genetic structure; type, severity, and duration of the training; etc.) can affect levels of vitamin D use. Though sufficient levels of vitamin D intake allow for the maintenance and development of athletic performance, the reduction of said vitamin levels inhibits muscle relaxation required after training and increases muscle pain while also triggering injury risk and stress fractures by causing muscle power loss and reducing bone mineral density [21].

In the findings of a study comparing elite athletes who were found to have insufficient vitamin D levels (n = 61) and the control group (n = 30) (eight weeks/daily vitamin D intake of 5000 IU), it was found that 25(OH)D levels observed in serum improved exercise performance (sprint, vertical jump) (Close et al., 2013), whereas another similar study reported that serum 25(OH) D concentrations measured on dancers (n = 98) were low (73%) due to vitamin D deficiency [43]. Additionally, in another study conducted on lab rats, it was observed that vitamin D reduced injury risk in anaerobic training loads by increasing the volume and number of type II muscle fibers, making positive contributions to maximal sprint and power capabilities [44].

In another study conducted by Allison et al. [45] (n = 506), vitamin D levels and structural coronary parameters were compared within the framework of the effects of vitamin D on the cardiovascular system and it was found that the structural parameters of the athletes below the 25(OH)D levels observed in serum were lower compared to the others [45].

VDR is among the first polymorphisms studied in the determination of the link to athletic performance. In the studies carried out within this framework, it was observed that the Fok1 polymorphism had a greater impact on bone mineral density along with exercise and nutrition in individuals with the CC homozygous genotype [46], and increased the risk of sarcopenia [9,47]. In addition to the impact of the related polymorphism on changes in bone mineral density linked with endurance training, it was reported to be related to environmental factors such as calcium intake [48,49]. Polymorphisms in different genes, including *VDR*, can impact muscle strength and alter responses to training stimuli [33].

Certain previous studies reported that the rs2228570 polymorphism increased the risk of stress fractures in athletes performing heavy exercise and that it is necessary to analyze the said genetic variable to create healthy training programs [50]. In a similar study conducted with young soldiers consisting of males and females (n = 385), 268 single nucleotide polymorphisms (SNP), including the *VDR* gene and 17 gene regions, were analyzed, and it was stated that the *VDR* gene is an important determinant in the onset of stress fractures [51]. In the findings of another similar study conducted on military personnel, it was found that patients with stress fractures were much more likely to possess the rs2228570 polymorphism [52]. In a previous study investigating the impact of VDR gene polymorphisms on the longitudinal changes in hormones related to bone mass, bones, and calcium in adolescent football players (n = 46; 11.8–14.2 years old), it was stated that the *VDR* gene polymorphism affected bone mass in adolescent football players (41.3% on the CC genotype rate, 47.3% on the CT genotype rate, and 10.9% on the TT genotype rate) and that its impact on bone mineralization likely occurred in early stages of puberty during bone maturation [53]. As in chronic obstructive pulmonary disease (COPD) patients, it was observed that individuals with systemic diseases had weaker quadriceps muscles in the CC alleles compared to the CT and TT genotypes [54].

Similarly, the study conducted by Eken et al. [6] stated that CC genotypes and C alleles were more dominant in elite athletes [6]. In contrast with the said study, in the present study, although the frequency of the T allele (n = 48) was higher compared to the C allele among all participants, no statistically significant intragroup and intergroup difference was observed (Table 1; $p > 0.05$). By contrast, another study (n = 206 men and women; 50–81 years old) reported that the Fok1 polymorphism had a greater effect on bone mineral density in lengthy and low-intensity aerobic and strength training applications, although the said interaction was unrelated to the development of aerobic resistance training levels [55]. Supporting the findings of the present study, in the study conducted by Morucci et al. [56] on gymnasts (n = 80), it was stated that there was no relationship between *VDR* genotypes and athletic performance in terms of genotype [56]. Similarly, in another study conducted by Gavin et al. [57] on female athletes (n = 62), the relationship between *VDR* and muscle strength was examined; however, no relation was found [57].

In the vertical jump performance tests carried out by Massidda et al. [58] on professional male football players (n = 90), the percentage of participants with the CC genotype (46.3%) was found to be higher compared to the CT (38.8%) and TT (14.8%) genotypes [58]. Differently from the results of the present study, in another study similar to that of Micheli et al. [59], the rate of young male football players (n = 125) with the CC genotype was found as 52%, whereas this percentage was 34% for the CT genotype and 14% for the TT genotype, stating that these genotype results can serve as important indicators in terms of the predisposition towards football [59].

The fact that the rate of football players with the CC genotype was found to be higher in both of the aforementioned studies compared to the competitive performance findings in the present study can be interpreted as notable differences between sports disciplines in terms of performance level, individual features, and the rate of use of energy sources. Though both studies show similarities with the analyses performed in the present study, this is not adequate to explain the reasons why the frequencies of the C or T alleles were high or low. The present study found no significant intragroup or intergroup difference (speed/throw/jump, long distance, and control) between the CC, TC, and TT genotypes.

This was also the case in genotype or allele distributions within the framework of both genders and sports experiences, and in the light of the findings obtained within this scope, no correlation was found between the *VDR* rs2228570 polymorphism and the personal bests (PBs) in the athlete groups (Table 2; $p > 0.05$). The fact that the number of studies and data on the cellular effects of this polymorphism are insufficient may lead to the conclusion that the T or C alleles provide a predisposition toward the status of elite athletes. However, the impact of either allele on a molecular level is not clearly known, which will cause their contribution or ineffectiveness toward individual performance to remain unclear.

With approximately three billion nucleotides in the human genome, the number of nucleotide combinations that can affect gene activity is essentially infinite. Beyond the aforementioned studies, many studies show that genes play an important role in the determination of athletic performance. Considering the improvements achieved in the past 20 years, it can be foreseen that the genetic elements affecting athletic performance phenotypes and training stimuli will become more comprehensible in the following years with multigene and multifactorial approaches. On the other hand, based on the available data, it is certain that the effects of the candidate genes involved in athletic performance development as the desired phenotype are not very clear, and that further studies are needed.

Although the present study underlined the insignificance between the related polymorphism and athletic parameters, there were some limitations that could affect the results. The single gene and polymorphism approach may alter the results of the associations in particular pathways where more than one protein is involved. Thus, a cumulative evaluation study for all genes and SNPs is under progress by our research group. In our study, biochemical measurements such as the athletes' vitamin D levels were not measured, and it was not checked whether the athletes took vitamin D before the study. In future studies, more comprehensive studies can be conducted by taking these factors into consideration. Nevertheless, such approaches are planned as further studies. Furthermore, the number of the participants was relatively lower, which could influence the results, but this criterion was restricted to the number of elite athletes in the population.

5. Conclusions

The unclarity of the *VDR* Fok1 rs2228570 polymorphism on a molecular level, as well as the limited number of studies on athletic performance in the literature, stand out as limiting factors in the interpretation of the outcomes of the present study. Within this framework, it is thought that the present study will contribute to the developments in the field of sports sciences along with the limited number of other similar studies.

In conclusion, the genetic profile in the selected gene was similar in elite endurance, sprint athletes, and in controls, suggesting that rs2228570 polymorphism does not determine competitive performance in the analyzed athlete cohort. Additionally, it is foreseen that conducting similar studies in the future with athlete groups in different sports branches will allow for the effects of the *VDR* gene on athletic performance to be identified more clearly.

Author Contributions: Conceptualization, C.B. and I.B.; methodology, H.H.K. and D.S.Y.; software, E.Z. and O.A.; validation, M.A.E. and M.C.; formal analysis, K.U. and Ö.E.; investigation, C.B. and H.İ.C.; resources, G.B., R.M. and W.R.G.; data curation, I.B. and H.H.K.; writing—original draft preparation, C.B. and D.S.Y.; writing—review and editing, G.B., R.M. and H.İ.C.; visualization, M.A.E. and M.C.; supervision, G.B. and H.İ.C.; project administration, E.Z. All authors have read and agreed to the published version of the manuscript.

Funding: This research received no external funding.

Institutional Review Board Statement: This study was conducted in accordance with the Declaration of Helsinki, and approved by the Ethics Committee from the Gazi University Non-Interventional Clinical Research Ethics Committee with the decision dated 5 April 2021 and numbered 09.

Informed Consent Statement: Participants gave written informed consent to participate in the present study and to its publication.

Data Availability Statement: The data presented in this study are available on request from the corresponding authors.

Acknowledgments: The authors are grateful to all participants who kindly provided their samples for DNA analysis. This study was supported by Gazi University Rectorate (Scientific Research Projects Coordination's Unit, project number: TCD-2021-7116).

Conflicts of Interest: All authors declare having no conflict of interest.

References

1. Cerit, M. Evaluation of the Soldier's Physical Fitness Test Results (Strength Endurance) in Relation to Ace Genotype. *Int. J. Sport Sci. Health* **2018**, *5*, 123–135.
2. Papadimitriou, I.D.; Lucia, A.; Pitsiladis, Y.P.; Pushkarev, V.P.; Dyatlov, D.A.; Orekhov, E.F.; Artioli, G.G.; Guilherme, J.P.L.F.; Lancha, A.H.; Ginevičienė, V.; et al. ACTN3 R577X and ACE I/D Gene Variants Influence Performance in Elite Sprinters: A Multi-Cohort Study. *BMC Genom.* **2016**, *17*, 285. [CrossRef] [PubMed]
3. Cerit, M. *The Unknown of Human Metabolism, Genetic and Athletic Performance*; State University of Tetowa: Tetowa, North Macedonia, 2021; ISBN 978-608-217-090-9.
4. Montgomery, H.E.; Clarkson, P.; Dollery, C.M.; Prasad, K.; Losi, M.-A.; Hemingway, H.; Statters, D.; Jubb, M.; Girvain, M.; Varnava, A.; et al. Association of Angiotensin-Converting Enzyme Gene I/D Polymorphism with Change in Left Ventricular Mass in Response to Physical Training. *Circulation* **1997**, *96*, 741–747. [CrossRef] [PubMed]
5. Papadimitriou, I.D.; Papadopoulos, C.; Kouvatsi, A.; Triantaphyllidis, C. The ACE I/D Polymorphism in Elite Greek Track and Field Athletes. *J. Sports Med. Phys. Fitness* **2009**, *49*, 459–463. [PubMed]
6. Eken, B.F.; Gezmiş, H.; Sercan, C.; Kapıcı, S.; Chousein, Ö.M.; Kıraç, D.; Akyüz, S.; Ulucan, K. Türk Atletlerde D Vitamini Reseptör Geni Fok1 (Rs2228570) ve Bsm1 (Rs1544410) Polimorfizmlerinin Analizi. *İstanbul Gelişim Üniv. Sağlık Bilim. Derg.* **2018**, 561–572. [CrossRef]
7. Norman, A.W. Vitamin D Receptor: New Assignments for an Already Busy Receptor. *Endocrinology* **2006**, *147*, 5542–5548. [CrossRef] [PubMed]
8. Ulucan, K.; Akyüz, S.; Özbay, G.; Pekiner, F.N.; İlter Güney, A. Evaluation of Vitamin D Receptor (VDR) Gene Polymorphisms (FokI, TaqI and ApaI) in a Family with Dentinogenesis Imperfecta. *Cytol. Genet.* **2013**, *47*, 282–286. [CrossRef]
9. Zmuda, J.M.; Cauley, J.A.; Ferrell, R.E. Molecular Epidemiology of Vitamin D Receptor Gene Variants. *Epidemiol. Rev.* **2000**, *22*, 203–217. [CrossRef]
10. Audí, L.; García-Ramírez, M.; Carrascosa, A. Genetic Determinants of Bone Mass. *Horm. Res. Paediatr.* **1999**, *51*, 105–123. [CrossRef]
11. Baker, A.R.; McDonnell, D.P.; Hughes, M.; Crisp, T.M.; Mangelsdorf, D.J.; Haussler, M.R.; Pike, J.W.; Shine, J.; O'Malley, B.W. Cloning and Expression of Full-Length CDNA Encoding Human Vitamin D Receptor. *Proc. Natl. Acad. Sci. USA* **1988**, *85*, 3294–3298. [CrossRef]
12. DeLuca, H.F.; Schnoes, H.K. Vitamin D: Recent Advances. *Annu. Rev. Biochem.* **1983**, *52*, 411–439. [CrossRef] [PubMed]
13. Wang, T.J.; Pencina, M.J.; Booth, S.L.; Jacques, P.F.; Ingelsson, E.; Lanier, K.; Benjamin, E.J.; D'Agostino, R.B.; Wolf, M.; Vasan, R.S. Vitamin D Deficiency and Risk of Cardiovascular Disease. *Circulation* **2008**, *117*, 503–511. [CrossRef]
14. Laplana, M.; Sánchez-de-la-Torre, M.; Aguiló, A.; Casado, I.; Flores, M.; Sánchez-Pellicer, R.; Fibla, J. Tagging Long-Lived Individuals through Vitamin-D Receptor (VDR) Haplotypes. *Biogerontology* **2010**, *11*, 437–446. [CrossRef] [PubMed]
15. Dusso, A.; González, E.A.; Martin, K.J. Vitamin D in Chronic Kidney Disease. *Best Pract. Res. Clin. Endocrinol. Metab.* **2011**, *25*, 647–655. [CrossRef] [PubMed]
16. Velluz, L.; Amiard, G. Chimie Organique-Equilibre de Reaction Entre Precalciferol et Calciferol. *Compt. Rend Assoc. Anat.* **1949**, *288*, 853–855.
17. Holick, M.F. Vitamin D Deficiency. *N. Engl. J. Med.* **2007**, *357*, 266–281. [CrossRef]
18. Pfeifer, M.; Begerow, B.; Minne, H.W. Vitamin D and Muscle Function. *Osteoporos. Int.* **2002**, *13*, 187–194. [CrossRef] [PubMed]
19. Annweiler, C.; Bridenbaugh, S.; Schott, A.-M.; Berrut, G.; Kressig, R.W.; Beauchet, O. Vitamin D and Muscle Function: New Prospects? *BioFactors* **2009**, *35*, 3–4. [CrossRef]
20. Bordelon, P.; Ghetu, M.V.; Langan, R.C. Recognition and Management of Vitamin D Deficiency. *Am. Fam. Physician* **2009**, *80*, 841–846.
21. Chatterjee, S. Vitamin D, Optimal Health and Athletic Performance: A Review Study. *Int. J. Nutr. Food Sci.* **2014**, *3*, 526. [CrossRef]
22. Houston, D.K.; Cesari, M.; Ferrucci, L.; Cherubini, A.; Maggio, D.; Bartali, B.; Johnson, M.A.; Schwartz, G.G.; Kritchevsky, S.B. Association Between Vitamin D Status and Physical Performance: The InCHIANTI Study. *J. Gerontol. Ser. A Biol. Sci. Med. Sci.* **2007**, *62*, 440–446. [CrossRef] [PubMed]
23. Windelinckx, A.; De Mars, G.; Beunen, G.; Aerssens, J.; Delecluse, C.; Lefevre, J.; Thomis, M.A.I. Polymorphisms in the Vitamin D Receptor Gene Are Associated with Muscle Strength in Men and Women. *Osteoporos. Int.* **2007**, *18*, 1235–1242. [CrossRef] [PubMed]
24. Lappe, J.; Cullen, D.; Haynatzki, G.; Recker, R.; Ahlf, R.; Thompson, K. Calcium and Vitamin D Supplementation Decreases Incidence of Stress Fractures in Female Navy Recruits. *J. Bone Miner. Res.* **2008**, *23*, 741–749. [CrossRef] [PubMed]

25. Young, A.; Edwards, R.H.T.; Jones, D.A.; Brenton, D.P. Quadriceps Muscle Strength and Fibre Size during the Treatment of Osteomalacia. *Mech. Factors Skelet.* **1981**, *12*, 137–145.
26. Larson-Meyer, D.E.; Willis, K.S. Vitamin D and Athletes. *Curr. Sports Med. Rep.* **2010**, *9*, 220–226. [CrossRef] [PubMed]
27. Ogan, D.; Pritchett, K. Vitamin D and the Athlete: Risks, Recommendations, and Benefits. *Nutrients* **2013**, *5*, 1856–1868. [CrossRef] [PubMed]
28. Rejnmark, L. Effects of Vitamin D on Muscle Function and Performance: A Review of Evidence from Randomized Controlled Trials. *Ther. Adv. Chronic Dis.* **2011**, *2*, 25–37. [CrossRef]
29. Ward, K.A.; Das, G.; Berry, J.L.; Roberts, S.A.; Rawer, R.; Adams, J.E.; Mughal, Z. Vitamin D Status and Muscle Function in Post-Menarchal Adolescent Girls. *J. Clin. Endocrinol. Metab.* **2009**, *94*, 559–563. [CrossRef] [PubMed]
30. Grant, W.B.; Soles, C.M. Epidemiologic Evidence for Supporting the Role of Maternal Vitamin D Deficiency as a Risk Factor for the Development of Infantile Autism. *Derm.-Endocrinol.* **2009**, *1*, 223–228. [CrossRef]
31. Willis, K.S.; Broughton, K.S.; Larson-Meyer, D.E. Vitamin D Status and Immune System Biomarkers in Athletes. *J. Am. Diet. Assoc.* **2009**, *109*, A15. [CrossRef]
32. Cannell, J.J.; Hollis, B.W.; Sorenson, M.B.; Taft, T.N.; Anderson, J.J.B. Athletic Performance and Vitamin D. *Med. Sci. Sports Exerc.* **2009**, *41*, 1102–1110. [CrossRef]
33. Willis, K.S.; Peterson, N.J.; Larson-Meyer, D.E. Should We Be Concerned about the Vitamin D Status of Athletes? *Int. J. Sport Nutr. Exerc. Metab.* **2008**, *18*, 204–224. [CrossRef] [PubMed]
34. Dubnov-Raz, G.; Livne, N.; Raz, R.; Rogel, D.; Cohen, A.H.; Constantini, N.W. Vitamin D Concentrations and Physical Performance in Competitive Adolescent Swimmers. *Pediatr. Exerc. Sci.* **2014**, *26*, 64–70. [CrossRef] [PubMed]
35. Spiriev, B.; Spiriev, A. IAAF Scoring Tables of Athletics, Revised Edition. *Int. Assoc. Athl. Fed.* **2014**, *2*, 12–20.
36. Van der Auwera, G.A.; Carneiro, M.O.; Hartl, C.; Poplin, R.; del Angel, G.; Levy-Moonshine, A.; Jordan, T.; Shakir, K.; Roazen, D.; Thibault, J.; et al. From FastQ Data to High-Confidence Variant Calls: The Genome Analysis Toolkit Best Practices Pipeline. *Curr. Protoc. Bioinform.* **2013**, *43*, 11.10.1–11.10.33. [CrossRef] [PubMed]
37. Wang, K.; Li, M.; Hakonarson, H. ANNOVAR: Functional Annotation of Genetic Variants from High-Throughput Sequencing Data. *Nucleic Acids Res.* **2010**, *38*, e164. [CrossRef]
38. Solé, X.; Guinó, E.; Valls, J.; Iniesta, R.; Moreno, V. SNPStats: A Web Tool for the Analysis of Association Studies. *Bioinformatics* **2006**, *22*, 1928–1929. [CrossRef]
39. Yıldırım, D.S.; Erdoğan, M.; Dalip, M.; Bulğay, C.; Cerit, M. Evaluation of the Soldier's Physical Fitness Test Results (Strength Endurance) In Relation to Genotype: Longitudinal Study. *Egypt. J. Med. Hum. Genet.* **2022**, *23*, 114. [CrossRef]
40. Bonder, M.J.; Kurilshikov, A.; Tigchelaar, E.F.; Mujagic, Z.; Imhann, F.; Vila, A.V.; Deelen, P.; Vatanen, T.; Schirmer, M.; Smeekens, S.P.; et al. The Effect of Host Genetics on the Gut Microbiome. *Nat. Genet.* **2016**, *48*, 1407–1412. [CrossRef]
41. Grundberg, E.; Brandstrom, H.; Ribom, E.; Ljunggren, O.; Mallmin, H.; Kindmark, A. Genetic Variation in the Human Vitamin D Receptor Is Associated with Muscle Strength, Fat Mass and Body Weight in Swedish Women. *Eur. J. Endocrinol.* **2004**, *150*, 323–328. [CrossRef]
42. Sercan, C.; Yavuzsoy, E.; Yuksel, I.; Can, R.; Oktay, S.; Kirac, D.; Ulucan, K. Importance of Vitamin D and Its Receptor in Sportsman Health and Athletic Performance. *J. Marmara Univ. Inst. Health Sci.* **2015**, *5*, 259. [CrossRef]
43. Constantini, N.W.; Arieli, R.; Chodick, G.; Dubnov-Raz, G. High Prevalence of Vitamin D Insufficiency in Athletes and Dancers. *Clin. J. Sport Med.* **2010**, *20*, 368–371. [CrossRef] [PubMed]
44. Kraemer, W.J.; Fleck, S.J.; Evans, W.J. Strength and Power Training: Physiological Mechanisms of Adaptation. *Exerc. Sport Sci. Rev.* **1996**, *24*, 363–397. [CrossRef]
45. Allison, R.J.; Close, G.L.; Farooq, A.; Riding, N.R.; Salah, O.; Hamilton, B.; Wilson, M.G. Severely Vitamin D-Deficient Athletes Present Smaller Hearts than Sufficient Athletes. *Eur. J. Prev. Cardiol.* **2015**, *22*, 535–542. [CrossRef] [PubMed]
46. Nakamura, O.; Ishii, T.; Ando, Y.; Amagai, H.; Oto, M.; Imafuji, T.; Tokuyama, K. Potential Role of Vitamin D Receptor Gene Polymorphism in Determining Bone Phenotype in Young Male Athletes. *J. Appl. Physiol.* **2002**, *93*, 1973–1979. [CrossRef] [PubMed]
47. Gross, C.; Eccleshall, T.R.; Malloy, P.J.; Villa, M.L.; Marcus, R.; Feldman, D. The Presence of a Polymorphism at the Translation Initiation Site of the Vitamin D Receptor Gene Is Associated with Low Bone Mineral Density in Postmenopausal Mexican-American Women. *J. Bone Miner. Res.* **1996**, *11*, 1850–1855. [CrossRef]
48. Gennari, L.; Becherini, L.; Falchetti, A.; Masi, L.; Massart, F.; Brandi, M.L. Genetics of Osteoporosis: Role of Steroid Hormone Receptor Gene Polymorphisms. *J. Steroid Biochem. Mol. Biol.* **2002**, *81*, 1–24. [CrossRef]
49. Hamilton, B. Vitamin D and Athletic Performance: The Potential Role of Muscle. *Asian J. Sports Med.* **2011**, *2*, 211. [CrossRef]
50. McClung, J.P.; Karl, J.P. Vitamin D and Stress Fracture: The Contribution of Vitamin D Receptor Gene Polymorphisms. *Nutr. Rev.* **2010**, *68*, 365–369. [CrossRef]
51. Yanovich, R.; Friedman, E.; Milgrom, R.; Oberman, B.; Freedman, L.; Moran, D.S. Candidate Gene Analysis in Israeli Soldiers with Stress Fractures. *J. Sports Sci. Med.* **2012**, *11*, 147–155.
52. Chatzipapas, C.; Boikos, S.; Drosos, G.I.; Kazakos, K.; Tripsianis, G.; Serbis, A.; Stergiopoulos, S.; Tilkeridis, C.; Verettas, D.-A.; Stratakis, C.A. Polymorphisms of the Vitamin D Receptor Gene and Stress Fractures. *Horm. Metab. Res.* **2009**, *41*, 635–640. [CrossRef] [PubMed]

53. Diogenes, M.E.L.; Bezerra, F.F.; Cabello, G.M.K.; Cabello, P.H.; Mendonça, L.M.C.; Oliveira Júnior, A.V.; Donangelo, C.M. Vitamin D Receptor Gene FokI Polymorphisms Influence Bone Mass in Adolescent Football (Soccer) Players. *Eur. J. Appl. Physiol.* **2010**, *108*, 31–38. [CrossRef] [PubMed]

54. Hopkinson, N.S.; Li, K.W.; Kehoe, A.; Humphries, S.E.; Roughton, M.; Moxham, J.; Montgomery, H.; Polkey, M.I. Vitamin D Receptor Genotypes Influence Quadriceps Strength in Chronic Obstructive Pulmonary Disease. *Am. J. Clin. Nutr.* **2008**, *87*, 385–390. [CrossRef] [PubMed]

55. Rabon-Stith, K.M.; Hagberg, J.M.; Phares, D.A.; Kostek, M.C.; Delmonico, M.J.; Roth, S.M.; Ferrell, R.E.; Conway, J.M.; Ryan, A.S.; Hurley, B.F. Vitamin D Receptor FokI Genotype Influences Bone Mineral Density Response to Strength Training, but Not Aerobic Training. *Exp. Physiol.* **2005**, *90*, 653–661. [CrossRef] [PubMed]

56. Morucci, G.; Punzi, T.; Innocenti, G.; Gulisano, M.; Ceroti, M.; Pacini, S. New Frontiers in Sport Training. *J. Strength Cond. Res.* **2014**, *28*, 459–466. [CrossRef] [PubMed]

57. Gavin, J.P.; Williams, A.G. No Association of α-Actinin-3 (Actn3) and Vitamin d Receptor (Vdr) Genotypes with Skeletal Muscle Phenotypes in Young Women. *Sport Sci. Pract. Asp.* **2010**, *7*, 5–11.

58. Massidda, M.; Scorcu, M.; Calò, C.M. New Genetic Model for Predicting Phenotype Traits in Sports. *Int. J. Sports Physiol. Perform.* **2014**, *9*, 554–560. [CrossRef]

59. Micheli, M.L.; Gulisano, M.; Morucci, G.; Punzi, T.; Ruggiero, M.; Ceroti, M.; Marella, M.; Castellini, E.; Pacini, S. Angiotensin-Converting Enzyme/Vitamin D Receptor Gene Polymorphisms and Bioelectrical Impedance Analysis in Predicting Athletic Performances of Italian Young Soccer Players. *J. Strength Cond. Res.* **2011**, *25*, 2084–2091. [CrossRef] [PubMed]

healthcare

Article

Impact of HIIT Sessions with and without Cognitive Load on Cortical Arousal, Accuracy and Perceived Exertion in Amateur Tennis Players

Vicente Javier Clemente-Suárez [1,2], Santos Villafaina [3,4,*], Tomás García-Calvo [3] and Juan Pedro Fuentes-García [3]

1. Faculty of Sports Sciences, Universidad Europea de Madrid, 28670 Madrid, Spain; vctxente@yahoo.es
2. Grupo de Investigación en Cultura, Educación y Sociedad, Universidad de la Costa, Barranquilla 080002, Colombia
3. Faculty of Sport Sciences, University of Extremadura, Avenida de la Universidad s/n, 10003 Cáceres, Spain; tgarciac@unex.es (T.G.-C.); jpfuent@unex.es (J.P.F.-G.)
4. Departamento de Desporto e Saúde, Escola de Saúde e Desenvolvimento Humano, Universidade de Évora, 7004-516 Évora, Portugal
* Correspondence: svillafaina@unex.es

Abstract: The aim of the present study was to investigate the effects of high-intensity interval training (HIIT) exercises, with and without cognitive load, on the accuracy, critical flicker fusion threshold (CFFT), and rating of perceived exertion (RPE) on recreational tennis players. A total of 32 players of tennis at recreational level (25 men and 7 women) were enrolled in this cross-sectional the study. Participants had to perform, randomly, two HIIT sessions. In one of them, cognitive load was induced by conducting an incongruent Stroop during rests. After training accuracy of tennis serve, CFFT, and RPE were measured. Results showed that accuracy after baseline and HIIT without cognitive load were significantly higher than after HIIT with cognitive load. RPE significantly increased (p-value < 0.001) after HIIT sessions in both, with and without cognitive load. However, significant differences were not observed between the two sessions in the RPE (p-value = 0.405). Furthermore, differences were not obtained in the CFFT neither within nor between sessions (p-value > 0.05). Therefore, HIIT with and without cognitive load increased the RPE in recreational tennis players. Furthermore, HIIT sessions with cognitive load significant altered tennis serve accuracy.

Keywords: high-intensity interval training; sport; mental load; Stroop

Citation: Clemente-Suárez, V.J.; Villafaina, S.; García-Calvo, T.; Fuentes-García, J.P. Impact of HIIT Sessions with and without Cognitive Load on Cortical Arousal, Accuracy and Perceived Exertion in Amateur Tennis Players. *Healthcare* **2022**, *10*, 767. https://doi.org/10.3390/healthcare10050767

Academic Editors: João Paulo Brito and Rafael Oliveira

Received: 30 March 2022
Accepted: 19 April 2022
Published: 21 April 2022

Publisher's Note: MDPI stays neutral with regard to jurisdictional claims in published maps and institutional affiliations.

1. Introduction

Tennis is an intermittent sport which combines intermittent anaerobic exercise bouts of varying intensities and a multitude of rest periods where players have to perform the technical action with power and accuracy [1]. A previous study indicated that the match result is highly influenced by the "breaks", or broken services, as well as the percentages of successes to errors, double faults, or errors in returns due to services with good speed and accuracy [2]. Thus, previous studies have analyzed speed [3,4] or accuracy [5] during a tennis serve. Furthermore, participation in recreational tennis may provide benefits in the physical [6,7], psychological, and social spheres [7].

Intermittent exercise has shown to cause a decline in the performance of athletes [8]. The complex phenomenon of fatigue in racquet sports can involve impairments in neural and contractile processes [9]. According to previous studies, this could manifest as mistimed shots, altered on-court movements, and incorrect cognitive choices [10–13]. In the same line, previous research in the field of tennis showed that fatigue has been shown to reduce the accuracy of returns by 81% [14], groundstrokes by 69% [10], and service by 30% [10]. Therefore, a major goal of tennis training should be to avoid the onset of fatigue during competition and training [1].

The critical flicker fusion threshold (CFFT) has been used to study fatigue and cognitive function in different sport events [15]. This technique is focused on the relationship of arousal level with central nervous system (CNS) [16,17]. When CFFT increases, it suggests an increase in cortical arousal and sensory sensitivity. Nevertheless, when this value decreases, it could mean that the efficiency of the system to process information [18] is reduced and, therefore, could be considered as a symptom of CNS fatigue [19,20].

A previous study [21] investigated the impact of high intensity interval training (HIIT) and intermittent interval training (IIT) on the forehand and backhand shots. Results showed that during the HIIT protocol, the number of errors was significantly higher (a total of 76.20 vs. 51.93 during HIIT and IIT, respectively). Furthermore, tennis performance depends on the interaction between technical, tactical, physiologic, and psychologic skills that often have to be sustained in hostile environmental conditions [9]. Thus, some stressors, such as cognitive load, might be included into training situations in order to simulate real conditions.

However, the impact of conducting HIIT with cognitive load on tennis serve accuracy has not been previously study. This is relevant since, as noted above, tennis players are often under hostile environmental conditions [9], which may cause cognitive load or anxiety. Therefore, the aim of the present study was to investigate the effects of HIIT exercises, with and without cognitive load, on tennis serve accuracy, CFFT, and rating of perceived exertion (RPE) on recreational tennis players. In order to induce cognitive load, an incongruent Stroop test was included during the HIIT rests. This test has been previously used to add cognitive load during physical activities in healthy [22,23] and special populations [24]. The primary hypothesis was that the inclusion of cognitive load in the HIIT session would produce a decrease in cortical arousal and accuracy as well as an increase in the RPE of players. Results and protocols of the present study could be useful for physical trainers and coaches to simulate real conditions (where players have to manage some cognitive stimulus) as well as to design motivating training. Furthermore, results could provide an idea of what happens to serve performance when cognitive load increases.

2. Materials and Methods

A total of 32 recreational tennis players (25 men and 7 women) were enrolled in this cross-sectional the study. Participants had a mean age of 21.40 (1.52) and an average experience in tennis practice of 0.84 years (0.80) with 3.26 (0.78) hours of weekly tennis training. The participants weighed an average of 72.18 (11.95) kg, with a mean height of 1.75 (0.8) m and a body mass index mean of 23.48 (2.55). Among the participants, 30 were right-handed and 2 were left-handed. All participants were enrolled in the Faculty of Sport Sciences, Cáceres (Spain).

Procedures were approved by the university ethic committee (approval number: CIPI/18/093), and participants gave written informed consent prior to participation in the study.

2.1. Procedures and Materials

Participants conducted a standardized warm-up composed of 2 min of joint mobility, 5 min of light aerobic running (50−60% of their maximum heart rate calculated with the Tanaka's Formula "208–0.7 × age" [25] and controlled using the V800, Polar Electro, Kempele, Finland), two series of 20 m of progressive running intensity [26], and five services. Tennis players had to perform seven first services in three different situations: (1) at baseline, (2) after a HIIT training session without cognitive load, and (3) after a HIIT training session with cognitive load. Participants had to serve in the area highlighted in Figure 1 at maximum power. This is an adaptation of a procedure followed by a previous investigation [2].

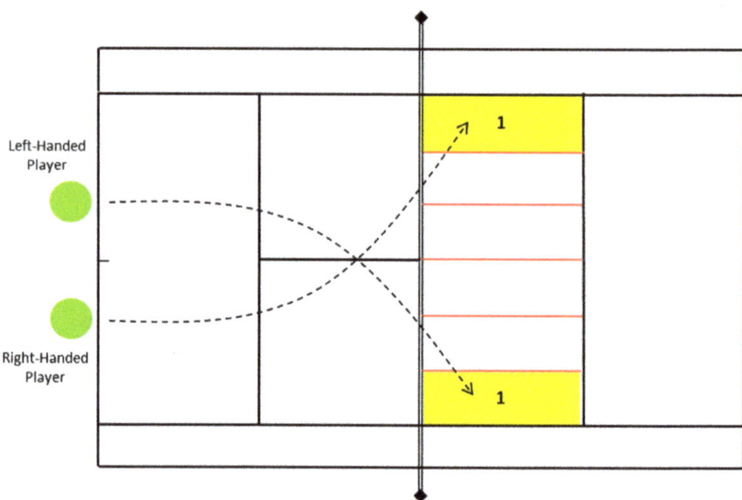

Figure 1. Graphical representation of the area where participants were required to perform the services.

We assessed if participants were able to serve in this area in either of the two attempts they had. Players were informed about the protocol and service area before starting the procedures.

Two researchers were needed for the data collection, one to provide the ball for the services and another to record if the service impacted in the area. These researchers did not participate in data analyses.

Participants were randomly divided into two groups. Whereas one group started the HIIT session with cognitive load, the other group started the HIIT session without cognitive load. Each of the HIIT sessions (with and without cognitive load) were performed with 48 h of rest between them.

HIIT training sessions consisted of the following:

(1) With cognitive load: Participants had to perform 12 repetitions of 30 s of push-ups, squats, and lateral displacements. After these exercises, participants had to conduct the incongruent Stroop in a validated mobile application (UMH-MEMTRAIN, Elche, Spain) for 30 s. The incongruous condition of the Stroop test consisted of selecting the name of a color, where the color word is printed in an incongruous color ink (i.e., the green word is printed in blue ink). Thus, in this incongruous condition, participants are asked to name the ink color instead of reading the word.

(2) Without cognitive load: 12 repetitions of 30 s of push-ups, squats, and lateral displacements. After these exercises participants rested for 30 s.

2.2. Outcomes

The main outcomes of the present study were as follows: (1) The tennis serve accuracy counting with yes/no if participants were able to conduct a tennis serve in the selected area (see Figure 1). (2) The cortical arousal (CFTT) using a Lafayette Instrument Flicker Fusion Control Unit (Model 12021) using the average of 5 incremental test (20 to 100 Hz) as performed in a previous research [27]; and (3) the RPE, on a 6–20 scale [28].

2.3. Statistical Analysis

The Statistical Package for the Social Sciences (SPSS) version 25.0 (SPSS Inc., Chicago, IL USA) was used to analyze the data. The Shapiro-Wilk test was conducted to examine the normality of the data. Taking into account the results, non-parametric statistic tests were conducted. The Friedman test was performed to explore the impact of HIIT sessions in the tennis serve accuracy. Moreover, Wilcoxon signed rank tests were performed to explore differences between pre-post assessments in the different sessions as well as to analyze pairwise comparisons in accuracy. In addition, the difference between post and pre values of CFFT and RPE was calculated. This allowed exploration of the impact, through Wilcoxon signed rank tests, of HIIT with cognitive load and HIIT without cognitive load sessions in the CFFT and RPE. The [r] effect size was calculated for Wilcoxon signed rank tests and classified as follows: ≥ 0.5 is a large effect, 0.5 to 0.1 is a medium effect, and ≤ 0.1 is considered as a small effect [29,30]. For the Friedman test, the partial η^2 was calculated and classified as follows: $\eta^2 = 0.01$ indicates a small effect; $\eta^2 = 0.06$ indicates a medium effect; and $\eta^2 = 0.14$ indicates a large effect [31]. The significance level was set at 0.05.

The G*Power software (version 3.1.9.7.) [32] was used to calculate the estimated power. Due to the study design (one group with three measurements), repeated measures ANOVA, within factor statistical test was selected in the G*Power software. This statistical test was chosen because the Friedman test (non-parametric equivalence) was not available in the G*Power software. The means of accuracy values (considered as the main outcome of the study) for baseline, HIIT without cognitive load, and HIIT without cognitive load (see Table 1) and the partial eta squared (effect size) were employed to calculate the statistical power in the G*Power software. Regarding partial eta squared, G*Power automatically transformed it into Cohen's f effect size. A power equal to 1 (100%) was achieved. The parameters included in the G*Power software were: a total sample size of 32, a Cohen's f(U) = 2.14 calculated from a 0.814 obtained in the partial eta squared effect size (with effect size specification as in SPSS), an alpha p-value = 0.05, one group, three measurements, and a non-sphericity correction of 1.

Table 1. Effects of HITT sessions, with and without cognitive load, in tennis serve accuracy.

Variable	Baseline	HITT without Cognitive Load Mean (SD)	HITT with Cognitive Load Mean (SD)	Z	p-Value	Partial η^2	Significant Pairwise Comparisons
Accuracy (%)	19.64 (16.53)	17.86 (13.58)	11.16 (10.73)	7.828	0.020	0.812	A > C B > C

A: Baseline; B: HITT without cognitive load; C: HITT with cognitive load.

3. Results

Table 1 shows the impact of HIIT training with and without cognitive load in the accuracy of the tennis serve. Friedman Test shows significant differences between the three conditions in the tennis serve accuracy. Pairwise comparisons show that accuracy was significantly higher at baseline than after HIIT with cognitive load (p-value = 0.034). Moreover, accuracy after HIIT without cognitive load was significantly higher than after a HIIT with cognitive load (p-value = 0.046).

Table 2 shows the impact of HIIT training (with and without cognitive load) on the CFFT and the RPE. Results showed that RPE significantly increased (p-value < 0.001) after HIIT sessions for both with and without cognitive load. However, significant differences were not observed between the two sessions in the RPE (p-value = 0.405). Furthermore, differences were not obtained in CFFT neither within nor between sessions (p-value > 0.05).

Table 2. Impact of HITT sessions, with and without cognitive load, in the CFFT and RPE.

Variables	HITT without Cognitive Load Mean (SD)			HITT with Cognitive Load Mean (SD)			Between Training Comparison		
	Pre	Post	*p*-Value	Pre	Post	*p*-Value	Z	*p*	Effect Size
CFFT (Hz)	34.98 (2.87)	34.26 (3.25)	0.278	33.88 (3.30)	34.15 (3.26)	0.135	−0.701	0.483	0.124
RPE	8.38 (2.22)	14.53 (1.90)	<0.001	9.06 (2.34)	14.44 (2.31)	<0.001	−0.833	0.405	0.147

CFFT: cortical arousal; RPE: Rating of perceived exertion.

4. Discussion

This study aimed to investigate the impact of two HIIT sessions, with and without cognitive load, on the tennis serve accuracy, CFFT and RPE on recreational tennis players. Cognitive load was induced by using an incongruent Stroop test during the rests. The primary hypothesis was that the inclusion of cognitive load in the HIIT session would decrease the cortical arousal and the tennis serve accuracy as well as increase the RPE of recreational tennis players. Main findings indicated that accuracy was significantly impacted by HIIT with cognitive load protocol, and that HIIT training (with and without cognitive load) significantly increased CFFT and the RPE.

Results showed that tennis serve accuracy was significantly affected by HIIT sessions, specifically, after a HIIT session with cognitive load. In this regard, it was observed that accuracy decreased from 19.64% in the baseline condition to 17.86% and 11.16% in the HIIT without and with cognitive load, respectively. Furthermore, according to our results, RPE significantly increased after HIIT protocols. Previous studies have reported that fatigue could reduce tennis hitting accuracy of returns by 81% [14], groundstrokes by 69% [10], and service by 30% [10]. Also, in our study it can be observed how accuracy significantly differs between HIIT conditions, showing significantly lower values after HIIT with cognitive load when compared to HIIT without cognitive load. This is consistent with previous studies, which indicated that cognitive load significantly decreased performance [33,34]. This is relevant since tennis performance depends on multiple variable, such as technical, tactical, physiologic, and psychologic skills [9]. In addition, concentrating attention is one of the most important psychological abilities for success in competitive tennis [35]. In relation with that, Pačesová et al. [36] reported that tennis players had better performance on the Stroop test, including in the incongruous condition, where information processing speed, selective attention in the visual system, and inhibitory control are required [37].

Regarding CFFT, results did not show significant changes in cortical arousal after HIIT sessions, with or without cognitive load. Previous studies in military population showed a decrease in cortical arousal [38,39] which has been considered a symptom of fatigue in CNS, reflected by the increase in CFFT values [40]. Furthermore, similar results have been observed after other high intensity activities such as simulated combat or tactical parachute jumps, or even activities with high cognitive requirement [40,41] such as chess [42,43]. A previous investigation hypothesized this could be due to the increase in sympathetic nervous system activation produced in the HIIT that can induce a greater number of cortex efferences to muscles [44].

The RPE showed that both HIIT protocols increased this perception. However, significant changes between the protocols were not observed. The results are similar to those observed in HIIT protocols in military population [38]. Furthermore, previous studies showed that motivation could counteract fatigue-induced performance decrements [45]. However, another study reported that verbal feedback (every 5 s during the 30 s of work) increased intensity, performance, and physical enjoyment during on-court drills [46]. Thus, the inclusion of cognitive elements into HIIT protocols, apart from mimicking real conditions (where players have to manage some cognitive stimulus), can be used as a way to increase motivation during training. In this regard, a previous study in elite youth padel players showed that high levels of motivation could increase players' mental effort and fatigue during padel training matches [47]. Moreover, HIIT protocols have been used in tennis players in order to improve endurance. Previous research showed that after six weeks of training, the HIIT induced greater improvements in tennis-specific endurance (HIIT 28.9% vs. repeated-sprint ability 14.5%) [48]. In the same line, Kilit and Arslan [49] showed that tennis-specific on-court drills training was more effective in improving agility and technical ability with greater physical enjoyment, whereas HIIT may be more appropriate for speed-based conditioning in younger tennis players. Therefore, cognitive load using incongruous Stroop test could complement on-court training.

This study has some limitations that should be acknowledged. First, the relatively small sample size might cause only the largest differences to reach a level of significance. Second, the sample was comprised of recreational tennis players. This fact means that result extrapolation with elite tennis players or other populations should be made with caution. Third, men and women recreational tennis players were included in the study. However, comparison regarding gender has not been conducted, since only seven females were included. Future studies could explore the differences between genders in the impact of HIIT with cognitive load on accuracy, RPE, and cortisol arousal. Fourth, the accuracy was registered as a binary variable (yes–accurate/no–not accurate). This was selected due to the nature of tennis where if the ball does not land in the area, it is not considered as valid. However, this binary nature does not allow calculation of the variability of the service. Therefore, future studies should explore how cognitive load could influence the variability in precision as previous studies have done with dart-throwing [50]. Despite these limitations, this article has some strengths. For instance, this is the first study exploring the impact of a HIIT session with cognitive load on the accuracy, RPE, and cortical arousal of tennis players. The results will help researchers, coaches, and physical trainers to design sessions that simulate real conditions (in order to improve accuracy in hostile conditions), as well as motivate players. Therefore, future studies and interventions protocols should include activities which combine physical and cognitive activities.

5. Conclusions

HIIT with and without cognitive load increased RPE in recreational tennis players. Furthermore, HIIT sessions with cognitive load significantly altered tennis serve accuracy. This is the first study that has examined CFFT after HIIT session with cognitive load in recreational tennis players. The results will help researchers, coaches, and physical trainers to design sessions that simulate real conditions, as well as motivate players. Therefore, future studies and intervention protocols should include activities which combine physical and cognitive activities. Future studies should confirm these results with elite tennis players.

Author Contributions: Conceptualization, V.J.C.-S., T.G.-C. and J.P.F.-G.; formal analysis, V.J.C.-S. and S.V.; investigation, V.J.C.-S., S.V., T.G.-C. and J.P.F.-G.; methodology, V.J.C.-S.; supervision, T.G.-C.; validation, J.P.F.-G.; writing—original draft, V.J.C.-S. and S.V.; writing—review and editing, T.G.-C. and J.P.F.-G. All authors have read and agreed to the published version of the manuscript.

Funding: The author S.V. was supported by a grant from the Universities Ministry of Spain and the European Union (NextGenerationUE) "Ayuda del Programa de Recualificación del Sistema Universitario Español, Modalidad de ayudas Margarita Salas para la formación de jóvenes doctores" (MS-03). This study has been made possible thanks to the contribution of the International Tennis Federation as well as the Department of Economy and Infrastructure of the Junta de Extremadura through the European Regional Development Fund: A Way to Make Europe (GR21094).

Institutional Review Board Statement: The study was conducted in accordance with the Declaration of Helsinki, and approved by the Ethics Committee of the University of Extremadura (protocol code CIPI/18/093).

Informed Consent Statement: Informed consent was obtained from all subjects involved in the study.

Data Availability Statement: Data are available upon reasonable request to corresponding author.

Conflicts of Interest: The authors declare no conflict of interest.

References

1. Kovacs, M.S. Tennis physiology. *Sports Med.* **2007**, *37*, 189–198. [CrossRef] [PubMed]
2. Menayo, R.; Fuentes-García, J.P.; Moreno, F.J.; Clemente, R.; García-Calvo, T. Relación entre la velocidad de la pelota y la precisión en el servicio plano en tenis en jugadores de perfeccionamiento. *Eur. J. Hum. Mov.* **2008**, *21*, 17–30.
3. Fett, J.; Ulbricht, A.; Ferrauti, A. Impact of physical performance and anthropometric characteristics on serve velocity in elite junior tennis players. *J. Strength Cond. Res.* **2020**, *34*, 192–202. [CrossRef]
4. Wang, L.-H.; Lo, K.-C.; Su, F.-C. Skill level and forearm muscle fatigue effects on ball speed in tennis serve. *Sports Biomech.* **2019**, *20*, 419–430. [CrossRef]
5. Menayo, R.; Sabido, R.; Fuentes, J.P.; Moreno, F.J.; Garcia, J.A. Simultaneous treatment effects in learning four tennis shots in contextual interference conditions. *Percept. Mot. Ski.* **2010**, *110*, 661–673. [CrossRef] [PubMed]
6. Chao, H.-H.; Liao, Y.-H.; Chou, C.-C. Influences of Recreational Tennis-Playing Exercise Time on Cardiometabolic Health Parameters in Healthy Elderly: The ExAMIN AGE Study. *Int. J. Environ. Res. Public Health* **2021**, *18*, 1255. [CrossRef] [PubMed]
7. Fernandez-Fernandez, J.; Sanz-Rivas, D.; Sanchez-Muñoz, C.; Pluim, B.M.; Tiemessen, I.; Mendez-Villanueva, A. A comparison of the activity profile and physiological demands between advanced and recreational veteran tennis players. *J. Strength Cond. Res.* **2009**, *23*, 604–610. [CrossRef]
8. Morin, J.-B.; Dupuy, J.; Samozino, P. Performance and fatigue during repeated sprints: What is the appropriate sprint dose? *J. Strength Cond. Res.* **2011**, *25*, 1918–1924. [CrossRef]
9. Girard, O.; Millet, G.P. Neuromuscular fatigue in racquet sports. *Phys. Med. Rehabil. Clin. N. Am.* **2009**, *20*, 161–173. [CrossRef]
10. Davey, P.R.; Thorpe, R.D.; Williams, C. Fatigue decreases skilled tennis performance. *J. Sports Sci.* **2002**, *20*, 311–318. [CrossRef]
11. Vergauwen, L.; Spaepen, A.J.; Lefevre, J.; Hespel, P. Evaluation of stroke performance in tennis. *Med. Sci. Sports Exerc.* **1998**, *30*, 1281–1288. [CrossRef] [PubMed]
12. Mitchell, J.B.; Cole, K.J.; Grandjean, P.W.; Sobczak, R.J. The effect of a carbohydrate beverage on tennis performance and fluid balance during prolonged tennis play. *J. Strength Cond. Res.* **1992**, *6*, 96–102. [CrossRef]
13. Girard, O.; Lattier, G.; Micallef, J.-P.; Millet, G.P. Changes in exercise characteristics, maximal voluntary contraction, and explosive strength during prolonged tennis playing. *Br. J. Sports Med.* **2006**, *40*, 521–526. [CrossRef] [PubMed]
14. Davey, P.R.; Thorpe, R.D.; Williams, C. Simulated tennis matchplay in a controlled environment. *J. Sports Sci.* **2003**, *21*, 459–467. [CrossRef] [PubMed]
15. Clemente-Suárez, V.J. Cortical arousal and central nervous system fatigue after a mountain marathon. (Activación cortical y fatiga del sistema nervioso después de una maratón de montaña). *Cult. Cienc. Deporte* **2017**, *12*, 143–148. [CrossRef]
16. Görtelmeyer, R.; Wiemann, H. Retest reliability and construct validity of critical flicker fusion frequency. *Pharmacopsychiatry* **1982**, *15*, 24–28. [CrossRef]
17. Simonson, E.; Brozek, J. Flicker fusion frequency: Background and applications. *Physiol. Rev.* **1952**, *32*, 349–378. [CrossRef]
18. Li, Z.; Jiao, K.; Chen, M.; Wang, C. Reducing the effects of driving fatigue with magnitopuncture stimulation. *Accid. Anal. Prev.* **2004**, *36*, 501–505. [CrossRef]
19. Costa, G. Evaluation of workload in air traffic controllers. *Ergonomics* **1993**, *36*, 1111–1120. [CrossRef]
20. Saito, S. Does fatigue exist in a quantitative measurement of eye movements? *Ergonomics* **1992**, *35*, 607–615. [CrossRef]
21. Rodríguez, D.S.; del Valle Soto, M. A study of intensity, fatigue and precision in two specific interval trainings in young tennis players: High-intensity interval training versus intermittent interval training. *BMJ Open Sport Exerc. Med.* **2017**, *3*, e000250. [CrossRef] [PubMed]
22. Killeen, T.; Easthope, C.S.; Filli, L.; Lőrincz, L.; Schrafl-Altermatt, M.; Brugger, P.; Linnebank, M.; Curt, A.; Zörner, B.; Bolliger, M. Increasing cognitive load attenuates right arm swing in healthy human walking. *R. Soc. Open Sci.* **2017**, *4*, 160993. [CrossRef] [PubMed]
23. Gwizdka, J. Using Stroop task to assess cognitive load. In Proceedings of the 28th Annual European Conference on Cognitive Ergonomics, Delft, The Netherlands, 25–27 August 2010; pp. 219–222.
24. Janssen, S.; Heijs, J.J.A.; van der Meijs, W.; Nonnekes, J.; Bittner, M.; Dorresteijn, L.D.A.; Bloem, B.R.; van Wezel, R.J.A.; Heida, T. Validation of the Auditory Stroop Task to increase cognitive load in walking tasks in healthy elderly and persons with Parkinson's disease. *PLoS ONE* **2019**, *14*, e0220735. [CrossRef] [PubMed]
25. Tanaka, H.; Monahan, K.D.; Seals, D.R. Age-predicted maximal heart rate revisited. *J. Am. Coll. Cardiol.* **2001**, *37*, 153–156. [CrossRef]
26. Tornero-Aguilera, J.F.; Gil-Cabrera, J.; Fernandez-Lucas, J.; Clemente-Suárez, V.J. The effect of experience on the psychophysiological response and shooting performance under acute physical stress of soldiers. *Physiol. Behav.* **2021**, *238*, 113489. [CrossRef]
27. Belinchon-deMiguel, P.; Clemente-Suárez, V.J. Psychophysiological, body composition, biomechanical and autonomic modulation analysis procedures in an ultraendurance mountain race. *J. Med. Syst.* **2018**, *42*, 32. [CrossRef]
28. Borg, G. *Borg's Perceived Exertion and Pain Scales*; Human Kinetics: Champaign, IL, USA, 1998.
29. Fritz, C.O.; Morris, P.E.; Richler, J.J. Effect Size Estimates: Current Use, Calculations, and Interpretation. *J. Exp. Psychol.-Gen.* **2012**, *141*, 2–18. [CrossRef]
30. Coolican, H. *Research Methods and Statistics in Psychology*; Psychology Press: East Sussex, UK, 2017.

31. Cohen, J. *Statistical Power Analysis for the Behavioral Sciences*; Routledge: Oxfordshire, UK, 2013.
32. Faul, F.; Erdfelder, E.; Lang, A.-G.; Buchner, A. G* Power 3: A flexible statistical power analysis program for the social, behavioral, and biomedical sciences. *Behav. Res. Methods* **2007**, *39*, 175–191. [CrossRef]
33. Villafaina, S.; Collado-Mateo, D.; Cano-Plasencia, R.; Gusi, N.; Fuentes, J.P. Electroencephalographic response of chess players in decision-making processes under time pressure. *Physiol. Behav.* **2019**, *198*, 140–143. [CrossRef]
34. Fallahi, M.; Motamedzade, M.; Heidarimoghadam, R.; Soltanian, A.R.; Miyake, S. Effects of mental workload on physiological and subjective responses during traffic density monitoring: A field study. *Appl. Ergon.* **2016**, *52*, 95–103. [CrossRef]
35. Crespo, M.; Miley, D. *ITF Advanced Coaches Manual*; International Tennis Federation: London, UK, 1998.
36. Pačesová, P.; Šmela, P.; Kraček, S.; Kukurová, K.; Plevková, L. Cognitive function of young male tennis players and non-athletes. *Acta Gymnica* **2018**, *48*, 56–61. [CrossRef]
37. Stroop, J.R. Studies of interference in serial verbal reactions. *J. Exp. Psychol.* **1935**, *18*, 643. [CrossRef]
38. Curiel-Regueros, A.; Fernández-Lucas, J.; Clemente-Suárez, V.J. Effectiveness of an applied high intensity interval training as a specific operative training. *Physiol. Behav.* **2019**, *201*, 208–211. [CrossRef] [PubMed]
39. Tornero-Aguilera, J.F.; Fernandez-Elias, V.E.; Clemente-Suárez, V.J. Ready for Combat, Psychophysiological Modifications in a Close-Quarter Combat Intervention after an Experimental Operative High-Intensity Interval Training. *J. Strength Cond. Res.* **2020**, *36*, 732–737. [CrossRef]
40. Clemente-Suárez, V.J. The application of cortical arousal assessment to control neuromuscular fatigue during strength training. *J. Mot. Behav.* **2017**, *49*, 429–434. [CrossRef]
41. Clemente-Suárez, V.J.; Delgado-Moreno, R.; González-Gómez, B.; Robles-Pérez, J.J. Respuesta psicofisiológica en un salto táctico paracaidista HAHO: Caso de Estudio. *Sanid. Mil.* **2015**, *71*, 179–182. [CrossRef]
42. Fuentes, J.P.; Villafaina, S.; Collado-Mateo, D.; de la Vega, R.; Gusi, N.; Clemente-Suárez, V.J. Use of biotechnological devices in the quantification of psychophysiological workload of professional chess players. *J. Med. Syst.* **2018**, *42*, 40. [CrossRef]
43. Fuentes-García, J.P.; Pereira, T.; Castro, M.A.; Santos, A.C.; Villafaina, S. Psychophysiological stress response of adolescent chess players during problem-solving tasks. *Physiol. Behav.* **2019**, *209*, 112609. [CrossRef]
44. Delgado-Moreno, R.; Robles-Pérez, J.J.; Clemente-Suárez, V.J. Combat stress decreases memory of warfighters in action. *J. Med. Syst.* **2017**, *41*, 124. [CrossRef]
45. Barte, J.C.M.; Nieuwenhuys, A.; Geurts, S.A.E.; Kompier, M.A.J. Motivation counteracts fatigue-induced performance decrements in soccer passing performance. *J. Sports Sci.* **2019**, *37*, 1189–1196. [CrossRef]
46. Kilit, B.; Arslan, E.; Akca, F.; Aras, D.; Soylu, Y.; Clemente, F.M.; Nikolaidis, P.T.; Rosemann, T.; Knechtle, B. Effect of coach encouragement on the psychophysiological and performance responses of young tennis players. *Int. J. Environ. Res. Public Health* **2019**, *16*, 3467. [CrossRef] [PubMed]
47. Díaz-García, J.; López-Gajardo, M.Á.; Ponce-Bordón, J.C.; Pulido, J.J. Is Motivation Associated with Mental Fatigue during Padel Trainings? A Pilot Study. *Sustainability* **2021**, *13*, 5755. [CrossRef]
48. Fernandez-Fernandez, J.; Zimek, R.; Wiewelhove, T.; Ferrauti, A. High-intensity interval training vs. repeated-sprint training in tennis. *J. Strength Cond. Res.* **2012**, *26*, 53–62. [CrossRef] [PubMed]
49. Kilit, B.; Arslan, E. Effects of high-intensity interval training vs. on-court tennis training in young tennis players. *J. Strength Cond. Res.* **2019**, *33*, 188–196. [CrossRef] [PubMed]
50. Sherwood, D.E.; Lohse, K.; Healy, A. The effect of an external and internal focus of attention on dual-task performance. *J. Exp. Psychol. Hum. Percept Perform* **2020**, *46*, 1–29. [CrossRef] [PubMed]

Article

Comparison between Olympic Weightlifting Lifts and Derivatives for External Load and Fatigue Monitoring

Joaquim Paulo Antunes [1,2,3,*], Rafael Oliveira [1,2,4,*] , Victor Machado Reis [4,5], Félix Romero [1,2], João Moutão [1,2,4] and João Paulo Brito [1,2,4]

1 Sports Science School of Rio Maior—Polytechnic Institute of Santarém, 2040-413 Rio Maior, Portugal
2 Life Quality Research Centre, 2040-413 Rio Maior, Portugal
3 Federação de Halterofilismo de Portugal, 2835-104 Moita, Portugal
4 Research Centre in Sports Sciences, Health Sciences and Human Development, 5001-801 Vila Real, Portugal
5 Sport Sciences Department, University of Trás-os-Montes e Alto Douro, 5001-801 Vila Real, Portugal
* Correspondence: joaquimantunes@esdrm.ipsantarem.pt (J.P.A.); rafaeloliveira@esdrm.ipsantarem.pt (R.O.)

Abstract: Load management is an extremely important subject in fatigue control and adaptation processes in almost all sports. In Olympic Weightlifting (OW), two of the load variables are intensity and volume. However, it is not known if all exercises produce fatigue of the same magnitude. Thus, this study aimed to compare the fatigue prompted by the Clean and Jerk and the Snatch and their derivative exercises among male and female participants, respectively. We resorted to an experimental quantitative design in which fatigue was induced in adult individuals with weightlifting experience of at least two years through the execution of a set of 10 of the most used lifts and derivatives in OW (Snatch, Snatch Pull, Muscle Snatch, Power Snatch, and Back Squat; Clean and Jerk, Power Clean, Clean, High Hang Clean, and Hang Power Clean). Intensity and volume between exercises were equalized (four sets of three repetitions), after which one Snatch Pull test was performed where changes in velocity, range of motion, and mean power were assessed as fatigue measures. Nine women and twelve men participated in the study (age, 29.67 ± 5.74 years and 28.17 ± 5.06 years, respectively). The main results showed higher peak velocity values for the Snatch Pull test when compared with Power Snatch ($p = 0.008$; ES = 0.638), Snatch ($p < 0.001$; ES = 0.998), Snatch Pull ($p < 0.001$, ES = 0.906), and Back Squat ($p < 0.001$; ES = 0.906) while the differences between the Snatch Pull test and the derivatives of Clean and Jerk were almost nonexistent. It is concluded that there were differences in the induction of fatigue between most of the exercises analyzed and, therefore, coaches and athletes could improve the planning of training sessions by accounting for the fatigue induced by each lift.

Keywords: Clean and Jerk; load monitoring; Olympic exercises; Power Clean; power; Snatch; Squat; weightlifting derivatives

Citation: Antunes, J.P.; Oliveira, R.; Reis, V.M.; Romero, F.; Moutão, J.; Brito, J.P. Comparison between Olympic Weightlifting Lifts and Derivatives for External Load and Fatigue Monitoring. *Healthcare* **2022**, *10*, 2499. https://doi.org/10.3390/healthcare10122499

Academic Editors: Herbert Löllgen and Daniele Giansanti

Received: 11 October 2022
Accepted: 7 December 2022
Published: 10 December 2022

Publisher's Note: MDPI stays neutral with regard to jurisdictional claims in published maps and institutional affiliations.

1. Introduction

Olympic Weightlifting (OW) is a dynamic strength and power sport in which two complex lifts/exercises are performed in competition: the Snatch and the Clean and Jerk (C&J). During these lifts, weightlifters have achieved some of the highest peak power outputs reported in the literature [1,2].

The Snatch requires a weighted barbell to be lifted from the floor (usually using a wide grip) to an overhead position in one continuous movement [3]. The C&J is divided into two main phases, in which the first requires the barbell to be raised from the floor (using a shoulder-width grip) to the front of the shoulders in one continuous movement [4], and the second phase consists of a jerk, in which the barbell is propelled from the shoulders to arm's length overhead by forces produced primarily by the hips and thighs [5].

Considering that weightlifting is used in strength development in most sports, some questions arise: how to train with weightlifting efficiently and how it reflect in the performance of other sports that use weightlifting exercises and their derivatives (i.e., variations that omit part of the full lift such as the Hang Clean, Hang Snatch, Power Clean, Power Snatch, and High pull). The answer to these questions may require enhancing the testing and training methods of weightlifting with a combination of the main exercises and their derivatives [6]. Weightlifting exercises and their derivative exercises have become a popular training modality to improve high strength and power expressions throughout the whole force–velocity spectrum during movement, across a range of sports [7–13].

Monitoring, planning, and periodizing training loads are critical factors when it comes to the athlete's development and progression. There has been an attempt by researchers to increasingly identify the variables of training and to control them. In fact, in the past, even the successful Bulgarian methodology tried to reduce some variables by reducing the variety of exercises used [14]. Therefore, there are still some factors that remain unknown. In OW training, load variables such as volume (number of repetitions multiplied by the number of sets) are often manipulated. On the other hand, intensity is expressed relative to the maximal load (kilograms) obtained in the main exercises. Another variable that is also commonly used is the total load, which is characterized by the number of sets multiplied by the number of repetitions multiplied by the kilograms lifted, also known as tonnage [15].

The magnitude of force production and the capacity to perform a given amount of work as rapidly as possible are often suggested as the primary underpinning qualities of sports skills. Thus, developing strength, power, and speed capabilities is frequently the primary aim of many athletic development programs [16]. Despite the variables that define the load being described in terms of intensity by volume [17], there are several parallel factors that may still be associated with this quantification, namely, the type and exercise selection [18]. More recently, other algorithms for the grouping and selection of exercises have been proposed, in some cases based on technical efficiency [19]. Factors such as the number and type of muscle fibers involved, either because of the complexity of the movement or because of the amount of force developed in a given unit of time, can vary in each exercise [11], thus creating an unknown amount of additional fatigue. Nonetheless, in strength training, the external load is related to the external resistance (load) lifted, but it can also be related to the work completed or the velocity achieved during exercises [20].

Several researchers [21–23] have highlighted strong relationships between load and movement velocity fatigue, with the assessment of strength qualities being load–velocity specific. In fact, previous studies have confirmed that the speed of movement provides a determinant of the level of effort during resistance training as well as a variable of the degree of fatigue [24,25]. Therefore, it is particularly important to know the fatigue induced by the different OW derivatives when programming the training load. It is essential to know which exercises induce greater fatigue and its magnitude. High-power outputs and the rate of force development expressed in weightlifting movements and derivatives [2], in conjunction with the motor control and coordination demands on the trunk and lower body muscles to stabilize and transmit forces [26], can effectively impact and compromise various aspects of an athlete's load–velocity profile [16].

This is a topic that has been scarcely addressed in the literature, which lacks results that could improve coaching, both in terms of exercise selection and the distribution of exercise along microcycles, mesocycles, and macrocycles. Several attempts have already been made to try to organize the various exercises into the clusters approach [15]. However, exercise-induced fatigue has never been investigated.

Thus, the aim of the present study was to compare the external load and fatigue prompted by the Clean and Jerk, the Snatch, and their derivative exercises (Snatch, Snatch Pull, Muscle Snatch, Power Snatch, and Back Squat; Clean and Jerk, Power Clean, Clean, High Hang Clean, and Hang Power Clean) among male and female participants, respectively. It was hypothesized that when volume and intensity are equated, there are differences in external load and fatigue induced by performing the different OW derivative exercises.

2. Materials and Methods

2.1. Design

This was a cross-sectional study conducted over two separate days, set apart by a minimum of three days and a maximum of five days. All procedures were recorded for future consultation at the protocols.io website (accessed on 1 January 2020.), and the sample represents more than 10% of the OW Portuguese population [27].

2.2. Participants

A priori power analysis using G*Power (Statistical Power Analyses software for Windows—RRID: SCR_013726) was completed [28]. A sample size calculation was made for the difference between two dependent means (paired sample t-test), an effect size of 0.8, an alpha of ≤0.05, and a beta of 0.95. It was determined that at least 19 participants were needed. Twenty-one Caucasian participants, twelve males and nine females, volunteered to participate in the study.

The inclusion criteria were to be aged between 18 and 40 years; having more than 2 years of OW training; competing at the national level; and having between 61 and 96 kgs of bodyweight for the male group and between 49 and 71 kgs for the female group. The characteristics of the participants are presented in Table 1.

Table 1. Participants' characteristics.

	Age (Years)	Height (cm)	Weight (kg)	BF (%)	FFM (kg)
Female	29.7 ± 5.7	158.8 ± 6.7	60.8 ± 7.3	17.8 ± 7.6	48.9 ± 7.7
Male	28.1 ± 5.0	174.5 ± 6.0	79.5 ± 5.3	17.0 ± 5.1	65.9 ± 5.0
Total	28.8 ± 5.3	167.8 ± 10.1	71.5 ± 11.2	17.3 ± 6.2	58.6 ± 10.6

BF, body fat; FFM, fat-free mass.

Data collection took place at each participant's usual training gym. Prior to their participation, each participant was familiarized with all procedures. Moreover, they read and signed a written informed consent form, in accordance with the university's institutional review board, before data collection. This study was designed according to the recommendations of the World Medical Association's Declaration of Helsinki of 1975, as revised in 2013, for human studies and approved by the Institutional Ethics Committee (approval number: 07A-2021ESDRM).

2.3. Exercise Selection

The rationale for the exercise selection was based on three factors: its ability to enhance the force–velocity profile of athletes [12]; the ability of each derivative to serve as a foundational exercise that enables the progression to more complex weightlifting movements [11,12]; and the exercise frequency applied by OW coaches [29]. The selected exercises were the Snatch and its derivative exercises (Muscle Snatch; Power Snatch; Snatch; Snatch Pull; Back Squat) and the Clean and Jerk (C&J) and its derivative exercises (Power Clean; C&J; Clean; High Hang Clean; Hang Power Clean).

2.4. External Load and Fatigue Assessment

Usually, the isometric mid-thigh pull test (IMTP) is a reliable and popular way to test maximal strength in adult athletes. Administering a partial movement test is a safer and more time-efficient method than traditional 1RM testing. The IMTP produces relatively little fatigue and has a low potential for injury [30], but it proved to be less effective in predicting the competitive performance of OW than other tests [31]. When considering the concept of neuromuscular fatigue, it is important to note that isometric versus dynamic measurements do not provide the same results. Additionally, the bar's range of motion (ROM) also plays an important role in OW, and it seems to be an important factor when assessing fatigue [32]. Therefore, we opted for the Snatch Pull test (SPT) as a reference measure, which has been correlated with the personal record (PR) of the Snatch exercise (r = 0.99) [33]. In all OW derivatives, the external load variables of mean velocity, peak velocity, mean power, and ROM were measured using the Isoinertial Dynamometer Vitruve (Vitruve encoder; Madrid, Spain) (previously, Speed4Lifts) [34]. Both mean and peak velocities were considered based on the measurement of fatigue according to previous references [24,25]. Moreover, this type of test can regularly be applied during weightlifting training as a valid alternative to the personal record Snatch test to assess individualized progression in weightlifting performance over time [33].

Since all these lifts have correlation intensity with each other, and Muscle Snatch is referenced as 60 to 65% of the Snatch PR, an intensity load of 60% was chosen. Therefore, setting it as the baseline intensity, the volume chosen (4 sets of 3 repetitions) was the amount of load that is usually performed by lifters within the intensity already settled upon [29].

2.5. Procedures

On the first day of data collection, participants started early in the morning for an anthropometric assessment, namely, height, weight, and body composition using bioimpedance analysis.

2.5.1. Anthropometric and Body Composition Assessment

The anthropometric and body composition measurements were obtained with the subjects dressed in light clothing without shoes following previous recommendations [35] using a stadiometer with an incorporated scale (Seca 220, Hamburg, Germany) according to standardized procedures [36]. The body composition data were obtained with bioelectrical impedance analysis using Inbody S10 (model JMW140, Biospace Co, Ltd., Seoul, Korea), according to the manufacturer's guidelines [37,38]. Eight electrodes were placed on eight tactile points (thumbs, middle fingers of both hands, and the ankles of both feet) to perform a multi-segmental frequency analysis. The parameters collected were body fat mass (BFM) and fat-free mass (FFM).

The measurements were carried out in the morning in a room with an ambient temperature and relative humidity of 22–23 °C and 50–60%, respectively, after a minimum of 8 h of fasting and after the bladder was emptied, following previous suggestions [35,39]. The participants adopted a supine position with their arms and legs abducted at a 45° angle; the skin was cleaned with ethyl alcohol and hydrophilic cotton at the eight electrode placement sites. After a 10 min rest in a room without noise, eight electrodes were placed on the cleaned surfaces, and the measurements were performed.

Before data collection, participants did not exercise or ingest caffeine or alcohol during the 12 h prior to the assessment. In addition, participants removed all objects that could interfere with the bioelectrical impedance assessment.

Female participants were only assessed if they were in the luteal phase of ovulatory menstrual cycles. Otherwise, they waited until they were in the luteal phase. All the assessments were performed by the same evaluator to minimize possible measurement errors [40].

2.5.2. Test Protocol

After anthropometric and body composition assessments, an explanation of the protocol was provided. A 10 min warm-up, including mobility exercises, OW repetitions, and jumps, was carried out before the beginning of each training session. To minimize the risk of injury, there were always two assistants to monitor exercise execution.

Participants started their personal warm-up exercises/specific-for-training session: up to 60% of the Snatch 1RM followed by two 50%, one 70%, and one 100% Snatch 1RM and

SPT attempts separated by 1 min of recovery [41]. Before each SPT, verbal feedback cues were given by coaches in a standardized form, namely, "Pull hard and fast".

On the first test day, the Snatch and derivative exercises protocol took place. After the warm-up, the baseline SPT evaluation occurred (Figure 1), making a 1RM Snatch of personal record, after which, data were collected. Then, participants rested for 1 min, followed by a Muscle Snatch protocol of 4 sets of 3 repetitions at 60% of the Snatch 1RM (1 min rest between sets). After the protocol, participants then took a 1 min rest before the post Muscle Snatch SPT evaluation (1RM).

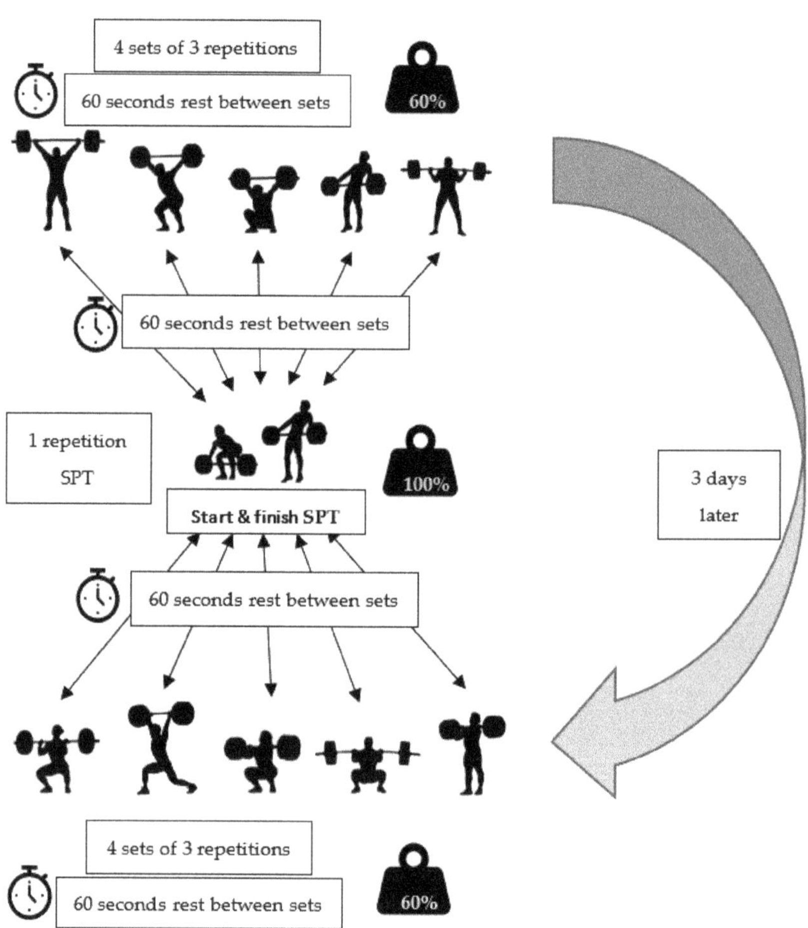

Figure 1. Testing protocol schematic.

This was followed by a Power Snatch protocol of 4 sets of 3 repetitions at 60% of the Snatch 1RM (1 min rest between sets). After the protocol, participants then took a 1 min rest before the post Power Snatch SPT evaluation, followed by one repetition at 100% Snatch 1RM, after which, data were collected; participants would then rest 1 min.

The same protocol was used for the Snatch, Snatch Pull, and Back Squat. On the second test day, three days after the tests were performed on the Snatch derivatives, the C&J and derivative exercises protocol was performed.

The same protocol used in the Snatch derivatives was used for all C&J derivatives in the following order: Power Clean, C&J, Clean, High Hang Clean, and Hang Power Clean. In this protocol, 60% of the C&J 1RM was used.

2.6. Statistical Analysis

All statistical analyses were performed using IBM SPSS for Windows (IBM Corp. Released 2020., Version 28.0. Armonk, NY, USA). The data are described as mean $\pm$ standard deviation (SD), standard error of the mean (SEM), and mean difference (MD) with a 95% confidence interval (CI). The Shapiro–Wilk test was used for testing normality. Differences between exercises were examined using a paired samples t-test (velocity, range of motion, and mean power within each exercise monitored using the isoinertial dynamometer). An a priori level of significance was set at $p < 0.05$ and a percentage change with a 95% CI. The effect size (ES) was calculated to determine the magnitude of the effects through Cohen's d (by the difference of two pairs of means, which are then divided by the standard deviation from the data), and the following thresholds were applied: large d, > 0.8; moderate d, between 0.8 and 0.5; small d, between 0.49 and 0.20; trivial d, < 0.2 [42].

3. Results

3.1. Snatch Derivative Protocols

Analyzing the mean power for the entire sample ($n = 21$), it was found that, after the Muscle Snatch protocol, there were no significant differences while, post-Power Snatch, -Snatch, -Snatch Pull, and -Back Squat showed a significant difference (Table 2). However, when considering the gender group analyses separately, the female group ($n = 9$) reveals no difference after the Muscle Snatch and Power Snatch protocols, whereas Snatch, Snatch Pull, and Back Squat manifested a significant difference. The male group ($n = 12$) did not reveal significant differences in mean power for any exercise.

Table 2. Baseline and post-values of the Snatch Pull test for the Snatch derivatives (♀♂ = 21).

Parameter	Weightlifting Derivative		Mean $\pm$ SD	SEM	MD (95% CI)	p (ES)
ROM (cm)	SPT		Baseline			
			106.49 $\pm$ 7.49	1.64		
			Post			
	Pair 1	Muscle Snatch	107.33 $\pm$ 7.75	1.69	−0.85 (−2.65; 0.95)	0.338 (−0.214)
	Pair 2	Power Snatch	105.15 $\pm$ 7.93	1.73	1.34 (−0.36; 3.04)	0.116 (0.358)
	Pair 3	Snatch	104.19 $\pm$ 7.85	1.71	2.30 (0.87; 3.73)	0.003 * (0.731)
	Pair 4	Snatch Pull	102.82 $\pm$ 8.63	1.88	3.67 (1.97; 5.36)	<0.001 ** (0.986)
	Pair 5	Back Squat	103.97 $\pm$ 9.41	2.05	2.52 (0.42; 4.62)	0.021 * (0.547)
Mean Power (W)	SPT		Baseline			
			706.55 $\pm$ 187.58	40.93		
			Post			
	Pair 1	Muscle Snatch	701.93 $\pm$ 189.80	41.42	4.61 (−18.41; 27.64)	0.680 (0.091)
	Pair 2	Power Snatch	681.19 $\pm$ 181.14	39.53	25.36 (0.93; 49.79)	0.043 * (0.472)
	Pair 3	Snatch	677.11 $\pm$ 183.49	40.04	29.44 (0.32; 58.55)	0.048 * (0.460)
	Pair 4	Snatch Pull	664.41 $\pm$ 180.76	39.44	42.14 (15.84; 68.44)	0.003 * (0.729)
	Pair 5	Back Squat	671.32 $\pm$ 190.58	41.59	35.22 (10.03; 60.42)	0.009 * (0.636)

Table 2. *Cont.*

Parameter		Weightlifting Derivative		Mean ± SD	SEM	MD (95% CI)	*p* (ES)
Peak Velocity (m/s)		SPT		Baseline			
				1.81 ± 0.17	0.04		
				Post			
	Pair 1	Muscle Snatch		1.78 ± 0.18	0.04	0.04 (−0.01; 0.09)	0.125 (0.350)
	Pair 2	Power Snatch		1.76 ± 0.19	0.04	0.06 (0.02; 0.10)	0.008 * (0.638)
	Pair 3	Snatch		1.73 ± 0.17	0.04	0.08 (0.05; 0.12)	<0.001 ** (0.998)
	Pair 4	Snatch Pull		1.72 ± 0.15	0.03	0.09 (0.05; 0.14)	<0.001 ** (0.906)
	Pair 5	Back Squat		1.73 ± 0.18	0.04	0.08 (0.04; 0.13)	<0.001 ** (0.906)
Mean Velocity (m/s)		SPT		Baseline			
				0.94 ± 0.13	0.03		
				Post			
	Pair 1	Muscle Snatch		0.93 ± 0.12	0.03	0.01 (−0.02; 0.04)	0.508 (0.147)
	Pair 2	Power Snatch		0.91 ± 0.13	0.03	0.03 (0.00; 0.06)	0.050 (0.455)
	Pair 3	Snatch		0.90 ± 0.13	0.03	0.04 (0.00; 0.07)	0.030 * (0.509)
	Pair 4	Snatch Pull		0.92 ± 0.15	0.03	0.02 (−0.06; 0.10)	0.604 (0.115)
	Pair 5	Back Squat		0.89 ± 0.13	0.03	0.05 (0.02; 0.08)	0.003 * (0.727)

SPT, Snatch Pull test; ROM, range of motion; *, $p < 0.05$; **, $p < 0.001$; SD, standard deviation; SEM, standard error of the mean; MD, mean difference; CI, confidence intervals; ES, effect size.

Mean velocity evidenced a significant difference in the Snatch and Back Squat (Table 2) for the total sample. When the gender groups were analyzed, the female group showed differences after the Snatch, Snatch Pull, and Back Squat (Table 3). No differences were found in the male group.

Table 3. Baseline and post-values of Snatch Pull test for the Snatch derivatives based on gender (♀= 9; ♂= 12).

Parameter			Weightlifting Derivative		Mean ± SD	SEM	MD (95% CI)	*p* (ES)
ROM (cm)	**Female**	SPT			Baseline			
					105.22 ± 8.25	2.75		
					Post			
		Pair 1	Muscle Snatch		105.16 ± 9.00	3.00	0.07 (−2.18; 2.32)	0.947 (0.023)
		Pair 2	Power Snatch		104.58 ± 9.29	3.10	0.64 (−1.95; 3.23)	0.585 (0.189)
		Pair 3	Snatch		102.04 ± 9.17	3.06	3.18 (1.17; 5.18)	0.006 * (1.218)
		Pair 4	Snatch Pull		100.40 ± 10.44	3.48	4.82 (2.73; 6.91)	0.001 * (1.776)
		Pair 5	Back Squat		100.03 ± 10.91	3.64	5.19 (2.48; 7.90)	0.002 * (1.474)
	Male	SPT			Baseline			
					107.43 ± 7.10	2.05		
					Post			
		Pair 1	Muscle Snatch		108.97 ± 6.59	1.90	−1.53 (−4.45; 1.38)	0.272 (−0.334)
		Pair 2	Power Snatch		105.57 ± 7.16	2.10	1.87 (−0.71; 4.44)	0.139 (0.460)
		Pair 3	Snatch		105.79 ± 6.66	1.92	1.64 (−0.55; 3.84)	0.128 (0.475)
		Pair 4	Snatch Pull		104.63 ± 6.91	2.00	2.80 (0.11; 5.49)	0.042 * (0.663)
		Pair 5	Back Squat		106.91 ± 7.23	2.09	0.52 (−2.28; 3.32)	0.692 (0.117)

Table 3. *Cont.*

Parameter			Weightlifting Derivative	Mean ± SD	SEM	MD (95% CI)	*p* (ES)
Mean Power (w)	Female		SPT	Baseline			
				557.79 ± 128.94	42.98		
				Post			
		Pair 1	Muscle Snatch	540.79 ± 113.25	37.75	17.00 (−0.81; 34.81)	0.059 (0.734)
		Pair 2	Power Snatch	536.32 ± 121.34	40.45	21.47 (−4.20; 47.13)	0.090 (0.643)
		Pair 3	Snatch	521.19 ± 113.48	37.83	36.60 (13.67; 59.53)	0.006 * (1.227)
		Pair 4	Snatch Pull	518.99 ± 121.16	40.39	38.80 (19.07; 58.53	0.002 * (1.512)
		Pair 5	Back Squat	518.82 ± 128.90	42.97	38.97 (21.13; 56.81)	0.001 * (1.679)
	Male		SPT	Baseline			
				818,12 ± 142.13	41.03		
				Post			
		Pair 1	Muscle Snatch	822.79 ± 137.79	39.78	−4.68 (−45.08; 35.73)	0.804 (−0.074)
		Pair 2	Power Snatch	789.84 ± 137.47	39.68	28.28 (−13.89; 70.44)	0.188 (0.428)
		Pair 3	Snatch	794.05 ± 130.53	37.68	24.07 (−28.01; 76.14)	0.331 (0.294)
		Pair 4	Snatch Pull	773.47 ± 135.84	39.21	44.65 (−2.79; 92.08)	0.063 (0.598)
		Pair 5	Back Squat	785.70 ± 143.73	41.49	32.42 (−13.22; 78.05)	0.146 (0.451)
Peak Ve-locity (m/s)	Female		SPT	Baseline			
				1.88 ± 0.17	0.06		
				Post			
		Pair 1	Muscle Snatch	1.84 ± 0.15	0.05	0.04 (−0.01; 0.09)	0.102 (0.615)
		Pair 2	Power Snatch	1.86 ± 0.16	0.05	0.02 (−0.04; 0.08)	0.422 (0.282)
		Pair 3	Snatch	1.78 ± 0.15	0.05	0.10 (0.05; 0.16)	0.002 * (1.469)
		Pair 4	Snatch Pull	1.76 ± 0.17	0.06	0.12 (0.05; 0.20)	0.005 * (1.258)
		Pair 5	Back Squat	1.74 ± 0.21	0.07	0.14 (0.09; 0.20)	<0.001 * (2.058)
	Male		SPT	Baseline			
				1.76 ± 0.16	0.05		
				Post			
		Pair 1	Muscle Snatch	1.73 ± 0.19	0.05	0.04 (−0.05; 0.12)	0.378 (0.265)
		Pair 2	Power Snatch	1.68 ± 0.18	0.05	0.09 (0.03; 0.14)	0.009 * (0.910)
		Pair 3	Snatch	1.69 ± 0.17	0.05	0.07 (0.01; 0.13)	0.025 * (0.745)
		Pair 4	Snatch Pull	1.69 ± 0.14	0.04	0.07 (0.00; 0.14)	0.039 * (0.675)
		Pair 5	Back Squat	1.73 ± 0.17	0.05	0.04 (−0.01; 0.09)	0.134 (0.467)
Mean Velocity (m/s)	Female		SPT	Baseline			
				0.99 ± 0.14	0.05		
				Post			
		Pair 1	Muscle Snatch	0.97 ± 0.13	0.04	0.03 (−0.00; 0.06)	0.063 (0.719)
		Pair 2	Power Snatch	0.97 ± 0.14	0.05	0.02 (−0.01; 0.06)	0.144 (0.540)
		Pair 3	Snatch	0.93 ± 0.14	0.05	0.06 (0.02; 0.10)	0.006 * (1.228)
		Pair 4	Snatch Pull	0.92 ± 0.11	0.04	0.07 (0.03; 0.11)	0.003 * (1.372)
		Pair 5	Back Squat	0.92 ± 0.14	0.05	0.07 (0.04; 0.10)	0.001 * (1.660)
	Male		SPT	Baseline			
				0.89 ± 0.12	0.03		
				Post			
		Pair 1	Muscle Snatch	0.90 ± 0.10	0.03	−0.01 (−0.05; 0.04)	0.806 (−0.073)
		Pair 2	Power Snatch	0.86 ± 0.11	0.03	0.03 (−0.02; 0.08)	0.174 (0.419)
		Pair 3	Snatch	0.87 ± 0.12	0.03	0.02 (−0.04; 0.08)	0.412 (0.246)
		Pair 4	Snatch Pull	0.91 ± 0.18	0.05	−0.02 (−0.17; 0.13)	0.800 (−0.075)
		Pair 5	Back Squat	0.86 ± 0.13	0.04	0.03 (−0.02; 0.09)	0.174 (0.420)

SPT, Snatch Pull test; ROM, range of motion; *, *p* < 0.05; SD, standard deviation; SEM, standard error of the mean; MD, mean difference; CI, confidence intervals; ES, effect size.

Peak velocity did not show a significant difference in the Muscle Snatch, while the remaining derivatives showed significant differences (Table 2). The female group did not report differences after the Muscle Snatch and Power Snatch protocols (Table 3). No differences were found for the male group in the Muscle Snatch and Back Squat.

For the total sample, only the Muscle Snatch protocol revealed differences in the range of motion (Table 2). In the gender analysis (Table 3), the female group revealed that, after the Muscle Snatch protocol, the Snatch, Snatch Pull, and Back Squat exercises presented a significant difference, while in the male group, only the Snatch Pull showed differences.

In the assessment of the Snatch variables, it was verified that the ROM and post-Snatch Pull protocol, as well as the peak velocity, post-Snatch, -Snatch Pull, and -Back Squat, showed differences when the total sample was analyzed.

3.2. Clean and Jerk Derivative Protocols

Differences in the C&J mean power and mean velocity were only found when considering the whole sample (Table 4) and when considering the male group (Table 5).

Table 4. Baseline and post-values of Snatch Pull test for the Clean and Jerk derivatives (♀♂ = 21).

Parameter		Weightlifting Derivative	Mean ± SD	SEM	MD (95% CI)	*p* (ES)
ROM (cm)			Baseline			
		SPT	106.01 ± 8.00	1.75		
			Post			
	Pair 1	Power Clean	105.77 ± 7.91	1.73	0.24 (−0.91; 1.39)	0.671 (0.094)
	Pair 2	Clean and Jerk	103.91 ± 8.88	1.94	2.10 (0.46; 3.73)	0.015 * (0.582)
	Pair 3	Clean	104.67 ± 8.77	1.91	1.34 (−0.27; 2.96)	0.098 (0.378)
	Pair 4	High Hang Clean	105.03 ± 8.98	1.96	0.98 (−0.43; 2.38)	0.164 (0.316)
	Pair 5	Hang Power Clean	104.92 ± 8.41	1.83	1.09 (−0.81; 2.99)	0.245 (0.261)
Mean Power (w)			Baseline			
		SPT	699.81 ± 176.31	38.47		
			Post			
	Pair 1	Power Clean	700.49 ± 183.15	39.97	−0.68 (−16.10; 14.74)	0.928 (−0.020)
	Pair 2	Clean and Jerk	675.26 ± 170.43	37.19	24.55 (1.65; 47.44)	0.037 * (0.488)
	Pair 3	Clean	679.59 ± 180.17	39.32	20.22 (−5.70; 46.14)	0.119 (0.355)
	Pair 4	High Hang Clean	690.40 ± 178.72	39.00	9.41 (−16.66; 35.48)	0.460 (0.164)
	Pair 5	Hang Power Clean	687.63 ± 176.81	38.58	12.18 (−13.46; 37.82)	0.334 (0.216)
Peak Velocity (m/s)			Baseline			
		SPT	1.75 ± 0.16	0.03		
			Post			
	Pair 1	Power Clean	1.75 ± 0.17	0.04	−0.01 (−0.05; 0.04)	0.809 (−0.054)
	Pair 2	Clean and Jerk	1.74 ± 0.18	0.04	0.01 (−0.02; 0.04)	0.456 (0.166)
	Pair 3	Clean	1.74 ± 0.18	0.04	0.01 (−0.03; 0.04)	0.712 (0.082)
	Pair 4	High Hang Clean	1.74 ± 0.16	0.03	0.01 (−0.02; 0.04)	0.511 (0.146)
	Pair 5	Hang Power Clean	1.75 ± 0.15	0.03	0.00 (−0.03; 0.04)	0.819 (0.051)

Table 4. *Cont.*

Parameter		Weightlifting Derivative	Mean ± SD	SEM	MD (95% CI)	*p* (ES)
Mean Velocity (m/s)			Baseline			
		SPT	0.93 ± 0.11	0.02		
			Post			
	Pair 1	Power Clean	0.93 ± 0.11	0.02	0.00 (−0.02; 0.02)	0.846 (0.043)
	Pair 2	Clean and Jerk	0.90 ± 0.12	0.03	0.03 (0.00; 0.06)	0.050 (0.478)
	Pair 3	Clean	0.90 ± 0.11	0.02	0.03 (−0.00; 0.06)	0.071 (0.415)
	Pair 4	High Hang Clean	0.91 ± 0.11	0.03	0.01 (−0.02; 0.04)	0.358 (0.205)
	Pair 5	Hang Power Clean	0.91 ± 0.11	0.02	0.02 (−0.01; 0.05)	0.227 (0.272)

SPT, Snatch Pull test; ROM, range of motion; *, $p < 0.05$; SD, standard deviation; SEM, standard error of the mean; MD, mean difference; CI, confidence intervals; ES, effect size.

Table 5. Baseline and post-values of Snatch Pull test for the Clean and Jerk derivatives (♀= 9; ♂= 12).

Parameter			Weightlifting Derivative	Mean ± SD	SEM	MD (95% CI)	*p* (ES)
ROM (cm)	Female			Baseline			
			SPT	102.14 ± 6.68	2.23		
				Post			
		Pair 1	Power Clean	102.77 ± 7.98	2.66	−0.62 (−2.78; 1.54)	0.525 (−0.222)
		Pair 2	Clean and Jerk	101.93 ± 8.16	2.72	0.21 (−2.16; 2.58)	0.843 (0.068)
		Pair 3	Clean	102.18 ± 8.70	2.90	−0.03 (−3.03; 2.96)	0.980 (−0.009)
		Pair 4	High Hang Clean	102.03 ± 8.87	2.96	0.11 (−2.42; 2.64)	0.922 (0.034)
		Pair 5	Hang Power Clean	101.47 ± 7.26	2.42	0.68 (−1.42; 2.77)	0.477 (0.248)
	Male			Baseline			
			SPT	108.91 ± 7.91	2.28		
				Post			
		Pair 1	Power Clean	108.03 ± 7.38	2.13	0.88 (−0.51; 2.28)	0.192 (0.401)
		Pair 2	Clean and Jerk	105.40 ± 9.44	2.73	3.51 (1.35; 5.67)	0.004 * (1.033)
		Pair 3	Clean	106.53 ± 8.72	2.52	2.38 (0.46; 4.29)	0.020 * (0.786)
		Pair 4	High Hang Clean	107.28 ± 8.74	2.52	1.63 (−0.22; 3.47)	0.079 (0.559)
		Pair 5	Hang Power Clean	107.51 ± 8.55	2.47	1.40 (−1.84; 4.64)	0.362 (0.275)
Mean Power (w)	Female			Baseline			
			SPT	536.97 ± 100.78	33.59		
				Post			
		Pair 1	Power Clean	539.52 ± 125.88	41.96	−2.56 (−27.66; 22.55)	0.820 (−0.078)
		Pair 2	Clean and Jerk	539.18 ± 132.05	44.02	−2.21 (−33.10; 28.63)	0.873 (−0.055)
		Pair 3	Clean	533.22 ± 130.91	43.64	3.74 (−26.22; 33.70)	0.781 (0.096)
		Pair 4	High Hang Clean	542.37 ± 125.61	41.87	−5.40 (−27.63; 16.83)	0.591 (−0.187)
		Pair 5	Hang Power Clean	528.48 ± 120.75	40.30	8.49 (−15.44; 32.41)	0.437 (0.273)
	Male			Baseline			
			SPT	821.94 ± 105.66	30.50		
				Post			
		Pair 1	Power Clean	821.21 ± 111.21	32.10	0.73 (−22.23; 23.69)	0.945 (0.020)
		Pair 2	Clean and Jerk	777.33 ± 116.69	33.68	44.62 (13.47; 75.76)	0.009 * (0.910)
		Pair 3	Clean	789.37 ± 126.03	36.38	32.58 (−9.58; 74.73)	0.117 (0.491)
		Pair 4	High Hang Clean	801.42 ± 123.38	35.62	20.53 (−24.59; 65.64)	0.338 (0.289)
		Pair 5	Hang Power Clean	806.99 ± 99.85	28.82	14.95 (−30.17; 60.07)	0.481 (0.211)

Table 5. *Cont.*

Parameter			Weightlifting Derivative	Mean ± SD	SEM	MD (95% CI)	*p* (ES)
Peak Velocity (m/s)	Female		Baseline				
			SPT	1.80 ± 0.13	0.04		
			Post				
		Pair 1	Power Clean	1.80 ± 0.15	0.05	−0.01 (−0.10; 0.05)	0.795 (−0.090)
		Pair 2	Clean and Jerk	1.82 ± 0.12	0.04	−0.02 (−0.06; 0.02)	0.231 (−0.081)
		Pair 3	Clean	1.81 ± 0.15	0.05	−0.01 (−0.07; 0.05)	0.725 (−0.121)
		Pair 4	High Hang Clean	1.79 ± 0.15	0.05	0.01 (−0.06; 0.07)	0.849 (0.066)
		Pair 5	Hang Power Clean	1.77 ± 0.16	0.05	0.03 (−0.04; 0.10)	0.377 (0.312)
	Male		Baseline				
			SPT	1.72 ± 0.17	0.05		
			Post				
		Pair 1	Power Clean	1.72 ± 0.18	0.05	−0.00 (−0.08; 0.07)	0.903 (−0.036)
		Pair 2	Clean and Jerk	1.68 ± 0.20	0.06	0.04 (−0.01; 0.08)	0.089 (0.539)
		Pair 3	Clean	1.70 ± 0.19	0.05	0.02 (−0.03; 0.07)	0.437 (0.233)
		Pair 4	High Hang Clean	1.70 ± 0.16	0.05	0.01 (−0.03; 0.05)	0.459 (0.222)
		Pair 5	Hang Power Clean	1.73 ± 0.15	0.04	−0.02 (−0.06; 0.03)	0.492 (−0.205)
Mean Velocity (m/s)	Female		Baseline				
			SPT	0.96 ± 0.12	0.04		
			Post				
		Pair 1	Power Clean	0.96 ± 0.12	0.04	0.00 (−0.04; 0.04)	0.901 (0.043)
		Pair 2	Clean and Jerk	0.96 ± 0.12	0.04	0.01 (−0.04; 0.06)	0.773 (0.099)
		Pair 3	Clean	0.95 ± 0.11	0.04	0.02 (−0.04; 0.07)	0.493 (0.239)
		Pair 4	High Hang Clean	0.96 ± 0.11	0.04	−0.00 (−0.04; 0.04)	0.947 (−0.023)
		Pair 5	Hang Power Clean	0.94 ± 0.11	0.04	0.02 (−0.03; 0.07)	0.322 (0.352)
	Male		Baseline				
			SPT	0.90 ± 0.10	0.03		
			Post				
		Pair 1	Power Clean	0.90 ± 0.10	0.03	0.00 (−0.02; 0.03)	0.890 (0.041)
		Pair 2	Clean and Jerk	0.86 ± 0.10	0.03	0.05 (0.01; 0.08)	0.011 * (0.876)
		Pair 3	Clean	0.86 ± 0.10	0.03	0.04 (−0.01; 0.08)	0.091 (0.535)
		Pair 4	High Hang Clean	0.88 ± 0.11	0.03	0.02 (−0.02; 0.07)	0.282 (0.326)
		Pair 5	Hang Power Clean	0.89 ± 0.11	0.03	0.01 (−0.03; 0.06)	0.489 (0.207)

SPT, Snatch Pull test; ROM, range of motion; *, $p < 0.05$; SD, standard deviation; SEM, standard error of the mean; MD, mean difference; CI, confidence intervals; ES, effect size.

Regarding peak velocity, no differences were found in any of the exercises or in either group. For the total sample, ROM only showed a significant difference in the C&J (Table 4). In the group analysis, only the male group showed differences in the C&J and Clean (Table 5).

4. Discussion

The aim of the present study was to compare the external load and fatigue prompted by the Clean and Jerk, the Snatch, and their derivative exercises (Snatch, Snatch Pull, Muscle Snatch, Power Snatch, and Back Squat; Clean and Jerk, Power Clean, Clean, High Hang Clean, and Hang Power Clean) among male and female participants, respectively. The majority of these exercises are used in OW, as well as in general strength and conditioning training programs for various sports [9,43–46]. It was hypothesized that when volume and intensity are equated there are differences between external load and fatigue induced by different OW exercises.

The main results showed that, for the total sample, significant differences were found in the Snatch Pull, Snatch, and Back Squat ROM and on the C&J ROM. Regarding the mean power, significant differences were found in the Power Snatch, Snatch, Snatch Pull, Back Squat, and C&J. Regarding peak velocity, significant differences were found in the Power Snatch, Snatch, Snatch Pull, and Back Squat. Regarding the mean velocity, significant differences were found in the Snatch Pull and Back Squat.

When genders were analyzed separately, the female group showed significant differences in the Snatch ROM, Snatch Pull, and Back Squat, while in the male group, differences were found in the ROMs of the Snatch Pull, C&J, and Clean. Regarding mean power, the female group presented significant differences in the Snatch, Snatch Pull, and Back Squat, while the male group showed significant differences in mean power in the C&J. The female group also revealed significant peak velocity differences in the Snatch, Snatch Pull, and Back Squat, while the male group revealed significant differences in the Power Snatch, Snatch, and Snatch pull. In addition, the female group showed significant differences in mean velocity in the Snatch, Snatch Pull, and Back Squat, while the male group only showed significant differences in the C&J. The fact that women can perform a greater number of intermittent contractions than men, even when the two groups are matched for strength, has been reported before [47], and the same effect may occur in OW training.

Considering the whole sample, almost all variables presented significant differences, as well as moderate-to-large effect size values. Peak velocity seems to present the most significant differences in both groups; however, in the female group, Snatch derivatives seem to show significant differences in every variable studied. This effect might be related to better technique proficiency and consistency in female lifters. On the other hand, the male group only showed significant differences in peak velocity. The fatigue induced by each exercise may be related to the individualized load–velocity relationship and to the specific characteristics of the participant [48,49]. Some studies [49,50] have reported that intersubject variability seems to be reduced when the loads are prescribed based on the individual load–velocity relationship. However, some coaches prefer to prescribe the loads to match a specific number of repetitions rather than using a prescription method based on bar velocity. Still, there is high intersubject variability between the number of repetitions performed and neuromuscular fatigue [51].

Some authors [11,12,17,21,23,33,49] have described a theoretical relationship between force and velocity with special consideration for weightlifting derivatives. The high-force end of the force–velocity curve features weightlifting derivatives that develop the largest forces due to the loads that can be used. As Suchomel et al. [52] point out, the proper implementation and progression of resistance training exercises throughout training facilitate the development of an athlete's force–velocity profile [52,53], which has been cited as an important aspect of athletic performance [54,55]. Specifically, the biomechanical and physiological characteristics of each weightlifting derivative may indicate that certain derivatives should be prescribed during certain training phases to meet the training goals of each phase. Thus, information that may assist practitioners when it comes to programming exercises to optimally develop these characteristics would be beneficial. In the present study, the only exercise that did not show any difference in any variable was the Muscle Snatch, and this exercise was the one with the highest ROM.

A higher barbell ROM has a direct relationship with the subject's height [11,12,56,57], meaning that if the lifter is taller, the barbell needs to have a higher displacement than if the lifter is shorter. OW is a competitive sport that requires athletes to lift a maximal amount of weight in the Snatch and C&J. OW's main distinction in sports training is velocity, meaning that other sports mostly involve training to develop more speed, maintaining the load of the athlete (bodyweight in most cases). However, OW aims to maintain the ideal velocity for each exercise according to its height. Therefore, it is also correlated with the lifter's height [56], and manipulating barbell weight could also indicate that OW lifters are more resistant to velocity loss than other kinds of athletes.

Recent research has also reported that different individual physical characteristics lead to different fatigue levels and recovery [57], and this could have led to greater variability in the study results. More than half of the participants showed increases in most variables instead of an expected decrease induced by fatigue. The post-activation potential effect might be involved in these findings, as this effect is a possible result of muscle contractions, and, utilized during subsequent explosive activity, it could potentially enhance power and, therefore, performance. However, while a previous effort might also induce fatigue, it is the balance between the post-activation potentiation effect and fatigue onset that will determine the effect of a previous effort on performance in an explosive movement. This relationship is affected by several variables, including volume and intensity and subject characteristics, as well as others [47,58]. Thus, it can also be inferred that some athletes probably did not quite induce this effect during their warm-ups. The fact that the warm-up was not standardized can be considered a limitation herein. In future studies, the warm-up should be controlled and also equalized among subjects since it may affect performance in explosive movements [59].

Additionally, we can speculate that some types of exercises may contribute more to the better potentiation of muscle contractions due to the lifted load, the force–velocity curve, and the different levels of induced fatigue [47,57,58,60–62]. The neuromuscular adaptations induced by weightlifting training strongly depend on the manipulation of strength training variables, such as the exercise type and sequence, load magnitude, volume, interset and intraset rest periods, and lifting velocity [63,64]. A common concern for coaches is deciding how much weight their athletes should lift in a particular exercise, as resistance-training-induced adaptations are highly dependent on the intensity used [65].

In addition to the manipulation of variables intrasession, coaches program exercises within periodized programs to vary the intensity of the training stimuli. Regarding squat movements, the exercise stimulus may be varied based on the depth and variation of the squat [66], as well as the load that is prescribed. As a result, the force–velocity characteristics of the training stimulus will be modified, but the athlete's force–velocity profile may be fully developed. There was a report that certain weightlifting derivatives emphasize force or velocity more than others [12]. Thus, it seems that a sequential progression and combination of weightlifting derivatives can be beneficial to athletes when it comes to increasing force and power development rates. Moreover, techniques refined during earlier training phases may facilitate increases in the load used for each exercise.

Sports such as OW, along with its derivatives, require a single high-force or high-velocity effort. These movements typically involve a burst of concentric muscular activity in the agonist muscles, followed by a phase of relaxation, which, during the motion, continues due to stored momentum. This type of movement is also known as ballistic movement/action [67]. In voluntary muscle contractions, the total force output of a muscle depends primarily on the number of motor units and the firing frequency of those motor units, in which a higher force output will result in more motor units firing frequency [67]. In fact, motor unit recruitment is known to be a critical factor in maximal or ballistic contractions, as well as inducing fatigue. This principle—known to be the recruitment threshold of a motor neuron—can be directly related to the size of its axon. In other words, the larger the axon, the greater the amount of stimulation required [68]. In fact, there is some evidence of the selective activation of large motor units if the motor task readily demands those motor units [69]. Moreover, ballistic exercises elicit several acute and chronic neurological changes. The standard recruitment of motor units, according to the size principle, stays consistent at submaximal exercise intensities but appears to be violated in ballistic movements. It seems that the motor task, more than any other variable, determines the sequence of activation [67].

All OW exercises and their derivatives have relatively high motor recruitment. However, more complex exercises empirically require more units. Therefore, they are supposed to use more energy, leading to greater fatigue. The fact that some exercises did not show fatigue in the current study may be associated with the fact that the volume or intensity fatigue threshold was not met. Recording the bar velocity at which submaximal loads are lifted is a potential method of quantifying the load as a function of the fatigue it causes [24,70]. Researchers have reported the general relationship between lifting velocity and the %1RM in different exercises. Nowadays, it seems to be the consensus that the individualized load–velocity relationship allows for a better assessment of athlete fatigue, mainly because the %1RM–velocity relationship is subject-specific [51,71]. Unfortunately, little information exists regarding the possibility of predicting the number of repetitions from the recording of lifting velocity when powerlifting training (i.e., OW-derivative exercise) is used by strength and power athletes up until the final days prior to a competition. Therefore, understanding how different derivatives influence peak power performance is critical [31]. Recent evidence suggests that many coaches and support staff are taking an increasingly scientific approach to load monitoring [12,13,23,46,50,72].

Some limitations of the present study may be considered. The randomization of the sample could only be accomplished to a certain extent since the population to be studied is small in number by itself, and the inclusion criteria further narrow this choice. Therefore, the small sample could also be pointed to as a relative limitation because both males and females were analyzed as a group, and the samples were even smaller when the genders are separated. The fact that we compared parameters in women and men as one homogeneous group could be considered questionable and should be considered a study limitation. However, this methodological decision stems from a practical issue, in that trainers test both men and women together [6]. Moreover, specific warm-ups were not standardized, mainly because lifters have their own warm-up routines, which we choose not to interfere in. However, this may constitute another unaccountable variable that could have influenced the first and second SPT.

Future research should take the previous information into account and try to measure 1RM, for example, by determining the catch height of each lifter and then setting 1RM using their respective SPT height achievements.

Nonetheless, the main practical outcome of this study was adding the various relative fatigue values to the overall load, whether in form of a percentage or a fraction. For example, considering peak velocity in the female group, Snatch had a 71% (0.10 m/s) difference in relative fatigue, and the Back Squat had the maximum relative fatigue difference (0.14 m/s), meaning that, when we multiply for the same load, let us say (100 kgs) × (0.71), Snatch = 71 in terms of the relative fatigue, while for (100 kgs) × (0.100), the Back Squat = 100 in terms of the relative fatigue, meaning that, for the same load (volume × Intensity), the Back Squat will fatigue the athlete 30% more than the Snatch.

5. Conclusions

This intervention confirmed the study hypothesis: when volume and intensity are equated, there are differences between fatigue induced by various OW exercises.

In Snatch derivatives, peak velocity was a good variable to quantify fatigue in both genders, while in all other variables, it was only sensible in females. In addition, females seem more sensitive to fatigue in Snatch derivatives. Snatch derivatives are well known for their velocity-developing capability; therefore, fatigue may be explained more effectively using a test that mimics the movement itself, such as the SPT.

In C&J derivatives, females did not present statistically significant results; therefore, they showed that more volume and or intensity are needed to induce measurable fatigue. Regarding the male group, ROM seems to be the variable that we can better rely on, and, in addition, C&J derivative exercises are less velocity-dependent; this could explain ROM's ability to quantify fatigue.

The ten exercises studied showed different external load and fatigue levels between them. However, it was not possible to quantify the magnitude of the different variables. This is likely the consequence of individual physiological adaptations and responses to exercise.

Coaches may plan according to these findings, specifically, as to C&J variables, by using higher relative loads in the exercises where fatigue was not found. Furthermore, using peak velocity in the Snatch and its derivatives plus ROM in the C&J and its derivatives seems to be best for training control in OW.

Author Contributions: Conceptualization, J.P.A., R.O. and J.P.B.; methodology, J.P.A., R.O. and J.P.B.; software, J.P.A., F.R. and J.P.B.; validation, R.O. and J.P.B.; formal analysis, J.P.A. and F.R.; investigation, J.P.A., R.O., V.M.R., F.R., J.M. and J.P.B.; resources, J.P.A., R.O., V.M.R., F.R., J.M. and J.P.B.; data curation, J.P.A.; writing—original draft preparation, J.P.A., R.O., V.M.R., F.R. and J.P.B.; writing—review and editing, J.P.A., R.O., V.M.R., F.R., J.M. and J.P.B.; visualization, J.P.A., R.O., F.R. and J.P.B.; supervision, R.O. and J.P.B.; project administration, J.P.A., R.O. and J.P.B.; funding acquisition, R.O. and J.P.B. All authors have read and agreed to the published version of the manuscript.

Funding: This research was funded by the Portuguese Foundation for Science and Technology, I.P., Grant/Award Number UIDP/04748/2020. Victor Reis was also funded by the Portuguese Foundation for Science and Technology (FCT), I.P., project number UIDB/04045/2020.

Institutional Review Board Statement: The study was conducted according to the guidelines of the Declaration of Helsinki and approved by the Ethics Committee of the Polytechnic Institute of Santarém (07A-2021ESDRM).

Informed Consent Statement: Written informed consent was obtained from the participants to publish this paper.

Data Availability Statement: The data presented in this study are available upon request from the corresponding author.

Acknowledgments: The authors would like to thank the team's coaches and players for their cooperation during all data collection procedures.

Conflicts of Interest: The authors declare no conflict of interest. The funders had no role in the design of the study; in the collection, analysis, or interpretation of the data; in the writing of the manuscript; or in the decision to publish the results.

References

1. Garhammer, J. A Review of power output studies of olympic and powerlifting: Methodology, performance prediction, and evaluation tests. *J. Strength Cond. Res.* **1993**, *7*, 76–89. [CrossRef]
2. Garhammer, J.; Mclaughlin, T. Power output as a function of load variation in olympic and power lifting. *J. Biomech.* **1980**, *13*, 198. [CrossRef]
3. Derwin, B.P. Sports Performance Series: The Snatch: Technical description and periodization program. *Strength Cond. J.* **1990**, *12*, 6–15. [CrossRef]
4. Al-Khleifat, A.I.; Al-Kilani, M.; Kilani, H.A. Biomechanics of the clean and jerk in weightlifting national Jordanian team. In Proceedings of the Journal of Human Sport and Exercise, 2019, Summer Conferences of Sports Science, Budapest, Hungary, 11–13 September 2019. [CrossRef]
5. Grabe, S.; Widule, C. Comparative biomechanics of the jerk in olympic weightlifting. *Res. Q. Exerc. Sport* **1988**, *59*, 1–8. [CrossRef]
6. Králová, T.; Gasior, J.; Vanderka, M.; Cacek, J.; Vencúrik, T.; Bokůvka, D.; Hammerová, T. Correlation analysis of olympic-style weightlifting exercises and vertical jumps. *Stud. Sport.* **2019**, *13*, 26. [CrossRef]
7. Ebben, W. Hamstring activation during lower body resistance training exercises. *Int. J. Sport. Physiol. Perform.* **2009**, *4*, 84–96. [CrossRef] [PubMed]
8. Ebben, W.P.; Feldmann, C.R.; Dayne, A.; Mitsche, D.; Alexander, P.; Knetzger, K.J. Muscle activation during lower body resistance training. *Int. J. Sport. Med.* **2009**, *30*, 1–8. [CrossRef] [PubMed]
9. Simenz, C.; Dugan, C.; Ebben, W. Strength and conditioning practices of National Basketball Association strength and conditioning coaches. *J. Strength Cond. Res. Natl. Strength Cond. Assoc.* **2005**, *19*, 495–504. [CrossRef]
10. Hori, N.; Newton, R.; Nosaka, K.; Stone, M. Weightlifting exercises enhance athletic performance that requires high-load speed strength. *Strength Cond. J.* **2005**, *27*, 50–55. [CrossRef]
11. Suchomel, T.J.; Sole, C.J. Power-time curve comparison between weightlifting derivatives. *J. Sport. Sci. Med.* **2017**, *16*, 407–413.
12. Suchomel, T.; Comfort, P.; Lake, J. Enhancing the force–velocity profile of athletes using weightlifting derivatives. *Strength Cond. J.* **2017**, *39*, 10–20. [CrossRef]
13. James, L.P.; Haff, G.G.; Kelly, V.G.; Connick, M.J.; Hoffman, B.W.; Beckman, E.M. The impact of strength level on adaptations to combined weightlifting, plyometric, and ballistic training. *Scand. J. Med. Sci. Sport.* **2018**, *28*, 1494–1505. [CrossRef] [PubMed]
14. Garhammer, J.; Takano, B. Training for weightlifting. In *Strength and Power in Sport*; John Wiley & Sons, Ltd.: New York, NY, USA, 2003; pp. 502–515. [CrossRef]
15. Takano, B. *Weightlifting Programming: A Winning Coach's Guide*; Catalyst Athletics: Sunnyvale, CA, USA, 2012.
16. Morris, S.J.; Oliver, J.L.; Pedley, J.S.; Haff, G.G.; Lloyd, R.S. Comparison of weightlifting, traditional resistance training and plyometrics on strength, power and speed: A Systematic review with meta-analysis. *Sport. Med.* **2022**, *52*, 1533–1554. [CrossRef] [PubMed]

17. González-Badillo, J.; Izquierdo, M.; Gorostiaga, E. Moderate of high relative training intensity produces greater strength gains compared with low and high s in competitive weightlifters. *J. Strength Cond. Res. Natl. Strength Cond. Assoc.* **2006**, *20*, 73–81. [CrossRef]
18. Badillo, J.J.G. *Libros Tecnicos Deportivos: Halterofilia*; Comite Olimpico Espanol: Madrid, Spain, 1991.
19. Flores, F.; Redondo, J.C. Proposal for selecting weightlifting exercises on the basis of a cybernetic model. *Int. J. Adv. Res.* **2020**, *8*, 906–914. [CrossRef] [PubMed]
20. Scott, B.R.; Duthie, G.M.; Thornton, H.R.; Dascombe, B.J. Training monitoring for resistance exercise: Theory and applications. *Sport. Med.* **2016**, *46*, 687–698. [CrossRef]
21. De Ruiter, C.J.; Jones, D.A.; Sargeant, A.J.; De Haan, A. The measurement of force/velocity relationships of fresh and fatigued human adductor pollicis muscle. *Eur. J. Appl. Physiol. Occup. Physiol.* **1999**, *80*, 386–393. [CrossRef]
22. Sakamoto, A.; Sinclair, P.J. Effect of movement velocity on the relationship between training load and the number of repetitions of bench press. *J. Strength Cond. Res.* **2006**, *20*, 523–527. [CrossRef]
23. Hughes, L.J.; Banyard, H.G.; Dempsey, A.R.; Peiffer, J.J.; Scott, B.R. Using load-velocity relationships to quantify training-induced fatigue. *J. Strength Cond. Res.* **2019**, *33*, 762–773. [CrossRef]
24. González-Badillo, J.J.; Sánchez-Medina, L. Movement velocity as a measure of loading intensity in resistance training. *Int. J. Sport. Med.* **2010**, *31*, 347–352. [CrossRef]
25. Sánchez-Medina, L.; González-Badillo, J.J. Velocity loss as an indicator of neuromuscular fatigue during resistance training. *Med. Sci. Sport. Exerc.* **2011**, *43*, 1725–1734. [CrossRef]
26. Eriksson Crommert, M.; Ekblom, M.M.; Thorstensson, A. Motor control of the trunk during a modified clean and jerk lift. *Scand. J. Med. Sci. Sport.* **2014**, *24*, 758–763. [CrossRef]
27. Federação de Halterofilismo de Portugal. Federação de Halterofilismo de Portugal—Relatórios de Atividades e Contas. 2021. Available online: https://www.halterofilismo.net/documentos/organização-efuncionamento/relatórios-de-atividades-e-contas (accessed on 10 May 2022).
28. Faul, F.; Erdfelder, E.; Buchner, A.; Lang, A.G. Statistical power analyses using G*Power 3.1: Tests for correlation and regression analyses. *Behav. Res. Methods* **2009**, *41*, 1149–1160. [CrossRef] [PubMed]
29. Everett, G. *Olympic Weightlifting: A Complete Guide for Athletes & Coaches*, 2nd ed.; Catalyst Athletics: Sunnyvale, CA, USA, 2009.
30. Stone, M.; O'Bryant, H.; Hornsby, G.; Cunanan, A.; Mizuguchi, S.; Suarez, D.; Marsh, D.; Haff, G.; Ramsey, M.; Beckham, G.; et al. Using the isometric mid-thigh pull in the monitoring of weightlifters: 25+ Years of Experience. *Prof. Strength Cond.* **2019**, *54*, 19–26.
31. Travis, S.K.; Goodin, J.R.; Beckham, G.K.; Bazyler, C.D. Identifying a test to monitor weightlifting performance in competitive male and female weightlifters. *Sports* **2018**, *6*, 46. [CrossRef] [PubMed]
32. Cheng, A.; Rice, C. Fatigue and recovery of power and isometric torque following isotonic knee extensions. *J. Appl. Physiol.* **2005**, *99*, 1446–1452. [CrossRef]
33. Sandau, I.; Chaabene, H.; Granacher, U. Validity and reliability of a Snatch pull test to model the force-velocity relationship in male elite weightlifters. *J. Strength Cond. Res.* **2021**, *36*, 2808–2815. [CrossRef]
34. Pérez Castilla, A.; Piepoli, A.; Delgado García, G.; Garrido, G.; García Ramos, A. Reliability and concurrent validity of seven commercially available devices for the assessment of movement velocity at different intensities during the bench press. *J. Strength Cond. Res.* **2019**, *33*, 1258–1265. [CrossRef]
35. Arazi, H.; Mirzaei, B.; Nobari, H. Anthropometric profile, body composition and somatotyping of national Iranian cross-country runners. *Turk. J. Sport Exerc.* **2015**, *17*, 35–41. [CrossRef]
36. Lohman, T.G.; Roche, A.F.; Martorell, R. *Anthropometric Standardization Reference Manual*; Human Kinetics Books: Champaign, IL, USA, 1988.
37. Buckinx, F.; Reginster, J.Y.; Dardenne, N.; Croisiser, J.L.; Kaux, J.F.; Beaudart, C.; Slomian, J.; Bruyère, O. Concordance between muscle mass assessed by bioelectrical impedance analysis and by dual energy X-ray absorptiometry: A cross-sectional study. *BMC Musculoskelet. Disord.* **2015**, *16*, 60. [CrossRef]
38. Yang, E.M.; Park, E.; Ahn, Y.H.; Choi, H.J.; Kang, H.G.; Cheong, H.I.; Ha, I.S. Measurement of fluid status using bioimpedance methods in Korean pediatric patients on hemodialysis. *J. Korean Med. Sci.* **2017**, *32*, 1828–1834. [CrossRef] [PubMed]
39. Rahmat, A.J.; Arsalan, D.; Bahman, M.; Hadi, N. Anthropometrical profile and biomotor abilities of young elite wrestlers. *Phys. Educ. Stud.* **2016**, *20*, 63–69. [CrossRef]
40. Marfell-Jones, M.J.; Stewart, A.D.; de Ridder, J.H. *International Standards for Anthropometric Assessment*; International Society for the Advancement of Kinanthropometry: Wellington, New Zealand, 2012. Available online: https://repository.openpolytechnic.ac.nz/handle/11072/1510 (accessed on 20 April 2022).
41. Parcell, A.C.; Sawyer, R.D.; Tricoli, V.A.; Chinevere, T.D. Minimum rest period for strength recovery during a common isokinetic testing protocol. *Med. Sci. Sport. Exerc.* **2002**, *34*, 1018–1022. [CrossRef] [PubMed]
42. Cohen, J. *Statistical Power Analysis for the Behavioral Sciences*, 2nd ed.; Academic Press: Cambridge, MA, USA, 2013. [CrossRef]
43. Ebben, W.P.; Carroll, R.M.; Simenz, C.J. Strength and conditioning practices of National Hockey League strength and conditioning coaches. *J. Strength Cond. Res.* **2004**, *18*, 889–897. [CrossRef] [PubMed]
44. Ebben, W.P.; Blackard, D.O. Strength and conditioning practices of National Football League strength and conditioning coaches. *J. Strength Cond. Res.* **2001**, *15*, 48–58. [PubMed]

45. Gee, T.I.; Olsen, P.D.; Berger, N.J.; Golby, J.; Thompson, K.G. Strength and conditioning practices in rowing. *J. Strength Cond. Res.* **2011**, *25*, 668–682. [CrossRef]

46. Weldon, A.; Duncan, M.; Turner, A.; Sampaio, J.; Noon, M.; Wong, D.P.; Lai, V. Contemporary practices of strength and conditioning coaches in professional soccer. *Biol. Sport* **2020**, *38*, 377–390. [CrossRef]

47. Tillin, N.A.; Bishop, D. Factors modulating post-activation potentiation and its effect on performance of subsequent explosive activities. *Sport. Med.* **2009**, *39*, 147–166. [CrossRef]

48. García-Ramos, A.; Haff, G.G.; Pestaña-Melero, F.L.; Pérez-Castilla, A.; Rojas, F.J.; Balsalobre-Fernández, C.; Jaric, S. Feasibility of the two-point method for determining the one-repetition maximum in the bench press exercise. *Int. J. Sport. Physiol. Perform.* **2018**, *13*, 474–481. [CrossRef]

49. Sánchez-Moreno, M.; Rendeiro-Pinho, G.; Mil-Homens, P.V.; Pareja-Blanco, F. Monitoring training through maximal number of repetitions or velocity-based approach. *Int. J. Sport. Physiol. Perform.* **2021**, *16*, 527–534. [CrossRef]

50. Hernández-Belmonte, A.; Courel-Ibáñez, J.; Conesa-Ros, E.; Martínez-Cava, A.; Pallarés, J.G. Level of effort: A reliable and practical alternative to the velocity-based approach for monitoring resistance training. *J Strength Cond. Res.* **2021**, *36*, 2992–2999. [CrossRef] [PubMed]

51. Gonzalez-Badillo, J.J.; Yanez-Garcia, J.M.; Mora-Custodio, R.; RodriguezRosell, D. Velocity loss as a variable for monitoring resistance exercise. *Int. J. Sport. Med.* **2017**, *38*, 217–225. [CrossRef] [PubMed]

52. Suchomel, T.J.; Beckham, G.K.; Wright, G.A. Effect of various loads on the force-time characteristics of the hang high pull. *J. Strength Cond. Res.* **2015**, *29*, 1295–1301. [CrossRef] [PubMed]

53. DeWeese, B.H.; Hornsby, G.; Stone, M.; Stone, M.H. The training process: Planning for strength–power training in track and field. Part 2: Practical and applied aspects. *J. Sport Health Sci.* **2015**, *4*, 318–324. [CrossRef]

54. Suchomel, T.J.; Beckham, G.K.; Wright, G.A. The impact of load on lower body performance variables during the hang power clean. *Sport. Biomech.* **2014**, *13*, 87–95. [CrossRef]

55. Suchomel, T.J.; Comfort, P.; Stone, M.H. Weightlifting pulling derivatives: Rationale for implementation and application. *Sport. Med.* **2015**, *45*, 823–839. [CrossRef]

56. Roman, R.A. *The Training of the Weightlifter R. A. Roman*; Sportivny Press: Moscow, Russia, 1974; Volume 1, p. 4. Available online: https://www.dynamicfitnessequipment.com/product-p/sp109.htm (accessed on 5 February 2022).

57. Helland, C.; Midttun, M.; Saeland, F.; Haugvad, L.; Schäfer Olstad, D.; Solberg, P.; Paulsen, G. A strength-oriented exercise session required more recovery time than a power-oriented exercise session with equal work. *PeerJ* **2020**, *8*, e10044. [CrossRef]

58. Sale, D.G. Postactivation potentiation: Role in human performance. *Exerc. Sport Sci. Rev.* **2002**, *30*, 138–143. Available online: https://journals.lww.com/acsmessr/Fulltext/2002/07000/Postactivation_Potentiation__Role_in_Human.8.aspx (accessed on 5 May 2022). [CrossRef]

59. Silva, L.M.; Neiva, H.P.; Marques, M.C.; Izquierdo, M.; Marinho, D.A. Effects of warm-up, post-warm-up, and re-warm-up strategies on explosive efforts in team sports: A Systematic Review. *Sport. Med.* **2018**, *48*, 2285–2299. [CrossRef]

60. Ebben, W.P. A brief review of concurrent activation potentiation: Theoretical and practical constructs. *J. Strength Cond. Res.* **2006**, *20*, 985–991. [CrossRef]

61. Robbins, D.W. Postactivation potentiation and its practical applicability: A brief review. *J. Strength Cond. Res.* **2005**, *19*, 453–458. [CrossRef] [PubMed]

62. Stone, M.; Sands, W.; Pierce, K.; Ramsey, M.; Haff, G. Power and power potentiation among strength-power athletes: Preliminary study. *Int. J. Sport. Physiol. Perform.* **2008**, *3*, 55–67. [CrossRef] [PubMed]

63. Bird, S.; Tarpenning, K.; Marino, F. Designing resistance training programmes to enhance muscular fitness: A review of the acute programme variables. *Sport. Med.* **2005**, *35*, 841–851. [CrossRef]

64. Kraemer, W.J.; Ratamess, N.A. Fundamentals of resistance training: Progression and exercise prescription. *Med. Sci. Sport. Exerc.* **2004**, *36*, 674–688. [CrossRef] [PubMed]

65. Suchomel, T.J.; Nimphius, S.; Bellon, C.R.; Hornsby, W.G.; Stone, M.H. Training for muscular strength: Methods for monitoring and adjusting training intensity. *Sport. Med.* **2021**, *51*, 2051–2066. [CrossRef]

66. Hartmann, H.; Wirth, K.; Klusemann, M.; Dalic, J.; Matuschek, C.; Schmidtbleicher, D. Influence of squatting depth on jumping performance. *J. Strength Cond. Res.* **2012**, *26*, 3243–3261. [CrossRef]

67. Wallace, B.; Janz, J. Implications of motor unit activity on ballistic movement. *Int. J. Sport. Sci. Coach.* **2009**, *4*, 285–292. [CrossRef]

68. Henneman, E. Relation between size of neurons and their susceptibility to discharge. *Science* **1957**, *126*, 1345–1347. [CrossRef]

69. Gillespie, C.A.; Simpson, D.R.; Edgerton, V.R. Motor unit recruitment as reflected by muscle fibre glycogen loss in a prosimian (bushbaby) after running and jumping1. *J. Neurol. Neurosurg. Psychiatry* **1974**, *37*, 817–824. [CrossRef]

70. García-Ramos, A.; Torrejón, A.; Feriche, B.; Morales-Artacho, A.J.; Pérez-Castilla, A.; Padial, P.; Haff, G.G. Prediction of the maximum number of repetitions and repetitions in reserve from barbell velocity. *Int. J. Sport. Physiol. Perform.* **2018**, *13*, 353–359. [CrossRef]

71. Pestaña-Melero, F.L.; Haff, G.G.; Rojas, F.J.; Pérez-Castilla, A.; García-Ramos, A. Reliability of the load-velocity relationship obtained through linear and polynomial regression models to predict the 1-repetition maximum load. *J. Appl. Biomech.* **2018**, *34*, 184–190. [CrossRef] [PubMed]

72. Suchomel, T.J.; McKeever, S.M.; Comfort, P. Training with weightlifting derivatives: The effects of force and velocity overload stimuli. *J. Strength Cond. Res.* **2020**, *34*, 1808–1818. [CrossRef] [PubMed]

healthcare

Case Report

The Influence of Aerobic Type Exercise on Active Crohn's Disease Patients: The Incidence of an Elite Athlete

Konstantinos Papadimitriou 

Physical Education and Sport Science, Aristotle University of Thessaloniki, 54636 Thessaloniki, Greece; kostakispapadim@gmail.com; Tel.: +30-69-8026-5800

Abstract: A lifestyle factor which contributes to the remission of Crohn's disease (CD) is physical activity. The effect seems to positively impact the disease's symptoms, improving the quality of life, especially on patients in remission. Due to the lack of clinical studies about the effects of physical activity on active CD patients, the purpose of the present case study was to record the influence of swimming training (aerobic type of exercise) on an athlete with active CD. In this study participated a 22-year-old male, who is an elite swimmer and who was diagnosed in 2019 with CD. The research was conducted over the last three years (2019 2022). Both the athlete and doctor consented to the clinical examinations by the author. According to the present study, immediate medical examination and the prescription of anti-TNF-α therapy is probably the most appropriate solution for someone who is diagnosed with CD symptoms. Moreover, patient participation in any sport activity is discouraged because of the potential danger of exacerbation of the symptoms. Therefore, for the sake of patient safety, physical activity should only be encouraged when the disease is in remission.

Keywords: inflammation; bowel; malnutrition; training; physical activity

Citation: Papadimitriou, K. The Influence of Aerobic Type Exercise on Active Crohn's Disease Patients: The Incidence of an Elite Athlete. *Healthcare* **2022**, *10*, 713. https://doi.org/10.3390/healthcare10040713

Academic Editors: João Paulo Brito and Rafael Oliveira

Received: 22 March 2022
Accepted: 10 April 2022
Published: 12 April 2022

Publisher's Note: MDPI stays neutral with regard to jurisdictional claims in published maps and institutional affiliations.

1. Introduction

Inflammatory Bowel Disease (IBD) is an autoimmune disease which mainly affects the gastrointestinal tract, especially in young adults though rarely on children [1]. IBD is categorized into ulcerative colitis (UC) and Crohn's disease (CD). In CD, the symptoms are diarrhea, abdominal pain, urgency, fatigue, weakness, anorexia, and malnutrition, altering body composition [2–4]. A lifestyle factor that possibly contributes to the remission of the disease is physical activity.

The research was conducted by the author supported the hypothesis that physical activity contributes to the disease's remission [5]. More specifically, moderate to intense aerobic and/or resistance exercise, (60–80%, of VO2max or 1 RM), with an interval of 15–30 s and 2–3 min after each exercise, and between exercises respectively, reduces CD's symptoms, while improving the patient's quality of life, especially on the disease's remission [5–7]. However, a lack of clinical studies about the effect of physical activity on active CD patients is observed. As a result, doctors and physical activity instructors avoid the prescription of any type of exercise on active CD patients because of the belief that it could lead to the exacerbation of the disease [5,8–12].

Therefore, the purpose of the present case study was to monitor the influence of swimming (aerobic type of exercise) on an active CD's patient via the incidence of an adult male elite athlete who was diagnosed with CD. The research was conducted over the last three years (2019–2022) from when he was diagnosed with CD. Both athlete and doctor consented to the clinical exams carried out by the author.

2. Case Report

The present study is concerned with an elite 22-year-old male swimmer who was diagnosed with CD in 2019. This affected his performance and the will to participate at the training sessions in a sport which he begun in 2005 and continues still. To start with, in 2018, at his senior year as a high school student and during a tough training period, he manifested symptoms of diarrhea which ceased after three days of recovery, improving his condition. Thus, the doctor suggested that probably it was gastroenteritis. Since the appearance of these symptoms, he followed his schedule and training with under normal conditions.

In January 2019, after his dinner, which was a slice of pizza, he felt intense abdominal pain and a few hours later, manifested symptoms of diarrhea. Over the following two weeks since the first manifestation of diarrhea, his dietary and training schedule were performed normally. However, the symptoms of diarrhea continued, disturbing his daily routine. According to his clinical situation, the doctor concluded that the diarrhea was probably due to gastroenteritis; therefore, a dietary schedule was recommended that avoided mainly high-fat foods, such as processed meat, pork products, and fast food, etc., which could have exacerbated the diarrhea. Moreover, he felt more confident to consume mainly carbohydrate foods and poultry, while avoiding any kind of leafy vegetables, fruits, fishes, or legumes, etc. Despite the modifications to his diet, the manifested symptoms (abdominal pain and diarrhea) remained, and he continued training, although his body composition was unaffected.

In the February of the same year, his clinical situation had not improved, and he suffered from strong abdominal pain which affected his bodyweight, resulting in the loss of 4 kg. The doctor recommended a hematological analysis to examine the value of C—Reactive Protein (CRP), a biomarker which is used for the detection of possible inflammation [13]. According to the analysis, the athlete had mild inflammation, with the CRP value at 6.2 mg/dL. Thus, the doctor hypothesized that this inflammatory condition originated from the intestines. Therefore, an antibiotic treatment was administrated, with three tabs per day of Metronidazole.

In March, a month later, the athlete did not show any improvement since the beginning of the treatment. His body weight decreased more than 12 kg since January (body weight 69 kg) and the CRP was raised to 13.87 mg/dL. As a result, it was difficult for him to continue training at the same frequency and intensity as before despite his intentions to do so. According to the literature [5], this was due to the exacerbation of abdominal pain and diarrhea. Thus, in April, he completely stopped training and any other physical activity due to the weakness that he was feeling.

In the same month (April), he did an endoscopy of the colon. The results showed mild inflammation and mucosal atrophy. Therefore, the doctors observing his clinical situation, and in accordance with the histological examination, prescribed him a new treatment combining Metronidazole with the anti-inflammation Mesalazine (two tabs per day).

The two medications (Metronidazole and Mesalazine), a month later (May), inhibited the value of CRP to 0.14 mg/dL, improving his health significantly. As a result, he was able to resume the training, though at a relatively low frequency and intensity. Although his clinical situation improved, his performance was still lower than before, and the exacerbations continued. In the middle of the month, Mesalazine's treatment was discontinued and replaced with the antibiotic Ciprofloxacin (6 tabs per week) which was administrated until the end of May.

In June, the athlete's body weight and CRP value returned to almost normal levels (75 kg and 0.38 mg/dL, respectively). However, because of some exacerbations (abdominal pain and diarrhea) the doctor decided to examine the fecal calprotectin (CAL), which is a highly effective index to detect possible endoscopic ulcerations in CD [14]. Furthermore, a colonoscopy was performed and biopsies were taken from the colon.

According to the analysis, the excreta's inflammation index was 780 µg/g. Likewise, from the colonoscopy examination, the sigmoid and ileum colon showed very red parts with the presence of exudate ulcers. Finally, the results from the biopsies showed long-term active ileitis. Connecting the results of colonoscopy (Figure 1A,B), CAL, and the clinical situation of the athlete, the doctor's hypothesis of a mild CD infection was confirmed. Despite the athlete's situation, the doctor recommended that he observe his nutrition intake and to continue with Salofalk at a dose of 500 mg (4 tabs per day). Moreover, his training continued at a lower frequency and intensity, although it remained difficult to sustain the training for more than 30 min (Figure 2).

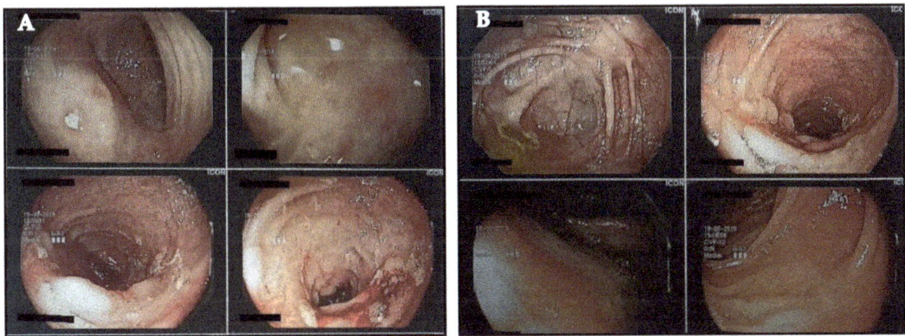

Figure 1. (**A**,**B**) Depiction of colon via colonoscopy.

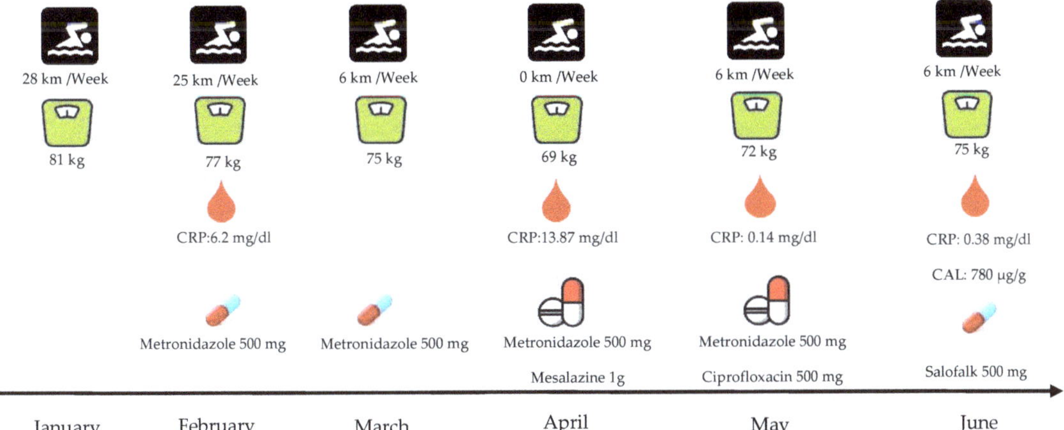

Figure 2. Patient's clinical situation during the first six months of CD manifestation. CRP: C—Reactive Protein, CAL: Fecal Calprotectin.

The athlete, after four months (July–October) of an unstable health condition, was, in November of 2019, hospitalized. After a sequence of hematological analysis and in accordance with his medical history, it was decided to begin the injection of anti-TNF-α (biological therapy) with Infliximab and Azathoprine. The first three therapies were conducted in a period of three months (from November of 2019 until January of 2020). From the March of 2020, until the present time, his therapy continues to take place every two months. In November of 2020, another colonoscopy was performed in which a physiological depiction of the colon instead of a mild inflammation in the sigmoid colon was found, which did not concern the doctors. Since the athlete started the therapy, his CRP value reached normal levels (<0.50 mg/dL) (Figure 3). The only exception was in August of 2021, when the CRP was raised above normal values. However, the athlete did not manifest any symptoms of abdominal pain or diarrhea. The doctor indicated that this raise was probably due to another factor, because CRP is an index which is associated with a variety of etiologies, ranging from sleep disturbances to periodontal disease [15]. Finally, his swimming performance was improved significantly when competing again in races.

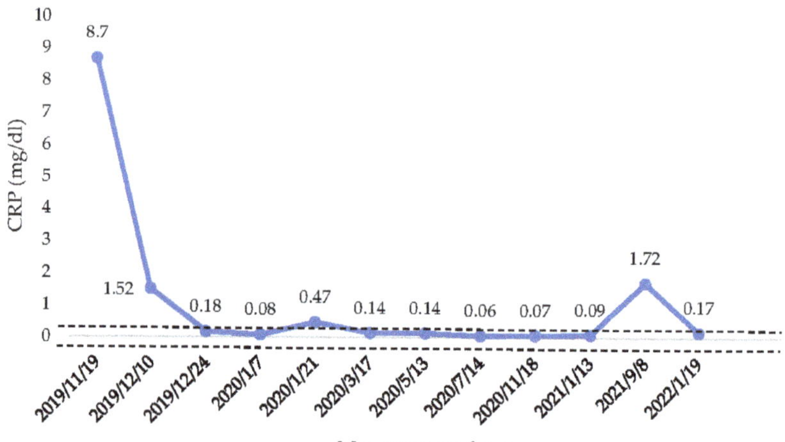

Figure 3. CRP concentration from the beginning of anti-TNF—α therapy until January 2022. CRP: C—reactive protein, ----: Range of normal values 0–0.50 mg/dL.

3. Discussion

This novel case study was performed with no limitations or difficulties. Both the athlete and the doctors contributed, giving reports about the hematological analysis and the clinical situation during the three-year period.

According to the athlete's situation, it is assumed that in the condition of an active disease, aerobic types of exercise exacerbate the disease; however, in remission, aerobic types of exercise do not cause exacerbation. Thus, the results of the present case are in accordance with the literature [5–7]. Specifically, many authors regard that aerobic or resistance type of exercise effectively mitigate CD's symptoms, especially when the patient is in remission. However, there is not any observation in the research of a case in which someone is in active disease's condition. Thus, there is only the hypothesis that any kind of exercise must be avoided when the symptoms are active [5].

Moreover, the immediate performance of a hematological analysis and colonoscopy after any suspicious CD symptom seems to be important for any patient. Specifically, since the first manifestation of the disease, the doctor administrated to the patient Metronidazole and Mesalazine two months later. For the athlete, these two months were a period which caused an extensive malnutrition and resulted in difficulty in the participation of his training or indeed of any other activity. Therefore, to deal with those types of symptoms, immediate examination, and offensive therapy with anti-TNF-α injection (Infliximab and Azathoprine) are the most appropriate actions [16]. In case of the athlete of this study, the doctors prescribed him medications at the beginning of symptoms' manifestation as a unique therapy. However, the most effective and appropriate way to deal with CD symptoms is the combined prescription of medication and anti-TNF-α therapy.

The present study is the first which recorded the incidence of an active CD athlete who participates in aerobic activity. Therefore, further studies are essential to inform doctors and physical activity instructors about the potential benefits of the prescription of exercise on CD patients.

4. Conclusions

According to the present study, immediate medical examination and prescription of anti-TNF-α therapy are probably the most appropriate solution for someone who is diagnosed with CD symptoms. Moreover, patient participation in any sport activity is discouraged because of the potential danger of exacerbation of the symptoms. Therefore, for the sake of patient safety, physical activity should only be encouraged when the disease is in remission.

Funding: This research received no external funding.

Institutional Review Board Statement: Not applicable.

Informed Consent Statement: Written informed consent was obtained from the athlete who signatured the statement about his participation, giving the permission to publish his clinical reports.

Data Availability Statement: Not applicable.

Acknowledgments: The author would like to thank the doctors who contributed giving athlete's clinical reports and the athlete.

Conflicts of Interest: The author declares no conflict of interest.

References

1. Aniwan, S.; Harmsen, W.S.; Tremaine, W.J.; Loftus, E.V., Jr. Incidence of inflammatory bowel disease by race and ethnicity in a population-based inception cohort from 1970 through 2010. *Therap. Adv. Gastroenterol.* **2019**, *12*. [CrossRef] [PubMed]
2. Legeret, C.; Mählmann, L.; Gerber, M.; Kalak, N.; Köhler, H.; Holsboer-Trachsler, E.; Brand, S.; Furlano, R. Favorable impact of long-term exercise on disease symptoms in pediatric patients with inflammatory bowel disease. *BMC Pediatr.* **2019**, *19*, 297. [CrossRef] [PubMed]
3. Bilski, J.; Mazur-Bialy, A.I.; Brzozowski, B.; Magierowski, M.; Jasnos, K.; Krzysiek-Maczka, G.; Urbanczyk, K.; Ptak-Belowska, A.; Zwolinska-Wcislo, M.; Mach, T.; et al. Moderate exercise training attenuates the severity of experimental rodent colitis: The importance of crosstalk between adipose tissue and skeletal muscles. *Mediat. Inflamm.* **2015**, *2015*, 605071. [CrossRef] [PubMed]
4. Elia, J.; Kane, S. Adult inflammatory bowel disease, physical rehabilitation, and structured exercise. *Inflamm. Bowel. Dis.* **2018**, *24*, 12. [CrossRef] [PubMed]
5. Papadimitriou, K. Effect of resistance exercise training on Crohn's disease patients. *Intest. Res.* **2021**, *19*, 275. [CrossRef] [PubMed]
6. Wiroth, J.B.; Filippi, J.; Schneider, S.M.; Al-Jaouni, R.; Horvais, N.; Gavarry, O.; Bermon, S.; Hébuterne, X. Muscle performance in patients with Crohn's disease in clinical remission. *Inflamm. Bowel. Dis.* **2005**, *11*, 296–303. [CrossRef] [PubMed]
7. Cabalzar, A.L.; Oliveira, D.J.F.; Reboredo, M.M.; Lucca, F.A.; Chebli, J.M.F.; Malaguti, C. Muscle function and quality of life in the Crohn's disease. *Fisioter. Mov. Curitiba.* **2017**, *30*, 337–345. [CrossRef]
8. Robinson, R.J.; Krzywicki, T.; Almond, L.; Al-Azzawi, F.; Abrams, K.; Iqbal, S.J.; Mayberry, J.F. Effect of a low-impact exercise program on bone mineral density in Crohn's disease: A randomized controlled trial. *Gastroenterology* **1998**, *115*, 36–41. [CrossRef]
9. Candow, D.; Rizzi, A.; Chillibeck, P.; Worobetz, L. Effect of resistance training on Crohn's disease. *Can. J. Appl. Physiol.* **2002**, *27*, S7–S8.
10. Pérez, C.A. Prescription of physical exercise in Crohn's disease. *J. Crohn's Colitis* **2009**, *3*, 225–231. [CrossRef] [PubMed]
11. De Souza Tajiri, G.J.; de Castro, C.L.; Zaltman, C. Progressive resistance training improves muscle strength in women with inflammatory bowel disease and quadriceps weakness. *J. Crohn's Colitis* **2014**, *8*, 1749–1750. [CrossRef] [PubMed]
12. Cronin, O.; Barton, W.; Moran, C.; Sheehan, D.; Whiston, R.; Nugent, H.; McCarthy, Y.; Molloy, C.B.; O'Sullivan, O.; Cotter, P.D.; et al. Moderate-intensity aerobic and resistance exercise is safe and favorably influences body composition in patients with quiescent inflammatory bowel disease: A randomized controlled cross over trial. *BMC Gastroenterol.* **2019**, *19*, 29. [CrossRef] [PubMed]
13. Wu, Y.; Potempa, L.A., El Kebir, D., Filep, J.G. C-reactive protein and inflammation. Conformational changes affect function. *Biol. Chem.* **2015**, *396*, 1181–1197. [CrossRef] [PubMed]
14. Buisson, A.; Mak, W.Y.; Andersen, M.J.; Lei, D.; Pekow, J.R.; Cohen, R.D.; Pereira, B.; Rubin, D.T. Fecal calprotectin is highly effective to detect endoscopic ulcerations in Crohn's disease regardless of disease location. *BMC Gastroenterol.* **2018**, *154*, 6. [CrossRef]
15. Nehring, S.M.; Goyal, A.; Patel, B.C. C Reactive protein. In *Stat Pearls*; StatPearls Publishing: Treasure Island, FL, USA, 2022. Available online: https://www.ncbi.nlm.nih.gov/books/NBK441843/ (accessed on 1 January 2022).
16. Peyrin-Biroulet, L.; Sandborn, W.J.; Panaccione, R.; Domènech, E.; Pouillon, L.; Siegmund, B.; Danese, S.; Ghosh, S. Tumour necrosis factor inhibitors in inflammatory bowel disease: The story continues. *Therap. Adv. Gastroenterol.* **2021**, *14*. [CrossRef] [PubMed]

Case Report

Bone Turnover Alterations after Completing a Multistage Ultra-Trail: A Case Study

Carlos Castellar-Otín [1], Miguel Lecina [2,*] and Francisco Pradas [1]

[1] ENFYRED Research Group, Faculty of Health and Sports, University of Zaragoza, 22002 Huesca, Spain; castella@unizar.es (C.C.-O.); franprad@unizar.es (F.P.)
[2] Faculty of Health and Sports, University of Zaragoza, 22002 Huesca, Spain
* Correspondence: miglecina@gmail.com

Abstract: A series of case studies aimed to assess bone and stress fractures in a 768-km ultra-trail race for 11 days. Four nonprofessional male athletes completed the event without diagnosing any stress fracture. Bone turnover markers (osteocalcin (OC), serum C-terminal cross-linking telopeptide of type I collagen (CTX), bone-specific alkaline phosphatase (BALP), and serum turnover calcium (Ca^{2+})) were assessed before (pre) and after the race (post) and on days two and nine during the recovery period (rec2 and rec9), respectively. Results showed: post-pre-OC = −45.78%, BALP = −61.74%, CTX = +37.28% and Ca^{2+} = −3.60%. At rec2 and rec9, the four parameters did not return to their pre-run levels: OC, −48.31%; BALP, −61.66%; CTX, +11.93% and Ca^{2+}, −3.38%; and OC = −25.12%, BALP = −54.65%, CTX = +93.41% and Ca^{2+} = +3.15%), respectively. Our results indicated that the ultra-trail race induced several changes in bone turnover markers, uncoupling of bone metabolism, increased bone resorption: OC and BALP and suppressed bone formation: CTX and Ca^{2+}. Bone turnover markers can help determine the response of bone to extreme effort and might also help predict the risk of stress fractures.

Keywords: ultra-endurance; bone mass density; bone remodelling markers; bone formation; bone resorption

Citation: Castellar-Otín, C.; Lecina, M.; Pradas, F. Bone Turnover Alterations after Completing a Multistage Ultra-Trail: A Case Study. *Healthcare* **2022**, *10*, 798. https://doi.org/10.3390/healthcare10050798

Academic Editors: João Paulo Brito, Rafael Oliveira and José Carmelo Adsuar Sala

Received: 7 March 2022
Accepted: 23 April 2022
Published: 25 April 2022

Publisher's Note: MDPI stays neutral with regard to jurisdictional claims in published maps and institutional affiliations.

1. Introduction

Regular physical activity has been recognized to have health benefits in general and specifically in the musculoskeletal system, increasing muscle strength and bone mineral density (BMD) [1]. Regular physical activity prevents multiple bone diseases, such as osteopenia or osteoporosis by increasing BMD [2]. Regardless of the importance of exercise in maintaining bone health, there is still no consensus in the scientific literature regarding the volume and intensity of effort required to prevent bone damage [3]. If exercise does not exert a minimum load on bone tissue, it will not increase BMD, as has been proven in studies carried out on swimmers [4]. Benefits in BMD depend on the type of exercise undertaken; weight-bearing exercise increases BMD, particularly at load-bearing sites, independent of muscular activity alone [5]. In contrast, when exercise overloads bone tissue because of excessive force, it may result in stress fractures and weaken BMD [6].

Among all the weight-bearing endurance sports, ultra-trail races include the longest (e.g., any distance in excess of the standard marathon distance 42.195 km or at least 6 h of duration) [7] and a great amount of negative and positive accumulated elevation, which increase the mechanical stress and consequently overload the microstructure of bone tissue (especially lower limbs) [8]. In addition, multi-stage ultra-trail competitions are usually held in extreme environments lasting several days and, as a consequence, athletes must carry their own provisions, resulting in additional weight and increasing the stress on bone tissue and reducing BMD [9]. Despite all these characteristics, the number of stress fractures found in competitions is relatively low [10,11]. Hoffman et al., in a descriptive study including 1212 ultra-runners, who completed an ultra-trail race, only found 0.3% of stress hip fractures and 1.9% of stress fractures involving tibia or fibula [12]. However, Scheer et al. found a higher incidence when assessing the stress fractures in a 12-month following study (10.3%) [13]. Apart from the length of the race, these runners have to complete a high volume of training prior to the competition in order to acquire the physiological and biomechanical adaptations required to face this extreme effort. Average training loads are between 66-83 km/week in adults and around 57 km/week in youth athletes [14]. This great amount of training may result in overtraining syndrome and subsequently in BMD loss increasing the likelihood of suffering stress fractures [15]. Thus, predictive markers that reflect stress of bone are needed to prevent stress fractures and help runners and coaches plan their training routines effectively [16].

BMD is usually assessed by using Dual-energy X-ray absorptiometry (DEXA) at the femur and, in the current criteria, OP is defined as a BMD T-score of -2.5 or lower at any one location or presenting a previous fragility fracture [4]. However, BMD analysis is not always conclusive as a predictive factor of stress fractures in the sports field [3]. Due to this fact, many researchers have raised the importance of bone turnover (BT) by analysing bone turnover markers (BTMs), as an essential factor when assessing bone microarchitecture and, therefore, the state of bone tissue [15,17]. BT consists of two dynamic processes: bone formation (anabolism) regulated by osteoblasts (responsible for bone matrix synthesis) and resorption (catabolism) mediated by osteoclasts (responsible for the secretion of proteolytic enzymes which digest bone matrix) [18]. BTMs are biochemical products measured usually in blood or urine that reflect the metabolic activity of bone, but which themselves have no function in controlling skeletal metabolism [19]. They are traditionally categorized as markers of bone formation or bone resorption. It has been shown that during the practice of endurance exercise there is an increase in markers related to bone resorption (serum C-terminal cross-linking telopeptide of type I collagen (CTX)) [20,21] and a decrease in those of formation (osteocalcin (OC), alkaline phosphatase (AP) and serum procollagen type I N propeptide (s-PINP)) causing a net loss of bone [22–24]. The analysis of BTMs offer some advantages over DXA alone: they allow the assessment of bone metabolic activity at a specific time, there are a high number of selective markers and the techniques are easily applicable and minimally invasive [25]. The analysis of BTMs, therefore, allows the monitoring of the likelihood of suffering stress fractures during the practice of physical activity by using selective biomarkers, apart from BMD [26]. The relationship between stress fractures and BT seems predisposed by an acceleration in the destruction of bone tissue that precedes the remodelling phase, which may cause a weakening of the tissue during this period and, therefore, increased probability of suffering stress fractures [27].

The aim of this case study was to assess the alterations suffered by four runners after completing a 768 km extreme ultra trail race on bone turnover markers. We hypothesized that bone resorption markers (CTX and Ca^{2+}) would increase and bone formation markers (CTX and BALP) would decrease. The alteration suffered in BTM may persist for days after the activity; a fact that may result in an increase in the suffering of stress fractures.

2. Case Report

Four non-professional healthy ultra-runners (38.08 $\pm$ 4.11 years) accepted participation in this case report after being invited by email. The four subjects included were males with broad experience (5 $\pm$ 1.26 years), well trained (11.61 $\pm$ 2.22 h·week^{-1}) and had accumulated large amounts of elevation both positive and negative in the preparatory period (116,615 $\pm$ 37,462 m). It took them 154 h 43 min (SD $\pm$ 23 min) (equivalent to 51% of VO_{2max}) to complete the 11 stages. They ran at an average speed of 5.11 $\pm$ 0.46 km h^{-1} and an average pace of 11 min 46 s (SD $\pm$ 3 min 4 s). No runner suffered any stress fracture. All participants were non-smokers and were not receiving medical, pharmacological, or dietary treatment.

Body composition measurements included: height, weight, skin folds and body mass index. All subjects were measured 2 h prior to the start of the race. Height measurement was made to the nearest 0.1 cm using a wall-mounted stadiometer (Seca 220, Seca, Hamburg, Germany), body weight was measured barefoot to the nearest 0.01 kg on calibrated electronic digital scales (Seca 769, Seca, Hamburg, Germany), skin folds used a compass accurate to ±0.2 mm (Seca 212, Seca, Hamburg, Germany) and a tape with an accuracy of ±1 mm was employed. Six skin folds were taken: abdominal, suprailiac, subscapular, tricipital, thigh and leg and perimeters; arms and legs were in a relaxed $90°$ position. The equations of Yushaz were used to calculate the percentage of fat [28] and the equation according to Lee to determine the percentage of muscle [29].

A cardiopulmonary test assessed the following physiological outcomes: maximum oxygen consumption (VO_{2max}), heart rate maximum (HR_{max}) and maximal aerobic speed (MAS). The laboratory test was performed on a treadmill (Pulsar, h/p/cosmos$^®$, Nussdorf, Germany). The test was run on a 1% slope and the start speed was set to 8 km h^{-1}, which increased 1 km·h^{-1} every minute. To warm up the subjects ran for 5 min on the treadmill operating at a speed of 6 km·h^{-1}. Respired gases were collected with an Oxycon Proanalyzer (Erich Jaeger GmbH, Hoechberg, Germany). The gas analysis system was calibrated according to ambient temperature and humidity, air flow and VO_2 and VCO_2 concentrations. A pulsometer was used to evaluate the maximal heart rate (Vantage M, Polar, Finland). The participants' pre-race characteristics are listed in Table 1.

Table 1. Pre-race individual characteristics of the population included ($n = 4$).

Parameters	Subject 1	Subject 2	Subject 3	Subject 4
Age (years)	33	37	41	42
VO_{2max} (mL/kg/min^{-1})	58.28	70.6	67.1	50.71
HR_{max} (beats·min^{-1})	194	186	194	176
Maximal aerobic speed (km·h^{-1})	18	17	16.7	16
Height (cm)	180.7	176.1	172.3	173.9
Weight (kg)	79.1	64.9	60.8	77.3
BMI	24.2	20.9	20.5	25.6
Fat mass (%)	8.82	6.88	8.70	8.14
Muscle mass (%)	43.4	47.38	57.63	38.55
Experience (years)	6	6	4	7
Distance covered (h·week^{-1})	11	11	15	11
Annual slope accumulated (m)	140,655	120,404	156,000	70,000

BMI, body mass index; HRmax, heart rate maximum; VO_{2max}: maximum oxygen consumption.

The GR-11 route joins the Mediterranean and Atlantic coasts along the Pyrenees, covering 786 km in 11 stages. The average stage/day consisted of 71.49 km (SD $\pm$ 8.2), the average positive elevation was 4260.45 ± 1063.26, and the average negative elevation was 4258.63 ± 989.13. The race had a warm temperature, with values ranging from 13.1 to 17.6 °C, and the humidity was (60.1–70.9%). In-race hydration was provided ad libitum. The characteristics of the ultra-trail race are listed in Table 2. This table has been previously published [30]. This case report is part of a series of case studies aimed at studying the effects on runners' health after completing this unique ultra-trail challenge called "GR-11"; accordingly, the characteristics of the ultra-trail (i.e., duration, positive and negative elevation) are the same in both case studies.

Table 2. Characteristics of the extreme ultra-trail [30].

Stages	Distance (km)	Elevation (m+)	Elevation (m−)
1	78.5	3136	3024
2	72.3	3886	3458
3	72	4655	4044
4	68.1	5660	4581
5	72.6	5411	6336

Table 2. *Cont.*

Stages	Distance (km)	Elevation (m+)	Elevation (m−)
6	76.1	5344	4788
7	63.7	5492	5163
8	66	3641	4576
9	66.1	3361	3841
10	66.5	2958	2934
11	83	3321	4100
Total	784.9	46,865	46,845
Md	71.35	4260.45	4258.63
Sd	±6.00	±1063.26	±989.13

Twenty milliliters of venous blood (antecubital vein) were withdrawn from each participant pre- and post-race, rec2, and rec9 evaluations (90 min before and 10 min after finishing the race, two days and nine days in the morning). Blood samples were collected in two 5-mL Vacutainer tubes (Beliver Industrial State, Plymouth PL6 7BP, UK) without anticoagulant for serum isolation and in two 5-mL tubes containing ethylenediaminetetraacetic acid (EDTA) as an anticoagulant. Once collected, the blood samples were coagulated for 25–30 min at room temperature and then centrifuged at 2500 rpm for 10 min to remove the clots. Serum samples were aliquoted into Eppendorf tubes (Eppendorf AG, Hamburg, Germany), washed with diluted nitric acid, and stored at $-80\ ^{\circ}$C until biochemical analysis. To facilitate the interpretation of the data, the change in analytical parameters was measured as follows: post-race, 2, 9 days less pre-race respectively. Statistical analyses were carried out using the Statistical Package for The Social Sciences software (IBM SPSS Statistics for Windows, version 26.0, 64 bits Edition, IBM Corp., Armonk, NY, USA). Descriptive analysis was carried out on all variables, and average, median and standard deviations were calculated. Normal distribution of the variables was verified by using Kolmogorov-Smirnov and Shapiro-Wilk tests, but normality criteria were not met because of the low number of subjects. *p*-value was calculated, but due to low number of subjects included and the design of the study as a series of case studies, its value was not considered for final analysis. The BTM changes analyzed are listed in Table 3. Range values were expressed for OC, ALP, CTX and Ca^{2+} according to age, sex, and race [31]. All bone formation markers included (OC and ALP) decreased their values when comparing pre- and post-exercise. Conversely, all the bone resorption markers decreased after race completion. During the recovery period, OC and ALP values remained above the basal line, even at rec9 (OC = -45.78% and BALP = -54.65%). In contrast, the CTX values increased slightly at rec2 (+11.93%) but soared at rec9, with values close to +100%. (CTX = +93.41%). Serum calcium levels decreased slightly when comparing pre- vs. post-and pre- vs. rec2. However, the rec9 values exceeded the pre-race levels (Ca^{2+} = +3.15%) (See Table 3). The chronological sequence of BTMs is fully shown in four different figures included in Figure 1.

Table 3. Blood parameters before (baseline) and after race (post-exercise day 2 and post-exercise day 9).

Parameter Blood (Reference Values)	Before-Race Pre (Baseline) Value	Post-Race Post (Post-Exercise) Value (% Difference)	Post-Race Day 2 (rec2) Value (% Difference)	Post-Race Day 9 (rec9) Value (% Difference)
OC (ng/mL) (13.98–41.99)	22.20 ± 7.41	11.15 ± 3.14 ↓ (−45.78)	10.30 ± 2.29 ↓ (−48.31)	15.14 ± 5.73 ↓ (−25.12)
BALP (ug/L) (6–30)	23.03 ± 4.68	8.64 ± 1.63 ↓ (−61.74)	8.50 ± 2.37 ↓ (−61.66)	10.29 ± 2.30 ↓ (−54.65)

Table 3. *Cont.*

Parameter Blood (Reference Values)	Before-Race	Post-Race		
	Pre (Baseline) Value	Post (Post-Exercise) Value (% Difference)	Day 2 (rec2) Value (% Difference)	Day 9 (rec9) Value (% Difference)
CTX (µg/L) (0.23–0.94)	0.24 ± 0.02	0.32 ± 0.09 ↑ (+37.28)	0.26 ± 0.12 ↑ (+11.93)	0.46 ± 0.14 ↑ (+93.41)
Ca^{2+} (mg/L) (8.70–10.40)	9.35 ± 0.33	9.22 ± 0.32 ↓ (−3.60)	9.03 ± 0.34 ↓ (−3.38)	9.64 ± 0.12 ↑ (+3.15)

Data are expressed as absolute values and as ± percentages from baseline values. OC, osteocalcin; BALP, alkaline phosphatase; CTX, C-terminal cross-linking telopeptide of type I collagen; Ca^{2+}, calcium.

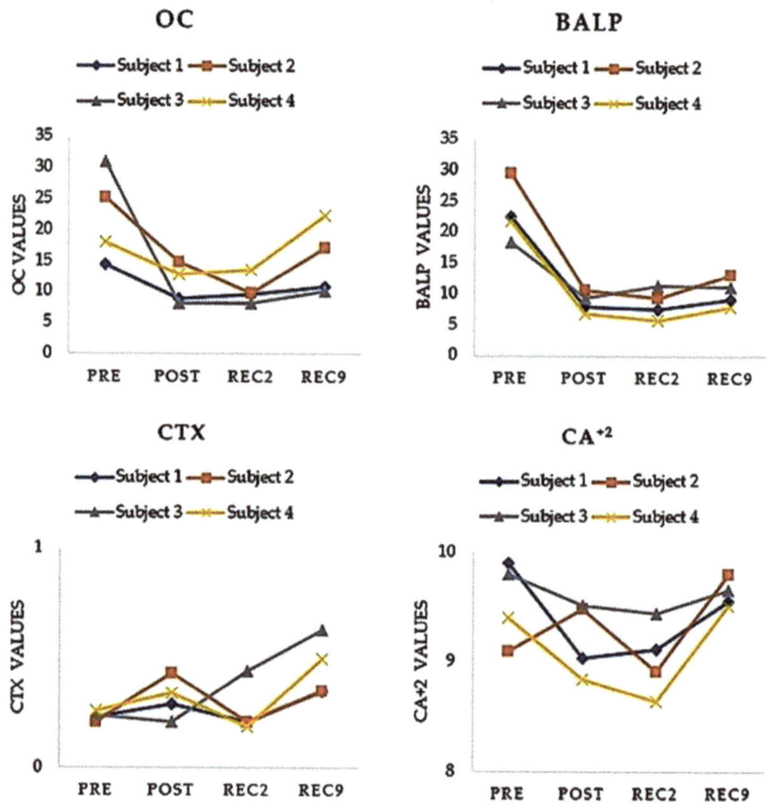

Figure 1. Chronological sequence of serum variables of bone metabolism of each of the 4 subjects. No statistically significant differences were found when comparing pre vs. post, rec2 and rec9 ($p > 0.05$).

3. Discussion

The objective of this case report was to assess the alterations suffered on BTMs after completing a multi-stage ultra-trail and in the recovery period. The main finding of our study was suppression in bone formation and an increase in the bone resorption process, not only after completing the race but also in the recovery period (2 and 9 days after), respectively. To the best of the authors' knowledge, this has been the first study to assess BTMs in such an extreme multi-stage ultra-trail after finishing the race and even nine days after in the recovery period. Only three previous studies have studied BTMs in ultra-endurance races so far, but the duration (from 245 to 308 km) and the elevation of these events were not as extreme as in the race included in this study [23,24,32]. To better understand the discussion, the text contains different points listed below.

3.1. Bone Formation Biomarkers

The results included in our study showed that bone formation markers reduced their values after completing an ultra-trail race competition and two and nine days after finishing it (See Table 3). These results fall in line with those previously reported by other authors [23,24,32]. However, these previous studies did not include such a long race nor assessed BTMs after nine days after completing the event. Despite the increase of new BTMs in recent years (e.g., Procollagen type I N-terminal propeptide (PINP) and procollagen type I C-terminal propeptide (PICP) [21], OC and BALP are still reliable markers of bone formation. Many of the most prestigious institutions related to bone health, including The National Health Allegiance (NBHA) [33], the International Osteoporosis Foundation (IOF) and the International Federation of Clinical Chemistry and Laboratory Medicine (IFCC), support the use of OC and BALP as clinical biomarkers for OP [19,34].

OC is a non-collagen protein synthesized exclusively by the osteoblasts and plays a pivotal role in osteogenesis [34]. Several studies have analyzed OC as the main BTM of bone resorption in ultramarathons [20,23,24,35]. Three out of the four found decreases in OC levels [20,23,24] but only one of them found no differences after finishing the race [32]. OC decrease during strenuous exercise has been explained by an increase in parathyroid hormone and cortisol [24]. The action of these hormones suppresses the activity of the osteoblasts or reduces osteoblast release as a consequence. Malm et al. found a four-fold increase in cortisol levels after completing a marathon [35]. In the same line, Knectle et al. found rises in cortisol and catecholamines and decreases in growth hormone, which shows the complexity of the alterations suffered by the hypothalamic-pituitary axis in these efforts [36]. In the recovery period, our study showed a relevant decrease of OC levels, even nine days after, which implies bone formation function remained partially suppressed days after completing the ultra-trail race. Other studies have also evaluated the activity of OC and have found similar results. Nizet et al. in a study that evaluated a marathon race, observed a decline in OC levels after the race (from 4.9 to 3.9 g/liter, −20%) and three days later. BALP is a homodimer anchored to the membrane of osteoblasts and matrix vesicles [37]. Although its exact function is not completely clear, the presence of alkaline phosphatase on the cell membrane is required for bone mineralization [38].

The results found in our study support the idea that repeated weight-bearing exercise may result in a suppression of the activity of BALP. However, BALP is mainly affected by hormonal levels altering its release. Malm et al., in a comparative study, only found significant decreases of BALP in the female group [39]. In this sense, the validity of BALP as a conclusive bone formation biomarker seems lower than OC or PINP [33]. Our study found higher decreases of BALP in the three measurements (−61.74%, −61.66% and −54.51%), despite the subjects of our study only being of male gender. This fact may be due to the excessive distance and the extraordinary elevation of the ultra-trail race in our study in comparison with previous research.

3.2. Bone Resorption Biomarkers

CTX is the result of osteoclastic bone resorption, and it is a type I breakdown product [38]. CTX has been proposed as the gold standard for assessing the bone resorption process [33]. Our study showed relevant decreases in CTX values after finishing the race and in the recovery period. Similarly, previous studies on marathons [40] and on ultra-trail races [20] have shown CTX increases ranging from +8% to +19%. The values found in our study were higher, especially at rec9 (+93.41%). The duration of the effort has been proposed as the main factor responsible for the increase of CTX [20]. According to many authors, [22,41] prolonged mechanical usage increases microdamage. It seems reasonable that higher values of CTX were found in our study because of the extreme duration of a 768 km run.

Ca^{2+} values slightly varied in the three measurements of our study (-3.60%, -3.38 and $+3.15\%$). Similar increases were reported by Nila et al. [42]. After the race Ca^{2+} increases from 9.2 ± 0.1 (mean $\pm$ SE) to 9.8 ± 0.1 mg/dL ($p < 0.01$). Another study that evaluated the alterations of Ca^{2+} in marathon runners (eleven men and seven women) found no increases in Ca^{2+} after the end of the race. The activity of Ca^{2+} has also been associated with duration and intensity of effort, apart from the adrenergic activation post-exercise [36]. The multi-stage here studied (768 km and 11 stages) exceeds the duration and the elevation of the races analyzed in previous research, so the higher increases in Ca^{2+} values at post and rec2 are mainly justified by these specific characteristics.

3.3. Stress Fractures and BTMs

The four subjects included in our study suffered no stress injury in the course of the ultra-trail race despite the BTM alterations found in our study. The incidence of stress fractures that the scientific literature has reported in ultra-endurance sports and ultra-trail, in particular, is relevant, oscillating their values from 0.3% in femur or hip to 1.2% tibia or fibula [12]. High rates of bone remodelling have been associated with an increase in suffering of stress fractures, regardless of BMD loss measured by DEXA or ultrasound [38]. Tian et al., in a systematic review. found positive relationships between CTX and risk fractures (1.20, 95%CI, 1.05–1.37). The ACSM has shown that the most influential factors for stress fractures are exercise mode, intensity and duration. Accordingly, stress fractures occur as a result of excessive training activity due to repetitive mechanical loading [16]. It is because of these microfractures that bone resorption activity increases. Vasikaran et al., in a systematic review, found several studies that associated BTMS changes and subsequent fractures [19].

Traditionally, the studies investigating hip fractures have usually focused on women, due to hormonal reasons behind the development of OP and, consequent higher incidence of hip fractures. Nevertheless, studies including men have also found BTM changes prior to suffering a stress fracture [16]. Studies that have analyzed the relationship between BMD loss and the likelihood of stress fracture in running activities have found significant associations [16,27,36].

Bennell et al., in a 12-month prospective study, found no differences in OC values between athletes who suffered a stress fracture and did not ($p = 0.010$) [15]. On the contrary, Sayaka et al., in another study including young athletes, found an incidence of stress fractures higher than in other similar studies (11.4% of 316 athletes), but there was no statistical difference in BTMs [16]. We can conclude that the value of BTM alterations as a tool for predicting stress fractures is contradictory. This discrepancy in the results obtained is due to the variety of BTMs analyzed, as well as the different characteristics in the population included and differences in running activities studied. The characteristics of the ultra-trail races (i.e., duration and elevation) would require further investigation considering these characteristics.

3.4. Limitations

It must be considered that the sample size ($n = 4$) and the only male gender used in this study could be a limitation that had an impact on the results obtained. The main reason to justify the design of this study was due to the uniqueness and extreme conditions of the race (e.g., duration, number of stages and positive and negative elevation accumulated). As our study shows, there are several alterations on BTMs after finishing an ultra-trail race and at least nine days after in the recovery period. However, further investigation is still required in order to clarify the mechanisms involved in BT and its relation with BMD and/or BC loss, and, ultimately, the etiology of stress fractures in these efforts. More epidemiological studies, including analyses of BTMS and DXA or ultra sound measurements, are needed to better elucidate the mechanisms involved in BMD.

4. Conclusions

According to this study, during an ultra-trail race it appears that bone resorption is increased and, conversely, bone formation is suppressed, resulting in a transient uncoupling of BT. The levels of all BTMs analyzed remained altered when compared with pre-run levels, especially in CTX and OC, even nine days later. This study showed that a 768-km multi-stage ultra-trail induces changes in the $OC/Ca^{2+}/BALP/CTX$ interaction, which may result in an increase in the likelihood of stress fractures as a consequence of damage to bone tissue. The special preparation that these athletes had to carry out to face the race implies a great amount of training volume (e.g., kilometers accumulated, $n°$ sessions·week^{-1}) prior to the race, so the study of which BTMs reflect bone damage may help in preventing runners suffering from BMD loss and in avoidance of stress fractures. Considering the results of this case report, runners and coaches should analyze alterations in BTMs, not only immediately after the race but also in the recovery period. The analysis of the BTMs here presented offers valuable information about load in bone tissue during the training process and may help runners reduce the likelihood of suffering stress fractures. The training process of these races requires a structured program of scientific monitoring in physiological, biomechanical and performance areas. According to these findings, BTMs should be measured as part of the preparation routine for ultra-races to prevent ultra-runners suffering bone turnover alterations.

Author Contributions: Conceptualization, M.L.; methodology, M.L., C.C.-O. and F.P.; software, M.L.; validation, M.L.; formal analysis, C.C.-O. and F.P.; investigation, F.P. and C.C.-O.; data curation, C.C.-O., M.L. and F.P; writing—original draft preparation, M.L. and C.C.-O.; writing—review and editing, C.C.-O. and F.P.; visualization, M.L. All authors have read and agreed to the published version of the manuscript.

Funding: The present research was funded by a research grant from the Instituto de Estudios Altoaragoneses of the Diputación Provincial de Huesca (Spain), file number N° 139 (3600 €) and with public funds from the Dirección General de Investigación e Innovación del Gobierno de Aragón to the ENFYRED research group.

Institutional Review Board Statement: The study was conducted according to the guide-lines of the Declaration of Helsinki, and approved by the Ethics Committee of the Department of Health and Consumption of the Government of Aragon (Spain), (protocol code 18/2015; date: 11 November 2015).

Informed Consent Statement: Informed consents were obtained from the subject involved in the study.

Data Availability Statement: Information about the case report is available at http://gr11en11.org/ (accessed on 26 October 2021).

Acknowledgments: The authors thank to the four ultra-trail runners for their participation in this research work and the Centro de Medicina del Deporte del Gobierno de Aragón for its invaluable help and collaboration.

Conflicts of Interest: The authors declare no conflict of interest.

References

1. World Health Organization. GLobal Recommendations on Physical Activity for Health. 2018. Available online: https://apps.who.int/iris/bitstream/handle/10665/325147/WHO-NMH-PND-2019.4-eng.pdf?sequence=1&isAllowed=y%0Ahttp://www.who.int/iris/handle/10665/311664%0Ahttps://apps.who.int/iris/handle/10665/325147 (accessed on 24 February 2022).
2. Benedetti, M.G.; Furlini, G.; Zati, A.; Mauro, G.L. The Effectiveness of Physical Exercise on Bone Density in Osteoporotic Patients. *BioMed Res. Int.* **2018**, *2018*, 4840531. [CrossRef] [PubMed]
3. Kopiczko, A.; Adamczyk, J.G.; Gryko, K.; Popowczak, M. Bone mineral density in elite masters athletes: The effect of body composition and long-term exercise. *Eur. Rev. Aging Phys. Act.* **2021**, *18*, 7. [CrossRef]
4. Gómez-Bruton, A.; Gónzalez-Agüero, A.; Gómez-Cabello, A.; Casajús, J.A.; Vicente-Rodríguez, G. Is Bone Tissue Really Affected by Swimming? A Systematic Review. *PLoS ONE* **2013**, *8*, e70119. [CrossRef] [PubMed]
5. Etherington, J.; Harris, P.A.; Nandra, D.; Hart, D.J.; Wolman, R.L.; Doyle, D.V.; Spector, T.D. The effect of weight-bearing exercise on bone mineral density: A study of female ex-elite athletes and the general population. *J. Bone Miner. Res.* **1996**, *11*, 1333–1338. [CrossRef] [PubMed]

6. Scott, J.P.R.; Sale, C.; Greeves, J.P.; Casey, A.; Dutton, J.; Fraser, W.D. The effect of training status on the metabolic response of bone to an acute bout of exhaustive treadmill running. *J. Clin. Endocrinol. Metab.* **2010**, *95*, 3918–3925. [CrossRef] [PubMed]
7. Scheer, V.; Basset, P.; Giovanelli, N.; Vernillo, G.; Millet, G.P.; Costa, R.J.S. Defining Off-road Running: A Position Statement from the Ultra Sports Science Foundation. *Int. J. Sports Med.* **2020**, *41*, 275–284. [CrossRef] [PubMed]
8. Eston, R.G.; Finney, S.; Baker, S.; Baltzopoulos, V. Muscle tenderness and peak torque changes after downhill running following a prior bout of isokinetic eccentric exercise. *J. Sports Sci.* **1996**, *14*, 291–299. [CrossRef]
9. Scheer, V.; Krabak, B.J. Musculoskeletal Injuries in Ultra-Endurance Running: A Scoping Review. *Front. Physiol.* **2021**, *12*, 664071. [CrossRef]
10. Dawadi, S.; Basyal, B.; Subedi, Y. Morbidity Among Athletes Presenting for Medical Care During 3 Iterations of an Ultratrail Race in the Himalayas. *Wilderness Environ. Med.* **2020**, *31*, 437–440. [CrossRef]
11. Vernillo, G.; Savoldelli, A.; La Torre, A.; Skafidas, S.; Bortolan, L.; Schena, F. Injury and Illness Rates during Ultratrail Running. *Int. J. Sports Med.* **2016**, *37*, 565–569. [CrossRef] [PubMed]
12. Hoffman, M.D.; Krishnan, E. Health and exercise-related medical issues among 1212 ultramarathon runners: Baseline findings from the Ultrarunners Longitudinal TRAcking (ULTRA) Study. *PLoS ONE* **2014**, *9*, e83867. [CrossRef] [PubMed]
13. Scheer, B.V.; Murray, A. Al Andalus Ultra Trail: An observation of medical interventions during a 219-km, 5-day ultramarathon stage race. *Clin. J. Sport Med.* **2011**, *21*, 444–446. [CrossRef] [PubMed]
14. Nikolaidis, P.T.; Knechtle, B. Performance in 100-km Ultramarathoners-At Which Age, It Reaches Its Peak? *J. Strength Cond. Res.* **2020**, *34*, 1409–1415. [CrossRef]
15. Bennell, K.L.; Malcolm, S.A.; Brukner, P.D.; Green, R.M.; Hopper, J.L.; Wark, J.D.; Ebeling, P.R. A 12-month prospective study of the relationship between stress fractures and bone turnover in athletes. *Calcif. Tissue Int.* **2020**, *4*, 80–85. [CrossRef]
16. Nose-Ogura, S.; Yoshino, O.; Dohi, M.; Torii, S.; Kigawa, M.; Harada, M.; Hiraike, O.; Kawahara, T.; Osuga, Y.; Fujii, T.; et al. Relationship between tartrate-resistant acid phosphatase 5b and stress fractures in female athletes. *J. Obstet. Gynaecol. Res.* **2020**, *46*, 1436–1442. [CrossRef]
17. Adami, S.; Gatti, D.; Viapiana, O.; Fiore, C.E.; Nuti, R.; Luisetto, G.; Ponte, M.; Rossini, M. Physical activity and bone turnover markers: A cross-sectional and a longitudinal study. *Calcif. Tissue Int.* **2008**, *83*, 388–392. [CrossRef]
18. Marini, S.; Barone, G.; Masini, A.; Dallolio, L.; Bragonzoni, L.; Longobucco, Y.; Maffei, F. Current Lack of Evidence for an Effect of Physical Activity Intervention Combined with Pharmacological Treatment on Bone Turnover Biomarkers in People with Osteopenia and Osteoporosis: A Systematic Review. *J. Clin. Med.* **2021**, *10*, 3442. [CrossRef] [PubMed]
19. Vasikaran, S.; Eastell, R.; Bruyère, O.; Foldes, A.J.; Garnero, P.; Griesmacher, A.; McClung, M.; Morris, H.A.; Silverman, S.; Trenti, T.; et al. Markers of bone turnover for the prediction of fracture risk and monitoring of osteoporosis treatment: A need for international reference standards. *Osteoporos. Int.* **2011**, *22*, 391–420. [CrossRef] [PubMed]
20. Kerschan-Schindl, K.; Thalmann, M.; Sodeck, G.H.; Skenderi, K.; Matalas, A.L.; Grampp, S.; Ebner, C.; Pietschmann, P. A 246-km continuous running race signinficant changes in bone metabolism. *Bone* **2009**, *45*, 1079–1083. [CrossRef] [PubMed]
21. Roberts, H.M.; Law, R.J.; Thom, J.M. The time course and mechanisms of change in biomarkers of joint metabolism in response to acute exercise and chronic training in physiologic and pathological conditions. *Eur. J. Appl. Physiol.* **2019**, *119*, 2401–2420. [CrossRef]
22. Zhang, R.; Geoffroy, V.; Ridall, A.L.; Karsenty, G.; Tracy, T.; Bonner, A.S.; Duffy, J.B.; Gergen, J.P.; Gergen, J.P.; Karlovich, C.A.; et al. Bone Resorption by Osteoclasts. *Science* **2000**, *289*, 1504–1509.
23. Shin, K.-A.; Kim, A.-C.; Kim, Y.-J.; Lee, Y.-H.; Shin, Y.-O.; Kim, S.-H.; Park, Y.-S.; Nam, H.S.; Kim, T.; Kim, H.S.; et al. Effect of Ultra-marathon (308 km) Race on Bone Metabolism and Cartilage Damage Biomarkers. *Ann. Rehabil. Med.* **2012**, *36*, 80. [CrossRef] [PubMed]
24. Mouzopoulos, G.; Stamatakos, M.; Tzurbakis, M.; Tsembeli, A.; Manti, C.; Safioleas, M.; Skandalakis, P. Changes of bone turnover markers after marathon running over 245 km. *Int. J. Sports Med.* **2007**, *28*, 576–579. [CrossRef] [PubMed]
25. Tian, A.; Ma, J.; Feng, K.; Liu, Z.; Chen, L.; Jia, H.; Ma, X. Reference markers of bone turnover for prediction of fracture: A meta-analysis. *J. Orthop. Surg. Res.* **2019**, *14*, 4–13. [CrossRef] [PubMed]
26. Torres, E.; Mezquita, P.; De La Higuera, M.; Fernández, D.; Muñoz, M. Actualización sobre la determinación de marcadores de remodelado óseo. *Endocrinol. Nutr.* **2003**, *50*, 237–243. [CrossRef]
27. Bennell, K.L. Models for the pathogenesis of stress fractures in athletes. *Br. J. Sports Med.* **1996**, *30*, 200–204. [CrossRef]
28. Escrivá, D.; Caplliure-Llopis, J.; Benet, I.; Mariscal, G.; Mampel, J.V.; Barrios, C. Differences in adiposity profile and body fat distribution between forwards and backs in sub-elite spanish female rugby union players. *J. Clin. Med.* **2021**, *10*, 5713. [CrossRef] [PubMed]
29. Lee, R.C.; Wang, Z.; Heo, M.; Ross, R.; Janssen, I.; Heymsfield, S.B. Total-body skeletal muscle mass: Development and cross-validation of anthropometric prediction models. *Am. J. Clin. Nutr.* **2000**, *72*, 796–803. [CrossRef]
30. Lecina, M.; Castellar, C.; Pradas, F.; López-Laval, I. 768-km Multi-Stage Ultra-Trail Case Study-Muscle Damage, Biochemical Alterations and Strength Loss on Lower Limbs. *Int. J. Environ. Res. Public Health* **2022**, *19*, 876. [CrossRef] [PubMed]
31. Fowler, J.; Cohen, L.; Jarvis, P. *Practical Statistics for Field Biology*; John Wiley & Sons: Hoboken, NJ, USA, 2013; ISBN 1118685644.
32. Sansoni, V.; Vernillo, G.; Perego, S.; Barbuti, A.; Merati, G.; Schena, F.; La Torre, A.; Banfi, G.; Lombardi, G. Bone turnover response is linked to both acute and established metabolic changes in ultra-marathon runners. *Endocrine* **2017**, *56*, 196–204. [CrossRef] [PubMed]

33. Dolan, E.; Varley, I.; Ackerman, K.E.; Pereira, R.M.R.; Elliott-Sale, K.J.; Sale, C. The Bone Metabolic Response to Exercise and Nutrition. *Exerc. Sport Sci. Rev.* **2020**, *48*, 49–58. [CrossRef] [PubMed]
34. Rubert, M.; De la Piedra, C. Osteocalcin: From marker of bone formation to hormone; and bone, an endocrine organ. *Osteoporos. Metab Miner.* **2021**, *12*, 146–151. [CrossRef]
35. Malm, C.; Sjödin, B.; Sjöberg, B.; Lenkei, R.; Renström, P.; Lundberg, I.E.; Ekblom, B. Leukocytes, cytokines, growth factors and hormones in human skeletal muscle and blood after uphill or downhill running. *J. Physiol.* **2004**, *556*, 983–1000. [CrossRef]
36. Knechtle, B.; Nikolaidis, P.T. Physiology and pathophysiology in ultra-marathon running. *Front. Physiol.* **2018**, *9*, 634. [CrossRef]
37. Nizet, A.; Cavalier, E.; Stenvinkel, P.; Haarhaus, M.; Magnusson, P. Bone alkaline phosphatase: An important biomarker in chronic kidney disease—Mineral and bone disorder. *Clin. Chim. Acta* **2020**, *501*, 198–206. [CrossRef]
38. Civitelli, R.; Armamento-Villareal, R.; Napoli, N. Bone turnover markers: Understanding their value in clinical trials and clinical practice. *Osteoporos. Int.* **2009**, *20*, 843–851. [CrossRef]
39. Malm, H.T.; Ronni-Sivula, H.M.; Viinikka, L.U.; Ylikorkala, O.R. Marathon running accompanied by transient decreases in urinary calcium and serum osteocalcin levels. *Calcif. Tissue Int.* **1993**, *52*, 209–211. [CrossRef]
40. Larsen, E.L.; Poulsen, H.E.; Michaelsen, C.; Kjær, L.K.; Lyngbæk, M.; Andersen, E.S.; Petersen-Bønding, C.; Lemoine, C.; Gillum, M.; Jørgensen, N.R.; et al. Differential time responses in inflammatory and oxidative stress markers after a marathon: An observational study. *J. Sports Sci.* **2020**, *38*, 2080–2091. [CrossRef]
41. Lee, J.H. The effect of long-distance running on bone strength and bone biochemical markers. *J. Exerc. Rehabil.* **2019**, *15*, 26–30. [CrossRef]
42. Vora, N.M.; Kukreja, S.C.; York, P.A.J.; Bowser, E.N.; Hargis, G.K.; Williams, G.A. Effect of exercise on serum calcium and parathyroid hormone. *J. Clin. Endocrinol. Metab.* **1983**, *57*, 1067–1069. [CrossRef]

MDPI

St. Alban-Anlage 66

4052 Basel

Switzerland

www.mdpi.com

Healthcare Editorial Office

E-mail: healthcare@mdpi.com

www.mdpi.com/journal/healthcare